Theory of shell structures

Theory of
SHELL STRUCTURES

C. R. CALLADINE

*Reader in Structural Mechanics, University of Cambridge
and Fellow of Peterhouse*

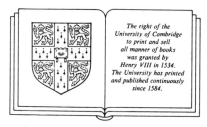

CAMBRIDGE UNIVERSITY PRESS
Cambridge
New York   New Rochelle
Melbourne   Sydney

Published by the Press Syndicate of the University of Cambridge
The Pitt Building, Trumpington Street, Cambridge CB2 1RP
32 East 57th Street, New York, NY 10022, USA
10 Stamford Road, Oakleigh, Melbourne 3166, Australia

© Cambridge University Press 1983

First published 1983
First paperback edition 1988

Printed in Great Britain at the University Press, Cambridge

Library of Congress catalogue card number: 82-4255

British Library Cataloguing in Publication Data

Calladine, C. R.
Theory of shell structures
1. Elastic plates and shells
2. Shells (Engineering)
I. Title
624.1'776   TA660.S5

ISBN 0 521 23835 8 hard covers
ISBN 0 521 36945 2 paperback

To Robert, Rachel and Daniel

Contents

Preface xiii   List of symbols xix

1.	**Introduction**	1
1.1	General remarks on shell structures	1
1.2	Theory of shell structures as a branch of structural mechanics	6
2.	**Generalised Hooke's law for an element of a shell**	12
2.1	Introduction	12
2.2	Shear distortion and Kirchhoff's hypothesis	13
2.3	Generalised Hooke's law for a simple flat shell element	17
2.4	Mechanical properties of curved elements	31
2.5	Non-uniform shells	36
2.6	Problems	37
3.	**Cylindrical shells under symmetric loading**	40
3.1	Introduction	40
3.2	Basic equations of the problem	42
3.3	The governing equation and its general solution	46
3.4	Semi-infinite cylindrical shell with edge-loading at $x = 0$	49
3.5	The ring-loaded long cylindrical shell	54
3.6	The band-loaded cylindrical shell	57
3.7	Cylindrical shell loaded sinusoidally along its length	59
3.8	Edge-loading of finite elastic cylindrical shells	67
3.9	Conclusion	72
3.10	Problems	73
4.	**Purely 'equilibrium' solutions for shells: the membrane hypothesis**	80
4.1	Introduction	80
4.2	A simple problem: the plane 'string'	83
4.3	Equilibrium equations for a doubly-curved shell	86
4.4	Equilibrium equations for cylindrical shells	90

viii   *Contents*

4.5	Axisymmetric loading of shells of revolution	96
4.6	Equilibrium equations for nearly-cylindrical shells	104
4.7	Boundary conditions in membrane analysis	112
4.8	Summary	117
4.9	Problems	118
5.	**The geometry of curved surfaces**	124
5.1	Introduction	124
5.2	The idea of a surface	125
5.3	General properties of plane curves	128
5.4	Curvature of a surface in terms of the geometry of cross-sections: principal curvatures, etc.	129
5.5	Gaussian curvature: an intrinsic view of surfaces	135
5.6	Inextensional deformation of surfaces	145
5.7	Nontriangular polygonalisation of surfaces	146
5.8	Summary	148
5.9	Problems	148
6.	**Geometry of distortion of curved surfaces**	151
6.1	Introduction	151
6.2	'Change of Gaussian curvature' in terms of surface strain	152
6.3	Connection between the two aspects of Gaussian curvature	156
6.4	Strain-displacement relations for a cylindrical shell	157
6.5	Inextensional deformation of a cylindrical surface	160
6.6	Strain-displacement relations for nearly-cylindrical surfaces	167
6.7	General remarks on the inextensional deformation of shells	170
6.8	Compatibility between surface strain and change of curvature in curvilinear coordinates	171
6.9	Symmetrical deformation of shells of revolution	178
6.10	Problems	183
7.	**Displacements of elastic shells stressed according to the membrane hypothesis**	188
7.1	Introduction	188
7.2	Testing the validity of the membrane hypothesis	189
7.3	Cylindrical shell with doubly-periodic pressure-loading	191
7.4	Use of the Airy stress function in the calculation of deflections in a cylindrical shell	201
7.5	A 'beam analogy' for 'quasi-inextensional' deformation of cylindrical shells	208
7.6	Discussion	212
7.7	Problems	214
8.	**Stretching and bending in cylindrical and nearly-cylindrical shells**	217
8.1	Introduction	217
8.2	The 'two-surface' idealisation: equilibrium equations	218
8.3	Response of the 'bending surface' to pressure-loading	224

8.4	Cylindrical shell subjected to a doubly-periodic pressure-loading	226
8.5	Donnell's equations for nearly-cylindrical shells	229
8.6	The improvement of Donnell's equations	242
8.7	Boundary conditions	249
8.8	Summary and discussion	264
8.9	Problems	265
9.	**Problems in the behaviour of cylindrical and nearly-cylindrical shells subjected to non-symmetric loading**	**269**
9.1	Introduction	269
9.2	A preliminary example: a cylindrical shell acting as a beam	270
9.3	Beam on elastic foundation	278
9.4	Long-wave solution: the formal 'beam' analogy	283
9.5	Cylindrical shell with one edge free and the other edge clamped	285
9.6	Cylindrical shell loaded by radial point forces	301
9.7	Concentrated load on a spherical shell	307
9.8	Conclusion	312
9.9	Problems	312
10.	**Cylindrical shell roofs**	**318**
10.1	Introduction	318
10.2	A simple cylindrical shell roof	321
10.3	The effect of edge-beams: a simple example	338
10.4	Concluding remarks	355
10.5	Problems	358
11.	**Bending stresses in symmetrically-loaded shells of revolution**	**361**
11.1	Introduction	361
11.2	Equations of the problem	364
11.3	The effects of an 'imperfect' meridian	374
11.4	Some pressure-vessel junction problems	387
11.5	Reconciliation of the present scheme with the 'two-surface' approach	403
11.6	General discussion	406
11.7	Conclusion	409
11.8	Problems	409
12.	**Flexibility of axisymmetric bellows under axial loading**	**413**
12.1	Introduction	413
12.2	Analysis of flexibility by an energy method	415
12.3	Comparison with previous work	422
12.4	Approximate analysis of strain in bellows	424
12.5	Discussion	426
12.6	Problems	426
13.	**Curved tubes and pipe-bends**	**429**
13.1	Introduction	429

13.2	A curved tube subjected to internal pressure	433
13.3	Pure bending of a curved two-flange beam	437
13.4	Pure bending of a curved tube	443
13.5	The effect of internal pressure	456
13.6	End-effects in the bending of curved tubes	461
13.7	Problems	470
14.	**Buckling of shells: classical analysis**	**473**
14.1	Buckling of structures	473
14.2	Eigenvalue calculations according to the 'two-surface' model	477
14.3	Cylindrical shell under uniform axial compression	484
14.4	Axial compression and 'side' pressure combined	495
14.5	Necessary corrections for small values of circumferential wavenumber n	504
14.6	The effect of clamped and other boundary conditions	512
14.7	Buckling of cylindrical shells in torsion	521
14.8	Experimental observations	532
14.9	A simple design problem	532
14.10	Unidirectionally reinforced shells	535
14.11	A special boundary condition	542
14.12	Problems	544
15.	**Buckling of shells: non-classical analysis**	**550**
15.1	Introduction	550
15.2	A simple model for the study of buckling	556
15.3	A re-examination of the 'classical' calculation	562
15.4	A nonlinear analysis of buckling	564
15.5	Distribution of tangential-stress resultants during buckling	578
15.6	Nonlinear behaviour of the S-surface	581
15.7	Discussion	583
15.8	Axial buckling of pressurised cylindrical shells	588
15.9	Nonlinear effects in the buckling of cylindrical shells under pure torsion	591
15.10	Problems	592
16.	**The Brazier effect in the buckling of bent tubes**	**595**
16.1	Introduction	595
16.2	The Brazier effect in a simple beam	598
16.3	The Brazier effect in a tube of circular cross-section	605
16.4	Local buckling	609
16.5	The effect of interior pressure	612
16.6	The effect of finite length	614
16.7	An improvement on Brazier's analysis	621
16.8	Problems	625
17.	**Vibration of cylindrical shells**	**626**
17.1	Introduction	626

17.2	Vibrations of a simple ring	629
17.3	Vibration of a cylindrical shell in 'shallow' modes	638
17.4	Low-frequency approximations	642
17.5	Boundary conditions	646
17.6	Natural frequencies for cylindrical shells having different boundary conditions	650
17.7	Concluding remarks	653
17.8	Problems	654
18.	**Shell structures and the theory of plasticity**	**656**
18.1	Introduction	656
18.2	Cylindrical shell subjected to axisymmetric loading applied at an edge	662
18.3	Upper- and lower-bound theorems	674
18.4	Cylindrical shell subjected to axisymmetric band-loading	677
18.5	Lower-bound analysis of axisymmetric pressure-vessels	679
18.6	Use of lower-bound theorem to design reinforcement for pressure-vessels	690
18.7	An upper-bound method	695
18.8	Discussion	708
18.9	Problems	708
	Appendices	
1.	Theorems of structural mechanics	711
	I The principle of virtual work	711
	II The theorem of minimum complementary energy	714
	III The theorem of minimum strain energy	717
	IV The theorem of minimum total potential energy	719
2.	'Corresponding' load and deflection variables	721
3.	Rayleigh's principle	723
4.	Orthogonal functions	725
5.	Force-like and stress-like loads	726
6.	The 'static-geometric analogy'	728
7.	The area of a spherical polygon	735
8.	The 'sagitta' of an arc	737
9.	Rigidity of polyhedral frames	738
10.	Fourier series	741
11.	Suggestions for further reading	743
	Answers to selected problems	744
	References	747
	Index	758

Preface

The theory of shell structures is a large subject. It has existed as a well-defined branch of structural mechanics for about a hundred years, and the literature is not only extensive but also rapidly growing. In these circumstances it is not easy to write a textbook. The character of any book depends, of course, mainly on the author's conception of its subject matter. Thus it may help the reader if I set out my basic views on the theory of shell structures at the outset.

Most authors of books and papers on the theory of shell structures would agree that the subject exists for the benefit of engineers who are responsible for the design and manufacture of shell structures. But among workers who share this same basic aim, a wide variety of attitudes may be found. Thus, some will claim that they can give the best service to engineers by concentrating mainly on the form and structure of the governing equations of the subject, expressed with due rigour and in general curvilinear coordinates: for once the foundations have been laid properly (they say), the solution of all problems becomes merely a mathematical or computational exercise of solving the equations to a desired degree of accuracy; and indeed unless the foundations have been laid properly (they say), any resulting solutions are of questionable validity. Another group will argue, on the contrary, that they can serve engineers best by providing a set or 'suite' of computer programmes, which are designed to solve a range of relevant problems for structures (including shells) having arbitrary geometrical configuration; and indeed that the provision of such programmes renders obsolete, at a stroke, what was formerly called the theory of shell structures.

There is merit, of course, in each of these two views, and indeed in various less extreme versions of them. But in relation to the task of writing this book I have rejected both of them. In my view, the important thing is for engineers to understand how shell structures behave, and to be able to express this understanding in the language of physical rather than of purely mathematical ideas. The key to all engineering design is an understanding of the relevant physical phenomena; and it seems clear to me, on the basis of experience, that neither of the two schemes outlined above normally engenders either in students – or indeed in more senior people – a level of understanding which is appropriate for the purposes of engineering design.

The theory of shell structures is set firmly in the field of *structural mechanics*, for which the main ideas have been well established for many years. The mechanics of shell structures is in principle of the same kind as that of simpler structures such as beams and trusses, and indeed of three-dimensional continua. But the subject of shells is more complicated than any of these other examples because of difficulties which spring from the geometrical aspects of doubly-curved surfaces.

An important simplifying idea was introduced to the theory of shell structures in the early days by Love and others. This idea was to model the physical shell as if it were a *surface* of zero thickness, but endowed with mechanical properties in the form of elastic resistance to both stretching and bending actions within the surface. We shall use this model widely in the book. It is a generalisation of the familiar idealisation of a beam or a member of a truss as a *line* which is endowed with precisely the same sort of mechanical properties. This kind of theory of beams and of trusses is entirely satisfactory for dealing with a wide range of practical problems. Nevertheless there are some aspects of the behaviour of beams and trusses which are beyond the scope of this type of theory, and for which more thorough analysis is needed: examples include the detailed localised response of a beam to concentrated loads and the behaviour of joints in a truss. Similar remarks apply to theories of shell structures in which the shell is idealised as a surface endowed with certain mechanical properties. Such theories are incapable of describing the complicated three-dimensional states of stress which can occur in shells near boundaries and in regions of application of concentrated loads.

In the theory both of beams and of trusses there is virtually no interaction between bending and stretching effects. By contrast, in the theory of shell structures the interaction of bending and stretching effects is crucial. It is obviously important for engineers to understand clearly in physical terms how the bending and stretching effects combine to carry the loads which are applied to shell structures. This topic poses a major difficulty, for neither the 'computer package' approach to the subject nor the 'basic equations' approach produces a useful measure of understanding of this interactive behaviour. Thus, although it is true that the computer may be programmed to print out in a given case detailed information on bending and stretching over the entire structure, nevertheless it is clear that such information constitutes 'knowledge' rather than 'understanding'. And again, although in principle the form of the solution of a governing equation may be deduced from the *structure* of the equation without the necessity for *solving* the equation, nevertheless very little work along these lines seems to have been done in a way which is useful to engineers.

In this book I put forward two innovations which are useful in tackling this basic problem. The first of these is to separate the shell, conceptually, into two coincident but distinct surfaces labelled the bending (B) and stretching (S) surface, respectively. Any load which is applied to the shell is sustained in general partly by the stretching surface and partly by the bending surface; and the balance in load-sharing is regulated mainly by the 'interface pressure' between the two surfaces, which varies from point to point over the surface. This interface pressure becomes a prime variable of the problem. In terms of classical structural mechanics it is a variable 'redundancy'.

This device of separating the action of the shell into two parts affords the possibility of thinking separately about these two aspects of behaviour, and the interaction between them. An advantage of this scheme becomes clear in chapters 8 and 9: *dimensionless groups* which are appropriate for the discussion of various classes of shell problem emerge clearly from this type of analysis. These enable us to present solutions in the most economical way, by means of dimensionless plots. Diagrams of this sort, which set out in the form of a 'map' the various regimes of structural behaviour, are of great benefit to engineers, particularly for the purposes of preliminary design; and they are an important feature of this book.

The second innovation concerns the description and presentation of the geometry of curved surfaces. In his important paper on the curvature of general surfaces, Gauss (1828) pointed out that there are two distinct but complementary ways of thinking about the curvature of surfaces. I have found no book or paper on the subject of shell structures in which this point is exposed plainly; and yet it is absolutely fundamental to a clear understanding of the subject. Gauss's dual view of curvature fits precisely the 'two-surface' description, and it provides succinctly the geometrical conditions which are necessary if the deformation of the two surfaces is to match. The key variable in this connection turns out to be a *scalar* quantity; and paradoxically it seems that the conventional treatment of the geometry of curved surfaces in terms of general curvilinear coordinates is somehow too elaborate to reveal this crucial scalar quantity. My treatment of these fundamental geometrical matters in chapters 5 and 6 aims at explaining the key ideas before general coordinate systems are introduced.

xvi   *Preface*

The first part of the book (chapters 1-8) is mainly concerned with the development of equations, and the 'two-surface' idea, for representing the behaviour of shells. The second part (chapters 9-18) is devoted mainly to the solution of problems which are important in a wide range of engineering applications. In these chapters several methods will be used in the solution of various *ad hoc* problems: it is not necessarily the best policy to begin by writing down all of the equations. Occasionally, as in chapters 16 and 18, a novel geometrical idea unexpectedly simplifies the calculation. In most cases the method which is used boils down to making a speculation of a physical kind about the behaviour of the shell which makes the subsequent calculations easier, whether these calculations involve an assessment of various kinds of energy, or the transformation to a simpler analogous mechanical problem. In each of these cases a *hypothesis* is made, and subsequently examined in the light of the results. And the object is, of course, not only to simplify the calculations but also to clarify the mechanics and aid the determination of dimensionless groups whose values will characterise the structural behaviour. This kind of procedure is, of course, central to scientific method.

Many books and papers on the subject of shell structures are written on the implicit assumption that the aim of the theory of structures is to describe, for a given structure subject to prescribed loading, the displacements, stresses, etc. as functions of position in the structure. This frame of reference grew naturally out of the development of the theory of structures as a branch of the theory of *elasticity*. The development of the *plastic* theory of structures from the 1930s onwards was healthy not only because many real materials are still structurally useful when they have passed beyond the elastic range, but also because the new theory was able to shift the emphasis of the theory from the narrow routine described above to more profound questions about the kind of information which the engineer requires in order to design satisfactory structures. Thus, for example, questions about the relevance of initial stresses to the failure of structures could be faced squarely for the first time. I have included a chapter (18) on the plastic theory of shell structures precisely to counteract an impression about the nature of the subject which might be given by the omission of such a chapter. In particular, plastic theory can contribute to an understanding of the response of shells to localised loading.

The book is aimed primarily at those who wish to understand about the basic features of the behaviour of shell structures before undertaking more detailed studies by means of experiment or computation. I hope that students will find a clear introduction to the subject, that teachers will appreciate some new insights into old problems, and that practising structural engineers will find my treatment of behaviour in terms of the relevant dimensionless groups

Preface xvii

to be useful for the purposes of, at least, preliminary design.

Although the book is large, it is nevertheless only an *introduction* to the subject in the form of a set of basic ideas, together with examples of their application to some sample problems. Problems about shell structures which are faced by engineers in industry are often more complicated than those which are discussed in this book, by way of less simple geometry, the provision of various kinds of reinforcement, and more varied operating conditions. And it is now common practice for engineers to make use of more-or-less standard computer packages in solving the problems which confront them. I have already explained that the 'computer package' approach is not central to the development of this book; and in fact there is very little explicit reference to computational methods. Nevertheless, the book is relevant to computational work in several ways. First, it aims to provide a basic understanding of the subject which should be, in a broad sense, a prerequisite of any computational work. Second, it enables workers to appreciate that some useful problems have already been solved analytically in a general way, and that for these problems large-scale computation is unnecessary. Third, it provides some useful general ideas about patterns of behaviour which are likely to appear, and which may be a key to the presentation of computer results in terms of suitable dimensionless groups. And, fourth, it provides specific answers to some simple problems which are directly useful in the nontrivial exercise of checking-out the operation of numerical schemes of computation.

I have provided some problems for the reader to tackle (with answers) in connection with each chapter. These are intended primarily to reinforce the material in the text; most of them, particularly in the earlier chapters, involve relatively simple calculations.

I assume that the reader is familiar with the basic theory of structures as taught in first- and second-year university courses for engineering students. The level of mathematics which is required is usually achieved by engineering undergraduates.

I have attempted to be consistent in matters of notation and sign convention throughout the book; but there remain some clashes in various places. The more important ambiguities of notation are pointed out in the *list of symbols*, which follows this preface.

A word about references. I have already remarked that the literature on the subject of shell structures is very extensive. It is also daunting, and sometimes overwhelming. In these circumstances a book could easily be swamped by its references. I have therefore tried to keep the number of references as small as possible, consistent with an adequate underpinning of important ideas and a proper attribution of original work. I hope that I have given enough references to enable the reader to follow up any topic of particular interest. Many important early contributions are not readily available to the modern reader: in several cases I have referred to these articles indirectly, through papers or textbooks which themselves have ample references to the early work. I fear that in any case I have not done justice to the Russian literature. In appendix 11 I give some suggestions for further reading among papers etc. which are not all cited in the body of the book.

Many colleagues and students have discussed various aspects of shell structures with me in Cambridge and elsewhere at various times over a number of years. I am grateful for their help, and I wish to thank particularly Jim Croll, Wilhelm Flügge, Sidney Gill, Ian Goodall, Jim Greenwood, Jacques Heyman, John Hutchinson, Fred Leckie, Peter Lowe, Chris Morley, Alan Morris, Turan Onat, Andrew Palmer, Nadarajah Paskaran, Milija Pavlovic, David Payne, Frank Pitman, Daya Reddy, Steve Reid, Jeremy Robinson, Andrew Smith, Tibor Tarnai, Paul Taylor, Mike Thompson, John Wilson, Bill Wittrick, Lindsay Woodhead and Jim Woodhouse. And I am specially grateful to Aaron Klug, a fellow member of Peterhouse high table, for explaining Gaussian curvature to me. I also thank several distant colleagues who have generously answered my requests for help on particular points: Gerry Galletly, Norman Jones, Warner Koiter, Les Morley, John Morton and Geoffrey Warburton. I thank Peter Clarkson for helping with computations, Roger Denston for helping with experiments and Iradj Kani for checking the manuscript. Shirley Barry, Hazel Dunn, Janice Eagle, Rosemary Tullock, Sandra Valentine, Sue Venn and Sarah Wells did the typing and Dennis Halls traced the diagrams. I thank them all.

List of symbols

Symbols defined in the main list below are used more-or-less consistently throughout the book. Those listed by chapter are defined for the limited purposes of that chapter. The list is not complete, but the meaning of other symbols should be clear from the text. There are some unavoidable minor ambiguities of notation in various places; but I do not think that they will act as stumbling blocks to the reader.

a, a^* radius of shell
b circumferential half wavelength ($= \pi a/n$)
e eccentricity of effective meridian (figs. 11.12, 11.13)
f load per unit area
g (i) small change of Gaussian curvature
  (ii) gravitational acceleration
h effective thickness of shell ($= t/(1 - \nu^2)^{\frac{1}{2}}$) (equation (3.25))
k (i) foundation-spring stiffness (fig. 9.6)
  (ii) peak-stress attenuation factor (equations (3.38), (3.40))
l (i) length
  (ii) longitudinal half wavelength
n circumferential mode-number ($= \pi a/b$)
p pressure
q tangential surface traction
r radius (figs. 4.10, 6.9)
s arc-length coordinate
t thickness of shell
u, v, w components of displacement
x, y, z Cartesian coordinates
A area

List of symbols

- B (i) peak bending strain/peak stretching strain (equation (7.1))
  (ii) flexural stiffness of beam element ($= EI$)
- D flexural stiffness of shell element ($= Eth^2/12$)
- E Young's modulus of elasticity
- F (i) force per unit length (fig. 3.5a)
  (ii) shearing force in beam (figs. 9.6b, 10.16)
- G shear modulus of elasticity
- I second moment of area
- K Gaussian curvature
- L length
- M bending-stress resultant (fig. 2.2)
- N tangential-stress resultant (fig. 2.2)
- P force
- Q (i) normal shearing-stress resultant (fig. 2.2)
  (ii) shearing force in beam
- R principal radius
- T (i) tension
  (ii) thickness
  (iii) kinetic energy
  (iv) combined tangential-stress resultant (fig. 8.11b)
- W point load
- U strain energy
- Y dimensionless natural frequency (equation (17.37))
- α angle
- β solid angle (fig. 5.6)
- γ shearing-strain component
- δ deflection
- ϵ direct-strain component
- η dimensionless circumferential wavenumber ($= \lambda/b$); cf. ξ
- θ angular coordinate
- κ change of curvature
- λ standard length ($= \pi(ah)^{\frac{1}{2}}/3^{\frac{1}{4}}$); cf. η, ξ
- μ standard length ($= (4B/k)^{\frac{1}{4}}$ for beam, $(ah)^{\frac{1}{2}}/3^{\frac{1}{4}}$ for shell)
- ν Poisson's ratio (see h)
- ξ dimensionless longitudinal wavenumber ($= \lambda/l$); cf. η
- ρ (i) radius of curvature (equation (5.1))
  (ii) density
- σ direct-stress component
- τ shearing-stress component
- ϕ (i) meridional angle coordinate (figs. 4.10, 6.9)

List of symbols   xxi

           (ii) Airy stress function (equation (7.36))
ψ, χ small angle of rotation (figs. 3.5, 6.9)
ω angular frequency
Γ^2 differential operator (equation (8.19)); cf. ∇^4
Δ sagitta (appendix 8)
Λ dimensionless load
Σ stress-concentration factor
Φ load-sharing factor (equation (3.45))
Ω dimensionless length ($= Lh^{\frac{1}{2}}/a^{\frac{3}{2}}$)
∇^4 differential operator (equation (7.40)); cf. Γ^2

Chapter 3
f dimensionless form of F ($=\mu^3 F/2D$)
m dimensionless form of M (equation (3.29))
n dimensionless form of N (equation (3.29))
q dimensionless form of Q (equation (3.29))
ζ dimensionless length ($= l/\mu$)
ξ dimensionless coordinate ($= x/\mu$)
ψ dimensionless form of Ψ (equation (3.29))
Ψ small rotation ($= -\mathrm{d}w/\mathrm{d}x$)

Chapter 4
m slope of generators of hyperbolic shell ($= (a/-a^*)^{\frac{1}{2}}$)
β $= 1/(-aa^*)^{\frac{1}{2}}$
ζ dimensionless length ($= l/(aa^*)^{\frac{1}{2}}$)
ξ dimensionless coordinate ($= x/(aa^*)^{-\frac{1}{2}}$)

Chapter 5
$c, \bar{c}$ curvature, twist
θ, ϕ angle of rotation, twist (fig. 5.3)
ξ auxiliary angle on Mohr circle (fig. 5.4c)

Chapter 6
α_1, α_2 Lamé parameters
ξ_1, ξ_2 curvilinear coordinates (fig. 6.7a)
ζ as in chapter 4

Chapter 7
F panel flexibility factor ($= w/p$)
η load per unit length on beam (fig. 7.5)

xxii *List of symbols*

Chapter 8
U, V, u, v auxiliary variables
    α, β roots of auxiliary equation (equation (8.32))
        ζ stiffness of S-surface/stiffness of B-surface (equation (8.16))

Chapter 9
η as in chapter 7 (fig. 9.6)
Φ stiffness of rotation-spring (fig. 9.14)

Chapter 10
B, H width, rise of shell (figs. 10.3, 10.13)
   J dummy force (fig. 10.7)
   α area ratio (equation (10.63))
   β shell-geometry parameter $(= B^2/L(ah)^{\frac{1}{2}})$
   η height ratio (equation (10.63))
   μ slope (fig. 10.12)
   $\xi = \mu_2/\mu_1$
   ψ angle subtended by shell (fig. 10.13)

Chapter 11
H horizontal shearing-stress resultant (fig. 11.3c)
U total horizontal shearing force $(= Hr)$
ζ through-thickness coordinate (equation (11.49))
ξ imperfection in form of meridian (fig. 11.7)

Chapter 12
b radius of convolution (fig. 12.1)
e strain in comparison tube
F flexibility factor

Chapter 13
b radius of bend (fig. 13.1a)
f flexibility factor
H separation of flanges (fig. 13.4)
L equivalent extra length (equation (13.51))
M bending moment
ζ flexibility factor for 'cos 2ϕ' mode (equation (13.57))
Γ pipe-bend geometry factor $(= a^2/bh)$

List of symbols xxiii

Chapter 14

- r radial coordinate in ξ, η plane
- Z function of modeform, defined in equation (14.34)
- N number of buckling modes occurring at almost the same load
- α, β stretching, bending stiffness factors for stiffened shell (section 14.10)
- $\alpha = N_x/N_y$ (equation (14.50))
- β parameter defined in equation (14.92)
- $\zeta = l/b = \eta/\xi = \tan \phi$
- ρ dimensionless radial coordinate (equation (14.55))
- ϕ (i) $= \arctan \eta/\xi$
  (ii) torsion function (equation (14.121))
- ψ load per unit length acting on beam (same as η in chapter 9)
- Φ torsion function (equation (14.122))
- $\Psi = \psi - \kappa$ (equation (14.88))

Chapter 15

- X, Y dimensionless Cartesian coordinates (equation (15.21))
- α, β dimensionless imperfection amplitudes (equations (15.40), (15.46))
- η, ν coefficients in elastic law for nonlinear spring (equation (15.2))
- θ rotation of rod (fig. 15.5)
- μ angular imperfection (fig. 15.5)
- Γ, Δ dimensionless imperfection amplitudes (equation (15.21))

Chapter 16

- c dimensionless form of C (equation (16.12))
- C curvature of centre-line of tube
- H initial separation of flanges (fig. 16.3)
- ζ cross-sectional shape-change parameter (equations (16.1), (16.26))
- η component of displacement normal to neutral axis (fig. 16.5)
- Γ cross-sectional shape-change parameter (equation (16.82))

Chapter 17

- m mass per unit length of beam
- N wavenumber

Chapter 18

- A radius of spherical shell

$\dot{\alpha}$ rotation-rate (fig. 18.25)
γ thickness ratio (fig. 18.21)
η eccentricity factor (fig. 18.21)
ρ dimensionless hole size (equation (18.42))
Δ deflection

1

Introduction

1.1 General remarks on shell structures

This book is about thin shell structures. The word *shell* is an old one and is commonly used to describe the hard coverings of eggs, crustacea, tortoises, etc. The dictionary says that the word shell is derived from *scale*, as in fish-scale; but to us now there is a clear difference between the tough but flexible scaly covering of a fish and the tough but rigid shell of, say, a turtle.

In this book we shall be concerned with man-made shell structures as used in various branches of engineering. There are many interesting aspects of the use of shells in engineering, but one alone stands out as being of paramount importance: it is the *structural* aspect, and it will form the subject of this book.

Now the *theory of structures* tends to deal with a class of idealised or rarified structures, stripped of many of the features which make them recognisable as useful objects in engineering. Thus a *beam* is often represented as a line endowed with certain mechanical properties, irrespective of whether it is a large bridge, an aircraft wing or a flat spring inside a weighing machine. In a similar way, the theory of shell structures deals, for example, with 'the cylindrical shell' as a single entity: it is a cylindrical surface endowed with certain mechanical properties. This treatment is the same whether the actual structure under consideration is a gas-transmission pipeline, a grain-storage silo or a steam-raising boiler.

Before we enter this realm of theory, in which shells are classified by their geometry (cylindrical, spherical, etc.) rather than their function, it is desirable to give a glimpse of the wide range of applications of shell structures in engineering practice. Indeed, a list of familiar examples will be useful in enabling us to pick out some structural features in a qualitative way in order to provide an introduction to the main body of theoretical work.

2 Introduction

It is instructive to assemble a list of applications from a historical point of view, and to take as a connecting theme for a sequence of brief sketches the way in which the introduction of the *thin shell* as a structural form has made an important contribution to the development of several different branches of engineering. The following is such a list. It is by no means complete. For a more comprehensive treatment, see Sechler (1974).

Architecture and building. The development of masonry domes and vaults in the Middle Ages made possible the construction of more spacious buildings; and in recent times the development of reinforced concrete has stimulated interest in the use of thin shells for roofing purposes.

Power and chemical engineering. The development of steam power during the Industrial Revolution depended to some extent on the construction of suitable boilers. These thin shells were constructed from plates suitably formed and joined by riveting. In recent times the use of welding in pressure-vessel construction has led to much more efficient designs. Pressure-vessels and associated pipework are key components in thermal and nuclear power plant, and in all branches of the chemical and petroleum industries.

Structural engineering. An important problem in the early development of steel for structural purposes was to design compression members against buckling. A striking advance was the use of tubular members in the construction of the Forth railway bridge in 1889: steel plates were riveted together to form reinforced tubes as large as 12 feet in diameter, and having a radius/thickness ratio of between 60 and 180, according to function.

Vehicle body structures. The construction of vehicle bodies in the early days of road transport involved a system of structural ribs and non-structural panelling or sheeting. The modern form of vehicle construction, in which the skin plays an important structural part, followed the introduction of sheet-metal components, preformed into thin doubly-curved shells by large power presses, and firmly connected to each other by welds along the boundaries.

The use of the curved skin of vehicles as a load-bearing member has similarly revolutionised the construction of railway carriages and aircraft. In the construction of all kinds of spacecraft the idea of a thin but strong skin has been used from the beginning.

Boat construction. The introduction of fibreglass and similar plastic materials has revolutionised the construction of small and medium-sized boats, since the skin of the hull can be used as a strong, stiff, structural shell.

Miscellaneous. Other examples of the impact of shell construction on technology include the development of large economical natural-draught water-cooling towers for thermal power stations, using thin reinforced-concrete shells; and the development of various kinds of economical silos for the storage of grain, etc., by the use of thin steel shells.

This list can easily be extended to include mediaeval armour, cartridge shells, arch dams, etc.

1.1.1 Continuity and curvature

The essential ingredients of a shell structure in all of the examples above are *continuity* and *curvature*. Thus, a fibreglass hull of a boat is continuous in a way in which the overlapping planks of clinker construction are not. A pressure-vessel must obviously be continuous to contain a fluid at pressure, although the physical components may be joined to each other by riveting, bolting or welding. On the other hand, an ancient masonry dome or vault is not obviously continuous, in the sense that it may be composed of separate stone sub-units or voussoirs not necessarily cemented to each other. But in general, domes are in a state of compression throughout, and the sub-units are thus held in compressive contact with each other. The important point here is that shells are *structurally continuous* in the sense that they can transmit forces in a number of different directions in the surface of the shell, as required. These structures have a quite different mode of action from *skeletal structures*, of which simple examples are the braced frame of an electricity supply pylon, and a tree. These structures are only capable of transmitting forces along their discrete structural members.

We shall return to the question of *curvature*, and its effect on the strength and stiffness of shell structures in section 1.2.4.

1.1.2 The empirical approach

Many of the objects in the preceding list were constructed long before there was anything like a textbook on the subject of shell structures. The early engineers had a strongly empirical outlook; they could see the advantages of shell construction from simple small-scale models, and clearly understood the practical advantages of doing 'overload tests' on prototypes or scale models. Much the same brand of empiricism is practised successfully today in the design of motor vehicles, where the geometry of the structure is so complicated as to defy even simple description, let alone calculation. But in other areas of engineering, where precision is needed in the interests of economical design, and where the geometry is more straightforward, the theory of shell structures is an important design tool.

1.1.3 Closed and open shells

Before we begin to describe the main body of the theory it is useful to discuss qualitatively an important practical point which must have been clear to those who pioneered the use of shell structures in the various branches of engineering.

Anyone who has built children's toys from thick paper or thin card will be familiar with the striking fact that a *closed box* is rigid, while an *open box* is easily deformable. Similarly, a biscuit tin or a chocolate box with the lid open can easily be twisted, yet it is effectively rigid when the lid is closed. The same sort of thing applies to a tin can, which may be squashed far more easily after one end has been removed. Again, it is noticeable that a boiled egg will not normally fit snugly into a rigid egg-cup until the top of the shell has been removed: the closed egg-shell is so rigid that small aberrations from circularity are noticeable, but it becomes flexible enough to adapt to the shape of the egg-cup as soon as an opening has been made.

There seems to be a *principle* here that *closed surfaces are rigid.* This principle is used in many areas of engineering construction. For example, the deck of a ship is not merely a horizontal surface for walking on: it also closes the hull, making a box-like structure. It is easy to think of many other examples of this form of construction, including aircraft wings, suspension bridge roadway girders, etc.

Conversely, a boat without a deck – such as a small fishing or rowing boat – has little rigidity by virtue of its form, and must rely on the provision of ribs, etc. for what little rigidity it has.

In practice, of course, it is not usually possible to make completely closed structural boxes. In a ship, for example, there will be various cutouts in the deck for hatches, stairways, etc. It is sometimes possible to close such openings with doors and hatch-covers which provide structural continuity; but this is often not feasible, and compromise solutions must be adopted. The usual plan is to reinforce the edge of the hole in such a way as to 'compensate', to a greater or lesser degree, for the presence of the hole. The amount of reinforcement which is required depends on the size of the hole, and to what extent the presence of the hole makes the structure an 'open' one. Large openings are of course essential in some forms of construction, such as cooling towers. A more extreme example is provided by shell roofs in general. Here the shell is usually very open, being merely a 'cap' of a shell, and the provision of adequate edge ribs, together with suitable supports, is of crucial importance. A main aim in the design of shell roofs is to eliminate those aspects of behaviour which spring from the 'open' nature of the shell.

It will be obvious from the foregoing discussion that although the ideas of

1.1 General remarks on shell structures

'closed' and 'open' shells respectively are fairly clear, it is difficult to quantify intermediate cases into which, of course, the majority of actual shell structures fall. While the effect of a small cutout on the overall rigidity of a shell structure may be trivial, the effect of a large cutout can be serious. The nub of the problem is to quantify the ideas of 'small' and 'large' in this context. If only there were a simple way of doing this, there would be little need for textbooks on the subject. In fact, the problem is difficult, for it involves the interaction between 'global' and 'local' effects; and it is largely for this reason that the subject of shell structures generally is a difficult one.

1.1.4 A simple geometrical approach

The notion that a closed surface is rigid is well known in the field of pure Euclidian geometry. There is a theorem, due to Cauchy, which states that a convex polyhedron is rigid. The concept of rigidity is, of course, hedged around with suitable restrictions (see appendix 9) but will be an obvious one to anyone who has made cardboard cutout models of polyhedra. It is significant that the word *convex* appears in the theorem. Although it is possible to demonstrate by means of simple examples that some non-convex polyhedra (i.e. polyhedra with regions of non-convexity) are rigid, it is also possible to demonstrate special cases of non-convex polyhedra which are not rigid, and are capable of undergoing infinitesimal distortions at least (see appendix 9). This is a difficult area of pure mathematics. For our present purposes we note that convexity guarantees rigidity (in the present context), while non-convexity *may* produce deformability. We shall return to these remarks later.

1.1.5 A disadvantage of rigidity

While rigidity and strength are in many cases desirable attributes of shell structures, there are some important difficulties which can occur precisely on account of unavoidable rigidity. An example of this occurs frequently in chemical plant where two large pressure-vessels, firmly mounted on separate foundations, are connected by a length of straight pipe. Thermal expansion of the vessels can only be accommodated without distortion if the pipe contracts in length: if it also expands thermally very large forces can be set up on account of the 'rigidity' of the vessels. In cases like this it is often convenient to accommodate expansion by a device such as a *bellows* unit. Alternatively, when the interconnecting pipework has bends, it is sometimes possible to make use of the fact that the bends can be relatively flexible. In the case of bellows and bends the flexibility is to a large extent related to the geometry of the respective surfaces. It is perhaps significant that both are *non-convex*; nevertheless this of itself does not constitute a proper explanation of their flexibility.

1.1.6 *Catastrophic failure of thin-shell structures*

The idea that closed structures are rigid, or strong, is of great practical importance. But it should not be used in ignorance of a second broad principle which is also well known to those who have made paper models. It may be stated thus: *efficient structures may fail catastrophically.* Here I use the term 'efficient' to describe the consequences of employing the first principle. By designing a structure as a closed box rather than an open one we may be able to use thinner sheet material, and hence produce an economical, or efficient, design.

Now one of the main difficulties in the design of thin-walled structures which are to be loaded in *compression* is that such structures are prone to *buckling* of a particularly unstable kind. Indeed this is why a large proportion of current research work on shells is concerned with buckling problems. This is not the place to go into the specialised concepts and terminology of buckling theory (but see chapters 14 and 15): we may simply appeal to the well-known experience of the crumpling of thin-walled tubes under load, with an irretrievable loss of the initial geometry which previously had seemed so efficient. It is instructive to note that crumpling of a thin, convex shell is accompanied by the introduction of non-convexity to the surface: it is partly on this account that the post-buckled rigidity of the shell is poor.

1.2 Theory of shell structures as a branch of structural mechanics

The foregoing qualitative description of some general properties of closed and open structural surfaces provides a useful introduction to important ideas. But we shall obviously need some much more specific tools when we are faced with questions about the detailed design of any of the structures in question. These tools are provided by the branch of mechanics known as *structural mechanics.* This subject works within the framework of what may be called scientific method: it is concerned with the rational explanation of phenomena by means of suitable conceptual models. In this connection we should note that 'natural science' is concerned with the phenomena of nature, whereas engineering is concerned mainly with the phenomena of artefacts, i.e. man-made systems. If we can understand the phenomena we shall be in a strong position to make rational designs. Experimentation plays a large part in scientific method.

An important aspect of structural mechanics is the explanation of structural collapses when they occur accidentally. The development of advanced structural mechanics is signposted by a number of spectacular failures which have become the subject of urgent enquiry. Although many failures result from human in-

1.2 Shell structures and structural mechanics

competence there have been some striking examples in which essentially new or unforeseen phenomena have appeared.†

1.2.1 The role of mathematics

It will be obvious that the development of the theory of shell structures will be to a large extent mathematical in character: the pages of much of the literature are covered with equations. Precisely for this reason it is necessary to emphasise that the role of the mathematics in the theory is exactly the same as its role in other branches of engineering and science. The basic strategem in mechanics is to apply mathematics not to the structure (or whatever) itself but to a *conceptual model* of the structure. In any given case the investigator must begin by exercising judgement in the choice of a model. The general aim is to choose the simplest model that is adequate for the task in hand. For example, in our previous discussion of the rigidity of closed surfaces we were thinking in purely geometrical terms about a surface which is free to bend but not to extend. The idea is adequate to account for the observed rigidity of closed surfaces and the non-rigidity of open ones; but it gives no means of calculating, for example, how *stiff* is a given open shell.

If we wish to answer questions of this sort we shall certainly need to know about the mechanical stiffness of the material of which the actual shell is made. We shall also need to take the conceptual model out of the realm of geometry into that of mechanics. Here, characteristically, we shall immediately be confronted by a choice: should we use a simple model in which the surface is inextensional in stretching but has stiffness in bending, or a more complex one in which there is finite stiffness in both actions?

From this simple example, we can see that the same physical structure may be studied by means of quite different conceptual models. Which particular model we elect to use depends on the kind of question which we are asking. In general our choice will depend to some extent on intuition. Intuition enters at several points in the above description and indeed in some unstated ideas which we have used, namely the notion of treating a physical structure as a surface; and the separation of mechanical effects into the two classes of bending and stretching.

---

† Examples include the failure of the first Quebec Bridge in 1908 by buckling; the collapse of the Tacoma Narrows suspension bridge in 1940 by wind-induced torsional oscillation; and the progressive collapse of the system-built Ronan Point block of flats in 1968.

1.2.2 *Some difficulties in the testing of conceptual models*

A constant problem in all of this sort of work is how to determine whether the analysis based on a particular conceptual model is adequate or satisfactory or successful. An obvious proper answer is that the model is good if the results obtained by its use agree well with the results of experimental investigations. This is an important point, and is indeed one of the cornerstones of the scientific method. But there are difficulties which may prove to be a snare for the unwary, and which we therefore need to be clear about. Let us take as an illustrative example the basic model of a shell which we shall use throughout the book. It takes as its first step the replacement of a shell by a *surface*. Consequently, it is clear that we shall not be in a position to discuss directly questions about the details of stress distribution around such features as fillet welds at the junction between two shells. We shall of course deal with junctions, but only junctions between surfaces; and much of the detail of the stress distribution will be suppressed precisely in the step of shrinking the three-dimensional physical shell onto a zero-thickness surface. Consequently our theory can neither be confirmed nor disproved by the results of strain-gauge measurements on a particular fillet weld.

It is, of course, pertinent to ask here what is the point of a model which at a stroke destroys a lot of detail. The answer is that the 'surface' theory of shell structures is extremely simple in comparison with other theories which do not idealise the shell as a surface. There are undoubtedly many important practical problems where the theory presented in this book will be inadequate. In these cases resort should be made to direct experimentation and numerical stress analysis using three-dimensional finite elements. However, the regions in which the simple theory is inadequate are all highly localised, and there are very many practical problems in which, for one reason or another, the local details are either unimportant or else can be treated more or less in isolation.

All of this is closely analogous to the classical methods for analysing beam and frame structures. Here the problems of analysis are enormously simplified if the beams are regarded as a set of lines endowed with certain mechanical properties such as bending stiffness and strength. We do calculations about the distribution of bending moment within the arrangement; but we are aware that when it comes to the design of the details of the joints connecting the members we have to consider another set of 'local' problems. Similar questions arise where a concentrated load is applied to the top flange of a rolled steel I-beam: it may be necessary to add a reinforcing web to the beam, but the necessity for this is in no way revealed by our simple beam theory. This does not condemn the simple beam theory: but we must clearly be aware of its in-built limitations. To use an old-fashioned but useful distinction, the 'theory

1.2 Shell structures and structural mechanics

of structures' must be supplemented by analysis of detailed features within the framework of 'strength of materials'.

In this section I have tried to set the foundations of the subject firmly in the context of structural mechanics, which is a kind of scientific method. Some students may regard the discussion of conceptual models as altogether too sophisticated, and irrelevant to the needs of practical engineers. I have a good deal of sympathy with this view, but I think that it is erroneous. An engineer who has the instinct to look for fresh simple conceptual models when faced with difficult new problems is more likely to solve problems than his less adventurous colleagues. The advantage of a conceptual model devised for a particular situation is that it leads to relatively simple mathematical manipulations and clear conclusions which can be put to the test. Less specific models, which have not been shorn of unnecessary features, tend to produce lengthy calculations which all too often are not brought to a conclusion, and consequently are of relatively little practical use.

1.2.3 The classical model for shell structures

Before we begin to set out the details of our methods of analysis it is appropriate to give a brief sketch of the way in which the subject will develop. We have already stated that the first step in the analysis is to replace the shell by a surface which is endowed locally with certain mechanical properties. Some difficulties associated with the specification of these properties will be dealt with in the next chapter, but it is sufficient for present purposes to note that the process of transferring mechanical properties onto a line or surface is identical in principle with what is done in the elementary theory of beams, trusses and plates.

We have as a result a set of mechanical properties expressed (in the case of an elastic material) in the form of a generalised Hooke's law relating the deformation of a typical element to the stresses applied to it. The subsequent steps in the theory of shells are also the same as in the theory of these simpler structures: we seek *equilibrium equations* relating the stress resultants in the structure to the applied external forces, and *compatibility equations* expressing the geometrical connection between the strain in an element and the displacement of points on the structure. The equations of equilibrium are a special case of Newton's laws of motion which apply when dynamic effects may be disregarded, i.e. when a problem may be regarded as 'quasi-static'. Where dynamic effects are important (e.g. in vibration problems) it is of course necessary to use the equations of motion in place of the equilibrium equations: see chapter 17. In the present section I use the term equilibrium equation to include the equations of motion, as appropriate.

In the classical theory which we shall be using throughout the book – except in the chapters on buckling and one other section – we shall express the equilibrium and compatibility equations in terms of the initial geometry of the structure. This is of course a well-established practice in all branches of structural mechanics, but it will restrict our analysis to relatively small displacements of the structure. We shall discuss aspects of this limitation at various points in the book.

In principle these three sets of equations, together with appropriate boundary conditions, constitute the mathematical aspect of the problem, just as they do in the simpler branches of structural mechanics. What makes the shell problem in general more complex is the fact that the equations have to be set up with respect to a surface in three-dimensional Euclidian space: in the theory of flat plates, by comparison, the three sets of equations have to be set up merely for a plane surface. It would be fair to say that the geometry of the shell surface dominates the structural problem.

1.2.4 *The dominating effect of the geometry of the shell*

It is of interest to note that in the mechanics of deformable *solids*, where again the same three sets of equations appear in an appropriate form, the character of the resulting mathematical problem is sometimes determined by the form of the material properties. Thus, the principal difference between the theories of *elasticity* and *plasticity* is that while in elasticity the material properties enter as a generalised linear Hooke's law, in (perfect) plasticity we have instead a yield condition and an associated flow rule (see, e.g. Prager, 1959; Symonds, 1962). In consequence, although the governing differential equations in elasticity are always elliptic, in plasticity they are sometimes hyperbolic, and then demand quite different approaches to the solution (Hill, 1950). In the theory of shells it seems that the governing equations are never rendered hyperbolic by the material properties alone, but they can be hyperbolic in some cases as a consequence of *geometrical* properties of the shell surface. In this sense the problem of shell structures may be said to be dominated by the geometry of the surface of the shell. It is largely for this reason that I have written most of the book in terms of a simple elastic material, with a single chapter on plastic analysis of shell structures at the end. Although there are obviously major differences between the two theories, the main difficulties which have to be overcome in the plastic theory of shells are rather similar to those involved in the elastic theory.

1.2.5 *Interaction of bending and stretching effects*

The mechanical properties of a shell element describe its resistance to deformation in terms of separable stretching and bending effects. Loads

1.2 Shell structures and structural mechanics

which are applied to the shell are carried in general by a combination of bending and stretching actions, which vary over the surface. One of the leading difficulties in the theory of shell structures is to find a relatively simple way of describing the interaction between the two effects. This aspect of the theory has been troublesome from the beginning. Rayleigh (1881; 1894 Chapter 10A) argued that the deformations of a thin hemispherical bowl would be predominantly inextensional, and accordingly he developed a simple special method of analysis which took into account only the strain energy of bending in the shell. Love (1888), on the other hand, argued that for *thin* shells stretching rather than bending was the dominant effect. At this time Love had not grasped the strong contrast between the behaviour of open and closed shells. This controversy was resolved by Lamb (1890) and Basset (1890), who solved Love's general equations in the simple case of a cylindrical shell and demonstrated the possibility of a relatively narrow *boundary layer* in which there was a rapid spatial transition between bending and stretching effects, with the width of the boundary layer being determined by the interaction between these effects.

In this book the general problem of the interaction between bending and stretching effects has been brought out in physical terms by the introduction of a model of the shell as two coincident surfaces (chapter 8). One of these carries the stretching effects while the other sustains the bending effects. It is possible to describe the interaction between the surfaces in terms of a pair of simple force and displacement variables. As we shall see, this idea unlocks a wide range of relatively simple problems which can be solved and the results expressed in simple, dimensionless terms.

2

Generalised Hooke's law for an element of a shell

2.1 Introduction

We are now ready to establish the mechanical properties of a typical small element of a thin uniform elastic shell. We have already decided to replace the shell itself by a model consisting of a surface, and we must now furnish this surface with appropriate mechanical properties.

This task is equivalent to the well-known piece of work in the classical theory of beams in which the beam is shown to be equivalent to a 'line' endowed with a flexural stiffness EI, where E is Young's modulus of elasticity of the material and I is a geometrical property of the cross-section. But the present task is more complex than the corresponding one for the beam, in two distinct ways. First, an element of a shell is *two dimensional*, whereas an element of a beam is one dimensional. Second, an element of shell is in general *curved* rather than flat.

A basic idea, which was proposed in the early days of shell theory, is that in relation to the specification of the *mechanical properties* of an element of a shell it is legitimate to proceed as if the element were flat, and not curved. Legitimate, that is, as a 'first approximation'. Much work has been done by many authors on the question of the degree of inaccuracy which is introduced by this idea: see, for example, Novozhilov (1964), Naghdi (1963). We shall not attempt to justify this idea formally. Instead we shall first give a derivation appropriate to a flat element essentially 'borrowed' from the theory of plates, and then argue by means of order-of-magnitude considerations that this is a valid procedure as a first approximation provided the thickness of the shell is sufficiently small in comparison with a local radius of curvature of the surface. For a definitive discussion of the errors inherent in 'first approximation' theories of shells, see Koiter (1960).

It is important to realise that for some *other* purposes it is essential to consider the element as being curved. We must, for example, have curved elements for the purpose of setting up the basic equilibrium equations: otherwise we shall rule out one of the most important effects in shell structures, namely, that tension in a curved member demands a transverse pressure for the satisfaction of the equilibrium conditions. It may thus seem paradoxical that we propose to regard a characteristic small element as being flat or curved, according to circumstances. There is in fact no basic inconsistency: in mathematical terms it turns out that in the limit as the size of the element tends to zero, some quantities involving curvature remain finite while others become negligibly small.

The plan for the chapter is to examine first the question of the magnitude of the *shearing* distortion of beam and plate elements. This leads to a key concept known as *Kirchhoff's hypothesis*, which we shall use in the derivation of the mechanical properties of a flat-plate element. Then we consider two ways in which the curvature of a shell element could conceivably render the flat-plate idealisation unsatisfactory; and in each case we establish order-of-magnitude criteria for the circumstances in which the flat-plate idealisation is adequate. Lastly we consider the order of magnitude of 'through-thickness' stresses in comparison with the bending stresses in the element.

Throughout the chapter we seek to relate the distortion of an element to the state of stress which it sustains. However, distortion may also occur in other ways. For example, a temperature gradient through the thickness of an element of a shell must be expected to alter the curvature on account of differential thermal expansion; but this does not necessarily induce a state of stress. Straining of this sort is not considered in the present chapter. Thermal-strain effects are important in some circumstances. For most purposes it is satisfactory to make *ad hoc* changes to the equations, as appropriate: see, for example, problems 3.22, 3.23 and 8.9.

2.2   Shear distortion and Kirchhoff's hypothesis

The problem of shear distortion is well known in the classical theory of beams, and it is essentially the same problem here, in a slightly more complicated format. Fig. 2.1*a* shows a cantilever beam of solid rectangular cross-section which is loaded transversely at its tip. Onto the surface of the beam in its original unloaded straight configuration has been inscribed a rectangular grid, and it is the distortion of this grid which is shown in the diagram: the results come from standard linear-elasticity theory. Fig. 2.1*b* shows an idealisation of the beam as a line, initially straight, which is deformed by the action of a transverse load. It is obvious that the full two-dimensional solution in fig. 2.1*a* cannot be condensed into a line, as in fig. 2.1*b*, without some loss of detail. Nevertheless, the mathematical advantages of a one-dimensional theory of beams are so strong that the sacrifice of some detail may be worthwhile in many cases.

It was realised by Bernoulli and Euler that the main effect in the distortion of slender beams was a change in curvature in response to the application of bending moment, and this provided the simple law

Change of curvature = Constant x (Bending moment).

This equation implies that the effect of shearing force on the distortion of the beam is negligible. The calculation of the constant in the above equation in terms of the dimensions of the cross-section etc., proved historically to be a difficult problem. The crucial step, which is now generally taken as the starting point of the analysis, was made in 1850 by Kirchhoff (see, e.g., Timoshenko, 1953, §55) in his study of the elastic behaviour of plates. In terms of the bending of a beam, Kirchhoff's idea was that the cross-sections of the original beam, which are of course plane and normal to the axis of the beam, *remain plane and normal to the axis* when the beam distorts. This is clearly not always

Fig. 2.1. Various aspects of beam theory. (*a*) Distortion of a grid inscribed on the surface of a loaded cantilever. (*b*) Conventional idealisation of a beam as a flexible line. (*c*) Distortion of an element of a uniform beam due to pure bending. (*d*) Shearing distortion of the same element.

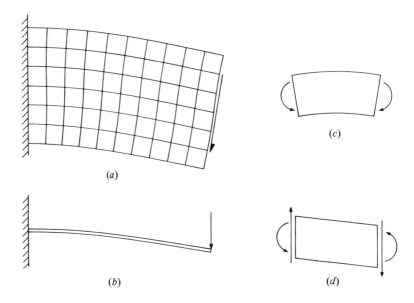

exactly true, as may be seen in fig. 2.1a and in experimental studies of beams loaded in shear. It is an *idealisation*, which must be justified by its fruits. In physical terms the idealisation involves suppression of shearing strains (which are there on account of the shear force and Hooke's law) which distort the cross-section into a gentle S-shape and give it a net rotation with respect to the centre-line. In other words, the beam, idealised in this way as a flexible line, is only allowed to distort by curving its centre-line in response to applied bending moment, and is not allowed to distort in shear in response to applied shear force. These ideas are shown pictorially in figs. 2.1c and d.

A consequence of this simplification is that the behaviour of the idealised beam *can* be expressed completely in the form used by Bernoulli and Euler: and moreover the constant of proportionality can easily be shown to be equal to EI (see p. 11).

The development of the 'simple elastic theory' of beams is of course well known. It gives results for the transverse displacement of beams which are in good agreement with those of the more complete theory, except in the case of *short* beams in which the effects of 'shear' distortion are not negligibly small in comparison with the 'bending' distortion. It is perhaps better to state this the other way round: for *long* beams the transverse displacements due to *bending* far exceed those due to *shearing*. The terms 'long' and 'short' as used here do not have self-evident meanings. For uniform beams of solid cross-section, bending effects predominate provided the length of the beam is greater than the depth. For typical I-beams in the standard rolled forms, bending effects predominate provided the length is greater than, say, twice the depth. But for some beams of 'sandwich' construction, consisting of surface layers separated by a 'core' of material having a much lower modulus of elasticity (e.g. plastic foam, 'honeycomb', etc.) bending effects predominate only if the length is many times the depth of the beam: that is, shearing effects ordinarily predominate for beams of this sort of construction. It is a straightforward matter to determine the distribution of shearing stress in the cross-section of a beam due to a transverse shearing force: see, e.g. Crandall, Dahl & Lardner (1972, §7.6). For an I-beam or a 'sandwich' beam the shearing force is carried almost entirely by practically uniform shearing stress in the web or core respectively; but for a uniform beam of rectangular cross-section the peak value of shearing stress is equal to 1.5 times the mean value.

Although these remarks have been expressed in the context of beams, we shall regard them as being relevant to shell structures also. In other words we shall express the general idea that deformations of shells due to transverse shear are unimportant by means of *Kirchhoff's hypothesis*. (This is sometimes known as the Kirchhoff-Love hypothesis, since Love (1888) was the first to apply it to curved shells.) We shall describe this hypothesis explicitly in section 2.3.2; it will enable us to describe the state of strain throughout the element in terms of the strains and changes of curvature at the mid-surface of the element. 'Hypothesis' is of course an apt term here, for it carries an implication that at the end of a piece of analysis a check or test on the validity of the hypothesis or idealisation should be carried out. We shall in fact make tests of this sort – involving order-of-magnitude calculations of various quantities – in the following chapters. In general we cannot make tests of this sort before we have solved specific problems.

Let us now interpret the above ideas about beams in the context of shells. Most shells are uniform in the sense that the material properties do not vary through the thickness, and it will therefore be reasonable to neglect transverse shear distortion, except perhaps in small regions close to points of application of localised loads. On the other hand there will not, *prima facie*, be a case for using the idealisation in relation to shells of sandwich construction of the type described: here a different idealisation is called for.

There are several common kinds of shell construction which may be suspected of being intermediate between 'uniform' and 'sandwich'. Reinforced concrete is normally regarded as sufficiently uniform for the Kirchhoff idealisation to be used. It is true that the moduli of the constituents (steel and concrete) differ by an order of magnitude, but the sandwich construction described above uses materials having elastic moduli differing by, say, four or five orders of magnitude.

It should be noted here that the 'sandwich-shell' idealisation which is sometimes used in the plastic theory of shell structures is of a different kind from that being described here. It is introduced in order to simplify certain calculations, but is intended to model uniform shells. Accordingly the core material is taken as being *rigid* in shear: see section 18.7.2.

Finally, we should add that the whole of the above discussion has presupposed that a typical short section of a beam is loaded both in bending and in shear. It is possible to arrange special examples in which there is zero shearing force, i.e. 'pure bending'. In this case, the cross-sections of the beam remain plane by virtue of symmetry (provided certain conditions are fulfilled: see problem 2.1), and the difficulty disappears. In practice, of course, it is unusual for shearing force to be absent. Nevertheless, one acceptable physical interpretation of the Kirchhoff hypothesis is that the elastic distortion of a small element is exactly as it would be if the local bending moment were a *pure* bending moment.

2.3 Generalised Hooke's law for a simple flat shell element

Here we consider the elastic properties of a typical small element of a shell. The element has thickness t and is made of an isotropic linear-elastic material having Young's modulus E and Poisson's ratio ν. The element is solid and uniform: there are no stiffening ribs, etc., attached to it. We have argued in the preceding section that it will be sufficient to analyse the elastic properties of a small flat element, and we take advantage of this fact in locating the centre of an x, y, z coordinate system at the centre of the element, with the z-axis normal to the natural surfaces of the element. Coordinates x, y, z define the position of a point in the element in the initial unstrained configuration.

2.3.1 Stress resultants

We shall describe the state of stress in a typical element by means of the *stress resultants* which it sustains. The basic idea here is the *concept of stress*, which is of course fundamental to the mechanics of all deformable bodies. The student may well have met the idea first with respect to simple structures in the form of tension, shear force and bending moment in a simple slender prismatic member, and then moved on to the idea of the tensor of stress in a continuum. The fundamental idea is the same in all of these cases. Basically, we are interested in how the structure (truss, beam, plate, etc.) carries some applied load. It does this by setting up internal forces in the material (which can in particular cause failure if they become excessive in magnitude). Our general tool for investigating these internal forces is to *cut* the structure into pieces and pretend that the state of stress existing in the structure is preserved, so that the forces formerly transmitted across an imaginary cut inside the structure are now balanced by external forces, i.e., forces external to the cut. For example, when a loaded beam is imagined to be cut into two parts at a certain cross-sectional plane, we must supply to each of the freshly-cut faces resultant forces equal to those which were previously transmitted across the plane. It is of course convenient to resolve these forces into normal and tangential components, which we call *tension* and *shear force* respectively. In general there will also be a couple, which may also be resolved into components about axes normal and tangential to the plane, which we shall call *twisting moment* and *bending moment* respectively.

Exactly the same idea applies when we cut out a small rectangular prism by means of six cuts from the interior of a loaded three-dimensional body. We can resolve the forces acting on the faces of the element in the normal and tangential directions. In this case we may regard the dimensions of the prism as vanishingly small in all directions, and we find, by arguments of elementary calculus, that in the limit as the size of the element tends to zero, the couples on the element become vanishingly small. The vanishing of couples on the faces of an infinitesimal element is in contrast to the case of a cut in a beam, cited above: the crucial distinction is that the cut face of a beam is *finite* in depth, and on that account there must be, in general, bending and twisting moments across it.

It is not difficult to extend the same ideas to a small (plane) element cut from a loaded shell. The labelling of the stress resultants is shown in relation to the chosen coordinate system in fig. 2.2. For the sake of clarity there are three diagrams showing the same element loaded by three sub-families of stress resultants. In one respect these diagrams are misleading. They represent a small element in the surface, in fact an infinitesimal element. The dimensions of such an element, dx, dy, are small in general, and if we think of them as being vanishingly small they will in particular be much smaller than the thickness of the shell. In fig. 2.2, however, the x and y dimensions of the element are shown as being somewhat larger than the thickness. The diagrams are *conventional* in this sense. Perhaps it is best to think of them as having been drawn at a late stage in the process of idealising the shell as a surface by steadily reducing its thickness to zero. It is important to realise that in these diagrams the cuts

Fig. 2.2. Stress resultants in a small element of a flat plate. (*a*) Tangential-stress resultants. (*b*) Out-of-plane shearing-stress resultants. (*c*) Bending- and twisting-stress resultants. (*d*) Coordinate system: the xy-plane is at the centre-surface of the element. The dimensions of the element in the x, y directions are dx, dy for some purposes and $2a$, $2b$ for others. Directions x, y, z are orthogonal.

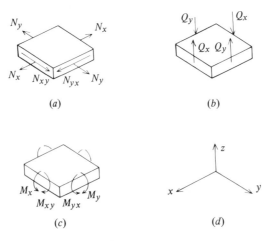

2.3 Generalised Hooke's law for a simple flat shell element

which isolate the element from the shell go through the *entire thickness* of the shell, even though this is shown, conventionally, at a much reduced scale.

In particular, the stress resultants include bending moments and twisting moments, just as they do in the case of a beam. Here, however, it is sensible to express the bending and twisting stress resultants in terms of 'moment per unit length'; thus the diagram of fig. 2.2c shows the bending-stress resultants, and not the actual couples acting on the face of the element, which are equal to the stress resultants multiplied by the length of the respective sides of the element.

The bending-stress resultants M_x, M_y (fig. 2.2c) which act on cuts normal to the x- and y-axes respectively, are analogous to bending moments in beams. The sign convention is that they are in the positive sense if they cause tension on the face of the element on the element on the positive z side. We shall use this sign convention consistently throughout the book.

The twisting-stress resultants M_{xy} and M_{yx} are represented by couples whose axes are the normals to the respective faces. They are shown in fig. 2.2c in the positive sense. We shall explain the sign convention later, and also discover that $M_{xy} = M_{yx}$ in general, for a flat element.

In general the forces transmitted across any cut in a structure have a resultant which may be expressed as a *force* passing through an arbitrary point, together with a *couple*. It is convenient to regard the resultant force on each face as acting through the centre of the face; then to resolve in the orthogonal x-, y- and z-directions, and finally to express the components as stress resultants having the units of force/length. Thus the force transmitted across an x-face (i.e. a face perpendicular to the x-axis) is expressed as three stress resultants N_x, N_{xy} and Q_x. The resultant N_x (fig. 2.2a) is a 'direct' stress resultant, and acts normally to the face. The other two are 'shear' stress resultants, acting tangentially, and we use the symbol N_{xy} for the component in the y-direction (thus N_x and N_{xy} may be called 'in-plane' or 'tangential' stress resultants) but symbol Q_x (fig. 2.2b) for the out-of-plane or normal or *transverse* shear-stress resultant. The symbol Q_x is really an abbreviation of Q_{xz}, but the double subscript is unnecessary on account of the use of the symbol Q as distinct from N. The sign convention is that used for stress resultants on a three-dimensional element: on the positive x-face, i.e. the one which overlooks the positive x-direction, the positive sense of Q_x is in the positive z-direction. Conversely, on the negative x-face, the positive sense of Q_x is in the negative direction of z. The same convention is used for N_{xy} and N_{yx}: on the positive x-face the positive sense of N_{xy} is in the positive y-direction, etc.

Various points should be made in connection with fig. 2.2.

First, there are no components in the z-direction of moment-stress resultants on the cut faces: see fig. 2.2c. This is because the x- and y-dimensions of the element are infinitesimal: there is negligible variation of normal stress along these short edges.

Second, the diagrams show *equal* stress resultants on the two x-faces and two y-faces respectively. When we come to study the *equilibrium* of the forces acting on an element in chapters 3 and 4, we shall be concerned closely with the way in which the stress resultants *vary* with x and y. But at present we are concerned merely with the elastic distortion of the element, which involves only mean values of the stress variables: consequently we have considered the various stress resultants as being *constant* over the element.

Those familiar with the equilibrium equations of beams will notice a paradox here. If there is to be uniform transverse shear (e.g. Q_x) in the element, equilibrium demands that M_x, etc. vary across the element. For this reason the transverse shear-stress resultants Q_x, Q_y are shown on a separate diagram, and their values will be taken as zero, as already argued, for purposes of finding the elastic response of the element. They have been included in fig. 2.2b for the sake of completeness, and in order to establish the sign convention which we shall use throughout the book.

Fig. 2.2 does not show any external loading, such as pressure or gravity force, acting on the element. In general, of course, there will be loadings of this sort, but for present purposes we shall ignore them on the grounds that the 'through-thickness' stress σ_z, and also any shearing stresses which may act on planes z = constant, are negligibly small in comparison with those produced by the stress resultants themselves: see section 2.4.2.

2.3.2   The required relations

Our immediate objective is to find a set of relations between the stress resultants acting on the element and the distortions which they produce. Since the material of which the element is made is linear-elastic we may call these relations a 'generalised Hooke's law'. We are concerned only with *small* distortions.

As we have stated earlier, we describe the distortion of the element in terms of the deformation of its centre-surface. There are 'stretching' strains ϵ_x, ϵ_y, γ_{xy} and 'bending' strains in the form of changes of curvature κ_x, κ_y, κ_{xy}. The strains ϵ_x, ϵ_y, γ_{xy} are defined just as in-plane elasticity: ϵ_x, ϵ_y are (small) elongations per unit length in the x- and y-directions respectively and γ_{xy}, an in-plane shear strain, is defined as the small change in angle (radian measure) at the corners of a distorted element which was originally rectangular in the unstressed condition.

The curvature changes κ_x, κ_y and κ_{xy} involve no distortion of the centre-surface in its own plane, but there are out-of-plane displacements which impart curvature. We shall define these quantities later on.

2.3 Generalised Hooke's law for a simple flat shell element

Consider first the special case in which $M_x = M_y = M_{xy} = M_{yx} = 0$, so that the only nonzero stress resultants are the in-plane ones N_x, N_y, N_{xy}, N_{yx}, as shown in fig. 2.2a. Since the element is made of uniform material and the stress resultants are applied on the central surface it is clear that the distorted centre-surface will remain in a plane, i.e. there will be no changes of curvature, for reasons of symmetry. Thus we seek to express $\epsilon_x, \epsilon_y, \gamma_{xy}$ in terms of the applied stress resultants.

There is just one preliminary before we do this. It was explained earlier that the stress resultants would be taken as uniform for present purposes, and that equilibrium equations were not involved. This is not quite true: moment equilibrium of the element shown in fig. 2.2a about the z-axis requires

$$(N_{xy} dy) dx = (N_{yx} dx) dy,$$

that is

$$N_{xy} = N_{yx}. \tag{2.1}$$

This relation holds good as the size of the element becomes vanishingly small. Throughout the book we shall use this identity, and write N_{xy} indiscriminately for either quantity.

Equation (2.1) would not be *strictly* true if the element were curved, and the stress resultants M_x, etc. were not zero; for then the twisting-stress resultants M_{xy} and M_{yx} would have small components about the normal to the centre of the small elements. Thus our proposed use of (2.1) as a general relation involves a certain degree of approximation. This kind of approximation was first proposed by Love (1927, §329). We shall discuss some of the issues raised by it in section 2.4.3.

Since we are neglecting the effect of stresses σ_z, τ_{zx} and τ_{zy}, we have in effect a state of plane stress throughout. The appropriate form of Hooke's law for a uniform isotropic material is (e.g. Timoshenko & Goodier, 1970, §6):

$$\begin{bmatrix} \epsilon_x \\ \epsilon_y \\ \gamma_{xy} \end{bmatrix} = \begin{bmatrix} \frac{1}{E} & \frac{-\nu}{E} & 0 \\ \frac{-\nu}{E} & \frac{1}{E} & 0 \\ 0 & 0 & \frac{2(1+\nu)}{E} \end{bmatrix} \begin{bmatrix} \sigma_x \\ \sigma_y \\ \tau_{xy} \end{bmatrix}. \tag{2.2}$$

Here the stresses $\sigma_x, \sigma_y, \tau_{xy}$ are defined as in fig. 2.3, and we have used the following identity relating the shear modulus G to Young's modulus E and Poisson's ratio ν:

$$G = \tfrac{1}{2}E/(1+\nu). \tag{2.3}$$

Since in the absence of bending moments the state of stress does not vary through the thickness t we have

$$\sigma_x = N_x/t, \quad \sigma_y = N_y/t, \quad \tau_{xy} = N_{xy}/t, \tag{2.4}$$

and consequently our required relation is

$$\begin{bmatrix} \epsilon_x \\ \epsilon_y \\ \gamma_{xy} \end{bmatrix} = \begin{bmatrix} \dfrac{1}{Et} & \dfrac{-\nu}{Et} & 0 \\ \dfrac{-\nu}{Et} & \dfrac{1}{Et} & 0 \\ 0 & 0 & \dfrac{2(1+\nu)}{Et} \end{bmatrix} \begin{bmatrix} N_x \\ N_y \\ N_{xy} \end{bmatrix}. \tag{2.5}$$

This is, of course, only a minor modification of the plane-stress relation (2.2).

The next step is to consider the special case $N_x = N_y = N_{xy} = 0$, so that the only nonzero stress resultants are the bending and twisting ones, M_x, M_y, M_{xy}, M_{yx}. We wish to express the changes of curvature in terms of these. By analogy with the previous case it would be convenient to invoke an equilibrium equation to show that M_{xy} and M_{yx} are equal. However, *no such equilibrium equation exists* since each of these two stress resultants, applied to opposite faces of the element, is precisely self-balancing.

It turns out to be much more straightforward to tackle this particular problem in the reverse direction, that is, to express the bending- and twisting-stress resultants in terms of arbitrary imposed curvatures. In order to do this we shall need to consider the distortion of the element as a three-dimensional body.

Fig. 2.3. A two-dimensional state of stress referred to an x, y coordinate system.

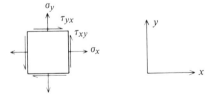

2.3 Generalised Hooke's law for a simple flat shell element

Since the reference plane lies in the centre of the element, the extent of the block in the z-direction is from $z = -\frac{1}{2}t$ to $z = +\frac{1}{2}t$. It is convenient, and it does not restrict generality, to define the other faces of the small block as $x = \pm a, y = \pm b$.

It was mentioned earlier that the position of every point within the block in its initial unstrained condition is described by coordinates x, y, z. In the course of deformation a typical point moves from its original position x, y, z to a nearby position $x + u, y + v, z + w$, with respect to the fixed coordinate system. Thus there is a displacement u, v, w of a typical point, and in general each of u, v and w is a function of x, y and z.

Consider a plane $z = $ constant $= z_0$ in the original configuration of the block, parallel to the central surface $z = 0$. We shall be particularly interested in the stretching of this plane in the x and y directions, even though it will also have displacement w in the z-direction. For such a plane $z = $ constant, and the displacements u, v (and w) are functions only of x and y. The strains $\epsilon_x, \epsilon_y, \epsilon_{xy}$ in such a plane are related to the displacements u, v by the following relations:

$$\epsilon_x = \frac{\partial u}{\partial x}, \quad \epsilon_y = \frac{\partial v}{\partial y}, \quad \gamma_{xy} = \frac{\partial v}{\partial x} + \frac{\partial u}{\partial y}. \tag{2.6}$$

These equations are precisely the same as for two-dimensional straining, and are readily derived: see, e.g., Timoshenko & Goodier (1970, §5).

If we can find u and v at a typical plane $z = z_0$ we can then use (2.6) and (2.2) to find $\sigma_x, \sigma_y, \tau_{xy}$ at this level, and finally do a suitable integration in order to determine the bending- and twisting-stress resultants. Therefore our next task is to study the geometry of distortion of the block, so that we can express u, v, w in terms of the imposed changes of curvature.

The coordinate system has been arranged so that the central surface of the block lies in the plane $z = 0$.

Consider first an example in which the central surface of the element is changed from its original plane form into a shallow cylindrical surface with generators parallel to the y-axis, as shown in fig. 2.4a. The centre of curvature has been put on the negative z side so that the bending moment induced will involve tension on the positive z face, i.e. will be in the positive sense, as already defined. In chapters 5 and 6 we shall study the question of curvature of surfaces in detail, but for the present we note that the example illustrated may obviously be regarded as 'pure' change of curvature in the x-direction, denoted by κ_x. It is clearly convenient to arrange that the displacement at the origin is zero, and also that at the origin there is no rotation of the surface about any axis.

The curvature κ of a plane curve $w(x)$ is expressed in general by

$$\kappa = \frac{d^2w}{dx^2} \bigg/ \left(1 + \left(\frac{dw}{dx}\right)^2\right)^{\frac{3}{2}}. \tag{2.7}$$

This may be simplified, when $dw/dx \ll 1$, to

$$\kappa = \frac{d^2w}{dx^2}, \tag{2.8}$$

as is well known in the simple theory of beams.

Using this expression (with a change of sign on account of our current sign convention) in the present circumstances, and integrating subject to the given conditions at the origin, we obtain the displacement function

$$w)_{z=0} = -\tfrac{1}{2}\kappa_x x^2. \tag{2.9}$$

This expression is sufficiently accurate for present purposes in a small region surrounding the origin. The magnitude of κ_x is restricted by the consideration that all strains associated with it are to be small. As the curvature is small and the resulting form of the central surface is 'shallow', and there is no rotation about the z-axis, we may also write

$$u)_{z=0} = v)_{z=0} = 0. \tag{2.10}$$

Having thus expressed the displacement of the central surface of the element we must next construct displacement functions for the entire block. We do this by applying Kirchhoff's hypothesis (Timoshenko, 1953, §55). Through the point $(x, y, 0)$ in the original configuration lay a normal to the central surface. According to Kirchhoff's hypothesis *this line in the material remains*

Fig. 2.4. Distortion of the centre-surface of an element on account of (*a*) pure bending in the *x*-direction, and (*b*) pure twist with respect to the *x* and *y* directions.

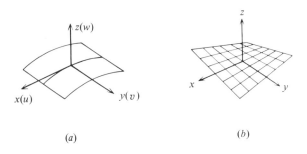

2.3 Generalised Hooke's law for a simple flat shell element

straight and normal to the central surface in the deformed configuration. The slope of the surface, from (2.9), is given by

$$\left(\frac{dw}{dz}\right)_{z=0} = -x\kappa_x, \tag{2.11}$$

and it follows that the component u of displacement of a point distant z from the central surface is given by

$$u = zx\kappa_x. \tag{2.12}$$

Further, since the normals tilt by rotation about the y-axis (in this example) there is no component of displacement in the y-direction; so

$$v = 0. \tag{2.13}$$

Expressions (2.12) and (2.13) describe the displacement field in the element, and we next use (2.6) to obtain expressions for the strains:

$$\epsilon_x = z\kappa_x, \quad \epsilon_y = 0, \quad \gamma_{xy} = 0. \tag{2.14}$$

As there are no through-thickness stresses (by assumption) the state of stress is one of plane stress throughout, for which the appropriate form of Hooke's law has been given in (2.2). Here we need the inverse form, which is (as may be verified easily),

$$\begin{bmatrix} \sigma_x \\ \sigma_y \\ \tau_{xy} \end{bmatrix} = \begin{bmatrix} \frac{E}{1-\nu^2} & \frac{\nu E}{1-\nu^2} & 0 \\ \frac{\nu E}{1-\nu^2} & \frac{E}{1-\nu^2} & 0 \\ 0 & 0 & \frac{E}{2(1+\nu)} \end{bmatrix} \begin{bmatrix} \epsilon_x \\ \epsilon_y \\ \gamma_{xy} \end{bmatrix}. \tag{2.15}$$

Consequently we find

$$\sigma_x = \left(\frac{E\kappa_x}{1-\nu^2}\right) z, \quad \sigma_y = \left(\frac{\nu E\kappa_x}{1-\nu^2}\right) z, \quad \tau_{xy} = 0. \tag{2.16}$$

This is the stress field in the element. It does not vary in the x- and y-directions, which is of course only to be expected for uniform cylindrical bending.

Our final task is to find, by means of an appropriate integration, the corresponding stress resultants. For an x-face of the element we take moments about the centre-line and equate the external and internal effects, integrating over the rectangular area:

$$2bM_x = \int_{-\frac{1}{2}t}^{\frac{1}{2}t} 2bz\sigma_x \, dz.$$

Thus
$$M_x = \kappa_x D, \tag{2.17}$$
where
$$D = Et^3/12(1 - \nu^2). \tag{2.18}$$

The constant D is known as the flexural rigidity of the plate element. It will appear in many places throughout the book. It is analogous to the constant EI used in the theory of beams, and like I is proportional to (thickness)^3. It does of course have different dimensions from EI because M_x (moment/length) has the units of force: D has units of force × length.

The factor $(1 - \nu^2)$ on the denominator comes from the inverse form of Hooke's law: when we apply a curvature change κ_x alone, Kirchhoff's hypothesis ensures that there is zero *strain* in the y-direction. Accordingly (see (2.15)) nonzero stresses σ_y are induced, and on the y-faces these constitute a bending-stress resultant. By integration, we find that this is given by

$$M_y = \frac{Et^3}{12(1 - \nu^2)} \nu\kappa_x = \nu D \kappa_x. \tag{2.19}$$

Since $\tau_{xy} = 0$ throughout there are certainly no twisting-moment resultants. Also the stress resultants N_x, N_y, N_{xy} are all zero, by direct integration.

Let us now consider another example, shown in fig. 2.4*b*, in which the originally plane central surface is given a pure *twist* with respect to the x and y axes. In this case the displacement function is

$$w)_{z=0} = -\kappa_{xy} xy, \tag{2.20}$$

and we shall call the quantity κ_{xy} the *twist* of the surface. The positive sense of κ_{xy} is defined so that the corresponding shearing stress τ_{xy} is positive (see fig. 2.3) on the positive z side of the element. Note that from (2.20) w is a linear function of y on lines x = constant and a linear function of x on lines y = constant; the displacement function involves zero change of curvature in the x- and y-directions. Also note that, as in the previous example, displacement and slope at the origin are zero. We shall also assume, as before, that there is no rotation about the z-axis; consequently, we have

$$u)_{z=0} = v)_{z=0} = 0. \tag{2.21}$$

Again we invoke Kirchhoff's hypothesis. In this example the normal through a given point on the central surface tilts by rotation about both the x- and y-axes, and by elementary geometrical considerations we find the following expressions for u and v.

$$u = zy\kappa_{xy}, \quad v = zx\kappa_{xy}. \tag{2.22}$$

2.3 Generalised Hooke's law for a simple flat shell element

Consequently we have, from (2.6), the strain field

$$\epsilon_x = \epsilon_y = 0, \quad \gamma_{xy} = 2z\kappa_{xy}. \tag{2.23}$$

Thus at each 'level' z = constant there is a pure shear strain, whose magnitude is proportional to z. Substitution into Hooke's law gives

$$\sigma_x = \sigma_y = 0, \quad \tau_{xy} = Ez\kappa_{xy}/(1+\nu). \tag{2.24}$$

This state of stress gives pure shear stresses on the cut surfaces as illustrated in fig. 2.5. It is clear that $M_x = M_y = 0$ in this example. To find the corresponding values of the twisting moments $M_{xy} = M_{yx}$ we must perform appropriate integrations. Fig. 2.5 shows not only the shearing stresses τ_{xy} acting on the edges, but also the statically-equivalent twisting-stress resultants M_{xy} and M_{yx} (both in their positive senses). On an x-face we have

$$2bM_{xy} = \int_{-\frac{1}{2}t}^{\frac{1}{2}t} 2bz\tau_{xy}\,dz.$$

Thus

$$M_{xy} = \left[Et^3/12(1+\nu)\right]\kappa_{xy} = D(1-\nu)\kappa_{xy}. \tag{2.25}$$

Similarly we find

$$M_{xy} = D(1-\nu)\kappa_{xy}. \tag{2.26}$$

Hence we obtain the important result, already mentioned:

$$M_{xy} = M_{yx}. \tag{2.27}$$

Fig. 2.5. Stresses and stress resultants in an element subjected to pure twist. The small arrows on the faces of the block represent shearing stresses whose magnitude is proportional to distance z from the central surface. The double-headed arrows represent the couples (right-hand screw rule) which are statically equivalent to the shearing stresses on the respective faces.

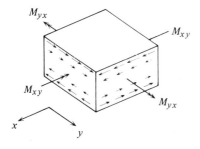

In future we shall use the same symbol M_{xy} indiscriminately for both quantities. It is important to realise that this result comes from a full analysis of the distortion of and stress in a typical element, in contrast to the other identity $N_{xy} = N_{yx}$ which came directly from an equation of equilibrium. Strictly, we have only demonstrated this result for the special case of pure twist, but it also holds for general distortions of the centre-surface of the element, as we shall see.

So far we have considered only two specially simple examples of the bending/twisting distortion of our flat element. The most general distortion of the element is specified, as we shall see in chapter 6, by three independent variables $\kappa_x, \kappa_y, \kappa_{xy}$. However, it may be shown, in the present context of linear elasticity and small strains and displacements, that the stress resultants on an element are the sum of those which would be obtained if each of the three components $\kappa_x, \kappa_y, \kappa_{xy}$ were applied separately (the 'principle of superposition'). We have already considered the separate cases of κ_x and κ_{xy}, and it is obvious that since the material is isotropic, we may transfer our results for κ_x to the component κ_y by simple alteration of subscripts. Thus, collecting results, we have, finally, the 'bending' part of the generalised Hooke's law:

$$\begin{bmatrix} M_x \\ M_y \\ M_{xy} \end{bmatrix} = \begin{bmatrix} D & \nu D & 0 \\ \nu D & D & 0 \\ 0 & 0 & D(1-\nu) \end{bmatrix} \begin{bmatrix} \kappa_x \\ \kappa_y \\ \kappa_{xy} \end{bmatrix}. \tag{2.28}$$

Although we have not done the integrations explicitly, it is clear that in each of these cases all three direct-stress resultants N_x, N_y, N_{xy} are zero: this is a consequence of the fact that all stress components are directly proportional to z, and in each case the integration is done over a rectangular area. It is straightforward to derive expressions for σ_x, σ_y and τ_{xy} as functions of z in an element on account of M_x, M_y and M_{xy}: see problem 2.2. In particular it is straightforward to derive the following simple expressions for the peak values of the stress components, which (in the absence of N_x, etc.) have the same absolute values at the two surfaces of the element:

$$\sigma_x = \pm 6M_x/t^2, \quad \sigma_y = \pm 6M_y/t^2, \quad \tau_{xy} = \pm 6M_{xy}/t^2. \tag{2.29}$$

The required generalised Hooke's law is thus given by the two independent relations (2.5), (2.28). For some purposes it is convenient to invert (2.28) so that κ_x etc. are given in terms of M_x etc.: then the two parts may be put into a 6 × 6 matrix expressing the six deformation quantities $\epsilon_x, \epsilon_y, \gamma_{xy}; \kappa_x, \kappa_y, \kappa_{xy}$ in terms of the six stress resultants $N_x, N_y, N_{xy}; M_x, M_y, M_{xy}$. We shall see,

2.3 Generalised Hooke's law for a simple flat shell element

however, when we come to solve problems concerning elastic shells, that the forms as given ((2.5), (2.28)) will be most convenient in general.

A loose end remains to be tidied up. In discussing the displacement field in the element we found general expressions for u and v, but not for w. It is clear that if there had been no strain in the z-direction, the w-component of displacement would have been constant along any normal to the central surface, i.e. would have been independent of z. But in fact it is the direct *stress* in the z-direction which is zero, and consequently the strain component ϵ_z, which is found from Hooke's law, is in general nonzero:

$$\epsilon_z = -\nu(\sigma_x + \sigma_y)/E. \tag{2.30}$$

This equation supplements (2.6). Its use in determining the w-displacements throughout the element is considered in problem 2.3.

2.3.3 Expressions for strain energy

In many of the problems to be solved in the following chapters we shall analyse the behaviour of elastic shells by first writing down the equations of equilibrium and compatibility and the generalised Hooke's law, and then solving them formally. On other occasions, however, (particularly in chapters 12, 13 and 16) it will be more convenient, for a variety of reasons, to obtain *approximate* solutions by means of *elastic energy methods*. In the application of such methods it is necessary to express the strain energy per unit area of shell in terms either of the local generalised strains $\epsilon_x, \epsilon_y, \gamma_{xy}; \kappa_x, \kappa_y, \kappa_{xy}$ or in terms of the stress resultants $N_x, N_y, N_{xy}; M_x, M_y, M_{xy}$. Since the derivation of these expressions involves only the generalised Hooke's law and some integration, it is convenient to do the calculations here.

Consider first an element of material in a state of plane stress which is achieved by the imposition of strain components $\epsilon_x, \epsilon_y, \gamma_{xy}$. At any stage the corresponding stress resultants $\sigma_x, \sigma_y, \tau_{xy}$ are given by (2.15), and it is therefore easy in principle to find the total work done by the loading agency per unit volume of material as it is taken from its original unstrained configuration to its current strained one. It is straightforward to demonstrate that this quantity of work depends only on the current strain, and not upon the *path* in $\epsilon_x, \epsilon_y, \gamma_{xy}$ space by which the current state was reached. We may therefore describe the specific work associated with the current state as the *strain energy*: it is an energy which is a function of the current *strain*.

It is easiest to evaluate the strain energy by following a path in which the three components of strain remain in constant proportions: thus we find from (2.15) that in plane stress the strain energy per unit volume is given by

$$\left[E/2(1-\nu^2)\right](\epsilon_x^2 + 2\nu\epsilon_x\epsilon_y + \epsilon_y^2) + E\gamma_{xy}^2/4(1+\nu). \tag{2.31a}$$

It is sometimes convenient to rearrange this expression as follows:

$$[E/2(1-\nu^2)]\{(\epsilon_x+\epsilon_y)^2+2(1-\nu)[-\epsilon_x\epsilon_y+(\tfrac{1}{2}\gamma_{xy})^2]\} \quad (2.31b)$$

(see problem 2.4).

When an element of shell is strained without curvature or twist, every 'slice' within the thickness experiences exactly the same pattern of strain. Thus expression (2.31b), when multiplied by the thickness t, gives the strain energy per unit area in terms of the components of strain at the centre-surface:

$$U_S = [Et/2(1-\nu^2)]\{(\epsilon_x+\epsilon_y)^2+2(1-\nu)[-\epsilon_x\epsilon_y+(\tfrac{1}{2}\gamma_{xy})^2]\}. \quad (2.32)$$

U_S is the strain energy of *stretching*, per unit area of shell surface.

Consider next an element of shell which is subjected to changes of curvature $\kappa_x, \kappa_y, \kappa_{xy}$ while its central surface remains unstrained. In this case we find, by Kirchhoff's hypothesis (cf. (2.14) and (2.23)) that

$$\epsilon_x = z\kappa_x, \quad \epsilon_y = z\kappa_y, \quad \gamma_{xy} = 2z\kappa_{xy}.$$

The ratios $\epsilon_x : \epsilon_y : \gamma_{xy}$ are thus independent of z; and we may therefore take advantage of (2.31) in integrating the strain energy through the thickness. In this way we obtain

$$U_B = \tfrac{1}{2}D[(\kappa_x+\kappa_y)^2+2(1-\nu)(-\kappa_x\kappa_y+\kappa_{xy}^2)]. \quad (2.33)$$

U_B is the strain energy of *bending* per unit area. In the important special case $\kappa_x \neq 0, \kappa_y = \kappa_{xy} = 0$, where the element is bent cylindrically, we have

$$U_B = \tfrac{1}{2}D\kappa_x^2. \quad (2.34)$$

In general, of course, an element of a shell will be subjected simultaneously both to stretching and bending distortion; and we may show without difficulty for our uniform, isotropic element, that the total strain energy U per unit area is given by

$$U = U_S + U_B, \quad (2.35)$$

where U_S, U_B are given by (2.32), (2.33): see problem 2.5. Love (1888) was the first to note this *separability* of the strain energies of stretching and bending, respectively, in an element of a uniform shell.

It is a straightforward matter to rearrange (2.32) and (2.33) in terms of *stress resultants* by means of (2.5) and the inverse of (2.28). For purposes of application of the complementary energy theorem (see appendix 1) we strictly need to compute an energy C per unit area which, though defined differently, is always equal to the strain energy U per unit area for linear-elastic material.

2.4 Mechanical properties of curved elements

Thus we find

$$C_S = (1/2Et)[(N_x + N_y)^2 + 2(1 + \nu)(-N_xN_y + N_{xy}^2)], \quad (2.36a)$$

$$C_B = (6/Et^3)[(M_x + M_y)^2 + 2(1 + \nu)(-M_xM_y + M_{xy}^2)], \quad (2.36b)$$

where

$$C = C_S + C_B. \quad (2.36c)$$

The subscripts S and B have the same meanings as before.

2.4 Mechanical properties of curved elements

We now turn to a qualitative justification of the use of the results of the preceding section in relation to an element of a curved shell. An element of a shell is a sort of two-way-beam element. If the shell is curved in both directions then this beam element (i) is curved and (ii) has a wedge-shaped or trapezoidal cross-section. Let us investigate what we might call the 'curved-beam' effect and the 'trapezoidal-edge' effect separately.

2.4.1 The 'curved-beam' effect

Consider the problem illustrated in fig. 2.6a and b. Two beams made from the same elastic material have the same uniform rectangular cross-section, and centre-lines of the same length; but one is straight in the unloaded condition while the other is curved, as shown. The two beams are now subjected to uniform bending moment M by the application of couples of the same magnitude but opposite sign at the ends.

The issue is whether or not the resulting *changes of curvature* in the two beams are different; and if so, by how much. If the difference turns out to be large, it will obviously not be appropriate to pretend that an element of shell is flat for the purposes of determining the generalised Hooke's law. In comparing the behaviour of the straight and curved beams of fig. 2.6a and b we shall of course assume that the bending moment is of magnitude which produces only small strains, and that the changes of curvature are correspondingly small. The problem as posed does not involve any shearing forces. These have been omitted for the reasons given above, but they may be reintroduced without too much difficulty: for a treatment of this problem see Timoshenko & Goodier (1970, § 33).

The essential difference between the pure bending of straight and curved beams can be grasped more easily with the help of a 'sandwich' model of the beam as shown in fig. 2.6c and d. In this model the couples are imposed at the ends in the form of tensile and compressive forces applied directly to the flanges.

When a sandwich beam is subjected to a bending moment the two flanges are put into tension and compression respectively. In the case of an initially *curved* beam the equilibrium of the flanges requires a state of compression in the core (for a bending moment in the sense of the diagram) in the radial direction: see fig. 2.6e. The stress in the flanges and in the core is accompanied by strain, due to the elasticity of the material: the flanges undergo small changes in length, and the core sustains a small change in thickness. The effect of these small changes on the overall configuration of the beam may be considered separately. In the absense of a change in thickness of the core, the strains in the flanges would produce a change of curvature, say κ_1, just as they would for a straight beam. On the other hand, a change in thickness of the core, in the absence of strain in the flanges, also produces a change of curvature, say κ_2. The ratio κ_2/κ_1 is independent of the value of M in the context of small strains. Thus we see that if in a given case the value of κ_2/κ_1 is sufficiently small, the response of the initially curved beam to bending moment will be substantially the same as that of an initially straight beam having the same centre-line length, to a first approximation.

It is straightforward to show that for a sandwich beam of the type illustrated

$$\kappa_2/\kappa_1 = (E/E')(TH/2R^2), \tag{2.37}$$

where E and E' are the Young's modulus of the flanges and core, respectively,

Fig. 2.6. Comparison of pure bending of an initially straight bar and an initially curved bar. (*a*) and (*b*) are straight and curved 'solid' bars, while (*c*) and (*d*) are corresponding bars of 'sandwich' construction. (*e*) shows the curved bar of (*d*), having been cut along its centre-line: equilibrium of each part requires a through-thickness compressive stress in the 'core'.

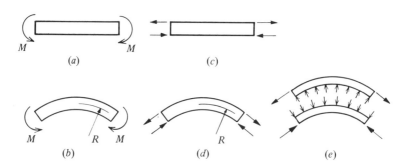

R is the original radius of curvature of the beam, H is the separation of the flanges and $T (\ll H)$ is the thickness of the flanges: see problem 2.6.

In the case of a curved beam of *solid* cross-section, the essential features of the behaviour are precisely the same, although the through-thickness compressive stress varies continuously across the thickness. The corresponding value of the curvature ratio may be shown to be given by

$$\kappa_2/\kappa_1 = \tfrac{1}{12}(t/R)^2 \tag{2.38}$$

where t is the thickness and R is the original radius of curvature of the beam: see problem 2.7. Note that there is now no 'modular ratio' since there is but one, uniform, material.

In many shell structures the value of t/R is of order 0.01, so the value of κ_2/κ_1 is of order 10^{-5}. Even when t/R is as large as 0.1, the value of κ_2/κ_1 is about 0.001, which is, of course, normally negligible. In a hypothetical extreme case in which the 'curved-beam' effect enhances the flexibility by 2%, we find from formula (2.38) that $t/R \approx 0.5$. This is of course a ridiculously large ratio of t/R in the context of thin shells, and no doubt for a beam having a geometry of these proportions the linearised analysis which we have used would be inadequate: see problem 2.8. What this calculation indicates positively is that the 'curved-beam' effect should be entirely negligible in the context of the bending of curved elements of thin shells.

On the other hand, some sandwich shells with 'soft' cores may in certain circumstances be subject to rather strong 'curved-beam' effects.

2.4.2 'Through-thickness' stresses

We have just seen that through-thickness stresses are developed in an initially curved beam when it is subjected to a pure bending moment, in order to satisfy the equations of equilibrium. Stresses of this kind may be required in a variety of other circumstances. The simplest example is a thin-walled spherical shell under internal pressure p. Everywhere in the shell there is a through-thickness stress varying from $-p$ at the inner surface to 0 at the outer surface, which is open to the atmosphere. These stresses are usually neglected in comparison to the in-plane stresses, since the latter are normally at least an order of magnitude larger.

But there are some circumstances in which it is not clear from the outset that through-thickness stresses are negligible. We can imagine, for example, a shell whose surface is loaded by a pressure acting over a small area; a 'localised' load or, in an extreme case a 'point' load. In a case like this the magnitude of the pressure – and consequently of the 'through-thickness' stress immediately under the load – might conceivably be of the same order as the main direct and bending stresses in the shell. Problems of this kind are not limited to shell structures, of course: they occur in beams and plates also. Thus, in the design of beams it is often necessary to make special provision for the accommodation of 'point' loads by the introduction of local pad reinforcement, web stiffeners, etc. These matters fall right outside the scope of the simple beam theory, which replaces the beam conceptually by an 'elastic line' of bending stiffness EI; and they must be considered separately. Similarly, in the design of plates and shells it is necessary to treat particular cases separately, in detail.

In some circumstances *plastic theory* is the appropriate tool with which to study the effects of highly localised loading on shells; and this topic provides part of the motivation for chapter 18. But we shall also be able to do some simple order-of-magnitude calculations on local stresses near regions of application of load in section 3.6.

A general problem which is raised by the idealisation of a thin shell as a surface concerns the precise location of any applied loading, since external loads are usually applied to the inner or outer surface of a shell. The obvious convention is to suppose that all external loading is applied at the mid-surface of a shell. This we shall do throughout the book. But we note here that it is possible to devise problems in which this idea is inherently unsatisfactory.

2.4.3 The 'trapezoidal-edge' effect

The second item on the list of effects possibly standing in the way of the legitimate use of a 'flat-element' expression for the generalised Hooke's law for an element of a shell is what might be called the 'trapezoidal-edge' effect. An exaggerated view of an element of a doubly-curved shell is shown in fig. 2.7. Its edges are normal to the mid-surface of the shell. Now suppose that mid-surface strains and changes of curvature have been prescribed, and that we have invoked Kirchhoff's hypothesis and Hooke's law in order to determine the various components of stress as functions of position in the element. In general these will vary linearly with z, the distance from the mid-surface. When we come to perform the integrations which are appropriate in order to determine the various stress resultants acting on the several faces, we find that they are more awkward on account of the fact that the faces are no longer rectangular. The key difference is that the edges vary in width with z; and we are, in effect, integrating over a trapezium rather than a rectangle. This complication does not disappear when the size of the element is steadily reduced to zero. These integrations are described clearly by Love (1927, §328), Novozhilov (1964, §6) and Kraus (1967, §2.5), and we shall not repeat them here. But we can see, even, without performing the integrations explicitly, that if σ_{xy} varies linearly with z, and the radii of curvature R_1 and R_2 are different,

2.4 Mechanical properties of curved elements

then the values of N_{xy} and N_{yx} will not be exactly equal. Neither, in fact, will the values of M_{xy} and M_{yx}. A careful analysis by Novozhilov (1964, §7) shows that these four quantities are related as follows:

$$N_{xy} - N_{yx} = (-M_{xy}/R_1) + (M_{yx}/R_2). \tag{2.39}$$

All of this is clearly contrary to our previous simpler results, namely $N_{xy} = N_{yx}$ (2.1) and $M_{xy} = M_{yx}$ (2.27) which were obtained by consideration of a small *flat* element. What are we to make of these discrepancies?

Much has been written on this and related topics. The crucial point (made clearly by Koiter, 1960, 1969a) is that anomalies of this sort are a direct consequence of the attempt to endow a *surface* with the mechanical properties of a curved plate of finite thickness. Such a theory is necessarily approximate, and too much insistence on rigour may lead us astray. We have already argued, on order-of-magnitude grounds, that it is reasonable to ignore distortion due to normal shearing stress and 'through-thickness' stress in many particular problems. Against this background we must also be prepared to ignore other anomalies which can and will arise in the course of our development of the subject. In the case of the contradictory equations (2.1), (2.27) and (2.39) we first remark that (2.39) is in fact the *equilibrium* equation of moments for the curved element about the normal axis − although, of course, it was derived by integration of stress resultants on the basis of the Kirchhoff hypothesis. The anomaly vanishes, of course for a flat plate, when $1/R_1 = 1/R_2 = 0$. In the so-called 'shallow-shell' equations the anomaly is tolerated, on the grounds that $1/R_1$ and $1/R_2$ are sufficiently small (in comparison with an appropriate quantity) for the mismatch to be insignificant in terms of numerical accuracy. Having regard to all of this, we shall adopt throughout this book the simple policy of disregarding (2.39) altogether. This will be amply justified in 'shallow-

Fig. 2.7. An element of a doubly-curved shell, showing (cf. fig. 2.5) the non-rectangular faces over which the stress must be integrated in order to give the various stress resultants. The x- and y-directions (not shown) are as in fig. 2.5.

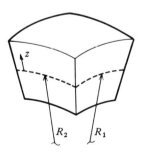

shell' situations; and we shall see, by means of occasional checks on solutions, that it is justified in other cases also. For example, we shall find in a rather extensive study (in chapters 8 and 9) of cylindrical shells under various kinds of nonsymmetric loading that in many practical problems the magnitude of the stress resultant M_{xy} is small in comparison with the magnitude of M_y or M_x: in these cases M_{xy} plays only a minor part in the resistance of the shell to loading.

It is possible, in fact, to get round the anomaly posed by equation (2.39) by defining alternative stress resultants which are a combination of N_{xy} and M_{xy}, etc: see, e.g., Novozhilov (1964 §8). But a multiplicity of slightly different equations emerges from studies of this sort by different authors: see Simmonds (1966) for an account in relation to circular cylindrical shells. In their search for a 'best' set of general equations, Budiansky & Sanders (1963) effect a redistribution of various anomalies in a way which satisfies both the basic conditions of classical mechanics and also some additional considerations of symmetry and neatness; but they concede that these cannot be defended on the grounds of strict logic. In particular, Budiansky & Sanders were not motivated by a search for *accuracy*; for, as they say, 'the comparative studies of Koiter (1960) clarify and substantiate the widely held impression that there is little difference, from the point of view of accuracy, among many of the existing sets of shell equations'.

Our general policy throughout the book will be to adopt the simple equations (2.1) and (2.27), acknowledging that this involves a measure of approximation; and then to search the subsequent solutions of particular problems for any evidence of substantial violation of (2.39) at the same time that we check on the magnitudes of the transverse shearing stresses and the through-thickness stresses in relation to those stresses which, according to the Kirchhoff hypothesis, are responsible for the elastic distortion of the shell.

2.5   Non-uniform shells

So far in this chapter it has been assumed that the element of the shell in question is isotropic. We may sometimes be concerned with a shell which, as shown in fig. 2.8, is stiffened by closely-spaced ribs. Orthotropic shells of this sort are widely used in structural applications. In cylindrical shells the stiffening ribs may take the form of circumferential rings or longitudinal stringers. The previous arguments about shear distortion and the 'curved-beam' effect hold good, but it is obviously not possible to generalise directly the previous equations (2.5) and (2.28) which together expressed the generalised Hooke's law for an isotropic element. In the previous derivation the obvious place for the reference surface ($z = 0$) was at the centre of the element; for then the application of (say) pure tension N_x caused no change of curvature and, reciprocally, the application of pure bending moment M_x caused no strain in the reference surface. In relation to the element of fig. 2.8 it is clear that a reference surface could easily be located in such a way that unidirectional bending and stretching effects were similarly uncoupled in each of the two principal directions separately: but the two reference surfaces so found would not coincide. Clearly we must have in general a *single* reference surface for any element; and consequently we must accept that bending and stretching effects are *coupled* in orthotropic shells. In this book we shall not be concerned systematically with orthotropic shells: but see Brush & Almroth (1975, §5.6) for a clear discussion of the problem of the derivation of an appropriate generalised Hooke's law. In some circumstances, however, it is possible to think about the mode of operation of orthotropic shells in relatively simple terms. Thus when the structural action of an orthotropic cylindrical shell involves primarily an interaction between bending in one principal direction and stretching in the other, it is possible to discuss the behaviour satisfactorily without an investigation of the details of the generalised elastic law: see section 14.10.

Fig. 2.8. An element of a shell with unidirectional reinforcing ribs attached to one side.

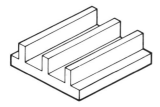

2.6 Problems

2.1 It may be shown that originally plane cross-sections of a uniform beam remain plane under pure bending, by means of an argument involving a comparison of the beam with its mirror-image in the plane of a cross-section. Demonstrate that this argument breaks down if the beam is *tapered*; if the modulus of elasticity varies along the length; or if there are other non-uniformities. (The argument presupposes, in a uniform beam, that there are no local instabilities: but see chapter 16.)

38   *Generalised Hooke's law for an element of a shell*

2.2 By superposing results like (2.16) and (2.24), express $\sigma_x, \sigma_y, \tau_{xy}$ in terms of z and $\kappa_x, \kappa_y, \kappa_{xy}$. Compare with expressions (2.28) for M_x, M_y, M_{xy} in terms of $\kappa_x, \kappa_y, \kappa_{xy}$. Hence find expressions for $\sigma_x, \sigma_y, \tau_{xy}$ in terms of M_x, M_y, M_{xy} and z (cf. (2.29)); and explain in particular why these do not involve the elastic constants of the material. Throughout this problem $\epsilon_x = \epsilon_y = \gamma_{xy} = 0$.

2.3 Use the expression $\epsilon_z = \partial w/\partial z$ together with (2.30) to determine the through-thickness displacement of the original centre-surface of an element with respect to the outer surfaces when the element is subjected to stress resultants M_x, M_y, M_{xy}.

2.4 Rearrange (2.31b) as the sum of two terms, one of which (by virtue of (2.3)) has the coefficients $\frac{1}{2}G$. Show by means of a Mohr circle construction that when $\epsilon_x + \epsilon_y = 0$ the state of strain may be expressed as a pure shear with respect to suitably inclined coordinate axes.

2.5 Derive expression (2.35) by considering the state of strain of an element whose centre-surface is subjected to $\epsilon_x^0, \epsilon_y^0, \gamma_{xy}^0$ and which is also subjected to curvature changes $\kappa_x, \kappa_y, \kappa_{xy}$. Use expression (2.31a) to determine the strain energy per unit volume in terms of z, and integrate. Note that $\int_{-\frac{1}{2}t}^{\frac{1}{2}t} z\,dz = 0$.

2.6 A curved beam has thickness H and radius of curvature R in its original configuration. The thickness now changes by a small amount δH in such a way that every longitudinal 'fibre' preserves its length exactly. Show by geometrical arguments that the corresponding small change of curvature of the beam is equal in magnitude to $\delta H/RH$.

Hence compute the change in curvature (κ_2) of the beam shown in fig. 2.6d on account of the change of thickness of the core due to the application of a pure bending moment; and thereby verify (2.37).

2.7 In the case of an initially *curved* beam of solid rectangular cross-section, having thickness t and radius R ($\gg t$) show that the condition of equilibrium of a thin curved slice parallel to the central surface is

$$\frac{ds}{dz} = \frac{\sigma}{R},$$

where $\sigma(z)$ is the 'bending stress' on account of the applied bending moment M per unit width and s is a 'through-thickness' compressive stress. Taking $\sigma(z)$ just as for an initially *straight* beam, and integrating between the inner and outer surfaces, show that the mean value of s is M/Rt; and that conse-

2.6 Problems

quently the corresponding small change of thickness of the beam is given by $\delta t = M/RE$. Hence, by using the first result of problem 2.6, verify (2.38).

2.8 According to the analysis of problem 2.7, the maximum 'through-thickness' stress in the bending of an initially curved beam is equal to $1.5 M/Rt$. In a more rigorous analysis of the same problem, using cylindrical polar coordinates Timoshenko & Goodier (1970, §29) show that in pure bending of a curved beam having $b/a = 2$, where b is the outer radius $(= R + \frac{1}{2}t)$ and a is the inner radius $(= R - \frac{1}{2}t)$, the maximum 'through-thickness' stress is equal to $1.070 M/a^2$. Compare the two results.

2.9 Verify that (2.36) gives $C_S = N_x^2(1 - \nu^2)/2Et$; $C_B = M_x^2/2D$ when $N_y = \nu N_x$ and $M_y = \nu M_x$.

3

Cylindrical shells under symmetric loading

3.1 Introduction

The subject of this chapter is the behaviour of thin elastic circular cylindrical shells when they are loaded by forces which are symmetrical about the axis of the cylinder. Cylindrical shells have structural applications in many fields of engineering, and the loading is often symmetrical, especially in pressure-vessel applications. Some of the results of this chapter will be directly useful and applicable in design. The behaviour of cylindrical shells when they are loaded by non-symmetric forces will be discussed later, particularly in chapters 8 and 9.

The main reason for the inclusion of the present topic early in the book is that it is uniquely *instructive*. Although it is obviously a particularly simple problem on account of the symmetry, it does nevertheless illustrate well some basic features of the behaviour of shell structures which reappear repeatedly, as we shall see, in much more complicated problems later in the book. In particular we shall be able to see clearly how the shell mobilises both stretching and bending effects in order to carry the applied loading. The problem will also illustrate how the choice of suitable dimensionless groups enables us to present useful results in the most economical way. Lastly we note that the symmetrically-loaded cylindrical shell provides a good introduction to the behaviour of symmetrically-loaded general shells of revolution, which will be discussed in chapter 11.

Our first problem (section 3.4) is the simplest of all: a semi-infinite shell which is loaded either by uniformly distributed radial shearing force or bending moment at its edge, as shown in fig. 3.1. By combining the two solutions in various ways – using the principle of superposition for small deflections of linear-elastic structures – we shall be able to analyse in sections 3.5 and 3.6 some practical problems involving both concentrated and distributed band-loading of the type illustrated in figs. 3.5*a* and 3.6*a*. In general

3.1 Introduction

we shall be concerned with questions about the distribution of stress and displacement throughout the structure in consequence of the applied loading, which are obviously important in engineering practice.

In section 3.7 we investigate a long cylindrical shell subjected to pressure-loading which varies sinusoidally with the axial coordinate, having a half wavelength l. In particular we find that the way in which bending and stretching effects combine to carry the loading depends strongly on the value of the dimensionless group $l/(at)^{\frac{1}{2}}$, where a and t are the radius and thickness, respectively, of the shell. This in turn leads to a description of the shell in terms of conceptually distinct but coincident 'stretching' and 'bending' surfaces. In more general terms this idea will have many fruitful applications throughout the book.

The *semi-infinite* shell is the first example to be considered because it is simplest. In section 3.8 we shall analyse shells of finite length. In principle these are no more difficult, but in practice they are more cumbersome on account of the doubling of the number of boundary conditions. The results of this section open up the solution of a wide range of practical problems.

For the sake of simplicity we shall restrict attention in this chapter to problems in which the shell is initially stress-free, so that all the stresses in the shell are a direct consequence of the applied loading. This is not to deny the importance of initial-stress and thermal-stress problems: it is relatively easy to extend the work of this chapter to include effects of this sort, and examples will be given in problems 3.22 and 3.23. Inelastic effects are not considered in this chapter: some aspects of inelastic behaviour of shells will be discussed in chapter 18.

In general, the solutions which are obtained in this chapter may be super-

Fig. 3.1. (*a*) Shearing force and (*b*) bending moment applied uniformly to the free edge of a semi-infinite circular cylindrical shell. The sense of each is positive when the positive sense of x is to the right: see also figs 3.2 and 3.3.

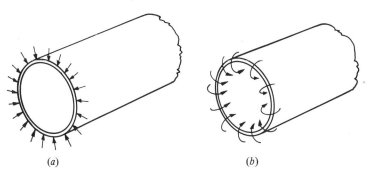

posed onto the uniform states of stress and strain which are present when, e.g., a cylindrical vessel with closed ends is subjected to interior pressure: the additional uniform circumferential and longitudinal stress makes no difference to the mode of action of the shell in resisting localised axisymmetric loads.

In fact the principle of superposition is not strictly correct in such cases. It is a general consequence of the 'small-deflection' assumption that the equations of equilibrium, etc., may be set up for the structure in its original undistorted configuration.

As an example consider a cylindrical shell which first sustains a ring-load as in fig. 3.5a and is then subjected to a second loading in the form of a uniform uniaxial tension. If the originally straight generators of the shell are appreciably *curved* by the action of the first load, the second loading may only be superposed without alteration of the stresses, etc. due to the first loading if it is accompanied by those pressures which are required to satisfy the conditions of equilibrium of curved tensile members. Since these pressures are not actually provided, we must conclude that the state of stress, etc. due to the first load is altered to some extent by the application of the second loading. This kind of *nonlinear* effect is outside the scope of the present chapter and of chapter 11. Nevertheless, effects of this sort can sometimes be important; and indeed they are of the essence in studies of the *buckling* of structures: see chapter 14. In order to discover something of the circumstances in which these effects are significant, we shall study briefly a simple nonlinear problem of this kind in section 3.7.3.

3.2 Basic equations of the problem

Let us begin by defining the coordinate system, the variables and the sign conventions which we shall use throughout the chapter.

The geometry of the shell is defined in fig. 3.2. The surface is a right circu-

Fig. 3.2. Coordinate system for cylindrical shell of radius a.

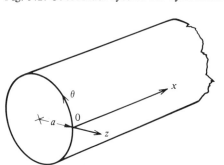

lar cylinder of radius a. We use an x, θ coordinate system in the surface, where x is an axial length coordinate and θ a circumferential angle. We also need to define the positive sense of z, a radial coordinate. Here (and indeed for cylindrical shells throughout the book) we shall take the positive sense of z as being directed away from the interior of the surface. The orthogonal system x, θ, z at any point on the surface is right-handed. The ends of the shell are defined by planes x = constant. Where the shell is semi-infinite it is convenient to arrange the coordinates so that $x = 0$ at the loaded end.

The stress resultants acting on a typical element of the shell are shown in fig. 3.3a. The notation follows generally the scheme used in chapter 2: see fig. 2.2 N_x, N_θ are positive when tensile. $N_{x\theta}$ is identically zero on account of the symmetry of the present problem. Bending moments M_x, M_θ have a positive sense when they tend to produce tension on the z-positive face of the shell. The positive sense of Q_x (strictly Q_{xz}) is shown: on the positive x edge of an element it points in the positive z direction, and on the negative x edge in the negative z direction. Stress resultants $Q_{\theta z}$ and $M_{x\theta}$ are zero by symmetry. In some cases we shall have a normal pressure-loading of intensity p per unit area, acting in the positive sense of z. In the diagram N_x has been shown, for the sake of completeness. In much of the chapter we shall have $N_x = 0$. However, N_x will play an important part in section 3.7.3. It is not clear at this stage whether or not it will be necessary to use M_θ as a variable: we shall discover later that it is a 'passive' variable in problems having symmetry about the axis.

Fig. 3.3. (a) Stress resultants in a typical element of the shell. (b) The same, but showing pressure-loading and changes in value of Q_x and M_x with x, for purposes of setting up the equilibrium equations.

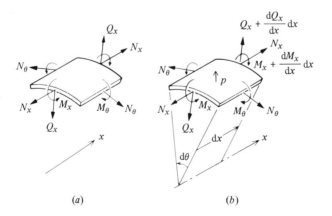

The next main group of variables describes the geometrical distortion of a typical element. When the shell deforms there will in general be (small) surface strains ϵ_x, ϵ_θ in the longitudinal and circumferential directions, respectively (positive for elongation), and shear strain $\gamma_{x\theta}$ in the surface. The symmetry of our present problem requires that $\gamma_{x\theta} = 0$. Deformation of the shell will also involve in general (small) changes of curvature κ_x, κ_θ in the longitudinal and circumferential directions, respectively, and change of twist $\kappa_{x\theta}$: by symmetry, however, $\kappa_{x\theta} = 0$ in the present problem. The positive sense of κ_x, κ_θ will be defined later.

The only remaining variable is the radial component, w, of the displacement of the shell. Its positive sense is outwards, i.e. the same as that of z. We are assuming that the circumferential component of displacement is zero, for there is no general reason why the shell should rotate (in contrast to, say, a problem involving torsional loading of the shell). The longitudinal component of displacement, u, plays a minor part in the analysis, as we shall see.

It is good practice in a problem of this sort to begin by writing down all of the equilibrium, compatibility and elastic-law equations. This first problem has been chosen, of course, so that the equilibrium and compatibility equations are particularly straightforward.

3.2.1 Equilibrium equations

While fig. 3.3a is adequate for the definition of stress resultants, it must be augmented by the indication of small changes in the variables, from one side of the element to the other as in fig. 3.3b, when we need to derive the equilibrium equations for the typical element. First let us resolve all the forces acting on the element in the z-direction. We have: (i) an applied pressure-loading of $pa\,\mathrm{d}x\,\mathrm{d}\theta$; (ii) an inbalance of shear force equal to $(\mathrm{d}Q_x/\mathrm{d}x)a\,\mathrm{d}x\,\mathrm{d}\theta$; and, (iii) a component from the two N_θ forces (each inclined at $\tfrac{1}{2}\pi + \tfrac{1}{2}\mathrm{d}\theta$ to the z-direction through the centre of the element) of $N_\theta\,\mathrm{d}x\,\mathrm{d}\theta$. Writing the force equilibrium equation and dividing throughout by $a\,\mathrm{d}x\,\mathrm{d}\theta$ we thus have

$$\frac{\mathrm{d}Q_x}{\mathrm{d}x} - \frac{N_\theta}{a} = -p. \tag{3.1}$$

The second equation is found by taking moments for the forces acting on the element about an axis which is tangential to the element in the circumferential direction. In the limit $\mathrm{d}x \to 0$ this becomes

$$\frac{\mathrm{d}M_x}{\mathrm{d}x} - Q_x = 0. \tag{3.2}$$

3.2 Basic equations of the problem

(It is precisely at this point that we exclude the nonlinear effect mentioned in section 3.1 by neglecting the couple associated with N_x (regarded as acting strictly parallel to the x-axis) by virtue of the slight difference in radial displacement at the two ends of the element.)

Equations (3.1) and (3.2) are the two equilibrium equations of the problem. It is sometimes convenient to combine them by the elimination of Q_x: this gives

$$\frac{d^2 M_x}{dx^2} - \frac{N_\theta}{a} = -p. \tag{3.3}$$

3.2.2 Compatibility equations

We shall now express the circumferential strain ϵ_θ and the changes of curvature κ_x, κ_θ all in terms of the displacement w. First consider ϵ_θ. At any plane x = constant, each point on the circumference is displaced outwards a small distance w, so that what was originally a circle of radius a is now enlarged to one of radius $a + w$. Since the circumferential strain is defined as change of circumference per unit original circumference, we have

$$\epsilon_\theta = w/a. \tag{3.4}$$

Here we are assuming – as indeed we shall do throughout the book – that $w/a \ll 1$; so there is no need to be fussy about the fact that we have used the original length in the definition of strain.

Next consider κ_x. In the undeformed configuration the generators are straight. The deformed generators have a curvature equal in magnitude to $d^2 w/dx^2$ (cf. equation 2.8). It is obviously convenient to have a sign convention for κ_x and κ_θ such that the coefficients on the leading diagonal of the Hooke's-law relation (2.28) are positive; and thus we must have

$$\kappa_x = -\frac{d^2 w}{dx^2}. \tag{3.5}$$

In particular, the positive sense of κ_θ is that which would increase the original curvature of the surface in the circumferential direction. This sign convention ensures that the positive sense of change of curvature corresponds to the positive sense of bending moment, as already defined, just as the positive sense of strain ϵ corresponds to the positive sense of stress resultant N.

Lastly, consider the change of curvature in the circumferential direction. A hoop of the shell had, originally, radius a, i.e. curvature $1/a$. In the distorted configuration the radius of curvature is $a + w$, and thus the curvature is $1/(a + w)$. Since $w/a \ll 1$ we can use the binomial theorem to write this, approximately, as $(1/a)(1 - w/a)$. Consequently, the change in curvature is given by

$$\kappa_\theta = -w/a^2. \tag{3.6}$$

In fact we shall neglect this, i.e. put $\kappa_\theta = 0$ in subsequent calculations. We make this hypothesis that κ_θ has an insignificant effect in comparison with κ_x on the grounds that we expect w to be a rapidly varying function of x, and that its successive derivatives with respect to x will make more important contributions to the behaviour. We ought, of course, to make a check on this hypothesis at the end of the calculation. This we shall do in section 3.5. We shall investigate this point more fully in chapters 8 and 9, when we consider non-symmetrical deformations of cylindrical shells.

3.2.3 Elastic law for the element

In this problem, as we have seen, $N_x = 0$ and $\kappa_\theta = 0$: consequently the elastic law for the element, which we developed in general in chapter 2 (equations 2.5, 2.28) reduces in the present problem to:

$$\epsilon_\theta = N_\theta/Et, \tag{3.7}$$

$$\kappa_x = M_x/D. \tag{3.8}$$

Here E is Young's modulus, t is the thickness of the shell and D is the flexural stiffness defined by (2.18). The elastic law also yields two other relations which are only indirectly involved in the analysis:

$$\epsilon_x = -\nu\epsilon_\theta, \tag{3.9}$$

$$M_\theta = \nu M_x. \tag{3.10}$$

The circumferential bending moment does not vary round the circumference, and is therefore *self-equilibrating* (cf. fig. 2.6b).

3.3 The governing equation and its general solution

The next task is to combine equations (3.3)–(3.5), (3.7) and (3.8) to give a 'governing equation' in a single variable. A convenient choice for this variable is w (but see problem 3.1 for the use of other variables). Starting with the compatibility equation (3.5) we have

$$\kappa_x = -\frac{d^2w}{dx^2}.$$

Then the elastic law (3.8) gives

$$M_x = -D\frac{d^2w}{dx^2}. \tag{3.11}$$

3.3 The governing equation and its general solution

Similarly, compatibility equation (3.4) may be combined with the elastic law (3.7) to give

$$N_\theta = (Et/a)w. \tag{3.12}$$

Lastly, we substitute into equilibrium equation (3.3) to obtain the required equation:

$$D\frac{d^4w}{dx^4} + \left(\frac{Et}{a^2}\right)w = p. \tag{3.13}$$

Now we shall mainly be concerned with problems involving edge-loads only, i.e. $p = 0$. In this case the governing equation simplifies to

$$\frac{d^4w}{dx^4} + \left[\frac{Et}{Da^2}\right]w = 0. \tag{3.14}$$

This is a kind of equation which we shall encounter frequently. It is sometimes known as the 'beam-on-elastic-foundation' equation: see section 9.3.

The properties of the shell enter only in the single constant enclosed in brackets, which has dimensions of (length)$^{-4}$. The constant may be rewritten, on account of (2.18):

$$Et/Da^2 = 12(1 - \nu^2)/a^2 t^2. \tag{3.15}$$

In order to solve the equation it is most convenient to rearrange the constant term by means of the substitution

$$Et/Da^2 = 4/\mu^4, \tag{3.16}$$

thereby defining a parameter μ, having the dimensions of length, as follows:

$$\mu = (at)^{\frac{1}{2}} / \left[3(1 - \nu^2)\right]^{\frac{1}{4}}. \tag{3.17}$$

The factor 4 has been used in (3.16) in order to make subsequent manipulations easier.

For a given shell the value of μ is determined primarily by the values of a and t; but it also depends weakly on the value of Poisson's ratio. For $\nu = 0.3$ (a typical value for metals)

$$\mu = 0.78 \, (at)^{\frac{1}{2}}; \tag{3.18}$$

but see the next section for a further discussion of the value of μ.

Note that in general μ is approximately equal to the geometric mean of a and t. For 'thin' shells in general $a \gg t$, and it follows that μ will normally be not only small in comparison with a, but also large in comparison with t.

Our governing equation may now be written

$$\frac{d^4w}{dx^4} + \left(\frac{4}{\mu^4}\right)w = 0. \tag{3.19}$$

It may be solved in the usual way by means of a trial solution of the form
$$w = Ae^{\alpha x}. \tag{3.20}$$
This yields the characteristic algebraic equation
$$(\alpha\mu)^4 = -4, \tag{3.21}$$
which has the four roots
$$\alpha\mu = \pm 1 \pm i. \tag{3.22}$$
The general solution of (3.19) may therefore be written
$$w = A_1 e^{-x/\mu} \cos(x/\mu) + A_2 e^{-x/\mu} \sin(x/\mu) + \\ A_3 e^{x/\mu} \cos(x/\mu) + A_4 e^{x/\mu} \sin(x/\mu). \tag{3.23}$$

Here $A_1, \ldots, A_4$ are four arbitrary constants, whose values will be determined in any given problem by boundary conditions.

Note that w is a function of the dimensionless variable x/μ: it is natural therefore to define a new dimensionless axial coordinate ξ, which becomes the working independent variable:
$$\xi = x/\mu. \tag{3.24}$$
There are several alternative equivalent ways of expressing (3.23), in terms of hyperbolic functions, etc. (see problem 3.2): the convenience of each depends mainly on the boundary conditions of particular problems, as we shall see in the following sections.

3.3.1 Definition of 'effective thickness' h

We saw in the preceding section that the convenient characteristic length μ depends mainly on the radius a and thickness t of shell, but also to some extent on the Poisson's ratio ν of the material through the factor $(1 - \nu^2)^{\frac{1}{4}}$. This slightly awkward function of ν will crop up again and again in the book, whenever there is an interaction between bending and stretching effects. It is tempting to decide once-for-all on a particular value of ν for use in numerical calculations. But it is also desirable to be able to put in different values of ν on occasion, if we wish. What we shall do, therefore, is to define an 'effective thickness' h by means of the relation
$$h = t/(1 - \nu^2)^{\frac{1}{2}}. \tag{3.25}$$
This enables us to write
$$\mu = (ah)^{\frac{1}{2}}/3^{\frac{1}{4}} = 0.760 \, (ah)^{\frac{1}{2}} \tag{3.26}$$
to three significant figures.

Thus for a shell of given thickness and made of a given material, our first step is to work out the 'effective thickness' h, and then use h in subsequent calculations. Of course, in many calculations it will be satisfactory to use

$$h \approx t$$

as a first approximation to (3.25); or, more accurately for materials having $\nu \approx 0.3$ (say),

$$h = 1.05t.$$

3.4 Semi-infinite cylindrical shell with edge-loading at $x = 0$

The semi-infinite shell with loads applied only at the edge $x = 0$ (with no pressure-loading) constitutes a special case of great practical importance. It is attractive for use as an example because the physically obvious requirement that w is finite as x (or ξ) $\to \infty$ immediately gives $A_3 = A_4 = 0$ in (3.23). We are then left with a general solution which decays rapidly in ξ, and is characterised by only two arbitrary constants, whose values will be determined by two conditions to be imposed at the edge $x = 0$.

In this particular problem, perhaps the most convenient way of proceeding is to rearrange the first two terms of (3.23) in the following form:

$$w = A e^{-\xi} \sin(\xi + \delta),$$

where A and δ (a phase angle) are two new constants, related to A_1, A_2 by

$$A_1 = A \sin\delta, \quad A_2 = A \cos\delta.$$

Now it is easy to show that

$$\frac{d}{d\xi}[e^{-\xi} \sin(\xi + \delta)] = -(2)^{\frac{1}{2}} e^{-\xi} \sin(\xi + \delta - \tfrac{1}{4}\pi);$$

and it follows that the nth derivative of this function is given by

$$\frac{d^n}{d\xi^n}[e^{-\xi} \sin(\xi + \delta)] = (-1)^n (2)^{n/2} e^{-\xi} \sin(\xi + \delta - \tfrac{1}{4}n\pi).$$

Thus the function w for the semi-infinite shell is closely related to all of its derivatives. In particular, differentiation with respect to ξ involves a simple *shift* of $\tfrac{1}{4}\pi$ in the abscissa in addition to a change of constant.

It is useful to construct a table showing how the various physically important quantities are related to the constants A and δ. For the sake of completeness we shall define another displacement variable Ψ, being the (small) rotation of the tangent to the generator:

$$\Psi = -\frac{dw}{dx}. \tag{3.27}$$

The positive sense of Ψ is determined by this definition. Using equations (3.27, 3.11, 3.2) we have:

$$w = Ae^{-\xi} \sin(\xi + \delta), \qquad (3.28a)$$

$$\Psi = -\frac{dw}{dx} = \frac{(2)^{\frac{1}{2}}}{\mu} Ae^{-\xi} \sin(\xi + \delta - \tfrac{1}{4}\pi), \qquad (3.28b)$$

$$M_x = -D \frac{d^2 w}{dx^2} = -\frac{2D}{\mu^2} Ae^{-\xi} \sin(\xi + \delta - \tfrac{1}{2}\pi), \qquad (3.28c)$$

$$Q_x = \frac{dM_x}{dx} = \frac{2(2)^{\frac{1}{2}}D}{\mu^3} Ae^{-\xi} \sin(\xi + \delta - \tfrac{3}{4}\pi). \qquad (3.28d)$$

We can also obtain easily the values of N_θ and κ_x from these expressions by means of the relations

$$N_\theta = wEt/a,$$

$$\kappa_x = M_x/D.$$

The form of (3.28) suggests that it will be convenient to define some normalised variables, as follows:

$$n = N_\theta a/Et = w, \qquad (3.29a)$$

$$\psi = \mu\Psi, \qquad (3.29b)$$

$$m = \mu^2 M_x/2D = \tfrac{1}{2}\mu^2 \kappa_x, \qquad (3.29c)$$

$$q = \mu^3 Q_x/2D. \qquad (3.29d)$$

Each of these variables has the dimensions of length. The new variables have been shorn of subscripts, since this is unlikely to cause confusion in the present problem.

Note that the normalised form n of N_θ is identical to w, and that similarly variable m does duty for both M_x and κ_x. The characteristic length parameter μ has been used in the definition of most of these variables, but it is a simple matter to express all variables in terms of the basic quantities of the problem: see problem 3.3.

For the remainder of the chapter we shall deal almost exclusively in terms of the dimensionally uniform variables w, n, ψ, m and q. In terms of the new variables, equations (3.28) simplify to:

$$w = Ae^{-\xi} \sin(\xi + \delta), \qquad (3.30a)$$

$$\psi = (2)^{\frac{1}{2}} Ae^{-\xi} \sin(\xi + \delta - \tfrac{1}{4}\pi), \qquad (3.30b)$$

$$m = -Ae^{-\xi} \sin(\xi + \delta - \tfrac{1}{2}\pi), \qquad (3.30c)$$

$$q = (2)^{\frac{1}{2}} Ae^{-\xi} \sin(\xi + \delta - \tfrac{3}{4}\pi). \qquad (3.30d)$$

Equation (3.19) is analogous to the governing equation for the transverse displacement of a *beam-on-elastic-foundation*. It is sometimes useful (see chapter 9) to express the results of the present chapter in terms of the analogous beam: see problem 3.4.

3.4 Cylindrical shell with edge-loading at $x = 0$

3.4.1 Some particular boundary conditions

We now give examples of some simple boundary conditions in which loading q, m is applied at edge $x = 0$. In this section we use subscript 0 to denote a value at $x = 0$.

(i) $q = q_0, m = 0$ (i.e. pure shear) at edge $x = 0$ (fig. 3.1a). The value of δ is found by substituting the boundary condition for m into (3.30c): hence $\delta = \frac{1}{2}\pi$. (Actually $\delta = \frac{1}{2}\pi + n\pi$ where n is an integer: we are, however, only interested in one value of δ.) Using this in (3.30d) we find $A = -q_0$. These values of A, δ, enable us to evaluate all of the quantities of interest at arbitrary values of ξ. In particular, we have displacement w_0 and rotation ψ_0 at the edge itself as follows:

$$w_0 = -q_0, \quad \psi_0 = -q_0.$$

(ii) $m = m_0, q = 0$ (i.e. pure couple) at edge $x = 0$ (fig. 3.1b). Proceeding as before we obtain

$$\delta = \tfrac{3}{4}\pi, \quad A = -(2)^{\frac{1}{2}} m;$$

and consequently

$$w_0 = -m_0, \quad \psi_0 = -2m_0.$$

(iii) $m = m_0, q = q_0$ at $x = 0$. For given numerical values of m_0, q_0 we can find the value of δ from (3.30c) and (3.30d). However, it is generally more convenient to regard the solution as the sum of solutions (i) and (ii), above. This is true for small deflections of this linear-elastic structure. In particular, the edge displacement and rotation are related to the applied shear force and bending moment by the following matrix relation, which combines the results of (i) and (ii):

$$\begin{bmatrix} w_0 \\ \psi_0 \end{bmatrix} = \begin{bmatrix} -1 & -1 \\ -1 & -2 \end{bmatrix} \begin{bmatrix} q_0 \\ m_0 \end{bmatrix}. \tag{3.31}$$

It is not surprising that the result can be expressed in terms of a matrix, as this problem is one in classical small-deflection elastic theory of structures. Indeed, since w_0, q_0 and ψ_0, m_0 are *corresponding* variables (see appendix 2 and problem 3.5), we should expect, as a consequence of Maxwell's reciprocal theorem, that the matrix will be symmetrical about the leading diagonal. And indeed it is. (But there might well have been an odd negative sign in consequence of the (arbitrary) sign convention which we have used.) The fact that the coefficients in the matrix are simple numbers is a consequence of our definition of normalised variables. It is a good exercise to rewrite (3.31) in the original variables.

An advantage of the use of constants A, δ instead of A_1, A_2 is that it is easy to plot the distribution of any required variable along the length of the shell. Suppose for example, we have $q_0 = 1$, $m_0 = 2$ and we wish to plot m, q, ψ and w along the shell. First we use (3.30c) and (3.30d) to obtain

$$w_0 = -3, \quad \psi_0 = -5.$$

Our next task is to find values of the constants A, δ, consistent with these edge-conditions. At $\xi = 0$, (3.30) can be simplified to:

$$w_0 = A \sin\delta, \tag{3.32a}$$

$$\psi_0 = (2)^{\frac{1}{2}} A \sin(\delta - \tfrac{1}{4}\pi) = A(\sin\delta - \cos\delta), \tag{3.32b}$$

$$m_0 = -A \sin(\delta - \tfrac{1}{2}\pi) = A \cos\delta, \tag{3.32c}$$

$$q_0 = (2)^{\frac{1}{2}} A \sin(\delta - \tfrac{3}{4}\pi) = A(-\sin\delta - \cos\delta). \tag{3.32d}$$

It follows that $\tan\delta = w_0/m_0 = -1.5$, so $\delta = 2.16$ or $124°$ (taking $0 < \delta < \pi$). The same value of δ could have been found, of course, from any pair of values of the edge variables.

Fig. 3.4a shows a graph of the function $e^{-\xi} \sin\xi$. Since the function decays so rapidly with x, a plot with a logarithmic ordinate is also given in fig. 3.4b. A disadvantage of this plot, of course, is that zeros are not shown clearly; but we know that they occur at $\xi = n\pi$, where n is an integer. The advantage of these graphs is that once the value of δ is known, any required function is easily found by means of an appropriate shift of abscissa. Here, for example, $\xi = 0$ for the w-function corresponds to point W on the curve, which has abscissa 2.16 on the diagram. The scale of the ordinate is arbitrary, but we know of course that $w_0 = -3$; consequently it is a simple matter to read off values of w at arbitrary values of ξ. A transparent overlay, made by transferring fig. 3.4b onto tracing paper, is obviously useful. The starting point for ψ is at point Ψ for this particular problem; this has been set back $\tfrac{1}{4}\pi = 45°$ from W in the ξ-direction. Similarly the starting points for m and q are marked M, Q. The resulting graphs are shown in fig. 3.4c.

An important point to emerge from this example is that the four functions w, ψ, m, q, are all essentially the same function, but starting at different points on account of the shift of $\tfrac{1}{4}\pi$ which corresponds to differentiation.

3.4.2 Other boundary conditions

In the preceding example two *force* boundary conditions were imposed at $x = 0$. It is easy to imagine problems in which two *displacements* (ψ_0, w_0) are specified (see problem 3.6), and it is required to find the corresponding forces. It is clear from an inspection of equations (3.30) that the

3.4 Cylindrical shell with edge-loading at x = 0

specification of any two of the four quantities m_0, q_0, ψ_0, w_0 is sufficient to determine the two constants A, δ, and hence fix the whole solution. In fact it is not necessary to do an *ad hoc* determination of A, δ in order to find the values of the two unspecified quantities. The information is contained in (3.31), and this may readily be rearranged in order to give any two of the four quantities on the right-hand side. There are altogether six cases, and the corresponding matrices are tabulated below.

Fig. 3.4. Graph of $e^{-\xi} \sin \xi$ on (a) linear and (b) log-linear scales. (c) Distribution of radial displacement w, etc. in a shell with particular boundary conditions, derived with the aid of (a) and (b). The points marked M, etc. in (a), (b) are explained in the text.

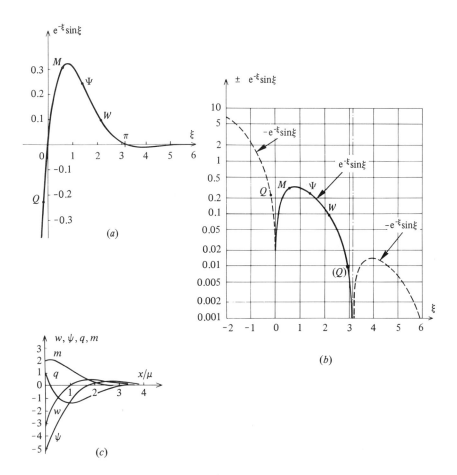

54   *Cylindrical shells under symmetric loading*

$$\begin{bmatrix} w \\ \psi \end{bmatrix} = \begin{bmatrix} -1 & -1 \\ -1 & -2 \end{bmatrix} \begin{bmatrix} q \\ m \end{bmatrix}, \quad \begin{bmatrix} q \\ m \end{bmatrix} = \begin{bmatrix} -2 & 1 \\ 1 & -1 \end{bmatrix} \begin{bmatrix} w \\ \psi \end{bmatrix},$$

$$\begin{bmatrix} \psi \\ m \end{bmatrix} = \begin{bmatrix} 2 & 1 \\ -1 & -1 \end{bmatrix} \begin{bmatrix} w \\ q \end{bmatrix}, \quad \begin{bmatrix} w \\ q \end{bmatrix} = \begin{bmatrix} 1 & 1 \\ -1 & -2 \end{bmatrix} \begin{bmatrix} \psi \\ m \end{bmatrix}, \quad (3.33)$$

$$\begin{bmatrix} q \\ \psi \end{bmatrix} = \begin{bmatrix} -1 & -1 \\ 1 & -1 \end{bmatrix} \begin{bmatrix} w \\ m \end{bmatrix}, \quad \begin{bmatrix} w \\ m \end{bmatrix} = \begin{bmatrix} -\tfrac{1}{2} & \tfrac{1}{2} \\ -\tfrac{1}{2} & -\tfrac{1}{2} \end{bmatrix} \begin{bmatrix} q \\ \psi \end{bmatrix}.$$

Note that in setting out these matrices we have dropped the subscript 0. The reason for this is that although we have so far thought of the linear relations between m, q, ψ, w as applying at the *edge* of the shell, they do in fact apply at *all values of x* in the shell. This idea can be explained in several ways. First, we can imagine a shell which is loaded at the edge $x = 0$ being subsequently cut at $x = l$, say, with the stress resultants q, m at this plane held at their previous values by externally applied forces. In other words, q, m at $x = l$ may be regarded as *edge*-loads for the *remainder* of the shell. Consequently, matrices (3.33) apply not only at the edges but also at the interior. This simple conclusion depends strongly on two features of a semi-infinite cylindrical shell. The first is that the *cylindrical* shell retains its characteristic geometry when a slice is cut off (unlike, e.g. a spherical shell: see chapter 11); and the second is that a *semi-infinite* shell remains semi-infinite when a slice is cut off (unlike a shell of *finite* length). An alternative explanation of the universality of the linear relations between m, q, ψ and w is simply that each of the matrices represents the two identities between the four variables which can be obtained by eliminating A and δ in various ways from the four relations (3.28), and is thus true for all values of x.

3.5   The ring-loaded long cylindrical shell

A problem which recurs frequently in the analysis of shell structures is the response of a long shell to a uniform radial ring-load of intensity F per unit length applied at a particular cross-section as shown in fig. 3.5a. If the cylindrical shell is sufficiently long we can regard it as being of infinite length for the purposes of stress analysis. We can then obviously regard the shell as consisting of two semi-infinite shells, back-to-back.

We must be careful when we apply our previous work to a shell extending to the *left* of the loaded edge. For this shell the x-axis (which we shall call x_1 to distinguish it from the x-axis of the other cylinder) extends with its posi-

3.5 The ring-loaded long cylindrical shell

tive sense to the left. Using precisely the same definitions as before, we find that the positive sense of the variables at the edge of this shell are as shown in fig. 3.5b. Here we have used the subscript 1 for all variables of the second shell. The diagrams for the two edges are thus complete mirror images of each other.

It is clear by symmetry, since the loading is in the plane of symmetry (cf. problem 3.7) that the corresponding variables at the two edges are equal:

$$Q = Q_1, \quad M = M_1, \quad w = w_1, \quad \Psi = \Psi_1. \tag{3.34}$$

Fig. 3.5. (a) A cylindrical shell under 'ring' loading. (b) Use of previous results for an edge-loaded shell and its mirror-image. (c) Distribution of the main variables. (d) The peak value of N_θ is equal to that which would be obtained if the ring-loading F were applied to a compact ring of radius a and cross-sectional area $2\mu t$. (e) The peak value of M_x would be obtained by applying a transverse load F to the centre of a simply-supported beam of unit width, length μ ($\approx 0.76\,(ah)^{\frac{1}{2}}$), and thickness t; $h(\approx t)$ is an effective thickness, defined in (3.25).

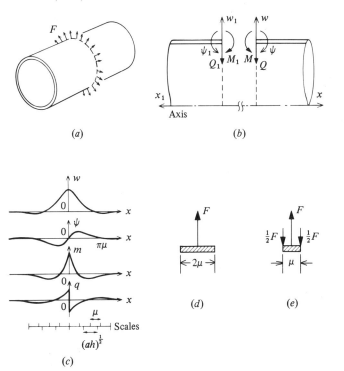

But there is also no discontinuity of slope at the junction; i.e. the clockwise (say) rotation of the two is identical. Thus $\Psi = -\Psi_1$; and it follows from (3.34) that $\Psi = 0$. Finally, the force F is carried by the shear forces at the two edges:

$$-F = Q + Q_1.$$

Therefore, from (3.34) we have

$$Q = -\tfrac{1}{2}F. \tag{3.35}$$

The two conditions imposed at the edge of the right-hand shell are thus that (i) the value of q is given, and (ii) $\psi = 0$. Defining a normalised ring-load f as follows (cf. (3.29)).

$$f = \mu^3 F/2D \; (= 2 \times 3^{\tfrac{1}{4}} \, Fa^{\tfrac{3}{2}}/Eth^{\tfrac{1}{2}}),$$

we have $q_0 = -\tfrac{1}{2}f$. From the lower right-hand matrix of (3.33) we thus obtain

$$w_0 = \tfrac{1}{4}f,$$
$$m_0 = \tfrac{1}{4}f.$$

Lastly, reverting to the original variables (see (3.29)) we find the following simple expressions for N_θ and M_x at the plane of the loading:

$$N_{\theta 0} = Fa/2\mu = 0.658F(a/h)^{\tfrac{1}{2}}, \tag{3.36a}$$
$$M_{x0} = \mu F/4 = 0.190F(ah)^{\tfrac{1}{2}}. \tag{3.36b}$$

The distribution of the main variables along the shell is shown in fig. 3.5c. Note that the diagrams for w and m are symmetric about the plane of loading, while those for ψ and q are 'antisymmetric'.

This is a convenient point at which to re-examine the validity of the step in the analysis at which we neglected κ_θ in comparison with κ_x. By (3.6) we find that κ_θ has a peak value of $f/4a^2$ at $x = 0$. The peak value of κ_x also occurs at $x = 0$: by (3.29) it is $f/2\mu^2$. Recalling that $\mu = 0.76(ah)^{\tfrac{1}{2}}$ we find that the peak-value ratio of κ_x to κ_θ is about $3a/t$. Our assumption that κ_θ is negligible is clearly justified for thin-walled shells in the present example.

We can also investigate the relative magnitudes of the various stresses induced by the given loading, with a view to checking on the assumption, introduced in chapter 2, that the level of shearing stress due to Q_x is so low in comparison with the level of tensile stress due to N_θ and M_x that distortion due to Q_x is negligible. For this purpose we can take $t \approx h$.

The largest value of Q_x has magnitude $\tfrac{1}{2}F$, and so the magnitude of the peak stress τ_{xz} is $3F/4h$. The peak hoop stress has magnitude $N_{\theta 0}/h = 0.66 \, Fa^{\tfrac{1}{2}}h^{-\tfrac{3}{2}}$, and the peak bending stress (see (2.29)) has magnitude $6M_{x0}/t^2 = 1.2 \, Fa^{\tfrac{1}{2}}h^{-\tfrac{3}{2}}$. We thus find that the hoop and bending stresses have the same order of magnitude $Fa^{\tfrac{1}{2}}h^{-\tfrac{3}{2}}$, while the magnitude of the shearing stress is about Fh^{-1}. Thus the latter has a negligible effect provided $(h/a)^{\frac{1}{2}} \approx (t/a)^{\frac{1}{2}} \ll 1$. This requirement will normally be satisfied for thin shells, but it could become stringent for less thick shells with $a/t < 20$, say.

In this analysis we have neglected the intense pressure stresses on the surface which are associated with a concentrated ring-load. We shall return to this point at the end of section 3.6.

Since the peak values of N_θ and M_x occur at the plane of loading, formulas (3.36) are useful in many design situations. It is sometimes helpful to endow them with physical meaning related to the characteristic length $\mu = 0.76(ah)^{\frac{1}{2}}$. If the ring-load were carried by a short ring of shell of length 2μ with *uniform* circumferential stress N_θ, the value of N_θ would be the same as the peak value in the infinitely long shell: see fig. 3.5d. Furthermore, M_{x_0} is equal to that which would occur at the centre of a beam of length μ, simply supported at its ends: see fig. 3.5e.

All of the above results were obtained, of course, for shells of uniform thickness. It is not difficult, however, to find corresponding formulae for a cylindrical shell having a step change in thickness at the plane of loading; see problem 3.8.

3.6 The band-loaded cylindrical shell

Consider next a related problem in which the long cylindrical shell is loaded by a uniform pressure p over a short length $2l$, as shown in fig.3.6a. It is convenient to regard this loading as a 'spread-out' version of the ring-load

Fig. 3.6. (a) As fig. 3.5a except that the ring-load F has been replaced by a band of uniform pressure $F/2l$ applied over an axial length $2l$. (b) The pressure load on a short section of shell may be treated as a ring-load. (c) N_θ at the plane of mirror symmetry may be found by an appropriate integration. (d) Reduction factors k_n, k_m for peak values of N_θ, M_ϕ on account of the change from ring-load to band-load.

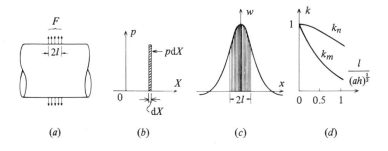

of the previous problem, and therefore to arrange for the total load on a longitudinal strip of the shell to be equal in both cases. Thus we put

$$p = F/2l. \tag{3.37}$$

We shall be interested to know, among other things, whether a 'spreading' of the load in this way over a short length of shell has an appreciable effect upon the magnitudes of the peak stress resultants N_θ and M_x.

We could solve the problem by thinking about a length $2l$ of uniformly loaded shell connected at each edge to a semi-infinite unloaded shell. It is much simpler, however, to make use of the previous results by means of the principle of superposition. For example, let us try to calculate the value of N_θ at the central plane in this way. Consider the effect of the pressure on the shell between cross-sections distant X and $X + dX$ from the central plane, as shown in fig. 3.6b. Its effect will be that of a ring-load of intensity pdX at a section distant X from the plane of loading; i.e. pdX multiplied by the ordinate of the curve of fig. 3.6c, which has been redrawn from a curve in fig. 3.5c. The value of N_θ at the central plane is thus equal to p times the shaded area in fig. 3.6c, from which the required answer may be calculated in principle. It is most convenient to express the answer in the form of (3.36) but incorporating an 'attenuation factor' k_n, which is a function of the length of the band of loading:

$$N_{\theta 0} = (Fa/2\mu) k_n. \tag{3.38}$$

It follows that the factor k_n is equal to the mean ordinate of the shaded area of fig. 3.6c divided by the peak ordinate. Thus, when $l = 0$, $k_n = 1$. The calculation for other values of l is straightforward, as the necessary integration is not difficult (see problem 3.9). The result is

$$k_n = (1 - e^{-\zeta} \cos\zeta)/\zeta, \quad \text{where } \zeta = l/\mu \tag{3.39}$$

The factor k_n is plotted in fig. 3.6d.

Similarly, we may define a peak bending-moment attenuation factor k_m by

$$M_{x0} = \tfrac{1}{4}\mu F k_m. \tag{3.40}$$

Clearly k_m is equal to the normalised mean value of m in fig. 3.5c, over the loaded region, and it follows (see problem 3.9) that

$$k_m = (e^{-\zeta} \sin\zeta)/\zeta. \tag{3.41}$$

This is also plotted in fig. 3.6d.

It is clear from this diagram that by spreading a ring-load over a *small* length of shell we may reduce the peak value of M_x considerably, while leaving the peak value of N_θ almost unchanged. For example, by having

$l = 0.5(ah)^{\frac{1}{2}}$ we reduce the peak value of M_x to about 0.5 of its former value, while reducing the peak value of N_θ only to about 0.9 of its former value.

In fact, for small values of ζ (say $\zeta < 0.5$ or $l/(ah)^{\frac{1}{2}} < 0.4$) the effect on M_x is almost exactly the same as redistributing uniformly the load on the simple 'equivalent' beam shown in fig. 3.5e; see problem 3.10. We shall return to this remarkable result at the end of the following section. A limiting case of the loading of fig. 3.6 is considered in problem 3.11.

The present example also enables us to make a rough estimate of the magnitude of the 'through-thickness' stresses σ_z in the vicinity of the loaded region of a ring- or band-loaded cylindrical shell: cf.section 2.4.2. In the shell immediately under the pressurised zone the magnitude of $|\sigma_z|$ varies between p at the loaded surface and zero at the unloaded surface. Let us therefore compare the magnitudes of p and the peak circumferential stress in order to assess the circumstances in which the magnitude of the through-thickness stress may be significant.

For these purposes it is adequate to put $k_n \approx 1$ in (3.38), and to write
$$\sigma_\theta = (N_{\theta 0}/h) \approx 0.66\,(F/h)\,(a/h)^{\frac{1}{2}}.$$
Putting $F = 2pl$ and rearranging, we have
$$(p/\sigma_\theta) \approx 0.7\,(h/a)\,[(ah)^{\frac{1}{2}}/l]. \tag{3.42}$$

Thus p/σ_θ reaches order 1 only if l is so small that $l/(ah)^{\frac{1}{2}}$ is of the same order as h/a. Typically, a band of pressure-loading might be applied through a member of thickness h; in which case $p/\sigma_\theta \approx 1.4(h/a)^{\frac{1}{2}}$, which will usually be small.

3.7 Cylindrical shell loaded sinusoidally along its length

Our next example, involving a shell loaded by axially varying pressure $p = p_0 \sin(\pi x/l)$ (see fig. 3.7) is important in two respects. First, it provides the basis of analysis of the shell under arbitrary periodic loading by means of Fourier series; and some useful results can be obtained in this way. Second, it furnishes a good example for a study of the way in which the shell carries the applied loading by a combination of bending and stretching effects. In this

Fig. 3.7. A pressure which varies sinusoidally with the axial coordinate x.

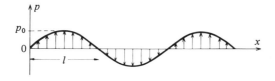

respect it provides an introduction to the ideas which are to be exploited in chapter 8. In particular, it enables us to understand more clearly the results of the preceding example concerning the effect of 'load-spreading' on the distribution of bending- and stretching-stress resultants. Third, it provides (in section 3.7.3) a simple means of studying the nonlinear consequences of a uniaxial tension in the shell.

In the absence of stress resultants N_x of significant magnitude, the governing equation for this problem is (see (3.13)):

$$\frac{d^4 w}{dx^4} + \frac{Et}{Da^2} w = \frac{p_0}{D} \sin\left(\frac{\pi x}{l}\right). \tag{3.43}$$

We shall assume that the shell is so long that the boundary conditions at the ends do not affect the portion under investigation. Thus, in contrast with the preceding problem, we shall be concerned only with the *particular integral* of the equation. This we now try to find. It is clear that the equation may be satisfied by a sinusoidal displacement function having the same half wavelength l, as the pressure distribution. Thus, putting

$$w = w_0 \sin(\pi x/l)$$

and substituting in (3.43), we find that

$$w_0 = (p_0 a^2/Et) \left[1 + (\pi^4 D a^2/l^4 Et)\right]^{-1}. \tag{3.44}$$

This may be rearranged in a more convenient form by the introduction of the length parameter μ and the definition of a function Φ of the dimensionless length l/μ as follows, with μ defined as before in (3.26):

$$\Phi(l/\mu) = \left[1 + \left((2)^{\frac{1}{2}} l/\pi\mu\right)^4\right]^{-1}. \tag{3.45}$$

Thus (3.44) may be written

$$w_0 = (1 - \Phi) p_0 a^2/Et. \tag{3.46}$$

Now imagine for the moment that the shell has *zero* bending stiffness in the longitudinal direction. In this condition it can only carry the applied loading by acting as a set of independent hoops. It is easy to calculate the radial displacement corresponding to the elastic distortion of these hoops, either from first principles or by using (3.43) with the first term (which corresponds to bending stiffness) omitted. In fact the deflection corresponds exactly to (3.46) with $\Phi = 0$. But inspection of (3.45) shows that if $l/\mu \gg 1$ the value of Φ will be negligibly small. Thus we conclude that for 'long-wave' loading $l/\mu \gg 1$ the shell carries the applied loading almost entirely by hoop action.

Consider on the other hand the case when $l/\mu \ll 1$, i.e. 'short-wave' loading. Here we see from (3.45) that the value of Φ is approximately equal to 1; and so the amplitude of radial displacement is, by (3.46), rather small. The shell now seems to be carrying the applied loading with negligible deflection. What is happening, in fact, is that the loading induces negligible hoop stress (which is proportional to w) but instead is carried almost entirely by beam-action along the generators. In this case the *second* term on the left-hand side of (3.43) has negligible effect.

The functions Φ and $1 - \Phi$ are plotted on both linear and logarithmic scales in fig. 3.8. At the two extremes which we have discussed Φ takes on values near 0 and 1 respectively; but it is also clear that there is an intermediate range of l/μ for which neither hoop stretching nor longitudinal bending dominates, and the load is carried by cooperative action between the two effects.

It is advantageous to endow the parameter Φ with a simple physical meaning. Φ is in fact the *fraction of the applied loading which is carried by longitudinal bending action*. The remaining fraction $1 - \Phi$ is thus carried by 'hoop' action.

3.7.1 A 'two-surface' interpretation of the behaviour of the shell

The picture of an interaction between bending and stretching effects which emerges from this problem can be clarified if the shell is imagined to be dissected into two surfaces which separately embody the 'hoopwise stretching' and 'longitudinal bending' aspects of shell stiffness, respectively. Now whenever we analyse a structure by making a 'cut', we must define as new 'external' force variables the stresses which were previously transmitted across the cut. In the present context we must have an 'interface' tensile

Fig. 3.8. Factors Φ and $1-\Phi$ expressed as functions of the dimensionless half wavelength $l/(ah)^{\frac{1}{2}}$; (a) on linear scales, (b) on doubly-logarithmic scales.

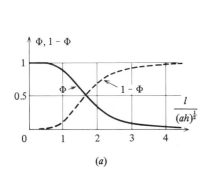
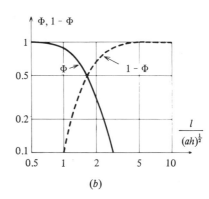

stress which pulls equally, but in opposite senses, on the two surfaces. These two surfaces and the interface stress p_B are shown in fig. 3.9a. In general p_B varies over the surface of the shell. At an arbitrary value of x the hoop stretching surface is required to carry a pressure $p - p_B$, while the longitudinal bending surface carries p_B itself. The interface tension p_B is in general an unknown function of x which must be determined so that the radial displacement of the two surfaces is equal for all values of x. The present example is particularly simple, of course: p_B turns out to be a constant fraction Φ of the applied pressure p, and the value of Φ depends only on the wavelength of the applied pressure function.

In the present problem it is instructive to evaluate the function Φ in a second way by considering the behaviour of the two surfaces separately. The governing equation of the stretching surface, written in terms of the radial displacement w_S of this surface is (cf.(3.43)):

$$\frac{Etw_S}{Da^2} = (1 - \Phi)\frac{p_0}{D}\sin\left(\frac{\pi x}{l}\right). \tag{3.47}$$

Similarly, the governing equation of the bending surface, in terms of its radial displacement w_B, is

$$\frac{d^4 w_B}{dx^4} = \Phi\frac{p_0}{D}\sin\left(\frac{\pi x}{l}\right). \tag{3.48}$$

Solving these separately we obtain

Fig. 3.9. (a) The shell under the loading of fig. 3.8 has been split into a 'stretching surface' S-S and a 'bending surface' B-B, here shown separately. There is an interface tension p_B; thus the S-surface carries the applied pressure less p_B, while the B-surface carries p_B. In this particular example p_B is a fraction Φ of p. (b) The S- and B-surfaces may be regarded as two springs in parallel, sharing an applied load unequally, according to their relative stiffness. (c) The interface pressure p_B exceeds $0.8p$ for $l/\mu < 1.5$, but is less than $0.2p$ for $l/\mu > 3.1$.

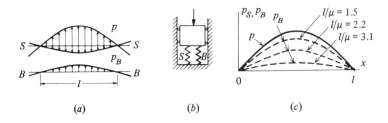

3.7 Cylindrical shell loaded sinusoidally along length

and
$$w_S = (1 - \Phi) \frac{p_0 a^2}{Et} \sin\left(\frac{\pi x}{l}\right) \tag{3.49}$$

$$w_B = \Phi \frac{l^4}{\pi^4} \frac{p_0}{D} \sin\left(\frac{\pi x}{l}\right). \tag{3.50}$$

Finally, equating w_S and w_B, we obtain the previous expression (3.45) for Φ.

In this analysis we are treating the bending and stretching surfaces as two 'parallel' structures, shown schematically in fig. 3.9b, which share the applied loading unequally. For short-wave loading the stiffness of the 'bending' surface is much greater than that of the 'stretching' surface, and so it dominates; but for long-wave loading the situation is reversed. The changeover of relative stiffness is due simply to the fact that the stiffness of a uniform elastic beam is inversely proportional to the fourth power of its span, here expressed as (3.50). This relatively high fourth power of l accounts for the relatively narrow range of $l/(ah)^{\frac{1}{2}}$ over which the function Φ changes between the extremes 0 and 1. Thus, a factor of $2^{\frac{1}{2}} \approx 1.4$ in l changes the relative stiffness by a factor of 4, with the results shown diagrammatically in fig. 3.9c.

It is, of course, not surprising that the parameter $\mu = 0.76(ah)^{\frac{1}{2}}$ should appear in this analysis. The half wavelength of the loading for which the load is shared equally between the two effects is given by

$$l = \pi\mu/2^{\frac{1}{2}} \approx 1.7(ah)^{\frac{1}{2}}. \tag{3.51}$$

3.7.2 The use of Fourier series

Let us now consider an example of the application of this result to a problem requiring the use of Fourier series.

A long cylindrical shell is loaded by a normal pressure which varies in the axial direction with a *triangular* waveform which has half wavelength L, and peak value p^*, as shown in fig. 3.10a. We require the value of M_x at values of x corresponding to $p = p^*$.

Fig. 3.10. (a) A pressure which has a 'saw tooth' variation with the axial coordinate x. (b) Dimensionless plot of the peak value M^* of M_x as a function of dimensionless wavelength $L/(ah)^{\frac{1}{2}}$.

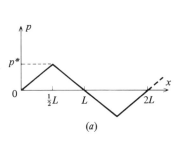
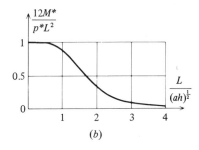

The first step is to do a Fourier analysis of the applied loading. This gives (see appendix 10)

$$p = \sum_{n=1,3,5\ldots} (-1)^{(n-1)/2} \, p_n \sin(n\pi x/L), \qquad (3.52)$$

where

$$p_n = (8/n^2\pi^2)p^*.$$

The value of coefficient p_n thus decreases sharply with n.

If the value of L is given as a multiple of $(ah)^{\frac{1}{2}}$ we can readily find from fig. 3.8 the fraction $\Phi = \Phi_n$ of the pressure-loading which is carried by beam action for any given term in the series. Thus the Fourier series for the interface pressure-load p_B applied to the beam-surface is:

$$p_B = \sum_{n=1,3,5\ldots} (-1)^{(n-1)/2} (8\Phi_n p^*/\pi^2 n^2) \sin(n\pi x/L). \qquad (3.53)$$

For a particular term in this series we can obtain the corresponding bending moment by using the equilibrium equation

$$\frac{d^2 M_x}{dx^2} = -p_B \qquad (3.54)$$

(cf.3.3). Hence, by performing a simple integration we have the following expression for M_x:

$$M_x = \sum_{n=1,3,5\ldots} (-1)^{(n-1)/2} (8L^2\Phi_n p^*/\pi^4 n^4) \sin(n\pi x/L) \qquad (3.55)$$

At the location in question, $x = \frac{1}{2}L$, the required result is given by the simple sum

$$M_x^* = p^*L^2 \, (8/\pi^4) \sum_{n=1,3,5\ldots} (\Phi_n/n^4). \qquad (3.56)$$

The summation is not difficult to perform numerically for any particular value of $L/(ah)^{\frac{1}{2}}$, and a graph showing the variation of M_x^* with $L/(ah)^{\frac{1}{2}}$ is shown in fig. 3.10b.

For sufficiently small values of this parameter, the factor Φ is approximately equal to unity for every term of the series. Putting $\Phi_n = 1$ (exactly) for all values of n in (3.56) gives $M_x^* = \frac{1}{12}p^*L^2$ (Jolley, 1961, p.64). In this case the 'hoop' action of the shell is entirely negligible; and the problem reduces to one of a long beam supporting the loading of fig. 3.10a, for which the above result may be obtained directly. At the other extreme, when $L/(ah)^{\frac{1}{2}}$ is large, the value of M_x^* may be determined by the use of the edge-response relations (3.31) for a semi-infinite shell: see problem 3.12.

Various other cases of periodic pressure-loading can be treated in a similar way by the use of Fourier series: see problem 3.13 for an example.

To conclude this section we give in fig. 3.11 some graphs showing how the applied loading is shared between the beam and hoop actions of the shell for the ring- and band-loading studied earlier (figs 3.5 and 3.6). For the range of cases shown an alteration in the width of the loaded band has virtually no effect on the pressure carried by the 'hoop' effect: the changes of loading occur over a short span of length and are consequently absorbed almost entirely by changes in loading carried by the 'beam-surface.' In this case, of course, most parts of the shell are free from surface loading; and consequently in these parts the pressure carried by the 'hoops' is of the same magnitude but of opposite sign from that carried by the longitudinal 'beams'.

The main aim of the analysis in this and the preceding section has been to introduce the idea of splitting the shell, conceptually, into two surfaces which carry load exclusively by 'stretching' and 'bending' action respectively. We shall find in chapter 8 that this idea is valuable in the analysis of more complicated problems. In the example which we have used – a symmetrically loaded cylindrical shell – the compatibility condition for the matching of displacements of the two deflected surfaces was easy to express: the radial displacements of the two surfaces were equal. The main problem in applying the idea to shells having more complicated surfaces is to develop a suitable compatibility condition. This will be done in chapter 6.

Fig. 3.11. Variation of applied pressure p and p_S ($= p - p_B$) for two particular uniform band-loadings having equal values of F. Two curves of p_S are practically the same. The shell responds to a change in local detail of loading almost entirely by means of changes in p_B ($= p - p_S$). These results and conclusions do not depend on the width of the bands, provided the wider band does not exceed by much the width indicated.

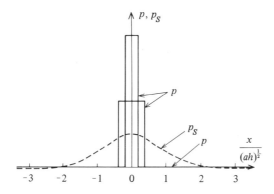

3.7.3 The nonlinear effect of axial tension

The contribution of N_x to the moment-equilibrium of a general element of the shell was explicitly excluded in the derivation of (3.2), and thus the nonlinear effects mentioned at the end of section 3.1 have been disregarded in all of the examples which we have considered in this chapter so far. It is clearly important for us to know under what circumstances this non-linear effect may make a significant difference to the mode of action of the shell. It is convenient to examine this problem at this point, in connection with a shell loaded by a pressure which varies sinusoidally along its length. The effect which has been ignored hitherto may be reintroduced into the problem in a particularly simple way in consequence of the remark that the applied pressure p is carried not only by a combination of circumferential stretching and longitudinal bending in the shell according to the 'two-surface' scheme, but also, in part, by the sinusoidally distorted generators acting as curved 'strings' under the action of N_x(= constant). The governing equation (3.43) may thus be altered in order to include this contribution:

$$D \frac{d^4 w}{dx^4} - N_x \frac{d^2 w}{dx^2} + \frac{Et}{a^2} w = p_0 \sin\left(\frac{\pi x}{l}\right). \tag{3.57}$$

The solution of this equation in the absence of boundary effects is, of course, very straightforward when w has the form $w_0 \sin(\pi x/l)$.

It is particularly instructive to enquire how the bending-stress resultant M_x varies along the shell in these circumstances. Solving the equation and using (3.11), we have

$$M_x = p_0 \sin\left(\frac{\pi x}{l}\right) \bigg/ \left[\frac{\pi^2}{l^2} + \frac{N_x}{D} + \frac{Et}{Da^2} \frac{l^2}{\pi^2}\right]. \tag{3.58}$$

Now when $N_x = 0$ we recover our previous solution. In this case the denominator on the right-hand side has a deep minimum with respect to l which occurs when (3.51) is satisfied, i.e. precisely when l is such that the bending and stretching surfaces share the applied loading equally. For this particular value of l the amplitude of M_x is thus maximum for a given value of p_0:

$$M_x = [ah/(4 \times 3^{\frac{1}{2}})] \, p_0 \sin(\pi x/l). \tag{3.59}$$

Suppose now that $N_x > 0$. The denominator on the right-hand side of (3.58) still has its minimum value at the same value of l as before, but the minimum is now somewhat less deep: consequently the amplitude of M_x will be somewhat attenuated. It is easy to show that for the value of l which maximises the amplitude of M_x, expression (3.59) still holds provided the right-hand side is divided by a denominator

$$(1 + N_x/N_0), \quad \text{where } N_0 = Eth/3^{\frac{1}{2}}a. \tag{3.60}$$

3.8 Edge-loading of finite elastic cylindrical shells

An axial tension N_x thus reduces the amplitude of M_x in general. This is precisely because N_x helps to carry part of the applied pressure by virtue of the curvature of the generators in the deformed state of the shell. Conversely, of course, a *negative* value of N_x – i.e. an axial compression – will *increase* the amplitude of M_x; and indeed, when $N_x = -N_0$ the denominator vanishes and so the amplitude of M_x formally becomes infinite. This corresponds, physically, to the *buckling* of the shell in an axisymmetric mode under a special value of uniform axial compression. We shall investigate buckling problems in detail in chapter 14. For present purposes we note that provided

$$|N_x| < 0.1 N_0, \qquad (3.61)$$

say, the axial loading will have relatively little effect on M_x even at the half wavelength l at which the amplitude of M_x is maximum, and *a fortiori* that the effect will be smaller at all other wavelengths. Thus, by invoking the idea that a general loading $p(x)$ may be expressed as an infinite Fourier series, we see that provided (3.61) is satisfied, the methods of the preceding sections will give satisfactory results. Wittrick (1963) studied in detail the influence of axial tension on the distribution of stress resultants in several axisymmetric problems involving discontinuities. He did not use Fourier series, but solved the governing equations directly. In particular he showed that in a cylindrical shell under ring-loading (fig. 3.5a) the presence of uniform axial tension reduces the peak values of both M_x and N_θ by the single factor $(1 + N_x/N_0)^{\frac{1}{2}}$.

3.8 Edge-loading of finite elastic cylindrical shells

We now return to the kind of problem which we studied earlier, involving the response of an otherwise unloaded shell to forces applied symmetrically to the edges. The previous examples were simplified, as we saw, by the fact that for a semi-infinite shell the complete solution of the governing differential equation involved only two constants. In this section we shall consider a shell of finite length l having two edges, each with its associated edge shearing force, bending moment, displacement and rotation, as shown in fig. 3.12. We use the same sign convention and notation as before, together with the convention that the value of x at the edge labelled 2 is greater than that for the edge labelled 1.

There are altogether eight edge variables. In general, the values of any four of these will determine the values of the four constants in expressions (3.23) and these in turn will fix the values of the remaining four edge variables. We expect therefore to be able to find a series of matrix equations corresponding to (3.33), but now with four variables on each side, and 4 × 4 coefficient matrices.

The algebra involved in the determination of these matrices is much heavier than in the previous problem. In the previous work it did not make much difference which two of the four variables were specified at the outset. In the present problem the choice of variables makes a big difference to the amount of algebra involved. It turns out that the algebra is minimised if we group the variables so that w and m appear on one side of the equation and ψ, q on the other. The results of the calculation (which is still lengthy) are given below.

$$\begin{bmatrix} q_1 \\ \psi_1 \\ q_2 \\ \psi_2 \end{bmatrix} = \begin{bmatrix} -a & -b & -c & d \\ b & -a & -d & -c \\ c & -d & a & b \\ d & c & -b & a \end{bmatrix} \begin{bmatrix} w_1 \\ m_1 \\ w_2 \\ m_2 \end{bmatrix}, \quad \begin{bmatrix} w_1 \\ m_1 \\ w_2 \\ m_2 \end{bmatrix} = \frac{1}{2} \begin{bmatrix} -b & a & d & c \\ -a & -b & -c & d \\ -d & -c & b & -a \\ c & -d & a & b \end{bmatrix} \begin{bmatrix} q_1 \\ \psi_1 \\ q_2 \\ \psi_2 \end{bmatrix}. \quad (3.62)$$

The normalised variables are related to the original variables exactly as before in (3.29). It is remarkable that only four distinct coefficients are involved both in the matrix and also in its inverse (except for the factor $\frac{1}{2}$). They are defined by the following expressions, in which the dimensionless length ζ of the shell is given by

$$\zeta = l/\mu = 3^{\frac{1}{4}} l/(ah)^{\frac{1}{2}} \quad (3^{\frac{1}{4}} \approx 1.32). \tag{3.63}$$

$$a = (\text{Sinh } 2\zeta - \sin 2\zeta)/k, \tag{3.64a}$$

$$b = (\text{Sinh } 2\zeta + \sin 2\zeta)/k, \tag{3.64b}$$

$$c = 2(\text{Cosh } \zeta \sin \zeta - \text{Sinh } \zeta \cos \zeta)/k, \tag{3.64c}$$

$$d = 2(\text{Cosh } \zeta \sin \zeta + \text{Sinh } \zeta \cos \zeta)/k, \tag{3.64d}$$

where

$$k = \text{Cosh } 2\zeta - \cos 2\zeta. \tag{3.64e}$$

Fig. 3.12. Sign convention for displacement, rotation, bending moment and shearing force at the ends of a cylindrical shell having finite length. Stress resultants are shown by solid-headed arrows, and displacements by open-headed arrows. Note that the positive senses of w_1 and q_1 are opposite, while those of w_2 and q_2 are the same.

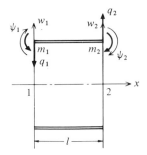

3.8 Edge-loading of finite elastic cylindrical shells

Note that coefficient a is distinct from the radius a of the shell. Here we have followed Hetényi's (1946) suggestion and have used capitals for the first letter of the hyperbolic functions. The same denominator appears in the expressions for all four constants.

The coefficients a to d are related by three identities, which are often useful in manipulations of (3.62) (e.g. problem 3.14):

$$ab + cd = 1, \qquad (3.65a)$$

$$a^2 + d^2 = b^2 + c^2, \qquad (3.65b)$$

$$(a^2 - c^2)(a^2 + d^2) = 2ab - 1. \qquad (3.65c)$$

Table 3.1. Coefficients a to d in (3.62).

$l/(ah)^{\frac{1}{2}}$	a	b	c	d
0.1	0.0877	7.5986	0.0439	7.5982
0.2	0.1755	3.8008	0.0877	3.7978
0.3	0.2631	2.5382	0.1314	2.5280
0.4	0.3503	1.9125	0.1748	1.8883
0.5	0.4366	1.5448	0.2173	1.4977
0.6	0.5213	1.3095	0.2583	1.2287
0.7	0.6033	1.1531	0.2965	1.0265
0.8	0.6810	1.0489	0.3308	0.8636
0.9	0.7531	0.9817	0.3595	0.7252
1.0	0.8177	0.9418	0.3811	0.6030
1.1	0.8736	0.9222	0.3944	0.4927
1.2	0.9198	0.9172	0.3985	0.3923
1.3	0.9560	0.9220	0.3933	0.3015
1.4	0.9825	0.9328	0.3792	0.2204
1.5	1.0004	0.9462	0.3576	0.1495
1.6	1.0112	0.9599	0.3301	0.0890
1.7	1.0164	0.9724	0.2986	0.0390
1.8	1.0178	0.9828	0.2647	-0.0011
1.9	1.0167	0.9908	0.2303	-0.0320
2.0	1.0143	0.9965	0.1965	-0.0547
2.1	1.0113	1.0003	0.1644	-0.0704
2.2	1.0083	1.0025	0.1347	-0.0803
2.3	1.0057	1.0035	0.1077	-0.0853
2.4	1.0035	1.0037	0.0837	-0.0866
2.5	1.0018	1.0035	0.0627	-0.0848
2.6	1.0007	1.0029	0.0447	-0.0809
2.7	0.9999	1.0023	0.0295	-0.0754
2.8	0.9995	1.0017	0.0170	-0.0689
2.9	0.9993	1.0012	0.0068	-0.0619
3.0	0.9992	1.0007	-0.0012	-0.0545

Values of the four coefficients are given in table 3.1 and are plotted against $l/(ah)^{\frac{1}{2}}$, on logarithmic scales, in fig. 3.13.

The coefficients a to d are given by the following simple asymptotic expression when ζ is either large or small. These are obtained formally by evaluating the limits of expression (3.64) as $\zeta \to \infty$ and $\zeta \to 0$, respectively (and using l'Hospital's rule as necessary):

$$\left.\begin{array}{l} a \approx 1 \\ b \approx 1 \\ c \approx 0 \\ d \approx 0 \end{array}\right\} \zeta \gg 1, \quad \left.\begin{array}{l} a \approx \tfrac{2}{3}\zeta \\ b = \zeta^{-1} \\ c = \tfrac{1}{3}\zeta \\ d = \zeta^{-1} \end{array}\right\} \zeta \ll 1. \tag{3.66}$$

When ζ is very large the matrices (3.62) thus decompose into two (2 × 2) matrices relating to the two edges separately: the edges then have practically no influence on each other, and we recover the 2 × 2 matrices (3.33) found already. On the other hand, when ζ is small the two edges interact strongly with each other.

This form of relationship (3.66) is useful for numerical calculations on shells having variable thickness in terms of a sequence of short, interconnected shells of different but uniform thickness; and a simplified version of it may be

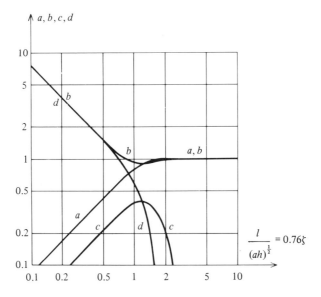

Fig. 3.13. Double-logarithmic plot of coefficients a to d in (3.62), describing the behaviour of the shell in fig. 3.12, as functions of $l/(ah)^{\frac{1}{2}}$: see table 3.1.

3.8 Edge-loading of finite elastic cylindrical shells

derived by the use of finite difference expressions, without the need for the introduction of the functions of (3.23); see problem 3.15.

The especially simple form of the 4 × 4 matrix relating the eight edge variables occurs only if the variables are grouped as shown in (3.62). It is of course possible to express (say) the vector w_1, ψ_1, w_2, ψ_2 in terms of vector q_1, m_1, q_2, m_2 by means of a 4 × 4 matrix; but the coefficients of this matrix do not repeat, and are not easy to express.

It turns out that although the grouping of variables in (3.62) may appear to be restrictive, it causes no difficulties when the matrices are used to solve a particular problem. To illustrate this, we shall consider some simple but representative examples in which end 2 (fig. 3.12) is supported in various ways (simple support, free, clamped) while arbitrary loadings are applied to edge 1. We wish to find expressions connecting w_1, ψ_1, q_1 and m_1 analogous to those already derived (3.33) for a semi-infinite shell.

The easiest case is when end 2 is *simply-supported*, i.e. $w_2 = m_2 = 0$. Then from the first of (3.62) we have, directly,

$$\begin{bmatrix} q_1 \\ \psi_1 \end{bmatrix} = \begin{bmatrix} -a & -b \\ b & -a \end{bmatrix} \begin{bmatrix} w_1 \\ m_1 \end{bmatrix} \quad \text{(end 2: s-s)}, \tag{3.67}$$

and this may easily be rearranged, if this is desired, in the manner of (3.33).

Next we consider edge 2 to be *free*, i.e. $q_2 = m_2 = 0$. The third equation in (3.62) gives us immediately an expression for w_2 in terms of w_1 and m_1, and this may be used in the first two equations of (3.62) to give

$$\begin{bmatrix} q_1 \\ \psi_1 \end{bmatrix} = \begin{bmatrix} -\frac{(a^2 - c^2)}{a} & -\frac{1}{a} \\ \frac{1}{a} & -\frac{a^2 + d^2}{a} \end{bmatrix} \begin{bmatrix} w_1 \\ m_1 \end{bmatrix} \quad \text{(end 2: free)}. \tag{3.68}$$

Similarly, when edge 2 is *clamped*, i.e. $w_2 = \psi_2 = 0$, the fourth equation of (3.62) enables us to express m_2 in terms of w_1, m_1 and we obtain:

$$\begin{bmatrix} q_1 \\ \psi_1 \end{bmatrix} = \begin{bmatrix} -\frac{(a^2 + d^2)}{a} & -\frac{1}{a} \\ \frac{1}{a} & -\frac{a^2 - c^2}{a} \end{bmatrix} \begin{bmatrix} w_1 \\ m_1 \end{bmatrix} \quad \text{(end 2: clamped)}. \tag{3.69}$$

By inspection of fig. 3.13 we see that when ζ is large each of the numerical coefficients in each of these matrices tends to unity, in agreement with our previous result (3.33) for an infinitely long cylindrical shell. Indeed, we find that the magnitude of all of the coefficients in the three matrices differ from unity by less than 2% provided $\zeta > 2.7$. We may thus say, as a rough guide, that if $l > 2(ah)^{\frac{1}{2}}$ the boundary conditions at the remote edge have practically no effect on the solution at the near edge.

As another example we consider the local effect of the application of a ring-load at the central plane of a cylindrical shell of finite length $2l$ which has free ends.

Using the same notation as before (section 3.5), we have boundary conditions as follows:

at edge 1: $\quad q_1 = -\frac{1}{2}f, \psi_1 = 0.$

at edge 2: $\quad q_2 = m_2 = 0.$

Thus we obtain from (3.62) the following expressions for the normalised quantities of interest:

$$m_1 = f/4b, \quad w_1 = (a^2 + d^2)(f/4b). \tag{3.70}$$

These expressions may readily be evaluated for a shell of given length. Equivalent expressions for shells with other end-conditions may easily be found (see problem 3.16).

The results expressed in (3.62) and fig. 3.13 may be used directly, if desired, in the solution of problems involving beams on elastic foundations, with the help of the table of problem 3.4. In this case the label $l/(ah)^{\frac{1}{2}}$ of the abscissa of fig. 3.13 should be disregarded.

3.9 Conclusion

The methods developed in this chapter may be used in the analysis of a wide range of problems concerning the symmetrical loading of thin-walled elastic cylindrical shells. Some examples are given in problems 3.17–3.23. These include cases where there are various pressure-loadings, some simple thermal-stress problems, and problems involving the calculation of changes in axial length of cylindrical shells.

The results of section 3.8 for the behaviour of finite-length shells may be used advantageously in *finite-element* formulations of problems concerning the axisymmetric behaviour of shells whose thickness varies discontinuously with the axial coordinate (Calladine, 1973a).

We have also seen in this chapter the emergence of an important idea which will be generalised in subsequent chapters to more complex situations. The idea is basically that the shell carries load by a combination of bending (in this case in the longitudinal direction) and stretching (in this case circumferentially), whose relative importance varies with position in the shell for a given loading. 'Slowly' varying loads and the 'low' terms in Fourier expansions are carried primarily by stretching action, while conversely, loads which vary rapidly with length, and the 'high' terms in Fourier expansions

3.10 Problems

are carried primarily by bending action. The basic reason for this is seen most clearly by splitting the shell, conceptually, into two surfaces which between them share the load. The 'bending' surface is very stiff for short-wave loading, and hence carries most of it. Conversely, it is very flexible for long-wave loading, which it thus sheds onto the 'stretching' surface. In this connection there is a characteristic length – in the present problem of order $(ah)^{\frac{1}{2}}$ – which enables us to distinguish between short-wave and long-wave effects.

Many of the results of the present chapter will be useful in approximate analyses of arbitrary shells of revolution under symmetric loading in chapter 11.

As we have remarked already, the fourth-order equation (3.13) on which the work of this chapter is based is often known as the equation of the *elastic beam on an elastic foundation*. In the present chapter the analogy is obvious: the 'beam' corresponds to a longitudinal strip of the shell, while the 'foundation' corresponds to the symmetric deformation of circumferential 'hoops' of the shell. The same equation will be useful in a different context in chapter 9.

The whole subject of beams on elastic foundations has been studied exhaustively by Hetényi (1946). Many of his results are directly useful in application to shell structures; and indeed he relates some of his work to the behaviour of arbitrary shells of revolution in a manner somewhat similar to that of chapter 11.

3.10 Problems

3.1 By differentiating (3.13) twice with respect to x and using (3.11), set up the following differential equation in terms of M_x:

$$D \frac{d^4 M_x}{dx^4} + \left(\frac{Et}{a^2}\right) M_x = -D \frac{d^2 p}{dx^2}.$$

3.2 Rearrange (3.23) into the two following equivalent forms, in which $B_1 \ldots B_4$, $C_1 \ldots C_4$ are sets of four arbitrary constants, and $\xi = x/\mu$.

$w = B_1 \text{Ch} \, \xi \cos \xi + B_2 \text{Ch} \, \xi \sin \xi + B_3 \text{Sh} \, \xi \cos \xi + B_4 \text{Sh} \, \xi \sin \xi;$

$w = C_1 \exp(1+i)\xi + C_2 \exp(1-i)\xi$
$\quad + C_3 \exp(-1+i)\xi + C_4 \exp(-1-i)\xi.$

3.3 Using (3.26), rearrange (3.29) as follows:

$\psi = (ah)^{\frac{1}{2}} \Psi / 3^{\frac{1}{4}}; \quad m = 2 \times 3^{\frac{1}{2}} a M_x / Eth;$
$m = ah\kappa_x / 2 \times 3^{\frac{1}{2}}; \quad q = 2 \times 3^{\frac{1}{4}} a^{\frac{3}{2}} Q_x / Eth^{\frac{1}{2}}.$

3.4 The governing equation for an unloaded elastic beam supported by an elastic foundation is

74   *Cylindrical shells under symmetric loading*

$$\frac{d^4w}{dx^4} + \frac{k}{B}w = 0,$$

where x is the lengthwise coordinate, w is the (small) transverse displacement and B, k are the appropriate stiffness factors for flexure of the beam and tension/compression of the foundation, respectively: for more details see section 9.3. By comparison of the above equation with (3.19), and inspection of the definitions in (3.29), set up the following table of equivalences between the variables of the present chapter (on the left) and those appropriate to the beam-on-elastic-foundation (on the right).

x	$=$	x
w	$=$	w
l	$=$	l
μ	$=$	$2^{\frac{1}{2}}(B/k)^{\frac{1}{4}}$
ζ	$=$	$l(k/B)^{\frac{1}{4}}/2^{\frac{1}{2}}$
m	$=$	$M/(Bk)^{\frac{1}{2}}$
q	$=$	$2^{\frac{1}{2}}Q/B^{\frac{1}{4}}k^{\frac{3}{4}}$
ψ	$=$	$2^{\frac{1}{2}}\Psi(B/k)^{\frac{1}{4}}.$

M, Q and Ψ are bending moment, shearing force and rotation, respectively, in the beam. ζ is a dimensionless length, defined in (3.39). (Note that in section 9.3 m and F are used in place of M and Q.)

3.5 Consider a cylindrical shell which exists in the region $x \leqslant x_0$. At the edge $x = x_0$, loading Q_0, M_0 is applied. If there is a displacement w_0, Ψ_0 at this edge, the external virtual work per unit length is $w_0 Q_0 + \Psi_0 M_0$; thus w_0, Ψ_0 and Q_0, M_0 are corresponding displacements and forces, respectively (see appendix 2). (If the loads had been applied at the edge of a shell extending in the positive x-direction, as in fig. 3.12, there would have been a negative sign in front of the entire expression.) By using the expressions given in problem 3.3, show that

$$wQ + \Psi M = (2 \times 3^{\frac{1}{4}} a^{\frac{3}{2}}/Eth^{\frac{1}{2}})(wq + \psi m);$$

and hence that w_0, ψ_0 and q_0, m_0 may also be regarded as corresponding normalised displacements and forces, respectively.

3.6 An open-topped water tank consists of a uniform vertical cylindrical shell of radius a, thickness t (effective thickness h) and height $l \gg (ah)^{\frac{1}{2}}$, set into a massive horizontal base which may be regarded as rigid. It is now filled with water. Show that if the shell were disconnected from the base the water would cause both displacement and rotation of the shell there; and that consequently the imposition of the proper boundary conditions involves a superposed case in which edge-displacement and rotation are both prescribed. Hence compute the stress resultant M_x at the base of the tank due to a depth l of water which has unit weight ρg. Sketch the distribution of M_x and N_θ in the tank, and compare the peak bending stress σ_x (see (2.29)) with the mean circumferential stress due to the fluid at the base of the tank with a 'free' lower edge. Under what circumstances is the contribution to M_x from the edge rotation negligible?

3.7 The ring-load F is now removed from the shell shown in fig. 3.5a, and replaced by a uniform couple $\bar{M}$ per unit circumference about the tangent to the cross-section. The sense of $\bar{M}$ is counterclockwise in the view of fig. 3.5b. Show by means of symmetry arguments that $Q = -Q_1$, $M = -M_1 = \frac{1}{2}\bar{M}$, $w = w_1 = 0$, and $\psi = -\psi_1$.

3.8 A long thin circular cylindrical shell of uniform material has radius a. It has two regions of uniform thickness t_1 and t_2 respectively, which join at a particular cross-section. At this cross-section a ring-load F per unit circumference is applied, as in fig. 3.5a. Show that the radial displacement w, bending-stress resultant M_x and rotation Ψ at this cross-section are related to F as follows:

$$F = \frac{Eth^{\frac{1}{2}}}{2 \times 3^{\frac{1}{4}} a^{\frac{3}{2}}} w \left(\alpha^3 + \alpha^{-3} + \frac{(\alpha + \alpha^{-1})^2}{(\alpha^5 + \alpha^{-5})} \right)$$

$$M_x = \frac{Eth}{2 \times 3^{\frac{1}{2}} a} w \frac{(\alpha + \alpha^{-1})}{(\alpha^5 + \alpha^{-5})}$$

$$\Psi = \frac{3^{\frac{1}{4}}}{(ah)^{\frac{1}{2}}} w \frac{(\alpha^4 - \alpha^{-4})}{(\alpha^5 + \alpha^{-5})}$$

where $t = (t_1 t_2)^{\frac{1}{2}}$, $h = (h_1 h_2)^{\frac{1}{2}}$ and $\alpha = (t_1/t_2)^{\frac{1}{4}}$.

(*Hint.* Start with the matrix in (3.33) which expresses ψ and q in terms of w and m. Express this in terms of the actual variables, using the substitutions $t_1 = t\alpha^2$, $t_2 = t\alpha^{-2}$. Imagine that w is imposed, and find the value of M_x at the junction so that the slopes match. Then find Q in the two sides, and hence F.)

3.9 Show that the definition of k_n in (3.38) is tantamount to

$$k_n = \int_{\frac{1}{4}\pi}^{\frac{1}{4}\pi+\zeta} e^{-\xi} \sin\xi d\xi \bigg/ \zeta e^{-\frac{1}{4}\pi} \sin(\tfrac{1}{4}\pi);$$

and verify (3.39) by noting (cf. (3.28)) that

76 *Cylindrical shells under symmetric loading*

$$-\frac{1}{2^{\frac{1}{2}}}\frac{d}{d\xi}[e^{-\xi}\sin(\xi+\tfrac{1}{4}\pi)] = e^{-\xi}\sin\xi.$$

Show also that k_m is equal to the mean value of the same function over the interval from $-\tfrac{1}{4}\pi$ to $-\tfrac{1}{4}\pi+\zeta$.

Show that $\int_{-\infty}^{\infty} w\,dx$ for $w(x)$ as in fig. 3.6c is equal to $2\mu w_0$, where w_0 is the peak value.

3.10 Show that the bending moment at the mid-point of the short beam shown in fig. 3.5e (which has unit width) is given precisely by (3.36b). Also show that when the central load F is uniformly distributed over a central length $2l$ of this beam, (where $2l < \mu$), the central bending moment is given by $M_{xo} = \tfrac{1}{4}\mu F(1-\zeta)$, where $\zeta = l/\mu$. Compare this with (3.40), where expression (3.41) for k_m is now replaced by the first few terms of a Taylor-series expansion in ζ.

3.11 A cylindrical shell is loaded by a uniform pressure over a band whose axial extent is $2l$, as in fig. 3.6. By putting $F = 2lp$ in (3.38) and (3.40) obtain expressions for $N_{\theta 0}$ and M_{xo}. Hence show that $N_{\theta 0} = pa$ and $M_{xo} = 0$ as $l \to \infty$. Show also, by inspection of fig. 3.5c, that $\Psi_0 = Q_{xo} = 0$. Check that these results agree with elementary considerations.

3.12 Consider the problem of fig. 3.10 by 'cutting' the shell at $x = \tfrac{1}{2}L, \tfrac{3}{2}L$, etc., and providing self-balancing bending-stress resultants M_x (and $Q_x = 0$) at the cut edges in order to make the edges join up without a discontinuity of slope. Show that the angle of rotation which must be eliminated in this way at each edge is equal to $2p^*a^2/EtL$; and by use of (3.31) or (3.33) (i.e. by regarding the length L as 'long') show that $12M^*/p^*L^2 = 2 \times 3^{\frac{1}{4}}/(L/(ah)^{\frac{1}{2}})^3$. Sketch this relationship on fig. 3.10b, and verify that the formula is satisfactory provided $L/(ah)^{\frac{1}{2}} > 2$, say.

3.13 A long cylindrical shell is subjected to ring-loads F as in fig. 3.5a, but applied at a series of cross-sections at a uniform spacing $2L$ along the axis. Express this loading as a Fourier (cosine) series, taking the origin of x at one of the planes of loading (see appendix 10). Hence express both w and M_x by means of Fourier series, and compute values of w, M_x at $x = 0$ and $x = L$ for the case $L = 2\mu$. Use the notation of section 3.7.2.

3.14 Verify the consistency of the two matrices in (3.62) by direct multiplication and use of (3.65).

3.10 Problems

3.15 Show that for a *short* length of a cylindrical shell subjected to edge-loads as in fig. 3.12, the following relations between the original variables are approximately true:

$$-Q_1 + Q_2 = \frac{Elt}{a^2}\left(\frac{w_1 + w_2}{2}\right)$$

$$Q_1 + Q_2 = (-M_1 + M_2)/l$$

$$-\Psi_1 + \Psi_2 = \left(\frac{l}{D}\right)\left(\frac{M_1 + M_2}{2}\right)$$

$$\Psi_1 + \Psi_2 = \frac{2}{l}(w_1 - w_2).$$

(The first relates the mean radial displacement to the load on the ring; the second is an approximate version of '$Q = dM/dx$'; the third relates the mean values of κ_x and M_x; and the fourth is a geometric relation.) Hence, by addition and subtraction of pairs of these equations, find the coefficients of matrices like those of (3.62); and finally put these in terms of the dimensionless form of the variables.

3.16 Show that the formulae corresponding to (3.70) are

$$m_1 = fb/2(a^2 + b^2), \quad w_1 = fa/2(a^2 + b^2)$$

when the ends of the shell are simply-supported; and

$$m_1 = f/4b, \quad w_1 = f(a^2 - c^2)/4b$$

when the ends are clamped.

3.17 Solve problem 3.13 by the use of (3.62) in connection with fig. 3.12, where edge 1 is at a plane containing a ring-load, and edge 2 is midway between ring-loads. Show that

$$w_1 = \tfrac{1}{4}fb, \quad w_2 = \tfrac{1}{4}fd, \quad m_1 = \tfrac{1}{4}fa, \quad m_2 = -\tfrac{1}{4}fc,$$

and by inspection of fig. 3.13 investigate the effect on these four quantities of changes in the value of l. The radial deflection of a long shell at the plane of application of a ring-load F was described in section 3.5 in terms of the deflection of an equivalent compact ring of length 2μ: see fig. 3.5a. Show that the same idea may be used in the present problem; and that if the length of the compact ring is taken as either $2l$ or 2μ, whichever is the smaller, the error in the deflection will not exceed 9%, whatever the value of l.

3.18 A uniform cylindrical shell has radius a and thickness t. It is reinforced by compact external rings of the same material, having cross-sectional area A and spaced regularly at a distance of $2l$ apart in the axial direction. The shell is now subjected to an interior gauge pressure p. By considering the force-interaction between the rings and the shell, and using the results of problem 3.17, show that the circumferential stress in the rings and throughout the shell on account of the pressure will be practically uniform provided $l < 0.75(at)^{\frac{1}{2}}$.

3.19 A long shell with unrestrained ends is subjected to a ring-load F as shown in fig. 3.5a. Show that the longitudinal strain in the shell is given by $\epsilon_x = -\nu w/a$. By using equilibrium equation (3.1) show that $Fa = \int N_\theta \, dx$, where the integration is performed over the entire length of the shell. Hence show that the overall axial shortening of the shell is equal to $\nu Fa/Et$. How is this result altered if F is now distributed over length $2l$ as shown in fig. 3.6a?

3.20 A uniform thin cylindrical shell of radius a, thickness t and length $L(\gg(at)^{\frac{1}{2}})$ has its ends firmly connected to rigid blocks. The shell is now loaded axially by compressive forces P applied at the centres of the end-blocks. On the assumption that P is much smaller than the critical value required for buckling, show that the clamping of the ends induces a local bending stress in the axial direction equal to $3^{\frac{1}{2}}\nu h/t$ times the mean axial stress; and that the loading causes an overall change of length equal to

$$(P/2\pi Eat) [L - \nu^2(ah)^{\frac{1}{2}}/3^{\frac{1}{4}}].$$

(Hint. Use the principle of superposition; imagine that the ends are put back-to-back; and use the result of problem 3.19.)

3.21 A pure shear loading ($q = q_0$, $m = 0$) is applied at the edge $x = 0$ of a long cylindrical shell. Show that m has a maximum value of $e^{-\frac{1}{4}\pi}q_0/2^{\frac{1}{2}}$ at $x = \frac{1}{4}\pi\mu$; and hence (see problem 3.3) that

$$M_x^{\max}/Q_{x0} = (ah)^{\frac{1}{2}}/(12)^{\frac{1}{4}} e^{\frac{1}{4}\pi}.$$

Also show that the edge-displacement w_0 is the same as that which would occur if the same loading were applied to a component ring of the same material, having the same radius and with a cross-sectional area Lt, where $L = \frac{1}{2}\mu \approx 0.38 \, (ah)^{\frac{1}{2}}$.

3.22 A long vertical cylindrical shell is stress-free at room temperature. Due to contact with a hot liquid the temperature of the portion $x > 0$ is raised by T, while the portion $x < 0$ remains at room temperature. The thermal conductivity of the material is such that although the temperature is uniform through the thickness at all values of x, there is an abrupt change of temperature at $x = 0$.

Imagine that the shell is cut at $x = 0$, and that the edges are re-aligned by the imposition of self-equilibrating shear forces Q_{x0}. Show that $M_{x0} = 0$, and use the results of problem 3.21 to show that the maximum longitudinal bending stress (cf. (2.29)) is approximately equal to $0.28\,E\alpha T$, independently of the value of a/h, where α is the coefficient of linear thermal expansion.

3.23 A thin elastic cylindrical shell with axially unrestrained ends is subjected not only to an interior pressure $p(x)$, but also to a temperature rise $T(x)$ above a datum at which the unloaded shell is stress-free. The temperature does not vary through the thickness of the shell. The coefficient of thermal expansion of the material is α. Re-examine (3.1)-(3.13) for this case, and show that the only required changes are:

$$\epsilon_\theta = N_\theta/Et + \alpha T, \tag{3.7'}$$

$$(\epsilon_x - \alpha T) = -\nu(\epsilon_\theta - \alpha T), \tag{3.9'}$$

$$N_\theta = (Et/a)w - Et\alpha T, \tag{3.12'}$$

$$D\frac{d^4w}{dx^4} + \left(\frac{Et}{a^2}\right)w = p + \left(\frac{Et\alpha}{a}\right)T. \tag{3.13'}$$

Hence show that temperature variations of this sort may be compounded with pressure-loading according to the right-hand side of (3.13′), and that any available solution for a pressure-load may be re-interpreted in terms of an equivalent temperature distribution provided N_θ is calculated according to (3.12′). Describe qualitatively the response of a shell to a temperature change

$$T = T_0 \sin(\pi x/l),$$

over a range of values of l.

4

Purely 'equilibrium' solutions for shells:
the membrane hypothesis

4.1 Introduction

Triangulated structural trusses of the kind used for bridges, electric power transmission towers, etc. carry the loads applied to them mainly by tensile and compressive stresses acting along the prismatic members. But the applied loads are also carried to a minor extent by transverse shear forces in the members, which are related to bending moments transmitted between members at joints of the frame. It is usual to begin the analysis of structures of this type by imagining that the joints are all made with frictionless pins, and also that the loads are applied only at these joints. In direct consequence of this idealisation there are no bending moments and transverse shear forces, and the analysis is much simpler than it would be otherwise (e.g. Parkes, 1974).

The displacements of the simplified structure are relatively easy to compute. But they involve, in particular, relative rotations of the members at the joints; and thus it is possible to use these computed rotations in order to assess the order of magnitude of the bending moments which were dispensed with at the outset. If these turn out to be substantial, the initial hypothesis that bending moments are negligible is clearly not justified, and the whole calculation must be abandoned in favour of one which pays proper respect to the bending effects.

There is a closely analogous state of affairs in the action and analysis of thin-shell structures. In general the applied loads are carried in shell structures by a combination of 'stretching' and 'bending' action, of which we have seen examples in chapter 3. But sometimes it seems clear that the bending effects are rather small. In these cases it is advantageous to have a procedure – analogous to the introduction of 'pin joints' in trusses – whereby the bending effects are discarded altogether, and the structure is analysed as if it carried the applied loading entirely by 'in-plane' stress resultants. This method is

4.1 Introduction

known as the *membrane hypothesis*. It is a hypothesis in the normally accepted sense of the word: if it should transpire in a given case that the deflected form of the shell calculated on this basis involves changes of curvature corresponding to substantial bending-stress resultants, then the hypothesis must be abandoned, and a new, more complete analysis must be attempted. In this respect it closely resembles the *pin-joint hypothesis* for the analysis of trusses.

In this chapter our task is to investigate the statical equilibrium of shell structures according to the membrane hypothesis, that is when all bending, twisting and normal shear-stress resultants (M_x, M_y, M_{xy}, Q_x and Q_y; see fig. 2.2) are zero throughout the shell. We shall find that the resulting problem is usually statically determinate, and we shall try to determine the values of the in-plane stress resultants throughout the shell by solving the equations of equilibrium set up for a typical infinitesimal element of the shell in its original, unloaded configuration. In particular, we shall not be concerned in this chapter with considerations of the *deflection* of shells. Calculations of this sort, which are, of course, crucial to the question of the validity of the membrane hypothesis in a given problem, are the subject of chapter 7. Nevertheless, it will be clear in some examples that the calculated distributions of stress resultants would produce strains and displacements leading clearly to a rejection of the hypothesis. Cases of this sort will be pointed out as they occur.

The meaning of *membrane* in the present context is 'a thin pliable skin'. The definition evokes pictures of thin toy rubber balloons, sails of ships, parachutes, envelopes of hot-air balloons, soap bubbles, etc. These pictures are helpful in emphasising the point that there are no bending moments, since the surfaces in question, being thin and pliable, can offer no resistance to bending. But these examples are less than helpful in a number of respects.

First, we are not concerned, in this chapter, with the analysis of structures of this kind. While their thinness makes them incapable of sustaining bending moments, it also renders them incapable of carrying compressive forces and indeed of preserving their shape under certain kinds of loading. Moreover, the deformation of such structures in response to stress often changes the effective geometry of the surface, so that the problem becomes *nonlinear*. For an example of nonlinear elastic membrane theory see Jordan (1962). In this book we are concerned almost entirely with shells which are capable of maintaining a given geometric form and sustaining compressive stresses. (The membrane hypothesis has recently been applied successfully by Heyman (1977) to the analysis of masonry construction: vaults, spires, etc.)

Second, we must emphasise that in this chapter we are concerned with satisfaction only of the equilibrium equations, and not with complete solutions of shell problems. We shall encounter examples where the isolated equilibrium equations can only be satisfied by concentrated 'rings of tension' and other singular distributions of stress. The moment we allow ourselves to think about how the material responds to such concentrations of stress, producing excessive strains which are geometrically incompatible, we shall see that the membrane hypothesis is untenable in these cases: cf. the problem of section 3.5. In the present chapter we shall ignore such 'realistic' consideration, and deal merely with the equilibrium equations for shells as if they were *rigid*. We do this because it turns out to be a very useful exercise. In thinking about the equilibrium equations as an *isolated* set of equations we are breaking away from the classical 'elasticity' point of view which has tended to dominate continuum mechanics since the mid-nineteenth century, and are following the ideas developed by applied mechanicians in the second half of the twentieth century. The essential point here is that the equations of structures are derived from three sources, namely: equilibrium of forces and stresses; geometric compatibility of displacements and strains; and the mechanical properties of the material. In the classical theory of elasticity a common procedure is to use Hooke's law in order to substitute the equilibrium equations into the compatibility relations (or vice versa) so as to obtain a single governing equation. And indeed some workers (e.g. Truesdell, 1945) have argued that it is advantageous to obtain the governing equations for thin-shell structures by a process of formal reduction from the general equations of the elastic theory of three-dimensional solids. It is now clear, however – particularly in relation to the mechanics of structures made from nonlinear inelastic materials – that the theory of structures benefits greatly if a somewhat higher status is accorded to the equilibrium and compatibility equations as such. One of the main themes of this book is that these benefits apply generally to the study of shell structures.

Lastly, we make the obvious remark that the equilibrium equations are concerned with external forces and internal stress resultants, and involve the shell only through the geometry of its (undistorted) surface. Such quantities as the thickness of the shell, the presence of reinforcing ribs and the properties of the material (except self-weight), simply do not enter into the equations.

The layout of the chapter is as follows. First we shall study the simple but illuminating equations of equilibrium of a plane 'string', and then the corresponding equations for a doubly-curved shell with reference to a simple 'local' Cartesian coordinate system. The special equations for a cylindrical shell are particularly straightforward to solve, and we investigate a sequence of simple examples. Another relatively simple case is provided by the equilibrium equations for a symmetrically-loaded shell of revolution having a meridian of arbitrary form, and we investigate some simple problems of this sort. Next, the response of shells of revolution to non-symmetric loading is studied through the device of 'nearly-cylindrical' shells, whose equations of equilibrium can be studied satisfactorily without the use of complicated curvilinear coordinates by virtue of the relatively small degree of curvature of the meridian. The character of solutions to the equilibrium equations depends much on the geometric form of these shells, and in particular on the sign of the 'Gaussian curvature' of the surface. Lastly, we discuss the question of boundary conditions according to the membrane hypothesis, mainly through the device of a 'framework analogy' for the shell.

The chapter is not intended to be exhaustive. In particular we do not consider special techniques which may be developed either for the study of equilibrium in shells of arbitrary form by means of generalised curvilinear coordinates or for particular kinds of problem involving, e.g. shell roofs: see Flügge (1973, Chapter 4). The aim of the chapter is to develop ideas which will be useful for subsequent work.

4.2   A simple problem: the plane 'string'

We begin with a simple problem which will illustrate several important points. An inextensible string is in the form of a circle of radius a (a plane figure) and is subjected to a uniform normally applied load of intensity p per unit length; see fig. 4.1a. What is the tension T in the string? The answer, $T = ap$, may be obtained in several ways.

First, by considering the equilibrium of a semicircle cut from the circle (see fig. 4.1a) we get $2T = 2ap$, and hence $T = ap$. Here we use the fact that the resultant of the loads p is a single force of magnitude $2ap$. This is seen most clearly by considering the equilibrium of the 'contents' of the semicircle, regarded as a weightless two-dimensional fluid, in fig. 4.1b. By an argument derived directly from a fundamental result in the most elementary hydrostatics we find that an arbitrary simply-closed area is in equilibrium under

Fig. 4.1. Various aspects of the determination of the tension T in a circular string of radius a which is subjected to a uniformly distributed normal load of intensity p per unit length. For details see text.

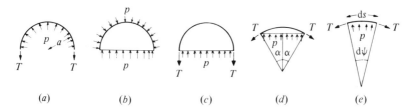

uniform normal loading p applied to the *whole* circumference. The restrictions implied by the words in italics are all crucial. It is often useful to consider as the system being investigated the string and the (imaginary) contents, as in fig. 4.1c, which is a superposition of figs. 4.1a and b.

Exactly the same result is found by considering the equilibrium of a piece of string subtending angle 2α at the centre, plus 'contents' see fig. 4.1d. The only nontrivial equation is found by resolving in the direction of the bisector of 2α. This gives $2T \sin\alpha = 2ap \sin\alpha$, and the same result as before.

Lastly, we consider the equilibrium of an infinitesimal section of string, as shown in fig. 4.1e (which for a non-uniform loading would be the only satisfactory element). We take a length ds and resolve normal to the chord. There is no need now to worry about the 'contents', since the length of the arc is a good approximation to the length of the chord for sufficiently small values of ds. Tangents to the end of the segment make angles $\frac{1}{2} ds/a$ to the chord. This comes from the fact that curvature of a plane curve may be expressed either as the reciprocal of the radius of curvature or as $d\psi/ds$, where ψ is the angle between the tangent to the curve at a point and a fixed reference direction. Here $d\psi/ds = 1/a$. Resolving normally to the chord we have $2T(\frac{1}{2} ds/a) = p ds$, so $T = ap$, as before.

A string of this sort under uniform loading (which might of course be difficult to arrange in practice) will adopt a circular form automatically. Suppose now that part of the circumference is loaded uniformly with normal load of intensity p_1 and the remainder is loaded with p_2, where $p_1 > p_2$ *and that the string remains circular.* (This last condition is a lead-in to shell surfaces, which have a distinct geometrical form.) From our previous analysis we find that in the two parts we have $T_1 = ap_1$ and $T_2 = ap_2$, respectively. Consequently, at the boundaries of the regions we must impose external forces equal in magnitude to $T_1 - T_2$, in the tangential direction: see fig. 4.2a. These forces obviously provide for the overall equilibrium of the string, but in general the necessity for forces of this kind cannot be deduced from considerations of overall equilibrium: see problem 4.1.

This example illustrates an important point in connection with calculations for plane strings in relation to the 'membrane hypothesis'. If we prescribe the *form* of the string and also arbitrary *normal* loading, then certain related *tangential* loading is required by the conditions of equilibrium. The tangential forces are 'reactions', analogous to those required at the ends of beams or trusses in order to keep arbitrary applied loads in equilibrium.

Note that we could if we wished turn the problem upside down by imposing tangential forces at various points of the string and asking what normal reactions p are required for equilibrium. This is like having the string wrapped

4.2 A simple problem: the plane 'string'

round a frictionless circular drum which is able to provide p as a reaction simply by preventing radial displacement. But note that, since the drum is frictionless, it is necessary for the tangentially applied loads to satisfy overall moment-equilibrium about the centre of the drum.

Consider next a more general case, illustrated in fig. 4.2b, where the string has a given form determined by the prescription of radius of curvature ρ as a function of arc-length s: $\rho = \rho(s)$. The curve is *smooth*. Normal and tangential loads of intensity $p(s)$ and $q(s)$ respectively per unit arc-length are applied to the string, and there is tension $T(s)$ in the string. By hypothesis there is neither bending moment nor transverse shear force in the string. We seek equilibrium relations between T, p and q.

Fig. 4.2c shows the forces acting on a small element of the string. In the limit $ds \to 0$ the forces are concurrent, so there are only two equations to be found. By resolving forces in the normal and tangential directions, respectively, we find in the limit $ds \to 0$,

$$T = \rho p, \tag{4.1}$$

$$q = \frac{-dT}{ds}. \tag{4.2}$$

The sign convention is that shown in fig. 4.2c.

There are two notable features of the equations, which will recur in subsequent 'shell' equations. First, the 'normal' equilibrium equation (4.1) is not a differential equation: it relates p and T directly at any point, and the constant of proportionality is the local radius of curvature. Second, the 'tangential' equilibrium equation is a differential equation in which the form of the string (i.e. $\rho(s)$) does not enter. Thus, if p is specified for a string of

Fig. 4.2. Aspects of the general problem of a string lying in a plane curve. (*a*) A circular string subjected to two different values of uniform normal loading p over different parts of its circumference requires concentrated tangential reactions at the change-points. (*b*) The general problem of a plane string subjected to normal and tangential loading $p(s)$, $q(s)$. (*c*) A small element of the string in (*b*). (*d*) Example of circular string with normal loading $p = p_n \cos n\theta$ (here $n = 2$); this requires also $q = np_n \sin n\theta$.

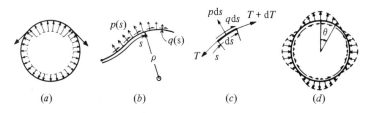

given form, T is determined directly by (4.1) and the tangential reaction q follows from (4.2).

For an example of this, consider a circular string ρ = constant = a, loaded by $p = p_n \cos n\theta$ where θ ($= s/a$) is an angular coordinate and n is an integer: see fig. 4.2d. Then by (4.1) and (4.2)

$$T = ap_n \cos n\theta, \tag{4.3}$$

$$q = \frac{-dT}{ds} = \frac{-1}{a}\frac{dT}{d\theta} = np_n \sin n\theta. \tag{4.4}$$

Note that the radius a does not appear in the final expression for q. This can be deduced independently by dimensional analysis, since q and p have the same dimensions and a (dimension: length) is the only other quantity. Radius a does of course occur in (4.3), by the same sort of argument. We should also remark that the amplitude of q is n times the amplitude of p; consequently rather 'extreme' tangential reactions are required when n is large. (As we shall see later (chapter 9) it is in these circumstances that a small amount of flexural stiffness in a ring makes a lot of difference.)

Finally we note (see problem 4.2) that a radial 'point' force carried by a string which retains its curved form (and has no flexural resistance) produces singularities in both T and q. The situation would be changed radically, of course, if the string were allowed to develop a slope discontinuity at the point of application of the force.

4.3 Equilibrium equations for a doubly-curved shell

Our next task is to find equilibrium equations corresponding to (4.1) and (4.2) for an element of a smooth *surface* according to the membrane hypothesis. While the only significant local feature of a plane curve is the (local) radius of curvature, a small region of a smooth surface is characterised by *two* radii of curvature. We shall consider the geometry of surfaces more fully in chapter 5 but for present purposes we may use a simplified exposition, as follows.

Fig. 4.3a shows a typical point P on a general smooth surface. A local Cartesian x, y, z coordinate system has been set up with origin at P, and the xy-plane tangential to the surface at P. The direction of the z-axis is uniquely determined with respect to the surface, but the x- and y-axes may be rotated about the z-axis. A sufficiently small region of the shell may be described adequately by z as a quadratic function of x and y (cf. appendix 8); and it may also be shown in general that the x- and y-directions may be chosen so that the quadratic expression contains no terms in xy. The equation of the surface may then be written

$$-z = x^2/2R_1 + y^2/2R_2. \qquad (4.5)$$

Here the constants have been arranged to include the two local 'principal' radii, R_1, R_2 of the surface. The idea of principal radii is explained briefly below, and at greater length in chapter 5. R_1 is the radius of curvature of the intersection of the surface with the xz-plane: this curve has the equation $-z = x^2/2R_1$, and since $dz/dx \ll 1$ in the region of interest, the curvature is equal to

$$\frac{-d^2 z}{dx^2} = \frac{1}{R_1}. \qquad (4.6)$$

Similarly R_2 is the radius of curvature of the intersection of the surface with the yz-plane. Here, and elsewhere in the book, the sign convention is that radius of curvature is positive if the centre of curvature is on the z-negative side of the surface. For example, if $R_1 = R_2 = R$; the surface is locally spherical with the positive sense of z in the direction of the outward-pointing normal.

Through any point P on a smooth surface we may draw a number of short line segments at various directions in the surface. It is convenient to describe these as the intersections of the surface and a small plane which can rotate about the z-axis. At various points on any one such line let us construct normals to the surface. As we move along the short curve the inclination of the normal will change (unless the surface is locally plane); that is, the normal will *rotate*. If in this process the rotation of the normal has a component about the forward axis of motion along the curve, as in fig. 4.3b, the surface is said to be *twisted*, but if there is no rotation of this kind, as in fig. 4.3c, the surface is not twisted for the direction in question. It may be shown in general that at a point on *any* smooth surface there are always two directions in the surface, which are mutually perpendicular, for which the twist is zero.

Fig. 4.3. (*a*) A Cartesian x, y, z coordinate system whose origin is at point P on a smooth surface. The xy plane is tangential to the surface at P, and x and y are the local *principal directions*. (*b*) shows that this surface is twisted in a diagonal direction, but (*c*) not in the x- (or y-) direction.

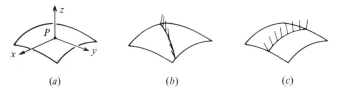

'Equilibrium' solutions: the membrane hypothesis

These directions are called *principal directions*, and the corresponding curvatures are called *principal curvatures*. In the present analysis we have arranged that $\partial^2 z/\partial x \partial y = 0$, i.e. there is no twist in the x- and y-directions; thus these are the principal directions of the surface, and it is for this reason that the constants in (4.5) were expressed in terms of the principal radii of curvature.

We are now ready to write down the equilibrium equations for a small element of the surface, depicted in fig. 4.4a. The element is bounded by planes at x, $x + dx$, y and $y + dy$, and the whole element is sufficiently close to the origin for the true lengths of the sides of the elements to be practically equal to the lengths dx, dy projected onto the plane $z = 0$. The element is loaded by surface tractions which may be resolved into components of intensity q_x, q_y and p per unit area, in the (positive) x-, y- and z-directions, respectively. These may be considered as uniformly distributed over the surface area of the small element. The stress resultants are essentially the same as in fig. 2.2a, but no distinction is made between N_{xy} and N_{yx} by reason of the condition of moment-equilibrium about the z-axis.

Consider first the equilibrium of the element in the x- and y-directions respectively. Fig. 4.4b shows a plan view of the element with the appropriate forces acting on each edge. The inclination of these forces to the xy-plane is so small that the cosine of the angles is unity for practical purposes. Resolving

Fig. 4.4. (a) A small element of the shell of fig. 4.3 showing normal and tangential loading of intensity p, q_x, q_y, together with tangential-stress resultants N_x, N_y, N_{xy}. (b) View normal to the element, showing changes in stress resultants across the element, in order to establish the two 'tangential' equilibrium equations.

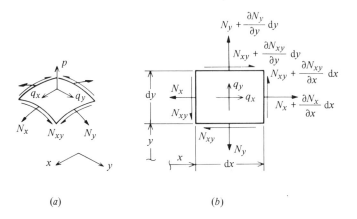

4.3 Equilibrium equations for a doubly-curved shell

forces in the x- and y-directions, and taking the limit as the size of the element vanishes, we have

$$\frac{\partial N_x}{\partial x} + \frac{\partial N_{xy}}{\partial y} = -q_x \tag{4.7a}$$

$$\frac{\partial N_y}{\partial y} + \frac{\partial N_{xy}}{\partial x} = -q_y. \tag{4.7b}$$

The third equation is found by resolving forces in the z-direction. The component of the force $N_x \mathrm{d}y$ on the left edge in the z-direction is $-N_x \mathrm{d}y\, (\partial z/\partial x)$, while the component of the force on the opposite side is

$$\left(N_x + \frac{\partial N_x}{\partial x}\,\mathrm{d}x\right)\mathrm{d}y\left(\frac{\partial z}{\partial x} + \frac{\partial^2 z}{\partial x^2}\,\mathrm{d}x\right)$$

in the z-direction. These forces are largely self-cancelling, and in the limit $\mathrm{d}x \to 0$ they have a resultant in the z-direction equal to

$$N_x \frac{\partial^2 z}{\partial x^2}\,\mathrm{d}x\mathrm{d}y.$$

There is also a corresponding contribution from the N_y forces on the other two edges equal to

$$N_y \frac{\partial^2 z}{\partial y^2}\,\mathrm{d}x\mathrm{d}y.$$

Lastly there is a contribution from N_{xy}, which, by similar arguments is equal to

$$2N_{xy}\frac{\partial^2 z}{\partial x \partial y}\,\mathrm{d}x\mathrm{d}y\,;$$

but this is identically zero, by our choice of coordinate system (cf. problem 4.3). Making use of (4.6), we have, finally,

$$N_x/R_1 + N_y/R_2 = p. \tag{4.8}$$

Equations (4.7) and (4.8) are the three equilibrium equations of the problem. Moment-equilibrium about the z-axis is automatically satisfied as we have already remarked; and the other two moment-equilibrium equations produce nothing, since in the limit the forces concerned are concurrent.

It is worthwhile to examine similarities with, and differences from, the equations given earlier for a curved string lying in a plane. In both cases the equations of *tangential* equilibrium are exactly as they would be for a *plane* element in Cartesian coordinates: the geometrical form of the surface or curve only enters in the *normal* equilibrium equation, which has been written in relation to an element orientated with respect to the principal directions of curvature. Furthermore, in both cases the tangential equilibrium equations are differential equations; but the normal equilibrium equation is not. There the similarities end. With the string there are two loadings (tangential and normal), one stress resultant (tension) and two equations of equilibrium: consequently one of the loadings must be regarded as a 'reaction'. With the shell element there are three loadings (two tangential, one normal) three stress resultants (N_x, N_y, N_{xy}) and three equations. In principle, therefore, the three loadings can be regarded as independent, and the three equilibrium equations may be solved for the stress resultants.

In practice, however, matters may not be so simple, and the nature of the solution may depend on the form of the shell surface and the nature of the boundary conditions. In the following sections we shall give a number of relatively simple miscellaneous examples to illustrate ways in which the equilibrium equations may be solved in particular cases, which will serve to illustrate the above remarks.

It must be emphasised that we have made no attempt in this section to write the equilibrium equations for an element in the most general form. Thus, having described the surface with respect to a locally tangential Cartesian coordinate system, we should not be surprised to recover the two 'quasi-plane' equilibrium equations with respect to the same x, y coordinate system. In the case of a cylindrical shell we may of course use a single Cartesian coordinate system over the entire surface; but for more complicated surfaces (e.g. general surfaces of revolution) we cannot expect that a Cartesian coordinate system will be at all satisfactory, except possibly over a strictly limited area. Thus we shall find in section 4.5 that the general small element of a surface of revolution is slightly wedge-shaped, like a small element in a plane (r, θ) coordinate system. Consequently we must work out the equivalent of (4.7) afresh for this case. On the other hand the equilibrium equation in the normal direction is insensitive to the choice of coordinate systems, provided that the x- and y-directions are in the locally principal directions.

We shall investigate some features of general curvilinear coordinate systems in chapter 5.

4.4 Equilibrium equations for cylindrical shells

It is convenient to use an x, θ coordinate system, as shown in fig. 4.5a. The coordinates are orthogonal, and may be transformed to an x, y system by means of the substitution $y = a\theta$. Clearly the generators lie in directions of zero twist, so the x- and θ-directions are principal directions for the entire surface. R_2, the principal radius of curvature in the θ-direction, is equal to a, and since the generators are straight, $R_1 \to \infty$. The equilibrium equations

4.4 Equilibrium equations for cylindrical shells

derived above apply everywhere on the surface. The stress resultants appropriate to the cylindrical shell are shown in fig. 4.5*b*. Equation (4.8) becomes

$$N_\theta/a = p, \tag{4.9}$$

and (4.7) are rewritten as

$$\frac{1}{a}\frac{\partial N_\theta}{\partial \theta} + \frac{\partial N_{x\theta}}{\partial x} = -q_\theta, \tag{4.10}$$

$$\frac{\partial N_x}{\partial x} + \frac{1}{a}\frac{\partial N_{x\theta}}{\partial \theta} = -q_x. \tag{4.11}$$

The loading component p is directed along the local outward normal, and q_x, q_θ act in the x- and θ-directions which are locally tangential to the surface. An immediate observation is that N_θ depends directly on p; if p is specified, N_θ can be evaluated. We can then substitute for N_θ in (4.10) and obtain an equation for $N_{x\theta}$, which can be solved with the help of a boundary condition. Lastly, (4.11) can be solved for N_x, with the help of another boundary condition.

4.4.1 Some simple examples

In all of the following examples we shall use the coordinate system shown in fig. 4.5*a*. In each case all edge and surface loadings which are not specified are zero. In most cases the shell has a free edge at the plane $x = 0$, and is suitably clamped at the other end, $x = L > 0$.

(1) *Arbitrary loading $N_x(\theta)$ at edge $x = 0$* (fig. 4.6). Since $p = 0$, (4.9) gives $N_\theta = 0$. Using this in (4.10) with $q_\theta = 0$ we have $\partial N_{x\theta}/\partial x = 0$. Therefore, as $N_{x\theta} = 0$ at $x = 0$, $N_{x\theta} = 0$ everywhere. In turn this gives, in (4.11), $\partial N_x/\partial x = 0$; consequently N_x does not vary with x; and for a given value of θ, N_x is equal to the value applied at the edge. In other words the shell carries the applied

Fig. 4.5. (*a*) Cylindrical shell of radius a and length L, showing x, θ coordinate system. (*b*) Typical small element, showing stress resultants.

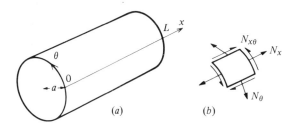

loading as if it consisted of a set of rods lying along the generators. In particular, the far edge must be capable of providing the necessary reactions; otherwise the membrane hypothesis is untenable. Notice that the solution imposes no requirements on the circumferential distribution of N_x, which can even be discontinuous. This example illustrates a kind of solution obtained when there are *straight lines* in the surface: the equilibrium equations allow forces to 'run' along these lines, without any 'diffusion' in the circumferential direction. Also notice, however, that the solution may well be 'unrealistic' in certain circumstances, since we would expect an ordinary elastic cylindrical shell to 'diffuse' concentrated loads applied axially at the ends, by St Venant's principle (see, e.g., Timoshenko, 1953, §32). For the present, however, we are concerned merely with the equilibrium equations, and in consequence St Venant's principle may appear to be broken. We shall investigate this point further in chapter 8.

(2) *Arbitrary loading $N_{x\theta}(\theta)$ applied at the edge $x = 0$* (fig. 4.7). As in the previous example, $N_\theta = 0$, from (4.9). Then (4.10) gives $\partial N_{x\theta}/\partial x = 0$; consequently the applied $N_{x\theta}$ is propagated undiminished along the generators

Fig. 4.6. Loading $N_x(\theta)$ applied to the free edge of the cylindrical shell.

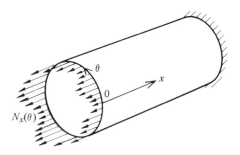

Fig. 4.7. Tangential shear loading applied to the free edge.

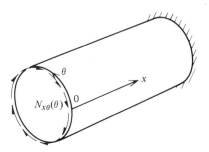

to the far end, where it must be resisted by reactions from the end fixture. Finally (4.11) gives, together with the edge-condition $N_x = 0$ at $x = 0$;

$$N_x = -\frac{x}{a}\left(\frac{\partial N_{x\theta}}{\partial \theta}\right). \tag{4.12}$$

Again, the remote boundary must provide the necessary reaction. Thus, for $N_{x\theta}$ = constant at $x = 0$ we have a state of uniform pure shear in the entire tube; but for more general distributions of the edge shear, axial forces N_x are induced, according to (4.12): see problem 4.4.

(3) *A horizontal uniform circular cylindrical shell loaded vertically by self-weight* (fig. 4.8*a*). Let the uniform weight per unit area of the shell be f. Taking $\theta = 0$ as the uppermost generator we may express the loading in terms of surface tractions:

$$p = -f\cos\theta, \quad q_\theta = f\sin\theta, \quad q_x = 0. \tag{4.13}$$

Equation (4.9) gives

$$N_\theta = -af\cos\theta. \tag{4.14}$$

Substituting this into (4.10) we obtain

$$\frac{\partial N_{x\theta}}{\partial x} = -2f\sin\theta.$$

At $x = 0$, $N_{x\theta} = 0$, so we find

$$N_{x\theta} = -2xf\sin\theta, \tag{4.15}$$

and when this is substituted into (4.11) we have

$$\frac{\partial N_x}{\partial x} = \frac{2xf}{a}\cos\theta.$$

Fig. 4.8. (*a*) Uniform 'self-weight loading' on a cantilever shell. (*b*) The stress resultants are unchanged in the lower part of the shell if the upper part is removed, provided appropriate edge-loads $N_{x\theta}$ are provided along the 'cut' edges. (*c*) These may be provided by reinforcing strips or 'strings'.

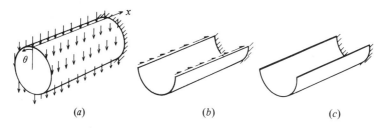

Since $N_x = 0$ at $x = 0$, this gives

$$N_x = \frac{x^2 f}{a} \cos\theta. \tag{4.16}$$

It is interesting to note that this solution corresponds exactly to that found by 'engineers' bending theory' when the shell is regarded as an elastic cantilever. But since that theory involves the concepts of Hooke's law and compatibility of strain, whereas the present equilibrium equations do not, the correspondence is fortuitous. What it tells us, however, is that, in contrast to example (1) the longitudinal strains computed from Hooke's law are compatible: but this kind of analysis, involving material properties and compatibility equations, is beyond the scope of the present chapter. It is of interest to note that since $N_\theta = 0$ at $\theta = \pm\frac{1}{2}\pi$ we could cut away the upper half (say) of the shell, leaving a cantilevered channel as in fig. 4.8b. But notice that $N_{x\theta}$ is not zero on these lines, so that a shear-stress resultant according to (4.15) must be provided externally. This can be done very simply by a 'string' along each of these edges, with a tension found by appropriate application of (4.2): we obtain a tensile force which is proportional to x^2. It is easy to check by an overall longitudinal force balance that the tensile forces in these two edge 'strings' exactly balance the longitudinal compressive force in the remainder of the shell. See problem 4.5 for a further example. Again we may observe that if we were to put in Hooke's law, we would find an incompatibility of strain which would oblige us to abandon the membrane hypothesis for this problem: except, of course, if we were to provide the tension by means of a prestressing tendon. But these are matters for a later chapter.

(4) *Arbitrary axially symmetric pressure-loading $p(x)$.* Here, (4.9) gives directly $N_\theta = ap$. As p does not vary with θ, and $q_\theta = 0$ (4.10) gives $\partial N_{x\theta}/\partial\theta = 0$; so if $N_{x\theta} = 0$ at $x = 0$ (a free end), $N_{x\theta} = 0$ everywhere. Similarly, by (4.11) $N_x = 0$ everywhere if $N_x = 0$ at $x = 0$. In other words, the applied loading is carried by the shell acting as a set of 'hoops'. There are no other restrictions on $p(x)$, which may even be discontinuous. This is the same kind of loading as was studied in sections 3.6 and 3.7, where bending effects were taken into account properly. Here the solution is much simpler, on account of the membrane hypothesis, but in general, of course, it is less realistic.

(5) *Arbitrary circumferentially varying pressure $p(\theta)$*, (which does not vary with x). By (4.9)

$$N_\theta = ap. \tag{4.17}$$

Substituting into (4.10) we have

$$\frac{\partial N_{x\theta}}{\partial x} = -\frac{dp}{d\theta},$$

and since $N_{x\theta} = 0$ at $x = 0$ we find

$$N_{x\theta} = -x\left(\frac{\mathrm{d}p}{\mathrm{d}\theta}\right). \tag{4.18}$$

Substituting this into (4.11) we obtain

$$\frac{\partial N_x}{\partial x} = \frac{x}{a}\left(\frac{\mathrm{d}^2p}{\mathrm{d}\theta^2}\right),$$

and since $N_x = 0$ at $x = 0$,

$$N_x = \frac{x^2}{2a}\left(\frac{\mathrm{d}^2p}{\mathrm{d}\theta^2}\right). \tag{4.19}$$

Again the remote end, $x = L$, must be supported so that the required edge reactions may be provided. Observe that this solution is much more complicated than that of example (4). In particular, if $p(\theta) = p_n \cos n\theta$ our results become

$$N_\theta = ap_n \cos n\theta \tag{4.20a}$$

$$N_{x\theta} = nxp_n \sin n\theta \tag{4.20b}$$

$$N_x = -(n^2x^2/2a)p_n \cos n\theta. \tag{4.20c}$$

Note that the magnitude of the shear- and longitudinal-stress resultants depends strongly on the value of the wavenumber n. See problem 4.6 for a further example. The function $p(\theta)$ need not necessarily be smooth. A discontinuity in $\mathrm{d}p/\mathrm{d}\theta$ produces a discontinuity in $N_{x\theta}$ across a generator, and a 'spike' in N_x, i.e. a line tension along a particular generator. See problem 4.7.

(6) *Normal shear $Q_n \cos n\theta$ applied at end $x = 0$* (fig. 4.9). This is paradoxical, since it is clear that the 'membrane' stresses within the shell can only be in equilibrium with edge-forces imposed *tangentially*. It is important to realise, however, that this kind of loading is admissible in the membrane hypothesis,

Fig. 4.9. (a) A normal shear loading applied to the edge is carried by an edge-string (b) whose necessary tangential shear reactions (cf. fig. 4.2a) provide tangential shear loading to the edge of the shell. In the example illustrated, $Q_x = Q_2 \cos 2\theta$ in (a); and in (b) the corresponding loading on the shell proper is $N_{x\theta} = -2Q_2 \sin 2\theta$.

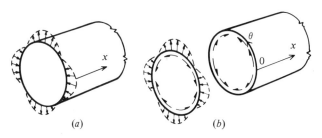

since it may readily be transferred to in-plane loading by means of a simple edge 'string'. This is the problem we studied in section 4.2, but with different notation and sign convention. Here, the edge-string (fig. 4.9b) is in equilibrium under radially directed loading $Q_n \cos n\theta$ if there is a string tension $T = aQ_n \cos n\theta$ and peripheral shear of $nQ_n \sin n\theta$ of the sense shown in the diagram. The latter is provided by an in-plane shear $N_{x\theta} = -nQ_n \sin n\theta$ in the shell itself, as shown. Thus, as far as the remainder $x > 0$ of the shell is concerned, there is an edge-load at $x = 0$ of $N_{x\theta} = -nQ_n \sin\theta$. This is precisely the case discussed as example (2) above, and the results are:

$$N_\theta = 0 \tag{4.21a}$$

$$N_{x\theta} = -nQ_n \sin n\theta \tag{4.21b}$$

$$N_x = (n^2 x/a)Q_n \cos n\theta. \tag{4.21c}$$

$N_{x\theta}$ and N_x depend on n in exactly the same way as in example (5). Indeed, this is not surprising because it should be possible (see problem 4.8) to obtain the same results (except for the 'string' at the edge) by taking a distribution of pressure as in example (5) but applied over a finite portion, say $0 < x < c$ of the shell, and then letting $c \to 0$. In effect we are saying here that if only an out-of-plane shear force $Q_n \cos n\theta$ could be displaced a small distance along the shell, we could treat it as the limiting case of a *surface* traction. It is simpler, however, to think of the end 'string' separately, and to use it as a device for turning radial loading into statically equivalent in-plane loading: see problem 4.9. Examples (5) and (6) are relevant to some problems of wind-loading on silo structures: see problem 4.10.

In all of these examples we have worked in terms of the in-plane stress resultants and the equilibrium equations as such. It is sometimes advantageous to reformulate the equilibrium equations in terms of *stress functions*. We shall investigate this method in chapter 7.

4.5 Axisymmetric loading of shells of revolution

Another class of problem for which the equilibrium equations have relatively simple solutions is provided by general shells of revolution loaded without overall torsion by surface tractions which are independent of the circumferential coordinate θ, i.e. are symmetrical about the axis.

First we must do some basic geometry. A surface of revolution is obtained by rotating an arbitrary curve, as a rigid body, about an *axis of revolution*. It is simplest if this generating curve lies in a plane containing the axis of revolution. Such a plane is called a *meridional plane*, and the intersection of this plane with the surface is called a *meridian*. A meridian is thus the most convenient generating curve of a given surface of revolution. It is often convenient to describe the meridian by means of intrinsic coordinates s, ϕ: s is arc-length measured from a suitable datum (often a 'pole', where the meridian intersects the axis) and ϕ is the angle between the tangent to the meridian and the normal to the axis, or, equivalently, the angle between the normal to the curve and the axis: see fig. 4.10. For the present we shall regard ϕ as a continuous function of s, but we shall later relax this condition so that we can study some elementary problems concerning the intersection of surfaces, e.g. cylindrical and spherical surfaces. Either s or ϕ may be used as a coordinate, but s is to be preferred in general as there will obviously be difficulties with the use of ϕ if the meridian has a *straight* segment or, indeed, if the curvature of the meridian changes sign.

The other coordinate, θ, gives the angular position of the meridional plane with respect to a fixed meridional plane.

Our first task is to find the direction of the local principal axes, and to obtain expressions for the principal radii of curvature R_1, R_2 at an arbitrary point on the surface. It follows from the foregoing description that any meridional plane is a plane of mirror symmetry of the surface, and thus that the (unique) normal to the surface at any point lies in the plane of the meridian. It is, of course, also normal to the meridian itself. Therefore, at any point on the surface the meridian passing through the point lies in a direction of principal curvature. The second principal direction is at right-angles; hence it is given locally by the tangent to the latitude circle which is obtained by rotating

Fig. 4.10. Meridian of an arbitrary surface of revolution showing the arc-length coordinate s, angular coordinate ϕ and various radii r, r_1, r_2, r_3 which are all functions of s (or ϕ). PU is normal to the meridian at P.

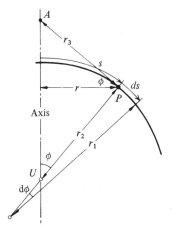

the point about the axis. One of the principal radii of curvature is thus the radius of curvature of the (plane) meridional curve; we shall call this radius r_1, adopting a lower-case r for the present problem. The second principal radius is obtained as follows. Consider a short length PP' of the latitude circle passing through a typical point P on the meridian, and let the distance PP' be dl. The normal to the surface at P passes through point U on the axis, as shown in fig. 4.10. The normal at P' also passes through the same point U. Let $PU = r_2$. When we pass from P to P' along the latitude circle the normal to the surface rotates through angle $d\psi = dl/r_2$. The corresponding curvature is defined as $d\psi/dl$, so the radius of curvature is, simply, r_2.

It is a common error among students to suppose that the radius of the *latitude circle*, r (see fig. 4.10) is the second principal radius. This seems plausible – as the latitude circle gives the second principal direction – but is wrong; the error is avoided if we think of curvature in terms of the rotation of the *normal to the surface* as we move a small distance along the latitude circle. The sphere is a good example to dispel possible doubt.

4.5.1   Equations of equilibrium

The three stress resultants acting on a typical small element of the surface are N_s, $N_{s\theta}$, N_θ, as shown in fig. 4.11a. Now in the present problem both the loading and the stress resultants are axially symmetric, and thus $N_{s\theta} = 0$; otherwise the stress distribution would not be a mirror image of itself

Fig. 4.11. (a) Portion of shell of revolution, showing circumferential coordinate θ and loads and tangential-stress resultants acting on a typical element. For symmetric loading $N_{s\theta} = 0$. (b) True view of element and auxiliary point A, in order to establish the equilibrium equation in the direction tangential to the meridian.

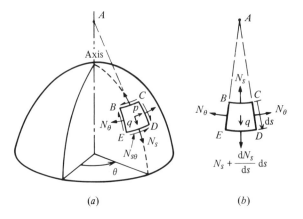

4.5 Axisymmetric loading of shells of revolution

in the plane of the meridian. The general surface loading consists of an outward-directed normal surface traction of intensity $p(s)$, and a tangential traction in the meridional direction of intensity $q_s(s)$. For reasons of symmetry we take $q_\theta = 0$: there is no *torsion* of the shell about its axis. All of the quantities N_s, N_θ, p, q are functions only of s, i.e. they are independent of θ.

There are three equations of equilibrium for an element, but one of these (tangential, in the θ-direction) is automatically satisfied by the assumed symmetry. We therefore seek the two equations of equilibrium in directions normal and tangential to the meridian, respectively.

Resolving normally to the surface we have, as before,

$$(N_s/r_1) + (N_\theta/r_2) = p. \tag{4.22}$$

When considering equilibrium in the direction tangential to the meridian we must take account of the fact that adjacent meridians are not parallel, unlike the previous cases. Thus fig. 4.11b shows a typical small element, viewed along its normal, together with the associated edge tractions. The projections of the meridional edges intersect at A on the axis of symmetry at distance r_3 from the element, where r_3 is the dimension indicated in fig. 4.10. The length BC of the element is equal to $rd\theta$, so the edges BE, CD make angle $rd\theta/r_3$ with each other. Consequently the sum of components of the N_θ forces in the direction of A is $(r/r_3)N_\theta \mathrm{d}s\mathrm{d}\theta$. The total force on edge BC is equal to $rN_s\mathrm{d}\theta$ towards A, and it follows that the total force on edge DE is equal to

$$rN_s\mathrm{d}\theta + \frac{\mathrm{d}}{\mathrm{d}s}(rN_s)\mathrm{d}s\mathrm{d}\theta,$$

acting away from A. The surface traction q_s contributes a force $rq_s\mathrm{d}s\mathrm{d}\theta$ away from A. Hence the required equilibrium equation is

$$\frac{\mathrm{d}}{\mathrm{d}s}(rN_s)\mathrm{d}s\mathrm{d}\theta + rq_s\mathrm{d}s\mathrm{d}\theta = (r/r_3)N_\theta\mathrm{d}s\mathrm{d}\theta.$$

In the limit as $\mathrm{d}s, \mathrm{d}\theta \to 0$ this reduces to

$$\frac{\mathrm{d}}{\mathrm{d}s}(rN_s) - \frac{1}{r_3}(rN_\theta) = -rq_s. \tag{4.23}$$

In this equation both r and r_3 are functions of s. In the special case of a cylindrical shell, $r = $ constant $= a$ and $r_3 \to \infty$: (4.23) then reduces to (4.7a), for here $N_{s\theta} = 0$.

It is not surprising, in view of our previous work, that the tangential equilibrium equation is a differential equation while the normal equilibrium equation is not. If we need to solve (4.23) we can in fact use various substitutions in order to make the equation easier to manipulate: this will be useful in chapter 11. It turns out, however, that there is a simple way of avoiding the

100 *'Equilibrium' solutions: the membrane hypothesis*

solution of a differential equation for the present family of problems. This involves consideration of the equilibrium of a 'cap' of the shell, obtained by cutting the shell around an arbitrary latitude circle, as shown in fig. 4.12. The only force transmitted across this cut is the meridional stress resultant N_s, and moreover this does not vary round the circumference. The axial component of the sum is equal to $2\pi r N_s \sin\phi$, and this must clearly balance the axial component of all the surface tractions acting on the cap. Hence we obtain

$$N_s = \frac{\text{Axial load on cap}}{2\pi r \sin\phi}. \tag{4.24}$$

Thus the only operation which we need to perform in order to obtain an explicit expression for N_s is a summation of the external forces acting on the cap. In general this is not difficult; and indeed it is particularly simple when there is a point load at the apex or a line load on a particular latitude circle.

It is not difficult to show formally that (4.24) is a direct consequence of (4.22) and (4.23): see problem 4.11. But (4.24) is particularly advantageous, as it avoids difficulties in cases where there are point or line loads. Some examples of the use of these equations are given in problems 4.12-19. They include two problems (4.17 and 4.18) which show that a relatively small change in the meridional slope of a shell can give disproportionately large changes in the stress resultant N_θ (but not in N_s); and an example (problem 4.19) in which the meridional shape is designed so that $N_\theta = 0$ everywhere.

4.5.2 Pressure-vessels

A particularly simple example of the use of these equations is provided by an axisymmetric simply-connected pressure-vessel loaded by a uniform internal (gauge) pressure p. Self-weight of the vessel is neglected. In this case the total force exerted by the pressure on a 'cap' is equal to the pressure

Fig. 4.12. 'Cap' of shell of revolution in equilibrium under total external load (here shown as single force F) and stress resultant N_s acting on the cut edge.

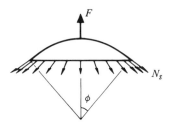

4.5 Axisymmetric loading of shells of revolution

multiplied by the cross-sectional area of the base circle of the cap, i.e. $\pi r^2 p$. Consequently, from (4.24),

$$N_s = \tfrac{1}{2} pr/\sin\phi. \tag{4.25}$$

Using the geometrical relationship (see fig. 4.10)

$$r_2 \sin\phi = r \tag{4.26}$$

we have, simply,

$$N_s = \tfrac{1}{2} pr_2. \tag{4.27}$$

The other equilibrium equation (4.22) can now be used to give an explicit expression for N_θ: hence

$$N_\theta = p(r_2 - r_2^2/2r_1). \tag{4.28}$$

These expressions may easily be evaluated for a shell having a meridian of given geometry.

Zick & St Germain (1963) have given a particularly useful graphical interpretation of (4.28). Consider two adjacent points B, C on the meridian, separated by distance ds, and the associated normals, as shown in fig. 4.13a. Let us find an expression for dA, the area enclosed between the meridian, the two normals and the axis. Area BCD is equal to $\tfrac{1}{2} r_1$ds. Area DEF is equal to this is multiplied by $(DF/DB)^2$, except for a small area near F which vanishes as d$s \to 0$. Now $DF = r_1 - r_2$ and $DB = r_1$; so

$$(DF/DB)^2 = (1 - r_2/r_1)^2 = 1 - (2r_2/r_1) + (r_2/r_1)^2.$$

Thus we find that

$$dA = (r_2 - r_2^2/2r_1)ds.$$

Fig. 4.13. Calculation of N_θ in a shell of revolution according to the membrane hypothesis by means of a graphical method. (a) Area dA enclosed between the meridian, two normals and the axis; (b) the area is evaluated algebraically. (c) A large area A is associated with a point on the meridian where there is a substantial discontinuity of slope.

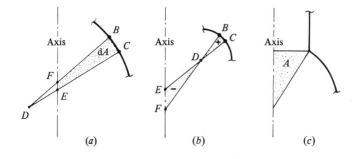

Comparing this with (4.28) we obtain, simply,

$$N_\theta = p\,dA/ds. \tag{4.29}$$

The diagram in fig. 4.13a has been drawn with $r_1 > r_2$, so dA is a positive area. The expression holds also if $r_1 < r_2$ provided the area dA is evaluated properly: see fig. 4.13b.

It is clear that the integral of dA taken from pole to pole of the meridian of a pressure-vessel, is simply equal to the area enclosed between the meridian and the axis. This corresponds to an overall equilibrium equation which can be obtained by cutting the vessel in two along a meridional plane: the overall force balance gives

$$\oint N_\theta\,ds = p\ (\text{area enclosed}). \tag{4.30}$$

In the case of a meridian which has an abrupt change of slope at a particular point, a simple limiting process indicates that the equilibrium solution involves a hoop of circumferential tension T. For example, the pressure-vessel shown in fig. 4.13c has a ring of tension at the junction given by $T = pA$, where A is the shaded area.

Fig. 4.14 shows the variation of N_s and N_θ with s for a 'torispherical' closure of a cylindrical pressure-vessel, for two different radii of the toroidal

Fig. 4.14. Stress resultants in a 'torispherical' pressure-vessel head, according to the membrane hypothesis. (a) Layout of the head, showing spherical, toroidal and cylindrical portions, with an alternative toroidal 'knuckle' shown in broken lines. (b) Scale plot of r_1 and r_2 (cf. fig. 4.10) for the two knuckles. (c) Scale plot of the stress resultants N_θ, N_s, showing how the strong compressive band of N_θ in the knuckle region is attenuated when the radius of the knuckle is enlarged.

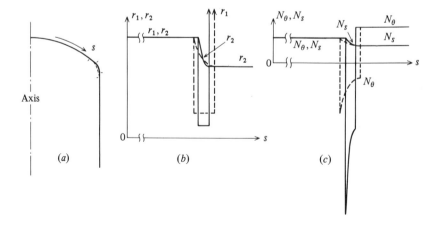

'knuckle'. Note that N_θ varies much more wildly than N_s, and that this may be seen as a direct consequence of (4.29). Note also that the sensitivity of N_θ to small changes in shape of the meridian (but the relative insensitivity of N_s) shown in fig. 4.14c is very similar to effects shown in problems 4.17-19. When the toroidal knuckle of fig. 4.14a is omitted entirely, there is an abrupt change in slope of the meridian at the junction between the spherical and cylindrical surfaces, and there is then a circumferential 'hoop of compression' at the junction: see problem 4.20.

These results are all based simply on the solution of the equilibrium equations alone, and it is clear even from qualitative thinking about the introduction of material properties and compatibility equations, that the membrane hypothesis will have to be abandoned in some cases. This forms the basis of the work in chapter 11.

4.5.3 Uniform ring-loads

Another problem is shown in fig. 4.15a. Here the shell is unloaded except for a uniform line load of intensity N_0 applied at a particular latitude circle of radius a. The load makes angle β with the tangent to the meridian, as shown.

From the previous analysis (cf. problem 4.13) the 'crown' of the shell, represented by the segment BC of the meridian, is stress-free (according to the membrane hypothesis), but beyond point C there will be nonzero stress resultants. It is easy to calculate N_s just beyond C by using (4.24) and to calculate

Fig. 4.15. (a) Shell of revolution loaded by a uniformly distributed line load of intensity N_0 applied in a given direction at a given latitude circle. (b) Calculation of the ring-tension T which is necessary for equilibrium according to the membrane hypothesis.

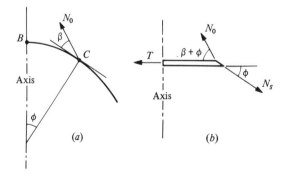

the circumferential hoop tension T at C by considering the equilibrium of half of the hoop shown in fig. 4.15b. The equations are, respectively,

$$N_s \sin\phi = N_0 \sin(\phi + \beta),$$
$$T = a[N_s \cos\phi - N_0 \cos(\phi + \beta)],$$

and they may be rearranged to give

$$N_s = N_0 \sin(\phi + \beta)/\sin\phi, \qquad (4.31a)$$
$$T = aN_0 \sin\beta/\sin\phi. \qquad (4.31b)$$

There are two specially simple cases. First, when $\beta = 0$, $N_s = N_0$ and $T = 0$: the tangentially applied load enters the shell without the need for a concentrated ring of tension. For all nonzero values of β (except $\beta = \pi$), equilibrium requires $T \neq 0$, i.e. a ring of tension. Second, when $\beta = \pi - \phi$, i.e. N_0 is in the plane of the latitude circle, $N_s = 0$ and $T = aN_0$. In this case the load is carried directly by a local hoop of tension and does not enter the shell proper at all.

The two directions (tangential to the meridian and normal to the axis) in which the solution of the equilibrium equations is specially simple, will be useful in the subsequent analysis of axially symmetric problems in chapter 11: in fact they will suggest the use of a simple nonorthogonal coordinate system.

4.6 Equilibrium equations for nearly-cylindrical shells

All of the equations of equilibrium which we have encountered so far in this chapter have been relatively simple. In both cylindrical shells and in shells of revolution we have found that one of the stress resultants can be calculated directly: N_θ for the cylindrical shell (see (4.9)) and N_s for the symmetrically-loaded shell of revolution (see (4.24)). To this extent, both problems are special cases. In the interests of completeness we ought, therefore, to give some more general examples. Shells of revolution with non-symmetric loading furnish some good instances, but the calculations can be involved. A typical example is a hyperboloidal cooling tower subjected to steady wind-loading (Martin & Scriven, 1961).

There are in fact some general features of the solution of the equilibrium equations which can best be studied by means of a shell whose geometry can be varied at will in certain respects. The simplest surface which illustrates these features is what may be called a 'nearly-cylindrical' shell, whose generators are not straight, but are slightly curved. Two examples, namely a 'barrel' shell and a 'waisted' shell are illustrated in fig. 4.16. It is clear that for surfaces of this sort we ought strictly to use a coordinate system appropriate to a general shell of revolution. Because the radius r of latitude circles is not constant, this coordinate system is more complicated than that used for the cylindrical shell.

4.6 Equilibrium eqns for nearly cylindrical shells

However, by confining our attention to a short length of shell near the point of maximum or minimum radius, respectively, and neglecting the variation of r, we can successfully use only a slightly modified version of the cylindrical shell equations. Of course, it can be misleading to use a not-strictly-proper coordinate system, and we must be cautious in interpreting the results. There is, however, a large and well-established branch of our subject known as 'shallow-shell' theory, and it is in the spirit of this work that we make our present study.

Let a be the (nearly constant) positive radius of the latitude circles of the 'nearly cylindrical' shell, and let a^* be the radius of curvature of the meridian, with $|a^*| \gg a$. Where the two principal radii have the same sense (the 'barrel') the values of a^* and a have the same sign, i.e. $a^* > 0$, but for the waisted shell $a^* < 0$. As we shall see, the character of solutions of the equilibrium equations will depend strongly on the sign of a^*: the cylindrical shell ($a^* \to \pm\infty$) will provide a special dividing case. For the sake of clarity it will be best to write certain equations in two different forms, depending on the sign of a^*.

In general, the three equations are (cf. (4.9)-(4.11))

$$\frac{N_\theta}{a} + \frac{N_x}{a^*} = p, \qquad (4.32)$$

$$\frac{1}{a}\frac{\partial N_\theta}{\partial \theta} + \frac{\partial N_{x\theta}}{\partial x} = -q_\theta, \qquad (4.33)$$

$$\frac{\partial N_x}{\partial x} + \frac{1}{a}\frac{\partial N_{x\theta}}{\partial \theta} = -q_x. \qquad (4.34)$$

We have retained the x, θ coordinate system from the cylindrical shell. The only change in the equations is the second term on the left-hand side of (4.32),

Fig. 4.16. 'Nearly-cylindrical' shells of (a) 'barrel' and (b) 'waisted' form, having almost constant radius $r \approx a$ (cf. fig. 4.10) and second principal radius $|a^*| \gg a$. The length of the shells is such that the variation in r does not exceed (say) 10%.

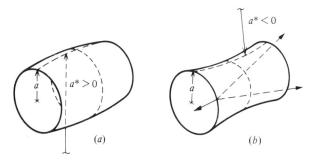

which enters because the generators are not straight; and as we have already implied, (4.33), (4.34) are not strictly accurate: see problem 4.21.

We shall consider the case of zero surface tractions ($p = q_\theta = q_x = 0$) and investigate cases in which the shell carries loading N_x applied at edges $x =$ constant. It is immediately clear that this problem is more complicated than it was in the case of a cylindrical shell: there we could solve the equations one by one, but here the presence of the N_x term in (4.32) makes that scheme impossible. Our first task therefore is to establish a differential equation for the problem in terms of one suitable variable. It will be most convenient to use N_x for this purpose.

From (4.32) we have

$$N_\theta = -(a/a^*)N_x; \tag{4.35}$$

and by eliminating $N_{x\theta}$ from (4.33) and (4.34) we obtain

$$\frac{\partial^2 N_x}{\partial x^2} - \frac{1}{a^2}\frac{\partial^2 N_\theta}{\partial \theta^2} = 0. \tag{4.36}$$

Substituting from (4.35) in (4.36) we have the required equation in N_x:

$$\frac{\partial^2 N_x}{\partial x^2} + \frac{1}{aa^*}\frac{\partial^2 N_x}{\partial \theta^2} = 0. \tag{4.37}$$

From this point it will be convenient to treat the two kinds of surface in different ways, so that their distinctive features may be studied separately.

4.6.1 'Barrel' shell, $a^* > 0$

First we consider the 'barrel' shell shown in fig. 4.16a with free edges at $x = \pm l$. It will serve our purposes best to consider an external loading

$$N_x = N_0 \cos n\theta.$$

applied to each of the two edges. In the case of a strictly cylindrical shell we would find, after a previous example, that N_x is the same function of θ for all values of x, and $N_{x\theta} = N_\theta = 0$. Here, however the governing equation (4.37) has a second term. We shall therefore hazard a solution of the form

$$N_x = N_0 \cos n\theta f(x). \tag{4.38}$$

Substitution into (4.37) immediately gives an ordinary differential 'auxiliary' equation for $f(x)$:

$$f''(x) - (n^2/aa^*)f(x) = 0, \tag{4.39}$$

where a prime denotes differentiation with respect to x. This equation is to be solved subject to the boundary conditions

$$f(-l) = f(l) = 1. \tag{4.40}$$

4.6 Equilibrium eqns for nearly cylindrical shells

The general solution of (4.39) may be written

$$f(x) = A \operatorname{Cosh}[n(aa^*)^{-\frac{1}{2}}x] + B \operatorname{Sinh}[n(aa^*)^{-\frac{1}{2}}x], \quad (4.41)$$

where A, B are constants to be determined by the boundary conditions (4.40): these give

$$A = (\operatorname{Cosh} n\zeta)^{-1}, \quad B = 0,$$

where ζ is a dimensionless length defined by

$$\zeta = l(aa^*)^{-\frac{1}{2}}. \quad (4.42)$$

The complete solution is thus:

$$N_x = N_0 \cos n\theta \ \operatorname{Cosh} n\xi / \operatorname{Cosh} n\zeta \quad (4.43a)$$

$$N_{x\theta} = -(a/a^*)^{\frac{1}{2}} N_0 \sin n\theta \ \operatorname{Sinh} n\xi / \operatorname{Cosh} n\zeta \quad (4.43b)$$

$$N_\theta = -(a/a^*) N_0 \cos n\theta \ \operatorname{Cosh} n\xi / \operatorname{Cosh} n\zeta \quad (4.43c)$$

where $\xi = x(aa^*)^{-\frac{1}{2}}$. The expressions for N_θ and $N_{x\theta}$ come from (4.35) and (4.34) respectively: in integrating (4.34) we have set an arbitrary constant equal to zero.

Since $N_{x\theta}$ is nonzero at the edges, it will be necessary to supply edge shear reactions accordingly. Having specified N_x at the two boundaries, we seem to have no option but to specify $N_{x\theta}$ as well. This may seem paradoxical, but it will be explained in section 4.7.

In general, (4.43) indicates that N_x dips from the prescribed values at the ends of the shell and is minimum at the central plane $x = 0$. The amount of dip depends on the values of n, l, a and a^*. By inspection of (4.43) there is no dip when $a^* \to \infty$, i.e. the cylindrical case is recovered; and then indeed also $N_\theta = N_{x\theta} = 0$. For a given geometry, i.e. a given value of ζ, the dip is larger for larger values of n. To obtain an idea of the magnitudes involved let us consider an example in which the radius at $x = 0$ is about 10% greater than the radius at the edges. The sagitta (see appendix 8) of the meridian is $l^2/2a^*$; putting this equal to $a/10$ gives $l^2/aa^* = 0.2$ i.e. $\zeta = 0.2^{\frac{1}{2}} \approx 0.45$. It is reasonable to suppose that our solution will be fairly good provided the variation in radius is not more than 10% (see problem 4.21), so the solution may become unreliable for larger values of ζ. For convenience we shall take $\zeta = 0.5$ in our numerical example. For this value of ζ the ratio of N_x at $x = 0$ to N_x at $x = l$ is equal to $1/\operatorname{Cosh}\tfrac{1}{2}n$. This is tabulated for several values of n below:

n	$1/\operatorname{Cosh}\tfrac{1}{2}n$		
1	0.887	5	0.163
2	0.648	6	0.099
3	0.425	7	0.060
4	0.266	8	0.037
		9	0.022

In this example there is thus a very striking die-away effect for N_x (and also for $N_{x\theta}$ and N_θ) as one moves away from the loaded edges towards the centre: a 'barrelling' of only 10% completely changes the character of the solution from that of a cylindrical shell.

It is instructive to consider a definite example, in which the edge-loading is represented by the following Fourier series:

$$N_x = N_0 \sum_n \cos n\theta, \quad n = 1, 3, 5 \ldots \qquad (4.44)$$

This represents (see appendix 10) a tensile point load of magnitude aN_0 at $\theta = 0$ in the meridional direction at each end, and a similar but compressive load at $\theta = \pi$. This produces a 'bending' effect in the shell as a whole. The circumferential distribution of N_x at the plane $x = 0$ is plotted in fig. 4.17a for various values of ζ. Note that there is a pronounced 'diffusing' effect due solely

Fig. 4.17. (a) Distribution of N_x around central cross-section of a 'barrel' shell which is loaded at its ends by concentrated tensile and compressive forces directed along the meridian at $\theta = 0$ and π respectively, as sketched in (b). The geometry of the shell is specified by the parameter $\zeta = l/(aa^*)^{\frac{1}{2}}$; the fractional change of radius along the shell is equal to $\frac{1}{2}\zeta^2$. Edge shearing reactions $N_{x\theta}$, shown schematically in (b), are required by the equilibrium equations.

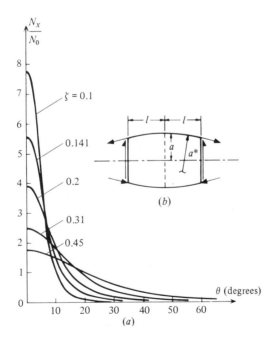

4.6 Equilibrium eqns for nearly cylindrical shells

to the (relatively small) barrelling of the shell, characterised by nonzero values of ζ.

Associated with each term in expression (4.44) is a shearing term, according to (4.43b): consequently the edge shear which must be provided at $x = -l$ is given by

$$N_{x\theta} = (a/a^*)^{\frac{1}{2}} N_0 \sum_n \mathrm{Tanh} n\zeta \sin n\theta, \quad n = 1, 3, 5 \ldots \quad (4.45)$$

The first term in this series has a resultant equal to $\pi(a/a^*)^{\frac{1}{2}} aN_0 \mathrm{Tanh}\zeta$ in the plane of the end of the shell. This force is necessary to preserve equilibrium of the shell as a whole: each of the point loads applied to the edge of the shell has a component normal to the axis, as shown in fig. 4.17b.

4.6.2 'Waisted' shell, $a^* < 0$

A 'waisted' shell is shown in fig. 4.16b, with two free edges. In this case it is more convenient to take the plane $x = 0$ at one end of the shell (not necessarily at the narrowest part) and to consider an edge-loading

$$N_x = N_0 \cos n\theta, \quad N_{x\theta} = 0 \quad (4.46)$$

applied at this edge. We shall see later why we do not choose to apply loads at both edges. The equations (4.32)–(4.34) are exactly the same as before, but it will be convenient, since a^* has a negative value, to rewrite (4.37) as

$$\frac{\partial^2 N_x}{\partial x^2} - \frac{1}{(-aa^*)} \frac{\partial^2 N_x}{\partial \theta^2} = 0. \quad (4.47)$$

Quantities $(-aa^*)$ and $(-a/a^*)$ (see later) are positive in value. Expressing a trial solution in terms of $f(x)$ as before (4.38) we have an ordinary differential equation

$$f''(x) + [n^2/(-aa^*)]f(x) = 0. \quad (4.48)$$

This has a general solution

$$f(x) = A' \cos[n(-aa^*)^{-\frac{1}{2}} x] + B' \sin[n(-aa^*)^{-\frac{1}{2}} x], \quad (4.49)$$

where A', B' are constants to be determined by the boundary conditions. Using (4.34) we find that these are

$$f(0) = 1, \quad f'(0) = 0. \quad (4.50)$$

Hence $A' = 1, B' = 0$ and the complete solution is

$$N_x = N_0 \cos n\theta \cos n\beta x, \quad (4.51a)$$

$$N_{x\theta} = (-a/a^*)^{\frac{1}{2}} N_0 \sin n\theta \sin n\beta x, \quad (4.51b)$$

$$N_\theta = (-a/a^*) N_0 \cos n\theta \cos n\beta x, \quad (4.51c)$$

where $\beta = (-aa^*)^{-\frac{1}{2}}$.

All of these expressions consist of products of trigonometric functions in the circumferential and longitudinal directions. Therefore each of the three stress resultants has zero value on a set of equally spaced meridians and latitude circles, and the signs of these variables form a 'chequerboard' pattern with respect to this grid. The dimensions of the 'panels' of this grid are $\pi a/n$ circumferentially and $(\pi a/n)(-a^*/a)^{\frac{1}{2}}$ longitudinally. The circumferential dimension depends, of course, on the arbitrary wavenumber n; but we see that the *proportions* of the panel are fixed only by the ratio of principal radii. In this connection it will be useful to define a geometrical parameter

$$m = (a/-a^*)^{\frac{1}{2}}. \tag{4.52}$$

For our 'nearly cylindrical' shells m^2 will be small in comparison with unity, so the 'panels' described above will be elongated in the meridional direction, with width/length = m.

In fact the solution (4.51) has an even more remarkable property, which can be seen most easily by writing (4.51a) as

$$N_x = \tfrac{1}{2} N_0 \cos n(\theta - \beta x) + \tfrac{1}{2} N_0 \cos n(\theta + \beta x). \tag{4.53}$$

The first part of this expression is constant along lines $\theta - \beta x$ = constant: along these lines there is neither growth nor decay of this part of the solution. And there is similar constancy of the second part of the solution along the lines $\theta + \beta x$ = constant. The two sets of lines make equal angles with the x-axis, and it is easy to show that this slope is $\pm m$. Note in particular that the slope is independent of the circumferential wavenumber n of the loading. This feature, which is quite different from anything in the solution for the barrel shell, makes it possible to write down an extremely simple solution of the given problem (after d'Alembert) when the applied loading at the edge is an arbitrary function of θ, say $N_x = N_0(\theta)$ at $x = 0$:

$$N_x = \tfrac{1}{2} N_0(\theta - mx/a) + \tfrac{1}{2} N_0(\theta + mx/a). \tag{4.54}$$

It is a simple matter to show by direct substitution that this satisfies all of the equations of the problem. There are no restrictions on the smoothness of $N_0(\theta)$: it could, for example, represent one or more point loads.

The families of lines along which parts of the solution neither grow nor decay are known as the *characteristics* of the solution. This feature is well known in the solution of *hyperbolic* partial differential equations, and the families of lines furnish a useful tool in the solution of problems involving this kind of equation. Hyperbolic equations occur in dynamics (e.g. the wave equation), in fluid mechanics (supersonic flow), in the theory of perfectly plastic solids, and in many other physical problems (Abbott, 1966). Flügge (1973, Chapter 4) gives a more complete account of the differences between

4.6 Equilibrium eqns for nearly cylindrical shells

solutions of the two kinds of equations. These special features have no counterpart in the solution (4.43) for the barrel shell, $a^* > 0$, for which the equations are *elliptic*.

The characteristic lines $a\theta = \pm mx$ have a striking geometrical significance: they are *straight* lines in the surface. This may be discovered by expressing the surface locally in Cartesian coordinates. Putting the origin of a Cartesian x, y, z coordinate system at a point on the cross-section of minimum radius, recalling that the principal radii are a and $-(-a^*)$, and using expression (4.5) we have, for a sufficiently restricted zone of the xy-plane

$$z = [x^2/2(-a^*)] - (y^2/2a). \tag{4.55}$$

The positive sense of z is directed outwards from the centre of the shell. For a characteristic $y = \pm mx$ (where $m = (-a/a^*)^{\frac{1}{2}}$) we find, simply, $z = 0$: in other words the characteristic lines are straight. This helps us to interpret physically the solution described by (4.54) for an arbitrary loading $N_x(\theta)$ applied at an edge $x = 0$. The load applied over any small interval $d\theta$ of the circumference may be regarded as a 'point' load, which is simply carried by two straight 'lines of tension' along the two characteristic directions passing through the point of application of the load, as sketched in fig. 4.16b. (Here we are tacitly assuming that the solution of the equilibrium equations is unique when we assert that *a* solution which obviously satisfies the equations is also *the* solution. Points of this sort are beyond the scope of this work.) This example also provides a connection with an earlier remark about an axial point load applied to the edge of a cylindrical shell and carried by the shell as a line of undiminished tension along a generator: we can see that the two families of characteristics of a waisted shell merge into a single family when the value of a^* is made so large that the shell becomes cylindrical.

Although our discussion of these points has been based on simplified, approximate equations, we have reached a point where we can see a simple physical interpretation of the solution of equilibrium equations for shells which possess *straight lines* in their surfaces, at least if these lines run from edge to edge of a shell. This property of nearly-cylindrical 'waisted' shells holds also for hyperboloids of revolution. Independently of the coordinate system or the merely local equations of the surface we can see that certain kinds of edge-loading will be carried in the shell by constant forces acting in the characteristic lines. At the risk of overemphasis, however, we remark that these conclusions apply only to the 'membrane' solution of the equilibrium equations alone. It is not difficult to see qualitatively that when elasticity, etc., are taken into account this kind of effect will in most cases (unless perhaps there are suitable prestressing wires) be sufficient to invalidate the membrane hypothesis and demand a full analysis. This is a subject of chapter 8.

The difference in kind between the solutions of the equilibrium equations for the 'barrelled' and 'waisted' shells, respectively may be attributed to the parity or non-parity of the coefficients of the two terms in (4.37). If the principal radii are of the same sign (and therefore, the Gaussian curvature of the surface $1/R_1 R_2$ is positive: see chapter 5) the governing equation is *elliptic*, and concentrated edge-loads are diffused by the shell. On the other hand, if the principal radii are of opposite sign (and so the Gaussian curvature is negative) the governing equations are *hyperbolic*, and at each point in the surface there are two real characteristic directions. If the characteristics run straight from edge to edge, loads applied (tangentially) at an edge are carried by undiminished tensions along these straight lines in the surface. In some respects this makes the specification of boundary conditions awkward, since a condition applied at one edge can impose itself on a remote edge. It was precisely for this reason that the boundary conditions for the two kinds of shell (barrelled, waisted) were specified differently for the two different cases: they were in fact chosen to give simple solutions.

4.7 Boundary conditions in membrane analysis

In the preceding section we have concentrated on differences between the solutions of the equilibrium equations for elliptic and hyperbolic problems. We shall now, for the sake of completeness, focus attention on the similarities between all the solutions we have encountered throughout this chapter, with particular reference to boundary conditions.

Consider a tubular shell (waisted, cylindrical or barrel-shaped) with two free edges, and a solution for the various stress resultants involving circumferential variations $\cos n\theta$, $\sin n\theta$. (This does not restrict the generality of the following argument; we can if we wish superpose Fourier components to build more general solutions.) By the nature of the equations the ordinary auxiliary equation (e.g. (4.39), (4.48)) is of second order, and its general solution therefore involves *two* arbitrary constants. But four 'boundary quantities' are necessary to specify N_x, $N_{x\theta}$ separately at the two edges. Consequently, by the elementary theory of algebraic equations, we are in general free to choose only two of these boundary quantities, and we must accept what the equations produce for the other two. We have seen several examples of this rule: for the cylindrical and hyperbolic shells we specified two quantities at one edge and provided the necessary values at the other edge by means of reactions from a 'foundation', while for the elliptic shell we specified one quantity at each end (maintaining overall equilibrium) and provided the other as a reaction. In principle this is straightforward, even in cases where the support system is arranged to provide a reaction of the form $N_{x\theta}$ = (constant)N_x. Nevertheless, there are some nagging problems. A good example is provided by the barrel shell, where we specified N_x at each end and then were obliged to provide $N_{x\theta}$ also. What would be the consequence of *not* providing the required $N_{x\theta}$?

4.7.1 A 'framework' analogy

In order to clarify such puzzles, and also questions of how 'reactions $N_{x\theta}$' might be provided, it is useful to set up a 'framework analogy' for equilibrium analysis of shells according to the membrane hypothesis. We shall use closely related ideas in chapter 5 to help to explain the geometry of curved surfaces. The idea is simply to replace the shell surface by a network of short straight *bars* connected by frictionless universal *joints*. Every bar is connected at its ends to two joints, and the entire surface is coverd by *triangles*. Within the limitations given above the scheme of triangulation is arbitrary as to the number and disposition of the joints; nevertheless it helps to avoid confusion if we adopt some common-sense rules like having a string of bars and joints around the edge of the shell and avoiding grossly elongated triangles. We also impose the restriction that loads are applied to the framework only at the joints.

This framework analogy clearly reproduces an important feature of the membrane hypothesis, since the framework, having frictionless joints, is only capable of carrying the loads applied to it by means of tensile or compressive forces in the bars lying in the surface. Bending moments and out-of-plane shear forces simply cannot be carried by the bars.

Now we have seen, earlier in this chapter, that according to the membrane hypothesis a shell is a statically determinate structure, at least in the sense that the equilibrium of a small element is governed by three equations in three unknowns (e.g. N_x, $N_{x\theta}$, N_θ). There may be exceptions, such as shells having *plane* regions, but they need not concern us here (see appendix 9). Consequently, the proposed framework analogy stands or falls according to whether or not it is statically determinate in an analogous sense.

The best way of investigating this problem is to consider first an arbitrary simply-connected surface, free in space, for which has been substituted a triangulated framework laid out as described above. The original surface and the derived framework are shown schematically in fig. 4.18*a* and *b* respectively. Let the framework have, in general, *j* joints and *b* bars.

The framework also defines a polyhedron having *j* vertices, *b* edges and (say) *f* triangular faces. Now there is a remarkable topological theorem of

Euler (see, e.g., Hilbert & Cohn-Vossen, 1952, §44), which states that for any simply-connected polyhedron:

Number of faces + Number of vertices = Number of edges + 2.

In the present problem, therefore,

$$f + j = b + 2. \tag{4.56}$$

But every face is triangular, and has three edges which are shared with two other faces: therefore

$$b = \tfrac{3}{2} f.$$

Eliminating f from these equations we have, finally

$$b = 3j - 6. \tag{4.57}$$

Now this relation is, precisely, Maxwell's general condition for a three-dimensional framework, free in space, to be 'stiff' (e.g. Parkes, 1974, §2.5). It is indeed also the general condition for the framework to be statically determinate. This may be demonstrated directly from the fact that there are $3j$ equilibrium equations and $b + 6$ unknowns (b bar tensions and 6 support reactions). For the (algebraic) equations to have a unique solution for arbitrary external loading, the number of equations and unknowns must be equal. Although there are some (important) exceptions to this rule (see appendix 9)

Fig. 4.18. (*a*) A smooth, simply-closed surface which is replaced (*b*) by an arbitrarily triangulated mesh of pin-jointed bars. (*c*) A cylindrical shell with closed ends and (*d*) a counterpart mesh. (*e*) As (*d*), but a hole has been cut from the shell. (*f*) As (*d*), but the ends are conical rather than plane.

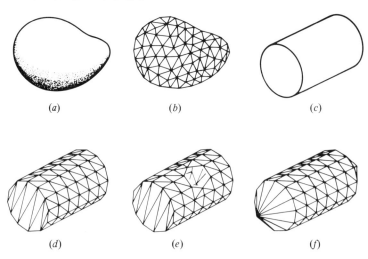

4.7 Boundary conditions in membrane analysis

we may conclude that, in general, the topology of the arrangement guarantees the statical determinacy of the framework irrespective of the geometric details of the layout. This is a remarkable and useful result.

Let us now apply this idea to an open-ended cylindrical shell, as shown in fig. 4.18c. In (d) the cylindrical surface has been replaced by an arbitrary network of triangles; and the ends have been closed in a triangulated fashion by the addition of bars connecting joints on the circumference. We have given the shell a closure of this sort so that, being a simply closed figure, it automatically satisfies the general requirement for statical determinacy.

Now it is a well-known feature of a three-dimensional framework which is just stiff, that the cutting of any bar converts the structure into a mechanism with one degree of freedom (in addition to the 6 rigid-body motion degrees of freedom); and that the cutting of each further bar imparts an extra degree of freedom to the mechanism. Consequently, in removing the extra bars at the two ends which are not part of our original cylindrical surface, we shall give to the assembly a large number of degrees of freedom. This is not altogether surprising, for an open tube constructed from bars and joints is obviously very 'floppy'. We are now in a dilemma, for our shell (fig. 4.18c) is only held in the required circular configuration by unwanted, extraneous, bars (fig. 4.18d). If we decide to remove these bars, the cylindrical surface is no longer a structure, but is now a mechanism; and it will only carry external loads if certain conditions between the loads themselves are satisfied (Calladine, 1978). We have already seen an example of this general idea in the case of a circular string carrying radial loads: equilibrium was only possible in the postulated circular configuration provided related tangential loads or 'reactions' were also supplied. The case of the end-loaded barrel shell, noted earlier, is explained in exactly the same way: the 'required' edge-loads $N_{x\theta}$ must be provided if the intended circular shape is to be maintained. If they are not provided, the surface behaves like a mechanism and distorts grossly. In algebraic terms, the equilibrium equations are insoluble: an equilibrium solution cannot be found.

But there is still a problem. How can we regard these $N_{x\theta}$ edge-loads as reactions? What can they react against? To answer this, we consider again the arrangement of fig. 4.18d, but this time we do *not* cut the bars at the ends. The cylindrical surface *plus* the end constraints does now constitute a proper *structure* to which arbitrary loads may be applied. The application of axial loading N_x to the ends will, in general, require tensions in these end-plane bars, just as it does in the other bars. It is these tensions which keep the ends of the shell circular. The $N_{x\theta}$ 'reactions' at the ends are sustained by the end-rings and the cross-bars in the end-planes acting as statically determinate two-dimensional structures. See problem 4.22 for an example.

We have just seen that if a shell is inadequately restrained at its edges it may be a *mechanism* rather than a true structure in the context of the membrane hypothesis; and that accordingly, it can only sustain applied loads without a change of configuration provided the loads satisfy certain requirements related to the modes of free distortion. This point can easily be overlooked, and therefore we give here a cautionary example, due to Novozhilov (1964, Fig. 11c). We have seen that a cylindrical shell which is restrained radially at its ends is statically determinate for arbitrary loads which satisfy the conditions of overall equilibrium. Suppose now that we cut a hole in the surface, and correspondingly remove bars over an appropriate area, as shown in fig. 4.18e. This action changes the system into a mechanism with several degrees of freedom, and in particular the arrangement in fact becomes incapable of carrying an arbitrary load such as a uniform axial tension N_x applied around the two ends. But actual cylindrical shells with cutouts of this sort may be required to carry uniform axial tension. Since the membrane hypothesis admits of no solution, the problem can only be analysed satisfactorily by means of 'full' shell theory. We shall encounter further examples of this type in chapter 9.

Let us now return to shells which are adequately supported, and attempt to interpret in terms of the framework analogy those cases in which a tubular shell was, when unloaded, free at one end and 'doubly' restrained at the other. In this case it is simplest not to start with Euler's theorem, but to imagine first a rigid base with n 'foundation points' arranged in a circle, as shown in fig. 4.19a. From each pair of points a 'Λ-frame' of two bars is erected, and finally the apices of the Λs are joined in pairs by a ring of bars, as shown in fig. 4.19b. As before, all joints are frictionless. In erecting this 'fence' or stage we have added n joints and exactly $3n$ bars; consequently the arrangement is in general statically determinate, and will continue to be so if another stage is built, using the top of the first stage as its foundation. In fact the arrangement will satisfy the general condition for static determinacy for an arbitrary number of stages. It follows that the structure is capable of supporting arbitrary load-

Fig. 4.19. (a) A rigid base with n mesh-points onto which (b) a 'fence' of $3n$ bars and n more joints may be built.

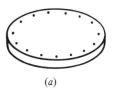
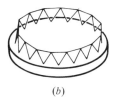

(a) (b)

ings applied at the joints, and in particular may withstand *independent* N_x and $N_{x\theta}$ loading applied as equivalent point loads to the upper edge. (In fact arrangements of this sort can, in certain circumstances, degenerate into 'exceptional cases' as far as the general rules for construction of frameworks are concerned (Tarnai, 1980). This does not affect adversely the general point which is being made in terms of the number of boundary conditions required.)

Lastly, we give in fig. 4.18*f* an example of a cylindrical shell held circular by 'spokes' forming a conical surface. As tension in any of these bars imposes both longitudinal and radial loading onto the cylindrical surface, the edge-conditions for the cylindrical part of the shell involve a relation between N_x and $N_{x\theta}$.

4.8 Summary

In this chapter we have been concerned with the solution of the equilibrium equations for shells which we obtain according to the membrane hypothesis, namely that bending moments and normal shearing-stress resultants are zero throughout the shell. In each case we have set up the equilibrium equations for the undistorted configuration of the shell. In general, the resulting problems are statically determinate, and we have seen several examples where the solution was perfectly straightforward. We have also seen examples in which there was no solution (because the shell constituted a *mechanism* rather than a structure) or the solution involved singularities such as concentrated lines or rings of tension. In all of these cases there are *prima facie* grounds for rejecting the membrane hypothesis as an oversimplification, and performing instead a 'full' shell analysis. The remainder of the book is largely devoted to such problems. In the present chapter, however, we have been concerned simply to *solve* the equations, singularities notwithstanding. In particular, we have not adopted the traditional practice of abandoning the task at the first sign of a singularity.

In the most elementary problems involving simple statically determinate trusses of the kind encountered at the beginning of courses on elementary mechanics, boundary conditions play an important part. The same is true for shell structures analysed according to the membrane hypothesis. In general there are two stress resultants N_n, N_{nt} at an edge of a shell (here n, t are directions normal and tangential to the edge in the plane of the surface), and on average one may be taken as an arbitrary loading and the other as a reaction: the average is in the sense that, as we have seen, we may sometimes have two reactions at one edge, and then will be allowed two arbitrary loadings at the other.

We have also seen that the solution of the equilibrium equations for surfaces of negative Gaussian curvature can involve effects being transmitted along the straight characteristic directions in the surface without 'diffusion' into the shell. Moreover we have seen that this has consequences for our freedom to impose boundary conditions. In contrast, shells of positive Gaussian curvature allow diffusion of loads from the edge to the interior. Cylindrical shells, which have zero Gaussian curvature, constitute a borderline case.

The membrane hypothesis is the best first step in the analysis of many shell structures: a statical analysis provides a good foundation for any subsequent work which may be necessary. In particular, solutions should be obtained even and especially when singularities emerge: as we shall see, they provide valuable clues to the nature of the results of 'full' shell theory.

4.9 Problems

4.1 A circular string of radius a is subjected to a uniform normal load of intensity p in regions $0 < \theta < \frac{1}{2}\pi$ and $\pi < \theta < \frac{3}{2}\pi$, and is otherwise unloaded: θ is an angular coordinate, as in fig. 4.2d. Show that four external tangential forces of magnitude ap are required for equilibrium; but that these cannot be deduced from the considerations of overall equilibrium of the ring. What form would the string adopt if these tangential forces were not provided?

4.2 A uniform normal load of intensity p is applied to a short section of the circumference of a circular string of radius a, subtending a small angle 2α at the centre. The magnitude of p is such that $2a\alpha p = P = $ constant. Tangential tractions are provided at the two ends of the arc, in order to preserve the circular form of the string. Show that the value of these tractions is $P/2\alpha$; which becomes infinite as $\alpha \to 0$ and the loading approximates a discrete normal force.

4.3 When the axes of principal curvature do not coincide with the x, y coordinate axes (cf. fig. 4.3a), show that (4.8) should be replaced by:

$$N_x \frac{\partial^2 z}{\partial x^2} + 2N_{xy} \frac{\partial^2 z}{\partial x \partial y} + N_y \frac{\partial^2 z}{\partial y^2} = -p.$$

4.4 Use equilibrium equations (4.9)–(4.11) to determine $N_x, N_\theta, N_{x\theta}$ in the shell shown in fig. 4.7, which is clamped at $x = l$ and sustains an edge-load $N_{x\theta} = N_0 \sin\theta$ at $x = 0$.

4.5 A semicircular cylindrical shell roof is defined by $-\frac{1}{2}\pi \leq \theta \leq \frac{1}{2}\pi, -l \leq x \leq l$ in the notation of fig. 4.8a. The cross-section $x = 0$ is a plane of mirror-symmetry, and the shell is supported at planes $x = \pm l$. The shell is loaded by uni-

4.9 Problems

form self-weight (see (4.13)). Solve the equilibrium equations (4.9)–(4.11) subject to the boundary conditions $N_\theta = 0$ at $\theta = \pm\frac{1}{2}\pi$, $N_x = 0$ at $x = \pm l$; and deduce what stress resultants $N_{x\theta}$ are necessary at these four edges. Along edges $\theta = \pm\frac{1}{2}\pi$ these shear-stress resultants are provided by tendons carrying pure tension, which has a value of zero at each end. Determine the tension in these tendons as a function of x; and describe the stress resultants which are transmitted across the central section $x = 0$ (including the tendons).

4.6 Solve equilibrium equations (4.9)–(4.11) for a complete cylindrical shell which is subjected to a normal pressure-loading $p = p_n \cos n\theta$, and using boundary conditions $N_x = 0$ at the edges $x = \pm l$. Determine the stress resultants $N_{x\theta}$ which are required at these edges in order to preserve the circular shape.

4.7 A complete circular cylindrical shell whose axis is horizontal has radius a and ends $x = \pm l$ at which $N_x = 0$ and which are held circular by reactions $N_{x\theta}$. The shell is half-full of a heavy liquid of density ρ, so that $p = 0$ on the upper half $-\frac{1}{2}\pi < \theta < \frac{1}{2}\pi$, while on the lower half $p = -\rho g a \cos\theta$. By using equilibrium equations (4.9)–(4.11), determine the state of stress throughout the shell; and in particular find the 'line tension' which is necessary along the generators $\theta = \pm\frac{1}{2}\pi$.

4.8 A shell as in fig. 4.7 is loaded by pressure $p = p_n \cos n\theta$ over the region $0 \leq x \leq h$, and is unloaded over the remaining part $h < x \leq l$. Use the equilibrium equations to determine the stress resultants in the shell. Now put $p_n = P_n/h$, and by steadily reducing the value of h obtain the stress resultants in a shell subjected to a load which is concentrated at the edge.

4.9 A shear loading $Q_{xz}(\theta)$ is applied to an edge $x = $ constant of a cylindrical shell of radius a. By considering the equilibrium of a circular 'edge-string', show that this is equivalent to a tangential shear loading $N_{x\theta} = dQ_{xz}/d\theta$ on the shell proper. Also show that, in consequence, the special loading $Q_{xz}(\theta)$ together with $N_{x\theta} = -dQ_{xz}/d\theta$ applied to an edge produces *zero* stress in the shell proper, but merely requires tension/compression in an 'edge-string' of the shell.

Pay particular attention to signs (the positive senses of $N_{x\theta}$ and $Q_{x(z)}$ are defined in figs 4.5 and 3.3, respectively; and cf. fig. 2.2), and show that the result holds whether the shell extends from the edge in the positive or in the negative direction of x.

4.10 In calculating the stresses in thin-walled grain-storage silos on account of wind-loading, it is useful to begin by regarding the silo as a cylindrical shell of radius a which is open at its top $x = 0$ and clamped at its base $x = l$. Tangential loading due to viscosity may be neglected (at large Reynolds' number) and it is usual to regard the pressure as a function only of the circumferential co-ordinate θ. Equations (4.18) and (4.19) then give $N_{x\theta}$ and N_x throughout the shell once $p(\theta)$ has been specified.

The following formula, due to Batch and Hopley (Martin & Scriven, 1961) and based on wind-tunnel tests, is often used in practice. For

$$0 < \theta < 47.6°, \quad p = -1.524 p_0 \cos 1.89\theta$$
$$47.6° < \theta < 100°, \quad p = 0.69 p_0 \sin[3.61(\theta - 47.6°)];$$
$$100° < \theta < 180°, \quad p = -0.21 p_0.$$

Here p_0 is the stagnation pressure of the wind (positive, by definition), and $\theta = 0$ is the windward generator. The main features of the distribution are an inward-directed pressure in the region around $\theta = 0$, and a region of outward-directed pressure which reaches a peak at about $\theta = 70°$. The pressure is uniform on the leeward side, in the region of turbulent flow beyond the generators at which the boundary layer breaks away. This particular distribution allows for 'suction' within the shell on account of flow past the open top.

Construct graphs, with suitable dimensionless ordinates, to show the variation of p, $N_{x\theta}$ and N_x with θ at the base of the silo.

It is sometimes more convenient to specify a pressure distribution of this sort by means of a Fourier series. For example, British Standard 4885, Part 4 (1975) (which is used in the design of water-cooling towers) gives

$$p/(-p_0) = \sum_{n=0}^{7} a_n \cos n\theta,$$

where

$$a_0 = -0.00071, \quad a_4 = +0.10756,$$
$$a_1 = +0.24611, \quad a_5 = -0.09579,$$
$$a_2 = +0.62296, \quad a_6 = -0.01142,$$
$$a_3 = +0.48833, \quad a_7 = +0.04551.$$

The strongest coefficients are for $n = 2$ and 3; and a relatively large number of coefficients is needed in order to give approximately constant pressure (here ≈ 0) in the leeward region.

Express the variation of $N_{x\theta}$ and N_x at the base of the silo as a Fourier series. (We shall discover in section 9.5 that the membrane hypothesis becomes invalid beyond a certain value of n, which is a function of the dimensions of the shell.)

4.11 Show that (4.24) is a direct consequence of (4.22) and (4.23). First use (4.22) to eliminate N_θ from (4.23); and verify that
$$\frac{1}{\sin\phi}\frac{d}{ds}(r\sin\phi N_s) = \frac{d}{ds}(rN_s) + \frac{rr_2}{r_1 r_3}N_s$$
by use of the identities (cf. fig. 4.10)
$$\sin\phi = r/r_2, \quad \cos\phi = r/r_3.$$

4.12 A conical shell has ϕ = constant = α, and it extends from the apex to a base-circle of radius b. It is suspended from its apex with its axis vertical. Show that when it is loaded by a self-weight of intensity f per unit area, the stress resultants according to the membrane hypothesis are given by
$$N_s = f(b^2 - r^2)/r\sin 2\alpha, \quad N_\theta = -fr/\sin\alpha.$$

4.13 A weightless hemispherical shell of radius a stands on a frictionless horizontal base. A uniformly distributed vertical line load of intensity q per unit length is applied downwards at a latitude circle ϕ_0. By solving the equilibrium equations show that
$$N_s = N_\theta = 0, \quad 0 < \phi < \phi_0;$$
$$N_s = -N_\theta = -q\sin\phi_0/\sin^2\phi, \quad \phi_0 < \phi < \tfrac{1}{2}\pi.$$

4.14 The shell of problem 4.13 is now loaded by a downward-directed point load P applied at the apex. Specialise the results of the problem to show that (according to the membrane hypothesis)
$$N_s = -N_\theta = -P/2\pi a \sin^2\phi.$$

4.15 The shell of problem 4.13 is now loaded by uniform self-weight of intensity f per unit area. Show that
$$N_s = -fa/(1 + \cos\phi), \quad N_\theta = fa((1 + \cos\phi)^{-1} - \cos\phi).$$
Sketch these functions, and verify that N_θ changes sign at $\phi \approx 52°$.

4.16 A spherical shell of radius a is subjected to a pressure $p(r)$ over a small region $r \ll a$ near the apex. By putting $\sin\phi \approx r/a$ and $\cos\phi \approx 1$ in the equations of equilibrium according to the membrane hypothesis, obtain the following expressions:
$$N_s/a = (1/r^2)\int_0^r rp(r)dr; \quad N_\theta/a = p(r) - (N_s/a).$$

122   *'Equilibrium' solutions: the membrane hypothesis*

A shell of this sort is loaded by pressure over a small area of radius $r = c$, but is otherwise unloaded. The pressure acting on the loaded region is non-uniform and is described as follows in terms of a dimensionless radial coordinate $\xi = r/c$:

$$p = p_0(1 - 3\xi^2 + 2\xi^3), \quad 0 \leq \xi < 1;$$
$$p = 0, \quad 1 \leq \xi.$$

Plot p, N_s/a and N_θ/a as functions of ξ for $0 \leq \xi < 3$.

4.17 The cap of a spherical shell of radius a has a horizontal base-circle of radius $a/2^{\frac{1}{2}}$. It forms a roof, with its apex uppermost. Show that the distribution of stress resultants N_s, N_θ in the shell on account of a uniform 'snow-load' of intensity f per unit area *in plan* are as follows, according to the membrane hypothesis:

$$N_s = -\tfrac{1}{2}fa, \quad N_\theta = -\tfrac{1}{2}fa\cos 2\phi.$$

Using the transformation $r = a\sin\phi$, plot these as functions of r/a in the range $0 < r/a < 2^{-\frac{1}{2}}$.

4.18 The shell of problem 4.17 is now replaced by a paraboloid of revolution $z = r^2/2b$, where z is an axial coordinate, measured vertically downwards from the apex.

For this surface show that $r = b\tan\phi$, $r_1 = b\sec^3\phi$, $r_2 = b\sec\phi$. Hence show that according to the membrane hypothesis the stress resultants due to the snow-load of problem 4.17 are as follows:

$$N_s = -\tfrac{1}{2}fb\sec\phi, \quad N_\theta = -\tfrac{1}{2}fb\cos\phi.$$

Show that if the two surfaces are to coincide at both the apex and the base-circle, $b = 0.854a$. Draw the meridians of the two surfaces to scale on the same diagram. Plot N_s, N_θ as functions of r in the range $0 < r/a < 2^{-\frac{1}{2}}$.

4.19 A shell of revolution with a vertical axis, and having uniform weight per unit area, is to be of such a shape that $N_\theta = 0$ everywhere, according to the membrane hypothesis, when it is loaded by its own weight. By putting this condition into (4.22) and (4.23), show that the profile of the meridian is determined by

$$rr_1\cos^2\phi = \text{constant}.$$

4.20 A cylindrical pressure-vessel of radius a is closed by a spherical cap of radius R in the manner of fig. 4.14, but without a toroidal knuckle. The vessel is loaded by an internal pressure p. Show that the equations of equilibrium demand a circumferential 'hoop of compression' of magnitude $\frac{1}{2}pa(R^2 - a^2)^{\frac{1}{2}}$ at the junction: (a) by considering the loading applied to the junction-ring by the spherical and cylindrical component shells, treated separately; and (b) by evaluation of the 'swept area' (fig. 4.13) associated with the junction.

4.21 A 'nearly cylindrical' shell (fig. 4.16) is subjected to a uniform tension along its axis of revolution. Investigate by the method of section 4.5.1 the variation of N_s and N_θ with s. By comparing the results with those obtained from a formal solution of (4.32)–(4.34) with $p = q_x = q_\theta = 0$ and $\partial/\partial\theta = 0$, show that for this problem the discrepancies will not exceed about 10% provided the radius of the shell does not vary by more than about 10%. Repeat the analysis for the same shell loaded by pure torque about the axis of revolution.

4.22 A cylindrical shell of length l has radius a. At end $x = 0$ a loading $N_x = N_0 \cos n\theta$ is imposed, while at the end $x = l$, $N_x = 0$. At each end reactions $N_{x\theta}$ are supplied in order to maintain a circular shape. Show that, according to the membrane hypothesis,

$$N_x = N_0(1 - x/l)\cos n\theta,$$
$$N_{x\theta} = (aN_0/nl)\sin n\theta,$$
$$N_\theta = 0.$$

(This is relevant to the behaviour of vertical cylindrical liquid-storage tanks built on foundations which settle differentially: see problem 7.7.)

5

The geometry of curved surfaces

5.1 Introduction

The geometry of curved surfaces plays an important part in the theory of shell structures. Many practical shell structures are made in the form of simple surfaces such as the sphere, the cylinder and the cone, whose geometry has been well understood for centuries. It has, therefore, been argued by some workers that sophisticated geometrical ideas are not needed for the analysis and design of a wide range of practical shell structures.

Gauss (1828) made a breakthrough in the study of *general* surfaces. He showed that there were two completely different ways of thinking about a curved surface, either as a *three*-dimensional or a *two*-dimensional object, respectively; and that the two different views had a very simple mathematical connection involving a quantity which is now known as *Gaussian curvature.* The ideas which Gauss described in his paper are of great importance for an understanding of the behaviour of all shell structures, however simple their geometrical form happens to be. The main object of the present chapter is to explain Gauss's work in relation to the geometry of curved surfaces. In chapter 6 we shall proceed further along these lines, with an investigation of the geometry of *distortion* of curved surfaces.

The basic geometrical ideas which Gauss discovered are not difficult to grasp by those who have at their disposal relatively modest mathematical tools. However, Gauss gave a very thorough treatment of the problem in his paper, and in particular he developed the use of *general curvilinear coordinates* for the description of surfaces. This was the beginning of the subject of *Differential Geometry*, and in due course Gauss' pupil Riemann used it to study non-Euclidean spaces. The sophisticated tensorial mathematics of four-dimensional space-time, which is vital in the theory of general relativity, is a direct descendent of this work.

Love, (1888), who was one of the early workers in the field of shell structures, saw the advantages of using differential geometry and its curvilinear coordinates in the description of general surfaces, and he employed them in deriving his famous general equations for thin elastic shells (Love, 1927, chapter 24).

Following Love's work the literature on shell structures has divided into two main streams, which still persist today. Workers of one school, following Love, emphasise the importance of general equations in the development of basic ideas, and consequently set up their work in curvilinear coordinates: they specialise these equations at a late stage if they ever have to deal with a shell having a particular geometric form. Those of the other school take an opposite view. According to them it is more convenient to set up the equations in relation to the particular geometrical form (sphere, cylinder, etc.) of the shell under investigation, and so avoid the complications of the notation of differential geometry: the resulting special equations are often relatively simple and easy to solve.

The purpose of this brief sketch of the methods used by these two schools of workers in shell structures is to point out that *neither* school has reaped the benefit of the basic ideas on curvature which were developed by Gauss. Ironically, many who follow Love have adopted the notation of differential geometry without apparently grasping the fundamental point which Gauss made; and many of those who use the simple specialised equations implicitly deny the need for any *general* geometrical ideas.

The aim of the present chapter is now plain: it is to explain the basic ideas of Gauss without getting side-tracked by the notation of differential geometry.

By far the most important geometrical idea which we need to grasp is that of *Gaussian curvature* of a surface. Since curvature is a *local* property of a surface it is possible to make considerable progress by considering only a small region of a general surface; and this is how we can avoid using a general coordinate system for our basic explanation. For this purpose we shall find that an elegant result of Maxwell is invaluable. However, for the sake of completeness, we shall give a short introduction to curvilinear coordinate systems in chapter 6.

5.2 The idea of a surface

Our starting point is the idea of a *surface*. A surface may be defined as the boundary between two distinct regions of three-dimensional Euclidean space. It is easiest to think first of a simply-*closed* surface which separates the interior from the rest of space, and then of an *open* surface as a portion of a closed surface. It will not be necessary to go into topological problems of the kind which centre round Klein's bottle (see, e.g. Hilbert & Cohn-Vossen, 1952, §46): in the surfaces which we shall use in the study of shell structures there will usually be no difficulty in labelling the two distinct sides of the surface. For quantitative work it is usually convenient to have a coordinate z normal to the surface, and to regard the side on which $z > 0$ as the 'positive' side of the surface.

The above definition implies that a surface has zero thickness, just as a mathematical line or curve has zero thickness. Of course, a physical representation of a surface, such as a soap bubble, will normally have a measurable thickness, just as a line drawn by a pencil on a piece of paper has measurable thickness. In this chapter and the next we shall be concerned exclusively with 'mathematical' surfaces, and not with physical realisations of them.

The definition given above puts surfaces securely in three-dimensional Euclidian space, and in this sense surfaces are *three dimensional*. But an equally important view is that surfaces are *two dimensional*, in the sense that the position of any point on a surface may be specified by only *two* coordinates. An obvious example of such a coordinate system is the arrangement of lines of latitude and longitude on the surface of the Earth. For purposes of navigation on the Earth's surface this is obviously a far more convenient system to use than a three-dimensional Cartesian coordinate system x, y, z, with its origin at the centre of the Earth.

The fact that surfaces possess, simultaneously, two- and three-dimensional features formed the basis of Gauss' paper; and it provides, as we shall see, a firm duality running right through the theory of shell structures.

Some aspects of the geometry of surfaces are best studied from a two-dimensional viewpoint. A simple example is the distance measured along an arc between two points on a surface. The length of the side of a triangle drawn on a piece of paper does not change when the paper is rolled into an arbitrary cylindrical or a conical surface, even though the surface itself changes its three-dimensional configuration considerably in the process. On the other hand, there are other aspects of the geometry of surfaces which can only be described satisfactorily from a three-dimensional standpoint. A simple example is the radius of a cylindrical surface.

Those properties of surfaces which depend only on measurements of length in the surface itself are known as *intrinsic* properties, while those which involve measurements in three-dimensional space are known as *extrinsic* properties. In describing surfaces, engineers frequently use the distinct ideas of intrinsic and extrinsic properties, without necessarily using these terms. For example, surfaces are often described by means of a series of cross-sections, cut according to a suitable predetermined scheme. This is the usual way of describing

5.2 The idea of a surface

the shape of the hull of a boat, as a sequence of longitudinal and transverse cross-sections, normally at equal spacings; it is an *extrinsic* description. On the other hand, surfaces are sometimes described by means of a *development* onto a plane. Of course, many surfaces cannot be flattened without stretching or tearing and the development then consists of a series of panels or 'gores'. Fig. 5.1a shows a development of the surface of a sphere made in this way: if the plane figure is cut out and the edges are joined in an obvious manner, a portion of a nearly-spherical surface is built. Of course, it is not possible to build a truly *smooth* surface in this manner by means of a finite number of gores, but old-fashioned cutout models of aeroplanes and ships, etc., made in thin card, achieved remarkably realistic representations of three-dimensional surfaces in this way. Other examples of the specification by means of developments are schemes for the construction of polyhedra – in this case non-smooth surfaces (see fig. 5.1b) – by cutting-out, folding and joining; and the manufacture of clothes by the joining at the edges of pieces of fabric cut out according to two-dimensional patterns. Sheet-metal workers are also familiar with the problem of making surfaces such as curved pipe ducts by joining together shaped pieces cut out from flat sheet. In all of these examples the form of the surface is described by means of its *intrinsic* properties.

One of the main tasks of this chapter is to establish a mathematical connection between these two distinct ways of thinking about curved surfaces. The connection is known as Gauss' theorem, and was first established in his famous paper. The theorem is a mathematical result, and it will, of course, be necessary to begin our analysis by expressing both intrinsic and extrinsic aspects of surfaces in precise mathematical language.

Fig. 5.1. (*a*) A string of gores which can be curved and joined into a practically hemispherical shape. (*b*) A set of squares which can be assembled into a cube.

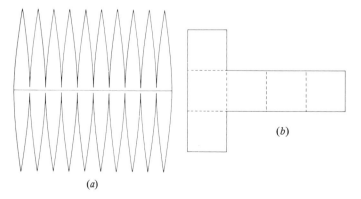

128   *The geometry of curved surfaces*

5.3   General properties of plane curves

At several places in the following analysis we shall need to refer to the elementary properties of a smooth curve in a plane; and indeed we have already touched on these properties in section 4.5. Fig. 5.2 shows a typical small region of a plane curve. The position of any point on the curve is determined by a single arc-length coordinate s, measured from a suitable datum point. A, B are two neighbouring points defining a short arc of length ds. At A and B normals are drawn to the curve (there is only one at each point, since the curve is smooth) and they are inclined with respect to a suitable datum direction in the plane at angles ψ and $\psi + d\psi$, respectively: ψ is in radian measure. The two normals intersect at 0, which is called the *centre of curvature* of the arc, by analogy with a small length of a circular arc passing through A and B and having the same normals as the curve. The centre of curvature is defined strictly in terms of a limiting process in which $ds \to 0$. The length $0A$ is called the (local) radius of curvature at A, and denoted by the symbol ρ. It follows by simple trigonometry from the above description that

$$\rho = \frac{ds}{d\psi} . \tag{5.1}$$

In practice it is more convenient to use *curvature c* as a variable, defined by

$$c = 1/\rho. \tag{5.2}$$

Thus a curve with smaller radius of curvature has larger curvature, in accordance with our intuitive ideas about imparting curvature to an originally straight

Fig. 5.2. A plane curve showing arc-length coordinate s, angle ψ and centre of curvature 0.

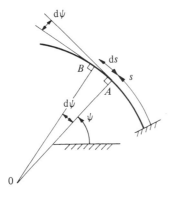

5.4 Curvature of a surface

line by bending it. From (5.1) and (5.2) we have a second definition of curvature, which is exactly equivalent to the first:

$$c = \frac{d\psi}{ds}. \quad (5.3)$$

The symbol c for curvature should not be confused with the symbol κ, used in chapter 2 and elsewhere in the book to denote *change* of curvature from an initial configuration.

Observe that the angle $d\psi$ is not only equal to the change of inclination of the *normal* in moving from A to B, but is also the difference in inclination of the *tangents* to the curve at the two points.

Finally we note that in order to fix the positive sense of curvature we may use the device of an auxiliary coordinate z to designate the positive side of a curve or surface. We shall define the curvature as positive when the centre of curvature lies on the negative z side of the curve. This is essentially the same sign convention as the one which we used for change of curvature in chapters 2 and 3.

5.4 Curvature of a surface in terms of the geometry of cross-sections: principal curvatures, etc.

In this section we consider the geometry of a typical small region of a smooth surface. By 'smooth' we mean that the surface is continuous, and contains no discontinuities of slope, i.e. no creases or vertices. The treatment repeats to some extent the simplified version already given in section 4.3. Fig. 5.3a shows a small region of such a surface, near a typical point P. There is a directed normal n to the surface at P. The normal to a surface at a point is defined as the normal to the tangent-plane at that point; and since the surface is smooth (as defined above) this plane and hence the normal are uniquely determined.

We shall now investigate the curvature of a family of plane curves which may be drawn on the surface to pass through point P. It is convenient to generate these curves by having a 'search plane' which includes the normal n and can be rotated about it; see fig. 5.3a. At any angular position this plane cuts the surface in a plane curve; and as the plane is rotated about the axis n the complete family of plane curves (or 'normal sections') is generated. Let the angular position of the plane be denoted by θ measured from a suitable datum. Fig. 5.3a shows the plane for an arbitrary value of θ, and the corresponding curve, where this plane intersects the surface.

By using the definition in the preceding section we may evaluate the curvature of the curve at P. If we take the positive direction of n as the positive (z)

side of the surface, we see that the curvature of the plane curve shown in fig. 5.3a is positive, according to our previous sign convention. Clearly the curvature defined in this way will depend on the value of θ: thus for a circular cylindrical surface of radius a, the curvature of the family of plane curves passing through a given point will vary between zero (when the plane curve is a straight generator) and $1/a$ (when the plane is perpendicular to the axis). Later on we shall investigate the general functional relationship between the curvature and angle θ, but before we do this we need to look a little closer at the notion of curvature of a *surface*.

In our discussion of curvature of a plane curve in the preceding section we used the idea of a normal to the curve, which rotated as a point moved along the curve. Therefore, in considering the curvature of a surface it is appropriate to think of the rotation of a *normal to the surface* as a point moves along a curved line through a given point P on the surface. In a 'true view' of the search plane (fig. 5.3b) we see the normal to the surface rotate, exactly as if we were dealing with a plane curve; but in general the normal to the surface will not lie in this plane (except of course at point P itself). These out-of-plane rotations of the normal cannot be seen in the view of fig. 5.3b, but they show

Fig. 5.3. (a) A small plane rotating about the normal n to a smooth surface at a point P. (b) The trace of all curves of intersection as θ increases by π. (c) An edge view of the plane along the direction tangential to the curve at P, showing rotation ϕ of the normal: cf. Fig. 4.3b. (d) Example of a twisted surface. (e) As (d), except that the two principal directions, of zero twist, are shown.

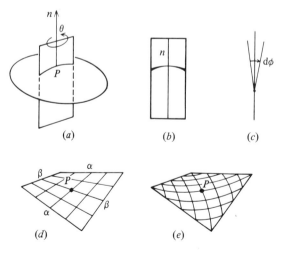

5.4 Curvature of a surface

clearly in a perspective view (cf. fig. 4.3*b*) or in an edge view of the plane in the direction of the tangent at *P*, as shown in fig. 5.3*c*. Let us denote by dϕ the rotation of the normal to the surface, corresponding to a movement of d*s* of a point along the curve, about an axis which is locally tangential to the curve. If dϕ/d*s* ≠ 0 at *P* for a given value of angle θ we say that the surface is *twisted* in this direction. It will be useful to define a twist variable $\bar{c}$ by the relation

$$\bar{c} = d\phi/ds. \tag{5.4}$$

This is analogous to the definition of curvature *c*, given in (5.3). We shall define the positive sense of $\bar{c}$ as giving a right-handed screw effect.

For example, the surface shown in fig. 5.3*d* is clearly twisted at point *P* in the sense that there is twist in the two directions $\alpha\alpha$, $\beta\beta$ which are marked; and indeed in these directions the curvature is zero. Along $\alpha\alpha$, $\bar{c} < 0$; but along $\beta\beta$, $\bar{c} > 0$. However, on two other lines through *P* shown in fig. 5.3*e* there is zero twist, but nonzero curvature.

Thus we see that for a given point on a given surface the values of curvature *c* and twist $\bar{c}$ depend on the direction of the curve through *P* along which we are moving; i.e. *c* and $\bar{c}$ are functions of θ. It is sometimes helpful to imagine that we are driving a toy car (with parallel axles) on a particular path over a surface. By analogy with the motion of a ship we define *roll* as angular rotation about the current direction of motion of the car, and *pitch* as angular rotation about the direction of the axles of the car. Then curvature is the rate of change of pitch with distance, and twist is the rate of change of roll with distance. For example, if we drive straight over an ordinary (cylindrical) humpback bridge we experience pitch but no roll, i.e. curvature but no twist. But on the other hand, if we drive our miniature car along the crest of the road surface of this bridge, perpendicular to the usual direction of traffic, we experience neither pitch nor roll, i.e. curvature and twist are both zero. Further, if we drive on a path making 45° with the usual direction of traffic we experience both pitch and roll: on this path curvature and twist are both nonzero.

This example illustrates the important point that curvature and twist of a surface are only defined for a particular *trajectory* through a given point on the surface; and thus they depend on the value of θ for the trajectory. (The definition of twist $\bar{c}$ in relation to a particular *direction* in the surface should not be confused with the definition of twist κ_{xy} in (2.20) in relation to an *x*, *y* coordinate system: see later.)

Our next task is to investigate in general the way in which *c* and $\bar{c}$ vary with θ. It is clear from the above description that *c*, $\bar{c}$ are *periodic* in θ, since obviously

$$c(\theta) = c(\theta + \pi), \quad \bar{c}(\theta) = \bar{c}(\theta + \pi).$$

Consequently we expect to find some relatively simple general relations.

The simplest way of tackling this problem is to use coordinate geometry. Let us define the surface locally in fixed Cartesian coordinates X, Y, Z by the relation $Z = Z(X, Y)$: see fig. 5.4a. It is obviously desirable to have the origin of coordinates at the point P of the surface in question, and it is also convenient to have the XY-plane touching the surface at the origin, so that the Z-axis is normal to the surface. As we shall be concerned only with a small region of the surface it is appropriate to express the function Z as a Taylor series. We thus put

$$-Z = \tfrac{1}{2} k_{11} X^2 + k_{12} XY + \tfrac{1}{2} k_{22} Y^2 + \text{terms of higher order.} \quad (5.5)$$

Terms of order 0 and 1 do not appear in this expression, since their coefficients vanish in consequence of the conditions imposed at the origin. There are three quadratic terms, as given, and the factors $\tfrac{1}{2}$ have been put in for the sake of convenience in later working. The negative sign has been inserted so that when $k_{12} = 0$ the two principal radii of curvature have the same parity as k_{11} and k_{22}, respectively; cf. (4.5). There is no point in writing down the higher-order terms because we shall consider only the small portion of the surface near the origin in which X and Y are sufficiently small for all terms of third and higher order to be negligible in comparison with the quadratic terms: cf. appendix 8.

The coordinate system X, Y, Z is fixed, and therefore the values of the three constants k_{11}, k_{12}, k_{22} completely define the surface in the region of interest. We shall also use another coordinate system x, y, z. This also has its origin at P, and has a common z-axis (normal to the surface) with the first co-

Fig. 5.4. (a) A portion of a smooth surface in relation to an X, Y, Z coordinate system centred at P. (b) An x, y coordinate system in the plane X, Y, obtained by rotation through θ. (c) Mohr's circle of curvature c and twist $\bar{c}$. Points on the circle correspond to directions in the X, Y plane. The positive sense of ξ is opposite from that of θ. The principal points are labelled 1, 2.

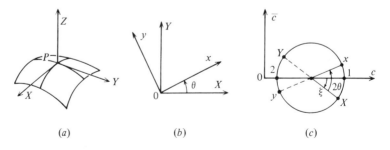

5.4 Curvature of a surface

ordinate system; but it has x- and y-axes which are inclined at angle θ to the X- and Y-axes, as shown in fig. 5.4b. The transformation corresponding to this rotation is expressed by the following relations, which may be written down by inspection:

$$X = x\cos\theta - y\sin\theta, \quad Y = x\sin\theta + y\cos\theta. \tag{5.6}$$

Hence we may express Z in terms of x and y by substitution, as follows:

$$\begin{aligned} Z = {\tfrac{1}{2}}x^2 [-{\tfrac{1}{2}}(k_{11} + k_{22}) - {\tfrac{1}{2}}(k_{11} - k_{22})\cos 2\theta - k_{12}\sin 2\theta] \\ + xy[{\tfrac{1}{2}}(k_{11} - k_{22})\sin 2\theta - k_{12}\cos 2\theta] \\ + {\tfrac{1}{2}}y^2 [-{\tfrac{1}{2}}(k_{11} + k_{22}) + {\tfrac{1}{2}}(k_{11} - k_{22})\cos 2\theta + k_{12}\sin 2\theta]. \end{aligned} \tag{5.7}$$

Let us now evaluate the curvature c_x and twist $\bar{c}_x$ in the x-direction of our new coordinate system. Near the origin we may write

$$c_x = \frac{-\partial^2 Z}{\partial x^2}. \tag{5.8}$$

Now in this small region angle ϕ is equal to $\partial Z/\partial y$, so we obtain

$$\bar{c}_x = \frac{\partial^2 Z}{\partial x \partial y}. \tag{5.9}$$

In these expressions the sign conventions are exactly as before: curvature c is positive if the centre of curvature is on the negative Z side of the surface, and the twist $\bar{c}$ is positive if the 'roll' is according to a right-handed screw rule.

From (5.7)–(5.9) we obtain

$$c_x = {\tfrac{1}{2}}(k_{11} + k_{22}) + {\tfrac{1}{2}}(k_{11} - k_{22})\cos 2\theta + k_{12}\sin 2\theta, \tag{5.10a}$$

$$\bar{c}_x = \phantom{{\tfrac{1}{2}}(k_{11} + k_{22}) +{}} {\tfrac{1}{2}}(k_{11} - k_{22})\sin 2\theta - k_{12}\cos 2\theta. \tag{5.10b}$$

This expression may be tidied up by the introduction of r and ξ, defined by the following expressions:

$$r = \{[{\tfrac{1}{2}}(k_{11} - k_{22})]^2 + k_{12}^2\}^{\tfrac{1}{2}}, \tag{5.11a}$$

$$\sin\xi = k_{12}/r, \quad \cos\xi = (k_{11} - k_{22})/2r. \tag{5.11b}$$

Then equations (5.10) become

$$c_x = {\tfrac{1}{2}}(k_{11} + k_{22}) + r\cos(2\theta - \xi), \tag{5.12a}$$

$$\bar{c}_x = \phantom{{\tfrac{1}{2}}(k_{11} + k_{22}) +{}} r\sin(2\theta - \xi). \tag{5.12b}$$

When $\theta = 0$ (5.12) give correctly the curvature and twist in the X-direction:

$$c_X = k_{11}, \quad \bar{c}_X = -k_{12}. \tag{5.13}$$

Similarly, when $\theta = {\tfrac{1}{2}}\pi$ we obtain expressions for curvature and twist in the Y-direction:

$$c_Y = k_{22}, \quad \bar{c}_Y = +k_{12}. \tag{5.14}$$

Thus the three constants as written in (5.5) correspond to physical quantities with reference to the original X- and Y-axes.

In general, as the value of θ varies, relations (5.12) describe a circle of radius r in a rectangular Cartesian c, $\bar{c}$ space. Here we drop the subscripts, but label points on the circle with the corresponding direction. This is indeed a *Mohr circle* of curvature and twist (cf. Timoshenko & Woinowsky-Krieger, 1959, §9), and is shown in detail in fig. 5.4c. As θ increases so that the x-axis rotates counterclockwise as viewed from the positive z-direction, the corresponding point on the circle also rotates counterclockwise, but through an angle of 2θ. Fig. 5.4c has been drawn with the following values of the constants:

$$k_{11} = 4, \quad k_{12} = 1, \quad k_{22} = 1.$$

The reader should verify this by the use of (5.11). Note in particular that the point marked X is obtained formally by putting $\theta = 0$, and that when $\theta = \frac{1}{2}\pi$ the x-axis coincides with the Y-axis and gives the point marked Y on the diagram. This illustrates the fact that if the twist of a surface has a certain value in a given direction, it has a twist of the same magnitude but opposite sign in a perpendicular direction: it is not simply a question of interchanging x and y in (5.9). Thus when $k_{12} > 0$ (as in fig. 5.4a) the surface droops more in the first and third quadrants of X, Y space than in the other two quadrants; and hence the twist is left-handed in the X-direction and right-handed in the Y-direction.

Various general results, as follows, may be written down by inspection of fig. 5.4. They will appear familiar to those who have encountered Mohr's circle in other connections.

(i) The circle describes the curvature of the surface at a particular point P.

(ii) Points on the circle corresponding to perpendicular directions on the surface lie at opposite ends of a diameter of the circle.

(iii) The circle is uniquely determined by data on curvature and twist in two perpendicular directions. The circle gives a complete description of the curvature of the surface in the region of P, and the three coefficients in (5.5) may be determined for an arbitrary inclination of the coordinates (see problem 5.1).

(iv) It is always possible to find two perpendicular directions on the surface such that there is zero twist. These directions are called *principal directions*, and the corresponding curvatures are called *principal curvatures*. It is clear from fig. 5.4c that the principal curvatures are maximum and minimum values, respectively, of c when θ varies. This result was first obtained by Euler. The principal directions and curvatures are usually given subscripts 1, 2. It is conventional to label the principal axes so that $c_1 \geqslant c_2$. Suppose that we know the values of curvature and twist at a given point on a given surface in two perpendicular directions a, b: namely c_a, $\bar{c}_a$, c_b, $\bar{c}_b$ ($= -\bar{c}_a$). It follows easily from fig. 5.4c that the principal curvatures are given by

$$c_1, c_2 = \tfrac{1}{2}(c_a + c_b) \pm \{[\tfrac{1}{2}(c_a - c_b)]^2 + \bar{c}_a^2\}^{\frac{1}{2}}. \tag{5.15}$$

Moreover, it is clear from the preceding analysis that the same answers are obtained irrespective of the choice of axes a, b.

(v) The inclination of the principal axes to the a- and b-axes is readily found from (5.11) since $\bar{c} = 0$ on the principal axes, by definition. The required condition is that $\sin(2\theta - \xi) = 0$, so $2\theta = \xi$ or $2\theta = \xi + \pi$. The two values of θ give the inclination of the principal axes to one of the axes a, b. We lose no generality by labelling the axes a, b so that a, b, z are a right-handed system and $c_a \geqslant c_b$. Then $\tan 2\theta = 2\bar{c}_a/(c_a - c_b)$; and taking the value of 2θ lying between $+\pi$ and $-\pi$ we may find the inclination θ of the principal axis 1 to the a-axis.

(vi) Note that the definition of change of twist κ_{xy} (see (2.20)) for use in the generalised Hooke's law (see (2.28)) corresponds by analogy to the value of the coefficient k_{12} in the quadratic expression (5.5). Unlike $\bar{c}_x$, it is not associated with a particular direction, but with a *pair* of perpendicular directions x, y. Although there are some obvious resemblances, the two definitions of twist are of two different kinds; and they are useful for different purposes.

Some exercises on the use of Mohr's circle of curvature are given in problems 5.2 and 5.3.

There are some important special cases, such as spherical and plane surfaces, in which all directions are principal directions: there is no twist along any direction, and the radius of the Mohr circle is zero.

It must be emphasised that the preceding analysis is *local* to the point P; but that we have been concerned with the curvature of a *general* smooth surface at this point.

In this section we have described the surface locally with respect to a three-dimensional x, y, z coordinate system; we have used in fact an *extrinsic* description of the surface. In the next section we shall begin to investigate the *intrinsic* geometry of surfaces. In particular we shall find that the product $c_1 c_2$ of the two principal curvatures – which is known as the *Gaussian* curvature of a surface at a point – is a specially significant quantity: see problem 5.4. There is nothing in the preceding derivation to suggest that the product $c_1 c_2$ has any remarkable properties.

5.5 Gaussian curvature: an intrinsic view of surfaces

We begin this section with a simple idea. In defining the curvature of a *plane* curve we found it useful to consider a small arc-length ds and an associated angle, dψ, subtended by the arc: we then found that curvature c defined as in (5.3) was equal to the reciprocal of the radius of curvature. This notion suggested to Gauss that in discussing the curvature of *surfaces* it would be useful to consider a small *area* dA of the surface, the *solid* angle dβ subtended by it, and to define a measure of curvature K by the following equation, in analogy with (5.3):

$$K = \frac{d\beta}{dA}.\tag{5.16}$$

This measure of curvature is known as *Gaussian* curvature, after its inventor. Equation (5.16) is sometimes known as the 'historical' definition, as distinct from the definition given at the end of the previous section in terms of principal curvatures.

Our main task in this section is to explore this definition of Gaussian curvature – which will involve an elementary investigation of the intrinsic geometry of surfaces – and to relate it to the second, or extrinsic, definition.

As we have already remarked, the relation between these two different views of curvature will prove to be very fruitful in the development of the theory of shell structures.

Our first task in investigating Gauss' definition (5.16) of curvature is to comprehend the idea that portions of surfaces subtend measurable solid angles.

The basic idea of a solid angle is most conveniently described in relation to a cone, as shown in fig. 5.5a. The solid angle subtended by a cone at its vertex is defined as A/r^2 where A is the area cut by the cone from a sphere of radius r whose centre is at the vertex of the cone. The solid angle so defined is dimensionless, and its value is obviously independent of the radius r of the sphere. Note that the cone need not have a circular cross-section; and indeed that two

Fig. 5.5. (a) The solid angle subtended by a cone is defined as A/r^2. (b) The plane angle subtended by two intersecting lines is defined as a/r.

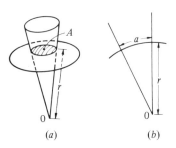

5.5 Gaussian curvature: an intrinsic view of surfaces

cones having different cross-sectional shapes may subtend the same solid angles. This definition is, of course, analogous to the ordinary definition of the angle between two lines indicated in fig. 5.5b: the angle is defined as a/r where a is the arc-length of a circle of radius r whose centre is at the point of intersection of the lines. The angle so defined is dimensionless, and its value is obviously independent of the radius r of the circle. Since the circumference of a circle is $2\pi r$, the *total angle* surrounding a point in a plane is 2π. Likewise, since the area of a sphere is $4\pi r^2$ the *total solid angle* surrounding a point in three-dimensional space is 4π.

Now although it is easy to envisage the angle subtended by an arc of a plane curve, it is not obvious how to think about the solid angle subtended by an area on an arbitrary surface. The basic idea which we need to employ, which is due to Gauss, is shown in fig. 5.6a. It involves the use of an *auxiliary surface* in the form of a *sphere of unit radius* (cf. Hilbert & Cohn-Vossen, 1952, §29).

Consider a simply-closed region Q of a curved surface S, bounded by a closed curve B and having an area A of curved surface. We wish to find the solid angle subtended by the region Q. The first step is to construct, on the auxiliary surface, the *spherical image* b of the curve B (sometimes known as the *horograph* of B (Thomson & Tait, 1879, §136)). The rule for mapping a point P on the original surface to its spherical image p on the auxiliary sphere is that the outward normal at p to the unit sphere (i.e. the radius passing through p) is *parallel* to the outward normal at P to the original surface. Since we are regarding the original surface as smooth (for the present, at least) there is a 1:1 correspondence between P and p, and therefore indeed between curve B and its image b on the unit sphere. Now let the surface area of the unit sphere enclosed by the curve b be β. Since the radius of this sphere is unity, β is equal to the solid angle subtended by the curve b at the centre of the sphere. We now define β also as the solid angle subtended by the region Q of the original surface. This definition satisfies several simple tests of self-consistency: see problem 5.5.

For every arbitrary simply-closed area A on the original surface there is thus a corresponding area β on the unit sphere; and for an infinitesimal area dA on the original surface there is a corresponding solid angle $d\beta$ on the unit sphere. In the limit as $dA \to 0$ we define the Gaussian curvature K by (5.16). In fact the value of the limit is unique at a point on a smooth surface, irrespective of the way in which the area is reduced to zero. A rigorous proof of this statement is beyond the scope of this book, but we give an intuitive proof later.

It is of interest to note that an analogous definition of curvature of a *plane* curve may be made by means of an *auxiliary circle* of unit radius, onto which the curve may be mapped by a rule of parallel normals, as shown in fig. 5.6b. The 'circular image' of an arc ds of the original curve has length dψ, since the radius of the circle is unity. Thus the previous definition of c (see (5.3)) may be interpreted as the ratio of the arc-length of the circular image to the original arc-length.

Some examples of the evaluation of K for some simple surfaces are illustrated in fig. 5.7.

The example illustrated in fig. 5.7a is a *plane*. Since the normals at all points on a plane are parallel, the entire plane maps onto a single point on the unit sphere. Thus $\beta = 0$ for an arbitrary region of the plane; and therefore $K = 0$.

Example (b) is a *circular cylindrical surface* of radius a. All points on any given generator have parallel normals, so each generator has a single point as its spherical image. The spherical images of generators labelled P, Q and R are shown as points p, q and r, and they lie on a great circle of the unit sphere on the plane through the centre of the sphere normal to the axis of the cylinder. Thus every point on the cylindrical surface maps onto a point on this single great circle. It is impossible therefore for the image b of any closed curve on the cylindrical surface to enclose a finite area: hence we conclude that $K = 0$.

Fig. 5.6. (a) The boundary curve B of a portion Q of a surface is mapped to curve b on a unit sphere by a rule of parallel normals. The solid angle subtended by Q is defined as the area β on the auxiliary sphere. (b) The end-points of a plane arc UV of length ds are mapped to points u, v on an auxiliary unit circle by a rule of parallel normals. The angle subtended by ds may be defined as the arc-length dψ on the auxiliary circle.

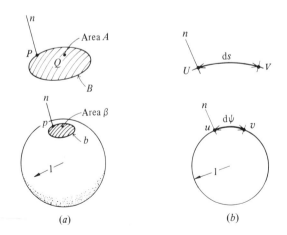

5.5 Gaussian curvature: an intrinsic view of surfaces

Example (c) is a *spherical surface* of radius R. The unit sphere has been drawn concentric for convenience, although this is not necessary for the mapping. It is clear that the image of any curve on the original sphere will be a similar curve on the unit sphere, with linear dimension $(1/R)$ times the original dimension. It follows that $\beta = A/R^2$, and consequently

$$K = 1/R^2.$$

The dimensions of K are thus length^{-2}, which is indeed only to be expected from the definition (5.16).

For the fourth example, (d), we consider a small region of a general smooth surface defined in Cartesian coordinates by the equation

$$Z = (X^2/2R_1) + (Y^2/2R_2). \tag{5.17}$$

Here R_1, R_2 are principal radii of curvature and X, Y are principal directions. We have dropped the negative sign from (5.5) for this particular example. We shall consider only a small region near the origin, in which the slope of the surface is small. The equation of a small portion of a unit sphere may be represented by the equation

$$z = \tfrac{1}{2}x^2 + \tfrac{1}{2}y^2 \tag{5.18}$$

Fig. 5.7. Examples in which various surfaces are mapped onto the unit sphere: see text for explanation.

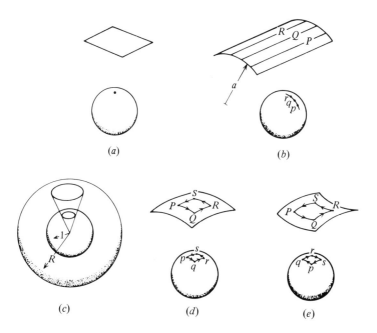

(since both principal radii are unity) in a second Cartesian system whose axes x, y, z are parallel to axes X, Y, Z.

At a typical point on the original surface the components of (small) slope in the X and Y directions are:

$$\frac{\partial Z}{\partial X} = \frac{X}{R_1}, \quad \frac{\partial Z}{\partial Y} = \frac{Y}{R_2}.$$

Similarly the components of slope of the unit sphere at point x, y are:

$$\frac{\partial z}{\partial x} = x, \quad \frac{\partial z}{\partial y} = y.$$

Now at corresponding points on the two surfaces the normals are to be parallel. In the present context this is easily expressed in terms of the condition that the tangent surfaces are parallel: thus

$$\frac{\partial Z}{\partial X} = \frac{\partial z}{\partial x}, \quad \frac{\partial Z}{\partial Y} = \frac{\partial z}{\partial y}.$$

It follows that the mapping from the original to the auxiliary surface is specified by

$$x = X/R_1, \quad y = Y/R_2. \tag{5.19}$$

Thus for a rectangular area $\mathrm{d}X\mathrm{d}Y$ on the original surface we have

$$\mathrm{d}A = \mathrm{d}X\mathrm{d}Y,$$

while on the corresponding spherical image we have

$$\mathrm{d}\beta = \mathrm{d}X\mathrm{d}Y/R_1 R_2.$$

It follows directly that

$$K = 1/R_1 R_2. \tag{5.20}$$

This establishes the required connection between our definition (5.16) of Gaussian curvature K and the work of the preceding section. It is a general result, originally discovered by Gauss. The proof given above seems to be due to Thomson & Tait (1879, §138).

Several points in this fourth example should be noticed.

First, the mapping consists of a simple (but non-isotropic) contraction in each of two orthogonal directions. It follows from this that the same result is achieved irrespective of the shape of the area $\mathrm{d}A$ or the way in which it is reduced to zero.

Second, (5.20) includes all of the previous examples as special cases. Observe that if either principal curvature $(1/R_1)$ or $(1/R_2)$ is zero, then $K = 0$. This is true not only for cylindrical and conical surfaces but also, as we shall see later, for plane surfaces with creases. (But this last result must be proved

5.5 Gaussian curvature: an intrinsic view of surfaces

separately, because we have used a Cartesian representation of a *smooth* surface in the proof set out above.)

Third, the sign of Gaussian curvature does not depend on the sign convention for principal curvatures, provided it is consistent: K is positive if R_1, R_2 are of the same sign, i.e. if the two centres of principal curvature lie on the same side of the surface; and negative if they are of opposite sign. It is interesting to note that in the case of negative Gaussian curvature the areas dA, dβ are of opposite sign: this corresponds to the cyclic senses of the curve B and its image b being opposite, as shown in fig. 5.7e.

5.5.1 The solid angle subtended by the vertex of a 'roof'

We begin this section by considering the spherical image and Gaussian curvature of several surfaces which involve creases in flat sheets, as shown in fig. 5.8. In diagram (*a*) a creased plane is shown, with its spherical image. Each of the two parts of the surface maps onto a single point on the unit sphere; and the crease itself maps onto the arc pq of a great circle lying in a diametral plane normal to the crease. At a crease, of course, the direction of the normal

Fig. 5.8. The mapping of plane surfaces containing creases onto the unit sphere: see text for explanation.

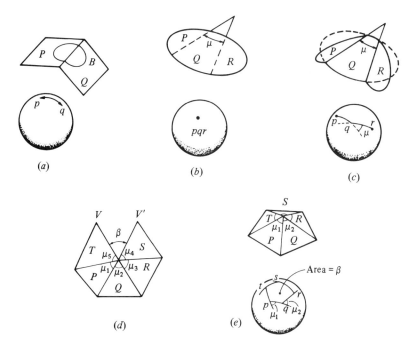

is not uniquely determined, and the range of possible directions is covered by the arc pq. Now any closed loop which crosses the crease maps onto the curve pq, which in particular encloses no area. Therefore, the Gaussian curvature of the creased plane is zero; and we conclude that the making of a simple fold in a flat sheet has no effect upon the Gaussian curvature. We could arrive at the same conclusion by considering the crease as a limiting process involving a portion of a cylindrical surface inserted in a plane.

A second example is shown in figs. 5.8b and c. Here a flat sheet, shown in fig. 5.8b, is to be creased on the two lines which are shown broken. The two potential creases include angle μ, and for the sake of convenience the creases intersect outside the boundary of the sheet. Before folding begins the three regions P, Q, R are coplanar, and their spherical image is a single point, pqr, as shown.

Now let the process of folding begin, with part Q of the sheet held rigid while flaps P and R are bent down, as shown in fig. 5.8c. Point q remains in its original position, but as the angle of the crease between Q and P is increased the point p moves away from q on the arc of a great circle which lies in the diametral plane normal to the axis of folding. Similarly when flap R is folded down, r moves along an arc of a great circle in a diametral plane normal to that crease. The planes of the arcs qp, qr respectively are normal to the respective hinge lines, and since the hinge lines make an angle μ, the arcs qp and qr make the same angle μ at their point of intersection q, as shown. This important result follows immediately from a view of the creased plane along the normal to Q together with a (parallel) view of the unit sphere along the radius through q.

The next step is to consider the spherical image of a 'roof' made by cutting out the sheet shown in fig. 5.8d, folding along the broken lines and joining the vertices marked V, V'. There are five radial creases in the completed roof shown in the diagram; however, this number is immaterial to the proof (provided it is at least three). Now a roof made in this way with hinges which are free to rotate is not a rigid structure. it is a *mechanism* with a number of degrees of freedom in addition to rigid-body motions. For present purposes we shall first consider the mechanism to be 'frozen' in an arbitrary configuration; then we shall investigate its image on the unit sphere. This image is shown for a particular configuration in fig. 5.8e. Each face of the roof maps to a single point, and each hinge maps to an arc of a great circle. The closed figure produced by these arcs has external angles which are equal to the angles between creases on the various faces, in consequence of the argument above. Thus the five external angles of the spherical image of the roof are precisely the angles $\mu_1 \ldots \mu_5$ marked in fig. 5.8d. Now the *lengths* of the arcs of this spherical

5.5 Gaussian curvature: an intrinsic view of surfaces

image depend on the values of the hinge-angles in the particular 'frozen' configuration; so it seems unlikely at first that we shall be able to obtain a general formula for the enclosed area of the spherical image. However, there is a remarkable theorem which enables us to do precisely this calculation. It is provided in appendix 7. It states that the area of a portion of a unit sphere bounded by arcs of great circles is equal to the 'angular excess' of the figure: angular excess is defined as the sum of the internal angles of the figure less the sum of the internal angles of a *plane* figure with straight edges having the same number of vertices.

Let us do the calculation for a spherical image having n vertices. The sum of the interior angles of a plane n-gon is equal to $(n-2)\pi$: this is a well-known result in plane geometry, and is easy to prove. When the exterior angle is as defined as in fig. 5.8e,

Interior angle = π - Exterior angle

therefore, for a spherical n-gon,

Sum of interior angles = $n\pi$ - Sum of exterior angles.

It follows that for a spherical n-gon

Angular excess = 2π - Sum of exterior angles.

But the exterior angles are equal respectively to the apex angles of the flattened roof, fig. 5.8d. Hence we see that the *angular excess of the spherical image* is exactly equal to the angle marked β in fig. 5.8d. It is convenient to call β the *angular defect of the flattened vertex*.

This is a very remarkable result, for it means not only that the solid angle subtended by a roof-vertex is trivial to calculate, but also that the solid angle subtended by the roof-vertex does not vary as the roof is distorted by changing the hinge-angles of the creases. Indeed, the solid angle does not change even if additional creases, passing through the vertex, are made in the faces of the roof; and this includes the possibility of inserting an infinite number of such creases and turning the roof into a conical surface. For a circular cone it is, of course, not difficult to calculate independently the solid angle subtended, and the result is, as expected, that the solid angle subtended is equal to the angular defect of the flattened conical surface: see problem 5.6. These results were first obtained by Maxwell (1854): see Hilbert & Cohn-Vossen (1952, §29).

It is important to realise that in constructing the spherical image of the roof we were strictly making the spherical image of a closed path B on the roof. The closed figure of fig. 5.8e is in fact the image of *every* such path which *encloses the vertex*. It is, of course, possible to construct closed paths which visit every face but do *not* enclose the vertex. In these cases the spherical image consists of the same arcs as in fig. 5.8e but it doubles back on itself and encloses zero area. Thus we conclude that *the solid angle subtended by the roof lies entirely in the vertex itself.*

5.5.2 The curvature of a polygonalised surface

The ease with which the solid angle subtended by a vertex of a polyhedron may be calculated suggests a simple and direct approach to the curvature of surfaces. Fig. 5.9a shows part of a smooth curved surface which has been divided, arbitrarily, into small cells. In the interior of each cell is marked a point on the surface. In fig. 5.9b these same points have been joined by straight lines which form a triangular grid and thus describe a many-faceted polyhedral surface. This polyhedral surface constitutes a crude approximate representation of the original surface, and it may be 'improved' in an obvious sense by repeating the process with a larger number of cells and vertices.

Consider now a typical cell on the original surface, having area A. It subtends a solid angle β which may be determined by constructing a spherical image of the cell boundary. In the polygonalised version of the surface it is only the *vertices* which subtend solid angle, and we can thus see that the process of polygonalisation is also one in which the distributed Gaussian curvature of the original smooth surface is concentrated into finite amounts at the vertices, and is zero elsewhere. Hence we discover that provided we can 'associate' an area of surface with each vertex of the polygonalised version, we may compute an approximation to the Gaussian curvature of a surface by means of the formula

$$K \approx \frac{\beta}{A} = \frac{\text{Angular defect of vertex}}{\text{Area associated with vertex}}. \tag{5.21}$$

Fig. 5.9. (*a*) A smooth surface onto which points have been marked. Each point is associated with a portion of the surface. (*b*) The same set of points, but now joined by straight lines to form a 'triangulated' version of the original surface.

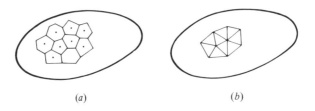

5.6 Inextensional deformation of surfaces

A good example of the usefulness of this equation comes from the construction of 'geodesic' domes, which are near-spherical enclosures constructed from plane triangular faces. It is relatively easy to decide upon a topological layout for the triangulation; but the accurate calculation of the lengths of the sides of the component triangles is usually regarded as a tedious problem in 'solid geometry'. The use of (5.21) simply removes this difficulty. As a first approximation, we can reckon that each vertex is associated with an equal area of the triangulated surfaces. Since the Gaussian curvature of the original sphere is uniform, there must be an equal angular defect at each vertex. Now the total solid angle enclosed by a sphere is 4π, and therefore the angular defect associated with each of the v vertices is just $720/v$ degrees. For example, a geodesic sphere having 180 faces and 92 vertices requires an angular defect of 7.83 degrees at each vertex. Once this point has been grasped it is a simple matter to work out the interior angles, and then the lengths of the sides, of the various triangles; see problem 5.7. The process may obviously be refined if it turns out that some triangles are smaller in area than others; this leads to a slightly unequal apportionment of the angular defects.

The fact that the total angular defect (summed over all vertices) is equal to 4π may also be deduced independently from Euler's theorem relating the numbers of faces, edges and vertices of an arbitrary but simply-connected polyhedron: see problem 5.8.

Examples of geodesic domes give a clear picture of the relation between angular defect and Gaussian curvature, but insofar as they involve specific geometrical calculations they do not relate to an important feature of the idea of a general triangulated surface. The mathematical connection between the original smooth surface and the polygonalised version is that as the number of faces is increased the polygonalised surface becomes closer to the original surface. The precise mathematical nature of the limiting process whereby the equation $K \approx A/B$ is transformed to $K = \mathrm{d}A/\mathrm{d}\beta$ is beyond the scope of this book. For our purposes it is satisfactory simply to regard the polygonalised surface, with sufficiently small triangles, as being 'practically equivalent' to the original surface.

5.6 Inextensional deformation of surfaces

A remarkable geometrical result follows immediately from the preceding description of surfaces in terms of approximate, polygonalised, surfaces. We have already seen that a 'cap' or roof consisting of a number of triangular facets meeting at a vertex subtends a solid angle which does not change when the vertex is deformed by folding. Such a mode of deformation is known as *inextensional*, since all distances measured along the surface are preserved, and there is no tearing.

Now we remarked in chapter 1 that, in general, closed surfaces are 'rigid' whereas open surfaces are deformable: we were in fact referring at this point to *inextensional* deformation. When a polygonalised approximation of a surface is permitted by its boundary conditions to undergo inextensional deformation, the angular defect is obviously preserved at every vertex, and thus there is *no change in Gaussian curvature* anywhere in the surface. This is a very remarkable property of curved surfaces. Paradoxically, it is given little, if any, emphasis in textbooks on shell structures.

Inextensional deformations are relevant to some problems in shell structures, and the invariance of Gaussian curvature in these deformations is an immediate aid to calculation. More usually, however, the deformation of a shell is not inextensional, and therefore the Gaussian curvature changes; in these cases we shall need a relation between the change of strain and the change of Gaussian curvature. The derivation of this sort of relation will be a task for chapter 6.

Before we return to our treatment of general curved surfaces, it is useful to mention an important special case, namely the *surface of zero Gaussian curvature*. In the polygonalised representation of such a surface there is precisely zero angular defect at every vertex; i.e. the sum of the interior angles of faces meeting at every vertex is exactly 2π. Each isolated vertex in turn may therefore be unfolded, without tearing, into a *plane* configuration; and provided restraints at the edges of the surface do not interfere, the entire surface may be deformed inextensionally into a plane. Such a surface is said to be *developable*. We have already seen that plane and cylindrical surfaces have zero Gaussian curvature, and that the creasing of a plane has no effect upon the Gaussian curvature: these are all examples of developable surfaces. Of course, the possibility of making a long straight crease in a plane polygonalised surface raises problems, for we cannot expect, in general, to find a straight row of vertices in any arbitrary direction. We may deal with difficulties of this sort by imagining the insertion of extra *ad hoc* vertices, as necessary.

5.7 Nontriangular polygonalisation of surfaces

The polygonalisation of a surface does not necessarily require the use of triangular facets. Quadrilaterals are sometimes useful, but then of course each set of four points on the original surface, which is to provide the vertices of one quadrilateral, must be chosen to lie in a plane. In this sense triangulation is both easier and more general, because three points always lie in a plane. Furthermore, the modes of deformation of a surface made of plane

5.7 Non-triangular polygonisation of surfaces

quadrilateral facets will be somewhat restricted; but these restrictions can easily be eliminated by the insertion of a potential hinge diagonally across each facet. It is instructive to consider some simple cases.

Fig. 5.10 shows the use of quadrilateral facets in the construction of a (nearly) spherical surface: cf. fig. 5.1a. Here it has been decided to use facets in the form of a trapezium, of uniform height a, as shown. The central facet of each gore is rectangular, of width b_0, and these rectangles will connect to form an equatorial band around the sphere. The radius of the sphere is R, and it is required to determine suitable dimensions $b_1 \ldots b_n$ for the edges of the successive trapezia. Consider the vertex labelled 0. As a first approximation we may associate it with surface area ab_0; hence we must insert an angular defect of ab_0/R^2 radians. This is easily done by inclining the edge of the next trapezium at an angle of $ab_0/2R^2$ to the edge of its neighbour on the gore, as shown. In this way the dimension b_1 is found. By the same argument the next edge must be set off at an angle $ab_1/2R^2$, as indicated. Of course, if ab/R^2 is very small, it may be desirable to do the calculations numerically; and indeed it is not difficult to see that if the step a is made small, it may be advantageous to use calculus to determine the smooth profile of the gore: see problem 5.9.

Fig. 5.10b shows another example of the way in which Gaussian curvature may be imparted to a region of an originally flat sheet by the insertion of easily calculated angular defects. Here a scheme has been adopted in which

Fig. 5.10. (a) The calculation of successive trapezoidal portions of gores like those of fig. 5.1a. (b) A scheme for making an approximation to a doubly-curved surface by means of rhomboidal tiles. The shear angle of each tile is marked: $q = Kc^2$ where K is the desired Gaussian curvature, here positive.

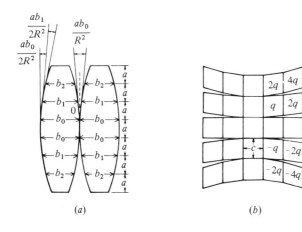

148   *The geometry of curved surfaces*

each facet is a rhombus, and the angular defects are inserted by adjusting the internal angles of the rhombuses. Let the side of each rhombus be c and let the required Gaussian curvature be K. Then the required angular defect at each vertex is the angle Kc^2: the size of c must be chosen so that this is a small quantity. The corresponding angles of shear are as marked on the diagram: we are assuming that the angles of shear are so small that the area of each rhombus may be taken as c^2. If $K < 0$ (as in fig. 5.3e) the angular defects are negative, i.e. there is an angular 'overlap' at each vertex.

5.8   Summary

In this chapter we have examined the curvature of an arbitrary smooth surface from two different points of view. The natural outcome of the *extrinsic* view is the expression of curvature by means of a Mohr circle of curvature and twist, and the associated ideas of principal axes, principal curvatures, etc. On the other hand the *intrinsic* view, most easily explored by means of a polygonalised version of the surface, brings out the importance of Gaussian curvature, which is equal to 'angular defect per unit area'. In particular we see that the Gaussian curvature of a surface is unchanged by any inextensional deformation of the surface.

These two quite different views of the curvature of surfaces are linked very simply by Gauss's result: Gaussian curvature is equal to the product of the principal curvatures. An appreciation of the two separate views of curvature, and their simple connection, is vital for a clear understanding of the deformation of shell structures.

5.9   Problems

5.1 A curved surface is described by the following equation in the vicinity of the origin of the right-handed Cartesian coordinate system X, Y, Z:

$$Z = 6X^2 + 2XY - 3Y^2.$$

Plot a Mohr circle of curvature/twist as in fig. 5.4c, and hence deduce the corresponding equation in terms of x, y, Z coordinates where (see fig. 5.4b) $\theta = 45°$.

5.2 Construct a Mohr circle of curvature/twist (*a*) for a cylindrical surface, (*b*) for a spherical surface.

5.3 Show that there are two lines of *zero* curvature passing through a point on a smooth surface if and only if the product $c_1 c_2$ of principal curvatures is negative or zero. In this case determine the inclination of the (locally) straight lines

5.9 Problems

in the surface to the principal direction corresponding to the principal curvature having the smaller absolute value. Discuss the special case of a cylindrical surface.

5.4 Show by use of (5.15) that
$$c_1 c_2 = c_a c_b - \bar{c}_a^2.$$

5.5 Verify that the definition of solid angle as an area after a mapping onto the unit sphere (fig. 5.6a) gives the correct result for the angle subtended by a cone (fig. 5.5a). Verify also that the total solid angle subtended at an interior point by a simply-connected closed surface is always 4π.

5.6 A right circular cone touches a sphere of radius R at a circle of radius r. Show by direct calculation that *both* the solid angle subtended by the circle at the centre of the sphere *and* the angular defect of the cone are both equal to $2\pi \{1 - [1 - (r/R)^2]^{\frac{1}{2}}\}$, and thus are equal to each other. (*Hint*: a theorem of Archimedes states that the surface area of a cap cut from a sphere of radius R, and having height h from base-circle to apex, is equal to $2\pi R h$.)

5.7 Consider the Geodesic sphere sketched below. It has 180 faces and 92 vertices. The twelve points (marked •) where 5 faces meet lie on the vertices of an icosahedron; and the shaded region of 9 faces corresponds to a simple subdivision of one of the 20 faces of that figure. Given that the angular defect at each vertex is to be $720/92 = 7.826$ degrees, and that the figure is to be as regular as possible, first calculate the angles marked + and then the angles marked △. Now suppose, arbitrarily, that the faces marked □ are to be congruent with

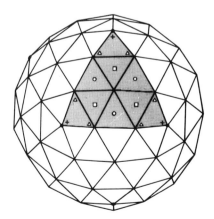

those already calculated, and hence find the angles of the faces marked ○. Lastly, determine the lengths of the edges as multiples of the length of the edges which meet at a 5-vertex. (Alternatively, suppose that the vertex angles of the faces ○ and □ are equal where these 6 faces meet; and hence compute the angles of all the faces.)

5.8 Euler's theorem for simply-closed polyhedra states that
$$v - e + f = 2,$$
where v, e and f are the numbers of vertices, edges and faces, respectively. When all the faces are triangles, $e = \frac{3}{2}f$ (since each edge is shared by two faces). Hence show that for a simply-closed but arbitrarily *triangulated* figure,
$$v = 2 + \tfrac{1}{2}f,$$
and that consequently the sum of all of the internal angles of all the faces, being equal to πf, is also equal to $2\pi v - 4\pi$. Thus show that the sum of the *angular defects* of all vertices is equal to 4π, irrespective of the pattern of triangulation.

5.9 With reference to the calculation of the shape of gores to cover a spherical surface of radius R (fig. 5.10a), let s be the meridional contour-length and $b(s)$ the (narrow) width of a gore. By considering successive small portions of the gore of length δs, and taking the limit $\delta s \to 0$, show that the required condition is
$$\frac{d^2 b}{ds^2} = -\frac{b}{R^2};$$
and that this leads to gores such that
$$b = b_0 \cos(s/R)$$
when the origin of s is at the equator.

6

Geometry of distortion of curved surfaces

6.1 Introduction

The main task of this chapter is to investigate some purely geometrical aspects of the distortion of curved surfaces. In general, if a given surface is distorted from its original configuration, every point on the surface will undergo a *displacement*; and at every point the surface will experience *strain* ('stretching') and *change of curvature* ('bending'). Clearly the components of strain and change of curvature are, in general, functions not only of the components of displacement but also of the geometry of the surface in its original configuration.

In the present chapter (and indeed throughout the book) we shall consider only the limited class of distortions in which displacements, strains and changes of curvature are regarded as *small*, just as they are in the classical theory of simpler structures. In consequence of this simplification, the functional relationships between strain, change of curvature, and displacement will be relatively simple, and indeed *linear*.

In chapter 3 we have already established *ad hoc* expressions for change of curvature and hoop strain in terms of radial displacement for symmetrical deformations of a cylindrical shell surface. Our present task includes not only the investigation of more general types of distortion of cylindrical shells but also the consideration of distortion of other kinds of surface.

The arrangement of the chapter is as follows. First we investigate some aspects of the distortion of initially plane and cylindrical surfaces: these have the advantage that they may be described in terms of simple Cartesian coordinates. In particular we analyse some simple 'inextensional' modes of deformation of these surfaces. Then we consider inextensional deformations in 'nearly-cylindrical' shells of the type which we considered in section 4.6. Next we discuss briefly some features of small deformations in general surfaces, in terms of curvilinear coordinates. Lastly we analyse the symmetrical distortion of an arbitrary surface of revolution, making use, to some extent, of general results from the preceding section.

The traditional approach to problems of this kind is to try and express the components of both strain and change of curvature in terms of arbitrary small displacements of the surface. In the present chapter we shall supplement calculations of this sort by means of an idea which springs directly from the work of chapter 5. There we saw that at an arbitrary point on a smooth surface the Gaussian curvature may be expressed *either* 'extrinsically' as the product of the two principal curvatures *or* 'intrinsically' in terms of angular defect per unit area, etc. It follows immediately that if the surface is deformed from its original configuration there will in general be a *change* of Gaussian curvature: and moreover that this may be expressed *either* in terms of changes of principal curvature *or* in terms of 'stretching' strains within the surface, since the latter will distort the constituent triangles of the surface and hence alter slightly the angular defects. It follows immediately that there must be a general relationship at a given point on a given surface between the change of curvature/twist on the one hand and surface strain on the other. Clearly a relationship of this sort can play an important part in the structural analysis of shells; and as we shall see, it is crucial to the development of the 'two-surface' idea. This concept was introduced in section 3.7.1 in the context of cylindrical shells under axisymmetric loading; and the proposed direct connection between change of curvature/twist and surface strain is in fact the key to the application of this concept to more general cases. A general relationship between change of curvature and surface strain may be found, in principle, by the elimination of the components of displacement from the expressions for strain and change of curvature in terms of displacements; but it seems likely that the relationship may also be established directly by means of an extension of the ideas of chapter 5.

6.2   'Change of Gaussian curvature' in terms of surface strain

Consider an assembly of small equal rectangular plane facets which originally fit precisely into a plane array. Now suppose that a small strain, which varies from place to place, is imparted to the assembly. How does the assembly distort in consequence? For the sake of simplicity let us suppose that each facet is distorted in such a way that the edges remain straight, and also that changes in length of touching edges of neighbouring facets are equal. The strains envisaged in this process may be provided in a variety of ways, which include the elastic response of the pieces to imposed stress in the plane of the facets; but it is perhaps simplest to think first in terms of thermal strain

6.2 'Change of Gaussian curvature' and surface strain

of the facets in response to imposed changes of temperature. In general, if the facets are not connected to each other when the straining takes place, a small *angular defect* will be introduced at each vertex of the assembly. Consequently we may conclude from the work of chapter 5 that in general the application of strain will impart some *Gaussian curvature* to the assembly. But it is equally clear that some simple patterns of straining – e.g. uniform strain in one direction – will not introduce angular defects, and so will not impart any Gaussian curvature. We must, therefore, try to find a general relationship between the strain in the assembly and the corresponding changes of Gaussian curvature.

It is convenient for this purpose to use a Cartesian x, y coordinate system to denote position in the originally plane assembly. The imposed pattern of small strains may be described by means of the components ϵ_x, ϵ_y of direct strain and γ_{xy} of ('engineering') shear strain. Each of these components will be assumed to be a smooth function of x and y. It will become clear later that the Gaussian curvature imparted by the strain will be the sum of the Gaussian curvatures which would be imparted if the three components of strain were applied separately. Therefore we may consider the effects of the three components ϵ_x, ϵ_y and γ_{xy} independently.

Fig. 6.1a shows four small rectangles in the xy plane which connect perfectly before the strains are imparted. The centres of the rectangles occupy positions $x, y; x + h, y; x, y + k$ and $x + h, y + k$, respectively. In due course we shall consider the limit as $h \to 0$ and $k \to 0$. We are particularly interested in the *angular defect* which the various components of strain impart to the vertex where the four rectangles meet; and indeed the local Gaussian curvature is equal to this angular defect divided by the area associated with the vertex, which is shaded in fig. 6.1a.

Fig. 6.1. Calculation of 'angular defect per unit area' imparted by surface strains. (a) Layout of four small contiguous rectangles. (b) Distortion due to shearing strain γ_{xy}. (c) Distortion due to tensile strain ϵ_x.

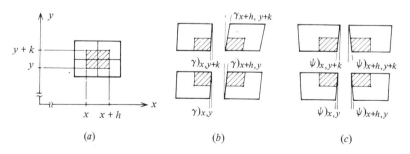

Geometry of distortion of curved surfaces

It is simplest to begin by examining the consequences of the application of the shearing-strain component γ_{xy} to the assembly. We assume that each rectangle shears uniformly with a strain equal to the value of γ at its centre point. Fig. 6.1*b* shows the four rectangles separately. It is clear that the angular defect at the common vertex when they are reassembled is

$$\gamma)_{x,y} - \gamma)_{x+h,y} - \gamma)_{x,y+k} + \gamma)_{x+h,y+k}.$$

The area associated with the vertex is hk. Thus, taking the limit as $h, k \to 0$ we obtain the required expression for the Gaussian curvature K:

$$K = \frac{\partial^2 \gamma_{xy}}{\partial x \partial y} \quad \text{(due to } \gamma_{xy}\text{)}. \tag{6.1}$$

Next suppose that the strain component ϵ_x is applied to the assembly. Each rectangle will change in width. But a *uniform* expansion or contraction does not change the corner angles. In order to calculate the change of angular defect we need to investigate the changes in *slopes* of the sides of the rectangles. Each rectangle will change into a slightly trapezoidal shape, as shown in fig. 6.1*c*. The slope of each side is equal to half the width of the rectangle multiplied by the rate of change of ϵ_x with respect to y, as shown in the diagram. The angular defect at the common junction is thus equal to

$$\tfrac{1}{2} h \left\{ \frac{\partial \epsilon_x}{\partial y} \bigg)_{x,y} + \frac{\partial \epsilon_x}{\partial y} \bigg)_{x+h,y} - \frac{\partial \epsilon_x}{\partial y} \bigg)_{x,y+k} - \frac{\partial \epsilon_x}{\partial y} \bigg)_{x+h,y+k} \right\}.$$

Dividing by the area hk and taking the limit as $h, k \to 0$ we obtain

$$K = \frac{-\partial^2 \epsilon_x}{\partial y^2} \quad \text{(due to } \epsilon_x\text{)}. \tag{6.2}$$

Here we have ignored the fact that the associated area changes slightly in the course of the straining; so our result is strictly only valid for sufficiently small strains. We shall make many approximations of this kind throughout the book; and indeed we have already done so in chapter 3 (e.g. (3.4)).

Equations (6.1) and (6.2) represent two different ways in which straining can locally impart Gaussian curvature to an initially flat sheet. It is clear that since angular defects are additive, the two methods may be used simultaneously and their effects superposed. Indeed, it is also clear that there is another method involving strains ϵ_y, giving a formula analogous to (6.2). Thus we obtain, finally, by superposition of the three cases:

$$K = -\frac{\partial^2 \epsilon_y}{\partial x^2} + \frac{\partial^2 \gamma_{xy}}{\partial x \partial y} - \frac{\partial^2 \epsilon_x}{\partial y^2}. \tag{6.3}$$

The piecemeal way in which (6.3) has been derived reflects strongly the picture of Gaussian curvature in terms of angular defect which has been built

6.2 'Change of Gaussian curvature' and surface strain

up in the preceding section. Now it may reasonably be objected that each of the four distorted rectangles in fig. 6.1c has nonzero shearing strain γ_{xy} at each corner. The important point here is that the *mean* value of γ_{xy} is zero in each trapezium. But if this explanation does not satisfy we must resort to other, more polished, ways of deriving the same equation. One of these will be given in section 6.8 in the context of more general coordinate systems.

In subsequent chapters we shall make use of (6.3) and variants of it in many problems: it is in fact a most important type of equation. At this stage it is useful to make some general remarks about it.

(i) The equation is purely *kinematic*, i.e. it is based on considerations of geometry alone. It applies whether the strains ϵ_x, ϵ_y, γ_{xy} are due either to elastic distortion in the material of the shell on account of changes of stress; or to thermal expansion; or indeed to any other cause.

(ii) If the straining of a plane sheet happens to leave the sheet plane, zero Gaussian curvature is imparted and (6.3) thus becomes

$$\frac{\partial^2 \epsilon_y}{\partial x^2} - \frac{\partial^2 \gamma_{xy}}{\partial x \partial y} + \frac{\partial^2 \epsilon_x}{\partial y^2} = 0. \tag{6.4}$$

This is in fact the well known *plane strain compatibility condition*. It is usually derived by expressing the strain components ϵ_x, ϵ_y, γ_{xy}, in terms of the components, u, v of small displacement in the x, y directions, respectively, and then eliminating these variables from the expressions: see problem 6.1.

(iii) We have derived the equation by considering the distortion of an originally plane sheet. In fact this is an unnecessarily restrictive view. The only critical requirement about the original configuration is that the original elementary rectangles fitted together exactly at their corners without any angular defect. This is, simply, a requirement that the original surface be *developable*. It follows that (6.3) applies not only to the distortion of plane surfaces (it is in fact the geometrical basis of 'large deflection' theory of flat plates) but also to cylindrical and other developable surfaces. In subsequent chapters we shall use it widely in the context of cylindrical shells, with the x-coordinate parallel to the axis of the cylinder and the y-coordinate in the circumferential direction. We could also use it for conical shells (except at the vertex), but in this case the Cartesian coordinate system would be inconvenient, of course. It is possible to recast the equation for a plane polar coordinate system.

(iv) Given that (6.3) is a special equation, applying only to surfaces which are initially developable, we may ask what form a more general version of the equation will take. The basic point, which was noted earlier, is that since an inextensional deformation of a surface causes no change to the Gaussian curvature, a more general deformation, involving surface strains, will be responsible for a *change* in Gaussian curvature. The general case therefore is concerned with a surface which initially has nonzero Gaussian curvature, and which is then deformed There is immediately a problem over coordinate systems, since the x, y Cartesian system is obviously only satisfactory for surfaces which originally have zero Gaussian curvature. Later in the chapter, in section 6.8, we shall give a derivation of the appropriate equation in terms of general curvilinear coordinates. This is included largely for the sake of completeness, as we shall need to make relatively little use of the general form in this book. In fact the special form already derived (6.3) applies as it stands to the important practical cases of cylindrical and other developable surfaces, but it is also a good approximation for what are known as *shallow shells*; for these nearly flat surfaces the simple Cartesian coordinate system, while not being exactly applicable, nevertheless is sufficiently accurate for many purposes: cf. section 4.6.

In general, in subsequent developments, we shall be concerned mainly with *changes* in Gaussian curvature, which we shall denote by g, in contrast with the Gaussian curvature of the original surface, which we shall write as K. In accordance with this notation we finally rewrite (6.3) as follows, for developable and shallow surfaces:

$$g = -\frac{\partial^2 \epsilon_y}{\partial x^2} + \frac{\partial^2 \gamma_{xy}}{\partial x \partial y} - \frac{\partial^2 \epsilon_x}{\partial y^2}. \tag{6.5}$$

6.3 Connection between the two aspects of Gaussian curvature

We have seen in chapter 5 that there are two precisely equivalent definitions of the Gaussian curvature K of a given surface. First, we have the *extrinsic* definition in terms of principal radii of curvature $R_1 R_2$, from (5.20):

$$K = \frac{1}{R_1 R_2}.$$

Second, we have an *intrinsic* definition, in terms of measurements on the surface itself, from (5.16):

$$K = \frac{d\beta}{dA}.$$

It is the connection between these two ideas of curvature which constitutes a powerful tool for the analysis not only of the geometry of undeformed surfaces but also for the analysis of the deformation of surfaces. In general the distortions of a shell surface will involve surface strains as a response to direct-stress resultants, temperature changes etc., and we can use (6.5) − or an appropriate generalisation of it – to express the local change of Gaussian curvature in terms of the local strains: but they will also involve changes of curvature in response to bending moments, and we naturally seek to express the change in Gaussian curvature of the surface in terms of these also, by using the first definition.

Let κ_1, κ_2 be the change of curvature in the principal directions 1, 2 respectively at a given point on a surface. Regarding these as small quantities we may obtain the required expression by differentiating (5.20) as a product. This gives immediately a general relation:

$$g = \frac{\kappa_1}{R_2} + \frac{\kappa_2}{R_1}. \tag{6.6}$$

Note that in general a distortion of the surface will also involve a local twisting κ_{12} with reference to the 1,2 directions (cf. section 2.3.2); for although directions 1,2 are the principal directions of the *original* surface there is no reason why they should in general also be principal directions for the *change* of curvature. However, although κ_{12} is in general nonzero, it does not appear in (6.6).

By matching equations (6.5) and (6.6) we have a powerful *compatibility* equation relating changes in surface strain to changes in curvature in the case of developable and 'shallow' shells:

$$g = \frac{\kappa_1}{R_2} + \frac{\kappa_2}{R_1} = -\frac{\partial^2 \epsilon_y}{\partial x^2} + \frac{\partial^2 \gamma_{xy}}{\partial x \partial y} - \frac{\partial^2 \epsilon_x}{\partial y^2}. \tag{6.7}$$

We shall find many uses for this equation in the solution of various problems. It forms the basis for the calculation of 'membrane' deformation of shells in chapter 7, in connection with other relations (to be developed in the following section) between displacement of the shell and changes of curvature. And it provides, in its more general form, an important geometrical relation in the deformation of general shells of revolution. Finally, and most importantly, it furnishes the main compatibility equation in the 'two-surface' theory of shell structures which we shall develop in chapter 8.

6.4 Strain–displacement relations for a cylindrical shell

The connection between surface strains and changes of curvature, established above, is important; but it is not in itself a sufficient description of the geometry of small distortion of surfaces. Thus, although it holds throughout the surface it is powerless to express boundary conditions in which *displacements* of the surface are specified. In general we also need to know the relationship between the *strains* of the surface (both extensional and bending)

158   *Geometry of distortion of curved surfaces*

and the *displacements* of the surface, just as we do in the analysis of simpler kinds of structure.

For the sake of definiteness we consider first the simple case of a cylindrical shell of radius a shown in fig. 6.2a. We use an orthogonal x, y coordinate system in the surface, with x in the axial direction. A general point (x, y) in the original undistorted configuration undergoes a small displacement whose components are u and v in the x- and y-directions respectively and w in the direction of the local (outward) normal z, as shown.

Our task is to express the components of strain and change of curvature, $\epsilon_x, \epsilon_y, \gamma_{xy}; \kappa_x, \kappa_y, \kappa_{xy}$ respectively, in terms of u, v, w.

In the special case $w = 0$ the deformation is confined to the xy-plane and accordingly the strains are given by the well-known 'plane strain' relations already given as (2.6):

$$\epsilon_x = \frac{\partial u}{\partial x}, \quad \epsilon_y = \frac{\partial v}{\partial y}, \quad \gamma_{xy} = \frac{\partial v}{\partial x} + \frac{\partial u}{\partial y}. \tag{6.8}$$

The distortion of generators and circumferential lines involves negligible change of curvature to a first order of approximation (see problem 6.2), so that we may write $\kappa_x = \kappa_y = 0$. There is, nevertheless, a non-negligible twist κ_{xy}, which may be worked out with the aid of fig. 6.2b. The edge AD of a general curved element has displacement v in the circumferential direction, and consequently it rotates about the axis of the cylinder through the small angle v/a, since the surface is exactly cylindrical. The edge BC rotates likewise, but through a slightly different angle

$$\frac{1}{a}\left(v + \frac{\partial v}{\partial x}dx\right).$$

There is thus a relative rotation of these two edges about the x-axis equal to

Fig. 6.2. (a) Coordinate systems for a cylindrical surface. Solid-headed arrows indicate components of (small) displacement in the x-, y- and z-directions. (b) A small element of the surface.

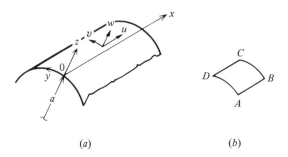

(*a*)   (*b*)

6.4 Strain–displacement relations: cylindrical shell

$$\frac{1}{a}\frac{\partial v}{\partial x}dx;$$

and by the previous definition this gives a twist

$$\kappa_{xy} = \frac{1}{a}\frac{\partial v}{\partial x}.$$

The sign corresponds to the convention used in chapter 2.

To all of these expressions must be added the effect of w, the normal component of displacement. This may be considered separately as a special case in which $u = v = 0$. Now a small radial displacement w has no effect upon ϵ_x and γ_{xy}, to a first approximation. But it has a definite effect on ϵ_y since the circumference of the surface is enlarged: it is straightforward to show (cf. section 3.2.2) that

$$\epsilon_y = w/a.$$

The effect of w on κ_x is exactly the same as for a straight beam lying along a generator:

$$\kappa_x = -\frac{\partial^2 w}{\partial x^2}.$$

Similarly w is responsible for a twist κ_{xy} just as for a plane:

$$\kappa_{xy} = -\frac{\partial^2 w}{\partial x \partial y}.$$

In the circumferential direction w has two separable effects on the curvature:

$$\kappa_y = -\frac{\partial^2 w}{\partial y^2} - \frac{w}{a^2}.$$

The first of these is analogous to κ_x, and the second represents the fact that a uniform expansion of a ring of radius a increases the radius to $w + a$ and thus produces a small decrease in the curvature: see problem 6.3. Collecting these results together and adding the various contributions, we have for a cylindrical surface:

$$\epsilon_x = \frac{\partial u}{\partial x}, \tag{6.9a}$$

$$\epsilon_y = \frac{\partial v}{\partial y} + \frac{w}{a}, \tag{6.9b}$$

$$\gamma_{xy} = \frac{\partial v}{\partial x} + \frac{\partial u}{\partial y}, \tag{6.9c}$$

$$\kappa_x = -\frac{\partial^2 w}{\partial x^2}, \tag{6.9d}$$

160 *Geometry of distortion of curved surfaces*

$$\kappa_y = -\frac{\partial^2 w}{\partial y^2} - \frac{w}{a^2}, \tag{6.9e}$$

$$\kappa_{xy} = -\frac{\partial^2 w}{\partial x \partial y} + \frac{1}{a}\frac{\partial v}{\partial x}. \tag{6.9f}$$

Note the significant difference between the expressions for ϵ_x, ϵ_y and κ_x, κ_y respectively.

It is easy to verify that these relations satisfy precisely the general compatibility equation (6.7): here, of course, x and y are principal axes, and on the left-hand side we have

$$\frac{\kappa_1}{R_2} + \frac{\kappa_2}{R_1} = \frac{\kappa_x}{R_y} + \frac{\kappa_y}{R_x} = \frac{\kappa_x}{a},$$

since $1/R_x = 0$. Hence we obtain the expression

$$\frac{1}{a}\frac{\partial^2 w}{\partial x^2} = \frac{\partial^2 \epsilon_y}{\partial x^2} - \frac{\partial^2 \gamma_{xy}}{\partial x \partial y} + \frac{\partial^2 \epsilon_x}{\partial y^2} ; \tag{6.10}$$

which is useful in the calculation of displacements of a cylindrical shell whose surface strains are given.

6.5 Inextensional deformation of a cylindrical surface

As a simple example, and as an introduction to the work of chapter 7, let us investigate the deformations of a cylindrical surface which are possible under the special restriction that the components $\epsilon_x, \epsilon_y, \gamma_{xy}$ of the surface strain are all equal to zero: these are called the *inextensional* deformations of the surface. The calculations are straightforward, but the complete solution involves a variety of important points.

The conventional approach to this problem is to use 6.9a, 6.9b and 6.9c to obtain differential equations for u, v and w; and these are then solved in relation to given boundary conditions. Here we shall begin instead by using (6.10), to obtain

$$\frac{\partial^2 w}{\partial x^2} = 0. \tag{6.11}$$

Hence we find immediately that

$$w(x, y) = f_1(y) + xf_2(y). \tag{6.12}$$

The two functions $f_1(y), f_2(y)$ depend on the boundary conditions.

Now although an x, y coordinate system is an obvious one to use for a developable surface, it is often more convenient to use an angular coordinate θ for a cylindrical shell, so that circumferential continuity can be ensured easily. The transformation is made by using the substitution

6.5 Inextensional deformation of a cylindrical surface

$$\theta = y/a, \qquad (6.13)$$

and we may then write (6.12) as

$$w(x, \theta) = f_3(\theta) + x f_4(\theta). \qquad (6.14)$$

Consider first the imposition of a particularly simple set of boundary conditions for a cylindrical shell of length l, with $0 \leqslant x \leqslant l$;

$$w(0, \theta) = w(l, \theta) = 0:$$

at each end there is zero radial displacement. Immediately we find $f_3(\theta) = f_4(\theta) = 0$, and consequently $w = 0$ throughout.

So far we have been concerned only with component w of displacement. Displacement components u, v can now be found from (6.9b) and (6.9c), suitably rewritten in terms of θ. Thus from (6.9b) we obtain

$$\frac{\partial v}{\partial \theta} = 0,$$

and therefore $v = f_5(x)$, while from (6.9a) we obtain

$$\frac{\partial u}{\partial x} = 0,$$

and therefore $u = f_6(\theta)$. Finally, from (6.9c) we have

$$\frac{\partial v}{\partial x} + \frac{1}{a}\frac{\partial u}{\partial \theta} = 0,$$

from which we find

$$\frac{df_5(x)}{dx} + \frac{1}{a}\frac{df_6(\theta)}{d\theta} = 0. \qquad (6.15)$$

The most general solution of this is given by

$$v = f_5(x) = C_2 + C_3 x, \qquad (6.16a)$$

$$u = f_6(\theta) = C_1 - C_3 a \theta, \qquad (6.16b)$$

where C_1, C_2 and C_3 are constants. The constant C_1 represents a uniform displacement in the axial direction, which is obviously not inhibited by our boundary supports which fix only the value of w. Similarly the constant C_2 represents an arbitrary rotation of the entire surface about the axis. The constant C_3 is the amplitude of the mode of deformation which is shown in fig. 6.3. Note in particular that since the u-component of displacement varies linearly with θ, the mode is only possible if the surface is slit along a generator, as shown. Normally, of course, we are concerned with complete, unslit cylindrical shells, for which we must have $C_3 = 0$. In general, the condition of continuity at a real or imaginary join on a given generator is that u is a *periodic* function of θ; and this constraint enables us to reject the term $C_3 \theta$ in the general expression for u.

The complete solution

$$u = C_1, \quad v = C_2, \quad w = 0 \tag{6.17}$$

thus corresponds to an open-ended paper cylindrical tube which is free to slide over two fixed and rigid circular discs at its ends, as shown in fig. 6.4a. But if a single drawing pin is pushed through the paper into one of the discs the values of C_1, C_2 are both determined and the entire surface is uniquely located. For this particular problem the solution is correct for arbitrarily large values of C_1 and C_2. In general we can only expect that solutions of (6.9) and (6.10) will be valid for *small* displacements.

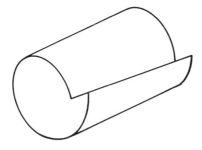

Fig. 6.3. A pattern of displacement with $w = 0$ which is only possible if the surface is slit along a generator.

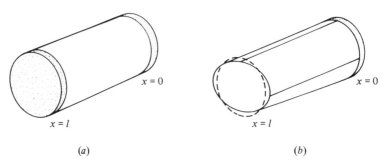

Fig. 6.4. (*a*) The two ends of a cylindrical paper tube are held circular by being slid over two fixed circular discs, which are shown here with exaggerated thickness. (*b*) When one end-disc is removed, modes of inextensional deformation become possible.

6.5.1 A second example

Consider now a second case in which there is a boundary condition $w(0, \theta) = 0$, but no restriction of displacement at the far end $x = l$: this corresponds physically to the removal of one of the discs in the previous problem. In this case we have, in (6.14),

$$f_3(\theta) = 0;$$

but now $f_4(\theta)$ is undetermined. The function $f_4(\theta)$ is *arbitrary* in the sense that it represents a freedom of the surface to deform when the boundary restrictions are too few to locate the surface completely. There are, however, certain restrictions on the function $f_4(\theta)$. First, since the displacement component w must be small over the entire surface – otherwise (6.9) would not be accurate – the magnitude of $f_4(\theta)$ is limited. Second, $f_4(\theta)$ must be periodic, since w itself is periodic in θ. Finally, there is a condition which corresponds to the fact that every 'hoop' of the surface at x = constant, is itself inextensional. From (6.9b) we have

$$w = -\frac{\partial v}{\partial \theta}.$$

Therefore, by integrating around a typical hoop from $\theta = 0$ to $\theta = 2\pi$ we find

$$\int_0^{2\pi} w \, d\theta = v_0 - v_{2\pi}.$$

But v is also a periodic function for this complete tube; hence $v_{2\pi} = v_0$ and we obtain

$$\oint f_4(\theta) d\theta = 0 \tag{6.18}$$

as a restriction on the function $f_4(\theta)$. This is satisfied, of course, for all functions $\sin n\theta$ and $\cos n\theta$ when n is an integer.

Let us proceed, as before, to solve for displacement components u, v; but for the sake of simplicity let us consider a specific radial displacement function

$$f_4(\theta) = \frac{\bar{w}_n}{l} \cos n\theta. \tag{6.19}$$

Thus at the edge $x = l$ there is a radial displacement of amplitude $\bar{w}_n$, with circumferential wavenumber n. This may, of course, be regarded as a term of a Fourier series representation of a more general function.

Substituting in (6.9b) we have

$$\frac{\partial v}{\partial \theta} = -\frac{x\bar{w}_n}{l} \cos n\theta,$$

and consequently

$$v = -\frac{x\bar{w}_n}{nl} \sin n\theta + f_7(x).$$

From (6.9a) we have
$$\frac{\partial u}{\partial x} = 0,$$
so
$$u = f_8(\theta).$$
Finally, substituting into (6.9c) we obtain
$$-\frac{\bar{w}_n}{nl}\sin n\theta + \frac{df_7(x)}{dx} + \frac{1}{a}\frac{df_8(\theta)}{d\theta} = 0. \tag{6.20}$$

This equation contains functions of x and θ respectively. The middle term can therefore only be constant. But a nonzero constant would give, on integration with respect to θ, a contribution to $f_8(\theta)$ which is linear in θ and, in particular, nonperiodic. Therefore, we conclude that the constant is zero, and hence $f_7(x)$ = constant = C_2, say.

Integrating (6.9c), we have
$$u = -\frac{a\bar{w}_n}{n^2 l}\cos n\theta + C_1. \tag{6.21}$$

As in the previous cases the constants C_1, C_2 in the expressions for u and v respectively represent simple rigid-body displacements.

Fig. 6.4b shows a sketch of the displacements for the case $n = 2$. The disc at $x = 0$ must be sufficiently thin, of course, not to interfere with the mode of deformation. Observe that both v and w are proportional to x, but that u is a function of θ only; and also that the displacements u and w are in phase circumferentially, while v is out of phase, having its maxima on the generators where both w and u are zero. Finally, observe the successive attenuation of the amplitudes of the three components of displacement at the end $x = l$:

Component	Amplitude
w	$\bar{w}_2$
v	$\bar{w}_2/2$
u	$a\bar{w}_2/4l$

For other values of n the numbers 2 and 4 in this table would be replaced by n and n^2 respectively. The relationship between w and v is determined entirely by the geometry of deformation of an inextensional hoop, and the radius cancels (cf. (6.9b) expressed in terms of θ). On the other hand, the amplitude of u depends also on the length/radius proportions of the cylinder: this becomes specially obvious in the simple case $n = 1$ (see problem 6.4).

The modes of deformation represented by the solutions of the present section may all be completely suppressed by the imposition of the boundary condition $u(0, \theta) = 0$. Physically, this corresponds to glueing the paper cylinder to the single end disc. Note that the boundary condition $w(0, \theta) = 0$ implies also $v(0, \theta) = 0$ for this inextensional surface: consequently the two boundary conditions which are needed at the end $x = 0$ to fix the surface, if the end $x = l$ is free, can be expressed in terms of w and u, or v and u, but *not* in terms of w and v.

The conclusion that a paper cylinder which is glued to a rigid disc at one end but is free at the other end cannot undergo inextensional deformations may seem surprising to those who are familiar with the fact that such a cylinder may readily be deformed inextensionally in the manner shown in fig. 6.5. In this diagram there is a curved (elliptical) crease ABC which lies in a plane; and the shaded portion is a mirror image of the original surface in this plane.

There are several points to be made about this paradoxical case. First, the displacements are obviously large, so the example is outside the scope of the equations which we have been using. And even if the plane ABC were arranged to make only a very small angle with the axis, so that displacements w were also small, we should need to introduce the possibility of certain kinds of discontinuity in the functional form of w. Our analysis so far has been limited in scope by the implicit assumption that all displacement functions are *smooth*; and although this is not a serious limitation for many problems, it will be wise to bear alternatives in mind. It may easily be demonstrated with paper models that the post-buckled form of some thin cylindrical shells has a 'concertina' appearance, with many almost-straight creases, like some kinds of pleated paper lampshade. Here we encounter the problem that an infinitely sharp crease in an actual shell of finite thickness is impossible, as it involves infinite local bending moments. Lastly, we note that if we wish to introduce the deformed pattern of fig. 6.5 into a perfectly cylindrical surface, we must begin

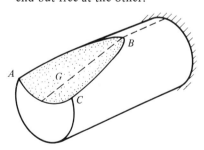

Fig. 6.5. An inextensional mode of large deflection, involving a curved, plane crease, in a cylindrical surface which is clamped at one end but free at the other.

at the free edge and 'roll' the crease down to the required position. To get from the original configuration to that shown in the diagram *without* moving the position of the crease requires either stretching or tearing of the surface: the final state is an inextensional transformation of the initial state, but intermediate states are *not* inextensional.

6.5.2 Remarks on boundary conditions

A feature common to both of the above examples on the inextensional deformation of cylindrical shells is that the displacement of the surface is determined – apart from the possibility of simple rigid-body motions – by the imposition of *two* boundary conditions, whether w is specified at each end or w and u are prescribed at one end. This rule can be generalised to include various other combinations, but it is clear from (6.9a) that the value of u must only be specified *once* on any generator.

We have seen that w is a convenient variable for use in these problems, by virtue of (6.11). Consequently, when boundary conditions in w are specified at both ends of the shell the solution is particularly straightforward. When on the other hand the boundary conditions involve both u and w (usually at one edge) it becomes desirable to express the condition on u in terms of w. This is easily done by elimination of v between (6.9b) and (6.9c):

$$\frac{\partial w}{\partial x} = \frac{a \partial^2 u}{\partial y^2} \ . \tag{6.22}$$

Two examples of the use of this expression are given in problems 6.5 and 6.6. These remarks about boundary conditions are curiously reminiscent of those made in chapter 4 in relation to the solution of equilibrium equations in cylindrical shells according to the membrane hypothesis. In fact there are some very close formal parallels between these two quite different problems. An obvious example is the result that the value of u is constant along any generator in any (small) inextensional deformation of a cylindrical shell. This corresponds directly, in terms of restrictions on boundary conditions, with the 'straight-through' transmission of axial point forces along the generators of a cylindrical shell, according to the membrane hypothesis.

In appendix 6 we shall investigate more fully the formal analogy between the 'membrane' and 'inextensional' solutions for shell surfaces; and we shall find strong analogies which will be helpful in the setting up of the full equations of elastic shells. In our present study it will be of interest – following an obvious parallel with chapter 4 – to study briefly the inextensional deformation of 'nearly cylindrical' surfaces, in order to check on the occurrence of elliptic and hyperbolic equations, which the above remarks lead us to suspect.

6.6 Strain–displacement relations for nearly-cylindrical surfaces

Let us now consider the inextensional deformation of a simple 'nearly-cylindrical' surface (fig. 4.16) whose generators have radius of curvature $|a^*| \gg a$. As in chapter 4, we hope to do the analysis by using a slight modification of the equations which we have already developed for our study of the distortion of cylindrical surfaces. Returning to the derivation of (6.9) we find that they must be augmented by the addition of terms containing w in (6.9a) and (6.9d), exactly analogous to the w-terms in (6.9b) and (6.9e); and also an extra term in (6.9f). The revised equations, with the new terms enclosed in brackets, are:

$$\epsilon_x = \frac{\partial u}{\partial x} + \left(\frac{w}{a^*}\right), \tag{6.23a}$$

$$\epsilon_y = \frac{\partial v}{\partial y} + \frac{w}{a}, \tag{6.23b}$$

$$\gamma_{xy} = \frac{\partial v}{\partial x} + \frac{\partial u}{\partial y}, \tag{6.23c}$$

$$\kappa_x = -\frac{\partial^2 w}{\partial x^2} - \left(\frac{w}{a^{*2}}\right), \tag{6.23d}$$

$$\kappa_y = -\frac{\partial^2 w}{\partial y^2} - \frac{w}{a^2}, \tag{6.23e}$$

$$\kappa_{xy} = -\frac{\partial^2 w}{\partial x \partial y} + \frac{1}{a}\frac{\partial v}{\partial x} + \left(\frac{1}{a^*}\frac{\partial u}{\partial y}\right). \tag{6.23f}$$

As the second principal radius is now nonzero, the strain–curvature compatibility equation becomes

$$\frac{\kappa_x}{a} + \left(\frac{\kappa_y}{a^*}\right) = \frac{\partial^2 \epsilon_y}{\partial x^2} - \frac{\partial^2 \gamma_{xy}}{\partial x \partial y} + \frac{\partial^2 \epsilon_x}{\partial y^2}. \tag{6.24}$$

In the case of the cylindrical shell ($a^* \to \infty$) the compatibility equation (6.24) is exactly satisfied by expressions (6.23), as we have seen. However, when the terms in brackets are included, substitution of (6.23) into (6.24) gives a term

$$\frac{w}{aa^*}\left(\frac{1}{a} + \frac{1}{a^*}\right) \tag{6.25}$$

on the left-hand side which has no counterpart on the right-hand side. This degree of inconsistency is a direct consequence of the use of a simple Cartesian coordinate system for a surface which is not strictly developable. For sufficiently large values of a^* the inconsistency is a small one. It can be removed entirely by omitting the second term on the right-hand side of each of (6.23d) and (6.23e). In the case of (6.23e) this step is justified even for a cylindrical shell for terms of the form $w_n \cos n\theta$ for sufficiently large values of n (see problem 6.7); this corresponds simply to a mode of displacement whose radial component varies rapidly round the circumference. On the other hand, it seems unlikely that inextensional deformation of nearly-cylindrical shells will also involve displacements varying rapidly in the x-direction, and so it does not seem possible to reject the second term of (6.25) on similar grounds. Note, however, the term a^{*2} on the denominator: for sufficiently large values of a^* this term will certainly be negligible.

6.6.1 Inextensional deformation of nearly-cylindrical surfaces

One of our intentions is to study the inextensional deformations of nearly-cylindrical surfaces. Making use of the approximations given above we find from the curvature–strain compatibility equation a governing equation in w:

$$\frac{\partial^2 w}{\partial x^2} + \frac{1}{aa^*} \frac{\partial^2 w}{\partial \theta^2} = 0. \tag{6.26}$$

This replaces (6.11), which was the appropriate form for a cylindrical shell. Any boundary conditions involving u or v will be taken into account by the use of (6.23a) and (6.23b).

Consider first a 'barrel' shell with $a^* > 0$. It will be convenient to have edges at $x = \pm l$ as in section 4.6.1. Let us impose the following simple set of boundary conditions:

$$w = \bar{w}_n \cos n\theta \quad \text{at } x = \pm l. \tag{6.27}$$

Assuming a solution of the form

$$w = \bar{w}_n f(x) \cos n\theta, \tag{6.28}$$

we obtain an auxiliary equation

$$f''(x) - (n^2/aa^*)f(x) = 0 \tag{6.29}$$

which must be solved subject to the boundary conditions

$$f(-l) = f(l) = 1.$$

Defining, as before, $\zeta = l(aa^*)^{-\frac{1}{2}}$ and $\xi = x(aa^*)^{-\frac{1}{2}}$ we find the solution

$$w = \bar{w}_n \cos n\theta \, \text{Cosh} n\xi / \text{Cosh} n\zeta. \tag{6.30}$$

When we put $a^* \to \infty$ we obtain a result appropriate to a cylindrical shell:

$$w = \bar{w}_n \cos n\theta \tag{6.31}$$

which corresponds (in accordance with (6.11)) to the generators remaining straight. But in general (6.30) indicates that the pattern of radial displacement imposed at the edges is attenuated as we move towards the centre of the shell.

6.6 Strain–displacement relations

This attenuation is a direct consequence of the barrel form of the shell.

Components u, v of the displacement may be calculated by substituting into (6.23a) and (6.23b) and solving these. The attenuating feature of the solution carries through to these components of displacement also.

Next consider a 'waisted' shell, $a^* < 0$. As in section 4.6.2 it will be convenient to impose two boundary conditions at one edge of the surface, which we put at $x = 0$. For simplicity, let the imposed boundary conditions be

$$w = \bar{w}_n \cos n\theta, \quad u = 0 \text{ at } x = 0. \tag{6.32}$$

As before, we consider a tentative solution of the form

$$w = \bar{w}_n f(x) \cos n\theta \tag{6.33}$$

which must satisfy the governing equation

$$\frac{\partial^2 w}{\partial x^2} - \frac{1}{(-aa^*)} \frac{\partial^2 w}{\partial \theta^2} = 0. \tag{6.34}$$

Here we have rearranged the equation in terms of the positive quantity $(-aa^*)$. The general solution of this equation is

$$f(x) = A \cos(n\mu x) + B \sin(n\mu x), \tag{6.35}$$

where $\mu = (-aa^*)^{-\frac{1}{2}}$, and our next task is to find the constants A, B in terms of the imposed boundary conditions. Now when both w and u are specified at an edge of a nearly cylindrical inextensional surface it is convenient to eliminate v between (6.23b) and (6.23c) to give the general expression

$$\frac{\partial w}{\partial x} = \frac{1}{a} \frac{\partial^2 u}{\partial \theta^2}. \tag{6.36}$$

This expression is precisely equivalent to that derived earlier for a cylindrical shell (see (6.22)).

In this case it gives, at $x = 0$,

$$\frac{\partial w}{\partial x} = 0.$$

It follows that $A = 1, B = 0$ and the normal component of displacement is thus given by

$$w = \bar{w}_n \cos n\theta \cos n\mu x. \tag{6.37}$$

This may be rearranged (cf. (4.53)) in the form

$$w = \tfrac{1}{2}\bar{w}_n \cos n(\theta - \mu x) + \tfrac{1}{2}\bar{w}_n \cos n(\theta + \mu x). \tag{6.38}$$

Thus, the solution may be expressed as the sum of two parts, each of which involves no attenuation in its *characteristic direction*, which is inclined at a slope of $\pm(-a/a^*)^{-\frac{1}{2}}$ to the x-axis. These directions correspond to *straight lines* in the surface, as we observed before. The propogation of w without diminution along these lines corresponds physically to the idea that we can insert a simple crease or fold in a surface in any straight-line direction, without any other straining or change of curvature in the surface being required, provided this is not prevented by boundary conditions. The coupling together of the two parts of the solution is required in fact to meet the circumferential connectivity of the surface. Thus, if a waisted surface is slit as shown in fig. 6.6, each part of the solution can exist separately: see problem 6.8.

The u, v components of displacement can readily be derived from expression (6.23) for w by substitution in (6.23a) and (6.23b) and solution of the resulting differential equations. (These solutions also satisfy (6.23c) exactly, since all of the equations are self-consistent once the w-terms in (6.23d) and (6.23e) are omitted.) The components u and v may be expressed in a form analogous to (6.38), thereby emphasising the role of the characteristic directions: see problem 6.9.

Various simple examples of the inextensional deformation of cylindrical and nearly-cylindrical shells are given in problems 6.10 and 6.11.

6.7 General remarks on the inextensional deformation of shells

The inextensional deformation of surfaces fits neatly into the present chapter as an example of the kinematics of deformation of curved surfaces in the specially simple case when there is zero strain in the surface. We have seen that if a shell is 'adequately' constrained at its edges (or indeed elsewhere) the condition of inextensibility prevents any kind of displacement. This feature is directly related to the 'rigidity' of closed surfaces described in chapter 1; but now we can see that the same rigidity is to be found in *open* surfaces provided they are adequately supported.

Inadequate support (see e.g. fig. 6.4b), on the other hand, allows unrestricted deformation of the surface in certain modes. Now in the case of a real thin

Fig. 6.6. A 'waisted' shell with straight inclined generators may be creased as shown if it has been slit longitudinally.

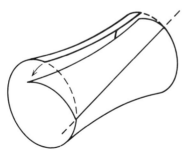

shell made of elastic material such deformation is not *completely* unrestricted, since the mode of deformation involves changes of curvature; and this can only occur, in a real shell, at the expense of bending moments which tend to resist the deformation. In the purely *geometric* calculations which we have been pursuing, we have discounted completely any bending stiffness which the surface may possess. In some cases it is possible to make a relatively simple analysis of the stiffness of an inextensional mode in a shell by considering the strain energy of bending which would occur in such a mode. This was indeed one of the earliest calculations ever done in the theory of shells: thus Rayleigh (1881) successfully estimated the lowest natural frequencies of vibration of a thin hemispherical bowl. The major problem in calculations of this sort is a worry that if there is a change of curvature there must be bending-stress resultants as a consequence of Hooke's law; and if there are bending-stress resultants there must also be direct-stress resultants, by considerations of equilibrium; and finally, if there are direct-stress resultants, there must also be surface strains as a consequence of Hooke's law. But this is contrary to the original hypothesis. In fact, inextensional deformation occurs only in consequence of a *hypothesis*, which may or may not turn out in a given problem to be tenable. In section 9.5.1 we shall see examples in which the hypothesis is examined in detail.

It should be noted that these cautionary remarks about the hypothetical nature of inextensional deformations are closely analogous to those which apply in the analysis of structures according to the *membrane* hypothesis: see chapter 4.

6.8 Compatibility between surface strain and change of curvature in curvilinear coordinates

Differential Geometry is the name of a branch of mathematics which deals with the general properties of curved surfaces. Love (1888) recognised that differential geometry could be used advantageously in the study of shell structures, and set up the general equations for elastic shells accordingly. It cannot be denied that the resulting long and complicated equations are beyond the grasp of many engineers; and the use of tensor notation makes the subject more compact but at the same time more elusive. The following exposition is an attempt to convey some (but not all) of the main ideas and results of the appropriate part of the subject of differential geometry in relatively simple terms. In particular we restrict the treatment to the use of *orthogonal* coordinates. Little of the remainder of the book depends strongly on this section, which may therefore be omitted at a first reading.

Fig. 6.7a shows a generalised representation in perspective of a curved surface onto which has been inscribed a set of right-handed orthogonal curvilinear coordinates ξ_1, ξ_2. The diagram shows a typical line ξ_1 = constant, and a neighbouring line ξ_1 = constant + $d\xi_1$; these are labelled $\xi_1, \xi_1 + d\xi_1$ respectively, and the same convention applies to the lines marked ξ_2 and $\xi_2 + d\xi_2$. A pair of values ξ_1, ξ_2 locates uniquely a point on the surface. Examples of coordinates of this sort are x, y and r, θ for planes; x, θ for cylinders; θ, ϕ for spheres (fig. 6.7b); and θ, s for general surfaces of revolution (figs. 4.10 and 4.11). Note that some of these coordinates have the dimensions of length while others are dimensionless.

It is obviously desirable to know the superficial dimensions of a typical element of a surface. Thus, in the cylindrical x, θ system the dimensions of an element $dx, d\theta$ are $(dx) \times (Rd\theta)$, where R is the radius of the cylinder; while in a spherical coordinate system the dimensions of an element $d\theta, d\phi$ are $(R \sin\phi \, d\theta) \times (R d\phi)$, where R is the radius of the sphere (see fig. 6.7b).

In general we may say that the dimensions of an element $d\xi_1, d\xi_2$ are $\alpha_1 d\xi_1 \times \alpha_2 d\xi_2$ where α_1, α_2 are functions of ξ_1, ξ_2 known as the *Lamé parameters*. In the above examples the Lamé parameters may be obtained by inspection, and the results are given below.

System	ξ_1	ξ_2	α_1	α_2
Plane Cartesian	x	y	1	1
Plane polar	r	θ	1	r
Cylindrical polar	x	θ	1	R
Spherical	ϕ	θ	R	$R \sin\phi$.

Fig. 6.7. (*a*) Perspective view of orthogonal curvilinear coordinates inscribed on an arbitrary smooth surface. (*b*) A spherical surface showing coordinates θ, ϕ and the dimensions of a typical small element.

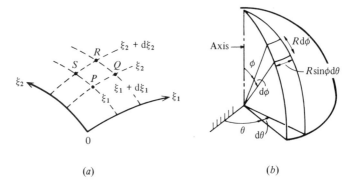

6.8 Compatibility relations in curvilinear coordinates

It is clear from these examples that the Lamé parameters themselves characterise the surface itself in some way; and indeed it may be shown in general that the Gaussian curvature and the principal curvatures of a surface at an arbitrary point are related to certain derivatives of the Lamé parameters.

6.8.1 Gaussian curvature of original surface in terms of Lamé parameters

We shall first establish the simplest of these expressions, which concerns the local Gaussian curvature of the surface. Our method will be to employ ideas described in section 6.2. Fig. 6.7a shows a small element lying between four coordinate lines $\xi_1, \xi_1 + d\xi_1, \xi_2, \xi_2 + d\xi_2$. The surface area of the small element is equal to $\alpha_1 \alpha_2 d\xi_1 d\xi_2$, and we can find the local Gaussian curvature if only we can find an expression for the *solid angle* which the element subtends. Now for a *pointed* vertex, we showed earlier that the solid angle subtended is exactly equal to the angular defect when the vertex is flattened inextensionally. For a region of a smooth surface it may be shown (see problem 6.12) that the corresponding solid angle may be found by the following procedure.

(i) Cut out a narrow ring of the surface around the edge of the region (fig. 6.8a).

(ii) Make a cut through this ring, shown as AB in fig. 6.8b.

(iii) Develop this ring, without stretching or tearing, onto a plane, as in fig. 6.8c.

(iv) Measure the angle β in the plane between the edges $AB, A'B'$ of the cut (fig. 6.8c): this angle is the required solid angle subtended by the region.

The idea lying behind this procedure is that a sufficiently *narrow* strip of any surface is developable. (The procedure just described seems easier to grasp than the exactly equivalent idea presented in some texts (e.g. Thomson & Tait, 1879, §137), whereby a tangent-plane is rolled, *without slip or 'spin'*, around the closed curve on the surface, and its change of angle measured after a complete circuit.)

In the case of a region of the kind shown in fig. 6.7a, the edge strip has four sides, as shown in fig. 6.8d, and as the corner angles are all right-angles, the value of β depends on the 'angle of turn' of each of the four sides.

Fig. 6.8f shows the calculation of this angle for the side PQ of the element. The side PQ is a line of constant ξ_2, and its ends are at the intersection with ξ_1 contours having a difference of $d\xi_1$, which we shall here call $\Delta\xi_1$. At present we shall regard $\Delta\xi_1$ as constant. $P'Q'$ is part of the contour $\xi_2 + d\xi_2$. We shall regard $d\xi_2$ as variable, and will seek a limit as $d\xi_2 \to 0$. For these reasons the proportions of $PQQ'P'$ have been drawn elongated, although $\Delta\xi_1$ and $d\xi_2$ are both small quantities of the same order.

174 Geometry of distortion of curved surfaces

Now the length PQ is equal to $\alpha_1 \Delta\xi_1$, by definition. The length of $P'Q'$ is slightly different, being equal to

$$\left(\alpha_1 + \frac{\partial \alpha_1}{\partial \xi_2} d\xi_2\right)\Delta\xi_1.$$

The angle of turn, $d\psi$, of side PQ, is found in exactly the same way as for the simple curved beam discussed in chapter 2, since the lines PP', QQ' are perpendicular to the curved edges: thus we have

$$d\psi = \frac{P'Q' - PQ}{QQ'}.$$ (6.39)

But $QQ' = \alpha_2 d\xi_2$, by definition; hence we obtain

$$d\psi = \frac{1}{\alpha_2} \frac{\partial \alpha_1}{\partial \xi_2} \Delta\xi_1.$$ (6.40)

Fig. 6.8. Calculation of Gaussian curvature for general surfaces.
(a) An area of surface, with a narrow peripheral band. (b) The band detached and about to be cut on AB. (c) The band flattened onto a plane and opening up by angle β. (d) and (e): As (b) and (c) but for an element bounded by coordinate lines. (f) Calculation of the curvature of the edge strip PQ. (g) Opening β due to curvature of PQ. (h) Corresponding change for curvature in RS. (i) Change of β due to change of corner angle at Q.

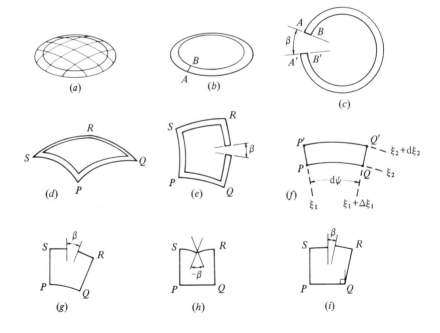

6.8 Compatibility relations in curvilinear coordinates

Now if PQ were the only edge to be curved in this way, the angular defect for the element would simply be equal to $d\psi$; see fig. 6.8g. On the other hand, if RS were the only edge to be curved, the angular defect would be $-d\psi$ calculated for this side: see fig. 6.8h. Consequently the angular defect due to sides PQ and RS taken together is

$$-d\psi)_{RS} + d\psi)_{PQ}.$$

Having obtained an expression for $d\psi)_{PQ}$ we can readily evaluate the change in this quantity across the element by differentiation: the required expression is

$$-\frac{\partial}{\partial \xi_2}(d\psi)\Delta\xi_2,$$

where $\Delta\xi_2$ is the coordinate difference between sides PQ and RS. Recalling that in (6.40) $\Delta\xi_1$ is a constant, we obtain finally the following expression for the angular defect on account of the curvature of the developed edges PQ, RS:

$$-\frac{\partial}{\partial \xi_2}\left(\frac{1}{\alpha_2}\frac{\partial \alpha_1}{\partial \xi_2}\right)\Delta\xi_1 \Delta\xi_2. \tag{6.41}$$

There is a similar expression (with subscripts 1 and 2 exchanged) for the contribution towards the angular defect of the element from the other two edges, QR and SP. The sum of these expressions, divided by the area of the element, gives the Gaussian curvature: thus we obtain, finally

$$K = -\frac{1}{\alpha_1 \alpha_2}\left[\frac{\partial}{\partial \xi_1}\left(\frac{1}{\alpha_1}\frac{\partial \alpha_2}{\partial \xi_1}\right) + \frac{\partial}{\partial \xi_2}\left(\frac{1}{\alpha_2}\frac{\partial \alpha_1}{\partial \xi_2}\right)\right]. \tag{6.42}$$

In the theory of surfaces this is known as the *equation of Gauss*. The original version, in Gauss's paper, is more complex as it is given for a general non-orthogonal coordinate system: and it also uses a different version of the Lamé parameters.

The reader should verify that the Lamé parameters tabulated at the beginning of this section satisfy this relation.

6.8.2 Change of Gaussian curvature due to surface strain

Equation (6.42) applies to the original undistorted surface. Our hope is to find how the Gaussian curvature *changes* when the original surface is now subjected to a general *surface strain*. Described in relation to the ξ_1, ξ_2 coordinate system the strains are $\epsilon_1, \epsilon_2, \gamma_{12}$: ϵ_1, ϵ_2 are tensile strains and γ_{12} is shear strain, as defined previously. We shall regard all strain magnitudes as small compared to 1. All three components of surface strain in general are functions of ξ_1 and ξ_2.

Geometry of distortion of curved surfaces

The previous analysis suggests a way of tackling the problem. For side PQ we need to recalculate the angle $d\psi$ in the deformed configuration of the element and use the change of value on account of the strains to compute the corresponding change of K.

Consider first the effect of strains ϵ_1, ϵ_2. The new length of PQ is $\alpha_1(1 + \epsilon_1)\Delta\xi_1$ and the new length of QQ' is $\alpha_2(1 + \epsilon_2)d\xi_2$. Thus instead of (6.40) we now have

$$d\psi = \frac{1}{\alpha_2(1 + \epsilon_2)} \frac{\partial(\alpha_1 + \epsilon_1\alpha_1)}{\partial\xi_2} \Delta\xi_1.$$

The term $(1 + \epsilon_2)$ on the denominator will be awkward, so we replace it by $(1 - \epsilon_2)$ in the numerator, by the binomial theorem. It corresponds exactly to the 'change of thickness' effect in section 2.4.1. Expanding the right-hand side, we have the following expression for $d\psi$:

$$\left(\frac{1}{\alpha_2} \frac{\partial\alpha_1}{\partial\xi_2} - \frac{\epsilon_2}{\alpha_2} \frac{\partial\alpha_1}{\partial\xi_2} + \frac{1}{\alpha_2} \frac{\partial(\epsilon_1\alpha_1)}{\partial\xi_2} - \frac{\epsilon_2}{\alpha_2} \frac{\partial(\epsilon_1\alpha_1)}{\partial\xi_2} \right)\Delta\xi_1.$$

The first term represents the angle before the strain was applied (cf. (6.40)); the next two terms give the effect of strain; and the last term is negligible as it involves a product of small quantities. Consequently, the first order change of $d\psi$ for edge PQ on account of strain ϵ_1, ϵ_2 is:

$$\left(-\frac{\epsilon_2}{\alpha_2} \frac{\partial\alpha_1}{\partial\xi_2} + \frac{1}{\alpha_2} \frac{\partial(\epsilon_1\alpha_1)}{\partial\xi_2} \right)\Delta\xi_1. \tag{6.43}$$

Next consider the effect of shear strain γ_{12}. By the principle of superposition we may do this in isolation, and add the result to the right-hand side of (6.43). Let us first imagine that the edge PQ is fixed. Then the effect of γ_{12} is to move point P' to the right a distance $\gamma_{12}\alpha_2 d\xi_2$ along the $\xi_2 + d\xi_2$ contour. Correspondingly, Q' moves a distance larger than this by the amount

$$\frac{\partial}{\partial\xi_1}(\gamma_{12}\alpha_2)d\xi_2 \Delta\xi_1. \tag{6.44}$$

Now the application of shear strain γ_{12} does not change the length of $P'Q'$ at all, so the result of keeping PQ in its original form is to give $P'Q'$ a bogus extension of the amount (6.44). PQ must therefore bend, and the change in the angle subtended is equal to the negative of expression (6.44) divided by $\alpha_2 d\xi_2$, i.e.

$$\left(-\frac{1}{\alpha_2} \frac{\partial}{\partial\xi_1}(\gamma_{12}\alpha_2) \right)\Delta\xi_1.$$

So far we have calculated the change in $d\psi$ on account of strain for the edge PQ. We now extend the calculation to edge RS, taking the difference, as before, and dividing by the area of the element in order to give the contribution to the change of K. Then we add the contributions from the other two sides, which we find simply by exchanging subscripts. This gives the following expression for the contribution to the change of Gaussian curvature from the changes of angle of the four sides:

$$\frac{1}{\alpha_1\alpha_2}\left[\frac{\partial}{\partial\xi_2}\left(\frac{\epsilon_2}{\alpha_2}\frac{\partial\alpha_1}{\partial\xi_2} - \frac{1}{\alpha_2}\frac{\partial(\epsilon_1\alpha_1)}{\partial\xi_2} + \frac{1}{\alpha_2}\frac{\partial(\gamma_{12}\alpha_2)}{\partial\xi_1}\right) \right.$$
$$\left. + \frac{\partial}{\partial\xi_1}\left(\frac{\epsilon_1}{\alpha_1}\frac{\partial\alpha_2}{\partial\xi_1} - \frac{1}{\alpha_1}\frac{\partial(\epsilon_2\alpha_2)}{\partial\xi_1} + \frac{1}{\alpha_1}\frac{\partial(\gamma_{12}\alpha_1)}{\partial\xi_2}\right)\right]. \quad (6.45)$$

There is one more effect to be reckoned with. When there are shearing strains γ_{12}, the angles at the *corners* of the element will no longer be exactly right-angles. Changes in corner angle contribute directly to the angular defect β, and it is easy to see that the contribution is positive at corners Q, S but negative at the other corners: see fig. 6.8i. A simple calculation shows that in the limit as the size of the element tends to zero the contribution to change of Gaussian curvature from this effect is:

$$-\frac{1}{\alpha_1\alpha_2}\frac{\partial^2\gamma_{12}}{\partial\xi_1\partial\xi_2}. \quad (6.46)$$

Putting this together with (6.45) and rearranging the terms containing γ_{12} we have, finally,

$$g = \frac{1}{\alpha_1\alpha_2}\left[\frac{\partial}{\partial\xi_1}\left(-\frac{1}{\alpha_1}\frac{\partial(\epsilon_2\alpha_2)}{\partial\xi_1} + \frac{\epsilon_1}{\alpha_1}\frac{\partial\alpha_2}{\partial\xi_1} + \frac{\gamma_{12}}{\alpha_1}\frac{\partial\alpha_1}{\partial\xi_2}\right)\right.$$
$$+ \frac{\partial}{\partial\xi_2}\left(-\frac{1}{\alpha_2}\frac{\partial(\epsilon_1\alpha_1)}{\partial\xi_2} + \frac{\epsilon_2}{\alpha_2}\frac{\partial\alpha_1}{\partial\xi_2} + \frac{\gamma_{12}}{\alpha_2}\frac{\partial\alpha_2}{\partial\xi_1}\right)$$
$$\left. + \frac{\partial^2\gamma_{12}}{\partial\xi_1\partial\xi_2}\right]. \quad (6.47)$$

This is the required general version of the expression given earlier (see (6.5)) for a rectangular Cartesian system. By putting $\alpha_1 = \alpha_2 = 1$ we recover the previous result (6.5) precisely. The expression, or a minor variant of it, is given in various texts (e.g. Novozhilov, 1964, §5). With suitable rearrangement (writing $\epsilon_{11}, \epsilon_{22}, 2\epsilon_{12}$ for our $\epsilon_1, \epsilon_2, \gamma_{12}$, etc.) equations of this sort can be expressed neatly in tensorial form: see, e.g. Sanders (1963).

In the special case of uniform isotropic expansion of a surface, $\epsilon_1 = \epsilon_2 =$ constant and $\gamma_{12} = 0$; then (6.47) gives $g = 0$. This is not exactly correct, since in the case of a sphere of radius R a uniform expansion by strain ϵ gives $g = -2\epsilon/R^2$ (see problem 6.13). The discrepancy arises at the point where, in calculating the change of angular defect per unit area we used the *original* area instead of the *current* area of the element, $\alpha_1\alpha_2(1+\epsilon_1)(1+\epsilon_2)d\xi_1 d\xi_2$. Our formula is thus entirely in accord with the traditions of small-deflection theory, in which changes are referred to the *original* geometry, and it corresponds to the neglect of terms in w in expressions (6.23d) and (6.23e), as we have noted earlier.

In the usual treatment of the geometry of curved surfaces the equation of Gauss (6.42) is associated with two other relations, due to Codazzi:

$$\frac{\partial}{\partial \xi_1}\left(\frac{\alpha_2}{R_2}\right) = \frac{1}{R_1}\frac{\partial \alpha_2}{\partial \xi_1}, \quad \frac{\partial}{\partial \xi_2}\left(\frac{\alpha_1}{R_1}\right) = \frac{1}{R_2}\frac{\partial \alpha_1}{\partial \xi_2}. \tag{6.48}$$

These apply to the original surface, and relate the principal radii. They only apply *when the coordinate lines are in the principal directions of the surface* ('lines of curvature' coordinates). The proof of the Codazzi relations requires the use of three-dimensional Euclidian geometry, and so is more difficult than the essentially two-dimensional proof of the Gauss equation. As we shall not use the Codazzi equations in this book, we do not give a formal proof of them. But the outline of a proof is given in problem 6.14.

Although (6.42) applies whether or not the coordinates lie in the principal directions, it is often a matter of simple convenience to use this kind of coordinate. In this case the change of Gaussian curvature is easily expressed in terms of curvature changes κ_1, κ_2 in the 1,2 directions:

$$g = \frac{\kappa_1}{R_2} + \frac{\kappa_2}{R_1}. \tag{6.49}$$

This equation, together with (6.47), provides a link, for a general surface, between surface strains and changes of curvature.

It might seem appropriate at this point to abandon the use of variable g, as it appears to be an unnecessary intermediate variable between strains and changes of curvature. There are perhaps some circumstances in which it is not useful to retain g as a separate variable; but in general we shall find that it has a useful role in clarifying the behaviour of shell structures.

6.9 Symmetrical deformation of shells of revolution

In chapter 4 we studied the equilibrium equations for symmetrically-loaded shells of revolution. Another ingredient of the complete solution for this kind of shell is the geometrical relationship between the strain, displacement and change of curvature variables. It is convenient to derive the relevant equations at this point, because they involve relatively small extensions of previous work.

Fig. 6.9 shows part of the meridian of a general shell of revolution. It defines angle ϕ between the normal to the meridian and the axis, and also the

6.9 Symmetrical deformation of shells of revolution

radii r, r_1, r_2, r_3, exactly as in fig. 4.10. (Subscripts 1, 2, 3 here have no connection, of course, with subscripts 1 and 2 of the previous section.) The positive sense of the auxiliary normal coordinate z is also indicated, primarily for the purposes of defining the positive sense of curvature and change of curvature. As drawn, r_1, r_2 are both positive at P. Radius r_2 is in fact a principal radius, and the other principal radius, r_1, is the radius of curvature of the meridian (see section 4.5). Coordinates s (contour length along the meridian) and θ (angular position of plane of meridian) correspond to ξ_1, ξ_2 respectively, and the Lamé parameters follow by inspection:

$$\alpha_1 = 1, \quad \alpha_2 = r. \tag{6.50}$$

As s is our chosen independent variable it will be necessary to eliminate the auxiliary angular variable ϕ by means of the following identities:

$$d\phi = \frac{ds}{r_1}; \quad \sin\phi = \frac{r}{r_2}; \quad \cos\phi = \frac{r}{r_3}. \tag{6.51}$$

Also, by inspection,

$$\frac{dr}{ds} = \cos\phi. \tag{6.52}$$

It is easy to verify by substitution that the equation of Gauss (6.42) is satisfied in general: here, of course, there is no variation of any quantity with $\xi_2 = \theta$.

Let us now express g, the change of Gaussian curvature, in terms of the

Fig. 6.9. (a) Geometry of the meridian of a surface of revolution, defining radii r, r_1, etc. (b) Components u, w of small displacement of point P, together with small rotation χ of the tangent at P.

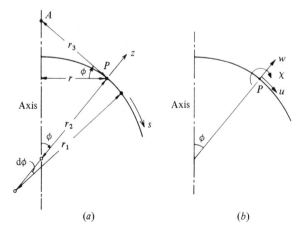

180   Geometry of distortion of curved surfaces

surface strains by specialising (6.47). In symmetrical deformation of a shell of revolution the shearing strain $\gamma_{s\theta}$ is zero by hypothesis. Also, there is no variation of any quantity with ξ_2. Therefore, we are left with only two nonzero terms of (6.47); and indeed these should no longer be expressed as partial derivatives. Thus we have

$$g = \frac{1}{r}\frac{d}{ds}\left(-\frac{d(r\epsilon_\theta)}{ds} + \epsilon_s \frac{dr}{ds}\right). \tag{6.53}$$

The expression enclosed in brackets may be rearranged as follows:

$$\left[-r\left(\frac{d\epsilon_\theta}{ds} + \frac{\epsilon_\theta - \epsilon_s}{r_3}\right)\right]. \tag{6.54}$$

We shall return to this expression after we have considered the other expression for g, in terms of changes of curvature.

In the Cartesian coordinate system which we used for a developable surface we found that it was convenient to express changes of curvature in terms of the normal component, w, of the displacement of the surface. This is shown in fig. 6.9b: its positive sense is the same as that of z. The tangential component u of displacement is also shown: its positive sense is the same as that of the coordinate s. For symmetrical deformation of general shells of revolution it turns out to be more convenient to use as the prime deformation variable the (small) rotation χ of a tangent to the meridian, which is also shown in fig. 6.9b. It follows immediately from the geometry of a plane curve that the change of curvature of the surface in the meridional direction is given by

$$\kappa_s = \frac{d\chi}{ds}. \tag{6.55}$$

The positive sense of rotation χ is marked in fig. 6.9b and the positive sense of κ_s is that which tends to produce tensile strain on the positive (z) side of the surface, in accordance with our previous sign convention.

The effect of a small rotation of the meridian on the circumferential curvature is readily found. Consider a small pure rotation χ about a typical point P on the meridian. It is equivalent to a small change in the angle ϕ. The original curvature in the circumferential direction is $1/r_2$, so we have a corresponding change in curvature equal to

$$d\left(\frac{1}{r_2}\right) = \frac{d}{d\phi}\left(\frac{1}{r_2}\right) d\phi. \tag{6.56}$$

Now $r_2 = r \csc\phi$, and since r is constant at P by supposition we have

$$\frac{d}{d\phi}\left(\frac{1}{r_2}\right) = \frac{1}{r}\cos\phi. \tag{6.57}$$

6.9 Symmetrical deformation of shells of revolutions

Eliminating $\cos\phi$ by means of (6.51) we obtain, finally,

$$\kappa_\theta = \frac{\chi}{r_3}. \tag{6.58}$$

The positive sense is defined just as for κ_s.

Let us now evaluate the change of Gaussian curvature in terms of χ:

$$g = \frac{\kappa_s}{r_2} + \frac{\kappa_\theta}{r_1} = \frac{1}{r_2}\frac{d\chi}{ds} + \frac{\chi}{r_1 r_3}. \tag{6.59}$$

This may be expressed as the derivative of a product (by means of (6.51), (6.52) and (6.57)) as follows:

$$g = \frac{1}{r}\frac{d}{ds}\left(\frac{r\chi}{r_2}\right). \tag{6.60}$$

By inspection of (6.54) and (6.60) we see that compatibility of strain and curvature is ensured if the contents of the brackets are equal in both cases. Hence our compatibility equation simplifies to

$$-\frac{\chi}{r_2} = \frac{d\epsilon_\theta}{ds} + \frac{\epsilon_\theta - \epsilon_s}{r_3}. \tag{6.61}$$

The variable χ is useful precisely because it gives such a simple equation, and is directly linked to both κ_s and κ_θ.

6.9.1 Strain-displacement relations

The next problem is to express the surface strains ϵ_s, ϵ_θ in terms of the components of the displacement of the surface.

It is most convenient to express the displacement of a point on the surface in terms of components u, v, w in the coordinate directions s, θ, z, respectively. For symmetrical deformations v is identically zero. Fig. 6.9b shows the components u and w. In the meridional direction we have, by the same considerations as those which we used for a cylindrical shell (cf. (6.9b)):

$$\epsilon_s = \frac{du}{ds} + \frac{w}{r_1}. \tag{6.62}$$

In the θ-direction the strain is simply equal to the component of displacement normal to the axis divided by the radius r: after some trivial manipulation we obtain

$$\epsilon_\theta = \frac{w}{r_2} + \frac{u}{r_3}. \tag{6.63}$$

And indeed it should not be surprising that the denominators of the two terms are, respectively, r_2 and r_3: see problem 6.15.

182   Geometry of distortion of curved surfaces

Lastly, we express the rotation χ in terms of u and w. When $w = 0$ the displacement is along the meridian; and since the meridian has a local radius of curvature r_1, the tangent rotates through a small angle u/r_1. Also, when $u = 0$ the displacement is normal to the meridian and there is a rotation of the tangent equal to $-dw/ds$. Combining these results we have

$$\chi = -\frac{dw}{ds} + \frac{u}{r_1}. \tag{6.64}$$

We now have expressions for ϵ_s, ϵ_θ and χ in terms of u and w. Elimination of u and w from the three expressions gives in principle a compatibility equation relating χ to ϵ_s and ϵ_θ. Now we have already obtained an equation of precisely this kind by specialisation of equation (6.47). We should therefore check whether equations (6.62) and (6.63) are consistent with equation (6.64); that is, whether they satisfy (6.61) identically. Detailed algebra shows that this is so. The calculation is straightforward but lengthy. In particular it makes use of the following identities which supplement relations (6.51) and (6.52), and which should be verified by the reader (also see problem 6.16):

$$\frac{dr}{ds} = \frac{r}{r_3} \; ; \quad \frac{dr_2}{ds} = \frac{r_2}{r_3}\left(1 - \frac{r_2}{r_1}\right); \quad \frac{dr_3}{ds} = 1 + \frac{r_3^2}{r_1 r_2}. \tag{6.65}$$

The precise self-consistency of the various equations is a consequence of the fact that the equations are exact in the context of small-deflection theory. The small inconsistency which we detected earlier in the equations of 'nearly-cylindrical' shells (section 6.6) was a consequence of the use of a slightly non-fitting coordinate system.

We obtained the compatibility equation in the first place by specialising the general equation (6.47), and then argued that we could have derived it instead by eliminating u and w between (6.62)–(6.64). In doing the elimination it is of course an advantage to know the answer.

The connection between χ, ϵ_s and ϵ_θ is purely a geometrical one, and as the distortion of the surface is symmetrical, the geometry can be done *ab*

Fig. 6.10. (a) Geometry of a small element PQ of the meridian. (b) Corresponding displacement diagram (origin not shown).

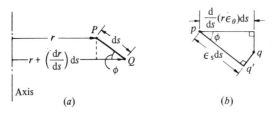

initio in the plane of the meridian. Fig. 6.10 shows (*a*) a general small element PQ of the meridian and (*b*) an associated displacement diagram. The component of displacement P normal to the axis is equal to $r\epsilon_\theta$, and that of Q is larger by the formally derived amount

$$\frac{d}{ds}(r\epsilon_\theta)ds,$$

where ds is the arc-length PQ. The other available information is that the arc-length PQ extends by $\epsilon_s ds$. This information is sufficient to solve the geometry of the displacement diagram, and the rotation χ of the element is equal to the displacement qq', divided by ds. In this way, using identities (6.51) we obtain

$$\chi = \frac{r_2}{r_3}\epsilon_s - \frac{r_2}{r}\frac{d}{ds}(r\epsilon_\theta), \tag{6.66}$$

from which (6.61) is obtained by rearrangement. See also problem 6.17.

6.10 Problems

6.1 Establish (6.4) in the case of a plane deformation, by eliminating u and v from the plane-strain relations (6.8) between the components of strain and (small) displacement.

6.2 A cylindrical surface of radius a distorts in such a way that the normal component w of displacement is zero; i.e. it remains a cylinder of radius a. Consider a line in the surface, which lay along a generator in the original configuration but is now inclined at a small angle $\partial v/\partial x$ to a generator of the current cylindrical shape. By constructing a Mohr circle of curvature/twist, show that the (change of) curvature of the line on account of this rotation is equal to $(1/a)(\partial v/\partial x)^2$.

6.3 A circle of radius a undergoes a small enlargement to a radius $a + w$. By use of the binomial theorem show that there is a small decrease in curvature (= radius^{-1}) of w/a^2.

Verify that in the case of a small displacement $u = 0$, $v = c\sin(y/a)$, $w = -c\cos(y/a)$ of a cylindrical surface (representing a rigid-body translation of the cylinder), equations (6.9) correctly make all components of strain zero.

6.4 A cylindrical surface of radius a lying between $x = 0$ and $x = l$ is given a small rigid-body rotation about the line $\theta = \frac{1}{2}\pi$ in the plane $x = 0$. Verify that the components of displacement are given by

$$u = -(ca/l)\cos\theta, \quad v = -(cx/l)\sin\theta, \quad w = (cx/l)\cos\theta,$$

184   Geometry of distortion of curved surfaces

where c is the magnitude of the small transverse displacement at $x = l$.

Verify that (6.9) again gives all components of strain zero (here $\theta = y/a$), and that the pattern of (6.21) (and the sentences following it) is satisfied.

6.5 A cylindrical shell of radius a and length l is free from geometric constraint at the end $x = l$. At end $x = 0$ the following components of small displacement are imposed:

$$u = u_n \cos n\theta, \quad w = 0 \quad (\theta = y/a).$$

If the shell undergoes purely inextensional deformation, show that the components of displacement are given by

$$u = u_n \cos n\theta \quad v = (nu_n x/a) \sin n\theta, \quad w = -(n^2 u_n x/a) \cos n\theta.$$

6.6 The shell of problem 6.5 is now subjected to edge-conditions $u = 0, v = v_n \cos n\theta$ at the edge $x = 0$. Show that the corresponding inextensional deformations are:

$$u = 0, \quad v = v_n \cos n\theta, \quad w = nv_n \sin n\theta.$$

6.7 Equation (6.23e) may be written

$$-a^2 \kappa_\theta = \frac{\partial^2 w}{\partial \theta^2} + w.$$

Examine the relative magnitude of the terms on the right-hand side for the mode $w = w_n \cos n\theta$; and hence show that the exclusion of the second term leads to an inaccuracy of less than about 6% in κ_θ for $n > 4$.

6.8 Equation (6.34) is a version of the *wave equation*. Show that, as an alternative to the solution (6.35), it is satisfied by (d'Alembert's solution)

$$w = f(\theta \pm \mu x),$$

where μ has the same meaning as before (6.35) and f is an *arbitrary* function. Verify that the situation depicted in fig. 6.6 may be represented by the solution

$$w = 0, \qquad \theta - \mu x \leqslant 0;$$
$$w = c(\theta - \mu x), \quad \theta - \mu x > 0$$

in the vicinity of the crease, provided the origin of the coordinate system lies on the crease and c is sufficiently small.

6.9 By using (6.23a), (6.23b) with $\epsilon_x = \epsilon_\theta = 0$, show that for the modeform (6.38) the u- and v-components of displacement are given by

6.10 Problems

$$u = (\bar{w}_n/2n\mu a^*)[\sin n(\theta - \mu x) - \sin n(\theta + \mu x)]$$
$$v = -(\bar{w}_n/2n)[\sin n(\theta - \mu x) + \sin n(\theta + \mu x)].$$

Verify also that when these expressions are put into (6.23c) they give $\gamma_{xy} = 0$.

6.10 A circular cylindrical surface lies between $x = -l$ and $x = l$. The radial component of displacement (w) is specified at each end; but the other two components of displacement at each end are unrestricted. Determine the inextensional displacements when

$$w = \pm w_n \cos n\theta \text{ at } x = \pm l.$$

6.11 Although the equations of section 6.6.1 have been derived for $|a^*| \gg a$, they may be used for sufficiently small regions of (say) a spherical shell $a^* = a$. Verify that in this case

$$w = c \cos(\pi y/L) \exp(-\pi x/L)$$

satisfies (6.26), where $y = a\theta$ and $x = 0$ is the equatorial edge of a hemispherical shell. Show that for sufficiently small values of the circumferential half wavelength L the displacement decays so rapidly in the x-direction that the deformation is confined to an equatorial band in which the radius r (fig. 6.9a) is practically constant.

6.12 Suppose that the region shown in fig. 6.8a is divided into a large number of small triangles, so arranged that the outer ring consists of a chain of facets. A view of such an arrangement in the flattened configuration − after sufficient cuts have been made − is shown schematically below. Demonstrate by direct geometrical arguments in relation to the shaded part of the diagram that the angle β is equal to the (algebraic) sum of the angular defects at all interior vertices, irrespective of the number and form of the triangles and the

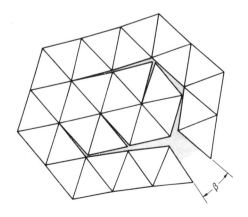

186   Geometry of distortion of curved surfaces

pattern of cutting; and thus that angle β of fig. 6.8c represents the *total solid angle* subtended by the region.

6.13 A sphere of radius R expands uniformly so that all linear dimensions are increased by a factor $1 + \epsilon$. Use the binomial theorem to show that to a first order of approximation the quantity $1/(\text{radius})^2$ changes by $-2\epsilon/r^2$.

6.14 The diagram below shows the portion $PQRS$ of the surface of fig. 6.7a, together with the normals at the vertices. These normals intersect in pairs because, in the present problem, the coordinate lines are lines of zero twist. Consider the angle marked $(\alpha_2/R_2)d\xi_2$ in the diagram, and the way in which it changes with respect to ξ_1 by virtue of the difference in length between QR and PS. Hence verify the first of (6.48).

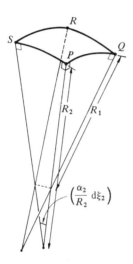

Check that (6.48) are satisfied by the four simple coordinate systems on p.172.

6.15 Set up (6.63) directly by considering the effect on the hoop passing through P in fig. 6.9 when components w and u of displacement are imposed separately.

6.16 Verify independently the special form of (6.65) for surfaces which have *straight* meridians, i.e. $(1/r_1) = 0$.

6.17 Let η be the component of displacement of the point P parallel to the axis of the surface of revolution in fig. 6.9. By means of the displacement diagram of fig. 6.10, establish the following expression for $d\eta/ds$:

$$\frac{d\eta}{ds} = \epsilon_s \operatorname{cosec}\phi - \cot\phi \frac{d}{ds}(r\epsilon_\theta).$$

7

Displacements of elastic shells stressed according to the membrane hypothesis

7.1 Introduction

We are now in a position to bring together the work of chapters 2, 4 and 6 in order to calculate the displacement of elastic shells which carry applied load, according to the membrane hypothesis, by direct-stress resultants only. In the case of a shell which is statically determinate according to the membrane hypothesis the procedure is straightforward, and consists of the same three steps which are used in the calculation of distortion of other kinds of statically determinate structure:

(i) Given the shell and its loading, and appropriate edge support conditions, use the equilibrium equations to find the direct-stress resultants, as in chapter 4.

(ii) Given the elastic properties of the material of which the shell is made (E, ν) and the thickness of the shell, use Hooke's law (chapter 2) to determine the surface strains in the shell.

(iii) Solve the strain-displacement equations, as in chapter 6, together with the appropriate boundary conditions, to determine the displacement of the shell.

Most of the problems which we shall investigate in the following chapters will involve interaction between stretching and bending effects in shell structures. It may seem odd therefore to wish to perform the sequence of calculations listed above, since in practice the membrane hypothesis will rarely be valid. And indeed, most of the results which will be obtained in the present chapter will reappear later as special cases of more general analyses, incorporating bending effects, which will be performed in subsequent chapters.

The motivation for the rather special work of the present chapter is fivefold. First, we need to be able to test the 'membrane hypothesis' which forms the basis of the calculations; and the proposed scheme provides a way of doing this. Having determined the displacements of the shell we shall be able to check on the magnitude of the peak 'bending strain' within the shell in comparison with that of the peak 'stretching strain', and use this as a rough first test of the validity of the hypothesis. We shall discuss this procedure in section 7.2. Second, it is useful for the engineer to be able to see clearly how shell structures *would* deform *if* they were to receive no help from bending effects. In particular it is useful to know under what circumstances a given cylindrical shell (say) is 'efficient', carrying the applied loads with relatively little displacement. Third, it will be a useful preparation for the work of subsequent chapters to investigate the role of *displacement boundary conditions* without the added complication of bending effects. Fourth, the calculations will provide the basis of a major part of the work of chapter 8, which deals in a particular way with the interaction of bending and stretching effects in shallow shells. And fifth, the work will lead to a simplified form of calculation which is valid in a wide range of circumstances, and which we may describe as 'quasi-inextensional' deformation. This in turn leads to a 'beam analogy' for cylindrical shells, which will play a useful and distinctive part in several subsequent chapters.

All of the examples in the present chapter involve cylindrical shells under loading which varies periodically in the circumferential direction. It would be possible also to do similar calculations for, say, the 'nearly-cylindrical' shells which we have met in chapters 4 and 6. However, the five points listed above can be illustrated adequately in the context of cylindrical shells; and once these points have been established, there is not a strong case for having more examples.

The problems to be investigated in the present chapter may be described briefly as follows. First (section 7.3) we consider a long uniform cylindrical shell, the surface of which is subjected to a normal pressure whose intensity varies sinusoidally with distance in both longitudinal and circumferential directions. Next (section 7.3.4) we extend the analysis to a statically indeterminate problem involving a shell of finite length whose two ends are 'clamped', being held both circular and plane. In section 7.4 we explore the possibility of using an Airy stress function to describe the state of membrane stress in a cylindrical shell. Lastly we point out in section 7.5 an approximate analogy between the equations governing a cylindrical shell, and those for an ordinary elastic beam.

7.2 Testing the validity of the membrane hypothesis

How can we decide, in a particular problem, whether or not the membrane hypothesis is justified? Let us suppose that the chain of calculations described in section 7.1 has been completed for a given shell subjected to given loads. In addition, let us calculate, from the known displacements, the changes of curvature over the entire surface of the shell. Now if any element of a shell having thickness t undergoes a change of curvature κ, there will be 'bending strains' of magnitude $\frac{1}{2}\kappa t$ at the exterior surfaces of the element; and if the material is elastic there will be corresponding stresses and, in particular, *bending-stress resultants*. But the presence of bending-stress resultants is explicitly ignored in the membrane hypothesis. Thus we may conclude that the membrane hypothesis is invalid unless the peak bending strain $\frac{1}{2}\kappa t$ is small in comparison with the peak 'stretching strain' ϵ corresponding to the peak tangential 'membrane' stress resultant. It is therefore useful to evaluate a dimensionless parameter

$$B = \kappa_{max} t / 2\epsilon_{max} \tag{7.1}$$

in any particular case, where κ_{max} and ϵ_{max} represent the magnitude of the peak change of curvature and tensile strain, respectively. Then, if $B \ll 1$ the membrane hypothesis is justified. For an order-of-magnitude calculation of this sort, of course, it is quite unnecessary to go into the details of principal axes, etc., when κ and ϵ are being determined, or to distinguish between t and h (cf. section 3.3.1).

We shall perform calculations of this sort at several points in the subsequent sections. At the present stage, however, we can immediately make a generalisation about the outcome of such calculations in all cases where κ is nonzero somewhere in the shell.

Suppose we have a shell of given geometry which carries given loading. Suppose also that the shell is somehow arranged in such a way that we can vary at will the value of the uniform thickness t without altering the elastic modulus of the material. By following the course of the calculations described in section 7.1 we can see that all strains, displacements and indeed changes of curvature are inversely proportional to t. Consequently, the peak 'bending strain' $\frac{1}{2}\kappa t$ will be independent of t; and it follows that the value of B will be directly proportional to t. Thus we may conclude that in any given problem for which $\kappa_{max} \neq 0$ the membrane hypothesis will cease to be justified if the thickness of the shell exceeds a certain critical value. This value will depend upon the other dimensions of the shell such as length and radius – and will also depend on the circumferential mode number for periodic loading – and we must therefore expect that the criterion will involve the achievement of a critical value by a certain *dimensionless group* which involves the thickness of the shell and other linear dimensions. We shall discover some groups of this sort in sections 7.3.2 and 7.4. In chapters 8 and 9 we shall abandon the mem-

7.3 Cylindrical shell with doubly-periodic pressure

brane hypothesis and investigate similar problems with the inclusion of bending effects; and we shall discover the dimensionless groups of the present chapter to re-emerge at the boundaries of the regime in which the membrane hypothesis is valid for practical purposes.

The sequence of calculations set out in section 7.1 seems to be straightforward in general, but there are some important cases in which stage (iii) breaks down completely because the strains determined by stages (i) and (ii) are plainly *incompatible*. In such cases the calculations simply cannot be completed; and on that account alone the membrane hypothesis must be discarded. An obvious example is a cylindrical shell carrying a radial load uniformly distributed on a circumferential line. As we saw in chapter 3, the equilibrium solution according to the membrane hypothesis gives a singularity in the form of a line tension around the circumference immediately under the loaded line. For a uniform shell this gives infinite stress, and the corresponding infinite strain (from Hooke's law) is clearly incompatible with the finite distortion of the adjoining parts of the shell. For this kind of problem it is necessary to do a calculation along the lines of chapter 3: the membrane hypothesis can be rejected *prima facie*.

7.3 Cylindrical shell with doubly-periodic pressure-loading

Consider first a simple case in which a long cylindrical shell of radius a sustains a pressure-loading of the form

$$p = p_0 \sin(\pi x/l) \sin(\pi y/b); \qquad (7.2)$$

see fig. 7.1. The positive sense of pressure is radially outwards, in the positive z-direction. It is convenient for this problem to use an x,y coordinate system,

Fig. 7.1. The doubly-periodic pressure-loading $p = p_0 \sin(\pi x/l) \times \sin(\pi y/b)$ has nodal lines as shown for the component w of displacement normal to the surface. Regions of outward/inward directed pressure (and displacement) form a shaded/blank chessboard pattern. Small circles and triangles mark points of maximum displacement in the x- and y-directions, respectively.

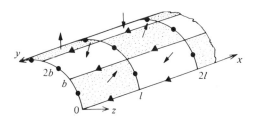

with the x-axis in the direction of the axis of rotational symmetry and the y-axis in the circumferential direction. The sign of the pressure is alternately positive and negative on a 'chessboard' pattern of 'panels' of length l and breadth b, as shown. The longitudinal half wavelength l may be chosen arbitrarily, but in the circumferential direction the condition of continuity requires that the number of whole waves around the circumference is integral; hence the circumferential half wavelength b must satisfy the relation

$$\pi a/b = n = \text{integer}: \tag{7.3}$$

n is the circumferential wavenumber, as in section 4.4.1. It will, however, be most convenient to work with b as if it were continuously variable, and introduce this restriction on b only at the end of the analysis.

It is clear that by having an effectively infinite shell we shall almost completely avoid boundary-condition problems: we shall simply require that the mean values of the components u and v of the displacement (see section 6.4) over the entire surface are zero.

Described in this way, the problem may seem to be singularly academic. In fact, it is extremely important in a wide range of applications, as we shall see: it forms the basis of Fourier-series methods of analysis (e.g. section 9.2), and it will play an important part in the study of buckling problems in chapter 14.

Our first step is to solve the equations of statical equilibrium. These are (see section 4.4)

$$\frac{N_y}{a} = p_0 \sin\left(\frac{\pi x}{l}\right) \sin\left(\frac{\pi y}{b}\right), \tag{7.4a}$$

$$\frac{\partial N_y}{\partial y} + \frac{\partial N_{xy}}{\partial x} = 0, \tag{7.4b}$$

$$\frac{\partial N_x}{\partial x} + \frac{\partial N_{xy}}{\partial y} = 0. \tag{7.4c}$$

Solving these in turn and putting all constants of integration to zero we obtain the following periodic expressions for the stress resultants:

$$N_y = ap_0 \sin\left(\frac{\pi x}{l}\right) \sin\left(\frac{\pi y}{b}\right), \tag{7.5a}$$

$$N_{xy} = \frac{l}{b} ap_0 \cos\left(\frac{\pi x}{l}\right) \cos\left(\frac{\pi y}{b}\right), \tag{7.5b}$$

$$N_x = \frac{l^2}{b^2} ap_0 \sin\left(\frac{\pi x}{l}\right) \sin\left(\frac{\pi y}{b}\right). \tag{7.5c}$$

Observe that N_x and N_y are of the same periodic form as p, and have the same sign; and that the shear-stress resultant N_{xy} is out of phase with N_x and N_y in

7.3 Cylindrical shell with doubly-periodic pressure

both x- and y-directions, having its maxima on the nodal lines of those functions. Also notice that $N_x/N_y = (l/b)^2$; consequently the direct-stress resultant with the larger magnitude is always that in the direction of the *longer* side of the 'panel' $l \times b$, whether that is in the axial or circumferential direction. The amplitude of N_{xy} is the geometric mean of the amplitudes of N_x and N_y.

Having solved the equilibrium equations, we now substitute into Hooke's law (2.5) and obtain expressions for the surface strains:

$$\epsilon_y = \frac{ap_0}{Et}\left(1 - \frac{\nu l^2}{b^2}\right) \sin\left(\frac{\pi x}{l}\right) \sin\left(\frac{\pi y}{b}\right) \qquad (7.6a)$$

$$\gamma_{xy} = \frac{ap_0}{Et} 2(1+\nu) \frac{l}{b} \cos\left(\frac{\pi x}{l}\right) \cos\left(\frac{\pi y}{b}\right) \qquad (7.6b)$$

$$\epsilon_x = \frac{ap_0}{Et}\left(\frac{l^2}{b^2} - \nu\right) \sin\left(\frac{\pi x}{l}\right) \sin\left(\frac{\pi y}{b}\right). \qquad (7.6c)$$

Next we substitute these expressions into the compatibility equation which involves change of Gaussian curvature (see (6.10)) to obtain, after rearrangement,

$$\frac{\partial^2 w}{\partial x^2} = -\frac{a^2 p_0}{Et} \frac{\pi^2}{b^2}\left(\frac{b}{l} + \frac{l}{b}\right)^2 \sin\left(\frac{\pi x}{l}\right) \sin\left(\frac{\pi y}{b}\right). \qquad (7.7)$$

It is interesting that Poisson's ratio does not appear in this expression: the reason for this will become clear in section 7.4, when we introduce a *stress function* as an aid to solving the equilibrium equations.

It is now a simple matter to integrate (7.7), setting constants of integration equal to zero: we thus obtain

$$w = \frac{a^2 p_0}{Et}\left(\frac{l}{b}\right)^2 \left(\frac{b}{l} + \frac{l}{b}\right)^2 \sin\left(\frac{\pi x}{l}\right) \sin\left(\frac{\pi y}{b}\right). \qquad (7.8)$$

Lastly, we use the compatibility equations (6.9a), (6.9b) and (6.9c) to obtain expressions for the axial and circumferential components of displacement:

$$u = -\frac{ap_0}{Et}\left(\frac{l}{\pi}\right)\left(\frac{l^2}{b^2} - \nu\right) \cos\left(\frac{\pi x}{l}\right) \sin\left(\frac{\pi y}{b}\right) \qquad (7.9)$$

$$v = \frac{ap_0}{Et}\left(\frac{b}{\pi}\right) \left\{\left(\frac{l}{b}\right)^2 \left[\left(\frac{b}{l} + \frac{l}{b}\right)^2 + \nu\right] - 1\right\} \sin\left(\frac{\pi x}{l}\right) \cos\left(\frac{\pi y}{b}\right). \qquad (7.10)$$

Note that the radial component of displacement, w, is in phase with the applied pressure: in particular it is zero all round the edge of each panel. Observe also that the u-component of displacement is greatest at the mid-point of the panel side of length b, while the v-component is greatest at the midpoint of the panel side of length l: see fig. 7.1.

It is instructive to study the relative amplitudes of the three components of displacement. Taking the amplitudes of u, v, w, as u_0, v_0, w_0, respectively, and putting $l = b$ and $\nu = 0.3$ for definiteness, we find that $u_0/w_0 \approx l/20a$ and $v_0/w_0 \approx b/4a$. These ratios depend on the panel size/radius ratio, but it seems clear that in many cases the u- and v-displacements will be small in comparison with the w-displacements. It is interesting to note that if the circumference had been *inextensional* the amplitude of v relative to that of w would have been $b/\pi a$, which is not much different. It is easy to show from (7.8) and (7.10) that the difference decreases as the ratio l/b (>1) increases.

7.3.1 Panel flexibility factor

Since w varies with x and y in the same way as p, it is convenient to think in terms of a 'panel compliance' or flexibility factor F equal to the amplitude of w divided by the amplitude of p. From (7.2) and (7.8) we find that

$$F = \frac{a^2}{Et} \left\{ 1 + \left(\frac{l}{b}\right)^2 \right\}^2. \tag{7.11}$$

This expression is plotted in fig. 7.2, on linear scales in (*a*) and on logarithmic scales in (*b*). When the right-hand side is expanded as a quadratic, the three terms $1 + 2(l/b)^2 + (l/b)^4$ correspond, respectively, to the effects of ϵ_y, γ_{xy} and ϵ_x; this may be verified easily. These three contributions to displacement are also shown separately in fig. 7.2. It is clear that for small values of l/b, ϵ_y (circumferential stretching) dominates, while for large values of l/b on the other hand it is ϵ_x (longitudinal stretching) which dominates. When $l/b = 1$ ('square' panels) the contributions to displacement from ϵ_x and ϵ_y are equal, and the shearing strain γ_{xy} contributes exactly half of the total flexibility. In general, the contribution from γ_{xy} is equal to twice the geometric mean of the contributions from ϵ_x and ϵ_y, and except for the range $0.5 < l/b < 2$ its contribution is intermediate between those from ϵ_x and ϵ_y.

The case $l/b = 0$ corresponds formally to an axially symmetric pressure distribution which varies sinusoidally with length: see problem 7.1. According to (7.11) the compliance is then equal to a^2/Et; which may easily be checked, since the pressure-loading is carried by hoop stress which is constant round any given circumference.

In general, the compliance does not depend on the *size* of the panels, but only on the *aspect ratio* l/b. When $l/b = 1$ the compliance is four times as much as the axisymmetric value, and it rises steeply for larger values of l/b. Panels with large values of l/b are highly compliant because the shell is obliged to carry the applied load 'inefficiently': the amplitude of N_x is greater than that of N_y and the effect of large circumferential variations in ϵ_x gives large displacements w by virtue of the compatibility equation (6.10). Note that the consequences of a high aspect ratio in the axial direction (l/b large) are quite different from those of a high aspect ratio in the circumferential direction (l/b small). This difference in behaviour is a consequence of the cylindrical form of the surface.

7.3.2 Criteria for validity of membrane hypothesis

Let us now investigate the validity of the membrane hypothesis in the crude way described in section 7.2 by examining the magnitude of the fictitious 'bending' strain in relation to that of the 'stretching' strains. As described already, it will be convenient to define a parameter B by (7.1) in terms of κ_{max}, the peak change of curvature, and ϵ_{max}, the peak tensile strain. There

Fig. 7.2. The 'panel flexibility factor' F for a cylindrical shell stressed according to the membrane hypothesis receives contributions from longitudinal and circumferential stretching ϵ_x and ϵ_y respectively and from shearing strain γ_{xy}; the magnitudes of these contributions are shown as functions of the panel aspect ratio l/b on (a) linear and (b) logarithmic scales. In (b) the curved line represents the total and the straight lines show the three separate contributions.

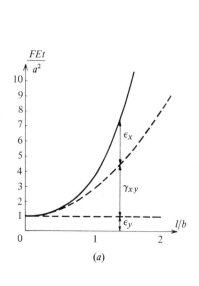
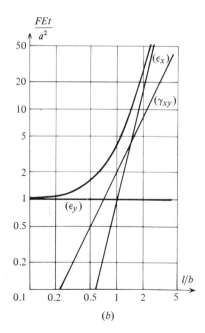

are two extreme cases to consider, depending on the value of l/b. When $l/b < 1$ the dominant surface strain is ϵ_y, as we have seen. Simplifying (6.9e) by omitting the second term on the right-hand side (which is a reasonable approximation in a basic, order-of-magnitude, calculation) we see that the dominant change of curvature is κ_x. Thus from (7.6) and (7.8) we have

$$B = \frac{\kappa_{max} t}{2\epsilon_{max}} = \frac{\pi^2 at}{2l^2}\left(1 + \left(\frac{l}{b}\right)^2\right)^2. \qquad (7.12)$$

On the other hand, when $l/b > 1$, ϵ_x and κ_y are the dominant surface strain and change of curvature, respectively. Unfortunately the calculation of κ_y is more complicated than that of κ_x, since the right-hand side of (6.9e) contains two terms in contrast to the single term of (6.9d). For present purposes, however, it is satisfactory to assume that the value of n is sufficiently high for the second term to be negligible; and in this way we find that (7.12) emerges as an approximate relation for the entire range of l/b. Thus, taking the very crude view that the membrane hypothesis should be reasonably satisfactory provided $B < 1$, we obtain a simple expression for the limiting value of t which is valid provided $n = \pi a/b > 3$, say:

$$\frac{at}{l^2}\left(1 + \frac{l}{b}\right)^2 < 0.2. \qquad (7.13)$$

Note that dimensions a, b and l are all involved, and that in contrast to expression (7.11) for F (with which (7.13) has some similarity) the formula cannot be expressed in terms of the ratio l/b.

When $l/b \ll 1$ (7.13) may be simplified to

$$at/l^2 < 0.2. \qquad (7.14)$$

This 'short panel' criterion (7.14) applies in particular to axisymmetric pressure-loading varying sinusoidally in the axial direction. We investigated this simple problem in the context of the 'bending' theory of shells in section 3.7; and we found that the proportions of the applied load carried by hoop stretching and longitudinal bending depended on a parameter of the form $l/(ah)^{\frac{1}{2}}$: see fig. 3.8. For large values of this parameter the hoop stretching carries most of the load – or in present terminology the membrane hypothesis is justified – and the hoop stretching carries more than half of the applied load provided (cf. section 3.7)

$$l/(at)^{\frac{1}{2}} > 1.8. \qquad (7.15)$$

Thus our present analysis certainly produces the governing dimensionless group discovered previously, and indicates the same broad conclusion.

7.3 Cylindrical shell with doubly-periodic pressure

On the other hand when $l/b \gg 1$ formula (7.13) simplifies to a 'long panel' version:

$$l^2 at/b^4 < 0.2. \tag{7.16}$$

This introduces a new dimensionless group which we shall find to be important in subsequent analysis. Rewriting (7.16) in terms of n we have, approximately,

$$n^2 < 4(a/l)(a/t)^{\frac{1}{2}} \tag{7.17}$$

as a condition for the validity of the membrane hypothesis. Relations of this sort will appear on many occasions in the following chapters.

Part of our motivation for studying a doubly sinusoidal distribution of pressure in the present example is that it leads directly to the study of more general pressure-loadings by means of Fourier series. If we can express the applied loading in this way it is an almost trivial operation to use the appropriate 'panel flexibility factor' for each term and sum the resulting displacements w. This calculation depends, of course, on the validity of the membrane hypothesis. It may be seen, by an examination of (7.14) and (7.17), that while the hypothesis may be valid for the lower terms in a Fourier-series representation, the terms with sufficiently high wavenumbers in either the circumferential or longitudinal directions will always involve violation of the membrane hypothesis.

We shall use Fourier series in this way in chapter 8, after we have done a 'full' analysis, complete with bending effects. It will then be instructive to make some comparisons with Fourier analyses based entirely on the membrane hypothesis.

7.3.3 Periodic tangential loading

It is straightforward to investigate the response of a long cylindrical shell to doubly sinusoidal *tangential* surface tractions in the circumferential direction by application of the same procedures. First, the equilibrium equations (7.4) are solved with the appropriate right-hand side; then the strains are found by means of Hooke's law, and finally the displacements are computed. A loading of the kind

$$q_\theta = q_0 \sin(\pi x/l) \cos(\pi y/b) \tag{7.18}$$

gives a particularly straightforward result, as may be seen from the following argument. Consider the equilibrium of a cylindrical shell under the simultaneous loading

$$p = -p_0 \sin(\pi x/l) \sin(\pi y/b), \quad q_\theta = p_0(\pi a/b) \sin(\pi x/l) \cos(\pi y/b). \tag{7.19}$$

198   *Displacements according to the membrane hypothesis*

The equilibrium equations are satisfied by

$$N_\theta = pa, \quad N_{x\theta} = N_x = 0; \tag{7.20}$$

in other words the radial and tangential loadings are self-balancing with the help of hoop tension alone, just as in the case of the circular 'string' studied in section 4.2. It follows that if we superpose the loading (7.19) onto the pressure-loading case studied earlier, we find that the stresses in the shell for the tangential loading

$$q_\theta = np_0 \sin(\pi x/l) \cos(\pi y/b) \tag{7.21}$$

(where $n = \pi a/b$ is the circumferential wavenumber) are precisely the same as for the normal pressure-loading

$$p = p_0 \sin(\pi x/l) \sin(\pi y/b)$$

(see (7.2)) except for a difference in the hoop-stress resultant N_θ. The significance of this is that in cases where the amplitude of N_θ is small in comparison with that of N_x and $N_{x\theta}$ - i.e. when $l/b > 2$, say - the strains and therefore the displacements in the shell will be practically the same for the two kinds of loading. In this restricted sense we may regard the two loading cases (7.20) and (7.21) as 'equivalent'.

7.3.4   *A different set of boundary conditions*

So far we have regarded the cylindrical shell as being indefinitely long. Let us now investigate a problem concerning a finite-length shell loaded according to (7.2) by 'cutting out' a section lying between cross-sections $x = 0$ and $x = l$. All of the previous results (7.5)-(7.12) continue to apply; and in particular they hold at the two ends of the shell. In other words, if we retain the previous solution we must accept whatever boundary conditions come with it. By inspection of (7.8)-(7.10) we find displacement boundary conditions $w = v = 0$ on both edges; and that $u \neq 0$. Thus the solution corresponds to a cylindrical shell whose ends are supported so that they remain *circular* but are not required to remain *plane*. Similarly, we find the force boundary conditions from (7.5): $N_x = 0$ but $N_{xy} \neq 0$ at both edges. The shear forces N_{xy} at the edges are required to prevent circumferential displacement, and the absence of axial forces N_x corresponds to the freedom of the shell to displace in the axial direction. Thus the force and displacement boundary conditions are self-consistent; and indeed they correspond closely to the closure of the ends of a cylindrical shell by plane diaphragms which are inextensional in their own plane but are practically free to distort out-of-plane.

Now suppose that we wish to apply the membrane hypothesis to a cylindrical shell of length l which is loaded according to (7.2) (we may plan to use

7.3 Cylindrical shell with doubly-periodic pressure

a Fourier series to represent a more general loading) but with its edges attached to massive abutments so that they remain not only *circular* but also *plane*.

Our previous solution is no longer appropriate, and since in particular it was a statically determinate one, we conclude that our new problem is *statically indeterminate*. Now in solving problems which involve statically indeterminate beams and frameworks we normally proceed by rendering the structure statically determinate by means of cuts and 'redundant' forces: the values of the redundant forces are then chosen so that corresponding displacement conditions are satisfied.

An analogous operation seems obvious in the present problem. Since our previous solution involved u-displacements at the edges proportional to $\sin(\pi y/b)$, we shall superimpose onto it a second solution in which the same shell has zero pressure-loading but supports self-equilibrating axial loads

$$N_x = N_0 \sin(\pi y/b) \tag{7.22}$$

applied at the two edges; and N_0 is a constant whose value is to be fixed so that the u-displacements at the edges in the previous solution are cancelled.

The general procedure for solving this second problem is exactly the same as before. The first step is to solve equilibrium equations (7.4) subject to the new loading conditions. As $p = 0$ the solution is trivial: $N_y = N_{xy} = 0$, and N_x is given throughout by (7.22). On substituting into Hooke's law we obtain

$$\epsilon_x = (N_0/Et) \sin(\pi y/b), \tag{7.23a}$$

$$\epsilon_y = (-\nu N_0/Et) \sin(\pi y/b), \tag{7.23b}$$

$$\gamma_{xy} = 0. \tag{7.23c}$$

The governing equation for w is obtained by substitution into (6.10):

$$\frac{\partial^2 w}{\partial x^2} = -\frac{\pi^2 a N_0}{b^2 Et} \sin\left(\frac{\pi y}{b}\right). \tag{7.24}$$

Hence, using the boundary conditions $w = 0$ at $x = 0$ and $x = l$ we obtain

$$w = \frac{\pi^2 a N_0}{2 b^2 Et} x(l - x) \sin\left(\frac{\pi y}{b}\right). \tag{7.25}$$

Finally, we substitute this into (6.9b) and (6.9c) and integrate to obtain expressions for v and u:

$$v = \frac{b N_0}{\pi Et} \left[\nu + \frac{\pi^2}{2 b^2} x(l - x) \right] \cos\left(\frac{\pi y}{b}\right), \tag{7.26}$$

$$u = \frac{N_0}{Et} \left(x - \tfrac{1}{2} l\right) \sin\left(\frac{\pi y}{b}\right). \tag{7.27}$$

Here we have fixed the constants of integration so that there is no rigid-body rotation of the shell about its axis, and $u = 0$ on the central plane $x = \frac{1}{2}l$, since these properties are possessed by the original solution.

Note that w and v vary *parabolically* with x. It is instructive to examine the case $n = 1$; here $b = \pi a$, and thus $\cos(\pi y/b) = \cos\theta$: see problem 7.2.

Returning to the main problem, we find N_0 by setting to zero the sum of the two expressions for u at $x = 0$ (or $x = l$):

$$-\frac{ap_0}{Et}\left(\frac{l}{\pi}\right)\left(\frac{l^2}{b^2} - \nu\right) - \frac{1}{2}l\,\frac{N_0}{Et} = 0;$$

hence

$$N_0 = -\frac{2ap_0}{\pi}\left(\frac{l^2}{b^2} - \nu\right). \tag{7.28}$$

Note that N_x in the second solution is of the same order as N_x in the first solution (see (7.5c)).

When the second solution, with the proper value of N_0, is added to the first, we obtain in particular the following expression for the radial component of displacement at the mid-section, $x = \frac{1}{2}l$:

$$w = \frac{a^2 p_0}{Et}\left[1 + 2\left(\frac{l}{b}\right)^2\left(1 + \tfrac{1}{8}\nu\pi\right) + \left(\frac{l}{b}\right)^4\left(1 - \tfrac{1}{4}\pi\right)\right]\sin\left(\frac{\pi y}{b}\right). \tag{7.29}$$

Observe that the addition of the second solution makes a substantial change to the term in $(l/b)^4$, reducing it to 0.21 of its value in the original solution (7.8). There is also a small increase (factor ≈ 1.1) in the coefficient of $(l/b)^2$. It follows that the 'flexibility factor' – now defined in terms of the pressure and radial displacement at the mid-plane $x = \frac{1}{2}l$ – is much reduced on account of axial restraint at the ends when l/b is large; but it is hardly affected when l/b is small.

These effects are not difficult to understand qualitatively. In the original solution (see (7.8)) the term $(l/b)^4$ represents the effect of ϵ_x while the other two terms represent the effect of γ_{xy} and ϵ_y respectively. The 'redundant' stress system (7.28) involves primarily strains ϵ_x, and these are the only ones to contribute to w, by virtue of the form of (6.10). Therefore, we expect the redundancy to affect only the term $(l/b)^4$. (Actually, this is not strictly true, as the u-deflection in the original solution (7.9) includes a term ν which derives from N_y.)

In fact, the effect of the second solution on the $(l/b)^4$ term is precisely the same as the effect of clamping the ends of a classical elastic beam carrying a sinusoidal load as shown in fig. 7.3: the parabolic form of (7.25) corresponds to the uniform bending of the beam under the action of equal redundant end-couples: see problem 7.3. We shall examine this analogy further in section 7.5; and we shall find in chapter 9 that it provides a useful interpretation of some simplifications to the 'full' equations of cylindrical shells which apply when l/b is large.

Returning to the present problem, we find that although the addition of the second part of the solution now gives boundary conditions $w = u = 0$ at both ends of the shell, there is also a nonzero displacement v at both edges:

$$v = -\frac{2}{\pi^2} \frac{abp_0}{Et} \nu \left(\frac{l^2}{b^2} - \nu\right) \cos\left(\frac{\pi y}{b}\right). \tag{7.30}$$

This is a consequence of the nonzero circumferential strain ϵ_y in the second solution, on account of the Poisson ratio effect.

Thus our solution illustrates the effect that a complete edge fixity, i.e. $u = v = w = 0$, inhibits circumferential strain which may be required at the edge by Hooke's law according to the membrane hypothesis. We shall return to this point in section 8.7.3.

7.4 Use of the Airy stress function in the calculation of deflections in a cylindrical shell

In this section we introduce the Airy stress function as an aid to the calculation of displacements in cylindrical shells which are stressed according to the membrane hypothesis. We illustrate the method in relation to a cylindrical shell which is free at one end and clamped at the other, under the action of a periodic shear load which is applied to the free edge, as shown in fig. 7.4b. We have already solved a problem of this type in section 4.4.1 in a relatively straightforward manner, but we shall find that the use of the Airy stress function produces a more 'tidy' calculation. In fact, this stress function will play an important part in more general analysis of shell structures, as we shall see in chapter 8.

The illustrative problem is described in fig. 7.4. The shell and its coordinate system are shown in fig. 7.4a. The axial coordinate is x and the planes $x = 0$ and $x = L$ correspond to the free and clamped edges of the shell, respectively.

Fig. 7.3. A clamped beam carrying a sinusoidal distribution of transverse load is formally analogous to the shell problem described in section 7.3.4.

For the circumferential coordinate it will be most convenient to use the angle θ (fig. 7.4a); but it will sometimes be advantageous to use y, as in the preceding section, where $y = a\theta$.

The shell is loaded at the edge $x = 0$ by a periodic tangential shearing force

$$N_{x\theta} = nQ_n \cos n\theta, \tag{7.31}$$

as shown in fig. 7.4b. The motivation behind this problem (which accounts for the use of nQ_n as the amplitude of $N_{x\theta}$) is that this loading would be applied to the shell proper by an 'edge-ring' which is itself loaded by a normal shearing force

$$Q_{xz} = Q_n \sin n\theta \tag{7.32}$$

as shown in fig. 7.4c: see section 4.4.1 and problem 4.9. And indeed, by assuming the presence of a suitable inextensional 'edge-ring' we may use the methods of the present chapter in order to analyse the displacement of a shell loaded by normal shearing force (7.32) at the free edge.

As we shall discover, the radial component of displacement at the loaded edge takes the form

$$w = w_n \sin n\theta, \tag{7.33}$$

and it is therefore convenient to define an edge-flexibility factor F_n as follows:

$$F_n = |w_n/Q_n| = -w_n/Q_n. \tag{7.34}$$

Fig. 7.4. (a) Dimensions and coordinate system for a cylindrical shell which is free at one end and clamped at the other. (b) Detail of periodic tangential shear loading $N_{x\theta} = nQ_n \cos n\theta$ at the free edge. (c) Equivalent normal shear loading, provided there is a suitable edge-ring: cf. fig. 4.9.

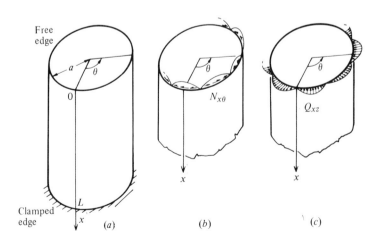

7.4 Use of the Airy function in the calculation of deflections

F_n is essentially positive, by definition: hence the negative sign on the right-hand side, since a loading of the sense of fig. 7.4c gives a radial displacement of negative amplitude. (Note, however, that the opposite would have been the case if the loading had been applied at the other end of the shell: this potential ambiguity is a consequence of using a *stress*-like quantity to describe a *force* loading (see appendix 5).) We expect F_n to be a function of the dimensions of the shell, the elastic constants of the material and the circumferential wavenumber n.

We now introduce the *Airy stress function* as a way of satisfying automatically two of the three simultaneous equations of equilibrium according to the membrane hypothesis. These equations are (see section 4.4)

$$\frac{N_y}{a} = p, \tag{7.35a}$$

$$\frac{\partial N_x}{\partial x} + \frac{\partial N_{xy}}{\partial y} = -q_x, \tag{7.35b}$$

$$\frac{\partial N_y}{\partial y} + \frac{\partial N_{xy}}{\partial x} = -q_y, \tag{7.35c}$$

where p is a surface load per unit area acting normal to the surface (an outward-directed pressure) and q_x, q_y are tangential surface tractions per unit area acting in the x- and y-directions respectively. Now *when the tangential loadings q_x, q_y are both zero* (7.35b) and (7.35c) are satisfied identically by stress resultants derived as follows from the Airy stress function ϕ:

$$N_x = \frac{\partial^2 \phi}{\partial y^2}, \quad N_y = \frac{\partial^2 \phi}{\partial x^2}, \quad N_{xy} = -\frac{\partial^2 \phi}{\partial x \partial y}. \tag{7.36}$$

(It is possible to adapt the Airy stress function to certain cases of tangential loading, but the result is somewhat cumbersome: see problem 7.4.) Consequently there remains the single equilibrium equation (7.35a), which may now be written

$$\frac{1}{a} \frac{\partial^2 \phi}{\partial x^2} = p. \tag{7.37}$$

In the present problem the only loads acting on the shell are applied at the edges, so $p = 0$ and the equilibrium equation is particularly simple:

$$\frac{\partial^2 \phi}{\partial x^2} = 0. \tag{7.38}$$

The next step is to express the change in Gaussian curvature, as computed from surface strains, in terms of ϕ by using (7.36) and Hooke's law (2.5). We find, by direct substitution that

$$g = -\frac{\partial^2 \epsilon_y}{\partial x^2} + \frac{\partial^2 \gamma_{xy}}{\partial x \partial y} - \frac{\partial^2 \epsilon_x}{\partial y^2} = -\frac{1}{Et} \nabla^4 \phi. \tag{7.39}$$

Here, $\nabla^4(\ldots) = \dfrac{\partial^4(\ldots)}{\partial x^4} + 2\dfrac{\partial^4(\ldots)}{\partial x^2 \partial y^2} + \dfrac{\partial^4(\ldots)}{\partial y^4}$ (7.40)

is the well-known bi-harmonic operator. Note in particular that Poisson's ratio does not appear in this expression: compare also (7.7).

The Airy stress function was first invented for the solution of elasticity problems under the conditions of plane stress. Now in *plane* problems there is no possibility of a change in Gaussian curvature; and we thus recover from (7.39) the well-known governing equation for certain problems of plane stress:

$$\nabla^4 \phi = 0. \tag{7.41}$$

In problems of this sort the distribution of stress is generally *statically indeterminate*, as is well known: the distribution of stress depends on the solution of (7.41) subject to appropriate boundary conditions.

Our present problem, concerning a cylindrical shell, is quite different. The stress resultants are *statically determinate* (at least, for the given boundary conditions) and the quantity $\nabla^4 \phi$ is a function which can be *evaluated* as soon as the equilibrium equations have been solved.

The sharp distinction between the role of $\nabla^4 \phi$ in the membrane-hypothesis theory of shells on the one hand and the elastic theory of plane stress on the other becomes somewhat blurred if we try to think of a portion of a rather *shallow* cylindrical shell, whose radius is very large. The paradox is resolved by observing that in these circumstances, the membrane hypothesis becomes untenable; and so it is necessary to work with the 'full' shell equations. Problems of this sort will be explored in chapter 8.

Returning to the present example we may readily express the change in Gaussian curvature in terms of the normal component of displacement:

$$g = -\frac{1}{a}\frac{\partial^2 w}{\partial x^2}. \tag{7.42}$$

Thus, from (7.39) and (7.42) we may write

$$\frac{1}{a}\frac{\partial^2 w}{\partial x^2} = \frac{1}{Et} \nabla^4 \phi. \tag{7.43}$$

Equations (7.37) and (7.43) express the problem of finding the displacements of the surface according to the membrane hypothesis. Our plan will be to solve the equilibrium equation (7.38) first, subject to appropriate force

7.4 Use of the Airy function in the calculation of deflections

boundary conditions, next to substitute the known function ϕ into (7.43), and then to solve for w subject to appropriate displacement boundary conditions.

The boundary conditions on force are

$$N_x = 0, \quad N_{x\theta} = nQ_n \cos n\theta \quad \text{at } x = 0. \tag{7.44}$$

Thus we have, from (7.36)

$$\frac{\partial^2 \phi}{\partial y^2} = 0, \quad \frac{\partial^2 \phi}{\partial x \partial y} = -nQ_n \cos n\theta \quad \text{at } x = 0. \tag{7.45}$$

The boundary conditions on displacement are

$$w = u = 0 \quad \text{at } x = L. \tag{7.46}$$

In order to express the second displacement condition in terms of w, we eliminate v between (6.9b) and (6.9c) to give a general relation

$$\frac{1}{a} \frac{\partial w}{\partial x} = \frac{\partial^2 u}{\partial y^2} + \frac{\partial \epsilon_y}{\partial x} - \frac{\partial \gamma_{xy}}{\partial y}. \tag{7.47}$$

At the boundary $x = L$ the first term on the right-hand side is zero since $u = 0$, and the second and third can be expressed in terms of ϕ, by means of (7.36) and Hooke's law.

It is clear by inspection that both ϕ and w will vary as $\sin n\theta$ in the circumferential direction. Consequently we can simplify the equations by putting

$$\phi = h(x) \sin n\theta, \quad w = f(x) \sin n\theta. \tag{7.48}$$

Thus the general equilibrium equation (7.38) becomes

$$h''(x) = 0, \tag{7.49}$$

where $'$ denotes differentiation with respect to x; and the boundary conditions on force become

$$h(0) = 0; \quad h'(0) = -aQ_n. \tag{7.50}$$

The solution of the equilibrium equation is therefore

$$h(x) = -aQ_n x; \tag{7.51}$$

and it is easy to verify that this agrees precisely with the solution given to example (6) in section 4.4.1, when allowance is made for the different origins of θ with respect to the loading in the two cases.

Equation (7.43) now becomes

$$f''(x) = \frac{-n^4 Q_n}{Eta^2} x, \tag{7.52}$$

while boundary conditions (7.46) transform to

$$f(L) = 0, \quad f'(L) = \frac{(2+\nu)n^2 Q_n}{Et}. \tag{7.53}$$

The solution for w is thus given by

$$f(x) = -\frac{n^4 Q_n}{6Eta^2}(x^3 - 3xL^2 + 2L^3) + \frac{(2+\nu)n^2 Q_n(x-L)}{Et}. \quad (7.54)$$

In particular, on putting $x = 0$ we find that the flexibility factor F_n may be written

$$F_n = \frac{n^4 L^3}{3Eta^2} + \frac{(2+\nu)n^2 L}{Et}. \quad (7.55)$$

It is not difficult to trace the second term on the right-hand side to the effect of shear distortion $\gamma_{x\theta}$ throughout the shell, together with a relatively small contribution from the circumferential component of displacement at the base: see problem 7.5. On the other hand, the first term on the right-hand side comes from the strain component ϵ_x throughout the shell. Formula (7.55) is thus somewhat analogous to expression (7.11) for the flexibility of a 'panel', as investigated in section 7.3.1, insofar as the different components of strain contribute different terms to the overall flexibility factor.

It is clear that in some circumstances F_n will be dominated by the first term on the right-hand side of (7.55). This requires

$$n^2 \gg 6(a/L)^2$$

or

$$n \gg 2.5(a/L). \quad (7.56)$$

This condition may be interpreted in terms of the proportions of a 'panel' whose length and breadth are, respectively, the length of the shell and $(1/2n)$ times the circumference of the shell.

It is of interest to note that if the loading at the edge of the shell had been specified in terms of $N_{x\theta}$ instead of Q_{xz}, it would then have been appropriate to define a *peripheral* flexibility factor in terms of the applied load and the peripheral displacement v. Let us therefore investigate the component v of the displacement at $x = 0$. Now at this edge both N_x and N_y are zero, according to the equilibrium solution; consequently, by Hooke's law $\epsilon_y = 0$ also (and in this sense the inextensional ring attached to the edge is compatible). Thus the component v of the displacement may be found from (7.33) by solving the equation

$$\frac{\partial v}{\partial \theta} = -w \quad (7.57)$$

at the edge. This gives, together with the condition that there is no rotation of the shell as a whole about its axis,

$$v = \frac{w_n}{n}\cos n\theta. \quad (7.58)$$

We now define a peripheral flexibility factor as follows:

$$F_n^p = \frac{-v}{N_{xy}} = \frac{-w_n/n}{nQ_n} = \frac{F_n}{n^2} . \qquad (7.59)$$

Thus we see that the peripheral flexibility factor is just $(1/n^2)$ times the radial flexibility factor. The fact that the amplitude of the $N_{x\theta}$ edge-loading is n times the amplitude of the Q_{xz} loading (see (7.31)) is related to the fact that the amplitude of the v displacement is $1/n$ times the amplitude of the w-displacement in the inextensional deformation of a ring by reason of the principle of virtual work: see problem 7.6.

Using (7.55) we may write

$$F_n^p = \frac{n^2 L^3}{3Eta^2} + \frac{L}{Et}(2+\nu). \qquad (7.60)$$

Bearing in mind remarks made in section 7.3.4 about a certain 'beam-like' quality of the membrane displacement of cylindrical shells, we may note a similarity between the first term on the right-hand side of (7.60) and the classical formula $\delta = PL^3/3EI$ for the deflection δ of the tip of an elastic cantilever due to a transverse point load P at the tip: the similarity certainly runs to the term L^3, and the factor 3 on the denominator. Indeed the second term on the right-hand side of (7.60) suggests a sort of 'shear deflection' effect. We shall return to this point in section 9.5.

In terms of engineering applications our main interest is in the 'radial' flexibility factor F_n. Henceforth we shall deal with this almost exclusively.

In subsequent work on cylindrical shells we shall drop the second term on the right-hand side of (7.55), in the interests of simplicity: in any given case it is easy to make a check on the magnitude of this term, and to insert it if appropriate. Our flexibility formula thus reduces to

$$F_n = \frac{n^4 L^3}{3Eta^2} . \qquad (7.61)$$

It is not surprising that the flexibility factor increases steeply with n, since we are dealing with effects which are similar to those discussed in section 7.3.1: there we found a flexibility factor proportional to $n^4 l^4$ (if we replace b in (7.11) by $\pi a/n$ and put $l/b \gg 1$), and the different power of l reflects the difference between line- and pressure-loading. It is also obvious why the flexibility factor is inversely proportional to Et: this is a direct consequence of our working with surface strains derived by the membrane hypothesis.

We must not, of course, expect the flexibility of an actual shell to increase indefinitely with n, since large flexibility implies high displacements and consequently large changes in curvature, which in turn invalidate the membrane hypothesis. In chapter 8 we shall investigate the same problem by solving the full equations, including bending terms in the equilibrium equations: in particular we shall see in what circumstances the membrane hypothesis gives satisfactory results. But it is worthwhile at this point to make a first investigation of the limits of validity of the membrane hypothesis by evaluating the parameter B as defined in (7.1), and setting its value to unity.

In the present problem the largest surface strain is in the x-direction at the base of the shell, and the largest change of curvature is at the edge $x = 0$ in the circumferential direction. It is straightforward to show that, for sufficiently large values of n,

$$B = \tfrac{1}{6} n^4 (tL^2/a^3).$$

Consequently, setting $B = 1$ we find that the membrane hypothesis is valid provided

$$n^2 < 2.2(a/L)(a/t)^{\frac{1}{2}}. \tag{7.62}$$

This is precisely the same kind of limit as (7.17), except of course that here L is the length of the shell whereas in (7.17) l is the longitudinal half wavelength of a doubly-periodic pressure-loading. In view of the order-of-magnitude nature of the calculations, the difference in numerical constants between the two expressions is hardly significant.

Take, for example, a shell having $L/a = 4$ and $a/t = 300$. We find that (7.62) gives $n < 3$. Up to about this value of n we may expect formula (7.61) to be reasonably accurate. As we shall see in chapter 8, the behaviour for larger values of n is dominated by bending effects in the circumferential direction, and the flexibility actually *decreases* for higher values of n.

7.5   A 'beam analogy' for 'quasi-inextensional' deformation of cylindrical shells

At various places in this chapter we have touched on apparent analogies between the behaviour of a cylindrical shell having stress resultants which are periodic in the circumferential direction, and the displacement of a simple beam. It is useful to investigate this further, because the analogy gives some interesting insights into the behaviour of cylindrical shells; and it also leads to some powerful approximate methods, which we shall develop in chapter 8, for taking into account the bending of cylindrical shells.

The most straightforward way of establishing the analogy is to set up a compatibility relation between the circumferential component v of displacement and the strains in the surface. By eliminating w and u between the strain–

7.5. 'Beam analogy' for deformation of cylindrical shells

displacement equations (6.9) for a cylindrical shell we obtain the equation

$$\frac{\partial^2 v}{\partial x^2} = -\frac{\partial \epsilon_x}{\partial y} + \frac{\partial \gamma_{xy}}{\partial x}. \tag{7.63}$$

This equation may be seen to describe a 'beam-like' effect along an arbitrary generator of the shell. The term on the left-hand side describes a change in curvature of a generator acting as a sort of beam in the tangent plane. The first term on the right-hand side can be directly associated with a change in curvature of a thin longitudinal strip in which the generator is embedded, since $\partial \epsilon_x / \partial y$ represents a relative rotation of lines originally normal to the generator, cf. (2.14). Finally, the term $\partial \gamma_{xy}/\partial x$ corresponds to an in-plane shear distortion of the strip. This equation is the basis of the analogy which we have perceived earlier.

We can make the analogy more concrete by dealing with a simple periodic variation of v, ϵ_x and γ_{xy} around the circumference. On the basis of previous examples we put

$$v = v_0 \cos n\theta, \quad \epsilon_x = \epsilon_0 \sin n\theta, \quad \gamma_{xy} = \gamma_0 \cos n\theta, \tag{7.64}$$

where, as before, $y = a\theta$.

Substituting in (7.63) and dividing throughout by $\cos n\theta$ we have

$$v_0'' = -(n/a)\epsilon_0 + \gamma_0', \tag{7.65}$$

where ' denotes differentiation with respect to x. Clearly we shall obtain an analogy with a 'classical' beam only by dropping the γ_0' term.

Now although the basis of a beam analogy lies in the circumferential component v of the displacement, it is more convenient in practice to have equations in which the *radial* displacement w is the variable. The reasons for this choice will become clear later on. We have already seen that there is a simple connection between w and v when the circumference of the shell is inextensional; and moreover if we seek a simple analysis it will be advisable to neglect the shearing strains γ_{xy}. Together these considerations suggest a simplified form of analysis in which we take

$$\epsilon_y = \gamma_{xy} = 0 \tag{7.66}$$

and allow only the component ϵ_x of strain to have nonzero value. We shall call this a *quasi-inextensional* theory, since we have suppressed all but one component of strain. Of course, there may be circumstances in which this kind of simplification is most unsuitable. But there are in fact many situations in which this quasi-inextensional approach is very satisfactory. The crux of the matter is that in the compatibility equation (6.10) it often happens that the right-hand side is dominated by the last term. It is easy to check if this is true in a particular case by taking the stress resultants from an equilibrium analysis and using Hooke's law. It is also possible to give more general criteria in certain circumstances: see problem 7.7. In cases where the last term *does* dominate, the quasi-inextensional approach is justified.

It is convenient to set up the equations of the 'quasi-inextensional' version of cylindrical shell theory for a general case in which the shell is loaded by normal and tangential surface tractions of the following form:

$$p = p^*(x) \sin n\theta, \quad q_y = q^*(x) \cos n\theta. \tag{7.67}$$

Here p^* and q^* are arbitrary functions of x. As we shall see, these two kinds of surface traction act together to give stress resultants and displacements of the same form.

When we solve the equilibrium equations (7.35) we find that the stress resultants are periodic in the circumferential direction, so it is convenient to put

$$N_x = N_x^*(x) \sin n\theta, \quad N_y = N_y^*(x) \sin n\theta, \quad N_{xy} = N_{xy}^*(x) \cos n\theta. \tag{7.68}$$

On substituting into the equilibrium equations we obtain:

$$N_y^* = ap^*, \quad N_{xy}^{*\prime} = -n(p^* + q^*/n), \quad N_x^{*\prime} = nN_{xy}^*/a, \tag{7.69}$$

where $'$ denotes differentiation with respect to x. Similarly, the strain and displacement variables are all periodic in the circumferential direction, and it is convenient to put

$$u = u^*(x) \sin n\theta, \quad v = v^*(x) \cos n\theta, \quad w = w^*(x) \sin n\theta,$$
$$\epsilon_x = \epsilon_x^*(x) \sin n\theta, \gamma_{xy} = 0, \qquad \epsilon_y = 0. \tag{7.70}$$

Substituting these into (6.9), (6.10) and (7.66) we obtain

$$v^* = w^*/n, \tag{7.71a}$$
$$v^{*\prime} = -nu^*/a, \tag{7.71b}$$
$$u^{*\prime} = \epsilon^*, \tag{7.71c}$$
$$w^{*\prime\prime} = -n^2 \epsilon^*/a. \tag{7.72}$$

Lastly, we have a simplified form of Hooke's law (2.5)

$$\epsilon_x^* = N_x^*/Et. \tag{7.73}$$

The reason for omitting Poisson's ratio ν here is as follows. As we have seen in section 7.3.1, the 'panel flexibility' for a doubly-periodic pressure-loading is independent of the value of ν for the material. Therefore it is reasonable to choose a particularly convenient value of ν for the purposes of a simplified, approximate analysis.

Equations (7.69), (7.72), (7.73), now constitute the mathematical problem. In fact, they are directly analogous to the equations of a uniform classical elastic beam. The beam variables are shown in fig. 7.5 in their positive sense:

7.5 'Beam analogy' for deformation of cylindrical shells

they are displacement w, rotation ψ $(=-w')$, change of curvature κ $(=-w'')$; bending moment m, shearing force F $(=m')$, and transverse load intensity η $(=-F')$. Bending moment is related to change in curvature by the flexural rigidity EI: $m = EI\kappa$.

The analogous quantities for the quasi-inextensional cylindrical shell are set out in table 7.1. Starting with the common axial coordinate x and the direct correlation between w^* for the shell and w for the beam, we find the quantities corresponding to ψ and κ by using (7.71b) and (7.72). Similarly, starting with the correspondence between beam load η and the corresponding shell loading p^*, q^*, we find quantities analogous to F and m by using (7.69).

Table 7.1. Equivalence between classical beam variables and quasi-inextensional shell variables (cf. fig. 7.5).

Classical beam variable	Equivalent variable for quasi-inextensional shell
x	x
w	$w^* = nv^*$
ψ	$n^2 u^*/a$
κ	$n^2 \epsilon_x^*/a$
η	$\pi a(p^* + q^*/n)$
F	$(\pi a/n)N_{xy}^*$
m	$(\pi a^2/n^2)N_x^*$
EI	$E\pi a^3 t/n^4$
P	$\pi a(\bar{p} + \bar{q}/n)$

Note: The shell variables are defined explicitly in the text as amplitudes of periodic functions $\sin n\theta$, $\cos n\theta$. If these are interchanged, some changes of sign occur, e.g. in the shell variables analogous to η and P.

Fig. 7.5. Load variables (solid-head arrows) and displacement variables for a beam formally analogous to a cylindrical shell which behaves according to the membrane hypothesis: see table 7.1.

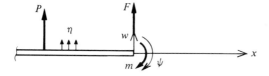

212   *Displacements according to the membrane hypothesis*

The factor πa in the quantity analogous to η gives it the proper dimensions of force per unit length, and also produces the tabulated equivalent to EI: the factor π ensures that when $n = 1$ we obtain the actual flexural rigidity for pure bending of the tube, which is convenient.

It is sometimes useful to be able to deal with radial and tangential line loads of intensity $\bar{p} \sin n\theta$ and $\bar{q} \cos n\theta$ per unit length applied at a particular cross-section of the shell. The positive sense of $\bar{p}$ is radially outwards, and the positive sense of $\bar{q}$ is in the direction of increasing θ. These together correspond to a transverse point load P applied to the equivalent beam, and the correspondence recorded in table 7.1 is deduced from that of η by a simple limiting process. Observe that κ and m are directly related to ϵ_x and N_x, respectively, which is a simple consequence of the form of (7.72) and (7.73). The radial and tangential circumferential loading combine into a single load variable, and there is a corresponding equivalence between the radial and circumferential components of displacement.

For large values of the circumferential wavenumber n the flexural rigidity is small, corresponding to large flexibility of a kind which we have seen already.

It should be emphasised that the whole of the present analysis is based on the membrane hypothesis. It applies only to problems in which all variables are periodic in the circumferential direction, and in which the displacements are due mainly to strains in the axial direction. In the present kind of problem we find that the changes of curvature of the shell are largest in the circumferential direction, and our parameter B (equation 7.1) is equal to $(n^2 - 1)w^*t/2a^2\epsilon^*$. The corresponding quantity for the analogous beam is $n^2(n^2 - 1)wt/2a^3\kappa$, which may be rewritten as

$$B = \left(\frac{n^2 - 1}{n^2}\right) \frac{\pi E t^2 w_{max}}{2 m_{max}}. \tag{7.74}$$

The factor $(n^2 - 1)$ comes from the correct form of the curvature–displacement relation (6.9e). Thus when $n = 1$ the formula indicates zero circumferential change of curvature; which is correct, since the cross-section of the tube is then displaced as a rigid body: see problem 7.2.

The whole analysis is invalid when $n = 0$: this gives the case of axially symmetric loading, and our restriction of circumferential inextensibility renders our method inappropriate.

7.6   Discussion

Most of the results of the present chapter in connection with cylindrical shells may be obtained to a close approximation by the use of this beam

7.6 Discussion

analogy. It will be sufficient to consider the problem illustrated in fig. 7.4 in terms of a simple cantilever loaded by a transverse force at the tip.

First we consider the various different ways in which the loading at the edge of the shell may be specified. If the edge-loading is of the form

$$N_{xy} = N_0 \cos n\theta$$

it may be conceived in terms of a shearing force F imposed on the beam, and the appropriate value of F may be read from the table: $F = \pi a N_0/n$. Alternatively we may consider that there is a tangential *line* load $\bar{q} \cos n\theta$ applied to the shell a short distance from the free edge, with $\bar{p} = 0$, where $\bar{q} = N_0$. According to the table this corresponds to a transverse load P applied to the beam a short distance from the edge, whose value is precisely the same as the value of F already determined. Furthermore, according to the table the same load P on the beam could be considered as being equivalent to a radial line load $\bar{p} \sin n\theta$, with $\bar{q} = 0$, where $\bar{p} = N_0/n$. And indeed, if this line load were applied at the edge itself, we could describe it as a transverse shear loading $Q_{xz} = (N_0/n) \sin n\theta$. Thus we see that a transverse force applied to the tip of the beam is analogous either to an edge traction

$$N_{xy} = N_0 \cos n\theta$$

applied to the shell, or to a transverse shear loading

$$Q_{xz} = \frac{N_0}{n} \sin n\theta;$$

or indeed to a linear combination of the two. This equivalence is directly attributable to the inextensibility of the shell in the circumferential direction which we assumed at the outset. In particular, the end 'ring' of the shell is capable of acting as the 'inextensional string' of section 4.2 in order to 'convert' an applied Q_{xz} loading into an equivalent N_{xy} loading.

By insisting on inextensibility in the circumferential direction we have thus blurred some of the detail surrounding boundary conditions and also the way in which loads are applied. We shall re-examine these points in chapters 8 and 9. Another example of the blurring of detail is provided at the clamped end of the shell. Here the conditions $w = u = 0$ imply also $v = 0$ according to this view, and lead directly to $w = w' = 0$ for the beam: i.e. the end of the beam is clamped, or encastré.

Returning to our particular problem, we observe first that the transverse displacement at the tip of the cantilever is equal to the transverse load multiplied by $L^3/3EI$. We have already determined the value of the load on the beam as $\pi a N_0/n$; and by reading from the table what shell quantities correspond to the beam stiffness EI we may write down the following expression for the amplitude of w at the loaded edge of the shell:

$$\left(\frac{\pi a N_0}{n}\right)\left(\frac{L^3}{3}\right)\left(\frac{n^4}{\pi a^3 t E}\right). \tag{7.75}$$

This agrees with the first term on the right-hand side of (7.55). The term corresponding to shear strain within the shell will obviously not be given by this quasi-inextensional method, which puts $\gamma_{x\theta} = 0$ explicitly in (7.66). There are in fact some problems in which the shearing strains play a dominant part in the deformation of the shell; and for these it is necessary to tackle the equations in the manner of section 7.3. Problems of this sort may usually be recognised from the fact that the value of the ratio l/b is smaller than, say, 1. This can occur, for example, in vertical oil storage tanks for low values of n, since these tanks commonly have $l/a \approx 1/3$: see problem 7.7.

Similarly, expression (7.74), gives, for sufficiently large values of n, the same result (7.62) as before.

Problems 7.7–7.9 give some simple examples of the use of this method, and demonstrate boundary conditions for equivalent beams which correspond to various practical force and displacement boundary conditions for cylindrical shells.

7.7 Problems

7.1 Show that the panel flexibility factor F is given by (7.11) for a pressure-loading of the form

$$p = p_0 \sin(\pi x/l) \cos(\pi y/b) \quad \text{(cf. (7.2))};$$

and that in this case an axially symmetric loading $p = p_0 \sin(\pi x/l)$ is recovered formally by putting $b \to \infty$.

7.2 Consider a cylindrical shell subjected to axial loads (7.22) at the two edges in the case $b = \pi a$, i.e. $N_x = N_0 \sin\theta$. Verify that the displacements w and u given by (7.25) and (7.27) correspond precisely to those which would be obtained by regarding the tube as a simple beam under a pure bending moment. In particular, verify that the deformed cross-sections of the tube remain plane.

7.3 Consider the uniform elastic beam shown in fig. 7.3 under a transverse load whose intensity varies sinusoidally with length. Derive expressions for the corresponding transverse displacements (a) when the ends are simply-supported and (b) when the ends are clamped, as shown. In particular, show that the transverse displacements at the centre of the beam are in the ratio $1 : (1 - \frac{1}{4}\pi)$.

7.7 Problems

7.4 Consider (7.35b) and (7.35c), in the particular case in which both q_x and q_y are derived from a single scalar potential function $V(x, y)$ as follows:

$$q_x = -\frac{\partial V}{\partial x}; \quad q_y = -\frac{\partial V}{\partial y}.$$

Show that the two equilibrium equations are satisfied identically by means of the modified Airy stress function (cf. (7.36)):

$$N_x = \frac{\partial^2 \phi}{\partial y^2} + V, \quad N_y = \frac{\partial^2 \phi}{\partial x^2} + V, \quad N_{xy} = \frac{-\partial^2 \phi}{\partial x \partial y}.$$

7.5 Show that if the term $\partial \epsilon_y / \partial x$ is omitted from (7.47), the second term on the right-hand side of (7.55) becomes $2(1+\nu)n^2 L/Et$; and hence that it is the shearing distortion γ_{xy} in the shell which is mainly responsible for the similar term in (7.55).

7.6 Consider a circular 'string' in equilibrium under radial and tangential forces of intensity $Q_n \sin n\theta$ and $nQ_n \cos n\theta$ respectively, as in fig. 4.9b. Now let the string undergo a small inextensional displacement of the form $w = w_n \sin n\theta$, $v = (w_n/n) \cos n\theta$. Verify that the sum of the external virtual work (appendix 1) performed by the two sets of loading is zero when due account is taken of the sense of the tangential loading. (Note that the origin of θ is not the same as in fig. 4.9b.)

7.7 Analyse the displacements of the shell of problem 4.22 by the techniques of section 7.3, taking $w = 0$ at both ends and setting the mean values of u and v equal to zero at $x = 0$; also take $N_x = N_0 \sin n\theta$ at $x = 0$. In particular show that

$$w = \frac{n^2 l^2 N_0}{6Eat} \left[\left(\frac{x}{l}\right)^3 - 3\left(\frac{x}{l}\right)^2 + 2\left(\frac{x}{l}\right) \right] \sin n\theta;$$

and that at $x = 0$

$$u = -\frac{N_0 l}{3Et} \left\{ 1 + \frac{3(2+\nu)}{n^2} \left(\frac{a}{l}\right)^2 \right\} \sin n\theta.$$

Verify that the second term in { } would not be present according to the 'quasi-inextensional' scheme; and that this term has a relatively small effect provided $nl/a > 8$, i.e. $l/b > 3$, approx.

Find expressions for κ_θ ($\approx a^{-2} \partial^2 w/\partial \theta^2$) and ϵ_x ($\approx N_x/Et$, cf. problem 4.22): hence determine κ_{max} and ϵ_{max} and evaluate B according to (7.1). By putting $B = 1$ show that the membrane hypothesis is valid for this problem provided $n^2 < 6(a/l)(a/t)^{\frac{1}{2}}$, approximately: cf. (7.17).

(This problem is relevant to the stresses induced in a cylindrical oil tank when the base $x = 0$ 'settles' unevenly by $u = u_0 \sin n\theta$: the second equation enables N_0 to be determined. In this analysis the top of the tank is held strictly circular.)

7.8 Use the 'beam analogy' (section 7.5) to investigate the displacements of the cylindrical shell of fig. 4.8a under uniform loading according to the membrane hypothesis. Start by comparing the loads of (4.13) with the standard variables of (7.67), and choose a new origin for θ in fig. 4.8a.

7.9 A uniform elastic cylindrical shell of radius a and thickness t is clamped at edge $x = 0$ and is free at edge $x = l$. It sustains a pressure $p = p_n \sin n\theta$ which does not vary with x. Use the 'beam analogy' to compute the maximum stress-resultants and displacements in the shell according to the membrane hypothesis; and by evaluating the factor B by means of (7.74) determine the circumstances in which the membrane hypothesis is valid.

Repeat the analysis for the same shell, but with the end $x = l$ 'held circular' instead of free.

(It is sometimes useful to investigate problems of wind-loading on grain-storage silos by describing the pressure distribution by means of a Fourier series: see problem 4.10. An important advantage of this approach is that it is relatively easy to analyse silos which are statically indeterminate, according to the membrane hypothesis, by virtue of a roof which holds the top edge circular.)

8

Stretching and bending in cylindrical and nearly-cylindrical shells

8.1 Introduction

In chapter 3 we studied the way in which stretching and bending effects combine to carry axially symmetric loads applied to a uniform elastic cylindrical shell. In chapter 4 we investigated the way in which applied loads of a more general kind are carried by in-plane stress resultants alone, according to the membrane hypothesis. It became clear when we examined the resulting deflections of the shells in chapter 7 that this hypothesis is untenable in certain circumstances, and that indeed in such cases the bending effects, explicitly neglected in the membrane hypothesis, might well sustain a major portion of the applied loading.

We have reached the point, therefore, where we must consider the more general problem of a shell which is capable of carrying the loads applied to it by a combination of bending and stretching effects. This is our task in the present chapter.

An important idea which we shall develop is that it is advantageous to regard the shell as consisting of two distinct surfaces which are so arranged to sustain the 'stretching' and 'bending' stress resultants, respectively; and indeed, the chapter as a whole explores various consequences which flow directly and indirectly from this idea.

In section 8.2 we introduce the 'two-surface' idea to the equilibrium equations, and show that we may treat the two surfaces separately provided we introduce appropriate force-interactions between them. The 'stretching surface' is identical to a shell analysed according to the membrane hypothesis, and we may therefore use directly the work of chapters 4–7 for this part. The behaviour of the 'bending surface' is closely related to that of a flat plate, and is analysed briefly in section 8.3. The simplest possible problem involving an interaction between the two surfaces is a cylindrical shell subjected to a doubly-periodic distribution of pressure; and this is analysed in section 8.4.

The two-surface method leads directly to the well-known 'Donnell' simplification of the equations of cylindrical shells. These are developed in section 8.5 for the 'nearly-cylindrical shell' which we have already encountered in chapters 4 and 6.

The next two sections give a comprehensive treatment of the roots of Donnell's equations for zero surface traction and with all variables periodic in the circumferential direction: some clear physical interpretations emerge, and these are helpful in the consideration of 'improved' versions of the equations.

Lastly, section 8.7 discusses in general terms the role of the stretching and bending surfaces in meeting various relatively simple boundary conditions which may occur in actual shells.

The chapter is concerned with general aspects of the two-surface idea. Various detailed applications to particular problems will be given in the following chapters, and in many cases direct benefit in the form of useful simplifications will be gained from physical ideas which are developed in the present chapter.

8.2 The 'two-surface' idealisation: equilibrium equations

Fig. 8.1a shows the stress resultants acting upon a typical small element of a general shell. In our earlier treatment of the equilibrium equations for a shell (in the absence of bending, twisting and out-of-plane shear-stress resultants) we used an x, y, z coordinate system arranged with its origin at the point of interest, the z-axis in the direction of the local normal, and the x- and y-axes inclined in the local principal directions of the surface; see fig. 4.4a. Here again we shall use the same element, but we must now introduce those stress resultants which we previously omitted in accordance with the demands of the membrane hypothesis. The resulting picture is much more complicated.

As in the earlier problem we must take into account small differences in the values of the stress resultants acting on opposite edges of the element (see fig. 4.4b). Thus, stress resultant N_x acts on edge CD, while on edge AB the corresponding quantity is

$$N_x + \frac{\partial N_x}{\partial x} dx,$$

where dx is the small length DA (or CB) of the element in the x-direction. In fig. 8.1a this quantity has been labelled N_x^+ for the sake of simplicity, and the

8.2 Two-surface idealisation: equilibrium equations

same procedure has been adopted for the other variables also. For the present analysis we call the principal radii of the surface R_1, R_2 respectively.

It is a straightforward matter to write down the various equilibrium conditions for the element, and to take the limit as $dx, dy \to 0$. In this way we obtain the following equilibrium equations. By resolution of forces in the z-direction we have:

$$\frac{N_x}{R_1} + \frac{N_y}{R_2} - \frac{\partial Q_x}{\partial x} - \frac{\partial Q_y}{\partial y} = p. \tag{8.1}$$

And by resolution in the x- and y-directions, respectively, we obtain:

$$\frac{\partial N_x}{\partial x} + \frac{\partial N_{xy}}{\partial y} + \left[\frac{Q_x}{R_1}\right] = -q_x \tag{8.2a}$$

$$\frac{\partial N_y}{\partial y} + \frac{\partial N_{xy}}{\partial x} + \left[\frac{Q_y}{R_2}\right] = -q_y. \tag{8.2b}$$

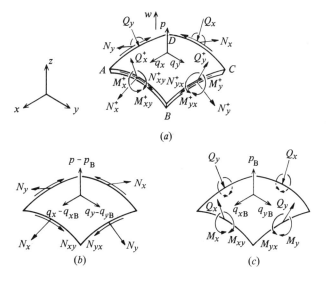

Fig. 8.1. A scheme for decomposing the stress resultants (a) acting on an element into distinct 'stretching' (b) and 'bending' (c) families, respectively. The necessity for unknown 'interface pressure', etc., is discussed in the text. The xy-plane is tangential to the element, and x and y are the local principal directions.

220   *Stretching and bending in shells*

Lastly, by resolution of moments in the x- and y-directions, respectively, we find:

$$Q_y - \frac{\partial M_y}{\partial y} - \frac{\partial M_{xy}}{\partial x} = 0 \tag{8.3a}$$

$$Q_x - \frac{\partial M_x}{\partial x} - \frac{\partial M_{xy}}{\partial y} = 0. \tag{8.3b}$$

In writing down these equations we have ignored differences between N_{xy} and N_{yx}, and between M_{xy} and M_{yx}, in accordance with the policy set out in section 2.4.3. We shall also disregard the strict requirement of moment-equilibrium about the z-axis (cf. (2.39)): see later.

These five equations are somewhat untidy. It will turn out to be useful to rearrange them as two separate sets of equations by means of the idea, illustrated in figs. 8.1b and c, of separating the stress resultants into two distinct families: cf. fig. 2.2. In this diagram we have split the element into two parts which we shall call the S-element (for stretching) and the B-element (for bending) respectively. We shall in fact regard the shell itself as consisting of two distinct surfaces – the S-surface and the B-surface – of which these sub-elements are parts. Later on we shall endow these separate surfaces with the in-plane and bending elastic stiffness, respectively, of the actual shell: but for the present we are concerned merely with the equilibrium arrangements.

Now when we separate the element into two pieces we must be careful to introduce force-of-interaction variables between the two parts, so that the equilibrium of each part is preserved. This is exactly analogous to the provision of bending moment and shearing force at a cross-section of a beam when we make an imaginary transverse cut – except, of course, that here we are not concerned with a *transverse* cut. At the new interface inside our shell element we need a normal and two tangential force interactions, which are designated p_B, q_{xB} and q_{yB}, respectively, in fig. 8.1c. All of these quantities have the dimensions of force per unit area. Thus, if we imagine that the external loads p, q_x, q_y are applied to the shell, the load which is transferred to the B-surface is p_B, etc. while the net load applied to the S-surface is $p - p_B$, etc., as indicated in figs. 8.1b and c. Now although we have thought of the B-surface as fitting 'beneath' the S-surface, the two surfaces are in fact coincident, and in particular, since they have zero thickness, the forces q_x and q_{xB}, etc. lie in the same plane: consequently these forces together do not apply a couple to the S-surfaces. Equations (8.1)–(8.3) may now be written in two distinct families:

$$\frac{N_x}{R_1} + \frac{N_y}{R_2} = p - p_B, \tag{8.4a}$$

8.2 Two-surface idealisation: equilibrium equations

$$\frac{\partial N_x}{\partial x} + \frac{\partial N_{xy}}{\partial y} = -q_x + [q_{xB}], \quad (8.4b)$$

$$\frac{\partial N_y}{\partial y} + \frac{\partial N_{xy}}{\partial x} = -q_y + [q_{yB}]. \quad (8.4c)$$

$$\frac{\partial Q_x}{\partial x} + \frac{\partial Q_y}{\partial y} = -p_B, \quad (8.5a)$$

$$\frac{\partial M_x}{\partial x} + \frac{\partial M_{xy}}{\partial y} = Q_x, \quad (8.5b)$$

$$\frac{\partial M_y}{\partial y} + \frac{\partial M_{xy}}{\partial x} = Q_y. \quad (8.5c)$$

$$\frac{Q_x}{R_1} = -q_{xB}, \quad (8.6a)$$

$$\frac{Q_y}{R_2} = -q_{yB}. \quad (8.6b)$$

The S-surface equations (8.4) are of exactly the same form as the membrane equilibrium equations. Indeed, the only difference is that an *unknown* amount of the applied load (p_B etc.) is now carried by the B-surface. The B-surface equations (8.5, 8.6) have the same 'interface' unknowns appearing as applied loads. Note, however, that (8.5a), (8.5b) and (8.5c) may be combined to give:

$$\frac{\partial^2 M_x}{\partial x^2} + \frac{2\partial^2 M_{xy}}{\partial x \partial y} + \frac{\partial^2 M_y}{\partial y^2} = -p_B. \quad (8.7)$$

This is precisely the equation of equilibrium of an element of a flat plate which is locally tangential to the curved surface of the shell. This feature is not altogether surprising, since we have removed from the B-surface any capacity for carrying loads by in-plane stress resultants: the B-surface is therefore obliged to act in a generally plate-like fashion.

There is, however, a small but important difference in the equilibrium conditions of a curved B-surface element and the corresponding flat-plate element. If we have an element of a flat plate, and allow it to sustain no in-plane stress resultants, it is clear that the element is incapable of supporting tangentially applied loads (acting at the centre-surface) even though it is capable of sustaining loads applied normally: any tangential loading must clearly be resisted by in-plane stress resultants. The only significant difference between the equilibrium equations of a *flat* plate and of our *curved* B-surface is that for the B-surface some tangential loading is *required* to balance the shearing forces Q: see (8.6). The key to the situation is that the shear forces Q are normal to the surface, so they demand for equilibrium a small tangential traction, as indicated in fig. 8.2.

Thus the interaction between the S- and B-surfaces involves a primary effect in the form of a normal force interaction p_B, together with some tangential interaction q_{xB}, q_{yB}, which depends simply on the disposition of shearing-stress resultants Q_x, Q_y within the B-surface.

At this point we shall make a simplifying assumption. We shall reckon that the only *significant* interaction between the two surfaces is the normal pressure p_B; and that consequently the terms q_{xB}, q_{yB}, enclosed in square brackets in (8.4), may be ignored.

A step of this sort must of course be put to the test before it can be accepted. In making this assumption we are not supposing that q_{xB}, q_{yB} are equal to zero – which would imply that Q_x, Q_y are also zero – but that their effect on the membrane equilibrium conditions for the S-surface, and furthermore on the resulting deformation of the S-surface, is negligible.

This kind of approximation is well known in shell theory, though it is usually presented in different terms. It is in fact a standard approximation of 'shallow-shell theory', in which the terms enclosed in square brackets in (8.4) (corresponding precisely to those in square brackets in (8.2)) are discarded (see, e.g. Reissner, 1960, §§ 7-10; Flügge, 1973, Chapter 7). The general justification of this is simply that the denominators, R_1, R_2 in terms Q_x/R_1, Q_y/R_2 are large for 'shallow' shells, and consequently give only a very weak coupling between the bending and stretching effects; a coupling which disappears entirely, of course, in the case of a flat plate. It is in precisely the same spirit that the equation of moment-equilibrium about the z-axis is disregarded in shallow-shell theory and elsewhere.

In this chapter we shall be concerned with both cylindrical and 'nearly-cylindrical' shells, and these are not necessarily obviously 'shallow' in the circumferential direction. Our hypothesis will lead to a simplified set of equations which we shall call *Donnell's equations*. We shall be able to put our

Fig. 8.2. The transverse shear-stress resultant Q in a *curved* element requires forces in the tangential direction for equilibrium.

8.2 Two-surface idealisation: equilibrium equations

assumption to the test in some simple cases, and to discover some limits on its validity for general use.

So far we have only discussed the idea of splitting the shell into two surfaces in the context of the equilibrium equations. What are the other steps in the analysis? We have explicitly excluded bending and transverse shear-stress resultants from the S-surface, and we may therefore use the membrane hypothesis for this part of the shell. There is, of course, the complication that the S-surface sustains the three unknown force interactions p_B, q_{xB}, q_{yB} at each point, in addition to the loading which is applied to the shell; but the effect of the assumption described above is to reduce these three unknown forces to one. In principle, therefore, we can use the methods of chapter 7 to determine the deflection of the S-surface, subject to appropriate force or displacement boundary conditions at its edges, and in terms of the unknown function p_B. Similarly we can analyse in a relatively simple way the behaviour of the B-surface subjected to the unknown loading p_B; and in particular we can express its displacement in terms of p_B, subject again to appropriate boundary conditions. Analysis of the B-surface will be the subject of the next section. Finally, we can choose the function p_B so that the displacements of the two surfaces match.

This method is essentially a generalisation of that which we used in sections 3.7 and 3.7.1 for the analysis of cylindrical shells under axially symmetric loading which varies sinusoidally in the axial direction: there the problem was treated as one in which the 'interface pressure' between what we would now call the S- and B-surfaces was the main variable.

There are many details of this process which we shall need to scrutinise later. Our plan will be to examine first some simple problems which avoid considerations of boundary conditions, etc., and then to examine various other features in turn.

The notion of separate S- and B-surfaces has been introduced to clarify the interaction between the 'stretching' and 'bending' effects which combine in the actual shell to resist the applied loading. It is tempting to think about a physical splitting of the material shell like the separation of two adjacent layers in an onion; but this temptation should be resisted. We have already decided to regard the material shell as a geometrical surface which is endowed with elasticity with respect to stretching and bending effects. It is this geometrical surface which we are now regarding as two separate surfaces, each of which is endowed with a distinct part of the elasticity of an element of the shell. Since the 'bending' and 'stretching' elasticity are both derived from the same elastic material in the actual shell, it is pointless to think in terms of splitting this *material*. The clarification of the structural action which comes from the two-surface idea follows from the separation of the general problem into two distinct and simpler problems, which can be recombined at a late stage.

8.3 Response of the 'bending surface' to pressure-loading

As we remarked above, the B-surface is capable of sustaining loads only by virtue of its bending stiffness, and it can offer no resistance to in-plane forces. In consequence, it is free to undergo surface strains while developing zero in-plane stress resultants. This is analogous to the way in which the S-surface is free to undergo changes of curvature without developing corresponding bending moments by virtue of the membrane hypothesis. It is not obvious how an arrangement of a 'free-to-extend' B-surface could be realised physically. Perhaps the best plan is to think first of a one-dimensional version in the form of a beam made of interlocking components which are free to slide relative to each other in the axial direction, as shown in fig. 8.3.

The well-known 'classical' small-deflection analysis of plates under transverse loading may be regarded as an example of a B-surface structure. If a real plate is displaced appreciably from its original flat configuration, second-order surface strains are set up which require, by Hooke's law, in-plane stress resultants; and these in turn are able to support a portion of the applied loading. The neglect of these 'membrane' effects in the classical analysis is simply tantamount to the idealisation of the plate as a B-surface.

In order to complete the analysis of the B-surface – which we began by writing down the equilibrium equations (8.5) and (8.6) – we proceed first to use Hooke's law in order to express the compound equilibrium equation (8.7) in terms of changes of curvature and twist $\kappa_x, \kappa_y, \kappa_{xy}$; then we shall express these changes of curvature and twist in terms of the displacements of the surface.

Here, for the sake of simplicity, we shall consider a 'shallow' B-surface, for

Fig. 8.3. An assembly of interlocking units of this kind constitutes a beam which is not capable of withstanding components of external loading in the tangential direction: the rollers are assumed to be frictionless. Similarly, the B-surface is incapable of withstanding tangentially applied loads.

8.3 Response of 'bending surface' to pressure

which the relations between curvature change and normal displacement w (see (6.23d)–(6.23f)) may be simplified to:

$$\kappa_x = -\frac{\partial^2 w}{\partial x^2}, \quad \kappa_{xy} = -\frac{\partial^2 w}{\partial x \partial y}, \quad \kappa_y = -\frac{\partial^2 w}{\partial y^2}. \tag{8.8}$$

This kind of simplification is one which we have used on several occasions already: we shall consider its justification and possible modifications of the equations in section 8.6.

Using Hooke's law (see (2.28)) and the above relations we find, simply:

$$D\nabla^4 w \equiv D\left(\frac{\partial^4 w}{\partial x^4} + \frac{2\partial^4 w}{\partial x^2 \partial y^2} + \frac{\partial^4 w}{\partial y^4}\right) = p_B. \tag{8.9}$$

Here D is the 'flexural rigidity' of the plate, which was derived in terms of Young's modulus, thickness, etc. in section 2.3.2: see (2.18). ∇^4 is the biharmonic operator.

Equation (8.9) is in fact none other than the governing equation for a flat isotropic elastic plate according to the classical theory. That the governing equation for the B-surface reduces to this simple form is partly a consequence of the use of a simplified version (8.8) of the relevant curvature–displacement relations. But this should not obscure the fact, which was pointed out earlier, that the role of the B-surface is almost precisely that of a (locally nearly flat) plate in bending and transverse shear.

It is instructive to associate each of the three fourth-order terms enclosed in the brackets in (8.9) with a physical quantity. Now the governing equation for a classical straight beam is

$$EI\frac{d^4 w}{dx^4} = p, \tag{8.10}$$

where $w(x)$ is the transverse displacement, EI is the flexural stiffness and p is (here) an intensity of transverse force loading (force $\times$ length^{-1}). We can therefore identify the first term in (8.9) as a 'beam-action' in the x-direction. Similarly the last term represents a beam action in the y-direction; and the central term may be associated with a 'twisting' effect with respect to the x- and y-axes. In particular note that the applied load is carried by the additive combination of three structural effects.

As a first example of the solution of (8.9), consider a B-surface in the xy-plane which sustains a transverse pressure loading of the form

$$p_B = p_{nB} \sin(\pi x/l) \sin(\pi y/b). \tag{8.11}$$

This is the kind of loading that we used in section 7.3 in an example concerning the membrane displacement of a cylindrical shell; and indeed our motivation is to use that previous result in relation to the S-surface, and thus determine the way in which a doubly sinusoidal pressure-loading is shared, in general, between the S- and B-surfaces of a shell.

It seems clear, as the equation involves only even derivatives, that the displacement will also be doubly sinusoidal. Putting, as a trial solution,

$$w = w_n \sin(\pi x/l) \sin(\pi y/b) \tag{8.12}$$

in (8.9) we find:

$$w_n = \frac{p_{nB}}{\pi^4 D} \bigg/ \left(\frac{1}{l^2} + \frac{1}{b^2}\right)^2. \tag{8.13}$$

This is the required relation for the 'shallow' version of the B-surface.

8.4 Cylindrical shell subjected to a doubly-periodic pressure-loading

We are now in a position to investigate the interaction of the S- and B-surfaces of a cylindrical shell in supporting a doubly-periodic pressure-loading. This is the simplest possible problem involving both stretching and bending effects in a shell apart from the axisymmetric version which we studied in section 3.7; and indeed this simpler problem may be extracted as a special case of the doubly-periodic loading, as we shall see.

The response of the B-surface to a doubly-periodic pressure-loading is described by (8.13). The corresponding relation for the S-surface, from (7.8) is, in the present notation,

$$w_n = (p_n - p_{Bn}) \frac{a^2 l^4}{Et} \left(\frac{1}{l^2} + \frac{1}{b^2}\right)^2. \tag{8.14}$$

Since the S- and B-surfaces must have the same normal displacement, (which is indeed why we have not used different symbols for the normal displacement of the two surfaces), we have

$$p_n = w_n \left[\frac{Et}{a^2 l^4} \left(\frac{1}{l^2} + \frac{1}{b^2}\right)^{-2} + \frac{\pi^4 E t h^2}{12} \left(\frac{1}{l^2} + \frac{1^*}{b^2}\right)^2 \right]. \tag{8.15}$$

The two terms on the right-hand side represent the load carried by stretching and bending action, respectively. In expressions of this sort, which we shall encounter frequently, the term corresponding to the S-surface may be recognised easily by the in-plane stiffness Et, and the B-surface term by the flexural stiffness $D = \frac{1}{12} Eth^2$. For present purposes the symbol 1^* may be read as 1; but in section 8.6 a 'correction' to this value will be discussed.

Equation (8.15) clearly suggests that the S- and B-surfaces are acting 'in parallel', like the two springs in the simple model shown in fig. 3.9b. The relative magnitude of the two terms clearly depends strongly on the dimensions a, t, l and b. Now we argued in chapter 7 that in some circumstances the membrane hypothesis would be justified for essentially this same problem. In terms of the present analysis, this would correspond to the second term on the right-hand side of (8.15) being much smaller than the first term.

Let us therefore investigate in general the way in which the applied loading is shared between the two surfaces. For this purpose it is useful to define a parameter ζ as the ratio of the stiffnesses of the S- and B-surfaces: thus

$$\zeta = \frac{p_{nS}/w_n}{p_{nB}/w_n} = \frac{\text{Stiffness of } S\text{-surface}}{\text{Stiffness of } B\text{-surface}} = \frac{12l^4}{\pi^4 a^2 h^2} \left[1 + \left(\frac{l}{b}\right)^2\right]^{-4}. \tag{8.16}$$

Consider first the circumstances in which $\zeta = 1$, i.e. the two surfaces share the applied load equally. From (8.16) we obtain, on taking a fourth root,

$$\frac{l}{(ah)^{\frac{1}{2}}} = \frac{\pi}{12^{\frac{1}{4}}} \left[1 + \left(\frac{l}{b}\right)^2\right]. \tag{8.17}$$

This expression is plotted in fig. 8.4a as the curve labelled $\zeta = 1$ in a graph of $l/(ah)^{\frac{1}{2}}$ against $b/(ah)^{\frac{1}{2}}$. The plot is double logarithmic. The curve is asymptotic to two straight lines of slope 2 and 0, respectively, when $l/b \gg 0$ and

Fig. 8.4. (a) A 'map' in relation to doubly-periodic pressure-loading of a cylindrical shell for sufficiently high values of the circumferential wavenumber n. Within the shaded region the applied loading is shared on more-or-less equal terms by the S- and B-surfaces; but outside this region the load is carried almost entirely by one or other of the two surfaces. (b) Another diagram in relation to the same problem, and using the same axes, but showing contours of 'panel stiffness'. (The six points marked in (a) are in connection with section 15.4.1.)

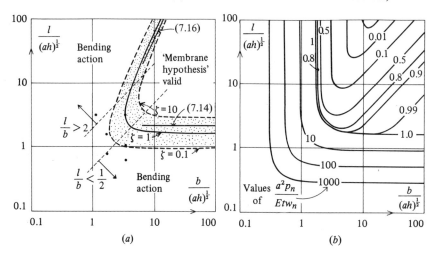

$l/b \ll 0$, and these lines (not shown) intersect at the point (1.69, 1.69): 1.69 ≈ $\pi/12^{\frac{1}{4}}$. In the region lying within the nearly V-shaped curve $\zeta = 1$ the B-surface carries less load than the S-surface; and vice versa in the region outside the curve.

In chapter 7 we investigated the validity of the membrane hypothesis by investigating the relative magnitude of peak values of bending and direct strain, respectively. As we remarked above, the present analysis furnishes an alternative and rather more satisfactory criterion in relation to the present problem. It is reasonable to regard the membrane hypothesis as satisfactory provided the B-surface supports less than, say, 10% of the applied load, i.e. $\zeta > 10$, roughly. The formula for this limit is the same as (8.17) except that the constant 1.69 is replaced by 2.92. This curve is also plotted in fig. 8.4a, together with its counterpart which marks the boundary of the region in which the S-surface carries less than 10% of the applied load, i.e. $\zeta < 0.1$. The entire diagram may thus be divided into three almost distinct regions. In two of these the 'membrane hypothesis' and what we might call the 'plate-bending hypothesis' are justified, respectively, while in the intermediate regime there is interaction between the S- and B-surfaces on more-or-less equal terms.

The limits on the validity of the membrane hypothesis which were determined by the 'simple method' of section 7.3.2 are also marked on fig. 8.4a by their equation numbers (7.14) and (7.16): they are seen to be satisfactory.

The simple form of expression (8.17) is due largely to the fact that the expression $(l^{-2} + b^{-2})$ occurs in the analysis for both the S- and the B-surfaces: it is in the numerator of one term and the denominator of the other. In the B-surface analysis we see this term as a result of using the operator ∇^4 on the displacement function $w_n \sin(\pi x/l) \sin(\pi y/b)$. On the other hand, in the S-surface analysis (section 7.4) we find that this term comes, in a similar way, from the expression $\nabla^4 \phi = -gEt$ for the change of Gaussian curvature in terms of the Airy stress function ϕ. In fact it is no accident that the same operators should appear in both calculations. We shall examine this connection in general terms in appendix 6.

The region $l/b > 2$ is important in many practical problems. From (8.16) we see that here $\zeta \propto b^8/l^4 a^2 h^2$, and consequently that relatively small changes in b (in the ratio $100^{\frac{1}{8}} \approx 1.8$) are needed to cross the transition zone between $\zeta = 10$ and $\zeta = 0.1$ in fig. 8.4a when the other dimensions are fixed.

Fig. 8.4b shows another plot of the same data (from (8.15)), using the same axes: it shows contours of 'panel stiffness' p_n/w_n. The panel stiffness is the simple sum of contributions from the S-surface and the B-surface respectively. Over most of the diagram either one or the other term dominates, according to the scheme of fig. 8.4a; so in these regions only the dominant term need be plotted. In the intermediate region (shaded in fig. 8.4a) contours of the sum are indicated. This diagram shows various interesting features. The nearly L-shaped contours of the B-surface stiffness correspond to the fact that when $l/b > 2$, almost all of the stiffness comes from bending across the shorter span, b, of the panel, while for $l/b < \frac{1}{2}$ the opposite is true, and the l dimension governs. As we saw in section 7.3, the deflection of the S-surface in the region $l/b > 2$ depends almost entirely on strains in the l-direction, and vice versa in the region $l/b < \frac{1}{2}$: however, this does not show up on the contours, which are all lines of l/b = constant, by virtue of the fact that the value of the first term on the right-hand side of (8.15) depends in fact only on the ratio l/b: see section 7.3.1. Notice, however, that in the region $l/b < 0.2$ the stiffness is almost *constant*, having the value corresponding to simple circumferential stretching, whereas in the region $l/b > 2$ the stiffness falls off rapidly with increasing values of l/b.

The interaction between bending and stretching effects in the upper arm of the shaded region of fig. 8.4a is characterised by the V-shaped contours in the same part of fig. 8.4b. Contours of stiffness of the S-surface are lines of slope 1, as remarked above, while those for the B-surface are lines of constant b, corresponding to beam-action across the minor dimension of the 'panel'. The V-shaped contours indicate a sloping, steep-sided valley. In consequence, for any given value of $l/(ah)^{\frac{1}{2}} > 10$, say, there is a particular value of $b/(ah)^{\frac{1}{2}}$ – and correspondingly of the circumferential wavenumber n – at which the panel stiffness is a minimum: see problem 8.1. An alternative way of plotting (8.17) is discussed in problem 8.2 and some useful special cases are considered in problem 8.3.

8.5 Donnell's equations for nearly-cylindrical shells

The example used in the previous section illustrates the way in which the S-surface and B-surface interact to carry the applied load in the particularly simple case in which pressure varies sinusoidally in both longitudinal and circumferential directions. The interaction was simple because the response of the two surfaces separately to the total applied load involved the same mode-form: consequently the fractional division of load between the two surfaces was uniform over the entire shell. Although the results of this simple case may be combined, with Fourier series, to give solutions to more complex problems (see e.g. section 9.2), it is desirable to study the equations of the problem in rather more general terms. In this section therefore we shall derive the general equations for a shell subject to the following conditions:

(i) the shell is cylindrical or 'nearly-cylindrical' in the sense of the analyses of sections 4.6 and 6.6.1;

(ii) the force interaction between the two surfaces is regarded as purely normal;

(iii) the loading consists of a normal pressure $p(x, y)$, i.e. there is no tangential surface traction; and

(iv) the load-deflection response of the B-surface is assumed to be that of a flat plate.

Condition (i) enables us to work in the xy-plane, and has been justified already. Condition (ii) is introduced in order to simplify the equations. The circumstances under which it is not justified have already been mentioned and will be examined in section 8.6. The prohibition of tangential surface tractions (condition (iii)) makes possible the use of the Airy stress function, as we shall see: this restriction will be relaxed at the end of section 8.6. Finally, use of the 'flat-plate' equations for the B-surface again simplifies the analysis: the validity of this approximation will also be examined in section 8.6.

As in section 4.6, we denote the radius of the cylindrical shell by a, and the radius of curvature of the meridian by a^*: in most cases we shall assume that $|a^*| \gg a$.

The main variables are ϕ, the Airy stress function for the in-plane stress resultants of the S-surface, and w, the normal component of displacement of the B-surface. Our third variable is p_B, the 'interface' pressure between the two surfaces. All three variables are functions of x and y.

The normal equilibrium equation (8.4a) for the S-surface may be written in terms of ϕ as follows:

$$\Gamma^2 \phi = p - p_B, \tag{8.18}$$

where

$$\Gamma^2 \equiv \frac{1}{a} \frac{\partial^2}{\partial x^2} + \frac{1}{a^*} \frac{\partial^2}{\partial y^2}. \tag{8.19}$$

The governing equation for the B-surface was derived in section 8.3 (see (8.9)):

$$D \nabla^4 w = p_B.$$

Clearly p_B may be eliminated by addition of these two equations, to give

$$\Gamma^2 \phi + D \nabla^4 w = p. \tag{8.20}$$

This is the general equilibrium equation in the sense that the two terms on the left-hand side represent the load-carrying contributions of the S- and B-surfaces respectively. The second term does of course embody Hooke's law and the compatibility of the B-surface.

We now seek a 'compatibility' equation to express the fact that the displaced forms of the S- and B-surfaces are coincident. In the examples of sec-

8.5 Donnell's equations for nearly-cylindrical shells

tions 3.7.1 and 8.4 we simply matched the normal components of displacement of the two surfaces. In general, however, the structure of the equations is better suited to a statement about the coincidence of the surfaces in terms of change of Gaussian curvature; and as we saw in section 6.3 the compatibility condition is expressed by the simple statement that *the change of Gaussian curvature of the two surfaces is equal.* This is only strictly true if we make the surfaces coincide at the boundary; and we shall assume throughout that this is so.

For the S-surface we have, from section 7.4, the change of Gaussian curvature, g_S, in terms of ϕ:

$$g_S = -\frac{\nabla^4 \phi}{Et}. \tag{8.21}$$

Also, for the B-surface we may easily express the change of Gaussian curvature in terms of w, from (6.6), (8.8) and (8.19):

$$g_B = \frac{\kappa_x}{a} + \frac{\kappa_y}{a^*} = -\frac{1}{a}\frac{\partial^2 w}{\partial x^2} - \frac{1}{a^*}\frac{\partial^2 w}{\partial y^2} = -\Gamma^2 w. \tag{8.22}$$

On elimination of $g_S = g_B$ from these expressions we have, finally,

$$(1/Et)\nabla^4 \phi - \Gamma^2 w = 0. \tag{8.23}$$

This is the general compatibility equation, stating that the change of Gaussian curvature, resulting from the deformation of the S- and B-surfaces under their respective loading, is precisely the same for both surfaces. The first term, expressing the change of Gaussian curvature for the S-surface, embodies Hooke's law and the equilibrium equations (through the Airy stress function).

Equations (8.20) and (8.23) are the field equations for the problem, and in general they must be solved subject to appropriate boundary conditions. They are coupled equations in terms of the two variables w and ϕ. The curious circumstances whereby both of the operators ∇^4 and Γ^2 appear in each equation will be examined in appendix 6.

It is a simple matter to obtain a single equation in either of the two variables by eliminating the other. Thus we can obtain the following alternative equations, each in terms of one variable:

$$D\nabla^8 w + Et\Gamma^4 w = \nabla^4 p \tag{8.24}$$

$$D\nabla^8 \phi + Et\Gamma^4 \phi = Et\Gamma^2 p. \tag{8.25}$$

Here,

$$\nabla^8 \equiv \nabla^4(\nabla^4), \quad \Gamma^4 \equiv \Gamma^2(\Gamma^2). \tag{8.26}$$

Note that in the special case of cylindrical shells,

$$\Gamma^4 \equiv \frac{1}{a^2} \frac{\partial^4}{\partial x^4} . \qquad (8.27)$$

In some circumstances it is most convenient to deal with a single eighth-order equation, (8.24) or (8.25), in one variable. On the other hand it is sometimes more convenient to deal with two coupled fourth-order equations in the two variables w and ϕ, possibly by means of complex-variable techniques. Which form is more convenient in a given case depends to some extent on the arrangement of the boundary conditions.

The equations which we have just developed are collectively known as the *shallow-shell* equations. They are obtained by making a series of approximations which are justified for 'shallow' shells. Approximations of this kind were made first by Donnell (1933) in his study of the buckling of cylindrical shells under torsional loading. The shallow-shell equations are sometimes known as *Donnell's* equations or the *Donnell–Mushtari–Vlasov* equations. Their derivation has been given by many writers, e.g. Flügge (1973, chapter 7), Reissner (1946), Vlasov (1964, chapter 7), Novozhilov (1964, §17). The present treatment, in terms of the two-surface idea and change of Gaussian curvature, appears to be the simplest and most direct way of obtaining the governing equations.

More 'exact' forms of the equations, obtained by cutting fewer corners, have been obtained by several workers, notably Flügge (1973, chapter 5); we shall discuss these briefly in section 8.6.

Donnell's equations pose a problem of a general kind which recurs in many branches of mechanics, and indeed in science in general. There is of course a pleasant prospect if we can obtain a particularly simple set of equations: but if they have been obtained only in consequence of a number of gross simplifications, can they be used with confidence?

Many attempts have been made to assess the accuracy of Donnell's equations in comparison with other, less simple, and more complete equations. In the main the procedure has been to postulate solutions to the differential equations of a certain simple kind, then to extract an algebraic 'characteristic' equation, and finally to compare the roots of this equation with those which are obtained when a different version of the governing equation is used. The general conclusion to emerge from studies of this kind is that in some circumstances the roots of the characteristic equation derived from Donnell's equation agrees closely with the roots derived from other equations, while in other circumstances they do not. The relevance of these various circumstances does not seem to have been examined in physical terms by the various investigators. This weak kind of conclusion is of little use to engineers. We therefore propose to resolve this question by constructing a 'map' of solutions, along the

8.5 Donnell's equations for nearly-cylindrical shells

lines of fig. 8.4, which shows different regimes of behaviour. Maps of this kind have been constructed for cylindrical shells by Holand (1959): see also Hoff (1954a). For example, it is clear from the way in which we established (8.20) and (8.23) that these equations contain, as distinct special cases, the membrane hypothesis and the equations of ordinary flat plates respectively. We may therefore expect that for shells having certain dimensions – or, rather, having the values of certain dimensionless groups lying in certain ranges – these, and possibly other, special cases will emerge as strong approximations. Indeed, one of the advantages of our scheme of building up the subject by means of relatively simple special cases is that we shall have a good idea of what to expect in a variety of special circumstances.

8.5.1 Roots of Donnell's equation

Let us investigate the general features of the solution of Donnell's equation when $p = 0$, i.e. when the shell is subjected only to edge-loadings.

The general eighth-order equation may be written in terms of w or ϕ, and indeed the form of the equation is exactly the same in either case. For most practical purposes it is more convenient to use w as the variable; nevertheless in some circumstances ϕ is more convenient. In the present section we shall use w, but we shall make some remarks about the use of ϕ in section 8.7.4. When $p = 0$ (8.24) may be written:

$$D\nabla^8 w + Et\Gamma^4 w = 0. \tag{8.28}$$

We shall consider a trial solution of the form

$$w = f(x) \sin(\pi y/b). \tag{8.29}$$

Thus we seek information about the general function $f(x)$ for a solution which is periodic in the circumferential direction. This kind of solution is, in practice, much more important than the complementary form

$$w = f(y) \sin(\pi x/l), \tag{8.30}$$

which we shall study only briefly: see problem 8.4.

Substitution of (8.29) into (8.28) leads to an ordinary eighth-order differential equation for $f(x)$. In these circumstances it is convenient to put

$$f(x) = Ae^{-kx}, \tag{8.31}$$

and to solve the resulting eighth-order algebraic 'auxiliary' equation in k to find for what values of this parameter the postulated solution is valid. The auxiliary equation has eight roots which are all complex, as we shall see. The algebra is bound to be heavy, but we can make it less oppressive by means of a long chain of substitutions designed to reduce the auxiliary equation to its simplest possible form; and notwithstanding the heavy algebra the results are relatively simple, as we shall see, especially when they are displayed in graphical form.

In general the eight roots of the auxiliary equation may be written

$$k_j = \alpha_j + i\beta_j, \quad j = 1, \ldots, 8, \tag{8.32}$$

where α_j, β_j are real coefficients and $i = (-1)^{\frac{1}{2}}$. Consider the first of these functions, namely

$$f(x) = \exp[(\alpha_1 + i\beta_1)x] = [\exp(\alpha_1 x)](\cos\beta_1 x + i\sin\beta_1 x). \tag{8.33}$$

This is a product of an exponential function and a trigonometric function of x. Now we have already characterised the assumed trigonometric function of y by means of the circumferential half wavelength b. It is therefore convenient to express the trigonometric function of x in terms of a longitudinal half wavelength l_1 given by

$$l_1 = \pi/\beta_1. \tag{8.34a}$$

Indeed, in view of the close connection between the exponential and trigonometric functions it will also be convenient to express the exponential function $e^{\alpha_1 x}$ in terms of a 'pseudo half wavelength' l_1^*, defined as follows, by analogy with (8.34a):

$$l_1^* = \pi/\alpha_1. \tag{8.34b}$$

Thus we seek solutions of (8.28) of the type

$$w = A \exp(\pi x/l_1^*) \sin(\pi x/l_1) \sin(\pi y/b). \tag{8.35}$$

The significance of the pseudo half wavelength l_1^* is that for a change in value of x equal to l_1^*, the value of the exponential function changes by a definite factor, namely

$$e^\pi \approx 23.1.$$

In view of these definitions it will be convenient to write our trial solution as

$$w = \exp(\pi x/H) \sin(\pi y/b), \tag{8.36}$$

where H ($= \pi/k$) is a parameter having the dimensions of length which replaces k as the main variable; and we shall thus expect to obtain eight complex values of

$$(1/H)_j = (1/l^*)_j + i(1/l)_j. \tag{8.37}$$

We now substitute (8.36) into (8.28) in order to obtain an auxiliary equation. It is convenient also to define the following four parameters, one of which (ψ) is a new variable while the others are constants:

$$\psi = b/H, \quad \gamma = b/2^{\frac{1}{2}}\lambda,$$
$$\lambda = (\pi/3^{\frac{1}{4}})(ah)^{\frac{1}{2}} \approx 2.39(ah)^{\frac{1}{2}}, \quad m^2 = -a/a^*. \tag{8.38}$$

8.5 Donnell's equations for nearly-cylindrical shells

Here λ is a 'standard length' which is convenient in the subsequent algebraic manipulation. The final results will involve various quantities being expressed in terms of b/λ, where b is the basic independent variable, namely the circumferential half wavelength. For purposes of final display, however, we shall use the ratio $b/(ah)^{\frac{1}{2}}$: by (8.38)

$$b/(ah)^{\frac{1}{2}} = 2.39 b/\lambda.$$

The parameter m, whose value is zero for a cylindrical shell, describes the geometry of the 'nearly-cylindrical' surface. When the shell is elliptical (barrel-shaped), m^2 is negative, so m has an imaginary value. On the other hand, m^2 is positive for hyperbolic ('waisted') surfaces, so m is real; and indeed it represents the inclination of the real, straight, characteristic lines of the surface to the x-axis, just as it did in sections 4.6.2 and 6.6.1.

All of these substitutions are made with the aim of expressing the auxiliary equation in a simple form, and the result is:

$$(\psi^2 - 1)^4 + 16\gamma^4(\psi^2 + m^2)^2 = 0. \tag{8.39}$$

It is now convenient to change to a new variable ξ^2 defined by

$$\xi^2 = (\psi^2 - 1)/(1 + m^2), \tag{8.40}$$

and a new constant G defined by

$$G = \gamma^2/(1 + m^2) = (3^{\frac{1}{2}}/2\pi^2) b^2/ah(1 + m^2). \tag{8.41}$$

The auxiliary equation thus becomes, finally:

$$\xi^8 + 16G^2(\xi^2 + 1)^2 = 0. \tag{8.42}$$

This form of the equation factorises into two complex equations:

$$\xi^4 \pm 4iG(\xi^2 + 1) = 0. \tag{8.43}$$

Taking first the case of the negative sign we obtain a quadratic in ξ^2, namely

$$(\xi^2)^2 - 4iG(\xi^2) - 4iG = 0, \tag{8.44}$$

which is readily solved to give

$$\xi^2 = 2iG \pm 2iG(1 - i/G)^{\frac{1}{2}}. \tag{8.45}$$

The two solutions may be written

$$\xi^2 = R + iI_S, \quad \xi^2 = -R - iI_L, \tag{8.46}$$

where

$$R = G[2(1 + G^{-2})^{\frac{1}{2}} - 2]^{\frac{1}{2}}, \tag{8.47a}$$

$$I_S = G[2(1 + G^{-2})^{\frac{1}{2}} + 2]^{\frac{1}{2}} + 2G, \tag{8.47b}$$

$$I_L = I_S - 4G. \tag{8.47c}$$

236 Stretching and bending in shells

The reason for subscripts S, L will become clear later. It is easy to show that the two solutions of equation (8.43) with the + sign are the complex conjugates of the roots given in (8.46). These four complex roots ξ^2 of (8.42) are shown on the Argand diagram in fig. 8.5. Values of $1/G$ are marked on the curves.

The last step in the analysis is to return to the variable $\psi = b/H$ by means of the substitution (cf. (8.40))

$$\psi^2 = 1 + (1 + m^2)\xi^2. \tag{8.48}$$

We thus find that the eight roots for ψ are in two families of four;

Fig. 8.5. Argand diagram representation of the four complex roots ξ^2 of the auxiliary equation (8.42), which describes circumferentially periodic solutions of Donnell's equation ((8.24) or (8.25)) when there is no surface traction. The variable ξ and the parameter G are defined in order to produce the simplest form of the auxiliary equation.

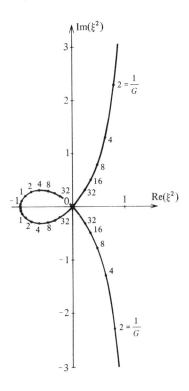

8.5 Donnell's equations for nearly-cylindrical shells

$$b/H = \pm u_S \pm iv_S, \quad b/H = \pm u_L \pm iv_L, \tag{8.49}$$

where

$$u_j + iv_j = (U_j + V_j)^{\frac{1}{2}}, \quad j = S, L, \tag{8.50a}$$

and

$$U_S = 1 + (1 + m^2)R \tag{8.50b}$$
$$V_S = (1 + m^2)I_S \tag{8.50c}$$
$$U_L = 1 - (1 + m^2)R \tag{8.50d}$$
$$V_L = (1 + m^2)I_L. \tag{8.50e}$$

In evaluating $(U + iV)^{\frac{1}{2}}$ we may use the following expressions, which are easy to verify:

$$2^{\frac{1}{2}}u = ((U^2 + V^2)^{\frac{1}{2}} + U)^{\frac{1}{2}}, \quad 2^{\frac{1}{2}}v = ((U^2 + V^2)^{\frac{1}{2}} - U)^{\frac{1}{2}}. \tag{8.51}$$

The eight roots of the auxiliary equation corresponding to (8.28) thus involve, for a particular pair of values of m and G, only four distinct real numbers, u_S, v_S, u_L, v_L.

Although these formulas seem complicated, we may determine the four numbers in any given case by means of a relatively simple chain of substitutions. For a particular pair of values of $b/(ah)^{\frac{1}{2}}$ and m we first evaluate G, by (8.41); then R, I_S, I_L by (8.47); then U_S, V_S, U_L, V_L by (8.50), and lastly u_S, v_S, u_L, v_L by (8.51). Finally we have

$$\frac{l_S^*}{c} = \frac{b/c}{u_S}, \quad \frac{l_S}{c} = \frac{b/c}{v_S}; \quad \frac{l_L^*}{c} = \frac{b/c}{u_L}, \quad \frac{l_L}{c} = \frac{b/c}{v_L}, \tag{8.52}$$

where (here) $c = (ah)^{\frac{1}{2}}$.

8.5.2 Physical interpretation of the roots

These four functions are plotted logarithmically in fig. 8.6 for a cylindrical shell, $m = 0$. The inset diagram enables one to read off the value of $b/(ah)^{\frac{1}{2}}$ directly for given values of n and a/h.

It can be seen that all four functions are asymptotic to very simple expressions when the value of $b/(ah)^{\frac{1}{2}}$ is either large or small. This may seem surprising in view of the detailed work necessary to extract the roots. The explanation for the simplicity of the results lies in the fact that ξ^2 quickly becomes asymptotic to simple formulae when G is either large or small; and indeed in the fact that for *either* large *or* small values of γ, the expanded form of the basic equation (8.39) is dominated by two terms: see problem 8.5.

What is the significance of the curves of fig. 8.6? First, and in practice most important, we notice that for large values of $b/(ah)^{\frac{1}{2}}$ there are two distinct

238   *Stretching and bending in shells*

asymptotic solutions, which it is convenient to designate 'short-wave' and 'long-wave' respectively: hence subscripts S and L in the preceding analysis.

For the short-wave solution

$$l_S/(ah)^{\frac{1}{2}} \approx l_S^*/(ah)^{\frac{1}{2}} \approx 2.39. \tag{8.53}$$

Thus the half wavelength of the sinusoidal term and the quasi half wavelength of the exponential term are both equal to $2.39(ah)^{\frac{1}{2}}$ ($= \gamma$). This is in fact precisely the result which we obtained in chapter 3 for the response of an axially symmetric shell to arbitrary edge-loading, and it is not difficult to trace this result to the fact that the behaviour is dominated by circumferential stretching and longitudinal bending. It is important to realise that this solution stands for four distinct terms in the complementary function of (8.28): two

Fig. 8.6. The solutions of Donnell's equation for a cylindrical shell with zero surface traction may be expressed as

$$w \text{ [or } \phi\text{]} = A e^{\pm \pi x/l^*} \sin[\text{or cos}] \, (\pi x/l) \sin[\text{or cos}] \, (\pi y/b)$$

(cf. fig. 7.1). The diagram expresses l and l^* in terms of b (or circumferential wavenumber n by means of the inset scale). There are two pairs of l, l^* for each value of b: they are shown by continuous and broken curves, respectively.

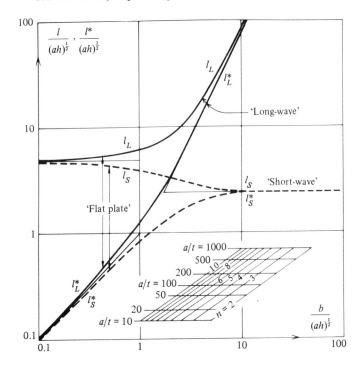

8.5 Donnell's equations for nearly-cylindrical shells

decay exponentially and two grow exponentially, just as they did in section 3.3.

The long-wave solutions (likewise four in number) are asymptotic to the lines

$$l/(ah)^{\frac{1}{2}} = l^*/(ah)^{\frac{1}{2}} = 0.838\, b^2/ah \tag{8.54}$$

when $b/(ah)^{\frac{1}{2}} \gg 1$. The fact that $l \approx l^*$ indicates that again the solution is analogous to a beam-on-elastic-foundation, with specific 'beam' and 'foundation' constants, which we shall determine in due course. This all follows from the fact that the solution involves interaction between longitudinal stretching and circumferential bending; and indeed, we may see the analysis of section 7.5 as just the 'beam' part of this analysis. We shall explore this analogy in section 9.4 in order to obtain some simple solutions to certain problems. This type of approximate analysis was first explored by Vlasov (1964, Chapter 11): cf. Novozhilov (1964, §49). It is often described as the 'semi-membrane theory'.

On the left-hand side of fig. 8.6, i.e. in the range of small values of $b/(ah)^{\frac{1}{2}}$ we observe a different kind of pattern. The first difference is that the two families of curves become close to each other, so that there are four repeated roots rather than eight distinct ones; and the second is that the quasi half wavelength of the exponential term becomes smaller than the half wavelength of the sinusoidal term. This means that the solution is dominated by exponential decay from the edge; and in fact the quasi half wavelength of the exponential factor is simply equal to the half wavelength of the imposed solution in the circumferential direction. All of these features are found in the bending of semi-infinite *flat plates* with periodic loading applied at the edge; and indeed the only effect which the curvature of the shell has on the solution is to impose small differences between the two families of solutions. This may readily be understood in terms of the domination of (8.39) by the first term on the left-hand side. We shall return to this region in section 8.7.4.

In the remaining region of fig. 8.6, where the value of $b/(ah)^{\frac{1}{2}}$ is in the region of 2, the curves cannot be explained in terms of simple special cases: the various terms of (8.39) represent non-negligible effects; and consequently the form of the solution will not be so simple as in the other two regions.

It is interesting to note that there are some strong similarities between figs. 8.6 and 8.4. Since fig. 8.4 represents the response of a shell to periodic *surface* loading, while fig. 8.6 is related to *edge*-loadings in the absence of surface tractions we cannot expect to find a direct correlation between the two diagrams. Nevertheless, it may be seen that the line of slope 2 in fig. 8.4, which corresponds to the most flexible modes under pressure-loading, seems to emerge also in fig. 8.6 to give the preferred modes for edge-loaded shells.

240   *Stretching and bending in shells*

Under doubly-periodic pressure-loading the *S*- and *B*-surfaces cooperate in resisting the applied loading, whereas under edge-loading the shell responds by an antagonism between the two surfaces. It is easiest to study the detailed relationship between the lines of slope 2 on the two diagrams in terms of a general 'beam-on-elastic-foundation' equation: see problem 8.5.

The line of slope 1 in fig. 8.6 appears to have no counterpart in fig. 8.4, but it does in fact correspond to a steeply sloping 'valley' in the stiffness of the bending surface which dominates in the lower left-hand part of fig. 8.4*b*. As we have already remarked, this part of fig. 8.4 represents almost pure plate-like behaviour.

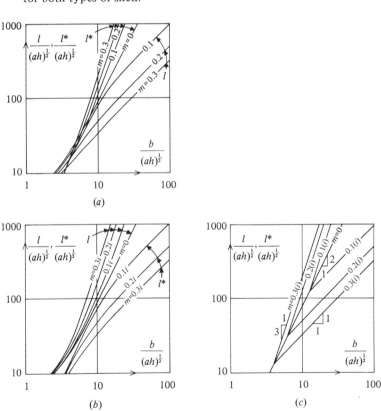

Fig. 8.7. The upper right-hand part of fig. 8.6 is the only region which needs to be modified appreciably when the shell is nearly cylindrical (cf. fig. 4.16), with $a/a^* = -m^2$. (*a*) is for hyperbolic shells ($a/a^* < 0$); (*b*) is for elliptic shells ($a/a^* > 0$) and (*c*) gives asymptotes for both types of shell.

8.5 Donnell's equations for nearly-cylindrical shells

Fig. 8.6 describes the modes of response of a long *cylindrical* shell to edge disturbances. Parts of the corresponding diagrams for hyperbolic and elliptic 'nearly-cylindrical' shells are shown in fig. 8.7. The curves have been plotted in the way described above, for various values of the geometrical parameter m. A comparison of the curves over the entire range of $b/(ah)^{\frac{1}{2}}$, $l/(ah)^{\frac{1}{2}}$ reveals that the effect of having a slightly curved meridian instead of a straight one can only be discerned for practical purposes in the region which is plotted in fig. 8.7. Detailed analysis shows that for hyperbolic shells (fig. 8.7a) the long-wave formula for l_L is asymptotic to

$$l_L = b/m. \tag{8.55}$$

This is exactly the result obtained by the membrane hypothesis in section 4.6.2. Here, of course, there is also an exponential decay factor, expressed in terms of a quasi half wavelength l_L^*; the fact that this becomes much larger than l_L in this region indicates a very weak exponential decay, representing only a small modification to the membrane-hypothesis solution. In the case of elliptical shells (fig. 8.7b) these effects are reversed: there is a strong exponential decay, corresponding to the membrane-hypothesis solution, together with a very weak oscillatory factor. It is interesting to see that the lines corresponding to the membrane hypothesis lie precisely in that region of fig. 8.4 in which S-surface effects dominate the stiffness of the shell.

Thus we may conclude that for a hyperbolic shell, the special features of the membrane-hypothesis solution which correspond to the presence of straight lines within the surface, only exert themselves for modes which have relatively *long wavelengths in the circumferential direction*. Specifically, we see from the diagram that the straight characteristics of the shell only exert an appreciable effect when

$$b > 5(ah)^{\frac{1}{2}}/m.$$

Hence we reach the important conclusion that for some purposes shells with curved meridians behave in a similar way to those with straight meridians. In particular, the 'short-wave' response, which involves only a trivial generalisation of the axisymmetric solution, is not affected by the curvature of the meridian. In this connection it should be emphasised that the object of broad generalisations of this kind is to help us to obtain a clear picture of different regimes of structural behaviour. The dividing lines between the regimes are somewhat blurred, but provided this fact is recognised it can hardly lead to erroneous conclusions.

The 'maps' shown in figs. 8.6 and 8.7 clarify some features of the response of shells to edge-loading, but they do not, of course, give a complete picture of the behaviour. Thus there is no indication of the *flexibility* of the various modes on the diagram; nor do the diagrams show how boundary conditions may be brought in. Indeed, these matters are related, and we shall investigate them in section 8.7.

8.6 The improvement of Donnell's equations

The analysis of section 8.5, which led to the remarkably simple 'Donnell' equations and the simple maps of behaviour, was based on a set of somewhat forbidding simplifications. It is now appropriate to investigate the consequences of these.

Conditions (ii) and (iv) in the list of simplifications in section 8.5 point to an investigation of the bending of cross-sections of the shell in the circumferential direction, for it is there that the effect referred to in fig. 8.2 is at its most severe. A similar effect occurs in the longitudinal direction for non-cylindrical shells on account of the curvature of the meridian: but since we are concentrating attention on shells having small absolute values of the principal radius ratio a/a^*, the effects in the circumferential direction will always dominate.

It is clear that we should begin by considering the small-displacement behaviour of a simple circular elastic ring under the action of a purely radial loading whose intensity varies sinusoidally around the circumference, as shown in fig. 8.8a. For simplicity we shall regard the ring as inextensional.

Fig. 8.8b defines the positive senses of tension T, shearing force Q and bending moment M, all of which are functions of the angle θ. There is a radial (outward) load of intensity p per unit length, and we shall take $p = p_n \sin n\theta$ as our given loading. It is easy to derive the three equilibrium conditions for an element in its original configuration. From fig. 8.8b, by resolving in the

Fig. 8.8. (*a*) A ring subjected to a circumferentially periodic normal loading. (*b*) Stress resultants etc. for a small element.

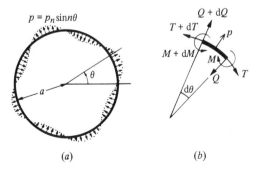

8.6 The improvement of Donnell's equations

radial and tangential directions, and taking moments, respectively, we obtain

$$\frac{dQ}{d\theta} - T = -p_n a \sin n\theta, \tag{8.56a}$$

$$\frac{dT}{d\theta} = -Q, \tag{8.56b}$$

$$\frac{dM}{d\theta} = aQ. \tag{8.56c}$$

Equation (6.9e) enables us to express the change in curvature κ in terms of the radial component w of displacement:

$$\kappa = -\frac{1}{a^2}\left(\frac{d^2w}{d\theta^2} + w\right). \tag{8.57}$$

Finally we have as the appropriate form of Hooke's law:

$$\kappa = \frac{M}{EI}, \tag{8.58}$$

where EI is the flexural rigidity of the member.

We now proceed to solve these equations. We can obtain from (8.56) a single second-order equilibrium equation in M, as follows. Eliminating Q from (8.56b) and (8.56c) we have

$$\frac{dM}{d\theta} = -a\frac{dT}{d\theta}$$

and therefore

$$M = -aT + \text{constant}. \tag{8.59}$$

We expect that both M and T will vary sinusoidally round the circumference, so we set the constant at zero, tentatively: this can be verified later on. Putting Q and T in terms of M, we have from (8.56a):

$$\frac{d^2M}{d\theta^2} + M = -a^2 p_n \sin n\theta. \tag{8.60}$$

The solution of this, taking into account the circumferential continuity of M is

$$M = \frac{a^2 p_n \sin n\theta}{n^2 - 1}. \tag{8.61}$$

Note that the term $n^2 - 1$ in the denominator is a consequence of the form of the left-hand side of the equation.

It is instructive to compare this result with that for a *straight* beam whose axial coordinate y is expressed in terms of a dimensionless variable θ by the substitution $y = a\theta$. The equilibrium equation for the beam is the same as (8.60) except that the term M on the left-hand side is missing. Consequently the variation of M along the beam is given by (8.61) except that the term $n^2 - 1$ in the denominator is replaced by n^2. We are, of course, overlooking here the question of boundary conditions of the beam. We thus obtain an important result: for the given loading the bending moment in the ring is *larger* than in the equivalent straight beam by the factor $n^2/(n^2 - 1)$. For high values of n this factor is close to unity, and there is an insignificant difference between the two solutions. On the other hand, for small values of n the difference can be large. Thus for $n = 2$ the factor is $\frac{4}{3}$. Indeed, for $n = 1$ the factor is infinite, which suggests that our analysis is inappropriate: the explanation is simply that the loading applied to the ring is not *self-balancing* in this special case.

Returning to our equations, we may now express κ in terms of the applied loading by using (8.58) and (8.61), which leaves us with a differential equation (8.57) for w. The solution of this equation is

$$w = \frac{a^4 \, Pn \, \sin n\theta}{EI \, (n^2 - 1)^2}. \tag{8.62}$$

In this expression there is yet another term $(n^2 - 1)$, which comes from the form of the right-hand side of (8.57): for a given change of curvature κ, varying sinusoidally round the circumference, the geometrically compatible radial component of displacement is larger than the corresponding displacement of a straight beam, by the factor $n^2/(n^2 - 1)$. We therefore conclude that, in terms of the relation between sinusoidally varying radial force and radial displacement, the ring is exactly equivalent to a straight beam whose flexural stiffness EI', say, is related to that of the ring by:

$$EI' = (1 - n^{-2})^2 \, EI. \tag{8.63}$$

What are the implications of this for our analysis of shells? So far in our discussion of the equations for shells we have treated the bending stiffness of the B-surface exactly as if that surface were flat. It is clear that, as far as bending in the *circumferential* direction is concerned, the 'flat-plate' version of the B-surface overestimates the stiffness.

It is not difficult to adapt the governing equation for the B-surface in a way which makes an appropriate modification to the bending stiffness in the circumferential direction. There are several ways of going about this, and we shall follow a suggestion of Morley (1959).

8.6 The improvement of Donnell's equations

The governing equation (8.9) for a uniform isotropic flat plate is:

$$\nabla^4 w = p/D.$$

The operator ∇^4 may be written $\nabla^2(\nabla^2)$, i.e. the Laplace operator applied twice. Now

$$\nabla^2 w \equiv \frac{\partial^2 w}{\partial x^2} + \frac{\partial^2 w}{\partial y^2},$$

so for the flat plate we may write in particular

$$-\nabla^2 w = \kappa_x + \kappa_y. \tag{8.64}$$

But for the cylindrical shell, following our analysis of the bending of a circular ring, we should write instead

$$-\kappa_y = \frac{d^2 w}{dy^2} + \frac{w}{a^2}.$$

This suggests that we should replace the operator ∇^2 by $\bar{\nabla}^2$, where

$$\bar{\nabla}^2 = \nabla^2 + a^{-2}. \tag{8.65}$$

Accordingly, the governing equation for the B-surface becomes

$$(\bar{\nabla}^2)^2 w = (\nabla^2 + a^{-2})^2 w = p/D. \tag{8.66}$$

In our earliest example (section 8.3) we considered a B-surface loading of the form

$$p_B = p_{nB} \sin(\pi x/l) \sin(\pi y/b).$$

Using (8.66) instead of (8.9) we find that the relationship between p_{nB} and the corresponding displacement w_n (see (8.13)) is:

$$p_{nB} = \pi^4 D w_n \left(\frac{1}{l^2} + \frac{1 - n^{-2}}{b^2} \right)^2, \tag{8.67}$$

where n is the circumferential wavenumber, as before. In modes of deformation in which the generators remain straight we have $l \to \infty$; and putting this in (8.67) we recover essentially the previous result (8.62) for the ring.

The final equation (8.15) must be modified accordingly: the term 1* must be replaced by the corrected version $1 - n^{-2}$. Notice that the term $1/b^2$ in the 'stretching' expression still stands. Although we have found a need to modify the operator ∇^4 in the governing equation of the B-surface, there is no reason to make a corresponding change in the operator ∇^4 in the S-surface equations: the relationship $\nabla^4 \phi = -gEt$, expressing the change of Gaussian curvature in terms of the Airy stress function, is correct as it stands, and must not be altered.

The corrected version of (8.16) is

$$\zeta = \frac{12l^4}{\pi^4 a^2 h^2} \left[1 + \left(\frac{l}{b}\right)^2\right]^{-2} \left[1 + (1-n^{-2})\left(\frac{l}{b}\right)^2\right]^{-2}. \tag{8.68}$$

It is not difficult to follow the effect of these alterations on the two diagrams of fig. 8.4. For high values of n there is virtually no change to either diagram. Indeed, the distortion is relatively minor even when the value of n is as low as 3 or 2 (see problem 8.6). But when $n = 1$ there are big changes; the circumferential bending effect, which was formerly dominant in some parts of the diagrams, now vanishes. It is usually best to think about this special case as a separate, simple problem, for which the membrane hypothesis is often valid: see section 4.4.1.

It is also a straightforward matter to examine the consequences of the correction of the 'circumferential bending' term with respect to the response of a cylindrical shell to edge-loading, expressed in fig. 8.6 in terms of the roots of Donnell's equation. In section 8.5.2 we picked out three practically distinct regimes of relatively simple behaviour: therefore it will be useful to examine the effect of a change in circumferential bending stiffness on each of these in turn.

First, in the 'short-wave' region, we found that the behaviour was dominated by bending in the longitudinal direction and stretching in the circumferential direction: therefore a correction to the circumferential bending stiffness can have, at most, a negligible effect.

Second, in the 'flat-plate' region, where the stretching effects are negligible, there is *prima facie* a case for considering the correction. However, the region corresponds to $b/(ah)^{\frac{1}{2}} < 2.4$, or, in terms of the circumferential wavenumber n

$$n > 1.3(a/h)^{\frac{1}{2}}. \tag{8.69}$$

Thus
$$n^2 > 1.7(a/h),$$

and the difference between $1 - n^{-2}$ and 1 is trivial for practical 'thin-shell' values of a/h. Therefore we conclude that in this region also the correction to the circumferential bending stiffness has a negligible effect.

Third, in the 'long-wave' region, we found that the behaviour was dominated by stretching in the longitudinal direction and bending in the circumferential direction. It is clear therefore that the correction to the circumferential bending stiffness can have an important effect here in some circumstances. As a first approximation it seems reasonable to take account of the correction only when the factor $(1 - n^{-2})^2$ differs from unity by more than (say) 10%. This gives $n < 4.5$ or, equivalently

8.6 The improvement of Donnell's equations

$b/(ah)^{\frac{1}{2}} > 0.7(a/h)^{\frac{1}{2}}.$

It is a straightforward matter to show that in general the asymptotic form of the long-wave roots should be

$$\frac{l}{(ah)^{\frac{1}{2}}} = \frac{l^*}{(ah)^{\frac{1}{2}}} = \frac{0.838}{(1-n^{-2})^{\frac{1}{2}}} \frac{b^2}{ah}. \tag{8.70}$$

Thus, for example, when $n = 2$ the corresponding line on fig. 8.6 (of slope 2) should be raised by the factor 1.15 or, equivalently, moved to the left by the factor 0.93. Thus, as far as this graph is concerned, the change on account of the circumferential bending stiffness correction is relatively small for $n \geq 2$ in the only region in which it can be argued that it is not entirely negligible! Again, $n = 1$ gives a simple special case.

The conclusion is clear. We may use Donnell's equations, i.e. treat the B-surface as if it were flat, with confidence for all calculations except those corresponding to 'long-wave' behaviour. In this regime we should multiply the circumferential bending stiffness by the 'correction factor' $(1 - n^{-2})^2$. The value of this will only differ appreciably from unity for $n < 5$, say.

There is one final point which we need to consider. It was shown in section 8.2 that the idea of separating the shell into two distinct surfaces, S and B, was strictly correct if the force interaction between these surfaces involved three components. We then argued that in many cases the normal component of this interaction would be far larger in magnitude than the two tangential components, and in consequence we built up our theory on the basis of the tangential components being zero. In order to study the case of circumferential bending with low circumferential wavenumbers we examined the bending of a circular ring under purely radial loading. As we saw, it was necessary to consider the circumferential tension T in the ring in addition to the bending moment M and shearing force Q which together would be sufficient for analysis of a straight beam. From the solution we may recover the following expression for the tension:

$$T = -\frac{ap_n \sin n\theta}{(n^2 - 1)}. \tag{8.71}$$

The tension is thus in phase with the applied loading, but is *opposite in sign* from what it would be if it carried the loading as a momentless string: cf. section 4.2. It is, of course, usually much smaller in magnitude also, on account of the term $(n^2 - 1)$ in the denominator. This tension constitutes an embarrassment to our theory, for we began with the clear idea that the B-surface could carry no in-plane stress resultants. This tension is a direct consequence of our decision to have zero tangential interaction between our two surfaces. In practical terms, however, the consequences are not serious. We can imagine that the tensions in the B-surface are actually carried by the S-surface as additional loading. In a cylindrical shell they are in the circumferential direction. But we have found that the 'circumferential bending correction' is only significant in the 'long-wave' behaviour of the shell. Now in this range the normal displacement of the S-surface is attributable almost entirely to strains due to *longitudinal* in-plane stress resultants. Hence we conclude that the 'feed-back' of tension from the B- to the S-surface should have a trivial effect on the displacement of the S-surface, and consequently on the interaction between the two surfaces.

It was mentioned in section 8.5 that several investigators have examined the roots of the characteristic equations corresponding to Donnell's equations and various other equations of the same sort but based on less drastic simplifications, in order to assess the accuracy of Donnell's equations. The most remarkable of these studies is by Morley (1959); but see also Simmonds (1966). Donnell's equation for a circular cylindrical shell having zero surface traction is:

$$D\nabla^8 w + \frac{Et}{a^2} \frac{\partial^4 w}{\partial x^4} = 0. \tag{8.72}$$

Morley showed that if this equation is modified simply by replacing the expression $\nabla^8 w$ by

$$\nabla^4(\nabla^2 + a^{-2})^2 w,$$

the roots agree extremely well with those of the more 'exact' equations over the entire range of geometric variables. In other words, he showed that a *simple* modification to Donnell's equation is all that is necessary to make it almost as accurate as the other more complicated equations. Morley's modification corresponds almost exactly to the introduction of our 'correction' term to the circumferential bending stiffness of the B-surface. We can therefore, if we wish, quote Morley's work in justification of the various approximations which are inherent in Donnell's equations and our simple modification of them.

We have not yet considered the restrictions on the validity of Donnell's equations which apply when the surface of the shell is loaded *tangentially*. Here the basic problem lies in the fact that the form of Airy's stress function which we have used (see (7.36)) is only strictly valid for an S-surface which is not subjected to tangential surface tractions. We may consider this matter further by the methods of section 7.3.3.

8.7 Boundary conditions

So far we have discussed solutions of Donnell's equation in general terms; and although the 'maps' presented in figs. 8.4, 8.6 and 8.7 show clearly some general trends, we cannot claim to have mastered the behaviour of cylindrical shells until we have solved some specific *problems*. This will form the basis of chapter 9. But before we can proceed to solve a particular problem we must investigate the general question of *boundary conditions*; for it is clear that these will play an important part in the solution of the differential equations, just as they do in the analysis of beams.

In the present section we shall be concerned only with boundary conditions at edges x = constant of a cylindrical shell, for the sake of definiteness. The shell is continuous in the circumferential direction, and continuity of the various stress resultant, strain and displacement variables is assured directly by having an integral number of waves in the circumferential direction. (Some problems of cylindrical shell roofs, with edges along generators, are discussed in chapter 10.)

In fact boundary conditions of this kind are more complicated than those for beams, because there are twice as many of them. This is a direct consequence of the fact that shell equations (e.g. (8.28)) are eighth-order whereas beam equations are fourth order. Thus, while the satisfaction of the boundary conditions for a beam may be reduced to the solution for four simultaneous algebraic equations, the corresponding problem for shells involves the solution of eight simultaneous equations; and this is of course very much more laborious, in general.

Fortunately, there are some special features of the class of problems which we are considering here which make it possible to avoid solving eight simultaneous equations. The first of these is that when the value of $b/(ah)^{\frac{1}{2}}$ is sufficiently large, the original eighth-order differential equation degenerates, as we have seen, into two practically separate fourth-order equations of the 'beam-on-elastic-foundation' type. In these circumstances the boundary conditions separate out into two sets of four. We shall concentrate our analysis on this special case because it leads to clear and instructive results; and also, as we shall see in chapter 9, the modes which give the most prominent displacements in practical structures usually lie in this range. But we shall also comment briefly on the position when the value of $b/(ah)^{\frac{1}{2}}$ is small in section 8.7.4.

The second special feature which simplifies the boundary-value problem is that for the 'short-wave' solution (at least) there are usually two boundary conditions to be met at each end of the shell, and it often happens that for practical purposes the conditions at one end do not interfere with those at the other end. We saw an example of this in section 3.8: only for particularly

250   *Stretching and bending in shells*

short shells under axially symmetric loading is it necessary to deal with all four boundary conditions simultaneously. In this context, and also for the 'short-wave' solution of the non-symmetric problem, 'short' means 'of the order of $(ah)^{\frac{1}{2}}$'. Even when we make these simplifications we find that the necessary algebra is rather long and tedious. Fortunately, there are some relatively simple conclusions at the end, and some of the detail may well be omitted at a first reading. The analysis has much in common with that of Reissner & Simmonds (1966).

8.7.1 Separation of boundary conditions for the long- and short-wave solutions

The problem of boundary conditions now revolves about the question of how the 'long-wave' and 'short-wave' parts of the solution dovetail into each other. For the sake of simplicity we shall consider a cylindrical shell which is completely clamped at one end, but is kinematically unrestrained at the other, where arbitrary forces may be imposed. We shall also consider only solutions in which all stress resultants and displacements, etc. vary periodically in the circumferential direction. At the free end, therefore, we shall wish to impose stress resultants of the following kind (see fig. 8.9):

$$N_x = N^* \sin(\pi y/b), \quad N_{xy} = N^+ \cos(\pi y/b),$$
$$Q_x = Q^* \sin(\pi y/b), \quad M_x = M^* \sin(\pi y/b). \tag{8.73}$$

(We shall consider later the application of twisting moment M_{xy} to this boundary.)

How can we match these four boundary conditions to the two pairs of two

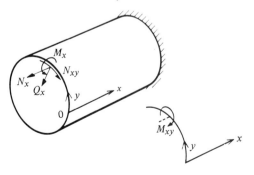

Fig. 8.9. Four stress resultants which may be applied at the free edge of a cylindrical shell, showing the positive sense in each case in relation to the given coordinate system. Inset: the positive sense of twisting moment M_{xy} (cf. fig. 2.5).

8.7 Boundary conditions

boundary conditions corresponding to the 'long-wave' and 'short-wave' beam-on-elastic-foundation equations respectively?

The key to the situation lies in the distinctive features of the two partial solutions. In the idealised 'long-wave' solution we have an interaction between longitudinal stretching and circumferential bending effects. Since bending in the longitudinal direction contributes negligibly to the carrying of pressure-loading, we conclude that at the edge of the shell we must have

$$M_x \approx 0, \quad Q_x \approx 0.$$

This therefore leaves the two in-plane stress resultants N_x, N_{xy} to be accommodated by the 'long-wave' part of the solution. There is certainly no problem over N_x ('longitudinal stretching') but it is not quite so clear how we can deal with N_{xy}. A closer examination of the 'long-wave' assumption shows that although N_x is the primary stretching effect, the magnitude of N_{xy} is not necessarily negligible: it is required to satisfy the in-plane equilibrium equation (8.4b). The crucial point about the 'long-wave' simplification is that N_{xy}, through γ_{xy}, makes only a trivial contribution to the change in Gaussian curvature in comparison to N_x, through ϵ_x.

On the other hand, the distinctive feature of the idealised 'short-wave' solution is that there is structural interaction between bending of the B-surface longitudinally and stretching of the S-surface circumferentially. So it seems clear that it will be possible to accommodate by means of the 'short-wave' solution the boundary conditions on M_x and Q_x which were effectively repelled by the 'long-wave' solution. In fact the situation is a little more complicated than this, as we shall see when we examine the equations carefully: the 'short-wave' solution also demands that Q_x be accompanied by a related stress resultant N_{xy}.

In order to proceed further we must write down explicitly the simplified governing equations of the problem. We may express the superposition of the two parts by writing the variables w, ϕ as follows:

$$w = [w_0(x) + w_1(x)] \sin(\pi y/b), \quad \phi = [\phi_0(x) + \phi_1(x)] \sin(\pi y/b). \tag{8.74}$$

Here subscripts 0 and 1 refer to 'short-wave' and 'long-wave' parts of the problem.

Our analysis of the roots of Donnell's equation was based on the assumption that there was no surface loading of the shell (the 'homogeneous case'). We shall use the same assumption here also. Surface loading will be introduced in the examples of chapter 9.

It is convenient to begin with the 'short-wave' part of the equations. Following the description given above we obtain these from (8.20) and (8.23) by retaining only that term of the operator ∇^4 which has the highest order derivative with respect to x (see problem 8.5). For a cylindrical shell

$$\Gamma^2(\ldots) = (1/a)\partial^2(\ldots)/\partial x^2$$

and we therefore obtain:

$$(1/a)\phi_0'' + Dw_0'''' = 0, \tag{8.75a}$$

$$(1/Et)\phi_0'''' - (1/a)w_0'' = 0. \tag{8.75b}$$

Here ' denotes differentiation with respect to x. The second of these equations is the condition for the compatibility of displacements of the S- and B-surfaces. We can obtain a single sixth-order equation in w_0 by eliminating ϕ_0 between the two equations. But it is tempting to reduce (8.75b) to a simpler equation by integrating twice:

$$(1/Et)\phi_0'' - (w_0/a) = C_1 + C_2 x. \tag{8.76}$$

Physically, the right-hand side with two arbitrary constants, corresponds to a family of *inextensional* deformations (cf. section 6.5) which may be superposed onto any given solution, and which we shall here discard for the sake of simplicity. They can be re-introduced later, if necessary. The compatibility equation thus becomes

$$(1/Et)\phi_0'' - (w_0/a) = 0, \tag{8.77}$$

and on substituting for ϕ_0'' in (8.75a) we obtain the governing equation

$$w_0'''' + (Et/Da^2)w_0 = 0. \tag{8.78}$$

This is, as we expect from the preceding work, not only a version of the well known 'beam-on-elastic-foundation' equation, but also precisely the same, formally, as the equation which we used in chapter 3 for the analysis of axially symmetric deformations of cylindrical shells in the absence of pressure-loading.

Let us now try to determine an expression for the out-of-plane shear-stress resultant Q_x in terms of w_0. Taking into account the negligible contribution of M_{xy} and M_y towards the carrying of load in the B-surface in the idealised 'short-wave' equations, we have, from the equilibrium equation (8.3b),

$$Q_x = \frac{\partial M_x}{\partial x}.$$

The appropriate form of Hooke's law is

$$M_x = -D\frac{\partial^2 w}{\partial x^2} \tag{8.79}$$

So taken together with (8.74) these give

$$Q_x = -Dw_0''' \sin(\pi y/b). \tag{8.80}$$

8.7 Boundary conditions

Indeed, this is what we should expect from the analogy with the equations for axially symmetric deformation.

Now although N_y is the predominant S-surface stress resultant in the 'short-wave' solution, the other in-plane stress resultants are required for the satisfaction of the in-plane equilibrium equations of the S-surface. Of these N_{xy} is more prominent than N_x (since it corresponds to a higher derivative of the stress function with respect to x); and it may be obtained from the equilibrium equation (8.2b)

$$\frac{\partial N_y}{\partial y} + \frac{\partial N_{xy}}{\partial x} = 0.$$

In the present case

$$N_y = \phi_0'' \sin(\pi y/b),$$

and so we obtain after integration

$$N_{xy} = -(\pi/b)\phi_0' \cos(\pi y/b). \tag{8.81}$$

Here we have set a constant of integration to zero on the grounds that all stress resultants decay rapidly to zero as x increases. Let us now express N_{xy} in terms of the variable w_0. Integrating (8.75a), and setting a constant at zero for the same reason as above, we obtain

$$\phi_0' = -aDw_0'''.$$

Therefore (8.81) may be written

$$N_{xy} = nDw_0''' \cos(\pi y/b). \tag{8.82}$$

Thus *both Q_x (8.80) and N_{xy} are directly related to w_0'''.* Hence Q_x and N_{xy} are directly related to each other as far as the 'short-wave' solution is concerned: if

$$Q_x = Q^* \sin(\pi y/b) \tag{8.83a}$$

then

$$N_{xy} = -nQ^* \cos(\pi y/b). \tag{8.83b}$$

Fig. 8.10 shows these two stress resultants acting on part of a shell cut off by an imaginary plane x = constant. The relationship between them has a very simple physical interpretation. In section 4.2 (cf. problem 4.9) it was shown that radial and tangential loading applied to a plane ring could be balanced by a varying pure tension (and no bending moment) provided a certain relationship held between the two loadings. The connection (8.83) between N_{xy} and Q_x satisfies precisely this relationship. Consequently, if a loading of the type shown in fig. 8.10 is applied to the edge of a cylindrical shell, it will be supported, *according to the membrane hypothesis*, by a simple line tension varying around the edge. In particular no stresses are required within the shell itself for satisfaction of equilibrium equations. Now we may recall that in the loading of the edge of a cylindrical shell by a shear-stress resultant Q_x which is *uniform* around the circumference, the membrane hypothesis leads to a stress distribution consisting only of a ring of uniform tension round the edge of the shell. Then, when the membrane hypothesis is replaced by a proper analysis which takes into account bending of the shell, we have instead the solution described in section 3.5. This pattern of behaviour provides a direct link with our present analysis of the 'short-wave' part of the non-symmetric edge-loading of a cylindrical shell. We find that the only boundary shear force conditions which can be accepted by the 'short-wave' part of the solution are those which, according to the membrane hypothesis, would cause a 'tension ring' loading at the edge.

We may also recall that in our calculations of the membrane displacements of shells in chapter 7 we studied an example with Q_x loading at the edge (section 7.4), which we could only deal with by inserting a suitable edge-ring. We can now see that the role of the 'short-wave' solution is to redistribute the 'ring of tension' required by the membrane hypothesis into a relatively narrow 'boundary layer' near the edge of the shell.

Thus we conclude that the role of the 'short-wave' solution is essentially a simple one: it provides a 'boundary-layer' effect to compensate for the fact that an actual shell is incapable of providing the concentrated 'rings of tension' which are one of the demands made by the membrane hypothesis.

It may seem odd that an applied edge shearing load $Q_x = Q^* \sin(\pi y/b)$ should require a companion in-plane shear loading N_{xy} according to (8.83). It should be emphasised that this demand is made only if the loading is to be carried exclusively by the 'short-wave' solution. If the Q_x-related N_{xy} is not in fact provided by the given force boundary conditions, the balance must be

Fig. 8.10 A combination of normal and tangential shear loading at an edge which is carried entirely by 'short-wave' effects: see (8.83).

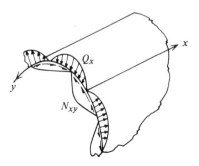

8.7 Boundary conditions

distributed somehow between the 'short-wave' and 'long-wave' solutions, as we shall see.

We discovered the need for the companion N_{xy} loading by studying the equations of the 'short-wave' solution. A simple qualitative explanation of the same idea may be derived from (8.77). In terms of N_y this states that

$$N_y = Etw/a, \qquad (8.84)$$

or, in consequence of Hooke's law

$$\epsilon_y = w/a. \qquad (8.85)$$

This is therefore a simplified form of the compatibility condition which we used implicitly in the 'short-wave' approximation. In fact this is precisely the form of the compatibility equation which is obtained from (6.9b) by putting $v = 0$; it corresponds to the displacement of the surface being purely *radial*. This provides a physical link with the axially symmetric problem, where the displacement is purely radial on account of symmetry. Now when we apply shear loading $Q_x = Q^* \sin(\pi y/b)$ to the edge of a cylindrical shell we may expect that some of the radial component of the displacement at the edge is due to tangential displacements, after the manner of the 'circumferentially inextensional' solutions of section 7.5. Such modes of displacement play no part in 'short-wave' solutions, where the displacement is essentially radial. It is precisely to ensure that there are no tangential displacements that the in-plane Q_x-related shear forces N_{xy} must be provided.

In conclusion we may write down the boundary condition for shear force on the assumption that (8.83) is satisfied: from (8.82) we have

$$Q^* = -Dw_0'''. \qquad (8.86)$$

Since the short-wave approximation corresponds to strong bending effects in the longitudinal direction, the edge of the shell is capable of sustaining bending moments M_x which are applied to it. With the present arrangement of variables (8.73) we may have at the edge

$$M_x = M^* \sin(\pi y/b);$$

and the corresponding boundary condition is, by (8.79),

$$M^* = -Dw_0''. \qquad (8.87)$$

We must now turn to the equations for the 'long-wave' part of the solution and any limitations on boundary conditions which may be implicit in them. The general line of attack is the same as before.

In the 'long-wave' behaviour the dominant effects are stretching of the S-surface longitudinally and bending of the B-surface circumferentially. Both of these effects are picked out in (8.20) and (8.23) by keeping only the term

$\partial^4/\partial y^4$ in the operator ∇^4. Using the present form (8.74) of the variables we obtain for the cylindrical shell in the absence of surface loading,

$$(1/a)\phi_1'' + D(\pi/b)^4 w_1 = 0 \tag{8.88a}$$

and

$$(1/Et)(\pi/b)^4 \phi_1 - (1/a)w_1'' = 0. \tag{8.88b}$$

Again, ϕ_1 may be eliminated, to give the following governing equation in terms of w_1:

$$w_1'''' + (Da^2/Et)(\pi/b)^8 w_1 = 0. \tag{8.89}$$

If necessary, ϕ_1 may be expressed in terms of w_1 as follows:

$$\phi_1 = (Et/a)(b/\pi)^4 w_1''. \tag{8.90}$$

Equation (8.89) is of the familiar 'beam-on-elastic-foundation' type, and its solution presents little difficulty, provided we can find appropriate expressions for the boundary conditions. The restrictions on boundary conditions which are intrinsic in the simplified equation (8.89) are simpler to investigate than for the 'short-wave' solution. As we have already remarked, the 'long-wave' solution is incapable of sustaining boundary forces Q_x or M_x; so we are left to investigate the way in which N_x and N_{xy} are accommodated.

A boundary condition

$$N_x = N^* \sin(\pi y/b)$$

may readily be expressed in terms of the Airy stress function ϕ_1:

$$N^* = -(\pi^2/b^2)\phi_1$$

or, indeed in terms of w_1 by use of (8.90):

$$N^* = -(Et/a)(b/\pi)^2 w_1''. \tag{8.91}$$

Suppose that a boundary condition on N_{xy} is also applied:

$$N_{xy} = N^+ \cos(\pi y/b).$$

In terms of ϕ_1 this gives

$$N^+ = -(\pi/b)\phi_1'$$

or, in terms of w_1

$$N^+ = -(Et/a)(b/\pi)^3 w_1'''. \tag{8.92}$$

In terms of the beam-on-elastic-foundation equation (8.89) we thus see that boundary conditions N_{xy} and N_x in the shell correspond to a shearing force and bending moment, respectively in the beam. The connection between N_x and a bending moment is precisely that which we discovered in the calculation of membrane displacements in section 7.5; and indeed the 'elastic foundation' term in (8.89) is precisely the effect of circumferential bending which we

8.7 Boundary conditions

concluded in section 7.3.2 would interfere with the membrane hypothesis in certain circumstances. Some quantitative calculations will be done in chapter 9.

The results of this section may be summarised, as follows. For an edge-loading of the following general form;

$$N_x = N^* \sin(\pi y/b), \quad N_{xy} = N^+ \cos(\pi y/b),$$

$$Q_x = Q^* \sin(\pi y/b), \quad M_x = M^* \sin(\pi y/b),$$

the boundary conditions may be expressed in the following way:

short-wave part long-wave part

$$\left.\begin{array}{ll} M^* = -Dw_0'' & N^* = -(Et/a)(b/\pi)^2 w_1'' \\ Q^* = -Dw_0''' & N^+ + a\pi Q^*/b = -(Et/a)(b/\pi)^3 w_1'''. \end{array}\right\} \quad (8.93)$$

Note that the shearing force N_{xy} received by the long-wave part of the solution includes a term contributed by the reaction from the response of the short-wave part of the solution to the transverse shear loading Q_x.

8.7.2 Twisting moments M_{xy} applied at an edge

We must now consider what happens if a twisting-moment stress resultant is applied to an edge of the shell. So far the boundary conditions have worked out neatly: broadly speaking N_x, N_{xy} are boundary conditions for the 'long-wave' solution and M_x, Q_x are boundary conditions for the 'short-wave' solution. Certainly we need *two* conditions at each edge for each of the two separate beam-on-elastic-foundation problems. But clearly we could also apply a loading M_{xy} at the boundary if we wished, thus making apparently a third condition. This kind of problem is almost as old as the theory of plates, and was resolved in 1850 by Kirchhoff for flat plates (see, e.g., Timoshenko, 1953, §55) and Lamb (1890) and Basset (1890) for shells. The problem is most easily grasped in relation to flat plates. If the Kirchhoff hypothesis is made (see section 2.3) the plate may be reduced to a two-dimensional surface and its governing partial differential equation established in terms of the transverse displacement w. It is a fourth-order equation, demanding in general two boundary conditions at each edge. But the force boundary conditions on (say) an edge x = constant of a physical plate involve three quantities which may obviously be specified independently:

$$M_x, M_{xy}, Q_x.$$

The discrepancy is resolved by observing that the rate of change of twisting moment with respect to y along the edge is *statically* indistinguishable from a transverse shearing force Q_x: see fig. 8.11a. Indeed, this may also be seen in the equilibrium equations (8.5); and see problem 8.7. Now we may protest that twisting moments M_{xy} may in fact be applied to the edge of a plate by pairs of forces acting *parallel* to the xy-plane, whereas shearing force Q_x acts *normal* to this plane; so the effects may clearly be distinguished. However, the crucial point is that by invoking Kirchhoff's hypothesis we have reduced everything to the xy-plane, and that in this *idealised* situation the two effects cannot be distinguished. For a real plate with finite thickness it may be clear that we are applying M_{xy} as distinct from Q_x; and we reconcile the two views by invoking a narrow edge strip whose width is the same order as the thickness. Beyond this strip, as a consequence of St Venant's principle, the effects are indistinguishable.

In this discussion we have established the connection between M_{xy} and Q_x in terms of *statical equivalence*. This is also the key to the more complicated situation which obtains when we are dealing with the curved edge of a shell. On the edge of a cylindrical shell there are three distinct quantities involving shear-stress effects:

$$M_{xy}, N_{xy}, Q_x.$$

A simple analysis (see fig. 8.11b) shows that these are statically equivalent to 'effective' tangential and normal shear forces T_{xy}; V_x, respectively, where

$$T_{xy} = N_{xy} + \frac{M_{xy}}{a}, \qquad (8.94a)$$

$$V_x = Q_x + \frac{\partial M_{xy}}{\partial y}. \qquad (8.94b)$$

Fig. 8.11. (a) The statical equivalence of Q_x and dM_{xy}/dy on a straight edge: the couples M_{xy} per unit length are represented here by opposing forces parallel to Q_x. (b) Along a curved edge the arrangement is more complicated: M_{xy} is statically equivalent to components of both normal and tangential shear.

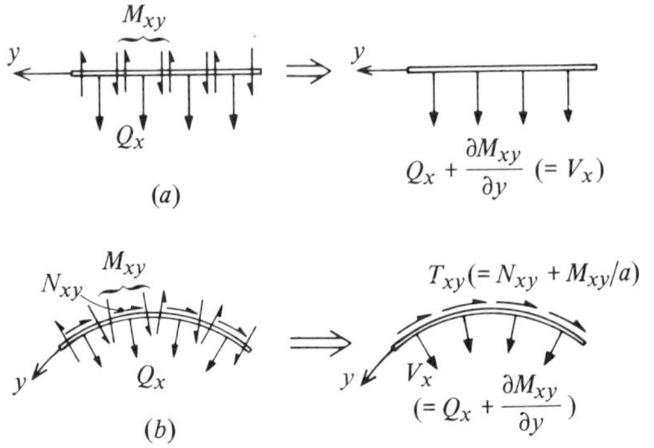

8.7 Boundary conditions

Thus, if twisting moments M_{xy} are applied to an edge, (8.94) should be used to convert the loading into equivalent normal and tangential shear loading: then (8.93) may be used as before. Note that, according to this scheme, the application of an M_{xy} boundary condition would involve both the 'short-wave' and 'long-wave' parts of the solution. It is easy to rewrite (8.94) in more general coordinates if this should be necessary. In the small-deflection behaviour of uniform isotropic plates it is straightforward to express Q_x, M_{xy} and V_x in terms of the derivatives of transverse displacement w: see problem 8.8.

8.7.3 Displacement boundary conditions

Lastly we consider the simple case of a fully-clamped boundary as an illustration of a *kinematic* or *displacement* boundary condition. As before, we shall consider a case in which the value of $b/(ah)^{\frac{1}{2}}$ is sufficiently large for the separation of the governing equation into 'short-wave' and 'long-wave' parts to be reasonable.

In considering the various kinds of *forces* which could be applied to the free end of our shell, we found that the displacement of that edge of the shell may be expressed as the sum of the displacements for two separate problems. Therefore, if the *displacement* is specified at an edge of the shell, there must be, in effect, two separate force boundary conditions which interact in a way which gives the imposed displacement as the sum of two parts. This may seem to be a more awkward problem than for a free edge loaded by forces, but in fact it is not difficult to solve.

The key to the situation is that, as we have seen, the displacements corresponding to the idealised short-wave solution are *purely radial*. Therefore, although a short-wave solution in the vicinity of an edge has the power to change the values of w and dw/dx at the edge, it cannot affect the values of the displacement components u and v. Our policy is therefore clear. First we must use the 'long-wave' part of the solution to satisfy displacement conditions on the axial and tangential components u and v respectively of the displacement at the edge, and then we must superpose a 'short-wave' solution which will bring w and dw/dx to their prescribed values.

For an edge which is completely clamped the four kinematic conditions are:

$$u = v = w = \frac{\partial w}{\partial x} = 0. \tag{8.95}$$

As before, all of the variables are periodic in the circumferential direction. At the edge we may put

$$\left. \begin{array}{lll} u = u^* \sin(\pi y/b), & v = v^* \cos(\pi y/b), & w = w^* \sin(\pi y/b), \\ N_x = N^* \sin(\pi y/b), & N_{xy} = 0, & N_y = 0. \end{array} \right\} \tag{8.96}$$

Here we have set stress resultants N_{xy} and N_y to zero in accordance with the basic simplifying assumption of the 'long-wave' equations, although this is a more drastic simplification for N_{xy} than the one we used earlier. In general N_{xy} is not zero in the long-wave solution, in the sense that it must satisfy the equilibrium equation (8.2a); but at present our main interest is in the strains and displacements, and we wish to exclude shearing strains γ_{xy} as a first approximation for the reasons set out in section 7.4. By Hooke's law (2.5) therefore,

$$\epsilon_x = (N^*/Et)\sin(\pi y/b), \tag{8.97a}$$

$$\epsilon_y = -(\nu N^*/Et)\sin(\pi y/b), \tag{8.97b}$$

$$\gamma_{xy} = 0. \tag{8.97c}$$

Substitution into the strain displacement equations (6.9) gives, in general,

$$u^{*\prime} = N^*/Et, \tag{8.98a}$$

$$w^* - nv^* = -\nu a N^*/Et, \tag{8.98b}$$

$$v^{*\prime} + (\pi/b)u^* = 0. \tag{8.98c}$$

We now impose the boundary conditions

$$u^* = v^* = 0. \tag{8.99}$$

The first gives, by (8.98c), $v^{*\prime} = 0$, which we can use by differentiating the general expression (8.98b) with respect to x: thus we have

$$w^{*\prime} = -\nu a N^{*\prime}/Et. \tag{8.100}$$

The second boundary condition gives, by (8.98b)

$$w^* = -\nu a N^*/Et. \tag{8.101}$$

We plan to solve the 'long-wave' equations by using the beam-on-elastic-foundation analogue, and we can see at once that for a shell made of a special material having $\nu = 0$, the boundary conditions would be particularly simple:

$$w^* = w^{*\prime} = 0.$$

The equivalent beam would be simply *clamped*. Indeed, since the actual boundary conditions for the shell are

$$w^* = w^{*\prime} = 0,$$

where w is the sum of a 'long-wave' and a 'short-wave' component, i.e.

$$w^* = w_0 + w_1,$$

we see that the boundary condition for the 'short-wave' partial solution in this case is simply

$$w_0 = w_0' = 0. \tag{8.102}$$

In other words, when $\nu = 0$ the 'short-wave' solution is not required at the

8.7 Boundary conditions

clamped edge. This simple result depends strongly on our assumption that the shear strains γ_{xy} are negligible, and is thus not strictly correct. But it is satisfactory for many purposes. Note that the specially simple boundary conditions in the 'quasi-inextensional' theory of section 7.5 were partly a consequence of having $\epsilon_y = 0$, which corresponds to $\nu = 0$ in the present analysis.

In general, of course, we have to reckon with nonzero values of Poisson's ratio, and it thus seems that we shall not be able to use the particularly convenient boundary conditions (8.102). Fortunately, this is not so.

Consider the following argument. Suppose we have solved the 'long-wave' equations with $\nu = 0$ in the in-plane part (2.5) of Hooke's law (but not in the bending part), subject to boundary conditions (8.99). Now let us gradually increase the value of ν, making sure that we keep the distribution of N_x the same throughout. The governing equation (8.24) is unchanged by this variation of ν, and the only changes which we need to make are in the boundary conditions. Since N_x does not change, we simply need to adjust w^* and $w^{*\prime}$ in accordance with (8.100) and (8.101) as the value of ν changes, for whatever values of N^* and $N^{*\prime}$ were found from the original 'beam-on-elastic-foundation' solution. Observe in particular that this prescription satisfies *both* boundary conditions (8.99) at once. This feature is a consequence of putting $N_{xy} = N_y = 0$ in (8.96), and is therefore not quite correct. But this simplification is hardly misleading. Having solved the basic long-wave equations for the simple boundary conditions (8.99), and then having made the adjustments described above, we must now simply add a short-wave solution which will bring w^*, $w^{*\prime}$ back to their prescribed values of zero. Thus we must superpose a 'short-wave' solution in the vicinity of the far edge for which

$$w_0 = \nu a N^*/Et, \tag{8.103a}$$
$$w_0' = \nu a N^{*\prime}/Et. \tag{8.103b}$$

As we have already shown, the short-wave equations are directly analogous to the equations for axially symmetric deformation of a cylindrical shell. Therefore it is easy to find the short-wave part of the solution by direct application of the results of chapter 3. Indeed, we can see that the short-wave solution that is required is of precisely the same kind as that which is required to take account of the 'Poisson ratio' effect when a cylindrical shell in axially symmetric longitudinal tension is connected to a perfectly rigid boundary (see problem 3.20); the results may be used directly, provided of course that the axially symmetric functions are multiplied by $\sin(\pi y/b)$.

Other displacement-type boundary conditions may be treated along the same lines. Conditions on u and v should be applied to the 'long-wave' part of the solution, but expressed in terms of w, and used with $\nu = 0$. The relaxation of this restriction on Poisson's ratio then gives nonzero edge displacements w, dw/dx for the 'long-wave' solution, which may then be cancelled by the superposition of a simple 'short-wave' solution in the vicinity of the edge. Provided the value of $b/(ah)^{\frac{1}{2}}$ is sufficiently large for the 'long-wave' and 'short-wave' solutions to be separable (cf. fig. 8.6), we may take over directly the results of chapter 3 to meet a variety of 'short-wave' boundary conditions.

8.7.4 'Flat-plate' region

The entire analysis of the last three sections, in terms of the superposition of 'long-wave' and 'short-wave' solutions, depends crucially on the assumption that the governing eighth-order characteristic equation can be separated into two independent fourth-order parts. This can only be justified, of course, for sufficiently large values of $b/(ah)^{\frac{1}{2}}$, which implies (for a given value of a/h) sufficiently small values of the circumferential wavenumber n. We have developed methods which give a clear picture of the physical behaviour of the shell in these circumstances; but what happens when the value of $b/(ah)^{\frac{1}{2}}$ is not particularly large?

As we saw in fig. 8.6, the region $b/(ah)^{\frac{1}{2}} < 1$ corresponds to behaviour which closely resembles that of a semi-infinite flat plate. Here the form of the shell deviates from a plane by so little in the order of one circumferential half wave that radial displacements induce relatively little change of Gaussian curvature, and consequently the applied loads are carried almost exclusively by bending actions.

The above explanation accords with the view which we have taken throughout the chapter, of behaviour seen in terms of the radial displacement w. We remarked earlier that the governing equation may alternatively be expressed in terms of ϕ: see (8.24), (8.25). For cases in which $b/(ah)^{\frac{1}{2}} > 3$, say, a reworking of the boundary conditions of the problem in terms of ϕ gives precisely the same results as those which we have already obtained: the eighth-order differential equation degenerates into two fourth-order equations representing the 'long-wave' and 'short-wave' modes respectively, and either may be worked in terms of w or ϕ.

The situation is quite different when $b/(ah)^{\frac{1}{2}} < 1$. In this case (8.20) and (8.23) degenerate into two fourth-order equations which may be written (when $p = 0$)

$$\nabla^4 w = 0, \quad \nabla^4 \phi = 0. \tag{8.104}$$

The first of these describes the pure bending of a flat plate, while the second describes the behaviour of a plate in conditions of *plane stress*. In these circumstances it is meaningless to attempt to use variables w, ϕ as alternatives.

8.7 Boundary conditions

The consequences of all this are straightforward in terms of boundary conditions. In the range $b/(ah)^{\frac{1}{2}} < 1$ periodic edge-loadings M_x, M_{xy}, Q_x are carried by the shell behaving essentially as a semi-infinite flat plate in bending, while periodic loadings N_x, N_{xy} are carried by the shell acting as a semi-infinite flat plate in plane stress. The interrelation between Q_x and N_{xy} described in section 8.7.1 does not apply here, because the circumferential wavenumber n is high.

All of these remarks hold only in an approximate sense, of course: the two equations (8.104) are not truly separate unless $b/(ah)^{\frac{1}{2}} \to 0$. Nevertheless, they provide a useful qualitative picture in the region $b/(ah)^{\frac{1}{2}} < 1$, say. Thus, for example, if the loading shown in fig. 8.12 (cf. Kildegaard, 1969) were analysed in terms of Fourier series in N_x we could say that the terms having *low* circumferential wavenumbers would be carried by means of long-wave behaviour in a way which will be described in chapter 9, while the terms having *high* circumferential wavenumber would be carried according to the well-known Boussinesq solution for a flat plate (Timoshenko & Goodier, 1970, §36). The wavenumber at which the crossover between these two modes occurs is given by $b/(ah)^{\frac{1}{2}} \approx 3$, i.e. $n \approx (a/h)^{\frac{1}{2}}$. Note particularly that the behaviour of the shell is substantially different from that which we described in section 4.4.1 on the basis of the membrane hypothesis. Similar remarks apply to the problem of a 'waisted' shell subjected to 'point' loads as in fig. 4.16b. According to the membrane hypothesis the load is carried across the shell by straight 'lines of tension'. But by studying the full eighth-order equation we can now see that the response of the shell in the immediate vicinity of the edge force is to 'diffuse' the stress in practically the same way as a flat plate. The 'hyperbolic' nature of the membrane-hypothesis solution is only recognisable for Fourier terms which have sufficiently large circumferential half wavelengths: see fig. 8.7a.

Fig. 8.12. A type of edge-loading for which the low Fourier terms are carried by 'long-wave' behaviour of the shell, while the high Fourier terms are carried by 'Boussinesq' plate-like action: cf. fig. 4.6.

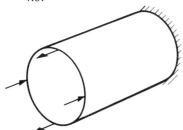

8.8 Summary and discussion

The division of a shell, conceptually, into coincident but distinct S- and B-surfaces enables us to grasp clearly the way in which the shell carries applied loading by a combination of in-plane and bending-stress resultants. The force interaction between the surfaces usually consists mainly of a normal interface pressure (but see section 11.5) and the corresponding compatibility condition can sometimes be expressed in terms of common normal displacements (as in the example of section 8.4) or, more generally, in terms of common change of Gaussian curvature on account of the distortions produced by the respective stress resultants.

The two-surface idea leads directly to the Donnell equations for cylindrical shells and a variant for 'nearly-cylindrical' elliptic and hyperbolic shells. It is useful to determine the roots of the characteristic equation for the case in which all quantities vary sinusoidally in the circumferential direction, and to map out the results on a dimensionless plot. This reveals important simplifying features of the solution in certain circumstances: in particular the solution degenerates into two independent beam-on-elastic-foundation systems in a wide range of practical problems. The 'Donnell' equations are based on a number of simplifications. Of these, the one which must be taken most seriously concerns the circumferential bending stiffness of the B-surface for low circumferential wavenumbers, say $n \leq 4$. A single correcting term gives a sufficient improvement in the equation for all practical purposes. The boundary conditions for two simplified equations are relatively straightforward, and in particular we can see clearly the role of the 'short-wave' set of solutions in providing a boundary layer effect which performs a simple and specific task.

Throughout the chapter we have emphasised the circumstances in which the governing eighth-order differential equation 'degenerates' into more-or-less 'uncoupled' fourth-order equations. The key to this uncoupling is the value of $b/(ah)^{\frac{1}{2}}$ or, equivalently, the value of n if a/h is given: see fig. 8.6. In particular, we have paid almost no attention to those circumstances in which the equations do not uncouple; say

$$1 < b/(ah)^{\frac{1}{2}} < 3.$$

It is certainly possible to invent problems which lie in this range, and indeed it will sometimes be necessary to work in this range: see section 14.6.

The application of the ideas of the present chapter to practical problems almost always involves the decomposition of the behaviour of a shell by means of a Fourier series into a set of components which are periodic in the circumferential direction. When we come to tackle practical problems in the following chapter, we shall almost always find that the behaviour is dominated by only a *few terms* of the Fourier series; and moreover that the values of n for these terms depend on the dimensions of the shell and the particular boundary conditions. It is only at this stage in the analysis that we can discover what values of n we should use in locating the problem on the 'maps' of figs. 8.4, 8.6 and 8.7. We shall find in almost all cases that the dominant Fourier terms are operating in the range $b/(ah)^{\frac{1}{2}} > 3$, for which the governing equation does degenerate into two simpler equations. In other words, it is impossible to check on the validity of the simple solution in a given case until the solution has been worked out in detail.

The adaptation of Donnell's equation to a problem involving thermal expansion is considered in problem 8.9.

8.9 Problems

8.1 Rearrange (8.15) to give a measure of the panel stiffness $(a^2/Et)(p_n/w_n)$ in terms of $l/(ah)^{\frac{1}{2}}$ and $b/(ah)^{\frac{1}{2}}$. Evaluate this factor as a function of $b/(ah)^{\frac{1}{2}}$ for $l/(ah)^{\frac{1}{2}} = 10, 20, 50$ and 100, and plot the results as four curves on double-logarithmic paper.

When $b/l < \frac{1}{3}$, say, l^{-2} may be neglected in comparison with b^{-2}. Simplify (8.15) in this way and hence show that $\partial(p_n/w_n)/\partial b = 0$ when $b^2/(ah) \approx 1.69$ $l/(ah)^{\frac{1}{2}}$; or, equivalently, $n^2 \approx 5.85(a/l)(a/h)^{\frac{1}{2}}$.

8.2 Show that if (8.17) is rearranged as a relationship between $(ah)^{\frac{1}{2}}/l = X$ (say) and $(ah)^{\frac{1}{2}}/b = Y$ (say), it represents a circle in a Cartesian X, Y space, having its centre on the X-axis and passing through the origin.

8.3 Show that when (8.17) is satisfied, (8.15) may be written

$$p_n = w_n \frac{\pi^2}{3^{\frac{1}{2}}} \frac{Eth}{al^2}.$$

Hence show that for the axisymmetric mode $(b/l \to \infty)$ satisfying (8.17), $l = l_0 = (\pi/12^{\frac{1}{4}})(ah)^{\frac{1}{2}}$, $p_n/w_n = 2Et/a^2$. Confirm from first principles that half of p_n comes from simple axisymmetric stretching of the shell. Also show that the mode having $l = b = 2l_0$ satisfies (8.17), and that the value of p_n/w_n is one-quarter of the value corresponding to the above axisymmetric mode.

8.4 Investigate the roots of (8.28) of the type (8.30) for a cylindrical shell (i.e. $m = 0$) by the methods of section 8.5.1. Let the solutions be

$$w = Ae^{\pi y/b^*} \sin(\pi y/b) \sin(\pi x/l), \text{ etc.}$$

where b and b^* are functions of the (given) longitudinal half wavelength l.

Define $\psi = l/b^* + il/b$, and $\gamma = l/2^{\frac{1}{2}}\lambda$ (cf. 8.37, 8.38), and show that the counterpart of (8.39) (with $m = 0$) is the simpler equation $(\psi^2 - 1)^4 + 16\gamma^4 = 0$. Verify that the roots of this are

$$\psi^2 - 1 = 2^{\frac{1}{2}}\gamma(\pm 1 \pm i) \text{ and that}$$
$$b/c = (2^{\frac{1}{2}}l/c)[(1 \pm 2R + 2R^2)^{\frac{1}{2}} - (1 \pm R)]^{-\frac{1}{2}}$$
$$b^*/c = (2^{\frac{1}{2}}l/c)[(1 \pm 2R + 2R^2)^{\frac{1}{2}} + (1 \pm R)]^{-\frac{1}{2}}$$

where $R = (3^{\frac{1}{4}}/\pi)(l/c)$. Here, the ambiguous signs correspond to two families of solutions, which are labelled 1, 2 in the plot below.

Verify that when $\gamma \to 0$ the equation degenerates into that of a flat plate; and that in the lower left-hand corner the curves match those of fig. 8.6, when due account is taken of the fact that l is now the independent variable. Also confirm that when $\gamma \gg 1$ the solutions, corresponding to the roots of $\psi^8 + 16\gamma^4 = 0$, are:

$$\frac{b_2}{c} \approx \frac{b_1^*}{c} \approx \left(\frac{l}{c}\right)^{\frac{1}{2}} \left\{\frac{2\pi}{(2^{\frac{1}{2}} + 1)3^{\frac{1}{4}}}\right\}^{\frac{1}{2}},$$

$$\frac{b_1}{c} \approx \frac{b_2^*}{c} \approx \left(\frac{l}{c}\right)^{\frac{1}{2}} \left\{\frac{2\pi}{(2^{\frac{1}{2}} - 1)3^{\frac{1}{4}}}\right\}^{\frac{1}{2}}.$$

Note that in the region $l/(ah)^{\frac{1}{2}} > 10$ (say) the solutions, like the 'long-wave' solution of fig. 8.6, lie broadly in the shaded zone of slope 2 in fig. 8.4; but that there is no counterpart of the short-wave solution in this case. Verify also

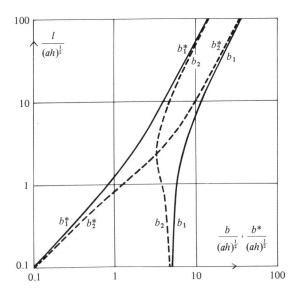

that in this region the roots satisfy $b_1/b_1^* \approx b_2^*/b_2 = 1 + 2^{\frac{1}{2}}$: and that there is thus a disparity between the half wavelength b and the 'pseudo half wavelength' b^*

8.5 (a) Show that when $G \ll 1$ (i.e. $b/(ah)^{\frac{1}{2}} \ll 1$) (8.47) reduces to $R \approx I_S \approx I_L \approx (2G)^{\frac{1}{2}}$; and that when $G \gg 1$ (i.e. $b/(ah)^{\frac{1}{2}} \gg 1$) (8.47) reduces to $R \approx 1$, $I_S \approx 4G$, $I_L \approx 1/4G$. Relate these results to fig. 8.5.

(b) Consider the expanded polynomial form of (8.39) for $m = 0$ when $\gamma \gg 1$, i.e. when $b/(ah)^{\frac{1}{2}} \gg 1$. Show that for roots in which $|\psi| \gg 1$ the equation reduces to $\psi^4 + 16\gamma^4 = 0$, of which the roots are $\psi = (\pm 1 \pm i)2^{\frac{1}{2}}\gamma$; and that for roots in which $|\psi| \ll 1$ the equation reduces to $\psi^{-4} + 16\gamma^4 = 0$, of which the roots are $\psi = (\pm 1 \pm i)/2^{\frac{3}{2}}\gamma$. Relate these results to fig. 8.6 by using (8.37) and (8.38).

(c) Investigate which terms in the original differential equation (8.28), (8.29) are retained in the two special cases in (b), thereby giving two distinct 'beam-on-elastic-foundation' equations; and show that in 'short-wave' solutions ($|\psi| \gg 1$) circumferential stretching and longitudinal bending are the dominant effects, whereas in 'long-wave' solutions ($|\psi| \ll 1$) longitudinal stretching and circumferential bending effects dominate.

8.6 Rearrange (8.68) to give relationships between $l/(ah)^{\frac{1}{2}}$ and $b/(ah)^{\frac{1}{2}}$ for which ζ = constant in cases (i) $(l/b) \ll 1$, (ii) $(l/b) \gg 1$. Hence show that the correction required to fig. 8.4a for small values of n ($\geqslant 2$) is practically equivalent to a shift of all curves to the left by the factor $(1 - n^{-2})^{\frac{1}{4}}$; which is equal to 0.93, 0.97 and 0.98 for $n = 2, 3$ and 4, respectively.

In relation to fig. 8.4b, show by examination of (8.67) that the circumferential bending correction for low values of n ($\geqslant 2$) involves a shift to the left by the factor $(1 - n^{-2})^{\frac{1}{2}}$ of all those lines for which circumferential bending is dominant; i.e. of all lines parallel to the $l/(ah)^{\frac{1}{2}}$ axis.

8.7 A rectangular plate is in a state of uniform twist under the action of uniform twisting-stress resultants M_{xy} ($= M_{yx}$) applied to the four edges (cf. fig. 2.5). Show that the loading applied to any one edge is statically equivalent to concentrated transverse forces of magnitude M_{xy} applied in opposition at the two ends of the edge; and that consequently the loading on the entire plate is statically equivalent to transverse forces of magnitude $2M_{xy}$ applied at each of the four corners.

8.8 Make use of (2.28), (8.5), (8.8) and (8.94b) to establish the following relations for the flexure of a uniform isotropic flat plate:

$$M_x = -D\left(\frac{\partial^2 w}{\partial x^2} + \nu \frac{\partial^2 w}{\partial y^2}\right), \quad M_{xy} = -D(1-\nu)\frac{\partial^2 w}{\partial x \partial y}$$

$$Q_x = -D\left(\frac{\partial^3 w}{\partial x^3} + \frac{\partial^3 w}{\partial x \partial y^2}\right), \quad V_x = -D\left(\frac{\partial^3 w}{\partial x^3} + (2-\nu)\frac{\partial^3 w}{\partial x \partial y^2}\right).$$

(The sign conventions for stress resultants are shown in figs. 2.2 and 8.11a.)

8.9 Donnell's equations have been derived in section 8.5 on the assumption that the shell does not suffer any strains due to thermal expansion in addition to these due to stress, on account of the elasticity of the material. In a particular case there are strains $\epsilon_x = \epsilon_y = \epsilon_0(x, y)$, $\gamma_{xy} = 0$ due to thermal expansion of the material. Writing a modified form of Hooke's law, as follows

$$\epsilon_x = \frac{1}{Et}(N_x - \nu N_y) + \epsilon_0,$$

$$\epsilon_y = \frac{1}{Et}(-\nu N_x + N_y) + \epsilon_0,$$

$$\gamma_{xy} = \frac{2(1+\nu)}{Et} N_{xy}$$

in place of (2.5), show that (8.21) must be replaced by

$$g_S = -\frac{\nabla^4 \phi}{Et} - \nabla^2 \epsilon_0;$$

and that consequently the right-hand side of (8.23), (8.24) and (8.25) must be replaced by $-\nabla^2 \epsilon_0$, $\nabla^4 p + Et\nabla^2 \Gamma^2 \epsilon_0$ and $Et\Gamma^2 p - DEt\nabla^6 \epsilon_0$, respectively.

9

Problems in the behaviour of cylindrical and nearly-cylindrical shells subjected to non-symmetric loading

9.1 Introduction

The ideas and equations which we have developed in chapter 8 provide tools for the solution of a wide range of practical problems. The aim of the present chapter is to give examples of some of these applications. The order follows roughly that of chapter 8.

The first example is concerned with a cylindrical tube which is partly full of a heavy fluid and acts as a beam between supports at its ends. We shall be able to make direct use of results from chapter 8 in determining the stresses and displacements in the shell, and we shall find that the pattern of behaviour depends strongly on the value of a certain dimensionless group Ω (to be defined in (9.9)) which involves the length, radius and thickness of the shell.

The next group of examples involves the response of a cylindrical shell to forces applied at one end while the other end is supported in various different ways which are met in engineering applications. The applied forces all vary periodically in the circumferential direction. The solutions will involve mainly the 'long-wave' behaviour of the shell, and it is advantageous to begin by setting out some standard 'beam-on-elastic-foundation' results which will be useful in the subsequent work. The last example in this group concerns the response of a cylindrical shell to a radial point load.

The final problem to be considered in this chapter is the response of a spherical shell to a radial point load: this may be discussed in terms of a 'nearly-cylindrical' shell.

An important idea which runs through the chapter is that it is advantageous in each particular problem to consider the response of the shell to the applied loading in terms of appropriate dimensionless groups which describe the geometry of the shell. This idea emerges naturally in the first example, and it carries through to all of the subsequent problems.

9.2 A preliminary example: a cylindrical shell acting as a beam

Cylindrical pipes and tubes are often required to act as beams carrying their self-weight and the weight of their contents over spans between points of support, in addition to their design requirements as pressure containers. The proportions of these structures vary over a wide range. At one end of the range are 'thick-walled' pipes having a radius/thickness ratio of order 3 and span/radius ratio of order 100, while at the other end there are boiler-drums having a radius/thickness ratio of order 100 and span/radius ratio of order 3. For a 'thick-walled' pipe, as described above, it seems reasonable to use simple beam theory, since the loading is unlikely to produce any appreciable change in the cross-sectional shape. But for a thin-walled boiler, possibly half full of water, we must clearly use shell theory; and in particular we must be prepared for the possibility that the membrane hypothesis will be inadequate (cf. section 7.3.2).

The problem which we shall consider here is shown schematically in fig. 9.1. A horizontal uniform cylindrical shell of length L, radius a and thickness t is simply supported at its ends. We shall mainly be concerned with the case where the shell is half full of a heavy fluid; but we shall also discuss briefly the much simpler case where the shell is completely full of a pressurised heavy fluid, as in fig. 9.2. In both cases the shell is subjected to (normal) pressure, which varies with position; and it will be possible to make use of various results obtained in chapters 7 and 8. In practice, loading of this sort will be combined with gravitational self-weight loading of the shell itself. We have already considered a closely related problem in section 4.4.1: except possibly at the ends, the membrane hypothesis is satisfactory for this loading case since the circumferential variation of all relevant quantities is of the kind $n = 1$.

Fig. 9.1. A cylindrical shell, half-full of a heavy fluid, acts as a beam between end-supports (for details of which, see text).

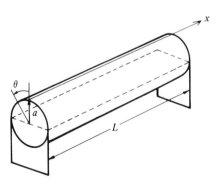

9.2 Example: cylindrical shell acting as a beam

For this particular problem we shall not be concerned with the details of stresses in regions of the shell adjacent to the supports. For the sake of simplicity we shall assume that the ends are held circular, but that no restraint is provided in the axial direction. Moreover we shall imagine that an axial 'expansion joint' is provided so that the force exerted by fluid pressure on the ends is not transmitted to the shell, but is sustained by external abutments. We shall be concerned mainly with the stresses near the mid-point of the span, where the overall bending moment in the shell, when it is treated as a beam, is largest. Essentially the same problem has been treated by Schorer (1936), and in a different way by Flügge (1973, p.258).

It will be convenient to use an x, θ coordinate system, as shown in fig. 9.1, with the generator $\theta = 0$ at the top of the cylinder.

When the pipe is half full of a fluid of density ρ, the pressure exerted on the shell wall varies around the circumference as follows:

$$0 \leqslant \theta \leqslant \tfrac{1}{2}\pi, \; \tfrac{3}{2}\pi \leqslant \theta \leqslant 2\pi: \quad p = 0 \tag{9.1a}$$

$$\tfrac{1}{2}\pi \leqslant \theta \leqslant \tfrac{3}{2}\pi: \quad p = -(\rho g a)\cos\theta. \tag{9.1b}$$

This is shown in fig. 9.3.

Fig. 9.2. The same shell, but full of the fluid, which is now under excess pressure.

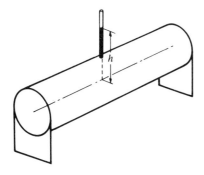

Fig. 9.3. Circumferential distribution of fluid pressure for the loading of fig. 9.1, showing also an approximate representation by means of the first three terms of a Fourier series.

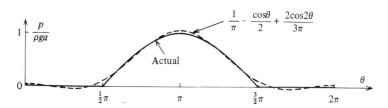

This function may be expressed as a Fourier series by means of standard procedures, as follows (see appendix 10):

$$p/\rho g a = (1/\pi) - \tfrac{1}{2} \cos\theta - (2/\pi) \sum_{n=2,4,6\ldots}(-1)^{n/2}(n^2 - 1)^{-1} \cos n\theta. \quad (9.2)$$

The sum of the first three terms of this series is also shown in fig. 9.3.

The fluid pressure does not vary with x. But since the results from previous chapters which we intend to use concern doubly-periodic pressure distributions, we shall introduce periodic functions in the x-direction by noting that over the interval $0 < x < L$ we may write (see appendix 10)

$$1 = (4/\pi)(\sin(\pi x/L) + \tfrac{1}{3}\sin(3\pi x/L) + \tfrac{1}{5}\sin(5\pi x/L) + \ldots). \quad (9.3)$$

In all of the subsequent working we shall consider only the first term of this series, which is in practice the most important. It is not difficult to repeat the analysis for the higher-order terms: see later.

The pressure distribution with which we are concerned may thus be written

$$p = \sin(\pi x/L)(p_0 + p_1 \cos\theta + p_2 \cos 2\theta + p_4 \cos 4\theta + \ldots), \quad (9.4)$$

where

$$p_0 = 4\rho g a/\pi^2 \quad (9.5a)$$

$$p_1 = -2\rho g a/\pi \quad (9.5b)$$

$$\frac{p_n}{p_1} = \frac{(-1)^{n/2}(4/\pi)}{n^2 - 1}, n = 2, 4, 6 \ldots \quad (9.5c)$$

Here we have expressed the coefficients of the higher terms as multiples of the coefficient p_1. As we shall see, it will be useful to normalise our results with respect to the fundamental term $n = 1$.

In principle, all we now need to do is to apply previously obtained results to each term in turn, and sum the results. We shall be mainly interested in the distribution for N_x, M_θ and w in the shell, and particularly their distribution around the circumference at mid-span. The most straightforward way to proceed is to begin by evaluating the stiffness ratio ζ (see (8.16), (8.68)) for each Fourier term in turn. Since ζ is the ratio of the separate S- and B-surface stiffnesses, it follows immediately that the fraction $\zeta/(1 + \zeta)$ of the load is carried according to the membrane hypothesis by the S-surface, while fraction $1/(1 + \zeta)$ is carried by the B-surface. For the present problem we shall also assume that $L/a > 3$, approximately, so that $l/b > 2$ for every term $n > 2$. Consequently, we may simplify (8.68) by omitting $(b/l)^2$ in comparison with 1, and we may also reckon that circumferential bending is the only significant action in the B-surface (see section 8.4, and particularly fig. 8.4a).

9.2 Example: cylindrical shell acting as a beam

Thus we find, on collecting results from sections 7.3 and 8.6 that for the component

$$p = p_n \sin(\pi x/L) \cos n\theta \qquad (9.6)$$

we have

$$N_x = \frac{n^2 L^2 p_n}{\pi^2 a} \left(\frac{\zeta_n}{1 + \zeta_n}\right) \cos n\theta \sin\left(\frac{\pi x}{L}\right), \qquad (9.7a)$$

$$w = \frac{n^4 L^4 p_n}{\pi^4 a^2 E t} \left(\frac{\zeta_n}{1 + \zeta_n}\right) \cos n\theta \sin\left(\frac{\pi x}{L}\right), \qquad (9.7b)$$

$$M_\theta = \frac{a^2 p_n}{n^2 - 1} \left(\frac{1}{1 + \zeta_n}\right) \cos n\theta \sin\left(\frac{\pi x}{L}\right). \qquad (9.7c)$$

The required expression ζ_n is, by rearrangement of (8.68):

$$\zeta_n = \left(\frac{12\pi^4}{n^4(n^2 - 1)^2}\right)\left(\frac{a^6}{L^4 h^2}\right). \qquad (9.8)$$

From this point onwards, the main problem is the *organisation* of the calculations. Inspection of (9.8) suggests that it will be useful to define a dimensionless length Ω, as follows:

$$\Omega = L h^{\frac{1}{2}} / a^{\frac{3}{2}}. \qquad (9.9)$$

We may then write

$$\zeta_n = (c_n/\Omega)^4, \qquad (9.10)$$

where c_n is a numerical constant given by

$$c_n = 12^{\frac{1}{4}} \pi / n(n^2 - 1)^{\frac{1}{2}} \approx 5.85/n(n^2 - 1)^{\frac{1}{2}}. \qquad (9.11)$$

The dimensionless length Ω emerges naturally from our analysis once we have agreed to consider only cases for which $l/b > 2$. In these circumstances it is the only significant dimensionless group in the problem. The behaviour of the shell will therefore be characterised by the value of Ω. Now the thick-walled pipe described earlier had $a/t \approx 3$ and $L/a \approx 100$; thus $\Omega \approx 60$, which we must regard as an exceptionally large value of Ω. On the other hand the pressure-vessel having $a/t \approx 100$ and $L/a \approx 3$ gives $\Omega \approx 0.3$, which we shall regard as a relatively low value of Ω.

If we wish to include in the analysis terms for which $l/b < 2$, it will of course be necessary to use the full version of (8.68) for the calculation of ζ; and it will no longer be reasonable to consider the *B*-surface only in terms of bending in the circumferential direction. We shall return to this point at the end of the section.

Notice that when $n = 1$, (9.8) gives $\zeta_n \to \infty$. In this case the *S*-surface alone carries the applied load, since the mode of deformation involves no change of curvature in the circumferential direction: here we are neglecting the contribution of B-surface twisting and bending in the longitudinal direction: see problem 9.1.

The factors containing ζ_n in (9.7) will clearly play an important part in the analysis. They have particularly simple asymptotes, as follows:

$$\zeta_n/(1+\zeta_n) \to 1 \text{ when } \Omega \to 0; \to \zeta_n \text{ when } \Omega \to \infty, \quad (9.12a)$$

$$1/(1+\zeta_n) \to \frac{1}{\zeta_n} \text{ when } \Omega \to 0; \to 1 \text{ when } \Omega \to \infty. \quad (9.12b)$$

The asymptotes will provide a useful guide to future calculations. Expressions (9.12) hold for all values of n except $n = 1$: in this case $\zeta_1/(1+\zeta_1) = 1$ for all values of Ω; which corresponds to the fact that the B-surface carries precisely zero load when $n = 1$, irrespective of the value of Ω.

All of the expressions (9.7) involve the term $\cos n\theta \sin(\pi x/L)$. In doing the calculations it is convenient to work in terms of the coefficients alone. Indeed, it is convenient to express these coefficients as multiples of the corresponding coefficients N_{x1} and w_1 for the case $n = 1$, which are obtained by specialisation of (9.7):

$$N_x = N_1 \cos\theta \sin(\pi x/L), \quad (9.13a)$$

$$w = w_1 \cos\theta \sin(\pi x/L), \quad (9.13b)$$

where

$$N_1 = p_1 L^2/\pi^2 a, \quad (9.13c)$$

$$w_1 = p_1 L^4/\pi^4 a^2 E t. \quad (9.13d)$$

Therefore for general values of n we have

$$\frac{N_{xn}}{N_{x1}} = n^2 \left(\frac{p_n}{p_1}\right)\left(\frac{\zeta_n}{1+\zeta_n}\right), \quad (9.14a)$$

$$\frac{w_n}{w_1} = n^4 \left(\frac{p_n}{p_1}\right)\left(\frac{\zeta_n}{1+\zeta_n}\right), \quad (9.14b)$$

$$\frac{6M_{\theta n}}{N_{x1}t} = \frac{6\pi^2}{n^2 - 1}\left(\frac{p_n}{p_1}\right)\frac{1}{\Omega^2}\left(\frac{1}{1+\zeta_n}\right). \quad (9.14c)$$

In the last expression M_θ has been normalised with respect to N_{x1} because a suitable normalising bending-stress resultant does not occur in the analysis for $n = 1$: the point of the factors 6 and t is to give a comparison in terms of 'an extreme fibre hoop stress' (see section 2.3.2).

It is useful to plot these coefficient ratios as functions of Ω and n. The best

9.2 Example: cylindrical shell acting as a beam

plan is to begin by finding the asymptotes for the two cases $\Omega \to 0$ and $\Omega \to \infty$. The corresponding asymptotic formulae are tabulated below.

	Formula for small Ω	Formula for large Ω	'Interaction ordinate'
$\dfrac{N_{xn}}{N_{x1}}$	$n^2 \dfrac{p_n}{p_1}$	$n^2 \dfrac{p_n c_n^4}{p_1 \Omega^4}$	$n^2 \dfrac{p_n}{p_1}$
$\dfrac{w_n}{w_1}$	$n^4 \dfrac{p_n}{p_1}$	$n^4 \dfrac{p_n c_n^4}{p_1 \Omega^4}$	$n^4 \dfrac{p_n}{p_1}$
$\dfrac{6M_{\theta n}}{N_{x1} t}$	$\dfrac{6\pi^2}{n^2-1} \dfrac{p_n}{p_1} \dfrac{\Omega^2}{c_n^2}$	$\dfrac{6\pi^2}{n^2-1} \dfrac{p_n}{p_1} \dfrac{1}{\Omega^2}$	$\dfrac{6\pi^2}{n^2-1} \dfrac{p_n}{p_1} \dfrac{1}{c_n^2}$

In each case the asymptotes intersect at $\Omega = c_n$, and the (common) value of the asymptotic expressions at the intersection is as given in the last column of the table.

These coefficient ratios are plotted logarithmically against Ω for different values of n in fig. 9.4. In the construction of these diagrams the location of the intersection of asymptotes was found first, then the asymptotes were drawn in at the appropriate slope, and finally the curves corresponding to the proper formulae (9.14) were put in.

Various features emerge clearly from these graphs. First, the 'transition region' in which the curves cross over from one asymptote to the other is relatively narrow in all cases. For practical purposes, we need only use the full formula rather than the appropriate asymptotic one for values of Ω differing from c_n by a factor of 2 or 3 in either direction. The reason for this lies in the fact that ζ_n depends on the relatively high fourth power of Ω: see (9.10). We shall discuss similar effects in sections 9.3 to 9.5.

Let us now examine in detail the graphs shown in fig. 9.4. In relation to N_x, shown in (a), it is clear that for $\Omega > 5$, say, the only significant term is $n = 1$: shells of such proportions (including our thick-walled pipe example) behave essentially as simple beams, and in practice we need not invoke shell theory at all. However, when the value of Ω steadily decreases we find that terms $n = 2, 4 \ldots$ come into operation successively. Thus in the range $0.7 < \Omega < 3$ the terms $n = 1$ and $n = 2$ dominate the behaviour, but for $\Omega < 0.7$ it is necessary also to consider terms $n = 4$, etc. For very small values of Ω the higher terms obviously become important; and indeed all of their coefficients have roughly the same value. Now a Fourier series whose terms are of equal magnitude represents a concentrated load (see appendix 10); so our analysis suggests that concentrated lines of tension are building up at $\theta = \pm \frac{1}{2}\pi$ as $\Omega \to 0$.

This indeed corresponds to the analysis given by the membrane hypothesis: see problem 4.7. The corresponding graph for radial deflection, fig. 9.4b, indicates that these become large as the value of Ω decreases. This reflects the fact that the membrane hypothesis gives stress resultants N_x which change rapidly with θ, and the corresponding strains are only compatible by virtue of large radial displacements: see section 7.3.1. Finally, the bending-stress coefficients

Fig. 9.4. Amplitudes of Fourier components for (a) tangential-stress resultant N_x, (b) normal component w of displacement and (c) circumferential bending-stress resultant. In each case the variable is proportional to $\cos n\theta \sin(\pi x/L)$, and the coefficient depends only on the dimensionless length $\Omega = Lh^{\frac{1}{2}}/a^{\frac{3}{2}}$.

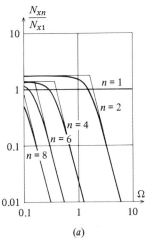

(a)

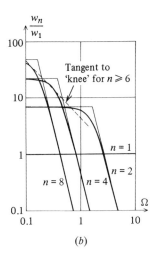

(b)

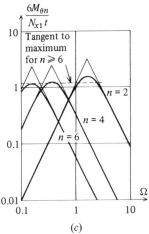

(c)

9.2 Example: cylindrical shell acting as a beam

(fig. 9.4c) show that the bending stress for each of the terms $n = 2, 4$ etc. reaches a *maximum* at a particular value of Ω; and moreover that these correspond to hoop bending stresses of the same order as the longitudinal stress for the case $n = 1$.

Perhaps the most interesting range of behaviour is found around $\Omega = 2$, where the term $n = 2$ is becoming prominent. Fig. 9.5 shows some results for $\Omega = 0.5, 1, 2$ and $\Omega > 3$; and it is not difficult to construct corresponding graphs for other values of Ω by reading the values of the coefficients from fig. 9.4 and summing. Note that the term $n = 1$ corresponds to a linear variation of N_x with depth (in the direction of gravity), just as in the calculation for a simple beam; and that the term $n = 2$ is parabolic in this plot: see problem 9.2.

An advantage of normalising the radial displacements with respect to the term $n = 1$ (fig. 9.4b) is that one can see, for example, that the magnitude of the $n = 2$ 'ovalisation' exceeds that of the $n = 1$ 'translation' for $\Omega < 2.6$. In these circumstances the uppermost generator of the shell deflects *upwards* on account of the heavy fluid.

All of our results agree closely with Flügge's analysis of the same problem, but by a different method. Flügge (1973, p.258) plots results for a shell having a particular geometry, which corresponds to $\Omega = 1$ in our notation.

Our use of Fourier series enables us to investigate shells having different geometries with equal ease. The plots of fig. 9.4 are especially useful for delineating ranges of values of Ω in which different simplifications may be used. For design purposes they are probably more useful than curves such as those of fig. 9.5 in which summations of the Fourier series have been made for different values of Ω.

Fig. 9.5. Distribution of N_x across the depth of the shell for several values of Ω, determined by use of fig. 9.4.

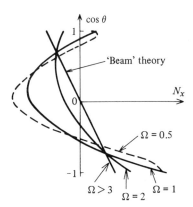

Our analysis of this problem is incomplete in the sense that we have taken account of only the *fundamental* term of the Fourier series (9.3) in the longitudinal direction. Contributions from the higher terms may be computed without difficulty, but before such a calculation is begun, it would be advisable to plot the longitudinal and circumferential half wavelengths onto the 'maps' of fig. 8.4: see problem 9.3. Terms $n = 1$ should not be included in such an exercise, for they fall directly within the scope of the membrane hypothesis, at least when $l/b > 2$. Self-weight loading (see section 4.4.1) and pressure-loading from a full pipe (fig. 9.2) also fall within this category: problems of 'ovalisation' come only from the $n \geqslant 2$ components of pressure distribution associated with a partially full tube.

Lastly we remark that the boundary conditions at the ends of the pipe were somewhat artificial, for the sake of simplicity. Problems of the same type, but with more 'realistic' boundary conditions, are best tackled by methods which will be described in the following sections.

9.3 Beam on elastic foundation

In the preceding example, it was possible to use directly the simple doubly sinusoidal solutions of the shell equations which we had developed in chapter 8. We were concerned only with 'long-wave' solutions which involve structural interaction between longitudinal stretching and circumferential bending. In the following sections we shall also be concerned with 'long-wave' solutions of the shell equations, but not those which involve simple sinusoidal functions in the longitudinal direction.

The most convenient way of handling these solutions, which are not in themselves particularly complicated, but which can involve untidy algebra, is to relate them to a set of standard solutions to the analogous equation for a uniform *beam-on-elastic-foundation*. We did of course encounter and solve equations of this type in connection with axially symmetric problems in chapter 3, but it will be most convenient here to develop the equations again

Fig. 9.6. Variables used in the analysis of a uniform elastic beam supported on a uniform elastic foundation. (*a*) General layout (*b*) forces acting upon a typical small element.

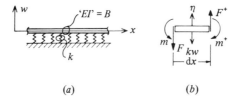

(*a*)   (*b*)

9.3 Beam on elastic foundation

from the beginning, using a notation which will suit better our present purposes.

The present section takes the form of an interlude, in which we explore the solution of a range of problems concerning a beam on an elastic foundation. We shall apply the results to shell problems, by analogy, in the following sections.

The general arrangement is shown in fig. 9.6a. A uniform slender straight beam has a flexural rigidity ('EI') of B. It is secured to an elastic foundation whose spring constant is k: if the transverse displacement of the beam at any point is w, the foundation exerts a normal restoring force there equal to kw per unit length. The dimensions of k are thus force x length^{-2}. Note that this kind of foundation is like a set of isolated springs, rather than a continuous elastic bed for which, of course, the surface displacement at x would depend on the surface loading over a region around x. This simple kind of foundation is known as a *Winkler* foundation, after its originator (see Timoshenko, 1953, §89).

The beam originally lies along the x-axis of any x, y Cartesian coordinate system. The only relevant displacement is w, measured in the positive sense of y; and throughout our analysis we shall assume that w is 'small'. The beam is subjected to a load of intensity η per unit length acting in the positive sense of y. The only other variables which we need are the bending moment m, and shearing force F, whose positive senses are defined in fig. 9.6b. (Symbols M and Q have been eschewed here in order to avoid confusion with the 'shell' variables.)

Force and moment equilibrium equations for a typical element may be written down as follows, by inspection

$$\frac{dF}{dx} + \eta = kw \tag{9.15}$$

$$\frac{dm}{dx} - F = 0. \tag{9.16}$$

The beam deforms, by hypothesis, only in response to bending moment and the appropriate form of Hooke's law is

$$B\frac{d^2w}{dx^2} = -m. \tag{9.17}$$

Eliminating variables m and F from (9.15), (9.16) and (9.17), we obtain a single fourth-order governing equation:

$$B\frac{d^4w}{dx^4} + kw = \eta. \tag{9.18}$$

This equation is to be solved subject to various end conditions at (say) $x = 0$ and $x = L$. Sometimes we shall have *kinematic* (or displacement) boundary conditions on deflection or slope, i.e.

$$w = \text{constant}, \quad \frac{dw}{dx} = \text{constant}.$$

For example, at a clamped end we have

$$w = \frac{dw}{dx} = 0,$$

or for a simply-supported end

$$w = 0.$$

On other occasions we shall have *force* boundary conditions, in which bending moment m and/or shearing force F are specified. These quantities are related to the displacement w by

$$m = -B\frac{d^2 w}{dx^2}, \quad F = -B\frac{d^3 w}{dx^3}. \tag{9.19}$$

For example, at a simply-supported ('hinged') end the bending moment is zero and the second boundary condition is therefore

$$\frac{d^2 w}{dx^2} = 0.$$

The general procedure for solution is as follows. First the governing equation (9.18) is rewritten as

$$\frac{d^4 w}{dx^4} + \frac{4}{\mu^4} w = \frac{\eta}{B}, \tag{9.20}$$

where

$$\mu = (4B/k)^{\frac{1}{4}} \tag{9.21}$$

is a convenient parameter having the dimensions of length. Compare the use of this symbol for a characteristic length in section 3.3, in the solution of essentially the same equation. Here μ is defined in terms of the 'beam' and 'foundation' constants pure and simple: cf. problem 3.4.

The part of the solution of (9.20) normally known as the *complementary function* has four terms of the same general kind, of which the function $A \exp(-x/\mu) \sin(x/\mu)$ is typical: cf. section 3.3. A is an arbitrary constant. When the loading has been specified the other part of the solution, normally known as the *particular integral*, may be found. The four arbitrary constants of the complementary function are fixed by four boundary conditions. All of this is described thoroughly by Hetényi (1946), and here we shall merely quote those results which are useful for our purposes.

9.3 Beam on elastic foundation

For example, a beam which is clamped at one end ($x = L$) and is free at the other ($x = 0$), where it sustains a transverse force P (see fig. 9.7a) suffers a displacement w given by

$$w = (-)\frac{2P}{\mu k} \frac{\text{Sh}[(L-x)/\mu] \cos(x/\mu) \text{Ch}(L/\mu) - \sin[(L-x)/\mu] \text{Ch}(x/\mu) \cos(L/\mu)}{\text{Ch}^2(L/\mu) + \cos^2(L/\mu)}. \tag{9.22}$$

(Here and later on Sh, Ch and Th stand for the hyperbolic functions Sinh, Cosh and Tanh respectively: capitals are used for the reason given by Hetényi (1946) in his Preface.) The displacement at the tip ($x = 0$) may be written

$$\frac{w_0 B^{\frac{1}{4}} k^{\frac{3}{4}}}{P} = -\frac{1}{2^{\frac{1}{2}}}\left(\frac{\text{Sh}\alpha - \sin\alpha}{1 + \frac{1}{2}\text{Ch}\alpha + \frac{1}{2}\cos\alpha}\right), \quad \alpha = \frac{2L}{\mu} = 2^{\frac{1}{2}}\frac{Lk^{\frac{1}{4}}}{B^{\frac{1}{4}}}. \tag{9.23}$$

Expression (9.23) is plotted logarithmically in fig. 9.7b against a dimensionless group $L(k/B)^{\frac{1}{4}}$, i.e. in terms of the (single) parameter specifying the main features of the arrangement. It is clear from the diagram that the solution becomes asymptotic to particularly simple formulae when L is either small or large, or, more precisely, when the dimensionless group $L(k/B)^{\frac{1}{4}}$ has either a small or a large value. In the former case, the tip deflection is given by the limiting form of (9.23) when $\alpha \to 0$, which may be written as

$$w_0 B^{\frac{1}{4}} k^{\frac{3}{4}}/P = (-)\tfrac{1}{3}\left(Lk^{\frac{1}{4}}/B^{\frac{1}{4}}\right)^3. \tag{9.24}$$

Fig. 9.7. (a) Cantilever beam loaded at its tip, and supported by an elastic foundation. (b) Dimensionless plot of tip deflection w_0 as a function of length. The two asymptotes are relatively simple to determine.

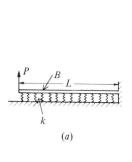
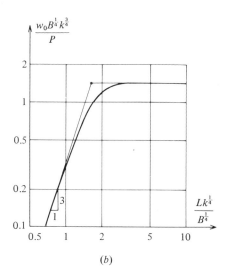

Note that the displacement w_0 is independent of the value of k; and indeed we can see by rearranging (9.24) that we have simply recovered the well-known result for a cantilever beam without a foundation:

$$w_0 = PL^3/3B. \tag{9.25}$$

Thus we see that when the foundation spring is sufficiently 'soft', it has practically no effect: the cantilever action of the beam is much stiffer overall than the foundation; see problem 9.4.

On the other hand, when the value of $L(k/B)^{\frac{1}{4}}$ is large, the expression (9.23) becomes asymptotic to

$$w_0 B^{\frac{1}{4}} k^{\frac{3}{4}}/P = (-)2^{\frac{1}{2}}. \tag{9.26}$$

The main feature of this result is that the length L of the beam is so large in comparison with the characteristic dimension μ that the boundary conditions at the end $x = L$ have virtually no effect on the solution near the end $x = 0$.

A particularly remarkable feature of the curve in fig. 9.7b is the compactness of the short 'transition zone' between the two simple asymptotic formulae, just as in fig. 9.4. This feature is somewhat similar to an effect which we noted in the curves of fig. 3.8. There we had a problem which may be described in present notation in terms of a simply-supported beam-on-elastic-foundation sustaining a transverse load η which varies sinusoidally from end to end. For 'short' beams the load is carried almost entirely by beam-action, but for 'long' beams it is the elastic foundation which carries the major part of the load. In the present example the behaviour is rather more complex, since for 'long' beams the shear load at the free end is carried by an interaction between beam and foundation effects. The same phenomenon of 'short transition zones' may also be seen in fig. 3.13; and indeed the curve of fig. 9.7b may be deduced from this diagram by the use of results from problem 3.4.

A convenient way of looking at the result shown in fig. 9.7b, which we shall use in later sections, is to focus attention on the two simple asymptotic forms (9.25), (9.26), and to regard the general result (9.23) as supplying the 'smoothing' function between them, which is important only within a limited 'transition zone'. We have already used this general procedure, in section 9.2.

The results of some other simple examples are shown in fig. 9.8. In every case it is relatively straightforward to obtain the asymptotes; and the transition curve has been plotted from the corresponding general formula given by Hetényi (1946), and which may also be obtained by specialisation of (3.62) and the use of problem 3.4.

In all of these diagrams we have plotted in dimensionless form the displacement at a convenient point of the beam against a suitable measure of the length of the beam. In any particular case we may of course be interested in other

9.4 Long-wave solution: the formal 'beam' analogy

variables such as shear force, bending moment, slope or deflection at various points of the beam. It is usually not difficult to obtain expressions for these variables, but the calculations may be tedious. If it is clear in a given case that the structural action is of a simple type (i.e. if we are outside the 'transition zone' in the appropriate curve of figs. 9.7 and 9.8) we can use this information to sketch the form of the variable of interest: see problem 9.5.

9.4 Long-wave solution: the formal 'beam' analogy

The basis of this analogy is the similarity in form between the beam-on-elastic-foundation equation (9.18) and the 'long-wave' form of Donnell's equation ((8.24) etc.), which we shall rewrite below. The variables for the 'beam' problem are defined in the preceding section, and we must now establish the connection between them and the corresponding variables for the

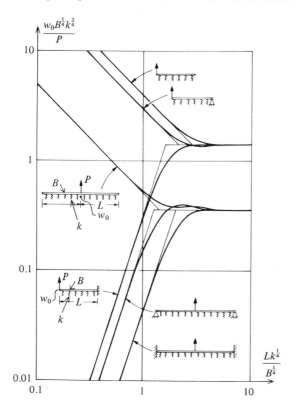

Fig. 9.8. Displacement of the point of application of a transverse load for beams on elastic foundation having various boundary conditions. Fig. 9.7 provides one of the curves. Results from Hetényi (1946).

shell. As before, we are dealing only with the nth harmonic of a Fourier series in the circumferential direction, and it is convenient to express the variables of the shell problem thus:

$$w = w^* \sin n\theta, \quad v = v^* \cos n\theta, \quad u = u^* \sin n\theta, \quad p = p^* \sin n\theta,$$
$$N_x = N_x^* \sin n\theta, \quad N_{x\theta} = N_{x\theta}^* \cos n\theta, \quad M_x = M_x^* \sin n\theta. \qquad (9.27)$$

Here the coefficients w^*, N_x^*, etc. are all functions of x. Using these variables, putting the 'long-wave' simplifications that only the dominant structural effects need be considered (namely, longitudinal stretching and circumferential bending), and inserting the circumferential bending correction described in section 8.6 we have, in place of (8.89), the governing equation

$$w^{*''''} + w^*[Dn^4(n^2 - 1)^2/Eta^6] = p^*(n^4/Eta^2). \qquad (9.28)$$

Here $'$ denotes differentiating with respect to x. This equation is clearly directly analogous to (9.18). It is convenient to make the two *deflections* directly analogous, i.e. to put

$$w \leftrightarrow w^*. \qquad (9.29)$$

We then have a degree of freedom in choosing the quantity analogous to B, but it is sensible to make use of the 'pure beam' analogy which we established in section 7.5 in relation to the displacement of shells according to the membrane hypothesis. Therefore we put

$$B \leftrightarrow E\pi a^3 t/n^4 ; \qquad (9.30)$$

and it follows immediately that

$$k \leftrightarrow \pi D(n^2 - 1)^2/a^3, \qquad (9.31a)$$
$$\eta \leftrightarrow \pi a p^*. \qquad (9.31b)$$

Equation (8.88b) gives the connection between the Airy stress function ϕ and displacement, and from it we find

$$m \leftrightarrow (\pi a^2/n^2)N_x^*. \qquad (9.32)$$

Again, from the in-plane equilibrium equation in the longitudinal direction we obtain

$$F \leftrightarrow (\pi a/n)N_{x\theta}^*. \qquad (9.33)$$

Lastly it follows from the equations of circumferential bending in section 8.6 that

$$w \leftrightarrow a^2 M_\theta^*/D(n^2 - 1). \qquad (9.34)$$

Alternatively, M_θ^* may be represented in terms of the force per unit length, kw, transmitted by the foundation to the beam.

Most of these analogies in the variables are similar to those established previously in the restricted analysis of the displacement of a cylindrical shell according to the membrane hypothesis, with inextensional behaviour in the circumferential direction: see table 7.1 (section 7.5). The main difference here is that we have now introduced bending effects in the circumferential direction in the shell, which add an elastic foundation to the analogous beam; consequently there are some new variables. Also, in the present context shear force in the beam is analogous with N_{xy} and not with a combination of N_{xy} and Q_x as in the previous analogy. If a periodic normal shear force Q_x is applied at the edge of a shell, it is 'converted' to an N_{xy} loading over a 'boundary layer' of shell by means of the short-wave solution of the eighth-order equations. Consequently, notwithstanding the above remarks, we may use the equivalence

$$F \leftrightarrow -\pi a Q_x^* \tag{9.35}$$

for an edge-loading

$$Q_x = Q_x^* \sin n\theta,$$

provided we subsequently add the corresponding short-wave solution. Similar remarks apply to the radial and tangential loads $\bar{p}, \bar{q}$ acting on a circumferential line, which we discussed in section 7.5. Provided an appropriate 'short-wave' solution is included on account of the radial loading $\bar{p}$, the previous equivalence of these loads carries through to the beam-on-elastic-foundation analogy. In the case of the pressure-loading p (see (9.27)) we may generally compound it with a circumferential tangential traction q as in table 7.1 (section 7.5) for the purposes of our new analogy, provided q^* does not vary so rapidly in the longitudinal direction as to induce short-wave effects; and there will also of course be small discrepancies in the circumferential stress resultant N_θ, which plays only a small part in the 'long-wave' solution.

Lastly, we need to express the tangential components of displacement in terms of w for purposes of satisfying boundary conditions. As we explained in section 8.7.3, the situation is almost the same as for the 'membrane-hypothesis' calculations, and we have from (7.71)

$$w \leftrightarrow nv^*, \tag{9.36}$$
$$w' \leftrightarrow -n^2 u^*/a.$$

9.5 Cylindrical shell with one edge free and the other edge clamped

Consider first the problem depicted in fig. 7.4, which we have already analysed in the absence of significant bending effects in chapter 7. A cylindrical shell of radius a, thickness t and length L is free at the edge $x = 0$ but clamped at edge $x = L$, i.e. restrained there so that all components of displacement are zero. At the free edge a normal shearing-stress resultant

$$Q_x = Q_n \sin n\theta \tag{9.37}$$

is applied. The problem is to express the radial displacement of this edge in terms of Q_n, i.e. to find an expression for the 'modal flexibility' of the shell for this particular kind of edge-loading. As we have argued already, the applied loading will be converted to a tangential shear-stress loading

$$N_{xy} = nQ_n \cos n\theta \qquad (9.38)$$

by a narrow edge zone, and this modified edge-load will be applied to the long-wave part of the solution. It will be necessary to add later an edge displacement corresponding to the short-wave effects.

The second force boundary condition at this end is

$$N_x = 0. \qquad (9.39)$$

Consequently, the boundary conditions for our analogous beam are:

$$m = 0, F = -\pi a Q_n \quad \text{at } x = 0. \qquad (9.40)$$

At end $x = L$ both u and v are held at zero; so we have, from section 8.7.3

$$w = 0, w' = 0 \quad \text{at } x = L. \qquad (9.41)$$

These four boundary conditions define a cantilever of the kind illustrated in fig. 9.7a, and we may therefore use the data of fig. 9.7b directly. There is a minor complication in that at end $x = 0$ of a beam which is clamped at $x = l$, the sense of F is opposite from that of w (fig. 9.6), whereas in fig. 9.7 it is implicit that P and w_0 have the same sense. In practice, however, we are concerned to evaluate the essentially positive 'mode flexibility', and the ambiguity of sign (see appendix 5) causes no difficulty.

It remains only to express the beam constants B and k in terms of shell variables, by means of (9.30) and (9.31a). It is most convenient to begin by considering the two asymptotic formulae separately.

For 'short' beams, i.e. for beams with

$$Lk^{\frac{1}{4}}/B^{\frac{1}{4}} < 1.62 \qquad (9.42)$$

(i.e. less than the 'crossover' value), we may use the ordinary cantilever formula as an approximation. Substituting for B and k from (9.30) and (9.31a) we find that inequality 9.42 becomes

$$Lh^{\frac{1}{2}}/a^{\frac{3}{2}} < 3.01/[n(n^2 - 1)^{\frac{1}{2}}] \qquad (9.43a)$$

or

$$n(n^2 - 1)^{\frac{1}{2}} < 3.01/\Omega, \qquad (3.43b)$$

where $\Omega = Lh^{\frac{1}{2}}/a^{\frac{3}{2}}$ is the parameter which we encountered in section 9.2. Substituting for B and F in the simple cantilever formula we find the 'mode flexibility' for a loading applied at the edge:

$$|w_{no}/Q_n| = n^4 L^3 / 3Ea^2 t. \qquad (9.44)$$

9.5 Cylindrical shell with free and clamped edges

This agrees exactly with the first part of (7.55); which is not surprising, for the circumstances in which the foundation stiffness of the analogous beam plays virtually no part are precisely those for which the membrane hypothesis is valid. We shall return to this point later.

Next, we find out what happens when inequality (9.42) is broken, and we must apply the second asymptotic formula. This gives, for the beam (cf. (9.26))

$$w_0/P = 2^{\frac{1}{2}}/B^{\frac{1}{4}}k^{\frac{3}{4}} \tag{9.45}$$

which becomes, in terms of the shell variables,

$$\left|\frac{w_{n0}}{Q_n}\right| = \frac{9.12}{E} \frac{n}{(n^2-1)^{\frac{3}{2}}} \frac{a^{\frac{5}{2}}}{th^{\frac{3}{2}}}. \tag{9.46}$$

In this range, the modal stiffness does not involve the length L of the shell; but neither is this surprising, since the remote boundary has a negligible effect on the solution in the vicinity of the loaded edge.

Formulae (9.44) and (9.46) have been plotted logarithmically in fig. 9.9, with the help of the simplification $(n^2 - 1) \to n^2$ in (9.46). The ordinate represents a dimensionless modal flexibility, while the abscissa is a dimensionless wavenumber n. The dimensionless groups used in this diagram are directly related to those of fig. 9.7b via the substitutions (9.30), (9.31a) and (9.35). But since the quantity corresponding to $w_0 B^{\frac{1}{4}} k^{\frac{3}{4}}/P$ is proportional to n^2, it has been divided by the second dimensionless group in order to eliminate n. The dimensionless length Ω emerges naturally from the manipulations.

It is clear from the diagram that for a shell of given dimensions, the modal compliance increases in proportion to n^4 (see (9.44)) in the range in which circumferential bending plays little part, but decreases as n^{-2} (see (9.46)) in the range in which the boundary conditions at the remote end play little part. We have not yet inserted the 'transition curve' between the two asymptotes, but it is clear nevertheless that the modal flexibility reaches a *maximum* around the value of n at which the two asymptotes cross. Denoting this by n^* we find from (9.44) and (9.46) that

$$n^* = 1.74/\Omega^{\frac{1}{2}} = 1.74 a^{\frac{3}{4}}/L^{\frac{1}{2}}h^{\frac{1}{4}}. \tag{9.47}$$

The value of n^* can be found readily for any given shell, and it immediately indicates the circumferential wavenumber of the *most flexible mode*; n^* itself, of course, will not normally be an integer.

From the same equations we may also determine the ordinate at which the two asymptotes cross:

$$|w_0/Q_n|(Eth/La) = 3.03. \tag{9.48}$$

It is perhaps slightly unfortunate that the thickness of the shell appears in this parameter in terms of both t and h. In practice, of course, we may use the two thicknesses indiscriminately for numerical work; but for the sake of completeness and in order to avoid the expression $(1 - \nu^2)$ we use the dimensionless group in fig. 9.9 which emerges unambiguously from the manipulations.

So far we have been concerned with the transformation of the two asymptotes of the 'beam' solution (fig. 9.7b) into the form appropriate for a cylindrical shell (fig. 9.9). We should also treat the transition curve in fig. 9.7b in the same way. The method of transfer is straightforward. For example, at the 'crossover' point in fig. 9.7b the true modal flexibility is 0.65 of the 'asymptotic' value; and this same factor applies to the crossover point in fig. 9.9. To obtain other points on the transition curve, we need to relate the parameter $Lk^{\frac{1}{4}}/B^{\frac{1}{4}}$ in fig. 9.7b to n in fig. 9.9. From (9.30) and (9.31a), and replacing $n^2 - 1$ by n^2 we find that

$$Lk^{\frac{1}{4}}/B^{\frac{1}{4}} = (\Omega/1.82)\,n^2. \tag{9.49}$$

Fig. 9.9. Dimensionless plot of the edge compliance of a cylindrical shell loaded as in fig. 7.4c. The edge at $x = L$ is supported in various ways: see text for details. The lines corresponding to short-wave and 'flat-plate' behaviour shift according to the value of $L/(ah)^{\frac{1}{2}}$, but the other curves do not.

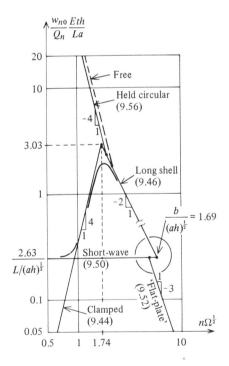

9.5 Cylindrical shell with free and clamped edges

Thus a value of $Lk^{\frac{1}{4}}/B^{\frac{1}{4}}$ in fig. 9.7b (say) 3 times the crossover value corresponds to a value of n in fig. 9.9 equal to $3^{\frac{1}{2}}$ times the crossover value n^*. Hence it is a straightforward matter to transfer the transition curve from fig. 9.7b to fig. 9.9: see problem 9.6. Note in particular that the transition zone covers only a small range of values of n.

Fig. 9.10 shows modal flexibilities determined experimentally by A. C. Smith for small deflections of a thin-walled cylindrical shell made from silicone rubber. The inverted V-shape can be seen clearly, and there is good agreement near the apex. For $n = 1$ the experimental flexibility is greater than the theoretical value. This may be due in part to 'short-wave' effects (see below). But more significant is the fact that for low values of n the 'beam analogy' for the membrane-hypothesis solution can become inaccurate on account of the explicit neglect of shear distortions due to N_{xy} in the basic analysis: we have already discussed this in section 7.4.

So far we have considered only the displacements due to the 'long-wave' part of the solution. The 'short-wave' part gives a relation between Q_n and edge deflection exactly as for the axially symmetric problem studied in chap-

Fig. 9.10. Experimental results of A. C. Smith in relation to a free/clamped thin cylindrical shell for which $L/a = 2.5$, $a/t = 36$, $\Omega = 0.45$ and $L/(ah)^{\frac{1}{2}} = 14$.

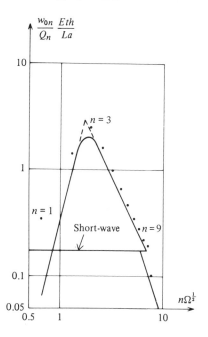

ter 3, with modal flexibility independent of n. From the results of section 3.4 we find

$$\left|\frac{w_{no}}{Q_n}\right| = \frac{2.63}{E} \frac{a^{\frac{3}{2}}}{th^{\frac{1}{2}}}. \qquad (9.50)$$

This result has been put onto figs. 9.9 and 9.10. Unlike the lines corresponding to (9.44) and (9.46), its location depends on the geometry of the shell. In particular its ordinate is found by dividing the ordinate of the crossover point by a factor of about $L/(ah)^{\frac{1}{2}}$, which will normally be of order 10, say.

The 'short-wave' flexibility must be added to that which is produced by the 'long-wave' part of the solution. It will only make a significant contribution in those cases where it exceeds (say) one half of the 'long-wave' flexibility. This occurs both at low and high values of $n\Omega^{\frac{1}{2}}$. It can be seen from fig. 9.9 that in a 'middle range' of values of n the contribution of the 'short-wave' solution to the deflection of the edge will be negligible. A short transition curve has been sketched in the lower left-hand region of fig. 9.9.

Now our analysis so far has been based on the hypothesis that the shell equations may be separated for practical purposes into two distinct parts. This is only reasonable (see fig. 8.6) provided $b/(ah)^{\frac{1}{2}} > 4$, say, where as before b is the circumferential half wavelength. We must therefore expect our solution to become inaccurate at large values of n. Now it is easy to show that the value of n at which the 'long-wave' line of slope -2 intersects the 'short-wave' line of slope 0 in fig. 9.9 is equal to $1.86(a/h)^{\frac{1}{2}}$, which corresponds to

$$b/(ah)^{\frac{1}{2}} = 1.69. \qquad (9.51)$$

Thus, as n increases, the separate short- and long-wave solutions *both* become questionable a little to the left of this intersection. It should not be surprising that the limit of validity of our calculations corresponds roughly to the intersection of the two lines, for we have a precisely similar effect in fig. 8.6.

Calculations are distinctly awkward in the 'transition region' indicated by the circle in fig. 9.9, but they become simple again for small values of $b/(ah)^{\frac{1}{2}}$ (or high values of n in the present context) because the governing equation then degenerates into separate 'flat-plate' and 'plane-stress' equations, as we pointed out in section 8.5.2. It is not difficult to determine the 'edge flexibility' of a semi-infinite flat plate subjected to an edge ('effective') shear of the form

$$V_x = Q_0 \sin(\pi y/b)$$

(see fig. 9.11). The answer is (see problem 9.7)

$$\left|\frac{w_0}{Q_0}\right| = \frac{24(1+\nu)b^3}{\pi^3(3+\nu)Et^3} \approx \frac{10.8}{n^3} \frac{a^3}{Eth^2}. \qquad (9.52)$$

9.5 Cylindrical shell with free and clamped edges

For the last expression, we have put $v \approx 0.3$ and have expressed b in terms of n by use of $bn = a\pi$. This is the only occasion in the present chapter on which Poisson's ratio v appears in a form other than $(1 - v^2)$.

This formula has also been plotted on fig. 9.9, in terms of circumferential wavenumber n. It gives a line of slope -3, which intersects the 'short-wave' line a little to the left (factor ≈ 0.86) of the intersection already discussed. The near-concurrence of the three lines corresponding to 'long-wave', 'short-wave' and 'flat-plate' behaviour suggests that a thorough analysis within the encircled region would produce a relatively smooth transition between the lines of slope -2 and -3. There is some support for this view in the experimental observations of A. C. Smith. see fig. 9.10.

It is clear that n^*, defined by (9.47) for a shell of given dimensions, is an important parameter: for practical purposes it is equal to the circumferential wavenumber n for which a shell with one end free and the other end clamped is *most flexible* with respect to loading applied at the free edge. For this reason a graphical chart for the rapid determination of n^* is presented in fig. 9.12. The scales are logarithmic and contours of n^* have been determined in terms of L/a and a/h by means of (9.47). An advantage of this chart over (9.47) itself is that it has been possible to include some 'cut-off' lines at various values of $L/(ah)^{\frac{1}{2}}$ of order unity. The point here is that for small values of this parameter the apex of the inverted V-shaped curve of fig. 9.9 falls *below* the curve corresponding to the 'short-wave' solution. In these circumstances (9.47) is not relevant. In the region of the diagram corresponding to small values of $L/(ah)^{\frac{1}{2}}$ the dominant effect is the bending of the B-surface, without help from the S-surface, just like a semi-infinite flat plate.

Before we consider other loadings and boundary conditions it is useful to make some general remarks about the form of the graph of fig. 9.9.

Basically the diagram is a plot of modal flexibility against circumferential wavenumber. It is plotted logarithmically because the various simple formulae

Fig. 9.11. Periodic transverse shear loading V_x applied to the edge of a semi-infinite flat plate.

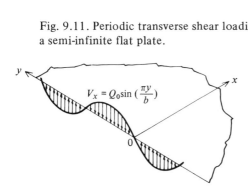

then appear as straight lines. In the form shown, it is not necessarily to scale. This is because the *three* independent geometrical dimensions L, a, h can only be specified by means of *two* dimensionless geometrical groups. We have found that the key group for the long-wave part of the solution is

$$\Omega = Lh^{\frac{1}{2}}/a^{\frac{3}{2}},$$

and we have seen earlier that for the short-wave part of the solution $L/(ah)^{\frac{1}{2}}$ is the appropriate group. Since the values of both groups are required to specify the geometry of the shell, a single diagram cannot be drawn to cover all possibilities. But this is not a serious disadvantage, since the various lines on the diagram have distinct slopes and the coordinates of the salient points can be expressed easily. However, when the value of $L/(ah)^{\frac{1}{2}}$ is of order unity, the form of the diagram must be re-examined: see problem 9.8.

A full version of the diagram has various 'transition curves' between the lines corresponding to the various simple formulae. But the diagram is almost more valuable without these when preliminary design work is being done. In addition to giving modal flexibility in terms of wavenumber n, the diagram shows clearly the several different regimes of behaviour. Thus in fig. 9.9 the line of slope +4 corresponds simply to the membrane hypothesis, and for the modes in this range it is reasonable to analyse the stresses in the structure by the equilibrium methods of chapter 4. Again, for larger values of n the action becomes progressively more localised near the free edge, and relatively simple

Fig. 9.12. 'Map' for the determination of the circumferential mode-number n^* of the most flexible mode for the clamped/free cylindrical shell under edge-loading. The lower region of the diagram corresponds to cases which are dominated by 'short-wave' and 'flat-plate' effects.

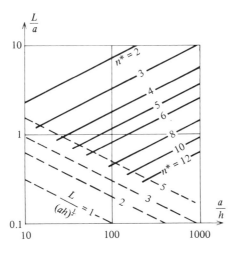

stress analysis calculations may be done. The diagram also shows clearly that there is a definite upper limit to modal flexibility for a shell with a clamped end, which may be calculated easily.

In the drawing of fig. 9.9 the 'circumferential bending factor' $(1 - n^{-2})$ has been omitted for the sake of simplicity. It is easy to examine the effect which inclusion of this factor would make to each line on the diagram in turn. Thus the line of slope +4, which corresponds to the membrane hypothesis, is not affected. On the other hand, the line of slope -2 depends significantly on the 'elastic foundation' effect which comes from circumferential bending. However, since the value of n in this region will usually be high, the correction is unimportant in most cases. We shall see, however, in the next example, that there are some circumstances in which the effect must be taken into account.

9.5.1 Other simple boundary conditions

It is interesting to extend the scope of the problem shown in fig. 9.9 by considering two more kinds of boundary condition at the end $x = L$, while retaining the same force conditions at end $x = 0$. So far we have considered the edge to be *clamped*, so that the edge is held both *circular* and also *in-plane*, i.e. is prevented from 'warping'. As we saw in section 8.7, the condition on circularity of the edge corresponds to $w = 0$ for the long-wave solution, while the prevention of warping implies $\partial w/\partial x = 0$, or $w' = 0$ for any given mode of the kind which we have been considering.

We shall now also consider a *free, unloaded* far edge, i.e. one with no kinematic constraints, and zero applied stress. This corresponds closely to the conditions imposed at the edge $x = 0$ in our previous study, but with both $N_x = 0$ and $N_{xy} = 0$. Consequently, in terms of the analogous beam-on-elastic-foundation we have a free end.

Lastly, we shall consider an arrangement at the far edge of the shell which holds the edge circular, but does not provide axial restraint against warping. This sort of edge condition would be furnished by an inextensible but flexible flat 'diaphragm' at the end. It is a 'mixed' displacement and force boundary condition in the sense that

$$w = 0, \quad N_x = 0. \tag{9.53}$$

In terms of the beam-on-elastic-foundation this gives a simply-supported end, with

$$w = 0, m = 0. \tag{9.54}$$

We shall refer to these three boundary conditions at the far edge as 'clamped', 'free' and 'held circular' respectively.

The results of a beam-on-elastic-foundation analysis corresponding to the 'free' and 'held circular' edge-conditions are given in fig. 9.8. For large values of the parameter $Lk^{\frac{1}{4}}/B^{\frac{1}{4}}$, the asymptote is the same as before, which is to be expected: when the beam is sufficiently long, the boundary conditions at $x = L$ have practically no effect on the behaviour at the other end. But for small values of this parameter the behaviour is quite different. When the far end is clamped the dominating effect is the stiffness of the cantilever as a beam. However, when the far end is either free or simply supported, the stiff beam can rotate as a rigid body, and it is now the *foundation* which provides the major resistance to the load. The asymptotic formulae are easy to derive, and we can readily see why the beam with the simply-supported end is not very much stiffer than the beam with the free end: see problem 9.9.

It is a straightforward matter to transfer these results to the diagram of fig. 9.9. For the 'held circular' case the end-deflection of the analogous beam is given by

$$w_0 = 3P/kL, \tag{9.55}$$

and this gives the corresponding formula for the shell:

$$\left|\frac{w_0}{Q_n}\right| = \frac{36}{(n^2 - 1)^2} \frac{a^4}{Eth^2L}. \tag{9.56}$$

For large values of n we may replace $n^2 - 1$ by n^2, and the modal flexibility is inversely proportional to the fourth power of n, i.e. is a line of slope -4 in fig. 9.9. This line intersects the asymptotic formula for long shells (see (9.46)) at $n = 1.99/\Omega^{\frac{1}{2}} = 1.14n^*$, as indicated. When the end $x = L$ is free, the short-beam formula corresponding to (9.55) is $w_0 = 4P/kL$; and there are corresponding changes, as shown, in the graph of fig. 9.9. It is interesting to note that all four of the lines in fig. 9.9 intersect, for a shell of given dimensions, in the region of $n = n^*$: the 'crossover' from one dominant mode to another occurs at about the same value of n for all of the various different boundary conditions which we have studied.

It is clear that these two boundary conditions give much more flexible modes for low wavenumbers than does the clamped edge. These very flexible modes involve practically *rigid-body* displacements of the analogous beam-on-elastic-foundation; and this corresponds simply to *inextensional* deformation of the cylindrical shell (cf. section 6.7). It is perhaps surprising that the 'held circular' boundary condition at $x = L$ gives almost as much flexibility as a completely unrestrained end. The important point is that both of these boundary conditions permit inextensional deformation; and the shell takes advantage of this when the circumferential wavenumber is low. On the other hand, when the edge $x = L$ is clamped, no inextensional deformation is possible, and for low wavenumbers the load is carried by stresses in accordance with the mem-

9.5 Cylindrical shell with free and clamped edges

brane hypothesis. In particular, high N_x forces are transmitted to the foundation.

These inextensional modes derive their entire stiffness from circumferential bending effects; and as the mode-number n is low, we must certainly investigate the 'circumferential bending correction'. It is clear from (9.56) that the inclusion of this effect, which was neglected in the plotting of fig. 9.9, will enhance the modal flexibility shown by the factor $n^4/(n^2 - 1)^2$. Thus for $n = 2$ the two curves of slope -4 should each be raised by a factor of 1.78. In the special case $n = 1$ the correction factor is infinite. This corresponds to the fact that neither the 'free' nor the 'held circular' boundary conditions can prevent a rigid-body displacement of the entire shell when an *unbalanced* force is applied at one edge.

It is clear from a comparison of these three cases that the ability of the foundation to prevent warping, i.e. to remain plane, plays a crucial part in the response of the shell to applied loading.

9.5.2 Shell with thick flat-plate closure

It is clear from fig. 9.9 that for low values of n, the compliance of the free edge of a cylindrical shell in response to shearing force $Q_x \sin n\theta$ may vary widely, depending on the nature of the boundary conditions at the far edge. For the sake of definiteness consider the lowest nontrivial mode, $n = 2$. The ratio of compliance for the far end 'held circular' to that for the far edge clamped is approximately equal to $(\frac{1}{2}n^*)^8$, where n^* is the value of n at the 'crossover', as indicated in fig. 9.12 and provided $n^* > 2$; if $n^* < 2$, of course, the edge-response for the mode $n = 2$ does not depend on the nature of the boundary condition at the far edge. The power of 8 comes from the fact that the relevant lines on fig. 9.9 have slopes of $+4$ and -4, respectively.

Now if the cylindrical shell is closed at the end $x = L$ by a flat plate of thickness T, and made from the same material as the shell itself, it is clear that for large values of T this plate will give a practically rigid foundation to the shell, while for small values of T it will constitute a thin diaphragm which merely holds the edge circular without providing appreciable axial restraint. It is therefore of great interest to know what is the minimum value of T for which a fully-clamped edge-condition is approximated, and what is the maximum value of T for which the plate provides effectively zero restraint against warping. Indeed, we would also like to know in general what happens if the value of T is intermediate between these two extremes.

Clearly the most important property of the end-plate is its ability to resist axial forces varying sinusoidally round the periphery. Suppose that edge-forces

$$N_x = N_n \sin n\theta$$

are applied to the edge of the plate as shown in fig. 9.13. The edge displacements in the axial direction may be written as

$$u = u_n \sin n\theta,$$

and we may call the ratio

$$K_n = N_n/u_n \tag{9.57}$$

the 'modal stiffness' of the edge of the plate. We have already quoted in (9.52) the modal stiffness for a load of this kind applied to the straight edge of a semi-infinite plate, and it may be shown that the error involved in using this formula (with the transformation $nb = \pi a$) for a circular plate is insignificant for high values of n, and not more than 10% in error for values of n as low as 2. We shall therefore write (putting $\nu \approx 0.3$)

$$K_n = 0.106 n^3 E T^3/a^3. \tag{9.58}$$

What effect does this have on the shell solution? We have already seen that N_x in the shell is analogous to bending moment m in our beam, and that displacement u is analogous to rotation of the beam. It seems clear therefore that the stiffness K_n will correspond to a spring, restraining the rotation of the end of the beam, whose stiffness, Φ, is the ratio of applied couple to rotation: see fig. 9.14a. Now from (9.32) and (9.36) we have the following expressions for bending moment in and rotation of the beam:

$$m = \pi a^2 N_n/n^2$$
$$w' = n^2 u_n/a.$$

Hence

$$\Phi = \frac{m}{w'} = \frac{\pi a^3}{n^4} \frac{N_n}{u_n} = \frac{\pi a^3 K_n}{n^4}. \tag{9.59}$$

This is our required expression. Now we could incorporate this feature in the classical beam-on-elastic-foundation analysis to reconstruct the boundary condition at $x = L$, and solve the governing equation accordingly. But we may proceed more simply and directly, as follows.

Fig. 9.13. A circular disc which is subjected to periodic transverse shear force applied at the edge.

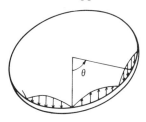

9.5 Cylindrical shell with free and clamped edges

Consider first the case of a *short* beam and a *small* value of Φ. Now for a vanishingly small value of Φ the end of the beam is effectively pinned, and the force P required for a given end-deflection w_0 is given by (9.55):

$$P = \tfrac{1}{3} w_0 k L. \tag{9.60}$$

For a given displacement w_0 the effect of the spring is to increase the transverse force. Since we are assuming here that the beam remains straight, the

Fig. 9.14. (*a*) Cantilever spring on an elastic foundation with a special rotation-spring at the root. (*b*) Logarithmic plot of compliance against length when the beam is effectively rigid. (*c*) Corresponding plot when the foundation springs have virtually no effect. (*d*) Composite diagram incorporating the results of (*b*) and (*c*), together with results for long cantilevers: only the asymptotes are shown. The family of lines of slope 2 corresponds to rotation-springs having a wide range of stiffness, and the numerical labels are values of $B^{3/4} k^{1/4} / \Phi$. (*e*) Diagram derived from (*d*) for the edge compliance of a thin cylindrical shell which is open at one end and closed by a flat plate at the other, cf. fig. 9.9. The numerical labels on the family of sloping lines are values of $L^{1/2} t^{7/4} a^{3/4} / T^3$.

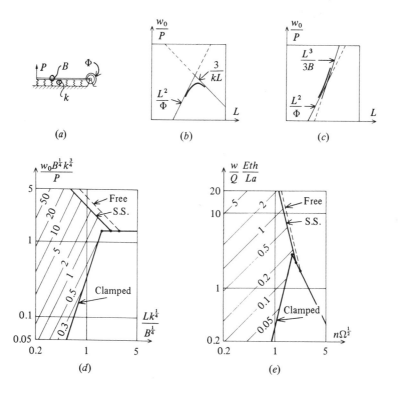

hinge rotation is equal to w_0/L; there is a couple $\Phi w_0/L$ at the hinge, which requires an extra transverse force P at the end of $\Phi w_0/L^2$. Hence (9.60) is altered to

$$P = w_0(\tfrac{1}{3}kL + \Phi/L^2).$$

Therefore,

$$\frac{w_0}{P} = \frac{1}{(\tfrac{1}{3}kL + \Phi/L^2)}. \tag{9.61}$$

This function is plotted schematically on double-logarithmic scales in fig. 9.14b. For very low values of L the Φ-term dominates, while for higher values the elastic foundation term governs: the required curve lies below the two curves corresponding to the separate effects, as shown.

Next consider a case in which Φ is so large that it almost provides a clamped end-condition to the beam. For a 'short' beam the foundation plays little part, and the deflection of the tip of the beam is equal to that for a clamped cantilever plus a term for the rotation of the root: thus

$$w_0 = PL^3/3B + PL^2/\Phi. \tag{9.62}$$

For sufficiently small values of L the Φ-term dominates, while for larger values the B-term is larger. Fig. 9.14c shows the summation on a logarithmic plot: the curve follows whichever of the two terms is higher.

Combining the curves of fig. 9.14b, c, we have the diagram shown in fig. 9.14d. We conclude from this diagram that we may use the formula

$$w_0/P = L^2/\Phi \tag{9.63}$$

in place of (9.61) and (9.62) throughout the region to the left of the lines inclined at +2 and −1, provided we bear in mind the need for suitable transition curves. Since the ordinate of the diagram is $w_0 B^{\frac{1}{4}} k^{\frac{3}{4}}/P$ when we change to the variables that are used in fig. 9.8, we rewrite (9.63) as

$$\frac{w_0 B^{\frac{1}{4}} k^{\frac{3}{4}}}{P} = \frac{(Lk^{\frac{1}{4}}/B^{\frac{1}{4}})^2}{(\Phi/B^{\frac{3}{4}} k^{\frac{1}{4}})}. \tag{9.64}$$

Lines have been drawn in fig. 9.14d for several values of the dimensionless group $\Phi/B^{\frac{3}{4}} k^{\frac{1}{4}}$. (It is easy to check by dimensional analysis that this is in fact the only dimensionless group containing the three stiffnesses Φ, B and k.)

In the corresponding diagram for a shell (fig. 9.14e cf. fig. 9.9) there will be a new family of lines lying in the region to the left of the existing lines of slope +4 and −4, and merging with them. Using (9.58) and (9.59) we may rewrite (9.63) as

$$\frac{w_0}{Q_n} = \frac{naL^2}{0.106ET^3}. \tag{9.65}$$

9.5 Cylindrical shell with free and clamped edges

Hence we obtain the lines of slope 1 which are shown in fig. 9.14e. For a given value of n the ordinate is inversely proportional to T^3.

There are various ways of assessing the significance of the value of T. Perhaps the most useful idea is to find at what values of T the line (9.65) intersects the line of slope +4 and −4 respectively at $n = 2$, for these will correspond closely to the limits on T described above. By inspection of fig. 9.14e, and making use of the parameter n^* (see (9.47)), the first of these intersections occurs when

$$\frac{2aL^2}{0.106ET^3} = 2.88 \frac{La}{Et^2} \left(\frac{2\Omega^{\frac{1}{2}}}{1.74} \right)^4,$$

which may be simplified to

$$\frac{T}{t} \approx 1.5 \left(\frac{a}{L} \right)^{\frac{1}{3}} \left(\frac{a}{t} \right)^{\frac{2}{3}}. \tag{9.66}$$

Here we have dropped the use of h, since the distinction between h and t is hardly warranted for these relatively crude calculations. The second intersection takes place when

$$\frac{2aL^2}{0.106ET^3} = 3.8 \frac{La}{Et^2} \left(\frac{4}{3} \right)^2 \left(\frac{1.74}{2\Omega^{\frac{1}{2}}} \right)^4,$$

where the factor $(4/3)^2$ represents the 'circumferential bending correction'. This reduces to

$$\frac{T}{t} \approx 1.7 \left(\frac{L}{a} \right). \tag{9.67}$$

Thus for given values of L/a and a/t, (9.66) indicates the value of T/t above which the base is effectively clamped for $n = 2$, while (9.67) gives the value of T/t below which the base is merely 'held circular'. These formulae are shown graphically in fig. 9.15; for given values of L/a and a/t the two extreme values of T/t may be read off by interpolation. The two formulas give the same value of T on a line which lies close to the line $n^* = 2$ as shown in fig. 9.12. Indeed, we may see from these two diagrams, that the *ratio* of the two extreme values depends only on the value of n^*, as indeed we would expect from the arguments above: see problem 9.10. Fig. 9.15 also shows the cut-off for short shells, corresponding to the line $L/(at)^{\frac{1}{2}} = 5$. For example, if a shell has $L/a = 4$ and $a/t = 300$, we find from the diagram that the two limits on the value of T/t are 7 and 43 respectively. For an intermediate value the modal stiffness for $n = 2$ lies between the two extremes indicated by the diagram of fig. 9.9; and can be worked out from (9.65).

So far in this section we have considered only the case $n = 2$. It is not difficult to find the corresponding relations for other values of n: see problem 9.11.

In the above example but for the higher mode $n = 3$, the two limits on T/t are 14 and 25, respectively. For $n = 4$ the two limits almost coincide, which is an indication that for this particular shell the edge-response is insensitive to the boundary conditions at the remote end at $n = 4$: from (9.47) $n^* \approx 3.6$ for this shell.

The above analysis assumes that the end-plate is made of the same material as the shell and is of radius a. These conditions may not be met in practice. For example, the foundation slab of a steel silo may well be made of reinforced concrete, and it may extend considerably beyond radius a. The bending stiffness of the slab is proportional to $E'T^3$ where E' is the Young's modulus of the material of the slab.

For such a silo, the value of E'/E is, typically, 0.1. The effect of this difference of moduli is to increase the value of T/t for which the slab is effectively rigid to $43 (10)^{\frac{1}{3}} = 93$. Similarly it is not difficult to allow for a base plate whose radius is larger than a. For a plate which is infinite in extent the value

Fig. 9.15. Map showing the values of T/t above which the plate is equivalent to a clamped end (solid lines) and below which it acts as a flexible diaphragm (broken lines) for edge-loading $n = 2$. In the empty upper region the shell is so long that the boundary conditions at the remote end have little effect: the boundary of this region approximates to the chain-dotted line corresponding to $n^* = 2$. In the lower left-hand region the behaviour of the shell is determined by short-wave effects.

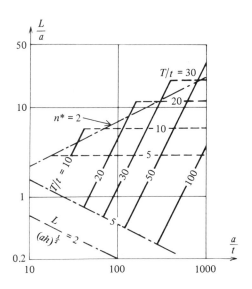

of K_n is about three times as large as that given by (9.58), and it is not difficult to carry through the effect of a change of this sort on the calculation.

It should be remembered that the above analysis is based on a relatively crude calculation on a beam-on-elastic-foundation having an elastic angular restraint at the end. If necessary, this analysis can be refined. However, the main benefit of the analysis is to give a relatively quick qualitative assessment of the kind of boundary support which is provided by a flat-plate closure of a given cylindrical shell.

9.6 Cylindrical shell loaded by radial point forces

In the previous example, we were concerned with the response of a cylindrical shell to an edge-loading in the form of a typical component of a Fourier series. It is appropriate now to consider a problem in which the loading involves the sum of such a series (cf. Calladine, 1977a). The shell and its loading are shown in fig. 9.16a. The shell is cylindrical, with radius a, thickness t and length $2L$. Its ends may be supported, as before, in a variety of ways – free, 'held circular', clamped – but in each case which we shall consider the same conditions are applied at both ends. The load consists of diametrically opposed point forces W applied at the central cross-section.

The application of localised loads to cylindrical shells constitutes a problem common to many branches of engineering, and particularly in the design of support arrangements for pressure-vessels and piping systems. It is perhaps unusual to find *diametrically* opposed loading in practice, but, as we shall see, it is not difficult to adapt the analysis to the case of single radial loads.

Fig. 9.16. (a) A thin-walled cylindrical shell which is subjected to diametrically opposed point loads applied at the central cross-section. (b) A typical component of the Fourier-series representation of the applied loading.

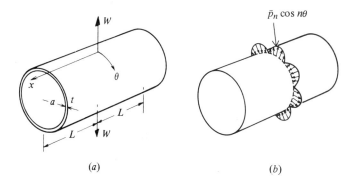

In practice, localised forces are not applied to vessels at points, but over small regions of the surface and through various kinds of fixture (see, e.g. Bijlaard, 1955 and Łukasiewicz, 1976). Some of the problems associated with localised loading fall properly within the scope of plasticity theory, and will be discussed in chapter 18. But there are nevertheless some important questions which can be answered by application of the methods of this chapter.

For example, if the loading shown in fig. 9.16a is applied to a tube with open ends, it is possible for the tube to deform by a purely inextensional mode, in which each generator remains straight and parallel to the axis. But this seems unlikely to occur for a very long tube, for which we must of course expect the deformation to be localised Therefore we shall seek, *inter alia*, a dimensionless group, whose value will indicate the general kind of deformation pattern which will occur. Our main study will be of the deflection of the shell immediately under the point of application of the load, but we shall be in a position to investigate in some detail the general deformation pattern and the disposition of stress resultants in the shell.

The first step of the analysis is to express the applied loading in terms of a Fourier series. Taking the origin of coordinates as shown in fig. 9.16a, we may express the load as the sum of a series of radial line loads applied at the cross-section $x = 0$, and varying as $\cos n\theta$ around the circumference:

$$\bar{p} = \frac{W}{\pi a} + \frac{2W}{\pi a} \sum_{2}^{\infty} \cos n\theta \quad (n = 2, 4 \ldots). \tag{9.68}$$

Here $\bar{p}$ is the intensity of radial load per unit circumference, defined in the spirit of section 7.5. The first term on the right-hand side is a uniformly distributed loading. The remaining terms of the series all have the same amplitude. The effect of the first term can be determined by the methods of chapter 3, and we shall return to it later. The response of the structure to a radial line load $\bar{p}_n \cos n\theta$, shown in fig. 9.16b, is obviously the most demanding question. This loading enters the shell in a way which is analogous to that of the preceding section. Except for high values of n there is a statically equivalent tangential loading of intensity $\bar{q}$ per unit circumference (cf. fig. 4.9), and given by

$$\bar{q} = -n\bar{p}_n \sin n\theta \tag{9.69}$$

which is transmitted directly to a 'long-wave' solution of the equations, while the remaining self-equilibrating loads

$$\bar{p} = \bar{p}_n \cos n\theta, \quad \bar{q} = n\bar{p}_n \sin n\theta \tag{9.70}$$

are carried by a localised 'short-wave' solution. (The negative sign in (9.69)

9.6 Cylindrical shell loaded by radial point forces

and the absence of one in (9.70) are a consequence of defining the radial line load as $\bar{p}_n \cos n\theta$ instead of $\bar{p}_n \sin n\theta$ as in sections 7.5 and 9.4.) For sufficiently large values of n these two solutions are replaced by a single 'flat-plate' solution of the equations for the normal loading $\bar{p}_n \cos n\theta$: cf. fig. 9.9.

Now since the sum of all the terms of the Fourier series, i.e. the actual applied loading, produces zero line load, except at $\theta = 0$ and $\theta = \pi$, we expect that the 'short-wave' and 'flat-plate' solutions give stresses which are effectively self-cancelling, except in the immediate vicinity of the points of application of load. It follows that except possibly in this small region of the shell, our main concern will be with the 'long-wave' solution.

The beam-on-elastic-foundation analogy for the long-wave part of the solution is very similar to that which has been described in section 9.5, except of course that the boundary conditions are different, and we are now dealing with quantities varying as $\cos n\theta$ instead of $\sin n\theta$, and vice versa. Corresponding to the tangential line load $n\bar{p}_n \sin n\theta$ acting on the shell is a transverse force

$$P = \pi a \bar{p}_n$$

acting on the beam (cf. (9.33). The expressions for beam and foundation stiffness, B and k respectively, are exactly the same as before (see (9.30) and (9.31a)), and the three boundary conditions on the shell, namely 'free', 'held circular' and 'clamped' correspond to free, simply-supported, and clamped edges of the beam respectively.

The central deflection of a centrally loaded beam on an elastic foundation is shown graphically in fig. 9.8 for the three different boundary conditions. The corresponding result for the shell, in terms of modal flexibility as a function of circumferential wavenumber n, is obtained from it by use of substitutions (9.30) and (9.31a). It is shown schematically in fig. 9.17, along the lines of fig. 9.9: see problem 9.12. For small values of n we obtain precisely the classical inextensional solution when the edges are free, but for the other two boundary conditions the behaviour is in accordance with the membrane hypothesis: in either case, the solution extends from the central plane to the ends of the shell. On the other hand, when the value of n is sufficiently large there is an interaction between longitudinal stretching and circumferential bending; the deformation is localised to the central portion of the shell and the various different end-conditions play virtually no part in the solution. These calculations are all based on the 'long-wave' part of the solution. The modal flexibility of the 'short-wave' part of the solution is also shown in fig. 9.17: it does not depend on the value of n. As in the previous case (fig. 9.9) the 'short-wave' and 'long-wave' division is not appropriate for sufficiently large values of n, and the 'flat-plate' solution supervenes, as shown.

For a shell of given dimensions, it is not difficult to determine from fig. 9.17 the modal flexibilities for $n = 2, 4, 6$, etc., (taking into account the shape of the transition curve between the asymptotes, as necessary) and to evaluate the sum of the various terms in order to obtain the deflection, δ, at the point of application of the load. This has been done in fig. 9.18 for shells having a range of values of the parameter Ω. Here the 'short-wave' effects have been ignored, because there is always (for large enough values of Ω) a term having much larger flexibility. For the shell with free ends the calculation recovers the classical inextensional result for sufficiently small values of Ω. It is clear that in general when $\Omega > 1$ the various boundary conditions have virtually no effect on the deflection. The curves corresponding to the ends being clamped and 'held circular' respectively have a wavy shape in the vicinity of $\Omega = 0.5$. This corresponds to the fact that for $\Omega < 1$, approximately, there is a particular mode which is most flexible; and although the sum of the various terms gives very nearly a straight line on the graph when the most flexible mode has

Fig. 9.17. Dimensionless graph showing the compliance of the shell under the loading of fig. 9.16b, for several types of boundary condition. The abscissa is the product of wavenumber n and the square root of the dimensionless length Ω.

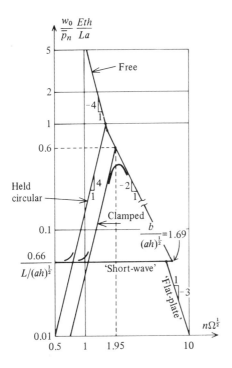

a circumferential wavenumber n of 6 or more, there are anomalies when this number is either 2 or 4: see problem 9.13.

Now if the length of the shell is steadily diminished, there comes a point where the diagram as drawn in fig. 9.17 becomes inappropriate. Just as in fig. 9.9, the abscissa and ordinate variables have been chosen so that the lines corresponding to the 'long-wave' solutions are located uniquely; and just as in fig. 9.9 the location of the 'short-wave' and 'flat-plate' lines depends on the value of the parameter $L/(ah)^{\frac{1}{2}}$. Thus as the value of $L/(ah)^{\frac{1}{2}}$ decreases, the 'long-wave' solutions become submerged beneath the rising level of the 'short-wave' solution. What should take the place of fig. 9.17 in these circumstances? The best way to proceed is to abandon the use of Fourier series and to think afresh of a point load W applied to the edge of a very short shell with clamped edges. It seems clear that if the shell is sufficiently short its deformation will be confined to a region whose circumferential length is of order L. Over such a region the shell is practically flat (if $L/(ah) \ll 1$) and we must therefore ex-

Fig. 9.18. Compliance of a cylindrical shell under the loading of fig. 9.16a, for several end-conditions. Various Fourier components are plotted separately in the lower right-hand part, but the three main curves show the summation of the several Fourier terms. At the lower left-hand corner the main curves are cut off, for short shells, by 'flat-plate' behaviour. δ is the radial displacement.

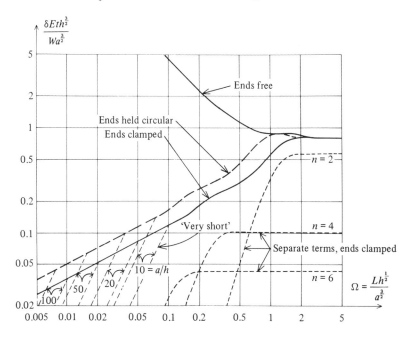

pect the shell to carry the applied load primarily by B-surface action. In other words the shell sustains the load by acting as a long plate of width $2L$ with clamped edges. This is a well-known problem in the theory of elastic plates, and the result is given, e.g., by Timoshenko and Woinowsky Krieger (1959, p.190). When the edges are clamped,

$$\delta = \frac{WL^2(1 - \nu^2)}{2.87Et^3}. \tag{9.71}$$

A similar argument applies to a short shell with simply-supported edges. In this case the corresponding displacement is given by

$$\delta = \frac{WL^2(1 - \nu^2)}{1.23Et^3}. \tag{9.72}$$

It is important to realise that these formulae represent a situation quite different from that corresponding to the lines labelled 'flat-plate' in figs. 9.9 and 9.17. In those cases the B-surface action was dominant, but the length of the shell was great in comparison with the circumferential half wavelength and the modal flexibility was independent of L. In the present problem it is precisely the *proximity* of the ends of the shell which provides the high stiffness of the arrangement; from (9.71) and (9.72) we see that the total flexibility is proportional to L^2. These two expressions have been plotted onto fig. 9.18. Since the dimensions L and h do not appear in a way which allows them to be put in terms of Ω, it is necessary to draw a family of lines, for different values of a/h, as shown: the parameter Ω, which was introduced as a convenient group for plotting the 'long-wave' part of the solution, is not particularly useful in the present context.

However, it is possible to replot the information in the lower left-hand part of fig. 9.18 in a way which gives a single curve, and this has been done in fig. 9.19b. The geometrical parameter which achieves this is $L/(ah)^{\frac{1}{2}}$, and the asymptotes intersect when it has a value of order unity. Here we have made no attempt to fill in the transition curve lying between the asymptotes: this task would require lengthy numerical work.

The information of fig. 9.18 has been redrawn twice in fig. 9.19 in a schematic way which couples the deflection of the shell with the dominant mode of action in the various regimes. In order to cover the entire range of behaviour it is necessary to have two diagrams using the two dimensionless groups Ω and $L/(ah)^{\frac{1}{2}}$ respectively. Although the curves only show explicitly the deflection δ of the shell under the point of application of the load, the appended information gives a qualitative indication of the corresponding mode of deformation. In particular, the value of the circumferential wavenumber n^* for the 'dominant' mode is indicated in fig. 9.19a. It is not difficult to make simple estimates of the stresses in the shell: see problems 9.5, 9.14 and 9.15.

Although this analysis has been done in terms of diametrically opposed loads W, it is not difficult to see that the results shown in figs. 9.18 and 9.19 may be adapted easily in relation to other patterns of loading, such as four loads W at $\theta = 0, \frac{1}{2}\pi, \pi$ and $\frac{3}{2}\pi$ or even a single load W: see problems 9.16 and 9.17.

9.7 Concentrated load on a spherical shell

In the previous section we have studied the response of a cylindrical shell to point loads applied radially. It is of interest to investigate finally the response of a spherical shell to a similar load, so that we can make a simple comparison of the behaviour of the two kinds of shell under this sort of loading. It seems clear that the most direct way of investigating the behaviour of a spherical shell loaded in this way is to use the equations for a shell of revolution. However, we can in fact employ the methods of chapter 8 to solve this problem, by writing the equations of a 'nearly-cylindrical' shell, and putting $a^* = a$. Previously we have insisted that $|a^*| \gg a$, so that even relatively long shells with curved meridians have more-or-less constant radius, as in fig. 4.16. But the equations developed previously are still valid when a^* is of the same

Fig. 9.19. (a) Schematic representation of fig. 9.18, omitting the detail of transition curves, etc., and indicating values of n^*, etc..
(b) Replot of lower left-hand part of (a) for clamped edges, so that the lines of slope 2 and $\frac{1}{2}$ each map to a single line. Also shown is the corresponding pair of lines for a spherical shell. The two points correspond to the calculations of Flügge & Elling (1972).

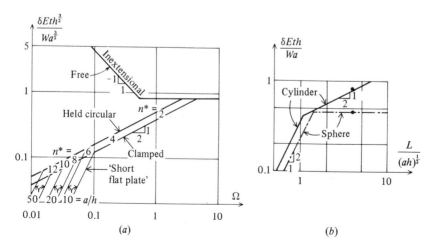

magnitude as a, provided we are content to deal with a sufficiently small zone of the shell. In the case of a spherical shell ($a^* = a$) we shall find that the main variables decay so rapidly with distance from the point of application of the load that this assumption is justified. In these circumstances the equations are known as the 'shallow-shell' equations: see section 8.2. We specialise (8.22) by putting

$$\Gamma^2(\ldots) = \frac{1}{a}\frac{\partial^2(\ldots)}{\partial x^2} + \frac{1}{a}\frac{\partial^2(\ldots)}{\partial y^2} = \frac{1}{a}\nabla^2 w. \tag{9.73}$$

Consequently (8.24) becomes

$$D\nabla^8 w + (Et/a^2)\nabla^4 w = 0. \tag{9.74}$$

The right-hand side is zero because there are no surface tractions on the shell: the load W will be accounted for by means of a suitable singularity at the origin.

It is clear that (9.74) will be satisfied by the solution of the lower-order equation

$$D\nabla^4 w + (Et/a^2)w = 0, \tag{9.75}$$

provided there are no anomalies at the boundaries. In (9.74), and therefore also in (9.75), the first term corresponds to the flexure of the B-surface as a flat plate, while the second term represents the behaviour of the S-surface. Equation (9.74) is a statement of the fact that the sum of the normal surface tractions on the two surfaces is zero. Now the first term of (9.75) is identifiable simply as the normal surface traction on the B-surface (see section 8.3); and it follows that the second term is equal to the pressure sustained by the S-surface. It is remarkable that the effect of the in-plane stiffness of the S-surface reduces to a simple multiple of the local displacement w in the present case of a spherical surface, but this is indeed true provided certain boundary conditions are satisfied: see problem 9.18. It follows that (9.75) may be considered simply as the equation of a uniform *plate on an elastic foundation*, where the plate represents the B-surface and the elastic foundation represents the S-surface. This is a well-known problem in the theory of elastic plates: see Timoshenko & Woinowsky-Krieger (1959, §58). The case which is easiest to solve is that in which a point load is applied to a plate of infinite extent resting on an elastic foundation: this is known as Hertz's problem. Since the arrangement is symmetrical about the point of application of the load, it is best to use as the basic variable

$$r = (x^2 + y^2)^{\frac{1}{2}}: \tag{9.76}$$

and the governing equation then becomes

$$\left(\frac{d^2}{dr^2} + \frac{1}{r}\frac{d}{dr}\right)\left(\frac{d^2 w}{dr^2} + \frac{1}{r}\frac{dw}{dr}\right) + \frac{12}{a^2 h^2}w = 0. \tag{9.77}$$

9.7 Concentrated load on a spherical shell

As usual, the factor D/Et has been expressed in terms of the 'effective' thickness h. It is convenient to put

$$\xi = r/b, \quad \text{where } b^4 = \tfrac{1}{12}a^2h^2, \tag{9.78}$$

so that the equation may be written

$$\left(\frac{d^2}{d\xi^2} + \frac{1}{\xi}\frac{d}{d\xi}\right)\left(\frac{d^2w}{d\xi^2} + \frac{1}{\xi}\frac{dw}{d\xi}\right) + w = 0. \tag{9.79}$$

This may be solved either by the use of power series or complex variables. With three boundary conditions

$$w \text{ and } \frac{dw}{d\xi} \to 0 \text{ as } \xi \to \infty, \quad \frac{dw}{d\xi} = 0 \text{ at } \xi = 0 \tag{9.80}$$

the solution is

$$w = C \text{ kei } \xi, \tag{9.81}$$

where kei is a Kelvin function tabulated by Olver (1965).

The remaining constant is determined by the condition that the total transverse shearing force across any cylindrical cut concentric with the axis is equal to the applied load W: this enables us to fix the value of C:

$$C = -Wb^2/2\pi D.$$

Therefore, the deflected form is given by

$$w = \frac{W \text{ kei } \xi}{2\pi (DEt/a^2)^{\frac{1}{2}}}$$

$$= -W\frac{3^{\frac{1}{2}}}{\pi}\frac{a}{Eth} \text{ kei}\left[\frac{12^{\frac{1}{4}}r}{(ah)^{\frac{1}{2}}}\right]. \tag{9.82}$$

At the origin, since $\text{kei}(0) = -\tfrac{1}{4}\pi$, we have

$$w_0 = \frac{3^{\frac{1}{4}}}{4}\frac{Wa}{Eth} \approx 0.32 \frac{Wa}{Eth}. \tag{9.83}$$

Fig. 9.20a shows the deflected form of the shell, and in (b) and (c) the loads imposed on the S- and B-surfaces respectively are shown separately. These diagrams show clearly that the point load is carried in the first instance by the B-surface, which is held in equilibrium by a pressure of interaction with the S-surface over a small area whose radius is of order $(ah)^{\frac{1}{2}}$. The loading on the S-surface is a pressure distribution which is statically equivalent to the point force, but is spread over the same small area. In other words, the B-surface acts locally as a device for 'spreading' the applied load over a small region of the S-surface. The size of this region is determined by the relative stiffness of the two surfaces in response to the applied loading.

It is clear from fig. 9.20 that provided the boundary of the curved surface lies at a radius r of more than about $3(ah)^{\frac{1}{2}}$, the precise form of the boundary conditions at the outer edge will have little effect on the local solution. In terms of the 'nearly-cylindrical' barrel-shaped shell with which we began, we find that the radius of the surface at this distance from the central plane is less than a by about $5h$: thus, for the usual range of values of a/h the equations are clearly satisfactory. If the shell actually has the form of, say, a hemispherical surface, we can easily extend the solution beyond the small region which we have studied: we can treat the entire shell as an S-surface to which is applied the pressure distribution shown in fig. 9.20c. It is then relatively straightforward to determine the distribution of the various stress resultants in the surface: see problem 9.19.

On the other hand, if the shell is merely a small shallow cap with (say) clamped edges, it is clear that the boundary conditions implicit in fig. 9.20 are inappropriate. It is possible to solve the equations for a cap having an arbitrary radius, but the algebra is somewhat complicated. However, when the radius of the cap becomes sufficiently small, the solution takes a particularly simple form, since the B-surface plays the dominant role in supporting the applied load as a flat circular plate of radius L (say) with clamped edges. This is a well-known problem in the theory of elastic plates, and the central deflection is given by Timoshenko & Woinowsky-Krieger (1959, §19):

$$w_0 = \frac{WL^2}{16\pi D} \approx 0.22 \frac{WL^2}{Eth^2}. \tag{9.84}$$

Equations (9.83) and (9.84) are plotted in fig. 9.19b. They provide asymptotes to an unknown transition curve whose form can be sketched: the asymptotes cross at $L/(ah)^{\frac{1}{2}} \approx 1.2$, and we have already argued that the boundary conditions should play no part beyond about $L/(ah)^{\frac{1}{2}} = 3$.

It is instructive to compare the two curves in fig. 9.19b, relating to the application of concentrated loads to cylindrical and spherical shells, respectively. There are some obvious similarities, but there is one important difference. For

Fig. 9.20. (a) Normal-displacement profile for a shallow spherical shell carrying a radial point load. (b) Distribution of external load and interface pressure on the B-surface. (c) Distribution of interface pressure on the S-surface.

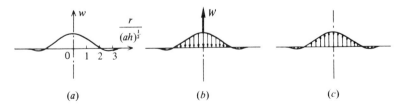

9.7 Concentrated load on a spherical shell

the spherical shell the region of bending action is localised in the sense that the deflection becomes *independent* of L when this variable exceeds a certain value; whereas for the cylindrical shell the central deflection increases steadily with L in both regimes. The difference springs from the fact that for a cylindrical shell the behaviour for low circumferential mode-numbers involves a predominantly 'membrane' mode in which load is carried by an analogous 'beam' action to the ends of the shell; and for this reason the deflection under the point of application of the load depends on L in a way quite different from that of a spherical shell.

This study of two simple kinds of shell gives results which may be considered as particular cases in a pattern of behaviour for a continuous range of shell geometries, including hyperbolic shells. Unfortunately, the equations are difficult to solve in general, when the two specially simple forms of the operator Γ^2 (see (8.27), (9.73)) do not apply; and solutions can only be found by means of numerical analysis. Flügge & Elling (1972) have studied the effects of a normal point load at the centre of a shell circular in plan and having a range of principal radii a, a^*, where a is constant and a^* varies between $-a$ and $+\infty$. The radius L of the shell (in plan) was equal to about $4(ah)^{\frac{1}{2}}$ for all of their examples.

The results of the present study are in good agreement with the calculations of Flügge and Elling, but although these writers consider a number of different shell geometries, it is not easy to draw any general conclusions as they quote results for only one value of $L/(ah)^{\frac{1}{2}}$. It does however seem reasonable to make the following remarks.

When the value of $L/(ah)^{\frac{1}{2}}$ is of order unity or less, the shell behaves essentially like a flat plate. When the value of $L/(ah)^{\frac{1}{2}}$ is greater than unity, the behaviour of the shell depends on the geometry of the surface. For a spherical surface (and other ellipsoidal surfaces?) the region in which there are appreciable bending stresses is localised, and in consequence the deflection of the shell at the point of application of the load is virtually independent of L. But for a cylindrical surface (and also other surfaces, e.g. hyperbolic, having straight generators?) the region in which there is appreciable bending extends out from the centre towards the edges along the straight generators; and the deflection increases as the value of L increases.

There is, however, one important aspect in which the behaviour is independent of the geometry of the surface. In all cases the applied point load is carried in the first instance by the B-surface acting locally as a plate. Flügge and Elling show that within a radius of about $0.4(ah)^{\frac{1}{2}}$ of the point of application of the load, the bending moments in the shell are virtually the same as they would be for a flat plate, independently of the value of a/a^*. This is a con-

312 *Nearly-cylindrical shells under non-symmetric load*

sequence of the fact that the bending moments in this region are dominated by a singularity in the plate-like solution of the bending equations of the *B*-surface.

9.8   Conclusion

In all of the problems which we have investigated in this chapter, we have been able to give results for a wide range of geometries specified principally in terms of Ω (= length(thickness)$^{\frac{1}{2}}$/(radius)$^{\frac{3}{2}}$), and to a lesser extent in terms of the group length/(radius × thickness)$^{\frac{1}{2}}$. Results for any particular geometry may be read from the appropriate curves. The key to the relatively simple analysis in each case has been the use of simple diagrams like those of figs. 9.9 and 9.17 which delineate almost distinct regions of structural behaviour: and these only interact in narrow zones near their boundaries. The narrowness of these zones is not an illusion deriving from the use of a logarithmic plot. In terms of integral circumferential mode-numbers, it is unusual for more than one mode to lie in a transition zone in a given problem: indeed, it is fairly common for modes with consecutive mode numbers to straddle almost entirely a transition zone (see problem 9.20). The diagrams of the kind shown in figs. 9.9 and 9.17 are based on the separation of the governing eighth-order differential equations into two fourth-order equations each representing a distinct beam-on-elastic-foundation analogue. Cases in which this separation is not appropriate can be considered separately.

In most of the examples the 'long-wave' fourth-order differential equation has played the dominant role – hence the importance of Ω – and the 'short-wave' equation has played a significant part only in the narrow 'boundary layers'.

Finally, we remark that although all of the examples in this chapter have involved uniform isotropic shells, it is not difficult to investigate some simple forms of anisotropy. As we shall see in an example to be given (in the context of *buckling*) in section 14.10, diagrams of the kind shown in figs. 9.9 and 9.17 are particularly useful in this respect: they can readily be modified to allow for anisotropy.

9.9   Problems

9.1   In the calculation of the deflection, according to the membrane hypothesis, of a shell such as that shown in fig. 4.8*a* (see problem 7.8), the load-carrying contribution of the flexure of shell elements in the longitudinal direction is implicitly ignored. Show that, in terms of the 'beam' analogy of section 7.5 this corresponds to the use of $I = \pi a^3 t$ for the second moment of area of the cross-section of a hollow tube of mean radius a and thickness t, rather than the

9.2 If $z = \cos\theta$, verify that
$$\cos 2\theta = 2z^2 - 1; \quad \cos 3\theta = 4z^3 - 3z; \text{ etc.}$$

exact expression $I = \pi a^3 t(1 + t^2/4a^2)$; and consequently that the longitudinal bending effect is normally negligible in such a case.

9.3 A cylindrical shell of radius a and length L is to be analysed by means of a double Fourier series having terms corresponding to half wavelengths $L(1, \frac{1}{2}, \frac{1}{3}, \frac{1}{4}$ etc.) and $\pi a(\frac{1}{2}, \frac{1}{3}, \frac{1}{4}$, etc.) in the longitudinal and circumferential directions, respectively. Mark a grid of these points, for use with fig. 8.4, on a piece of tracing paper; and verify in particular that changes in the dimensions of the shell correspond to translations of the entire grid of points.

9.4 Use the 'strain-energy' method of appendix 1 to obtain an approximation to the deflection at the tip of the cantilever of fig. 9.7a. Take for the mode of deflection the displacement of an unsupported cantilever. Plot the resulting curve onto fig. 9.7b.

9.5 An elastic beam on an elastic foundation has simply-supported ends. A transverse load P is applied at the centre. Hetényi gives the following formula for the central bending moment M_0:
$$M_0 = \frac{P}{2^{\frac{3}{2}}} \left(\frac{B}{k}\right)^{\frac{1}{4}} \left(\frac{\text{Sh}\xi + \sin\xi}{\text{Ch}\xi + \cos\xi}\right), \tag{i}$$
where $\xi = 2^{\frac{1}{2}} L(k/B)^{\frac{1}{4}}$, L is the half-length and B, k are the beam and foundation stiffnesses, respectively.

Show formally that
$$M_0 \approx \tfrac{1}{2} PL \quad \text{for } \xi \ll 1, \tag{ii}$$
and check this result by assuming that the beam is so short that the foundation support is negligible. Also show that
$$M_0 \approx P(B/k)^{\frac{1}{4}}/2^{\frac{3}{2}} \quad \text{for } \xi \gg 1. \tag{iii}$$
Verify that asymptotes (ii) and (iii) intersect at $L(k/B)^{\frac{1}{4}} = 2^{-\frac{1}{2}}$, i.e. $\xi = 1$; and compare the values of M given by the three formulae in this particular case.

9.6 Demonstrate that the process of transforming the curve of fig. 9.7 to its counterpart in fig. 9.9 involves (a) a contraction of the abscissa by a factor of two, giving a curve with asymptotes having slopes of $3 \times 2 = 6$ and 0, respectively; (b) a further transformation in which the ordinate is replaced by the

314   *Nearly-cylindrical shells under non-symmetric load*

ordinate divided by the square of the abscissa, giving a curve with asymptotes having slopes $6 - 2 = 4$ and $0 - 2 = -2$, respectively.

9.7 Obtain expression (9.52) for the edge-response of the plate shown in fig. 9.11 by solving the governing equation $\nabla^4 w = 0$ on the assumption that $w = f(x) \sin(\pi y/b)$, and subject to the boundary conditions $M_x = 0$ and $V_x = Q_0 \sin(\pi y/b)$ at $x = 0$. See problem 8.8 for expressions for M_x and V_x in terms of w.

9.8 Sketch the form of fig. 9.9 for a shell clamped at $x = L$ for values of $L/(ah)^{\frac{1}{2}} = 5, 2$ and 1 (cf. $L/(ah)^{\frac{1}{2}} \approx 10$ as shown); and hence show that the inverted V corresponding to the long shell (with asymptotes intersecting at $(1.74, 3.03)$) is submerged beneath the rising 'short-wave' horizontal at around $L/(ah)^{\frac{1}{2}} = 1$. Sketch the form of fig. 9.9 when $L/(ah)^{\frac{1}{2}} < 1$, retaining only the 'short-wave' and 'flat-plate' asymptotes.

9.9 A *rigid* beam of length L is connected to a uniform elastic foundation of stiffness k. One end is simply-supported, while the other is free. A transverse load P is applied at the free end. Show by first principles that the deflection w_0 at the free end is given by $w_0 = 3P/kL$, and reconcile this formula with the corresponding line in fig. 9.8 for sufficiently small values of $L(k/B)^{\frac{1}{4}}$. Also show that when both ends of the beam are free, the corresponding expression is $w_0 = 4P/kL$.

9.10 Use (9.66) and (9.67) to show that the ratio of the smallest end-plate thickness (T_1, say) for which the base is effectively clamped to the largest thickness (T_2) for which the base is merely 'held circular' is given by $T_1/T_2 \approx 0.2(n^*)^{\frac{8}{3}}$.

9.11 Equations (9.66) and (9.67) were developed for the case $n = 2$. Show that the corresponding equations for $n > 2$ are:

$$\frac{T_1}{t} \approx 1.5 \left(\frac{2}{n}\right)^{\frac{4}{3}} \left(\frac{a}{L}\right)^{\frac{1}{3}} \left(\frac{a}{t}\right)^{\frac{2}{3}}; \quad \frac{T_2}{t} \approx 2.1 \left(\frac{n}{2}\right)^{\frac{4}{3}} \left(\frac{L}{a}\right).$$

For this purpose, remove the factor $(4/3)^2$ from the precursor of (9.67). T_1 and T_2 in these expressions have the meanings assigned in problem 9.10.

Hence show that the consequence of increasing the value of n is to move all of the solid and broken lines in fig. 9.15 down and to the right at a slope of -1, until the intersection vertices bear the same relation to the line $n^* = n$ (fig. 9.12) as they do to the line $n^* = 2$ in fig. 9.15.

9.9 Problems

9.12 Working from the asymptotic formulae for beams on elastic foundations, and the transformations given in section 9.4, derive the following expressions for the 'long-wave' formulae plotted in fig. 9.17 when n is so large that $n^2 - 1 \approx n^2$.

When we write $w_0 Eth/\bar{p}_n La = Y$ and $n\Omega^{\frac{1}{2}} = X$, the expressions are:

$Y = 3^{\frac{3}{4}}/X^2$ (long shell);

$Y = 6/X^4$ (short shell, ends free);

$Y = \frac{1}{6}X^4$ (short shell, ends held circular);

$Y = \frac{1}{24}X^4$ (short shell, ends clamped).

Also determine the coordinates of the intersection of the long-shell and short-shell 'ends clamped' formulae, and thus obtain the following asymptotic expressions:

$Y = 0.60 \, (n/n^*)^4, \quad n < n^*$

$Y = 0.60 \, (n/n^*)^{-2}, \quad n > n^*$

where $n^* = 1.95/\Omega^{\frac{1}{2}}$.

9.13 In fig. 9.18 the curve for a shell with clamped ends has a slope of $\frac{1}{2}$ for $\Omega < 0.2$, approximately. Determine an approximate equation for this line in the following way, using the results of problem 9.12 on the assumption that $n^* \gg 2$ and that the transition curve in fig. 9.17 may be ignored.

Since $\bar{p}_n = 2W/\pi a$ (from 9.68), we may write

$\pi\delta Eth/2WL = 0.60 \, [(2/n^*)^4 + (4/n^*)^4 + \ldots (n < n^*)$
$+ (8/n^*)^{-2} + (10/n^*)^{-2} + \ldots (n > n^*)].$

Show that the contents of [ ] may be written approximately as

$(n^*/2)(\int_{2/n^*}^{1} z^4 dz + \int_{1}^{\infty} z^{-2} dz) \approx 0.6n^*$

(if the lower limit of the first integral is set to zero). Hence show that $\delta Eth^{\frac{3}{2}}/Wa^{\frac{3}{2}} \approx 0.45\Omega^{\frac{1}{2}}$.

Why does this line lie a little above its counterpart in fig. 9.18?

9.14 A cylindrical shell of length $2L$, radius a and thickness t is 'held circular' at both ends. At its central cross-section it carries a radial line load of intensity $\bar{p} = \bar{p}_n \cos n\theta$, as in fig. 9.16$b$. Use the results of problem 9.5 together with the transformations of section 9.4 to establish the following approximate formulae for the amplitude N_x^* of $N_x = N_x^* \cos n\theta$ at the central cross-section:

$N_x^* = \dfrac{n^2 L p_n}{2a}, \quad n < n^*;$ \hfill (i)

316   Nearly-cylindrical shells under non-symmetric load

$$N_x^* = \frac{3^{\frac{1}{4}} a p_n}{2(1-n^{-2})^{\frac{1}{2}}} \left(\frac{a}{h}\right)^{\frac{1}{2}}, \quad n > n^*; \qquad (ii)$$

where $n^*(n^{*2} - 1)^{\frac{1}{2}} = 3^{\frac{1}{4}}/\Omega$, i.e. $n^* \approx 1.1/\Omega^{\frac{1}{2}}$.

Verify that (i) agrees with an analysis according to the membrane hypothesis; and show by inspection of fig. 9.17 that (ii) is valid up to $n \approx 1.9(a/h)^{\frac{1}{2}}$.

9.15 A long elastic beam of flexural rigidity B is supported on an elastic foundation of stiffness k. It sustains a transverse load P at a point remote from the ends. Show that at this point the bending moment and displacement are given by $M = P(B/k)^{\frac{1}{4}}/2^{\frac{3}{2}}$, $w = P/2^{\frac{3}{2}} k^{\frac{3}{4}} B^{\frac{1}{4}}$. Thus show that the strain energy per unit length in *both* the beam *and* the foundation at this point is equal to $P^2/16(kB)^{\frac{1}{2}}$.

Hence show that for a shell loaded as in problem 9.14, and with $n^* < n < 1.9(ah)^{\frac{1}{2}}$ approximately, the amplitudes of the mean longitudinal stress and the circumferential bending stress will be roughly equal.

9.16 Show that the formulae of problem 9.12 transform to the following expressions in terms of the variables of fig. 9.18 when n is so large that $n^2 - 1 \approx n^2$. Writing $w_0 E t h^{\frac{3}{2}}/W a^{\frac{3}{2}} = Z$ (i.e. putting $\bar{p}_n = 2W/\pi a$, cf. (9.68)):

$Z = 1.45/n^2$,   (long shell);
$Z = 3.82/n^4 \Omega$ (short shell, ends free);
$Z = 0.106 n^4 \Omega^3$, (short shell, ends held circular);
$Z = 0.0265 n^4 \Omega^3$, (short shell, ends clamped).

9.17 Sketch the modified form of fig. 9.19 when the shell of fig. 9.16 is loaded at its central plane by *four* equally spaced radial outwards loads W.

First show that in this case

$$\bar{p} = \frac{2W}{\pi a} + \frac{4W}{\pi a} \sum_{4}^{\infty} \cos n\theta \quad (n = 4, 8, 12 \ldots);$$

i.e. that $\bar{p}_n$ is zero except when n is an integral multiple of 4, and that the nonzero values of $\bar{p}_n$ are double those of (9.68). Use the analysis of problem 9.13 to show that the lines of slope $\frac{1}{2}$ in fig. 9.19 are unchanged. Use the first result of problem 9.16 to show that the asymptote for $\Omega \to \infty$ is lowered by a factor of about 2; and similarly use the second result of problem 9.16 to show that the asymptote for the shell with free ends is lowered by a factor of about 8. Hence verify that the group of three asymptotes resulting from 'long-wave' solutions may be envisaged as shifting together on fig. 9.19 downwards and to the left at a slope of $\frac{1}{2}$ when the number of loads is raised from 2 to 4.

Explain the separate changes in position in the asymptotes in physical terms.

9.18 Verify the following equations in the case of a shallow uniform elastic spherical S-surface (cf. section 8.5):
$$\nabla^2 \phi = pa; \quad \nabla^4 \phi = -gEt; \quad \nabla^2 w = -ga.$$
Hence show that
$$\nabla^2 w = (a^2/Et)\nabla^2 p;$$
and that $w = a^2 p/Et$ if $w = p = 0$ around a closed contour surrounding the shallow region.

9.19 Use the results of problem 4.16 in connection with fig. 9.20 to sketch the distribution of N_θ and N_ϕ in the vicinity of a point load on a spherical shell. Show that at the point of application of the load, $N_\theta = N_\phi \approx W/6t$.

9.20 Let the extent of the 'transition zones' of the curves in fig. 9.8 be defined as the range of values of $Lk^{\frac{1}{4}}/B^{\frac{1}{4}}$ over which the ordinate of the curve differs by more than 5% from the closer relevant asymptote. By examination of the six curves show that this range involves a factor of between about 2 and 3. Invoke problem 9.6 to show that when these curves are transformed into diagrams such as figs. 9.9 and 9.17, in which the abscissa is proportional to the circumferential mode-number n, the corresponding transition zones cover a factor of between about 1.4 and 1.7 in n. Hence show that in a solution which involves only the *even* circumferential modes, it is not possible to have more than about 2 modes lying within the transition zone unless the value of n at the transition exceeds about 10.

10

Cylindrical shell roofs

10.1 Introduction

Reinforced concrete shells have been used in the construction of roofs for many large buildings such as airport terminals, exhibition halls and factories. From a structural point of view a shell is attractive for this purpose, since the continuity of surface which is required to keep out the weather is provided by the structural member itself. From an economic point of view, however, reinforced-concrete shell roofs cast *in situ* are less attractive, largely on account of the labour-intensive effort which is needed in the construction of the formwork.

According to chapter 5, it is easy to construct a surface having zero Gaussian curvature from rectangular plywood sheets, whereas the construction of other kinds of surface makes it necessary to cut the sheets individually into non-rectangular shapes. It is not surprising therefore that cylindrical shells have been popular for the roofing of relatively simple rectangular buildings according to the scheme shown in fig. 10.1 and extensions of it. Shells of this kind, simply supported at their ends, form the subject of the present chapter.

Several authors have written on the structural analysis of cylindrical shell roofs of this sort, and at least one conference has been devoted to this subject: see Timoshenko & Woinowsky-Krieger (1959, §126), Flügge (1973, §5.4.4.2), Gibson & Cooper (1954) and Witt (1954).

Almost all of the work which has been reported, however, is devoted to the analysis of particular examples having specific dimensions, and it cannot be said that any clear *design principles* have yet emerged from these studies.

The present chapter aims at general considerations rather than specific problems. We shall be concerned mainly with the overall structural action of a circular-cylindrical shell roof which is specified by its leading dimensions L, B, t and a for length, breadth, thickness and radius, respectively: see fig. 10.2.

10.1 Introduction

Most of this chapter will be devoted to a single section of the shell, as shown in fig. 10.1c, d, which is structurally separate from its neighbours. This isolated shell is conventionally taken to be an adequate representation of the end-shells of an array; and it is easy in fact to adapt the analysis to the case of a shell in the interior of such an array, which is connected to neighbouring shells at its edges, as we shall see. We shall study shells both with and without edge-beams (see fig. 10.1): in this way we shall be able to grasp the advantages associated with the provision of these auxiliary members, and to discuss ways of deciding on their dimensions.

In order to clarify the analysis we shall make certain simplifying assumptions throughout most of the chapter. First, we shall assume that the shell has uniform thickness, density and elastic modulus. These are not obviously justifiable in general for construction in reinforced concrete: after all, a prime advantage of this medium is the possibility of disposing the steel reinforcing bars non-uniformly over the surface of the shell, in order to give local strengthening; and this gives non-uniform elastic modulus and surface density. We shall return to this general question in section 10.4.

The second simplifying assumption which will be used in much of the chapter is that the shell is *shallow*, i.e. that the 'rise' or 'sagitta' H (fig. 10.3) is less than, say, 0.2 of the width B. One advantage of this assumption is that we may use the coordinate y either as a measure within the surface, as in fig. 10.2, or as a rectangular Cartesian coordinate as in fig. 10.3: very little error will spring from the discrepancy between these two definitions. Similarly, for a shallow shell there is not much numerical difference between the component

Fig. 10.1. Cylindrical shell roofs of the type studied in this chapter. (a) A set of parallel shells, connected along their edges. (b) As (a), but with edge-beams. (c) and (d) Isolated shell, without and with edge-beams, respectively.

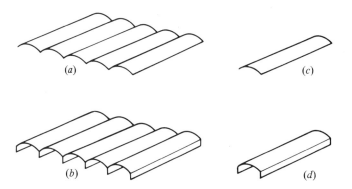

of displacement normal to the surface, and the component in the vertical direction, normal to the y-axis of fig. 10.3. It will also be convenient to observe that there is an insignificant difference between a shallow circular arc and a parabola: see appendix 8. Use of this scheme will enable us to establish rather clearly in section 10.2 some features of the behaviour of cylindrical shell roofs which might otherwise go unnoticed. Having clarified the picture in this way, we shall be in a position to relax to some extent the 'shallow-shell' restriction from section 10.3.1 onwards.

All of these simplifications must be abandoned, of course, if we ever wish to analyse a *deep* cylindrical shell, such as one with a semicircular cross-section.

In general we shall be concerned with a single 'panel' of a roof, and for most of our analysis we shall use an x, y coordinate system, as shown in fig. 10.2, having the origin at one corner. We shall consistently describe the curved arcs $x = 0$ and $x = L$ as the *ends* of the shell, and the straight lines $y = 0$ and $y = B$ as the *edges* of the shell. We shall refer to the x- and y-directions as *longitudinal* and *transverse* ($\equiv$ circumferential), respectively.

At the ends of the shell we shall assume supports in the form of diaphragms which retain the original form of the cross-section but provide negligible restraint in the x-direction. Thus they ensure that there is zero net axial tension in the shell, and zero restraint against warping at the ends.

Most of our analysis will be relevant only to shells having $L/B > 2$, say.

Fig. 10.2. Single cylindrical shell, showing dimensions and coordinate system.

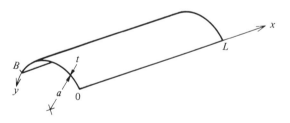

Fig. 10.3. Cross-section of a shallow circular cylindrical shell, here drawn as a parabolic arc.

10.2 A simple cylindrical shell roof

Let us consider first the behaviour of the simple shell shown in fig. 10.1c, as it carries its own weight between simple supports at the two ends. Normally, reinforced-concrete shells are constructed by casting onto a mould ('formwork') which is kept in place until the concrete has hardened. When the formwork is removed (or the shell is lifted off, in the case of a roof panel which is not cast *in situ*) the shell is obliged to carry its own weight. For the sake of definiteness in the subsequent analysis we shall imagine that the shell exists originally under conditions of effectively zero gravity in a perfectly cylindrical shape, and with zero stress throughout. Then, at a certain time, gravity is 'switched on', and the shell thereafter carries its own weight. Our task is to determine the corresponding stresses and displacements of the structure.

For an engineer not familiar with the theory of shell structures, the obvious first step in the analysis of the structure shown in fig. 10.1c is to consider the shell as a simple *beam* having a particular cross-sectional shape and carrying a uniform load per unit length of beam. On the other hand a student of shell structures may well regard it as obvious to begin the analysis by applying the membrane hypothesis. Now for this particular structure the membrane hypothesis leads almost immediately to a disqualifying contradiction, which may be explained as follows. We know that for a cylindrical shell there is, according to the membrane hypothesis, a direct equilibrium equation

$$N_y/a = p$$

connecting the normal surface loading intensity p and the circumferential direct-stress resultant at any point: see section 4.4. Here the self-weight of the (shallow) shell provides a roughly uniform distribution of p over the surface; and yet along the two unsupported edges there is an obvious force boundary condition $N_y = 0$. Consequently it is clear that the membrane hypothesis is inadequate: and hence there must in general be some bending action in the shell.

In chapter 9 we have used the methods of chapter 8 in order to solve a number of problems involving both bending and stretching action in cylindrical shells. Unfortunately, these methods are not directly applicable to the present problem on account of the 'stress-free' boundary conditions which may be imposed at the straight edges of the shell, although some ideas from chapter 8 will be useful later, in sections 10.2.4 and 10.3.3. For most of the chapter we shall use a different approach to the problem of shell structures, which makes use of the essential idea of S- and B-surfaces in a different way. The basic idea is that we start by considering a simple hypothetical mode of deformation of the shell. From this we first calculate the surface strains and then the tangential-stress resultants by means of Hooke's law. Lastly we use the equilibrium equations according to the membrane hypothesis in order to determine precisely what surface tractions are required by the assumed mode of deformation. In general these surface tractions will not be the same as the actual loading on the shell; and so we finally investigate the possibility that the difference is carried primarily by bending action in the shell.

It turns out to be useful to begin with some simple calculations made on the hypothesis that the shell acts as a simple beam spanning in the longitudinal direction between simple supports. As we shall discover, this sort of calculation bears a direct relationship with the one described above; and it enables some simple physical features of the behaviour to be delineated from the outset.

Our first aim is thus to study the behaviour of the simple shell shown in fig. 10.1c from two different points of view. In section 10.2.4 we consider the consequences of joining the edges of the shell to those of neighbouring shells. Then in section 10.3 we investigate the consequences of providing reinforcing beams along the straight edges. Lastly we discuss, with the aid of a simplified model of the behaviour of the shell, various factors which affect the choice of cross-sectional dimensions for edge-beams of this sort.

10.2.1 A preliminary analysis, treating the shell as a beam

We begin by analysing the behaviour of the shell according to classical beam theory.

An important parameter in the classical theory of elastic beams is the second moment of area, I, of the cross-section about the relevant centroidal axis. Now, provided the shell is sufficiently shallow, we may represent its circular cross-sectional profile of radius a by a parabolic arc as shown in fig. 10.3. Here, y, z is a rectangular Cartesian coordinate system in the plane of a cross-section, with the origin at one edge of the shell. The equation of the curve is:

$$z = (By - y^2)/2a. \tag{10.1}$$

The 'sagitta' or rise, H, of the cross-section is found by putting $y = \tfrac{1}{2}B$: hence

$$H = B^2/8a; \tag{10.2}$$

see appendix 8.

On the assumption that the thickness t is uniform in the vertical direction, it is a straightforward matter first to locate the centroid of the section at an

10.2 A simple cylindrical roof

elevation $z = z_0 = \frac{2}{3}H$, and then to evaluate the second moment of area I about the centroidal axis parallel to the z-axis:

$$I = \int_0^B (z - z_0)^2 t \, dy = \tfrac{4}{45} BtH^2 = \tfrac{1}{720} B^5 t/a^2 \tag{10.3}$$

(see problem 10.1).

It will also be convenient here, and throughout the chapter, to treat the uniform lengthwise distribution of weight on the shell in terms of a Fourier series; and indeed it will be sufficient, for most purposes, to consider the fundamental term in this series: cf. section 9.2 and problem 10.2. We shall therefore assume a displacement function of the following form:

$$w = w_0 \sin(\pi x/L). \tag{10.4}$$

Here w is a small displacement in the vertical direction; and in order to be consistent with our previous sign convention for cylindrical shells that w-deflection is positive when directed outward, we take the positive sense of w as upward. The displacement w varies with x but not with y, since the cross-sections preserve their original shape in simple beam theory. Equation (10.4) clearly satisfies the conditions of simple support at the ends of the shell/beam.

The steps in our analysis are first to determine the change in curvature κ of the beam, by using geometrical compatibility, $\kappa = -d^2w/dx^2$; then to use the appropriate form of Hooke's law, $M = EI\kappa$ in order to determine the bending moment M; and finally to use the equilibrium equation $q = -d^2M/dx^2$ in order to find the vertical load q per unit length of beam in terms of w_0. Then we may interpret q in terms of a surface loading on the original shell: at any value of x the *mean* vertical load per unit area, p_m is given by

$$p_m = q/B$$

By following this procedure we find that

$$p_m = (\pi^4/720)(Et \, w_0 \, B^4/a^2 L^4) \sin(\pi x/L). \tag{10.5}$$

This result has been expressed in terms of the radius a of the shell, rather than the rise H.

This computation of the relationship between the mean value of p and the displacement w_0 represents the highest level of detail about the applied loading which we can obtain by the use of the simple classical theory of beams. We may, however, use classical beam theory in order to determine the distribution of the longitudinal-stress resultant N_x in some detail; and in particular we must expect N_x to vary linearly with z (fig. 10.3) at any given cross-section (problem 10.3). We shall return to this calculation later.

10.2.2 A more refined calculation

Let us now attempt a more refined analysis of the same problem, by treating the structure as a *shell* rather than as a beam, in accordance with our declared aim. We shall, however, assume for the present that the mode of deformation has the special feature that each cross-section of the shell moves vertically, *without changing its shape*. We shall refer to this kind of deformation as one in which the cross-sections behave *monolithically* as they would, of course, if the shell were indeed a beam of fixed cross-section. We do not expect that this special 'monolithic' feature of the deformation will necessarily be found in practice. It has been introduced primarily to ensure some relatively simple calculations; but at the end of this somewhat artificial analysis we should be in a position to state under precisely what circumstances the mode of deflection would be 'monolithic'. Accordingly we now put

$$w = w_0 \sin(\pi x/L)$$

as a description of the mode of displacement (cf. (10.4)). Here we shall regard w as the radial component of small displacement, so that we may use directly the methods of previous chapters.

Our next task is to work out the surface strains, and then the corresponding stress resultants in the S-surface of the shell. As we have seen on previous occasions (see chapter 6) the first step in such a calculation is to evaluate the change of Gaussian curvature corresponding to the assumed mode of displacements. For small displacements of a cylindrical shell we have

$$g = -\frac{1}{a}\frac{\partial^2 w}{\partial x^2}.$$

Here, therefore,

$$g = (\pi^2/aL^2) w_0 \sin(\pi x/L). \tag{10.6}$$

Next we consider the governing equation for the uniform elastic S-surface, which may be written in terms of the Airy stress function ϕ (cf. section 7.4):

$$\nabla^4 \phi = -Etg. \tag{10.7}$$

We may solve this, and then evaluate the stress resultants N_x, N_y, N_{xy} in the usual way (see (7.36)) upon specification of suitable boundary conditions on stress. For this particular problem, however, it is more convenient to take a short cut by assuming that the proportions of the shell are such that

$$B/L < 0.5, \text{ say.} \tag{10.8}$$

Consequently we may use the approximate expression

$$\frac{\partial^2 N_x}{\partial y^2} = -Etg \tag{10.9}$$

10.2 A simple cylindrical roof

in place of (10.7): see section 8.4. From this we find, upon integration,

$$N_x = (Et\pi^2/aL^2) w_0 (C_1 + C_2 y - \tfrac{1}{2}y^2) \sin(\pi x/L). \tag{10.10}$$

The two constants of integration, C_1 and C_2, are determined by the following conditions:

(i) $\int_0^B N_x dy = 0$, since the support conditions at the ends ensure that there is no overall longitudinal tension in the shell; and

(ii) $N_x(y) = N_x(B-y)$, i.e. the distribution of N_x is symmetrical about the mid-plane $y = \tfrac{1}{2}B$, since vertical loading produces no overall bending moments about a vertical axis.

Thus we find

$$N_x = (Et\pi^2/aL^2) w_0 (-\tfrac{1}{12}B^2 + \tfrac{1}{2}By - \tfrac{1}{2}y^2) \sin(\pi x/L). \tag{10.11}$$

This expression agrees exactly with a subsidiary result which we may obtain by the 'beam' theory reported earlier: see problem 10.3. In particular note that, by virtue of (10.1), N_x varies linearly with z at any given cross-section.

The next step is to find expressions for the stress resultants N_{xy} and N_y in the surface by solving the following equations for the tangential equilibrium of the S-surface:

$$\frac{\partial N_x}{\partial x} + \frac{\partial N_{xy}}{\partial y} = 0 \tag{10.12a}$$

$$\frac{\partial N_y}{\partial y} + \frac{\partial N_{xy}}{\partial x} = 0. \tag{10.12b}$$

In putting the right-hand side of (10.12b) equal to zero we are assuming that the applied (vertical) self-weight loading has a negligible component of tangential surface traction in the circumferential direction: but see section 10.3.3, below.

Once we have solved these equations for N_y it is a simple matter to work out the corresponding value of p_S ($= p - p_B$) from the equilibrium equation (8.4a) of the S-surface in the normal direction:

$$p_S = N_y/a. \tag{10.13}$$

In this way we can determine the load distribution which is consistent with our assumed mode of deformation in more detail than was possible by the use of 'beam' theory.

Solving (10.12a) subject to the boundary condition $N_{xy} = 0$ at $y = 0$, we obtain

$$N_{xy} = (Et/a)(\pi^3/L^3) w_0 (\tfrac{1}{12}B^2 y - \tfrac{1}{4}By^2 + \tfrac{1}{6}y^3) \cos(\pi x/L). \tag{10.14}$$

Notice that the other boundary condition $N_{xy} = 0$ at $y = B$ is also satisfied:

this is a consequence of our having used two conditions in the solution for N_x in order to make N_x symmetrical about $y = \frac{1}{2}B$. Note that the shearing-stress resultant N_{xy} is zero on the transverse and longitudinal centre-lines

$$x = \tfrac{1}{2}L, \quad y = \tfrac{1}{2}B,$$

respectively. But along the end lines $x = 0$ and $x = L$, N_{xy} is not zero. According to the membrane hypothesis the only way in which the applied loading can be transferred to the supports is through N_{xy} at the ends of the shell. In an actual shell the transfer of load to the diaphragms will involve a combination of N_{xy} and Q_x; and there will be a narrow boundary layer (whose width is of order $(at)^{\frac{1}{2}}$) in which the membrane hypothesis is inadequate. We shall not discuss this point further: see section 8.7.1.

Next we use (10.14) in (10.12b) and solve for N_y, subject to the boundary condition $N_y = 0$ at $y = 0$. Thus we obtain, after some rearrangement,

$$N_y = \tfrac{1}{24} w_0 \, (Et/a) \, (\pi^4/L^4) \, y^2 \, (B - y)^2 \, \sin(\pi x/L). \tag{10.15}$$

Here again the second boundary condition $N_y = 0$ at $y = B$ is satisfied automatically. Lastly we use the equation of equilibrium in the direction normal to the shell to obtain

$$p_S = \tfrac{1}{24} w_0 \, (Et/a^2) \, (\pi^4/L^4) \, y^2 \, (B - y)^2 \, \sin(\pi x/L). \tag{10.16}$$

As we expected, this varies in the x-direction as $\sin(\pi x/L)$. In the y-direction the load varies as $[y(B-y)]^2$, as shown in fig. 10.4. This variation is far from uniform, and the applied load is heavily concentrated in the central region. In particular, it is quite different from the uniform distribution of the actual 'self-weight' loading of the shell. This special non-uniform loading is exactly what is required, according to the membrane hypothesis, to give the special 'monolithic' mode of deformation which we have assumed.

This completes the first stage of our analysis. The distribution of p_S according to (10.16) is far different from the uniform loading which self-weight

Fig. 10.4. Distribution of vertical load intensity which is required by the shell of fig. 10.3 in order that the cross-sections deform 'monolithically'.

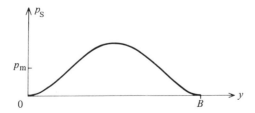

10.2 A simple cylindrical roof

actually applies to the shell. Now we have already argued on general grounds that bending effects play a significant role when the loading is uniform; and therefore it seems reasonable to suppose that the difference between p_S and a uniform load is carried by bending action. Our calculation of p_S was based, of course, on the assumption of a 'monolithic' mode of displacement; and it is therefore natural to make the conjecture that it might be possible to start with some other mode of deflection, as yet unknown, for which a calculation on the same lines would give a function p_S which is much nearer than (10.16) to a uniform distribution. In fact this would not be a useful line of enquiry, for it turns out that p_S can be made to approximate the uniform applied loading only if the assumed mode involves rather high changes of curvature in the y-direction. For the present, therefore, we shall proceed by invoking the idea that the shell consists not only of an S-surface but also of a B-surface. What we have called p_S in (10.16) is that part of the applied pressure which is carried by the S-surface. The difference between the applied pressure and p_S is equal to p_B, i.e. that part of the applied pressure which is carried by the B-surface. It is easy to verify that, as we expected, the mean value of p_S at any value of x in (10.16) is equal precisely to p_m as given by (10.5). Thus p_B may be written

$$p_B = \tfrac{1}{24} w_0 \, (Et/a^2)(\pi^4/L^4) \, [\tfrac{1}{30}B^4 - y^2(B-y)^2] \sin(\pi z/L). \quad (10.17)$$

For downward-directed uniform loading on the shell the sense of the p_B loading is downward at the edges and upward in the centre. We expect therefore that the B-surface will tend to deform by 'drooping' at the edges.

In order to quantify this effect we must first determine the distribution of M_x, M_{xy} and M_y in a B-surface which is subjected to the loading (10.17). Since we have already assumed that $L/B > 2$ in order to simplify the analysis of the S-surface, we make the corresponding approximation for the B-surface which is, simply, that the loading is carried almost entirely by bending moments M_y. In other words, the B-surface behaves like a set of simple strips running in the transverse or circumferential direction. We may therefore determine M_y from the equilibrium equation

$$\frac{\partial^2 M_y}{\partial y^2} = -p_B, \quad (10.18)$$

together with boundary conditions

$$M_y = 0 \quad \text{at } y = 0 \text{ and } y = B. \quad (10.19)$$

In this way we find

$$M_y = \frac{-Et}{a^2} \frac{\pi^4}{1440} \frac{w_0}{L^4} \{y^2(B-y)^2 \, [B^2 + 2y\,(B-y)]\} \sin\left(\frac{\pi x}{L}\right). \quad (10.20)$$

This is plotted in Fig. 10.5 for $x = \frac{1}{2}L$; and it is of broadly the same shape as the function p_S. The peak value of M_y is equal to $-\frac{3}{64} p_m B^2$.

It is now a straightforward matter to determine the deflected form of the transverse strips of the B-surface by means of the equation

$$\frac{\partial^2 w}{\partial y^2} = -\frac{M_y}{D}, \qquad (10.21)$$

where D is the flexural rigidity of the shell. The general tendency is for the transverse sections to 'droop' at the edges when the uniform loading on the shell is downward directed. The variation of w across the width is almost purely sinusoidal as a result of the double integration of the function shown in fig. 10.5; and the difference in values of w between the edges ($y = 0, y = B$) and the centre ($y = \frac{1}{2}B$) is found to be equal to

$$0.00676 \, (w_0 \, B^8/L^4 a^2 h^2) \sin(\pi x/L). \qquad (10.22)$$

Here we have introduced the 'effective thickness' h (see (3.25)) in order to eliminate Poisson-ratio terms from the expression.

The notion that transverse sections of the shell bend in this way is, of course, paradoxical, for it is strictly contrary to our original assumption about the mode of deflection, expressed in (10.4). It seems clear, however, that if the dimensions of the shell are such that the quantity above is sufficiently small in comparison with $w_0 \sin(\pi x/L)$, our entire solution should be moderately reliable. Thus, if we require the value of the above expression to be less than (say) $0.1 w_0 \sin(\pi x/L)$ we obtain a restriction on the geometrical parameters of the shell.

It is convenient at this point to introduce the dimensionless group

$$\beta = B^2/L(ah)^{\frac{1}{2}}. \qquad (10.23)$$

Fig. 10.5. Distribution of bending moment in transverse strips of the shell if the actual loading is distributed uniformly.

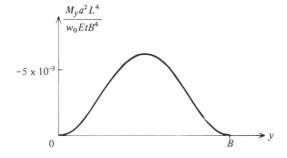

10.2 A simple cylindrical roof

The condition described above thus becomes

$$\beta < 2.0. \tag{10.24}$$

The dimensionless group β falls naturally out of this particular analysis. Indeed, we shall find later that it is relevant to several aspects of the behaviour of this kind of structure. The parameter β is a measure of the *width* of the shell in terms of its length, radius and thickness. Other things being equal, a wider shell 'droops' more. For a shallow shell, β may be expressed alternatively in terms of the rise, H, of the cross-section, by means of (10.2):

$$\beta = \frac{8H}{L}\left(\frac{a}{h}\right)^{\frac{1}{2}}, \quad (H/a < 0.2, \text{ say}).$$

Typical cylindrical shell roofs described in the literature have $L/H \approx 20$ and $a/h \approx 150$. Typical values of β are thus in the region of 5. This certainly does not satisfy inequality (10.24), and for these typical shells the 'droop' would exceed the (arbitrary) limit given above by a factor $(\frac{5}{2})^4 \approx 40$. Practical shells, of course, usually have edge-beams, which accounts for the discrepancy: inequality (10.24) expresses a condition for restricting the amount of 'droop' of an isolated shell *without* the help of edge-beams.

Throughout this chapter we shall use the term 'droop' to describe this kind of elastic deformation of the shell. It is distinct from the idea of 'rise' H of the cross-sectional shape, which refers to the geometry of the cross-section in its original configuration.

It is of interest to compare the relative magnitude of the peak transverse bending stress, σ_b^{max}, and the peak longitudinal tensile stress, σ_x^{max}. In general

$$\sigma_x^{max} = N_x^{max}/t, \quad \sigma_b^{max} = 6M_y^{max}/t^2. \tag{10.25}$$

For a shell loaded uniformly (e.g. by its own weight) N_x has its maximum value at the edges $y = 0$ and $y = B$ of the central cross-section, $x = \frac{1}{2}L$. Hence, from (10.11) and (10.20) we find

$$\sigma_b^{max}/\sigma_x^{max} = 0.046 \, B^4/L^2 at \approx \tfrac{1}{22}\beta^2. \tag{10.26}$$

The inequality (10.24) implies that

$$\sigma_b^{max}/\sigma_x^{max} < 0.2. \tag{10.27}$$

This kind of restriction is not altogether surprising, since a thickness of shell which is adequate to prevent appreciable 'droop' at the edges will also reduce the magnitude of the transverse bending stress. The comparison in (10.26) between the magnitudes of the 'bending' and 'tensile' stresses is reminiscent of the simple calculations which we did in sections 7.2 and 7.3 on the validity of the membrane hypothesis; and indeed the form of (10.26) is closely related

330   *Cylindrical shell roofs*

to that of (7.16) if we regard the width B of the shell as being roughly equivalent to the width b of a 'panel' in the earlier calculation. Thus we can see that the parameter β has close connections with ideas which we have explored previously.

Finally, we note that for self-weight loading $\sigma_x^{\max}$ is independent of t if t is the only quantity being varied. This reflects the fact that the load per unit area, and hence all tangential-stress resultants, are directly proportional to t. (In practice, of course, design calculations would be based on self-weight loading enhanced by a suitable 'load factor'. We are not concerned here with the choice of such factors.)

10.2.3 Another mode of displacement

It is clear from the above calculation that the assumed mode of 'monolithic' displacement (see (10.4)) is unlikely to occur under self-weight loading unless the shell is sufficiently thick to prevent 'droop' at the edges; and that in general the mode of deformation is likely to involve larger displacements at the edges than at the centre. As a step towards a more general analysis let us therefore consider the following mode of displacement as a replacement for (10.4):

$$w = [w_0 + w_1 \sin(\pi y/B)] \sin(\pi x/L). \tag{10.28}$$

We shall hope to determine the relative magnitudes of the two parameters w_0 and w_1 in terms of the dimensions of the shell later on. The term having amplitude w_1 represents crudely the anticipated 'droop' of the cross-section, and its precise form has been chosen in order to make the subsequent calculations easier. It is clear that we shall obtain the desired results by analysing the consequences of a displacement of the form

$$w = w_1 \sin(\pi y/B) \sin(\pi x/L) \tag{10.29}$$

and superposing the results onto those which we have obtained already.

Proceeding in exactly the same way as before, we obtain the following expressions in place of (10.6), (10.11) and (10.16). Note that we cannot simply use results from chapter 8, since there the boundary conditions on the edges $y = 0$ and $y = B$ involve substantial shearing-stress resultants N_{xy}, in contrast to the present case, in which these edges are stress-free.

$$g = \frac{\pi^2}{aL^2} w_1 \sin\left(\frac{\pi y}{B}\right) \sin\left(\frac{\pi x}{L}\right) \tag{10.30}$$

$$N_x = \frac{Et}{a}\frac{B^2}{L^2} w_1 \left[-\frac{2}{\pi} + \sin\left(\frac{\pi y}{b}\right)\right] \sin\left(\frac{\pi x}{L}\right) \tag{10.31}$$

10.2 A simple cylindrical roof

$$p_S = \frac{Et}{a^2} \frac{B^4}{L^4} w_1 \left[-\frac{\pi y(B-y)}{B^2} + \sin\left(\frac{\pi y}{B}\right) \right] \sin\left(\frac{\pi x}{L}\right). \tag{10.32}$$

It is easy to show by integration that the *mean* value of p_S at the cross-section $x = \frac{1}{2}L$ for this mode is given by

$$p_S = 0.1130 \frac{Et}{a^2} \frac{B^4}{L^4} w_1. \tag{10.33}$$

Therefore, from (10.5) we find that if the displacement is of the form (10.29) alone, and we put

$$w_1 = 1.197 w_0, \tag{10.34}$$

the two modes (10.4) and (10.29) separately will sustain exactly equal *total* load. Moreover, we can use (10.32) to determine the distribution of p_S across the central cross-section. This distribution should obviously be compared with that which was obtained from the previous analysis, and was plotted in fig. 10.4. This second curve appears not to have been drawn on fig. 10.4; but this is an illusion, for *the two curves are in fact almost identical.*

This observation has important consequences. It means that an extremely small redistribution of load from the arrangement shown in fig. 10.4 is required in order to alter the pattern of deflection of the S-surface from mode (10.4) to the considerably different mode (10.29). The transverse distribution of p_S is extremely *insensitive* to changes in mode between the two types which we have considered. Indeed, practically no change in the distribution of p_S is needed in order to give a *range* of modes in the form of the linear combination

$$w = w_0 [1 - k + 1.197 k \sin(\pi y/B)] \sin(\pi x/L), \tag{10.35}$$

where k is an indeterminate number. Displacements corresponding to several values of k are shown in fig. 10.6. The displacement at points $0.315B$ from the edges is independent of the value of k: these are 'pivot points' for the various modes.

It is worthwhile to draw attention to two aspects of this remarkable result before we examine its implications. First, we can see that the process of obtaining an expression for p_S from the original expression for w involves four integrations in the y-direction: we integrate twice in order to obtain an expression for N_x, and twice more to obtain an expression for p_S. It is because the expressions (10.11) and (10.31) for N_x have broadly the same form (a parabolic arc in one case and a sinusoidal one in the other) that the next two integrations give almost identical results if the relative values of w_0 and w_1 are adjusted appropriately.

Second, the fact that a small change of loading produces a disproportionately large change in displacement suggests that the changes from mode to mode are *almost inextensional*. They are obviously not exactly inextensional, since any inextensional deformation of a cylindrical shell requires the generators of the cylinder to remain straight in the course of deformation (see section 6.5). In chapter 4 we observed that it is sometimes useful, in discussing the membrane hypothesis, to replace the continuous surface of the shell by a triangulated array of bars which are freely pinned to each other at the ends. In particular, it is possible to discuss the number of kinematic constraints which are necessary at boundary nodes in order to satisfy - formally at least - the conditions for the assembly to be statically determinate. If we construct a model of this sort for the present shell we discover that, in comparison with the examples illustrated in fig. 4.18, we have omitted all of the constraints along the straight edges of the shell. We may therefore expect that the arrangement constitutes a mechanism with a large number of degrees of freedom. Consequently it should not surprise us to learn that a state of purely membrane stress in the original shell is only possible provided the distribution of load which is applied to the shell satisfies certain requirements: see fig. 10.4. This kind of argument must not be pressed too hard, however; for the fact that the generators of the shell are straight puts the analogous framework into a special category which is rather awkward: cf. appendix 9. Nevertheless it is worthwhile to make the point that the lack of *kinematic restraint* on the free edges of the shell makes certain demands on the pattern of loading which may be sustained according to the membrane hypothesis.

It follows from all of this that our previous calculation of the bending of cross-sectional arcs of the shell in the manner of transverse beams should be valid for larger displacements than those which we contemplated in section

Fig. 10.6. Distribution of vertical displacement w across the width of the shell. The curve marked $k = 0$ corresponds to a 'monolithic' displacement, while the other curves indicate greater or lesser 'droop' of the edges of the shell.

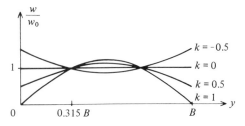

10.2 A simple cylindrical roof

10.2.2, since transverse bending of a simple kind (cf. fig. 10.6) has virtually no effect on the distribution of pressure between the S- and B-surfaces.

Lastly we note that to a good first approximation the stress resultants N_x associated with the various displacement modes of the form (10.28) vary linearly with z (cf. fig. 10.9) irrespective of the ratio w_1/w_0, within wide limits. Indeed, we observe that the distribution of N_x, being almost independent of the amount of 'droop' of the cross-section is effectively *statically determinate* in terms of p_m. Thus we may combine (10.5) and (10.11) to indicate that for a loading

$$p = p_0 \sin(\pi x/L)$$

the stress resultant N_x is given by

$$N_x \approx (720/\pi^2) L^2 (ap_0/B^4) (-\tfrac{1}{12} B^2 + \tfrac{1}{2} By - \tfrac{1}{2} y^2) \sin(\pi x/L). \quad (10.36)$$

It is sometimes necessary to calculate N_x in terms of an assumed mode of displacement. The simplest way of doing this – or, strictly, obtaining a close approximation to N_x which is valid for a wide range of values of the ratio w_1/w_0 – is to use the simple beam theory for the 'equivalent' monolithic displacement

$$w = (w_0 + 0.835 w_1) \sin(\pi x/L). \quad (10.37)$$

Finally we draw attention to the fact that our remarkable result indicates a high degree of 'uncoupling' of the behaviour of the S- and B-surfaces; and that this feature has a counterpart in the results of problem 8.4: see problem 10.4.

10.2.4 Continuity with adjacent shells

It is convenient at this point to investigate the change in behaviour of a shell without edge-beams when both of its edges are connected to the edges of equally loaded similar adjacent shells in the manner of fig. 10.1a. For the sake of simplicity we shall assume that the edges are freely hinged about their line of intersection, but are otherwise firmly joined, and that all of the shells are loaded uniformly.

Consider first the form of the new boundary conditions at the edges of the shell. It is clear from considerations of symmetry that the force-interaction between adjacent shells must be in a horizontal plane, as sketched in fig. 10.7a; and also that the displacement of the shell at these edges must be purely vertical.

The first mode of deformation which we considered (see (10.4)) was conceived as one in which cross-sections of the shell retained their original shape, and deflected vertically. On the other hand the part of the general displacement (see (10.35)) involving the factor k, which represents the 'droop' of the edges with respect to the centre, plainly involves an element of displacement of the edges towards each other. Therefore it seems clear that a major effect of the new boundary conditions is to induce the shell to deflect according to (10.4), i.e. to force the parameter k in (10.35) to have zero value. But it is also clear that the membrane analysis of the shell in this mode, which we considered in the preceding section, involves non-uniform loading p_S over the surface of the shell; so the full solution of the problem with the new boundary conditions must still involve some bending in the transverse direction of the shell. Since in our previous analysis the transverse bending was associated with the drooping of the edges with respect to the centre, we must obviously re-examine the mechanics of what we might call the 'transverse strip' effect.

Figure 10.7a,b shows a typical transverse strip of the shell, of unit width (in the x-direction), which is loaded firstly by p_B and secondly by the outward-directed horizontal forces J per unit width which are provided by the connection with adjacent shells. The strip is to be treated as a simple curved elastic beam, and the value of J is to be determined so that there is no displacement of the ends of the beam relative to each other. We shall assume that the strip is inextensional in the circumferential direction: cf. chapter 9.

Let the bending moment per unit width of the strip due to p_B alone be denoted by $M_0(y)$: this has already been calculated, and is plotted in fig. 10.5. Then the total bending moment, due to the combined effects of p_B and J, is given by

$$M_y = M_0 - Jz, \qquad (10.38)$$

where $z(y)$ describes the profile of the shallow arc of the cross-section. The change of curvature κ_y in the transverse direction, due to elastic distortion, is thus given by

$$\kappa_y = (M_0 - Jz)/D. \qquad (10.39)$$

This is compatible, in general, with an outward displacement, say δ, of the

Fig. 10.7. (a) The effect of attachment to neighbouring shells (as in fig. 10.1a, b) is to supply horizontal edge forces as shown here. (b) The distribution of pressure which is responsible for the bending of transverse strips. (c) 'Dummy' external forces and internal bending moment for use in a virtual work calculation.

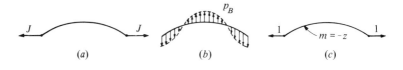

10.2 A simple cylindrical roof

ends relative to each other. We wish to establish the condition for the value of δ to be zero, which will enable us to calculate the appropriate value of J.

It is convenient to use the principle of virtual work. Fig. 10.7c shows the same strip of the shell, carrying unit outward-directed (dummy) loads at the ends, and therefore sustaining a bending moment

$$M_y = -z.$$

These forces and the corresponding bending moment provide an equilibrium set of (external) forces and (internal) bending moments, and we may therefore write the following equation of virtual work (appendix 1):

$$\int_0^B [(M_0 - Jz)/D] \, (-z) \, dy = \delta. \tag{10.40}$$

Here we are integrating with respect to y rather than arc-length s, since we are assuming the cross-sectional arc to be shallow.

The integrand on the left-hand side is the product of the actual elastic curvature change and the 'dummy' bending moment, while the right-hand side is the product of the actual end-displacement and the 'dummy' force. Since we require $\delta = 0$, we find that

$$\int_0^B M_0 z \, dy = J \int_0^B z^2 \, dy. \tag{10.41}$$

Since $M_0(y)$ and $z(y)$ are known functions, we can evaluate J; hence we can work out the final bending moment M_y from (10.38). Fig. 10.8 shows the distribution of bending moment M_y which has been calculated in this way for the section $x = \frac{1}{2}L$. The diagram shows both M_0 and Jz, and M_y as the difference between them. The integration of (10.41) shows that the peak value of Jz is equal to 0.825 (exactly, 52/63) of the peak value of M_0; and it is clear that the effect of J is to reduce considerably the magnitude of M_y. The peak

Fig. 10.8. Distribution of bending moment in a transverse strip when the edges of the shell are restrained by connection with neighbouring shells.

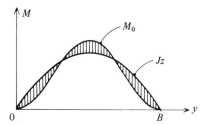

value of M_y is only 0.203 times the peak value of M_0, and it follows that (10.26) should be replaced by

$$\sigma_b^{\max}/\sigma_x^{\max} = 0.009\beta^2 \tag{10.42}$$

for the present problem.

It is interesting to note that the distribution of M_y is almost, but not exactly, of the form $\sin(3\pi y/B)$. This is in contrast with M_0, which obviously has a large component of the form $\sin(\pi y/B)$. The effect of the condition $\delta = 0$ is largely to eliminate the fundamental (Fourier-series) term from the distribution of M_y corresponding to free edges.

Hence we reach a general conclusion concerning the effect of connecting the edges of the shell to neighbouring shells: there is virtually no change in the distribution of N_x, but there is nevertheless a substantial attenuation of the transverse bending-stress resultants.

Now in our calculations on a shell with free edges, we concluded that in order effectively to eliminate 'droop' of the edges we must proportion the shell so that

$$\beta < 2$$

(see (10.23)). We also saw that this corresponds to the following restriction (cf. (10.27)) on transverse bending stress

$$\sigma_b^{\max}/\sigma_x^{\max} < 0.2.$$

On the other hand a shell which is connected to its neighbours has practically zero droop at the edges, but it will in general have some transverse bending. If we decide, arbitrarily but reasonably, to insist that

$$\sigma_b^{\max}/\sigma_x^{\max} < 0.5,$$

we find from (10.42) that

$$\beta < 7. \tag{10.43}$$

Now for given values of L, B and a, the value of β is proportional to $h^{-\frac{1}{2}}$. On this basis the change from free edges to connected edges permits an increase in β by a factor of 3.5, with a corresponding reduction in thickness by a factor in excess of 10. Now as far as self-weight effects are concerned, σ_x is independent of thickness, as we have shown. A clear lesson may thus be drawn from the above calculation. There is a tendency for the *free* edges of a shell to droop, which in general can only be counteracted by an increase in thickness. However, if the edges are connected to those of adjacent shells, the tendency to droop is immediately countered, and the shell may be made much thinner.

A feature of all of the above analysis has been the assumption that the shell carries the difference between the applied loading p and the 'S-surface' loading p_S by means of bending effects in the transverse direction alone. For the shell with free edges this is justified simply by noting that the mode of transverse bending of the B-surface due to M_y is of practically the same form as that which we showed to be 'almost inextensional' in section 10.2.3: so the S-surface offers virtually no resistance to the mode in question. On the other hand the shell with connected edges suffers a displacement essentially of the form

$$w = w_3 \sin(3\pi y/B) \sin(\pi x/L) \qquad (10.44)$$

as a result of p_B (see fig. 10.8), and we should therefore investigate the possibility (in view of chapter 8) that some appreciable resistance to this mode may in fact be provided by the S-surface. Here we can use directly the results of section 8.4. For a mode

$$w = w_0 \sin(\pi y/b) \sin(\pi x/l) \qquad (10.45)$$

the ratio ζ of S- to B-surface stiffness is given by

$$\zeta \approx \frac{12}{\pi^4} \frac{b^8}{l^4 a^2 h^2} \qquad (10.46)$$

when $l/b > 2$: see (8.16). In the present case we have

$$l = L, \quad b = \tfrac{1}{3}B \qquad (10.47)$$

and it follows directly from (10.46) that the S-surface will sustain less than 0.1 of the applied load provided

$$\beta < 8.5. \qquad (10.48)$$

In view of earlier remarks, it is not surprising that β is the relevant parameter in this case.

The fact that we have already adopted a more stringent limit (10.43) on β in order to limit the peak transverse bending stresses indicates that almost all of the loading which we have called p_B is in fact carried by means of transverse bending action in the shell.

However, since a small fraction of this loading is carried by the S-surface, we should check on the magnitude of stresses of the form

$$\sigma_x = \sigma_3 \sin(3\pi y/B) \sin(\pi x/L) \qquad (10.49)$$

which are thus induced in the S-surface and which we have hitherto neglected. It is easy to show by the methods of chapter 8 that for a deflection in mode (10.45) the ratio of peak longitudinal stress to peak transverse (bending) stress is approximately equal to

$$0.2 \, b^4/l^2 ah. \qquad (10.50)$$

Substituting for l and b from (10.47) we find that this expression is equal to 0.0025 β^2.

The design condition $\beta < 7$, which restricts the level of transverse bending stress to less than 0.5 of the peak longitudinal stress, thus also restricts the magnitude of the longitudinal stress σ_3 in (10.49) to less than 0.06 of the peak longitudinal stress. In other words, this is a minor effect. It should be noted, however, that the addition of a small component of kind (10.49) to the 'basic' distribution (10.14) makes the variation of σ_x in the shell slightly nonlinear with z, as shown in fig. 10.9.

10.3 The effect of edge-beams: a simple example

We have seen that an isolated shallow shell has a tendency for its edges to 'droop' under self-weight loading, and that this effect can be resisted by making the shell thicker. Also, the prohibition of horizontal displacements at the edges, which occurs when the shell is connected to its neighbours, interferes with the primary mode of drooping; and in this case the shell may be constructed an order of magnitude thinner without incurring excessive bending stress in the transverse direction.

It is obviously not always possible to connect a shell to an equal neighbouring shell, but it is clear that unless the main component of 'droop' can be eliminated, the shell may have to be thickened substantially. It seems obvious

Fig. 10.9. Distribution of longitudinal stress against coordinate z (fig. 10.3), at a typical cross-section. The continuous line corresponds to 'monolithic' behaviour of the cross-sections, while the broken curve shows the effect of the mode of distortion associated with the bending-moment distribution of fig. 10.8.

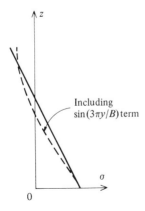

10.3 The effect of edge-beams

in these circumstances that the provision of some sort of *edge-beam* to the shell may be advantageous, for such a beam will clearly act so as to limit, to some extent, the vertical displacement of the edges.

The provision of edge-beams to an isolated shell will obviously complicate the structural analysis. The conventional treatment of this problem in terms of an interaction between the shell and the beams is particularly cumbersome: see Timoshenko & Woinowsky-Krieger (1959, §126) and Flügge (1973, §5.4.4.2). In this section we shall use our discovery that the shell has an 'almost inextensional' mode of deformation, in order to help in the setting up of a simplified analysis. But before we begin this work it will be useful to try to assess the general effects of edge-beams by considering a simple — and somewhat artificial — example, involving a minor extension of previous work, as follows.

Suppose we have a shell like to one in section (10.2.2) which is carefully loaded, as in fig. 10.4, so that the cross-sections behave monolithically. In these circumstances we can make an elementary analysis by considering the shell as a simple beam, and in particular we may use the well-known classical formulae for the stresses in and deflection of the beam under load in terms of the cross-sectional property $I = \int z^2 \, dA$, etc.

Let us now imagine a rearrangement of the material of this shell in which a central zone is left unaltered in the form of a shell, while the outer parts are rearranged so as to form edge-beams: see fig. 10.10a. This rearrangement of material is to be done so that the value of z associated with any arbitrary small portion dA of the cross-section does not change. An immediate consequence of this is that the value of the sectional property I remains exactly the same as for the original shell. This prescription for alteration of the cross-

Fig. 10.10. Transformation of a simple shell into one with edge-beams: a preliminary calculation on a shell with edge-beams which is assumed to deform 'monolithically'.

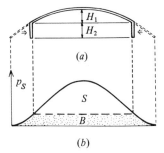

section produces, of course, edge-beams whose thickness varies with z. In this sense the design is artificial. But this variation of thickness will be rather small provided that only a relatively small zone at each edge is rearranged, and then it may be ignored for present purposes.

The process of rearrangement has no effect on the total weight per unit length of the beam, and consequently no effect on the distribution of overall bending moment. Furthermore, it has no effect on the distribution of longitudinal tension with respect to z, for

$$\sigma_x = Mz/I: \tag{10.51}$$

here, of course, the origin of z has been placed at the centroid of the cross-section. Therefore, in particular, the distribution of N_x in the (unchanged) central zone remains unaltered.

Now in section 10.2.2 we analysed the stress resultants N_{xy} and N_y by taking the distribution of N_x which could have been obtained by means of simple beam theory, and solving the equations of tangential equilibrium (10.12) with the help of appropriate boundary conditions. Since by symmetry $N_{xy} = 0$ at $y = \frac{1}{2}B$, we see from (10.12a) that the distribution of N_{xy} in the central zone is unaffected by our redistribution of material. In particular $N_{xy} \neq 0$ at the junction between the central zone and the edge-beams, and we must therefore suppose that the junction is strong enough to transmit this shear from the shell to the beams with no tangential 'slip'. The stress resultant N_{xy} is the main structural interaction between the edge-beams and the central zone.

The differential equation for N_y is also exactly the same as before; but the solution for N_y is different, since we can no longer use the boundary condition $N_y = 0$ at the edge of the beam. Instead, we shall use the condition $N_y = 0$ at the edge of the central zone. This is based on the supposition that the edge-beams are so much stiffer against vertical displacements than against horizontal displacements that they impose in practice only vertical loads on the shell in addition to the N_{xy} loads; and since the shell is shallow it is reasonable to assume that the component of loading N_y which is applied to the shell by the edge-beams is negligible. It follows that the distribution of N_y, and hence of p, in the central zone of an arbitrary cross-section is exactly the same as before except for a constant of integration; and therefore we may obtain the distribution of p by translating the curve of fig. 10.4 until $p = 0$ at the edges of the central zone. Thus the total load which is now carried by the central zone (i.e. the remaining shell) is proportional to the area marked S in fig. 10.10b; and it follows (since the total load on the entire structure is unchanged) that the area marked B is proportional to the load which is carried by the two edge-beams together.

10.3 The effect of edge-beams

It is now a straightforward matter to derive a general expression for the load carried by the edge-beams for an arbitrary position of the edges of the central zone. The result is shown in fig. 10.11 as a graph showing the fraction of the total load which is carried by the edge-beams, as a function of the fraction of the total cross-sectional area which is devoted to these beams: see problem 10.5.

In relation to shells which have edge-beams of rectangular cross-section, only the part of this curve in the lower left-hand corner is valid. Nevertheless it is clear that relatively small edge-beams do not manage to support even their own weight, and that only when the edge-beams occupy in the region of 0.2 of the total cross-sectional area (and hence in the region of 0.35 of the total depth of the cross-section) do they support a 'fair' share of the load. The particular designs studied by Timoshenko & Woinowsky-Krieger and Flügge have proportions which are in this range.

The preceding calculations are based on several questionable assumptions which include, in particular, the notion that the cross-sections behave monolithically when the load is applied. If the actual loading applied to the shell does not agree with the calculated distribution across the width of the section, this assumption cannot be expected to hold.

Furthermore, we have been considering here an *isolated* shell. If the shell is connected to neighbours, the analysis must take this factor into account.

Fig. 10.11. Results of the calculation with respect to fig. 10.10. H_1 and H_2 are the depths of the shell and edge-beams. A_1 and A_2 are the cross-sectional areas of the shell and the two edge-beams (together), while F_1 and F_2 represent the total vertical load carried by the shell and the edge-beams, respectively.

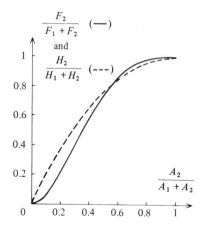

10.3.1 The effect of edge-beams: more general analysis

When we contemplate doing a general analysis of the behaviour of a shell with edge-beams of rectangular cross-section having arbitrary dimensions, we must make a decision about the way in which loading is applied to the structure. There are two aspects to this problem. The first is connected with the details of distribution of load over the shell itself. The actual load, of course, will normally be uniform, in contrast to the non-uniform distribution of p_S which emerges from the membrane hypothesis. The second concerns the gross distribution of load between the shell and the beams. We shall see later on that these two aspects are interrelated. For the next piece of analysis, however, we shall be concerned only with this gross distribution of load between the shell and the beams. Specifically, we shall assume that the load is distributed across the shell exactly in accordance with whatever distribution p_S is calculated for any given case. In other words, we shall make sure that the loading is applied in a way which does not require the B-surface to play a part: we shall leave until later the treatment of the B- surface which is required when the actual loading is considered.

The key to the calculation which follows is the observation, which we made earlier, that even if the displacement of a shallow shell involves 'droop' of the edges with respect to the centre, the distribution of N_x may be calculated satisfactorily by employing simple beam theory in relation to a specific postulated mode of deformation (see (10.4)) which involves monolithic behaviour of the cross-sections. In particular, the calculation relies on the fact that the distribution of N_x is linear with the coordinate z, regardless of the magnitude of the 'droop' of the cross-section.

Let the deflection of the shell be given by (cf. (10.28))

$$w = [w_0 + w_1 \sin(\pi y/B)] \sin(\pi x/L).$$

The dimensions of the shell and the coordinate system are exactly the same as in section 10.2.3. The variation of strain ϵ_x across the shell is exactly the same as for a simple beam undergoing a displacement (cf. (10.37))

$$w = (w_0 + 0.835 w_1) \sin(\pi x/L),$$

and in particular the variation of ϵ_x with z is given by

$$\frac{d\epsilon_x}{dz} = \frac{\pi^2}{L^2}(w_0 + 0.835 w_1) \sin\left(\frac{\pi x}{L}\right) = \mu_1 \sin\left(\frac{\pi x}{L}\right). \quad (10.52)$$

Here the central expression is simply the change of curvature of this 'equivalent' beam, and μ (the amplitude of $d\epsilon_x/dz$) is a convenient parameter.

On the other hand, the edge-beams undergo a vertical displacement of

10.3 The effect of edge-beams

$w_0 \sin(\pi x/L)$ (on the assumption that the shell is shallow), and so for them we have

$$\frac{d\epsilon_x}{dz} = \frac{\pi^2}{L^2} w_0 \sin\left(\frac{\pi x}{L}\right) = \mu_2 \sin\left(\frac{\pi x}{L}\right). \tag{10.53}$$

It follows from (10.52) and (10.53) that the distribution of ϵ_x in the cross-section is *bilinear*, as shown in fig. 10.12. In general the slopes of the two lines will be different, since in general there will be some 'droop' in the shell, i.e. $w_1 \neq 0$. If the edges of the shell deflect more than the crown, then $\mu_2 > \mu_1$.

This diagram provides the key to the subsequent analysis. We shall treat the combination of beams and shell as a sort of composite beam section, and we shall use ordinary beam theory except that we shall regard the relative slope of the two lines as a variable. In a given case we shall determine the relative slope by considering the way in which the load applied to the composite beam is distributed between the two parts.

The dimensions of the cross-section of the shell and associated beams are as shown in fig. 10.13. The shell has a circular profile and subtends angle 2ψ. Subscripts 1 and 2 refer to the shell and beams, respectively (cf. figs. 10.10 and 10.11). It will be convenient in this problem to have the origin $z = 0$ at the level of the junction between the shell and the beams. The cross-sectional area of the shell is A_1, and that of the two beams together is A_2; thus

$$A_1 = Bt_1, \quad A_2 = 2H_2 t_2. \tag{10.54}$$

In fig. 10.13, B is shown as the *arc-length* of the cross-section of the shell. For most purposes we shall regard the shell as shallow and use a parabolic approxi-

Fig. 10.12. Distribution of longitudinal strain with z in the shell ($z > 0$) and edge-beam ($z < 0$). The slopes μ_1, μ_2 are different if the vertical displacement of the edge-beams differs from that of the crown of the shell.

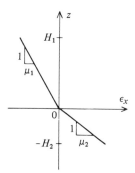

mation to the arc, as in fig. 10.3, in which B appears as the width of the shell. In the course of the following calculations we shall need to evaluate various integrals along the arc-length, and we shall perform these (in the spirit of (10.3)) with respect to the parabolic approximation, as follows:

$$\int_0^B z \, ds = \tfrac{2}{3} B H_1 \tag{10.55a}$$

$$\int_0^B z^2 \, ds = \tfrac{8}{15} B H_1^2. \tag{10.55b}$$

It turns out, however (see problem 10.6) that these formulae are accurate in relation to the actual cross-section to within about 1% provided we take B as the *arc-length* and H as the rise of the curve (indicated in fig. 10.3).

Consider the distribution of σ_x in the cross-section at $x = \tfrac{1}{2}L$. First suppose that $\sigma_x = 0$ at $z = 0$. Then in the shell

$$\sigma_x = -\mu_1 E z \tag{10.56}$$

while in the edge-beams

$$\sigma_x = -\mu_2 E z. \tag{10.57}$$

The quantities μ_1 and μ_2 depend on the assumed mode of deformation, as we have seen. We shall assume that they both have positive values, and the negative signs have been included in (10.56) and (10.57) so that we shall be considering a downward displacement of the shell. It will become clear that we shall be concerned mainly with the ratio

$$\xi = \mu_2/\mu_1. \tag{10.58}$$

The resultant force across the curved arc of the shell may readily be worked out with the aid of (10.55a), and the position of its line of action is at $z = \tfrac{4}{5} H_1$,

Fig. 10.13. Cross-sectional dimensions of a shell used for analysis of interaction with edge-beams; and some definitions.

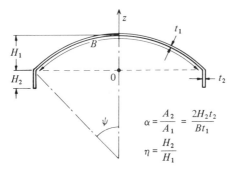

10.3 The effect of edge-beams

by (10.55a) and (10.55b). This is indicated in fig. 10.14 as a force T_1, and is given by

$$T_1 = \tfrac{2}{3} \mu_1 E H_1 A_1. \tag{10.59}$$

The sense of this force is compressive.

The resultant force in the beams, T_2, may be worked out similarly. Its line of action is at $z = -\tfrac{2}{3} H_2$, and T_2 is given by

$$T_2 = \tfrac{1}{2} \mu_2 E H_2 A_2 : \tag{10.60}$$

its sense is tensile; see fig. 10.14.

We shall now assume, as we have done elsewhere in this chapter, that there is zero net longitudinal force across any given cross-section, since the end-supports provide only vertical resultant reaction. Consequently we must superpose onto the distribution of strain, as assumed above, a strain which is constant across the entire cross-section. This additional strain is associated, by Hooke's law, with an additional tensile force T which must satisfy

$$-T_1 + T_2 + T = 0. \tag{10.61}$$

Since the elastic modulus is uniform, this force is divided between the shell and the beams in the ratio A_1/A_2. These forces are also shown in fig. 10.14, and it may be checked from this diagram that the total tensile force across the section is zero. The lines of action of the new forces pass through the centroids of the corresponding parts of the cross-section: in the shell this is at $z = \tfrac{2}{3} H_1$, by (10.55a) and in the beams at $z = -\tfrac{1}{2} H_2$.

Now one of our objectives is to find how the total shearing force on any vertical cross-section is shared between the shell and the beams. The easiest way of discovering this is to consider the equilibrium of one half of the struc-

Fig. 10.14. Resultant forces acting on the central cross-section of the shell: see text for details.

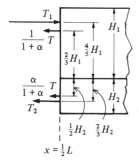

ture, separated into shell and beam components, as shown in fig. 10.15. The various forces which are shown may be described as follows.

Across the section $x = \frac{1}{2}L$ there are exactly the same forces as in fig. 10.14. There are no vertical shear forces, by symmetry, on this central cross-section for the loading under consideration. The forces marked S are shear forces transmitted across the junctions between the shell and the beams: they are the sum of stress resultants N_{xy} along the junction lines. There are no vertical force interactions between the two parts, since we are assuming, as before, that $N_y = 0$ at the edge of the shell. Lastly there are the vertical forces F_1, F_2. The upward-directed forces at the end $x = L$ are the reactions supplied to the shell and beams, respectively, by the end-supports, while the downward-directed forces represent the resultant total load which is applied to the two parts, respectively: and since the ratio of load applied to shell and beams is, we assume, independent of x, these two resultants have the same line of action.

The two 'free bodies' shown in fig. 10.15 are in equilibrium under the various forces. By taking moments about a point on the line $z = 0$ for each part we find that the ratio F_2/F_1 is equal to the ratio of the moments of the forces on the two parts of the cross-section about $z = 0$. Thus

$$\frac{F_2}{F_1} = \frac{\frac{2}{3}H_2T_2 + \frac{1}{2}H_2A_2\ (T_1 - T_2)/(A_1 + A_2)}{\frac{4}{5}H_1T_1 - \frac{2}{3}H_1A_1\ (T_1 - T_2)/(A_1 + A_2)}. \quad (10.62)$$

We may now substitute for T_1, T_2 from (10.59) and (10.60). It is convenient to define

$$\eta = H_2/H_1, \quad \alpha = A_2/A_1 \quad (10.63)$$

for this purpose (see fig. 10.13); and we find, after some manipulation,

$$\frac{F_2}{F_1} = \frac{1 + \xi\eta\,(1 + \frac{1}{4}\alpha)}{\xi + (\frac{4}{15} + \frac{8}{5}\alpha)/(\alpha\eta)}. \quad (10.64)$$

Fig. 10.15. Resultant forces acting on half of the shell and the associated edge-beams.

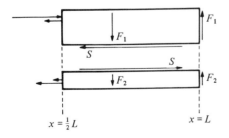

10.3 The effect of edge-beams

For a given geometrical design of cross-section, the ratios α and η are known; so (10.64) may be used to find the distribution factor F_2/F_1 as a function of the displacement-mode parameter ξ. The results of a calculation for a particular shell having $\alpha = \frac{1}{4}$, $\eta = \frac{1}{2}$ are shown in fig. 10.16. It is clear that larger values of ξ correspond to larger values of F_2/F_1. The distribution of stress σ_x against z is shown for three values of ξ, and we can see from these that as the loading ratio F_2/F_1 increases, the structure responds by a 'drooping' of the edges of the shell relative to the centre, so that the edge-beams deflect more than the central region of the shell. The stress-distribution diagrams have been drawn for a constant value of the total load.

When $\xi = 1$ we recover the stress distribution of classical beam theory, which is linear with z over the entire section: see fig. 10.16. It is possible to check this by treating the whole section monolithically, although the calculation is tedious. For this particular cross-section the neutral axis is nearer the top of the section than the bottom (it is about $\frac{1}{3}(H_1 + H_2)$ from the top), so the tensile stress at the bottom of the cross-section is larger in magnitude than the compressive stress at the top.

It is clear from fig. 10.16 that a relatively small alteration of the distribution of load between the beams and the shell can alter substantially the

Fig. 10.16. Relative proportions of total load carried by the shell and the beams, plotted against parameter ξ, which indicates the relative vertical displacement of the shell and the beams. Continuous curve: homogeneous shell and edge-beams (inset, distribution of strain with z). Broken curve: the result of concentrating the 'beam' material at the lower surface of the beam is to increase the proportion of the total weight which is carried by the beams (section 10.4).

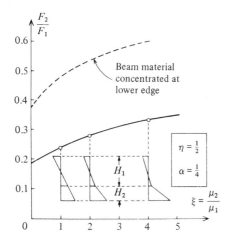

level of tensile stress at the bottom of the edge-beams. Nevertheless it is useful to introduce the problem of *design* of shells of this sort with reference to the properties of the composite beam in terms of the classical, 'monolithic' theory. For a given geometrical design this theory is only valid, of course, for a particular distribution of total load on the shell and beams. In general the actual distribution of both dead and live load will be different from this; and we shall return to the consequences in the next section.

On what basis should the cross-section of an isolated shell with edge-beams be proportioned? First we observe that for a very wide range of combinations of the geometrical ratios α and η the neutral axis lies above the mid-depth of the section. It is easy to show that this rule is violated only if

$$\eta < \alpha - \tfrac{1}{3} \tag{10.65}$$

(see problem 10.7), and this corresponds, broadly, to the cross-sections of the edge-beams being very 'compact'. Therefore in general we shall have to be content with the most highly stressed part of the section being the lowest part of the beams.

An obvious way to proceed is to argue that, roughly, the proportion of loading applied to the edge-beams and the shell will be the ratio of self-weights, i.e. $F_2/F_1 = A_2/A_1 = \alpha$. Putting this condition in (10.64), together with the requirement $\xi = 1$, we find that

$$\alpha = (\eta^2 + \eta - 0.2667)/(-\tfrac{1}{4}\eta^2 + \eta + 1.6). \tag{10.66}$$

In the range $0 < \alpha < 1$ the above relation between α and η is represented well by the simpler expression

$$\eta \approx 0.25 + \alpha. \tag{10.67}$$

This formula thus gives a scheme for design in which there remains one degree of freedom; and the design $\alpha = \tfrac{1}{4}, \eta = \tfrac{1}{2}$ which we have investigated previously fits this scheme. Once the values of the ratios α and η have been settled, it is a straightforward matter to fix the leading dimensions, etc.

10.3.2 Maximum tensile stress as a function of F_1 and F_2

In the next section we shall find, contrary to the supposition above, that the ratio, F_2/F_1, of applied loads is *not* normally equal to A_2/A_1; and we have already discovered that the tensile stress at the bottom of the cross-section appears to be sensitive to small changes in the ratio F_2/F_1. It is therefore appropriate to study next how the maximum tensile stress in the cross-section depends on the values of F_1 and F_2.

We shall do this by establishing, for a given cross-sectional design, the contours of maximum tensile stress in the cross-section in a Cartesian F_1, F_2

10.3 The effect of edge-beams

space. It will not be necessary to do much detailed calculation: fortunately the contours have a simple form and we shall be able to discuss the diagram in general terms.

Let us begin by observing that for any given cross-section there is a particular value of F_2/F_1 for which the calculation of maximum tensile stress is particularly easy, since the section behaves monolithically and we may thus use simple classical beam theory. For example, if $\alpha, \eta = \frac{1}{4}, \frac{1}{2}$ the condition $\xi = 1$ in (10.64) gives

$$F_2/F_1 = 0.242. \tag{10.68}$$

On this particular line in a plot of F_2 against F_1 (see fig. 10.17), therefore, it is easy to calculate those loads which are required in order to set up a given maximum tensile stress.

The simplest way of finding the contour of constant tensile stress which passes through a particular point on this line is to enquire what *changes* in F_1 and F_2 cause *zero* tensile stress at the bottom of the beam.

It is not difficult to derive an expression for the tensile stress at the bottom of the beam in terms of the quantities which we used in section 10.3.1: see fig. 10.14. We find that at $z = -H_2$,

$$\sigma_x = \mu_2 E H_2 + (T_1 - T_2)/(A_1 + A_2). \tag{10.69}$$

Fig. 10.17. Contours of longitudinal stress at the lower surface of the edge-beams (labelled $\sigma = 1$, etc.) in an F_1, F_2 load-space for a shell of a particular design. ($2F_2$ has been plotted for the sake of clarity.) The line AB corresponds to a redistribution of the total load between the shell and the edge-beams.

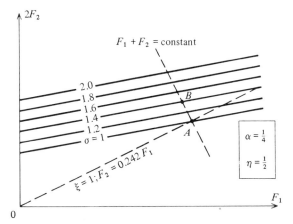

The first term on the right-hand side comes from (10.57) and represents the state of stress on the assumption that the neutral axis is at $z = 0$; and the second term represents the uniform stress in the cross-section corresponding to T in (10.61). Thus, putting $\sigma_x = 0$ in (10.69) and substituting for T_1 and T_2 from (10.59) and (10.60) we find, after some rearrangement,

$$\xi = -2/3\eta(1 + \tfrac{1}{2}\alpha). \tag{10.70}$$

Substituting this value of ξ into (10.64) we obtain the required relationship between these changes, ΔF_1 and ΔF_2, in load for which the change in tensile stress σ_x at the bottom of the beams is zero:

$$\Delta F_2/\Delta F_1 = \tfrac{5}{4}\alpha\eta/(1 + 3\alpha). \tag{10.71}$$

The value of the right-hand side is fixed for a particular design. Consequently the required contours are all straight and parallel. For $\alpha, \eta = \tfrac{1}{4}, \tfrac{1}{2}$ the right-hand side of (10.71) is equal to 0.0893; and the contours in fig. 10.17 have been drawn at this slope. The contour passing through point A on line (10.68) has been labelled arbitrarily, $\sigma = 1$, and the other contours correspond to $\sigma = 1.2, 1.4$ etc..

Line AB on the diagram corresponds to $F_1 + F_2 = $ constant. We can see that a change of load-point from A to B, involving the transfer of about 5% of the total load from the shell to the beams, increases the maximum tensile stress by 40%. This example illustrates the sensitivity of the maximum stress to changes in the value of ratio F_2/F_1; and it is clearly a straightforward matter to construct the contours for sections having different designs. We shall return to this diagram at the end of the next section.

A diagram of the same sort may also be constructed in relation to the compressive stress at the highest point in the cross-section. In this case the slope of the contours of constant stress is positive and considerably greater than the slope of the line for which the section behaves monolithically (see problem 10.8). Consequently the stress level depends primarily on the value of F_1, and is generally much lower than the maximum tensile stress at the bottom of the cross-section.

10.3.3 Sharing of applied load between shell and edge-beams

All of the preceding calculations have been based on the supposition that the loading on the surface of the shell is distributed in whatever way is required by statical equilibrium in the shell from the assumed 'beam-like' distribution of N_x in the shell. In other words, we have been using the membrane hypothesis for the shell; and in order to avoid the consideration of bending stresses we have artificially arranged that the loading which is applied to the shell is exactly equal to whatever pressure distribution p_S is acceptable to the

10.3 The effect of edge-beams

S-surface. In fact, of course, the actual loading on the shell which is imposed by dead weight, snow, etc. is unlikely to match p_S exactly, if at all: consequently there will usually be a nonzero distribution of pressure, p_B, which must be carried by bending of the shell, primarily in the transverse direction. We must now consider effects of this sort. The details can be complicated, but it is possible to study satisfactorily the general behaviour without too much attention to detail, as follows.

Suppose, for the sake of argument, that the distribution of p_S across the width of the shell is sinusoidal, say

$$p_S = p^* \sin(\pi y/B) \tag{10.72}$$

at the central cross-section, whereas the corresponding applied load p is constant, say p_0. What is the relationship between p and p_S? We can certainly expand $p = p_0$ as a Fourier series, as follows (appendix 10):

$$p = (4p_0/\pi) \, [\sin(\pi y/B) + \tfrac{1}{3} \sin(3\pi y/B) + \tfrac{1}{5} \sin(5\pi y/B)\ldots]. \tag{10.73}$$

Thus it is clear that

$$p_B \, (= p - p_S) \tag{10.74}$$

must necessarily involve the higher terms $\sin(3\pi y/B)$, etc. of this series. We shall argue that, provided the geometrical parameter β has a suitable value, the pressure distribution corresponding to each of these higher terms is carried entirely (or almost entirely) by B-surface action, precisely along the lines discussed in section 10.2.4. Now we can, if we wish, ensure that p_B contains *no* term in the fundamental component $\sin(\pi y/B)$, by making

$$p_0 = (\pi/4) \, p^*. \tag{10.75}$$

In this case the S-surface would carry all of the term $\sin(\pi y/B)$ in (10.73), while the B-surface would carry all of the other terms. This is certainly a simple way of dividing up the applied load, but can it be justified? The key to the situation lies in an observation which we made in section 10.2.3. There we explained that cylindrical shells of the kind which we are considering have a simple and 'almost inextensional' mode of deformation; and moreover we invoked this idea in the preceding section. Now if there were a component proportional to $\sin(\pi y/B)$ in the loading p_B, the bending moments M_y in the B-surface would produce changes of curvature in the shell corresponding to an excitation of this mode, or at least one generally similar to it. But this mode is so flexible that only negligibly small component of p_B of this kind would be necessary. Thus we argue that the Fourier series expansion for p_B contains a fundamental term whose magnitude is practically zero. In almost total contrast, however, we are implying that the pressure term containing $\sin(3\pi y/B)$, and all higher modes, are carried by the shell almost entirely by transverse bending of the B-surface. We have already seen that this is reasonable provided $\beta < 8.5$ (see (10.48)); and we shall assume that this inequality is fulfilled by any design under consideration.

Now the total vertical load per unit length of shell is equal to $p_0 B$. The total load carried by the first term in the series (10.73) is equal to $(2/\pi)[(4/\pi)p_0 B]$. Therefore the load carried by the S-surface is equal to $8/\pi^2 \approx 0.81$ of the total applied load; and it follows that about 0.19 of the applied load is carried by the B-surface. In particular, this latter fraction of the load is carried by beam-action in transverse 'strips' of the B-surface to the edge-beams; for unless the beams can provide the required normal reaction at the ends of transverse strips of the shell, the B-surface cannot be in equilibrium. Hence we may conclude in the present case, and on the basis of the contingent assumptions, that if the total uniformly distributed load which is applied to the shell is equal to P_1, while the loading applied to the beams is P_2, the loads which are actually sustained by the shell and the beams, according to the previous scheme, are

$$F_1 = 0.81 \, P_1$$
$$F_2 = 0.19 \, P_1 + P_2.$$

Consequently

$$F_2/F_1 = 0.23 + 1.23 \, P_2/P_1. \tag{10.76}$$

For example, for purely self-weight loading $P_2/P_1 = A_2/A_1 = \alpha$; so

$$F_2/F_1 = 0.23 + 1.23 \, \alpha. \tag{10.77}$$

It is important to realise that F_1 is the load carried by the S-surface of the shell, whereas P_1 is the actual load (self-weight plus live load, etc.) applied to the shell. It is because the S-surface only carries *part* of the load which is actually applied that the edge-beams are required to 'help out' to some extent. Now in the previous section we used

$$F_2/F_1 = \alpha$$

as a preliminary illustrative example; but we now see that this is conservative, and that the value of F_2/F_1 may be larger than α by, typically, 0.25.

An important conclusion which we drew from fig. 10.17 was that the tensile stress at the bottom of the beams is sensitive to the ratio F_2/F_1: consequently we must now be prepared for rather larger stresses than we were led to expect in the previous section.

Equation (10.67) provided a way of proportioning the cross-section of the shell and beams so that it behaved monolithically if $F_2/F_1 = \alpha$. In view of (10.77) it might seem more reasonable to design for monolithic response

10.3 The effect of edge-beams

when $F_2/F_1 = 1.5\,\alpha$, say. This can easily be done, by the use of (10.64). The result is, broadly, that η must now be determined, in relation to α, by

$$\eta \approx 0.3 + 1.5\,\alpha \tag{10.78}$$

instead of (10.67). Unfortunately this leads to designs with rather large values of η, i.e. with rather deep edge-beams. We shall suggest a way out of this difficulty at the end of the chapter.

There are several factors which make (10.78) perhaps unduly pessimistic, which may be enumerated as follows.

(1) The calculation above referred to purely dead (i.e. self-weight) loading. When the effects of 'live' loading are included, the situation is somewhat less severe. Suppose, for example, that the total dead + live load on the shell is equal to 1.5 times the dead load, and that there is zero live load on the beams. It follows that $P_2/P_1 = \frac{2}{3}$, so (10.77) must be replaced by

$$F_2/F_1 = 0.23 + 0.82\,\alpha. \tag{10.79}$$

But on the other hand, a change in live load will correspond to a change

$$\Delta F_2/\Delta F_1 = 0.23, \tag{10.80}$$

and since this is a considerably steeper slope than that of the contours in fig. 10.17, we must expect a striking response to changes in live load in terms of tensile stress at the bottom of the edge-beams.

(2) Expression (10.73) was based on the assumption that the self-weight loading could be considered to be equivalent to a uniform pressure. While this is satisfactory when $\psi < 15°$, say, (see fig. 10.13) it becomes less reliable for larger values of ψ, particularly in the popular range of about $30°$. The vertical loading which is applied to the shell should strictly be considered as the sum of normal and tangential loads. As we saw in section 7.3.3 (cf. section 4.2), the tangential loads are statically equivalent to a second (normal) pressure-load, if we can ignore some small changes in N_y. Consequently it may be shown (see problem 10.9) that for a vertical loading of f_0 *per unit area of surface* we have an 'almost equivalent' pressure-loading which is given by

$$p = -f_0 (2\cos\theta - \cos\psi). \tag{10.81}$$

Here θ is the circumferential angular coordinate, measured from the vertical. It is a straightforward matter to express this distribution as a Fourier series of the form

$$c_1 \cos(\pi\theta/2\psi) + c_3 \cos(3\pi\theta/2\psi) + \ldots,$$

then to pick out the fundamental term, and, lastly, to integrate the resultant vertical forces corresponding both to the given loading (10.81) and also to the first term of the Fourier series alone. In this way we find that the fraction of the applied vertical load which is carried by the fundamental term of the Fourier series for the pressure distribution is equal to

$$(8/\pi^2)\cos^2\psi\,[1 + 2\psi/\pi]^2/[1 - 2\psi/\pi]^2. \tag{10.82}$$

This factor rises from a value of 0.81 at $\psi = 0$ (as before), almost exactly in proportion to ψ^2 to a value of about 0.86 at $\psi = 30°$. Thus for a shell with $\psi = 30°$, the coefficient 0.81 in the earlier analysis must be replaced by 0.86, and (10.77) must be changed to

$$F_2/F_1 = 0.16 + 1.16\,\alpha. \tag{10.83}$$

This gives appreciably lower maximum tensile stresses, other things being equal.

(3) Lastly we should report a contrary tendency. The calculations above have assumed, without any real justification, that p_S is of the form $\sin(\pi y/B)$. In general, when the value of z at which $\sigma_x = 0$ in the shell is an appreciable fraction of H_1, p_S has a small but appreciable negative term of the form $\sin(3\pi y/B)$. This feature is similar to that which may be seen in the example of fig. 10.9. This effect reduces the load carried by the S-surface by a small amount, and gives a corresponding increase in the fraction of the load applied to the shell which is transferred to the beams.

The present section would not be complete without a brief discussion of transverse bending stresses in the shell, and indeed of any small longitudinal stresses with which they may be associated. For the sake of simplicity we shall consider only a shallow shell; but the results will be applicable to deeper shells in an approximate sense. As before, we shall consider a uniform downward loading f_0 per unit area to act on the shell, and as before we shall consider only the first term of a Fourier-series expansion in the longitudinal direction, of (outward-directed) pressure, namely,

$$-(4/\pi)f_0 \sin(\pi x/L). \tag{10.84}$$

We shall also assume, as above, that only the fundamental term of a Fourier series in the transverse direction, namely,

$$-(16/\pi^2)f_0 \sin(\pi x/L)\sin(\pi y/B)$$

is carried by the S-surface, which leaves the remaining terms

$$(16/\pi^2)f_0\,(\tfrac{1}{3}\sin(3\pi y/B) + \tfrac{1}{5}\sin(5\pi y/B) + \ldots)\sin(\pi x/L)$$

to be carried mainly by the B-surface. Here we shall consider only the first of these terms, namely,

$$p_3 = -(16/3\pi^2)f_0 \sin(3\pi y/B). \tag{10.85}$$

The response of the shell to loading of this form is precisely the same as in our earlier example involving a shell with no edge-beams, but with its edges connected to neighbouring shells: see section 10.2.4. We shall also assume that the shell is hinged to the edge-beams, in order to simplify the analysis.

As before, the condition that at least 0.9 of this load is carried by the B-surface is met by arranging the dimensions of the shell so that

$$\beta < 8.5.$$

Now the small remaining fraction of this load which is transmitted to the S-surface may be associated with appreciable S-surface stresses of the form

$$\sin(3\pi y/B)\sin(\pi x/L). \tag{10.86}$$

In the earlier example (section 10.2.4) we found that for $\beta < 7$ the magnitude of these stresses was restricted to 0.06 of the peak longitudinal stress, i.e. to 0.12 of the peak longitudinal compressive stress. In the previous example there were no edge-beams; and since the addition of edge-beams to a shell of given dimensions considerably reduces the level of the compressive stress in the cross-section, we must conclude that stresses of the form (10.86) may be of significant magnitude in comparison with the primary stresses σ_x. As before, their tendency is to make the distribution of σ_x with z somewhat non-linear in the shell (see fig. 10.9); and this effect may be seen clearly in the examples studied by Timoshenko & Woinowsky-Krieger and Flügge. Strictly, a term of form (10.86), if superposed directly onto the distributions of stress from the preceding sections, involves a violation of longitudinal equilibrium of the forces σ_x over the cross-section. Normally, however, this effect is small and can be neglected.

10.4 Concluding remarks

Certain broad conclusions may be drawn from the problems which we have studied in this chapter. But before we enumerate them we should draw attention to some of the simplifying assumptions on which the work has been based.

Throughout the chapter we have considered a loading which varies in the longitudinal direction as $\sin(\pi x/L)$, and which is the fundamental term of a Fourier-series representation of uniform load. It is not difficult to repeat the calculations in respect of the remaining terms in this series, and – since the structure is regarded as linear-elastic and the displacements are supposed to be small – the results may be superposed. In principle, therefore, the calculations can be refined in this sense. We have also regarded the material of which the shell is made as both uniform and perfectly elastic, and initially stress-free. All of these assumptions are questionable in the context of reinforced-concrete construction. But although concrete is well known to be subject to creep, shrinkage and cracking during its working lifetime, it is often regarded as linear-elastic for the purposes of calculation. In principle there is no difficulty in handling the different elastic moduli of the steel and the concrete, and the methods of 'transformation' into an equivalent structure made of homogeneous material are well known. Since these calculations lead to generally satisfactory designs we may conclude tentatively that our use of elasticity theory is not in itself altogether objectionable. Perhaps more questionable is our assumption about the homogeneity of the construction. For instance, what is the validity of our calculations of transverse 'extreme surface' bending stresses in the context of a shell whose steel reinforcement consists, perhaps, of a single-layer mesh of reinforcement at the centre-surface? Questions of this sort are of course well known in the application of standard methods of structural analysis to reinforced-concrete structures, and there are well known, if conventional, answers.

Lastly we should confess that our scheme for building up a final picture by the successive superposition of various effects may appear to some readers to be somewhat contrived: after all, at one stage (section 10.3.3) we were prepared to violate slightly the condition of longitudinal equilibrium. The success of the work hinges, in fact, on two circumstances. First is the fact of a simple 'almost inextensional' mode of deformation for the shell, and second is the assumption of a radically different response of the B-surface to the fundamental term in a Fourier-series representation of the distribution of pressure across the width of the shell on the one hand and to all higher harmonics on the other. The answer to these points is twofold. Firstly, the final results agree well with those based on more conventional methods, which have been reported by Timoshenko & Woinowsky-Krieger and Flügge (who use $\beta = 6.7$ and 8.3, respectively). Secondly, the entire calculation, including all of the effects at once could be set down systematically. But, thirdly, in this case we might well miss several of the illuminating features which we observed *en route* in our sequence of smaller problems.

In spite of all of these deficiencies in our analysis, it does seem reasonable to draw some firm general conclusions, as follows.

A single shell with no edge-beams and free edges acts mainly as a large beam, but is afflicted with a strong tendency for its edges to droop under self-weight loading. This tendency may be countered by thickening the shell, in order to inhibit transverse bending; or by connecting the edges of the shell to neighbouring shells; or by the provision of edge-beams. The main effect of the addition of edge-beams, other things being equal, is to deepen the cross-section of the shell (viewed as a beam) and thus to give the well-known corresponding benefits in the form of smaller bending stress, smaller deflections, etc.. The

10.4 Concluding remarks

tendency to droop at the edges is still present, however, even when edge-beams have been provided, in the sense that a relatively small increase in the load applied to the edge-beams causes disproportionately large stresses: see fig. 10.17. It is feasible to proportion the cross-section in such a way that the edge-beams support their own weight without appreciable 'droop' of the shell at the edges. But in general the edge-beams have to support, in addition to their own weight, a fraction (in the region of 0.1 to 0.2) of the load which is applied to the shell itself; and it is this extra factor which makes it difficult to design the shell and its associated edge-beams to act monolithically: see fig. 10.17.

Now throughout all of the calculations on which these conclusions are based it has been assumed that the edge-beams are of uniform rectangular cross-section, and are made of uniform material. This last assumption is particularly questionable, for it is obvious that in the design of reinforced-concrete structures, steel reinforcement is usually placed especially in regions of high tensile stress. Thus, although the shell itself may well be provided with a fairly uniform mesh of reinforcement – which should also indeed carry over into the beams to ensure the proper transmission of N_{xy} at the junction in particular – there must normally be some relatively massive rod-reinforcement towards the bottom edge of the beam sections. In consequence, we ought to question our treatment of the beams as having uniform rectangular cross-section. The most questionable part of the scheme shown in fig. 10.14 is therefore the magnitude of the resultant tension T_2 in the beam, and its line of action. In fact it is not difficult to repeat the whole analysis leading up to (10.64) for any arbitrary disposition of steel and concrete in the edge-beams. As a first shot at an appropriate modification we may consider the somewhat idealised arrangement shown in fig. 10.18, in which all of the beam material – as far as pure bending is concerned – is concentrated in a single 'rod' at $z = -H_2$ whose area is A_2 when expressed in terms of a material having the

Fig. 10.18. The diagram of fig. 10.14, modified for the 'extreme' case in which all of the material of the edge-beams is concentrated at the lower surface: see fig. 10.16.

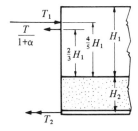

elastic modulus of concrete. There is also provision for sustaining shear forces N_{xy} in the beam without shear distortion, but this does not enter the analysis explicitly. The result for this particular cross-section, corresponding to (10.64), is

$$\frac{F_2}{F_1} = \frac{1 + \xi h(1.5)}{\xi + (\frac{4}{15} + \frac{8}{5}\alpha)/(2\alpha\eta)}. \tag{10.87}$$

This relation has also been plotted, as a broken line, in fig. 10.16, for $\alpha, \eta = \frac{1}{4}, \frac{1}{2}$. Its ordinate is roughly *twice* that of the original curve over the whole range of ξ. The beam is more effective in this configuration because its material is disposed at a larger 'lever arm' than formerly. The broken curve in fig. 10.16 provides an upper limit to what may be achieved by the use of reinforcement. In this diagram A_2 is always calculated in terms of an *equivalent* cross-sectional area of concrete. It is clear that the benefit which comes in consequence of a more realistic treatment of the reinforcement of the beam goes a long way to making it possible to design an arrangement which behaves monolithically.

10.5 Problems

10.1 The cross-section of a thin-walled beam has the form of a circular arc of radius a, which subtends an angle 2ψ at the centre (cf. fig. 10.13). The (uniform) thickness of the material is t. It may be shown that the second moment of area I of the section about the centroidal axis normal to the line of mirror symmetry is given by

$$I = ta^3 \{\psi + \tfrac{1}{2}\sin 2\psi - (2/\psi)\sin^2\psi\}.$$

By means of a power-series expansion of this expression show that, for sufficiently small values of ψ,

$$I \approx \tfrac{2}{45} ta^3 \psi^5 (1 - \tfrac{2}{7}\psi^2 + \ldots);$$

and that the two expressions in (10.3) are each equal to the first term of this expansion by virtue of the 'small-angle' approximation $\psi = B/2a$.

10.2 A simply-supported beam of length L and uniform flexural rigidity EI is subjected to a uniform transverse loading of q_0 per unit length. Express this loading as a Fourier series by means of (9.3), and hence find Fourier-series expressions for the bending moment M and the transverse displacement w. In particular, show that the first term of the series for M gives approximately 1.032 × the (total) bending moment at the centre, while the corresponding term in w gives approximately 1.004 × the central displacement.

10.5 Problems

10.3 Show that the bending moment M in a simply-supported beam subjected to a loading of Bp_m per unit length (see (10.5)) is given by

$$M = \frac{\pi^2}{720} \frac{EtB^5 w_0}{a^2 L^2} \sin\left(\frac{\pi x}{L}\right);$$

and that, according to classical beam theory the longitudinal 'bending' stress is given by

$$N_x = \frac{Mt(z - z_0)}{I} = \frac{\pi^2 Etw_0}{L^2}(z - z_0)\sin\left(\frac{\pi x}{L}\right).$$

By using (10.1) and (10.2) together with $z_0 = \tfrac{2}{3}H$, confirm that this agrees with (10.11).

10.4 A shell roof of the kind shown in fig. 10.2 has the following dimensions (in metres): $L = 36$, $a = 9$, $h = 0.09$, $B = 12$ (so $L/(ah)^{\frac{1}{2}} = 40$, $B/(ah)^{\frac{1}{2}} \approx 13$, $\beta \approx 4.3$).

Consider a mode $w = f(y)\sin(\pi x/L)$ as a solution of the (homogeneous) governing equation for the unloaded shell, and by reference to the roots diagram of problem 8.4 show that the characteristic half wavelength b of $f(y)$ is given by $b/(ah)^{\frac{1}{2}} \approx 20$. Hence deduce that the modes of bending which are considered in sections 10.2.2 and 10.2.3 have relatively *short* circumferential half wavelengths in comparison with the characteristic half wavelength; and thus that they involve only minor interaction with longitudinal stretching effects in the shell.

10.5 Derive the expressions which are plotted in fig. 10.11 in the following way. First put $\zeta = 2y/B - 1$, so that ζ is a dimensionless coordinate having values 0, 1 at the centre-line and edge, respectively of the cross-section. Noting that $p_S \propto (1 - \zeta^2)^2$ show that $F_2/(F_1 + F_2)$ = Area B/Area $(S + B)$ (in fig. 10.10) = $1 - 2.5\zeta^3 + 1.5\zeta^5$ by means of suitable integration. Also show that $A_2/(A_1 + A_2) = 1 - \zeta$ and that $H_2/(H_1 + H_2) = 1 - \zeta^2$.

10.6 The second moment of area I of the cross-section of problem 10.1 about the line through the end-points of the cross section is given by

$$I = ta^3 \{\psi - \tfrac{3}{2}\sin 2\psi + 2\psi \cos^2\psi\},$$

in contrast to the approximate form I' based on the integration of (10.55b):

$$I' = \tfrac{8}{15} BtH_1^2.$$

By putting $B = 2a\psi$ and $H_1 = a(1 - \cos\psi)$, show that I and I' differ by $\approx 0.2\%$ for $\psi = 30°$ and $\approx 1.5\%$ for $\psi = 45°$.

The corresponding (exact) expression for the *first* moment of area of the figure about the same axis is $2a^2(\sin\psi - \psi\cos\psi)$. Show that this differs from the approximate version (10.55a) by similar amounts when the above substitutions for B and H_1 are made.

10.7 By treating the cross-section shown in fig. 10.13 as that of a simple beam (and taking the centroid of the upper part as $\frac{2}{3}H_1$ above the level of the junction) show that the centroid of the entire section lies at a height of $H_1(4 - 3\alpha\eta)/6(1 + \alpha)$ above the level of the junction, where α is the area-ratio defined in the diagram. Hence show that (10.65) represents the condition that the centroid lies above the mid-depth of the section.

10.8 Show that the condition for zero change of stress at the *top* of the cross-section, analogous to (10.71) is
$$\Delta F_2/\Delta F_1 = \tfrac{5}{12}\eta(3\alpha + 4).$$

10.9 Consider a cylindrical shell of radius a with edges at $\theta = \pm\psi$ which is subjected to a vertical load of constant intensity f_0 per unit area. Resolve this load into normal (p) and tangential (q) components $p = -f_0\cos\theta$, $q = f_0\sin\theta$. Imagine that q is carried in the first instance by artificial circumferential strips which lie on surface of the shell without friction, and show by means of (4.2) that the tangential equilibrium of these strips requires a circumferential stress resultant $N = f_0\,a(\cos\theta - \cos\psi)$. Next consider the normal pressure between these strips and the shell which is required for normal equilibrium; and hence show that together with the loading p the shell is subjected to an equivalent normal pressure equal to $-f_0(2\cos\theta - \cos\psi)$.

11

Bending stresses in symmetrically-loaded shells of revolution

11.1 Introduction

The subject of this chapter is the behaviour of thin elastic shells of revolution which are subjected to loads applied symmetrically about the axis of revolution. In chapter 4 we studied the same problem, but there we worked under the simplifying conditions of the 'membrane hypothesis'; and the analysis involved only the equations of statical equilibrium. In the present chapter we shall not exclude in this way the possible occurrence of normal shear-stress and bending-stress resultants; and in consequence we shall need to consider not only the equations of equilibrium but also the conditions of geometric compatibility and the generalised Hooke's law. In the main we shall assume, for the sake of convenience, that the shell is stress-free in its initial, unloaded state. This is by no means always true in practice, and we shall discuss some important exceptions in section 11.6. However, it is always correct to regard our analysis as giving properly the *change* of stress resultants, displacements, etc. on account of a *change* of loading. Throughout the chapter we shall adopt the 'classical' assumption that displacements, strains and rotations are so small that the various equations may be set up in relation to the original, undeformed, configuration of the shell. Some remarks on the validity of this assumption are made in section 11.6.

On account of the symmetry of both the shell and its loading, the problem becomes *one dimensional*, in the sense that all of the relevant quantities are functions of a single variable which describes the position of a point on the meridian. This represents an enormous simplification in comparison with, say, the problems of chapter 9; and the resulting work turns out to be relatively simple, even for shells whose meridional shape is moderately complicated.

There are some important practical applications of the theory, particularly in the field of pressure-vessel engineering. Most of our examples will come from this source.

In the traditional approach to problems of this kind (cf. Flügge, 1973, chapter 6; Galletly, 1960; Kraus, 1967, §5.5; Bickell & Ruiz, 1967) a shell of revolution is first 'cut' into a number of sections, each of which has a meridian with a relatively simple geometrical form. Thus a pressure-vessel with a 'torispherical' end-closure (see fig. 11.1) is usually dissected into cylindrical, toroidal and spherical components, respectively. The various portions are then reconnected to each other by means of bending- and shearing-stress resultants, whose magnitudes are determined so that conditions of geometrical compatibility are satisfied. This process is essentially the same as the one which we touched upon in relation to cylindrical shells in chapter 3, but of course the response of the various non-cylindrical components to edge-loading is more complicated to work out.

In the present chapter we shall adopt a somewhat different approach, which has been described by Calladine (1972a) (but with different sign conventions), and is essentially an adaptation of older ideas which have been described, e.g. by Wells (1950). The main idea is not to *cut* the shell into components, but to regard it instead as a system whose behaviour is governed by a differential equation in which the coefficients and the 'loading' terms change – sometimes abruptly – from place to place along the meridian, in accordance with its geometrical form. The solution of equations of this sort is, of course, not trivial; but in many cases it is possible to make good approximate solutions by means of justifiable simplifications, as we shall see.

The method has several advantages over the traditional scheme in a wide range of problems. In the first place it avoids the introduction of 'internal boundary variables', and various complications which are associated with the corresponding equations. Second, the method emphasises the interaction between bending and stretching effects in the shell. This not only provides a useful qualitative picture of the behaviour of the shell, but it also gives clear

Fig. 11.1. A 'torispherical' end-closure to a cylindrical pressure-vessel.

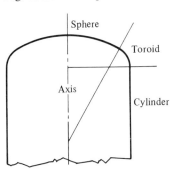

11.1 Introduction

indications of the circumstances in which the membrane hypothesis is valid. Moreover, it avoids certain conceptual difficulties which can arise in the traditional method and which actually led to erroneous solutions for some kinds of shell before certain features of the equations were clarified by Galletly (1959): the difficulties simply do not occur in the scheme to be presented here.

Third, it enables the necessary calculations to be simplified in many cases to such an extent that they are no more onerous than those corresponding to a simple cylindrical shell. Indeed, many problems which are by no means straightforward by the traditional methods become almost trivial under the present scheme.

Lastly, the method enables us to appreciate qualitatively the effect on the behaviour of the shell of local *changes* in the shape of the meridian. Information of this kind is of great help to the engineer for the purposes of preliminary design. The traditional method, on the other hand, is very cumbersome from this point of view.

The present method is ideally suited to a shell whose meridian is a single unbranched curve, as shown in fig. 11.2a. When the geometry is more complex, as in fig. 11.2b, c the traditional method has definite advantages: nevertheless some useful results can be obtained quickly by means of the present scheme. We include an example of the traditional approach in section 11.4.5.

The layout of the chapter is as follows. First we set up the governing equations in a general form for a shell having a meridian of arbitrary shape,

Fig. 11.2. Various pressure-vessel geometries. In (*a*) the meridional section is a simply-closed curve, whereas in (*b*) the topology is more complex. The vessel in (*c*) is connected to a foundation by two separate cylindrical 'skirts', and thus has a degree of external redundancy not found in (*b*).

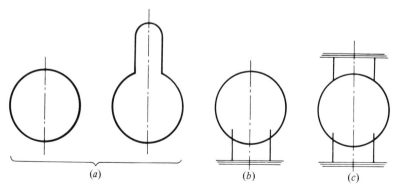

and subjected to arbitrary axisymmetric loading. We achieve a particularly simple set of equations by separating out the 'stretching' and 'bending' effects. For this particular problem we do not use directly the S- and B-surface idea which we have exploited in previous chapters, for reasons which we shall explain. Instead, we use a different scheme of separation which incorporates the equilibrium equations according to the membrane hypothesis (section 4.5) in a specially advantageous way. At the end of the chapter we examine the relationship between the present method and the S- and B-surface scheme.

Having set up the general equations, we next introduce and justify a particular simplification of them, due to Geckeler, which turns out to be valid in a wide range of circumstances. Next we apply these simplified equations to a range of 'junction' problems, beginning with some relatively mild discontinuities in the form of the meridian. We give results in a specially useful form, and include some simple approximate formulas for stress-concentration factors in structures such as those which are shown in figs. 11.1 and 11.2. We then discuss briefly a problem for which Geckeler's simplification is not valid. In section 11.5, as already mentioned, we relate the equations of the present chapter to previous work on the S- and B-surface description of shells.

Finally we discuss some practical limitations of the analysis, which spring from the underlying presuppositions of the work, and which have some important implications in design. For example, the behaviour of axisymmetric *bellows* does not fit conveniently into the scheme of the present chapter; it is discussed separately in chapter 12. The present chapter is concerned only with the classical analysis of elastic shells. Some problems of inelastic behaviour are considered in chapter 18. Throughout the chapter we use the same notation as in chapters 4 and 6.

11.2 Equations of the problem

Most of the necessary ingredients for the study of the problem under investigation have already been established elsewhere in the book. Thus in chapter 6 we studied the main geometrical features of surfaces of revolution, and the strains and curvatures which occur in the course of symmetrical deformations; and in chapter 4 we studied the equilibrium equations for these shells, at least in the context of the membrane hypothesis. The main geometric and kinematic variables for a general shell of revolution are defined in figs. 4.10 and 6.9.

11.2.1 *Equilibrium equations*

Our first task therefore is to investigate the general form of the equilibrium equations when bending-stress and normal shearing-stress result-

11.2 Equations of the problem

ants are not excluded by hypothesis. Fig. 11.3a shows the loads and stress resultants to which a typical small element of the shell is subjected. There is a normal surface traction of p per unit area, and a tangential traction of q_S per unit area in the meridional direction, as shown. Tangential surface tractions in the circumferential (θ) direction are not considered here: they lead to a twisting of the shell about its axis which is easy to analyse.

The edges of the element are subjected to tangential-stress resultants N_s, N_θ; bending-stress resultants M_s, M_θ; and the shearing-stress resultant Q_s. Here the subscripts s, θ denote, as before, the meridional and circumferential directions. The positive sense of each of these stress resultants is indicated in the diagram. Shearing- and twisting-stress resultants $N_{s\theta}$, Q_θ and $M_{s\theta}$ are not shown: they are identically zero, for reasons of symmetry.

It would be feasible to set up three equations of equilibrium for the small element of fig. 11.3a by resolving forces in the directions normal to the surface and tangential to the meridian, respectively, and by taking moments about an axis normal to the plane of the meridian. But the resulting equations can be expressed in a much simpler form if we first split the forces acting on the element in fig. 11.3a into two distinct sets, as shown in fig. 11.3b and c. This procedure is something like the scheme illustrated in fig. 8.1, in which an element of a shallow shell was decomposed into 'stretching' and 'bending' parts. However, in most previous applications we have been able to assume that the shearing-force interaction between the two surfaces (q_B) was negligible in comparison with the normal force interaction (p_B); and this has led to the 'shallow-shell' version of the governing equations. But for the present set of problems this simplification is not always justified, and it is more convenient to use a different approach, as follows.

Fig. 11.3. (a) Stress resultants acting on a typical small element of a symmetrically-loaded shell of revolution. (b) The same external loading is now carried according to the membrane hypothesis. (c) Stress resultants chosen so that $(a) \equiv (b) + (c)$. See the text for an explanation of H.

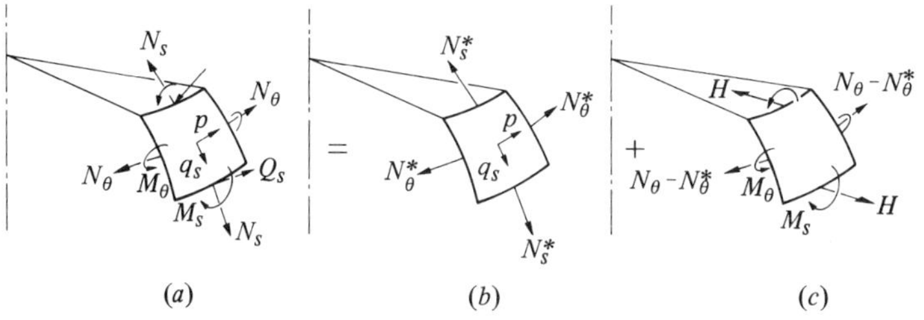

The element shown in fig. 11.3b carries *all* of the externally applied surface tractions p, q. Equilibrium is maintained by the stress resultants N_s^*, N_θ^*, which are calculated according to the *membrane hypothesis* by precisely the methods of section 4.5. Throughout the present chapter superior asterisks will indicate stress resultants calculated according to the membrane hypothesis in this way. The stress resultants shown in fig. 11.3c are simply those which must be superposed on those of fig. 11.3b in order to be statically equivalent to the full set of fig. 11.3a. They include the bending-stress resultants M_s, M_θ and the tangential hoop-stress resultant $(N_\theta - N_\theta^*)$. But observe that in place of a compound stress resultant involving Q_s, N_s and N_s^*, we have a single stress resultant H of a new kind, which acts radially *in the plane normal to the axis of revolution*. Note that there are no *surface tractions* in fig. 11.3c, since these have already been included in fig. 11.3b.

Why do the stress resultants H act in planes normal to the axis? The explanation is given in fig. 11.4, which shows the stress resultants acting at the edge of an arbitrary 'cap' of the shell: M_s has been left out for reasons of clarity. Fig. 11.4a shows N_s and Q_s, in accordance with fig. 11.3a, while fig. 11.4b shows N_s^* and H. Now N_s^* itself, by definition, is sufficient to maintain the equilibrium of the cap in the axial direction (see section 4.5.1); and therefore any additional stress resultant can have no component in the axial direction. Therefore H must act in the plane normal to the axis.

The resultant of N_s and Q_s must of course be identical to the resultant of N_s^* and H, since figs. 11.4a and b are merely different views of the same total forces. It follows immediately that

$$H = Q_s/\sin\phi = (N_s - N_s^*)/\cos\phi, \tag{11.1}$$

Fig. 11.4. The stress resultants acting at the edge of an arbitrary 'cap' may be expressed either (*a*) as the vector sum of Q_s and N_s (cf. fig. 11.3a) or, equivalently, (*b*) as the vector sum of N_s^* and H. Stress resultant M_s has been omitted from both diagrams.

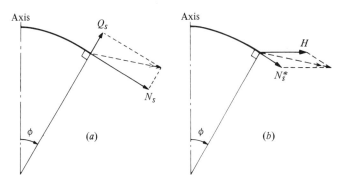

11.2 Equations of the problem

where $\phi(s)$ is the meridional angle defined in the usual way in fig. 11.4a. Thus H is closely related to Q_s; and we see moreover that H and Q_s are necessary only because the actual stress resultant N_s differs from that (N_s^*) which is computed according to the membrane hypothesis.

In fig. 11.3 stress resultants of the same magnitude have been shown at both ends of the element of meridional length ds, whereas of course all of the stress resultants vary with s. This must be taken into account when we consider the equilibrium of the small element. Since N_s^* and N_θ^* are known, in principle, from a membrane-hypothesis analysis, it is sufficient to consider the equilibrium of the element shown in fig. 11.3c. Equation (11.1) already guarantees equilibrium in the axial direction, so we need only consider force-equilibrium in the direction normal to the axis and moment-equilibrium about an axis normal to the plane of the meridian: all other conditions are fulfilled by symmetry.

By considering the force equilibrium of the element in the radial direction (perpendicular to the axis) we obtain

$$\frac{d}{ds}(rH) = N_\theta - N_\theta^*, \tag{11.2}$$

where r is the radius defined in fig. 4.10. Next, by considering the moment-equilibrium of the element about an axis normal to the meridian we find

$$\frac{dM_s}{ds} + \frac{M_s - M_\theta}{r_3} = H\sin\phi. \tag{11.3}$$

The radius r_3 enters because the tangents to the meridional edges of the element intersect on the axis of symmetry at a point which is distant r_3 from the element: cf. fig. 4.11b.

It is now convenient to define a new 'total shear' variable

$$U = rH\ (=r_2 Q_s), \tag{11.4}$$

and to collect together our three equilibrium equations, which may be written:

$$N_s - N_s^* = \frac{U}{r_3}, \tag{11.5}$$

$$N_\theta - N_\theta^* = \frac{dU}{ds}, \tag{11.6}$$

$$\frac{dM_s}{ds} + \frac{M_s - M_\theta}{r_3} = \frac{U}{r_2}. \tag{11.7}$$

These equations are precisely equivalent to those which may be found in many texts (e.g. Timoshenko & Woinowsky-Krieger, 1959 §127; Flügge, 1973, §6.1.3; Kraus, 1967, §5.1), but they are much simpler in form. This simplicity derives partly from the use of U as the 'shear' variable, and partly from the fact that the surface loading of the shell is not considered explicitly, but implicitly, through the calculation of the stress resultants N_s^*, N_θ^* according to the membrane hypothesis. Thus, when we come to assemble the governing equation for the shell, we shall find that the terms representing the applied loading on the shell enter in the form of functions of N_s^* and N_θ^*.

Equations (11.5) to (11.7) do of course have a direct physical significance. If the actual stress resultants N_s, N_θ happen to be equal to the stress resultants N_s^*, N_θ^* respectively, calculated according to the membrane hypothesis throughout the shell, then $U(s)$ is zero and no bending moments are necessary.

11.2.2 Equations of kinematics

The kinematic relations may be copied from chapter 6. We have

$$\epsilon_s = \frac{du}{ds} + \frac{w}{r_1}, \quad \epsilon_\theta = \frac{u}{r_3} + \frac{w}{r_2} \qquad (11.8)$$

$$\chi = \frac{u}{r_1} - \frac{dw}{ds} \qquad (11.9)$$

$$\kappa_s = \frac{d\chi}{ds}, \quad \kappa_\theta = \frac{\chi}{r_3}, \qquad (11.10)$$

where all quantities are as defined in fig. 6.9. Also, on eliminating u and w from (11.8) and (11.9) we find (cf. section 6.9)

$$\frac{d\epsilon_\theta}{ds} + \frac{\epsilon_\theta - \epsilon_s}{r_3} = -\frac{\chi}{r_2}. \qquad (11.11)$$

There is a formal similarity between (11.11) and (11.7), and indeed between (11.10), (11.6) and (11.5), respectively, if we omit the 'loading' terms N_s^*, N_θ^*: see appendix 6.

11.2.3 Generalised Hooke's law

The last set of basic equations which we need is the generalised Hooke's law for the element. From chapter 2 we have

$$M_s = D(\kappa_s + \nu \kappa_\theta) \qquad (11.12a)$$
$$M_\theta = D(\kappa_\theta + \nu \kappa_s) \qquad (11.12b)$$
$$\epsilon_s = (1/Et)(N_s - \nu N_\theta) \qquad (11.12c)$$
$$\epsilon_\theta = (1/Et)(N_\theta - \nu N_s), \qquad (11.12d)$$

where t is the thickness of the shell, and the other quantities are defined as usual.

11.2 Equations of the problem

11.2.4 Governing equations

It is now most convenient to proceed by putting (11.7) in terms of χ by using (11.12) and (11.10): thus we find

$$L(\chi) - \{\nu\chi/r_1\} - (U/D) = 0 \tag{11.13}$$

where the operator $L(\ldots)$ is given by

$$L(\ldots) = r_2 \left(\frac{d^2}{ds^2}(\ldots) + \frac{1}{r_3}\frac{d}{ds}(\ldots) - \frac{1}{r_3^2}(\ldots) \right). \tag{11.14}$$

Similarly, (11.11) may be put in terms of U by the use of (11.12), (11.5) and (11.6).

$$L(U) + \{\nu U/r_1\} + Et\chi = -Pr_2 \tag{11.15}$$

where

$$P(s) = \frac{d}{ds}(N_\theta^* - \nu N_s^*) + \frac{(1+\nu)}{r_3}(N_\theta^* - N_s^*). \tag{11.16}$$

The two simultaneous second-order equations (11.13), (11.15) constitute the mathematical description of the problem. They may be solved, at least in principle, subject to appropriate boundary conditions. These equations are not so 'transparent' as those which we derived previously for shallow shells in a local x, y coordinate system (see section 8.5), but they have some fundamental similarities. Thus (11.13) represents an equilibrium equation (although the variable χ is a displacement quantity) while (11.15) represents a compatibility equation ('in disguise') which relates the deformation due to bending and stretching, respectively. The function $P(s)$ on the right-hand side of (11.15) describes the 'stretching' distortions on account of the stress resultants calculated according to the membrane hypothesis: if these happen to be compatible without any change of Gaussian curvature (see problem 11.1) then $P(s) = 0$ and (11.13) and (11.15) have a trivial solution $\chi = U = 0$ except possibly near a boundary: see later. In section 11.5 we shall investigate further the relationship between the equations of the present chapter and those of 'shallow-shell' theory.

The convenience of the use of χ and a shearing-stress resultant (here U) as the variables for this problem was first noted by H. Reissner; and the operator $L(\ldots)$ which appears on both (11.13) and (11.15) is known as Meissner's operator (see, e.g. Timoshenko & Woinowsky-Krieger, 1959, §128). $L(\ldots)$ is often expressed in a much more complicated form than (11.14), but one which reduces identically to it: the usefulness of the geometrical parameter r_3 has not always been recognised. (The operator $L(\ldots)$ can be expressed in a more compact form through the identity

$$L(\xi) = r_2 \frac{d}{ds}\left[\frac{1}{r}\frac{d}{ds}(r\xi)\right] + \frac{\xi}{r_1};$$

but this will not be useful for our purposes. The almost-complete formal similarity between the left-hand sides of (11.13) and (11.15) is discussed in appendix 6.)

Equations (11.13) and (11.15) may be combined to give a fourth-order differential equation in χ. In general, of course, r_1, r_2 and r_3 are known functions of s, and the coefficients of the equation are therefore not constant. It is relatively straightforward to set up a general finite-difference numerical procedure for solving the equation, in which the details of shell geometry and loading may be inserted in relation to a given problem.

However, in many practical problems such a procedure turns out to be unnecessary, since in many cases the equations are dominated by only a few terms; thus the neglect of other terms makes little difference to the values of the stress resultants, etc.

Our aim in the next two sections will be to exploit the possibilities for simplification of the equations in this way. We shall see that the solutions of the simplified equations have some features which lead to a strongly 'physical' interpretation of the results. It may happen that in a given problem a numerical solution is advisable, since there is some doubt about the validity of the simplifications: nevertheless a physical interpretation of the results which is suggested by the simplified system provides a good guide to the interpretation of the numerical results.

Our procedure will therefore be firstly to examine the simplified form of the equations without giving a detailed justification of each simplifying step; and secondly to compare results obtained by means of the simplified equations with those obtained by numerical means from the original equations.

11.2.5 Geckeler's simplification

We know from chapter 3 that the application of an axisymmetric out-of-plane shear load to a thin cylindrical shell of radius a and thickness t yields a solution in which all relevant quantities vary in the axial direction in a strongly decaying oscillatory fashion with a characteristic length of order $(at)^{\frac{1}{2}}$.

Geckeler argued that in an expression of the form

$$\frac{d^2(\ldots)}{ds^2} + \frac{1}{r_3}\frac{d(\ldots)}{ds} - \frac{1}{r_3^2}(\ldots) \qquad (11.17)$$

the overall effect of the second term would be negligible in comparison with that of the first term provided the dimension r_3 were sufficiently large in comparison with the 'characteristic length' of the solution (see, e.g. Timoshenko & Woinowsky-Krieger, 1959, §130). As we shall see, the local characteristic

11.2 Equations of the problem

length for a general shell of revolution is of order $(r_2 t)^{\frac{1}{2}}$; so we may conclude tentatively that the first term of the expression will be dominant provided

$$r_3 \gg (r_2 t)^{\frac{1}{2}}. \tag{11.18}$$

The same reasoning shows that in these circumstances the effect of the third term is small in comparison with that of the second.

In order to assess the significance of this condition, consider the spherical cap shown in fig. 11.5, whose radius and thickness are a, t respectively. At a point on the meridian defined by the angle ϕ we have

$$r_3/(at)^{\frac{1}{2}} = (a/t)^{\frac{1}{2}} \tan\phi. \tag{11.19}$$

Let us now define an axial coordinate z, as shown, whose value is zero at the apex. For $0 < \phi < \frac{1}{2}\pi$ we have the following chain of self-evident expressions:

$$z = a(1 - \cos\phi) = 2a \sin^2(\tfrac{1}{2}\phi) < 2a \tan^2(\tfrac{1}{2}\phi) < \tfrac{1}{2} a \tan^2\phi.$$

Hence

$$\tan\phi > 2^{\frac{1}{2}} (z/a)^{\frac{1}{2}},$$

and it follows that

$$r_3/(at)^{\frac{1}{2}} > (2z/t)^{\frac{1}{2}}. \tag{11.20}$$

For this shell, therefore, a criterion for the validity of Geckeler's simplification which is given in terms of $r_3/(ah)^{\frac{1}{2}}$ may therefore be expressed alternately in terms of the parameter z/t. Detailed studies of the response of shallow spherical caps (and holes in shallow spherical shells) to edge-loading (Calladine & Paskaran, 1974) shows that the Geckeler approximation is satisfactory provided

$$z/t > 5 \tag{11.21}$$

Fig. 11.5. The 'cap' of a spherical shell, for the purpose of examining the validity of Geckeler's simplification.

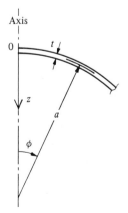

approximately. Thus we may conclude that it is only in the relatively *shallow* regions near the apices of thin shells of revolution that the Geckeler approximation is not satisfactory. (And it is generally the case that in such regions the 'shallow-shell' approximations are valid: see section 11.4.4.)

It follows immediately that the Geckeler approximation is valid for a very wide range of practical problems.

Returning to (11.13) and (11.15) we see that the operator $L(\ldots)$ (11.14) may normally be simplified to

$$L(\ldots) \approx r_2 \frac{d^2}{ds^2}(\ldots). \tag{11.22}$$

The terms enclosed { } in equations (11.13) and (11.15) may usually be discarded for similar reasons. When (11.18) is satisfied they are of the same order as the last of the three terms of $L(\ldots)$: see appendix 6.

If we are prepared to accept these simplifications, we find that (11.13) and (11.15) may be written, approximately,

$$\frac{d^2\chi}{ds^2} - \frac{1}{r_2 D} U = 0 \tag{11.23}$$

$$\frac{d^2 U}{ds^2} + \left(\frac{Et}{r_2}\right)\chi = -P. \tag{11.24}$$

If furthermore the geometry of a given shell has the feature r_2 = constant, then these equations have *constant coefficients* and the solutions are relatively simple. A *cylindrical* shell, of course, has r_2 = constant, and (11.23) and (11.24) are in this case precisely equivalent to those which we studied in chapter 3.

The *sphere* is the only other kind of surface in which r_2 is truly constant. Therefore there is a direct similarity between the governing equations for the symmetrical deformation of spherical and cylindrical shells, at least according to the Geckeler simplification. The simplest way of visualising this similarity is by means of the example illustrated in fig. 11.6 (cf. Pflüger, 1961, Fig. 39). The two shells carry uniform edge-loading in a plane normal to the axis of revolution, with no other loading. They have equal, constant, values of r_2. Therefore, according to (11.23), (11.24) they are governed by precisely the same differential equations in U and χ; and these are satisfied by identical functions $U(s)$, $\chi(s)$ where s is distance measured along the meridian from the loaded edge. In these circumstances $N_\theta(s)$ and $\kappa_s(s)$ are identical for the two shells; but note that $N_s(s)$ and $\kappa_\theta(s)$ are not, by (11.5), (11.10) and (11.12), since the shells do not share the same $r_3(s)$. However, in the context of the Geckeler simplification these differences are of little consequence; and we

11.2 Equations of the problem

may therefore conclude that the functions $M_s(s)$, $M_\theta(s)$ are approximately the same for the two cases. Note in particular that by (11.10) Geckeler's simplification implies $|\kappa_\theta| \ll |\kappa_s|$, and hence (by 11.12) that $M_\theta \approx \nu M_s$.

Note particularly that the practical equivalence of the two cases depends on the equality of the values of the shearing force U at the edge and not on the equality of the shearing force H per unit length.

The variable U has the dimensions of *force*, and since $U = Hr$ (see (11.4)) it may be regarded as the *total resolved force* across the radius r. The physical significance of this variable may be appreciated if we consider the equilibrium of half of the shell, cut along diametrically opposite meridians, when it is subjected to loading U, M applied at the edge, as in fig. 11.6. The value of U applied at the edge must be balanced by $\int N_\theta \, ds$ along the meridian; and this is equally true for both shells if $N_\theta(s)$ is the same. These considerations, of course, correspond exactly to (11.6): for edge-loading only $N_\theta^* = N_s^* = 0$.

The equivalence of the two shells considered in this example is subject to various obvious limitations; thus the cylindrical shell must be sufficiently *long* (see section 3.4) and the spherical cap must be sufficiently *deep*.

For all surfaces of revolution other than the cylinder and the sphere, the radius r_2 is a function of position along the meridian. Consequently (11.23), (11.24) have, in general, *variable* coefficients and cannot therefore be transformed into those of 'equivalent' cylindrical shells. Fortunately this feature

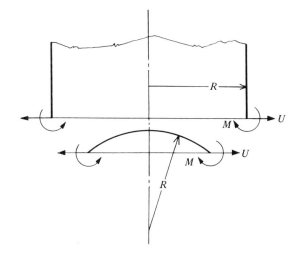

Fig. 11.6. The equivalence of cylindrical and spherical shells having the same radius and thickness, according to Geckeler's simplification.

of general surfaces of revolution is not as awkward as the above remarks may appear to imply. The key point is that the solutions of (11.23), (11.24) involve rapidly decaying oscillating functions of s whose half wavelength is of the order of $(r_2 h)^{\frac{1}{2}}$. If therefore the value of r_2 changes by relatively little over a segment of meridian whose length is *of this order*, it is reasonable, at least as a first approximation, to consider the value of r_2 to be effectively constant over the portion of meridian in question. Detailed numerical calculations on thin conical and other shells have demonstrated that the use of an equivalent constant value of r_2 gives excellent results in circumstances in which Geckeler's approximation is valid. There are, however, some important cases – see e.g. fig. 11.2a – in which the value of r_2 changes *discontinuously* precisely in a region where bending effects are important. Even here, as we shall see later, the use of a 'mean' value of r_2 can introduce remarkably little error.

So far we have not investigated the form taken by the function P, which constitutes the right-hand side of the governing equations (11.15) or (11.24). Clearly a lot will depend on the form of this function, and our next task will be to investigate it for some typical problems.

Our main interest is of course in so-called 'junction' or 'intersection' problems of the kind sketched in figs. 11.1 and 11.2. But it will be useful to begin with some problems which involve somewhat less drastic meridional profiles. These will provide a clear physical picture of the behaviour of shells with non-simple meridians; and we shall be able to build relatively easily onto these solutions when we come to consider problems such as those illustrated in figs. 11.1 and 11.2.

11.3   The effects of an 'imperfect' meridian

Consider the following problem (Calladine, 1972a). A shell of revolution has been designed in such a way that the axially symmetric loads which are applied to it are carried essentially according to the membrane hypothesis, without any significant assistance from bending and out-of-plane shear effects. Unfortunately, the shell has been constructed with a meridian of a slightly different shape: but it is nevertheless still truly axially symmetric. By how much do the stress resultants in the shell change on account of the unintended small change in the profile of the meridian? It will be assumed that the imperfection occurs in a region of the shell for which Geckeler's simplification is valid.

In solving this problem we shall describe the form of the shell as constructed in terms of the small normal displacement ξ of the meridian from its intended configuration, as indicated in fig. 11.7. Possible forms of imperfec-

11.3 The effects of an 'imperfect' meridian

tion in which we shall be interested include (small) kinks and jogs (see fig. 11.8), as well as periodic imperfections.

The motivation for this study is twofold. First, it may be important for the designer of a shell to understand the structural consequences of a small manufacturing imperfection. It is perhaps unlikely that such an imperfection would be truly axisymmetrical: nevertheless the present study can give some useful information. Second, as we have remarked already, some of the imperfections such as those illustrated in fig. 11.8 may be regarded as relatively 'mild' versions of junction problems; and some of the ideas which are to be developed in the present section will provide a useful introduction to junction problems in general.

What changes in (11.23) and (11.24) are brought about by the change in configuration of the meridian? There will certainly be changes in the radii r, r_1,

Fig. 11.7. Definition of an 'imperfection' in the smooth meridian of a shell of revolution.

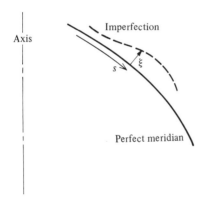

Fig. 11.8. Several types of 'localised' imperfection, plotted in Cartesian s, ξ space (see fig. 11.7).

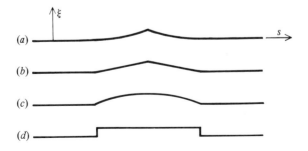

r_2 and r_3 which will affect the left-hand side of (11.13) and (11.15) (including the operator $L(\ldots)$) – or of (11.23) and (11.24) in circumstances where Geckeler's simplification is valid. Furthermore, the calculation of $P(s)$ on the right-hand side of (11.15) and (11.24) will be affected, since the changes of profile will also affect the calculation of stress resultants N_s^*, N_θ^*. We have already seen in section 4.5.2 that (according to the membrane hypothesis) relatively small changes in meridional profile can cause large fluctuations in N_θ^* while having little effect on N_s^*. But for present purposes we need to treat this question more systematically; and in particular we need to investigate the likely magnitudes of the several alterations.

The overriding factor is that for imperfections of a predominantly 'short-wave' type, the changes in $r_1(s)$ are by far the most significant in comparison with those of $r(s)$ or $r_2(s)$; and although there may also be relatively large changes in $r_3(s)$, these have little impact in the conditions for which Geckeler's simplification is appropriate.

Let us denote the changes in the several variables on account of the change in the meridional profile by the operator Δ. Since $-d\xi/ds$ is practically equivalent to the small-rotation variable χ we obtain immediately from (11.10) the relations

$$\Delta\left(\frac{1}{r_1}\right) \approx -\frac{d^2\xi}{ds^2}, \qquad (11.25a)$$

$$\Delta\left(\frac{1}{r_2}\right) \approx -\frac{1}{r_3}\frac{d\xi}{ds}. \qquad (11.25b)$$

Now the basis of Geckeler's simplification was that the effect of a term $(1/r_3)d(\ldots)/ds$ would be insignificant in comparison with that of a term $d^2(\ldots)/ds^2$ for sufficiently *short-wave* functions of s. In Geckeler's problem the functions χ and U have a characteristic wavelength of order $(r_2 t)^{\frac{1}{2}}$, whereas of course in the present problem the imperfection function is arbitrary. It is therefore clear that provided the character of ξ is sufficiently 'short-wave', it will be reasonable to neglect changes Δr_2 in comparison with changes Δr_1. Let us proceed, on this basis, to examine first the effect of these changes on the stress resultants N_θ^* and N_s^*.

Two equations of equilibrium may be written as follows (cf. section 4.5).

$$2\pi N_s^* r^2/r_2 = \text{Total axial load on a 'cap'}, \qquad (11.26)$$

$$(N_s^*/r_1) + (N_\theta^*/r_2) = \text{Normal component of surface traction.} \qquad (11.27)$$

We shall suppose that the external loading applied to the shell is unaltered by the change in meridional profile. Since we may ignore changes in r_2, and also in r, we find from (11.26) that

$$\Delta N_s^* \approx 0. \qquad (11.28)$$

11.3 The effects of an 'imperfect' meridian

Consequently, by differentiating the left-hand side of (11.27) we find

$$N_s^* \Delta (1/r_1) + \frac{1}{r_2} \Delta N_\theta^* \approx 0$$

and therefore, by (11.25),

$$\Delta N_\theta^* \approx r_2 N_s^* \frac{d^2\xi}{ds^2}. \tag{11.29}$$

This rather crude analysis thus confirms the remarks of section 4.5.2: although a small localised change in the meridional profile has relatively little effect upon the stress resultant N_s^*, there may be some striking changes in the stress resultant N_θ^*. This feature of the membrane-hypothesis solution for symmetrically-loaded shells of revolution has been noted by Flügge (1973, p.370) and Heyman (1967) after detailed calculations for shells with particular but slightly differing meridians. Expressions (11.28), (11.29) give remarkably good agreement with the cited calculations. The same effect may be seen in a comparison of the results of problems 4.17 and 4.18: see problem 11.2.

It is particularly straightforward to understand (11.29) in physical terms in the special case of a *cylindrical* shell to whose meridian has been imparted a small change $\xi(s)$: the change ΔN_θ^* balances the product of the axial stress resultant and the curvature of the meridian.

Now it is also easy to see that the change ΔN_θ^* is precisely the same as that which would occur if instead of varying the *profile* of the meridian of the cylindrical shell we had supplied an *additional normal pressure*, say p^*, given by

$$p^* = N_s^* \frac{d^2\xi}{ds^2}. \tag{11.30}$$

In other words, we may *mimic* the effect of a change in shape of the meridian on the stress resultants N_θ^*, N_s^* by means of an additional pressure-loading on the original shell, which is directly related to the change in the meridional profile. These remarks apply to an originally cylindrical shell, but they have a simple counterpart for general shells of revolution, as we shall see later.

So far we have assumed implicitly that $\xi(s)$ is a smooth function. The physical interpretation given above suggests that the counterpart of an abrupt change in the value of $d\xi/ds$ will be a concentrated ring-load of intensity q^* per unit length, where

$$q^* = N_s^* \left[\frac{d\xi}{ds} \right]. \tag{11.31}$$

Here, [ ] indicates the abrupt change in value. We shall consider the effect of a 'jog' (i.e. a step change in ξ itself) later.

Our aim in calculating the changes in N_θ^*, N_s^* on account of a change in the meridional profile has been of course the evaluation of the consequent changes in the right-hand side of the governing equation (11.24). But we now see – at least in the case of a cylindrical shell – that the small changes in the meridional profile are precisely equivalent to small changes in the pressure-loading on the original shell as far as the right-hand side of (11.24) is concerned. Now the given loads applied to the original shell were carried, according to the statement of the problem, by stress resultants in accordance with the membrane hypothesis. Consequently, the *changes* in stress resultant on account of the small *change* in profile are precisely the same as those stress resultants which would be set up in a shell loaded only by the additional equivalent pressure p^* and line load q^*. In particular, the bending moment M_s and shearing force U are a direct consequence of the additional loading p^*, q^*.

11.3.1 A 'change of slope' imperfection

It follows directly that the most basic case is that of a cylindrical (or nearly-cylindrical) shell loaded in simple axial tension whose meridian has a small-angle 'kink' at a particular location, as shown in fig. 11.9a. If such a shell is cut at the plane of the kink it is clear that for the membrane hypothesis to be fulfilled, some small inward-directed forces are required at the junction, as shown in fig. 11.9b. But since these are not in fact provided, we will have bending and out-of-plane shear-stress resultants in the shell on account of the outward-directed ring-loading as indicated in fig. 11.9c. Stated in these terms, the final result as given above seems self-evident. But our scheme has much more general application than is suggested by this simple example. Firstly, it is important to realise that the analysis covers a perturbation of a *general* state

Fig. 11.9. Physical argument whereby (a) an imperfect cylindrical shell under axial tension has 'bending' stresses on account of the ring-loading (c); the loading (b) is carried according to the membrane hypothesis.

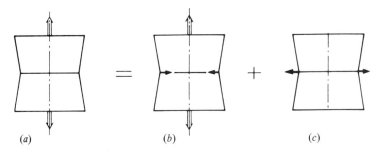

11.3 The effects of an 'imperfect' meridian

of membrane stress in the original shell. The result holds, for example, not only for the simple axial tensile loading shown in fig. 11.9, but also for internal pressure-loading of a shell with closed ends, or for self-weight loading, or indeed any combination of these kinds of loading. As far as the *perturbation* is concerned it is only the *local value* of N_s^* which is significant: the details of the loading arrangement are immaterial.

Secondly, the result applies not only to cylindrical shells but also to shells of arbitrary meridian provided the loading on the shell is such that for the original configuration the membrane hypothesis is justified: see fig. 11.10. In this more general case (11.28), (11.29) indicate that a perturbation of the meridian produces changes ΔN_θ^*, but almost no change in N_s^*. Consequently, if a small perturbation of the meridian is to be mimicked by small changes in the applied loads, these additional loads must all be self-equilibrating in planes normal to the axis of revolution. It follows from the statical analysis of section 4.5.2 that a small abrupt change in the value of $d\xi/ds$ at a point on the meridian produces (on the membrane hypothesis) a ring of tension there of value

$$T^* = r_2 N_s^* \left[\frac{d\xi}{ds} \right]. \qquad (11.32)$$

Exactly the same ring-tension would be produced by a uniform radial loading acting on the original shell in the plane of the slope discontinuity; and this is readily described as a loading causing a discontinuity of T^* in the shear variable U. The corresponding bending and shearing stresses in the shell are readily found by applying the same loading, in terms of U, to a cylindrical shell of radius r_2.

Let us now examine the local stresses in the shell on account of the ring-loading characterised by the hoop tension T^*. This problem was studied in

Fig. 11.10. The same argument applied to an imperfect version of a non-cylindrical shell. The ring-load in (c) acts in a plane normal to the axis of revolution.

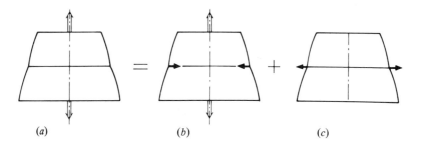

section 3.5, and we may quote results in terms of the present notation. Putting the origin of s at the point where the slope discontinuity occurs, we have, on account of the ring-loading, the following expressions in the region $s > 0$:

$$N_\theta = N_0 e^{-s/\mu} [\cos(s/\mu) + \sin(s/\mu)], \qquad (11.33a)$$

$$M_s = M_0 e^{-s/\mu} [\cos(s/\mu) - \sin(s/\mu)], \qquad (11.33b)$$

where

$$\mu = (r_2 h)^{\frac{1}{2}}/3^{\frac{1}{4}}, \qquad (11.33c)$$

$$N_0 = T^*/2\mu, \qquad (11.33d)$$

$$M_0 = T^*\mu/4r_2. \qquad (11.33e)$$

These expressions are plotted in fig. 11.11.

The most convenient scheme for displaying these results is one which was first used by Flügge (1973, Fig. 6.22). The basic idea is that the stress resultants N_s and M_s at any point on the meridian are statically equivalent to a single resultant N_s acting at an eccentricity e from the centre-surface of the shell (in the same sense as ξ), where

$$e = M_s/N_s. \qquad (11.34)$$

Thus, for example, at the point of discontinuity of slope, we have, from (11.32) and (11.33)

$$e_0 = \frac{\mu}{4} \left[\frac{d\xi}{ds} \right]. \qquad (11.35)$$

In fig. 11.12 the eccentricity determined from (11.34) has been 'set off' from a profile of the shell in the vicinity of the discontinuity. Note that the resulting curve is smooth. This is no accident, for in this particular scheme of

Fig. 11.11. Stress resultants N_θ and M_s in the vicinity of a radial ring-load which is carried by a ring of tension T^* according to the membrane hypothesis (cf. (11.33)).

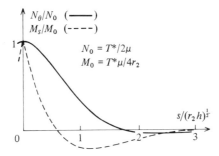

11.3 The effects of an 'imperfect' meridian

plotting, the discontinuity of slope of the meridian exactly balances the discontinuity of slope in the function $M_s(s)$ shown in fig. 11.11, which by equilibrium alone is directly related to T^*. We may call the resulting curve the 'effective meridian'. In fact this curve not only indicates the variation of M_s along the shell, but it also gives by means of the following expressions the difference, ΔN_θ, between the actual N_θ and N_θ in the original 'perfect' shell, on account of the presence of the imperfection in the meridian. The expression, which is easily demonstrated (see problem 11.3), is analogous to (11.29):

$$\Delta N_\theta = r_2 N_s^* \frac{d^2(\xi + e)}{ds^2} . \tag{11.36}$$

In physical terms this means that the stress resultant N_θ in the actual shell is precisely the same as the stress resultant N_θ^* calculated according to the membrane hypothesis for an artificial shell whose meridian is the same as the 'effective meridian', as defined above.

The relationship between the imperfect meridian and the effective meridian is shown for this problem in fig. 11.12. Note that the scale in the s-direction is set by the length μ, and that the scale in the ξ-direction is arbitrary.

Equation (11.36) is, unfortunately, only valid in circumstances where the eccentricity e is small. The idea of an effective meridian is therefore valid in the present problems, but it cannot be used directly for junction problems of the kind to be studied in the next section.

Perhaps the most striking feature to emerge from the above analysis is the severity of the meridional bending-stress resultant M_s. It is convenient to de-

Fig. 11.12. Results for a shell containing an imperfection in the form of a discontinuity of slope in the meridian, plotted in terms of an 'effective meridian' offset by $e = M_s/N_s$ from the actual meridian. Also shown is a localised 'smoothing arc' of overall length L, whose purpose is to moderate the severity of the original imperfection. The positive senses of e and ξ are the same.

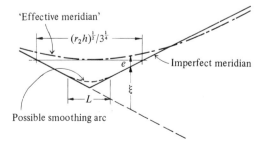

fine a 'peak bending stress' σ_s^b as the 'extreme fibre' stress due to the peak value of M_s: from (2.29), $\sigma_s^b = 6M_s/t^2$. It is straightforward to express σ_s^b as a fraction of the mean stress, N_s/t:

$$\frac{\sigma_s^b t}{N_s} = \frac{6e^{\max}}{t} = \frac{1.5}{3^{\frac{1}{4}}(1-\nu^2)^{\frac{1}{4}}} \left(\frac{r_2}{t}\right)^{\frac{1}{2}} \left[\frac{d\xi}{ds}\right]. \qquad (11.37)$$

Suppose that we decide, somewhat arbitrarily, that we can tolerate only a 30% increase in σ_s due to bending stress on account of a 'kink' imperfection of the meridian. From (11.37) we have, for $\nu = 0.3$,

$$\left[\frac{d\xi}{ds}\right] < 0.26 \left(\frac{t}{r_2}\right)^{\frac{1}{2}}. \qquad (11.38)$$

Thus, for $r_2/t = 50$, we can tolerate a kink of only about 2.1°.

We shall return to consider various developments of this analysis later.

11.3.2 Periodic imperfections

It is instructive to consider next the problem of a sinusoidal imperfection of the meridian, having the form

$$\xi = \xi_0 \sin(\pi s/l), \qquad (11.39)$$

where l is an arbitrary half wavelength. It is, of course, easiest to envisage such an imperfection to the meridian of a cylindrical shell having a constant value of N_s^*. Nevertheless our results, when expressed in terms of the relationship between the actual imperfect meridian and the effective meridian, should be applicable to more general shells of revolution provided Geckeler's simplification is valid and also that there is a reasonably long stretch of meridian over which the value of r_2 is approximately constant.

The analysis is straightforward. First we note that the imperfection in the meridian is mimicked by a sinusoidally varying pressure calculated from (11.30). The problem of a cylindrical shell under such a pressure was studied in chapter 3. The results are readily adapted. The profile of the effective meridian has the equation

$$\xi_e = (1 - \Phi)\, \xi_0 \sin(\pi x/l), \qquad (11.40)$$

where Φ is the function defined by (3.45) and plotted in fig. 3.8. For small values of l/μ the value of Φ is close to 1: in this case the effective meridian is almost straight, and the value of N_θ is almost constant; but the peak value of M_s is large, being almost equal to $N_s^* \xi_0$. On the other hand, for large values of l/μ the value of Φ is close to zero: in such cases there is little bending, since the effective meridian closely follows the actual imperfect meridian.

An important conclusion may be drawn from this piece of analysis. Rela-

11.3 The effects of an 'imperfect' meridian

tively 'gentle' changes in the meridian, having characteristic lengths which are large multiples of $(r_2 h)^{\frac{1}{2}}$, are accommodated simply by a readjustment of stress resultants according to the membrane hypothesis, which thus continues to be valid. On the other hand the response of the shell to 'rapid' changes in the line of the meridian, having short characteristic lengths, involves the introduction of bending and normal shearing-stress resultants, and hence a departure from the range of validity of the membrane hypothesis.

Earlier we saw that our expressions (11.28), (11.29) for changes in the stress resultants according to the membrane hypothesis were only valid for 'short-wave' changes in the form of the meridian. We can now see that the range in which these expressions become inaccurate is precisely the range in which the response does not involve the bending effects which it is our purpose to investigate.

We must, of course, use some caution in the application of a result which was obtained by consideration of a single sinusoidal perturbation to more general situations. But it is clear that our first example, concerning a small abrupt change of slope in the meridian, is precisely the sort of problem for which our method is suitable.

11.3.3 A 'change of curvature' imperfection

Another clear application of the method is to the case of a step change in the *curvature* of the meridian. Thus we can imagine a shell of revolution whose meridian is altered according to

$$\xi = 0, \qquad s \leq 0; \qquad (11.41a)$$
$$\xi = \tfrac{1}{2}\Delta(1/r_1)s^2, \qquad s > 0. \qquad (11.41b)$$

For sufficiently large values of s, of course, ξ becomes too large for our analysis to be valid; but this is in fact not important, because the 'bending' solution decays rapidly with $|s|$. The effect of this particular change in meridian is mimicked by the application of a uniform pressure to the original shell in the region $s > 0$ (on the assumption that N_s^* is constant over the relevant section of the meridian); and if we take r_2 as sensibly constant over the relevant section we have, in terms of the equivalent cylindrical shell, a particularly straightforward problem to analyse (see problem 11.4). The result is plotted in terms of an effective meridian, in fig. 11.13. By symmetry, the eccentricity of the effective meridian is zero at $s = 0$; and the diagram has a unique natural scale. The maximum eccentricity of the effective meridian has magnitude

$$r_2 h \Delta(1/r_1)/2 \times 3^{\frac{1}{2}}. \qquad (11.42)$$

This result may be applied directly to the junction between a cylindrical and a spherical shell having the same radius a and thickness t: in this case $r_2 = a$ and $\Delta(1/r_1) = 1/a$. The peak meridional bending stress is given by

$$\sigma_s^b t/N_s = 0.28 \, h/t \approx 0.29. \qquad (11.43)$$

Thus the bending stress nowhere exceeds about 0.3 of the mean meridional stress, irrespective of the value of a/t; so by comparison with a step change of *slope*, a step change of *curvature* seems mild indeed in terms of the meridional bending stress which it produces.

These results agree precisely with calculations done by Flügge (1973, Fig. 6.12) in the traditional way for a pressure-vessel of this form when it is subjected to internal pressure-loading. A pleasant feature of the present approach to the problem is that the result is seen to apply not only for pure pressure-loading but also for a wide range of other loading conditions.

11.3.4 Clusters of imperfections

Imperfections in the form of a meridian may occur in isolation, as in the preceding examples; but they may also occur in *clusters*. Thus, for example, in the fabrication of a welded cylindrical vessel, an isolated circumferential course of plates may be made accidentally to a slightly oversize radius; and in the resulting assembly the meridian then has two 'jogs' as shown in fig. 11.8d. Again, in the construction of a spherical vessel a circumferential 'ring' may be fabricated with an incorrect meridional curvature; in which case the imperfection-profile will be as shown in fig. 11.8c, provided there are no jogs in the assembly. It is not difficult to envisage manufacturing errors which can result in the imperfection profiles shown in the other diagrams of fig. 11.8.

Consider first the example of fig. 11.8b, which involves a succession of discontinuities of slope in the meridian. The stress analysis involves an appropriate superposition of the data of fig. 11.11. It is immediately obvious that

Fig. 11.13. An imperfection involving a discontinuity in the curvature of the meridian by an amount $\Delta(1/r_1)$. The 'effective meridian' is drawn to scale. In this drawing $\Delta(1/r_1)$ has a negative value.

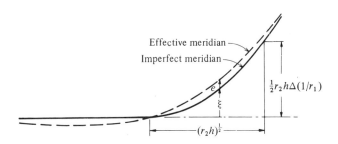

11.3 The effects of an 'imperfect' meridian

if the discontinuities of slope are widely separated, they may all be treated in isolation; whereas if they lie close to each other they may interact strongly. The results of these calculations are shown in fig. 11.14b by means of the 'effective meridian' for three different scales in the meridional direction. It is clear that for the scale which involves the largest distance between successive discontinuities in slope, the effective meridian in the vicinity of each discontinuity corresponds to that shown in fig. 11.12: and the single point shown in each case comes from that figure. Similar remarks apply to the diagrams of fig. 11.14a and c, which have also been constructed by superposition from fig. 11.11. On the other hand, the scale which involves the smallest distance between the successive features of the various imperfections gives effective

Fig. 11.14. Solutions in the form of 'effective meridians' for the imperfections shown in fig. 11.8. The diagrams have been standardised so that the imperfections are all of the same height and enclose the same area relative to the reference meridian. Effective meridians are shown for imperfections extending over three different lengths, as indicated. In each case the bending-stress resultants M_s were evaluated by superposition of data from fig. 11.11.

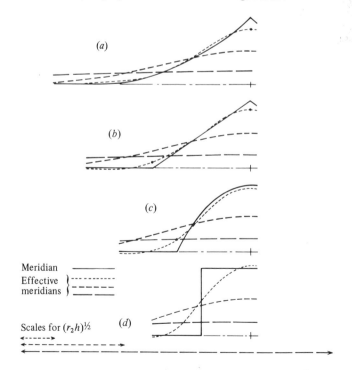

386   *Bending stresses in shells of revolution*

meridians which are almost straight as they pass through the cluster, with correspondingly large values of M_s. The imperfection shown in fig. 11.14d involves two 'jogs', i.e. discontinuous changes in the value of ξ. It is straightforward to show that the structural consequences of a 'jog' may be mimicked by externally applied couples of magnitude N_s^* multiplied by the amplitude of the jog. Imperfections of this sort can occur when two cylindrical shells are joined by a short sleeve to which both are connected. Similar effects can be found at the junction between shells reinforced with (say) external longitudinal stiffeners if the stiffeners are removed in the vicinity of the junction. Both of these examples, of course, are more complicated than the problem which we have studied, on account of changes in the properties of the shell element from point to point along the meridian. Nevertheless, the results given in fig. 11.14 can be useful in cases of this sort if they are applied with imagination.

11.3.5 *Moderation of a 'change of slope' imperfection*

In view of the potential severity of bending stress in the region of a discontinuity of slope in a meridian, and also in anticipation of a problem in section 11.4.1, it is instructive to investigate by how much the severity of a discontinuity of slope may be attenuated by 'smoothing' the kink. Suppose we insert a short, uniformly curved section of length L in place of the kink, as shown in fig. 11.12. It is clear from (11.30) and (11.31) that the pressure-loading which is required to mimic the perturbation of the original meridian must be altered from a concentrated 'ring' load to a uniformly distributed 'band' loading, of the same total magnitude. The variation of the peak bending moment M_s with the width L of the band is a problem which we studied in section 3.6, and an appropriate dimensionless graph is given in fig. 3.6d. Thus, in order to reduce the peak value of e to one-half of its former value, we must have

$$L \approx 0.9(r_2 h)^{\frac{1}{2}}. \tag{11.44}$$

It is interesting to note that for $L < 0.4(r_2 h)^{\frac{1}{2}}$, approximately, the 'effective meridian' is hardly altered by the presence of the smoothing length L: see fig. 3.11. In such cases the 'rounding' of the corner confers an immediate and direct diminution of the peak value of the eccentricity e.

11.3.6 *Discussion*

The analysis of the structural consequences of imperfections in the form of a meridian, as presented in the preceding six sections, constitutes a useful unifying tool for the study of symmetrically-loaded shells of revolution.

11.4 Some pressure-vessel junction problems

There are, however, some important restrictions on the method; and these should also be emphasised. We have assumed throughout that the shell has uniform thickness t over the relevant portion of the meridian. Consequently, we have been able to call upon results which were derived in chapter 3 for cylindrical shells of uniform thickness. In fact it is not particularly difficult to analyse a cylindrical shell having an abrupt change of thickness if we wish to do so; but the problem goes deeper than this. The starting point of the work is that in its original, perfect configuration, the shell carries the applied loading without any bending: that is, the strains in the shell when it is stressed according to the membrane hypothesis are geometrically compatible. Thus, for example, a cylindrical shell with a step change in thickness at a certain point on the meridian does not carry an applied uniform axial tension according to the membrane hypothesis in the vicinity of the junction, since there is a step change in ϵ_θ at this location on account of the 'Poisson ratio' effect. Since equations (11.13) and (11.15) were developed for shells of constant thickness, it is not possible to use the right-hand side of (11.15) as a detector of incompatibility in the solution according to the membrane hypothesis. The equations may easily be rewritten for shells of variable thickness $t(s)$, if this is felt to be desirable. In this connection we must point out that discontinuities of this sort are relatively *weak*: they fall in the general class of which the arrangement of fig. 11.13 is an example. There seems little point in investigating such weak effects in detail; and there is certainly no point if the much stronger effects which arise from discontinuities in *slope* of the meridian are present.

11.4 Some pressure-vessel junction problems

In the preceding sections we have studied the consequences of various discontinuities in the meridional profile which may be considered as *small* perturbations of what may be considered as originally 'smooth' meridians. Clearly the junctions illustrated in figs. 11.1 and 11.2 fall into a different category. Our next task is to investigate problems of this sort.

11.4.1 Spherical closure of a cylindrical vessel

Consider first the cylindrical vessel of radius a whose end is closed by a large-radius spherical cap of radius A, as shown in fig. 11.15a. The cylindrical and spherical parts have common thickness t. Under internal pressure-loading the stress resultants N_θ^*, N_s^* are uniform in the two portions; but there is a strong concentrated ring of compression around the junction itself: see section 4.5.2.

In order to analyse the bending stresses let us consider first a pressure-loading together with an outward-directed ring-loading at the junction whose magnitude is arranged so that, according to the membrane hypothesis, there is zero resultant ring-tension at the junction. This augmented loading case is shown in fig. 11.15b. Onto it must be superposed, in order to obtain the required loading (fig. 11.15a), the *inward-directed* ring-loading at the junction which is shown in fig. 11.15c.

For the loading case shown in fig. 11.15b there will be some mismatch of circumferential strain at the junction, but it seems safe to conclude, on the basis of the work of the preceding section, that the corresponding bending effects will be weak. In the present problem there is of course the additional novel feature of a step change in the value of r_2, but this will not in fact change the character of the mismatch; for the overriding consideration is that the bending effects which are associated with the loading case shown in fig. 11.15c are of a much higher order.

It remains only to study the response of the shell to the loading shown in fig. 11.15c. According to the Geckeler simplification, the governing equations (11.23) and (11.24) for the two component shells have exactly the same form, but different constant values of r_2. P is zero for both shells. The connecting conditions at the junction between the shells are as follows.

(i) There is a discontinuity in U equal to the applied junction force T^*.
(ii) The values of χ in the two component shells are equal.
(iii) The values of ϵ_θ in the two component shells are equal.

Now in Geckeler's simplification we ignore terms U/r_3 in comparison with dU/ds; and hence by (11.5), (11.6) and (11.12) – with $N_\theta^* = N_s^* = 0$ – we find that the continuity of ϵ_θ is practically equivalent to the continuity of dU/ds across the junction.

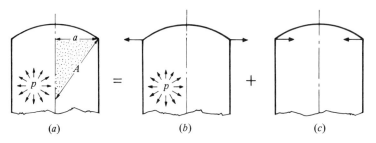

Fig. 11.15. Analysis of the stresses in the vicinity of the junction between a cylindrical vessel and its spherical end-closure, due to internal pressure. The actual loading (a) is the sum of two parts (b) and (c). The outward-directed additional loading in (b) is chosen so that the analysis according to the membrane hypothesis involves zero 'ring-tension' at the junction: cf. figs. 11.9 and 11.10.

11.4 Some pressure-vessel junction problems

The solution of this problem is particularly simple in the special case in which the two shells have a common thickness. It turns out in this case that $\chi = 0$ at the junction irrespective of the ratio a/A; that T^* is shared between the shells in proportion to $a^{\frac{1}{2}}$ and $A^{\frac{1}{2}}$; and that the peak values both of M_s and N_θ (which occur at the junction) are both precisely the same as they would be if the loading T^* had been applied instead to a uniform cylindrical shell of the same thickness and having a radius R such that

$$R^{\frac{1}{2}} = \tfrac{1}{2}(a^{\frac{1}{2}} + A^{\frac{1}{2}}):$$

see problem 11.5.

In this way we obtain the following approximate expressions for the (total) stress resultants N_θ and M_s at the junction, on account of an internal pressure p (for the loading shown in fig. 11.15a):

$$N_\theta \approx \frac{p}{2}(a + \tfrac{1}{2}A) - \frac{3^{\frac{1}{4}}pa}{2}\left(\frac{A^2 - a^2}{h}\right)^{\frac{1}{2}} \frac{1}{(A^{\frac{1}{2}} + a^{\frac{1}{2}})}, \quad (11.45)$$

$$M_s \approx -\frac{pa}{4 \times 3^{\frac{1}{4}}} \frac{(A^2 - a^2)^{\frac{1}{2}} h^{\frac{1}{2}}}{(A^{\frac{1}{2}} + a^{\frac{1}{2}})}. \quad (11.46)$$

In (11.45) the first term on the right-hand side represents the (tensile) hoop-stress resultant at the junction for the loading of fig. 11.15b: it has been assumed that the weak bending stresses at the junction roughly equalise the disparate values of N_θ in the two components. The other terms come direct from the loading of fig. 11.15c: T^* is equal to the pressure multiplied by the area which is shaded in fig. 11.15a: i.e.

$$T^* = \tfrac{1}{2}pa(A^2 - a^2)^{\frac{1}{2}}. \quad (11.47)$$

For shells having large values of a/h the second term on the right-hand side of (11.45) dominates. For example, if we take $A = 2a$, (11.45) and (11.46) reduce to

$$N_\theta \approx pa(1 - 0.47(a/h)^{\frac{1}{2}}),$$
$$M_s \approx -0.14\, pa\,(ah)^{\frac{1}{2}}.$$

The component of loading shown in fig. 11.15c clearly dominates the problem, and typically produces at the junction meridional bending stresses σ_s^b (= $6M_s/t^2$, see (2.29)) which are larger in magnitude than the hoop stresses in regions remote from the junction by a factor of about $(a/h)^{\frac{1}{2}}$. In other words, this particular kind of end-closure of a cylindrical pressure-vessel leads to rather high values of 'stress-concentration factor'. Indeed, these stress-concentration factors may be unacceptably high; and it is precisely for this reason that the so-called 'torispherical head' is introduced as a way of reducing extreme stress concentrations. This forms the subject of the next section.

11.4.2 Torispherical pressure-vessel heads

Fig. 11.1 shows the meridian of a torispherical pressure-vessel head or closure. The shell consists of cylindrical and spherical portions which are connected by a toroidal section of small radius which joins both parts smoothly. The entire shell has constant thickness, although in practice the spherical and toroidal sections (which are sometimes made in one piece which is then welded to the cylindrical shell) may be made from thicker material.

The arrangement may be considered as a variant of the closure shown in fig. 11.15, in which the cylindrical and spherical parts join with a discontinuity in the slope of the meridian; and indeed, as we shall see, it is useful to think of the insertion of the toroidal section as a way of mitigating the strong bending effects which are a consequence of a discontinuity of slope, in much the same manner as in section 11.3.5.

Consider first the distribution of stress resultants N_θ^*, N_s^*, according to the membrane hypothesis, when the shell is subjected to interior pressure p. These may be calculated in a given case from (4.22) and (4.24), and a typical result is shown in fig. 11.16. The most striking feature of this is the strongly negative (i.e. compressive) N_θ^* in the toroidal region.

It is a straightforward matter to show by the methods of section 4.5.2 that the total area of the region shaded in fig. 11.16 is almost precisely equal to the value of T^* calculated in the absence of a knuckle. Thus we see immediately that the magnitude of the compressive hoop stress N_θ^* decreases as the length of the knuckle region increases. This situation is much the same as that

Fig. 11.16. Stress resultant N_θ^* calculated according to the membrane hypothesis for pressure-loading on a torispherical end-closure of a cylindrical vessel: cf. fig. 4.14.

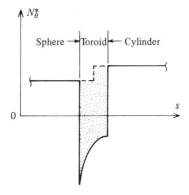

11.4 Some pressure-vessel junction problems

considered in section 11.3.5 in relation to a small change of slope of the meridian. It is clear, however, that some details are different, since there is a strong variation of N_θ^* within the knuckle region in fig. 11.16 in contrast to the uniform distribution of N_θ^* within the bands of loading in fig. 3.6 and its counterpart in fig. 11.12. This is attributable to the variation of r_2 within the region of the knuckle: see fig. 4.14.

In the absence of a knuckle it was convenient to regard the pressure-loading as the sum of two parts, as shown in fig. 11.15. We can adapt this scheme to the present problem by adding to the pressure-loading an outward-directed loading at the knuckle, as shown in fig. 11.17b, in order to 'cancel' the strong fluctuations in N_θ^* by direct 'hoop' action, and leave a shell having only 'weak' bending stresses. The major bending stresses are thus due to the inward-directed loading as shown in fig. 11.17c. Following the previous example we now transfer this loading to an 'equivalent' cylindrical shell; and our problem now resolves into an analysis of the behaviour of this 'transformed' cylindrical shell under the applied loading. This can be done by means of appropriate superposition of results obtained in chapter 3 for a cylindrical shell subjected to a concentrated 'ring' load. But before we proceed to do this, it is worthwhile to make some general observations.

The required calculation would certainly be simpler if the applied loading were to produce a band of *uniform* high negative hoop-stress resultant N_θ^*; and indeed we have already discussed the solution to this problem. The non-uniformity in the loading comes simply from the application of (4.28) for the given meridional profile. But we know already that relatively small changes in the meridional profile can produce disproportionately large changes in N_θ^*. It is therefore of interest to ask what *changes* in the actual meridional profile would be necessary in order to give a *uniform* value of N_θ^* in the short compression region. It is not difficult to show (see section 4.5.2) that the curva-

Fig. 11.17. A torispherical end-closure analysed by the same scheme as in fig. 11.15.

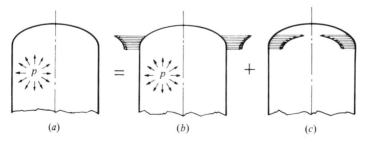

(a)   (b)   (c)

ture of the meridian needs to be decreased in the portion of the knuckle neighbouring the spherical section, but to be increased in the portion neighbouring the cylindrical section. The required changes can be accomplished by means of normal displacements which nowhere exceed about 0.02 of the radius of the knuckle. This appears to be within normal manufacturing tolerances. In other words, we may state that the distribution of N_θ^* shown in fig. 11.16 will be found if the meridian is made *exactly* to the pattern prescribed in fig. 4.14; but that relatively small imperfections in this profile – such as those which might occur in a typical manufacturing process – may produce appreciable changes in the distribution of N_θ^*. These changes do not, of course, affect the *mean* value of N_θ^* in the band. In view of these various uncertainties it seems reasonable to use the *uniform* distribution of N_θ^* as a first approximation. This can be refined, if required, in a case where the actual profile of the meridian has been determined with some accuracy.

It follows that the peak values of M_s and N_θ are given approximately by (11.45) and (11.46) provided the last terms on the right-hand side are multiplied by the respective attenuation factors which are shown in fig. 3.6d; here of course we take the radius of the equivalent cylindrical shell as $\frac{1}{4}(a^{\frac{1}{2}} + A^{\frac{1}{2}})^2$ (as in section 11.4.1) and the axial extent of the 'band' as the meridional length of the knuckle.

Calculations performed in this way agree well with the results of detailed computations which have been reported in the literature (e.g. Galletly, 1960): typically the discrepancy in the values of the stress resultants is no more than 5% (Calladine, 1972b, p. 261). See also Ranjan & Steele (1975). The sources of the small difference fall into two classes. First there is the complex of approximations in relation to the simplification of the shell equations into those relating to an equivalent cylindrical shell; and second there is our recent introduction of a simplified distribution of N_θ^*. The second class is likely to be the source of the larger part of the discrepancy. It is not difficult to study the effect of a change in the distribution of N_θ^*, simply by superposing the 'basic' solution (fig. 11.11) appropriately. It is clear by inspection that a change from a rectangular to a 'skew' distribution of N_θ^* will have a larger impact on M_s than on N_θ, both in relation to the magnitude of the peak value and the change of its location. As a guide to the order of magnitude of the changes corresponding to a redistribution of N_θ^* we can state that the effect of changing from a 'rectangular' to a 'triangular' distribution (see fig. 11.18) is, roughly, to give results for $N_\theta^{\max}$ and $M_s^{\max}$ corresponding to a rectangular distribution of width $\approx 0.8b$ (see Calladine, 1972b, p. 265). From fig. 3.6d we see that the corresponding changes in the peak values are in the region of 10%. The triangular distribution of N_θ^* differs, of course, from the actual distribution; but for a knuckle which subtends a meridional angle of 60° the peak values of N_θ^* do not differ by more than 20%: see problem 11.6.

11.4.3 Stress-concentration factors

We have seen that the main feature of structural behaviour in a vessel like that shown in fig. 11.15, on the application of internal pressure, is a concentration of both circumferential compressive stress and meridional bending stress in the region of the junction; and that the insertion of a short toroidal section moderates these effects to some extent. In situations like these it is convenient to describe the local peak stresses in terms of 'stress-concentration factors'. It frequently happens, of course, that material which according to the elastic theory lies in localised regions of high stress will in fact enter the plastic range; and when this happens it may be more appropriate to do the structural analysis within the framework of *plastic theory*. This will form the subject of chapter 18, where the same shell will provide an example. For the present, however, we shall assume that the behaviour of the shell is entirely in the elastic range; and it is therefore appropriate to discuss a stress-concentration factor Σ defined, broadly, as the peak stress at the junction divided by the hoop stress in (say) the cylindrical portion of the vessel, due to the application of internal pressure. Before we can evaluate Σ we must face various questions. First, since stress is certainly not a scalar quantity, we must choose a way of allocating a single number to the state of stress in a given element of material. Second, we have conducted our analysis almost entirely in terms of the variation of M_s and N_θ with s, and we shall need to consider variations of stress through the thickness of the shell.

We shall suppose that the vessel is made from a ductile metal or alloy.

Fig. 11.18. 'Rectangular' and 'triangular' approximations to the actual distribution of N_θ^* on account of the loading of fig. 11.17(c).

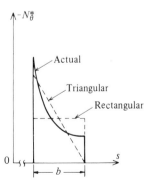

Consequently the most rational way of comparing different states of stress will be to use a 'yield criterion' for the material. For the sake of simplicity we shall adopt Tresca's yield condition (e.g. Prager, 1959), which states that under principal stresses $\sigma_1, \sigma_2, \sigma_3$ yield will occur when

$$\max\{|\sigma_2 - \sigma_3|, |\sigma_3 - \sigma_1|, |\sigma_1 - \sigma_2|\} = Y = \text{constant}. \tag{11.48}$$

Y is the yield stress of the material in simple tension or compression, i.e. with principal stresses $(Y, 0, 0)$. In the present problem the meridional and hoop directions through any point are the local principal axes. Furthermore the third principal stress, in the through-thickness direction, is typically two orders of magnitude smaller than the components of stress in the s- and θ-directions. Consequently we have essentially a two-dimensional stress system at a given location on the meridian and at a given position within the thickness.

In a two-dimensional σ_θ, σ_s space, Tresca's yield condition is represented by a hexagon, as shown in fig. 11.19a, whose leading dimension is Y. At a given point on the meridian we must now find at which position through the thickness the stress-point σ_θ, σ_s comes nearest to this boundary.

Now in general, as we have seen, an element of the shell is subjected to stress resultants $N_\theta, N_s, M_\theta, M_s$. According to the assumptions of classical 'thin-shell' theory (see section 2.3.2) the components of stress are given as a function of through-thickness position ζ by the relations

$$\sigma_\theta = (N_\theta/t) + (6M_\theta\zeta/t^2), \tag{11.49a}$$
$$\sigma_s = (N_s/t) + (6M_s\zeta/t^2). \tag{11.49b}$$

Fig. 11.19. (a) Yield condition in principal stress space σ_s, σ_θ: the third principal stress is zero. (b) Stress trajectory at the junction in fig. 11.15 at the pressure for which the yield condition is first reached. Line AB corresponds to the state of stress at different positions through the thickness: A, B correspond to the inner and outer surfaces, respectively.

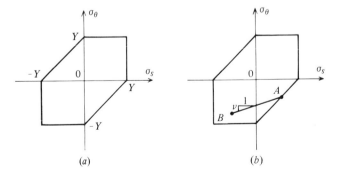

11.4 Some pressure-vessel junction problems

Here ζ is a 'through-thickness' coordinate, whose value varies linearly with distance from +1 at the outer surface to -1 at the inner surface of the shell. In consequence of Geckeler's simplification (see section 11.2.5) we find that $M_\theta \approx \nu M_s$.

Consider first the junction as illustrated in fig. 11.15. Clearly the 'worst' location on the meridian is at the junction itself. Let us take the stresses due to the two loading cases separately. Due to the loading of fig. 11.15c we have at the junction large negative values of both N_θ and M_s, with $N_s \approx 0$. The corresponding σ_θ, σ_s trajectory is thus as shown in fig. 11.19b: as ζ varies, the point traces out a line of slope ν (since $M_\theta = \nu M_s$); and since $N_s \approx 0$, the mid-point of this line is on the σ_θ-axis. It is clear that for $\nu < 0.5$ the 'worst' location is at $\zeta = -1$, i.e. on the inner surface: here the hoop compressive stress is compounded with a tensile stress in the meridional direction. In fig. 11.19b the hexagon has been drawn to pass through this point. Next we must add the effect of the loading shown in fig. 11.15b. In general we can say that at the junction N_s and N_θ are both tensile, and are smaller in magnitude in comparison with the hoop stress considered already. Moreover, at the junction their values will be roughly equal. It follows that the effect of this stress system is to move the 'worst' point a relatively small distance in approximately the same direction as the corresponding edge of the hexagon. In other words, as far as our chosen criterion of stress is concerned, we need not consider the second loading at all. (Strictly we must check that the second stress system does not move the stress point into the next quadrant of the σ_s, σ_θ space.) We therefore conclude that the required value of the equivalent tensile stress σ is given by

$$\sigma = |N_\theta^{\max}/t| + (1 - \nu)|6M_s^{\max}/t^2|, \qquad (11.50)$$

as a first approximation, where $N_\theta^{\max}$, $M_s^{\max}$ correspond to the loading of fig. 11.15c.

Our required stress-concentration factor is now defined as the ratio of σ to the hoop stress in the cylindrical portion of the vessel remote from the junction; i.e.

$$\Sigma = \sigma t/pa. \qquad (11.51)$$

For the junction of fig. 11.15 we find, from (11.45) and (11.46) that

$$\Sigma = \frac{(A^2 - a^2)^{\frac{1}{2}}}{h^{\frac{1}{2}}(A^{\frac{1}{2}} + a^{\frac{1}{2}})} \frac{3^{\frac{1}{4}}}{2} \left\{ 1 + \left(\frac{3(1 - \nu)}{1 + \nu}\right)^{\frac{1}{2}} \right\}. \qquad (11.52)$$

It is interesting to note that the contributions to this factor from N_θ and M_s respectively have a ratio which is a function only of ν. For $\nu = 0.5$ the contents of { } are $1 + 1$ exactly, while for $\nu = 0.3$ the corresponding expression is

1 + 1.27 – i.e. the bending contribution is about 27% larger than the stretching contribution.

Consider next a vessel having a toroidal knuckle at the junction, as shown in fig. 11.17. The preceding analysis of stress-concentration factors holds true provided we take account of the moderating effect of the knuckle on the peak values of M_s and N_θ. Indeed, equation (11.52) still holds provided the expression { } is replaced by

$$\left\{ k_n + \left(\frac{3(1-\nu)}{1+\nu} \right)^{\frac{1}{2}} k_m \right\}, \tag{11.53}$$

where k_n and k_m are the *moderating functions* plotted in fig. 3.6d.

It is straightforward to evaluate Σ for a given case by substituting in the values of A, a, t etc. But since (11.52) is in any case an approximate formula it is reasonable to simplify it by using a set of approximations which are set out in detail by Calladine (1972b). The final result is

$$\Sigma = \frac{3^{\frac{1}{2}}}{2} \left(\frac{A-a}{t} \right)^{\frac{1}{2}} \Big/ \left(1 + \frac{2^{\frac{1}{2}}}{3} \frac{r}{(at)^{\frac{1}{2}}} \right), \tag{11.54}$$

where a and A are the radii of the cylinder and sphere respectively and r is the radius of the knuckle. The formula is satisfactory provided

$r/(at)^{\frac{1}{2}} < 3$, approximately.

When $r = 0$ we recover very nearly the previous expression corresponding to a sharp junction. It is evident that the provision of a toroidal knuckle of suitably chosen dimensions can reduce the value of Σ appreciably. Nevertheless, it must be recognised that in general the value of Σ is of the order of $(a/t)^{\frac{1}{2}}$, and may prove on occasion to be unacceptably high.

There are several obvious remedies for this situation. First we may make the value of A/a smaller, and thereby reduce the value of the ring-load of fig. 11.15c. This will of course increase the axial height of the closure. According to (11.54), $\Sigma = 0$ when $A = a$, i.e. for a hemispherical closure, which is not quite correct: in concentrating on the loading case of fig. 11.15c we have ignored the 'weak' discontinuity corresponding to a hemispherical closure.

A second possible way of reducing the value of Σ is to increase the thickness of the vessel, or of the end-closure. Not much error is introduced if (11.54) is used for a closure in which the thickness of the spherical and toroidal portions exceeds that of the cylindrical portion, provided the greater thickness is used in the formula.

Some comments on the significance of elastic stress-concentration factors in the design of pressure-vessels will be made in section 11.6.

11.4 Some pressure-vessel junction problems

11.4.4 Cylindrical branches in spherical pressure-vessels

The methods of the present chapter may be used to study the behaviour of cylindrical branches in spherical pressure-vessels under conditions of symmetric loading. A typical case is illustrated in fig. 11.20, where a spherical vessel of radius A is connected to a cylindrical branch of radius a. We shall consider first the case of a pressure-loading. For the present we shall take the thickness to be uniform throughout. It is useful to begin by listing the obvious similarities to, and differences from, the problem depicted in fig. 11.15. In both cases we have a junction between cylindrical and spherical shells. The statical analysis according to the membrane hypothesis gives a ring-loading at the junction in each case, equal in magnitude to the product of pressure and the appropriate shaded area. But whereas in fig. 11.15 there is a ring *compression* (corresponding to the inward-directed loading of fig. 11.15c), in the present problem there is a ring-*tension*, corresponding to the outward-directed loading of fig. 11.20c. If the values of a, A and p are common to the two cases, the response of the shell – found through the response of the 'equivalent' cylindrical shell – is precisely the same in the two cases. And indeed, when this is recognised, we find that formula (11.52) is equally valid, subject to some small restrictions which we shall consider later.

Now for some differences. In the first place, the spherical shell is the main vessel in the present problem; whereas in the previous problem the cylindrical shell was the main vessel. Therefore in defining a stress-concentration factor we should normalise the stress for the present problem with respect to the stress in the body of the spherical shell. Secondly, while it was clear in the previous problem that conditions were suitable for application of the Geckeler simplification, there may well be cases where the radius a of the branch is so

Fig. 11.20. Analysis of the stresses at the junction between a spherical vessel and a cylindrical branch, according to the scheme of fig. 11.15.

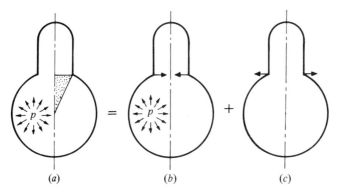

small in comparison with $(At)^{\frac{1}{2}}$ that this simplification is not valid. Thus we must be prepared for a somewhat more awkward piece of analysis. Thirdly, the insertion of a toroidal knuckle in the previous problem provided a suitable means of moderating a high stress-concentration factor; but in the design of branches for spherical vessels it is common practice to insert a thickened reinforcing ring or 'pad'. This raises in an acute form a question, which we have mentioned previously (cf. section 1.2.2), concerning the use of the simple 'shell theory' expressions (11.49) in the calculation of local stresses near a junction which involves an abrupt change of slope of the meridian. Not only does our previous expression for Σ become of somewhat dubious validity, but also an adaptation to the case of a reinforcing ring becomes questionable. We shall return to this point in chapter 18.

Bearing in mind these qualifications, let us apply equation (11.54) to the present problem. Putting $r = 0$ we find that if the value of a/A is steadily reduced, the value of Σ approaches a constant value of $(3^{\frac{1}{2}}/2)(A/t)^{\frac{1}{2}}$. But the normalising stress used in (11.54) was the hoop stress in the cylindrical vessel, namely pa/t. When we normalise instead with respect to the hoop stress in the spherical vessel (i.e. $pA/2t$) we have a new formula for small values of a/A:

$$\Sigma = 3^{\frac{1}{2}}a/(At)^{\frac{1}{2}}. \tag{11.55}$$

According to this relation, the stress-concentration factor is directly proportional to the radius of the branch, and depends only on the dimensionless group $a/(At)^{\frac{1}{2}}$. This should not be surprising, for the effect is due to a ring-tension (according to the membrane hypothesis) of value $\frac{1}{2}pa(A^2 - a^2)^{\frac{1}{2}}$ – which approximates $\frac{1}{2}paA$ for small values of a/A – and on account of bending effects this is carried over a meridional length of shell of order $(At)^{\frac{1}{2}}$. Equation (11.55) is important as a first-approximation relation, but it conceals an awkward paradox, since it is based on the Geckeler simplification which is valid only for sufficiently large values of, essentially, the dimensionless group $a/(At)^{\frac{1}{2}}$. Consequently we must investigate separately what happens when the Geckeler simplification is invalid. This constitutes our next task.

11.4.5 The analysis of branches in shallow shells

Consider the analysis of the loadings shown in fig. 11.21. Fig. 11.21a shows the actual pressure which acts on the junction; fig. 11.21b shows this loading together with an inward-directed loading H at the junction, which removes the ring-tension corresponding to the analysis according to the membrane hypothesis; and fig. 11.21c shows the reapplication of an outward-directed junction load so that figs. 11.21b and c together constitute the loading

11.4 Some pressure-vessel junction problems

shown in fig. 11.21a. Fig. 11.21c is somewhat more complicated than the comparable figs. 11.15c and 11.20c. The reason for this is that we cannot necessarily regard the case shown in fig. 11.21b as producing only *minor* discontinuity stresses, since that conclusion depended crucially upon the Geckeler simplification. Consequently figs. 11.21b and c have been drawn to show separately the constituent shells, together with their edge-loadings. There is an unknown meridional bending moment M at the junction, and the total outward loading is to be shared between the two shells in such a way that on account of their elastic distortion they have a common rotation and a common component of displacement normal to the axis. The common displacement is assured if there is a common value of hoop strain ϵ_θ: note particularly that this depends not only on the edge-loadings (fig. 11.21c) but also on the loadings of the case shown in fig. 11.21b. This problem constitutes an example of the 'traditional' method of tackling the structural analysis of junctions in shells.

Of the various sub-problems into which this problem degenerates, the only one which is at all awkward is the response of the spherical shell to the edge-loadings which are shown in fig. 11.21d. In general, we may relate the required hoop strain ϵ_θ and rotation χ to the imposed shear force U and moment M by means of the matrix equation

$$\begin{bmatrix} -\epsilon_\theta \\ -\chi \end{bmatrix} = \begin{bmatrix} \bar{A} & \bar{B} \\ \bar{B} & \bar{C} \end{bmatrix} \begin{bmatrix} U \\ M \end{bmatrix}, \tag{11.56}$$

whose coefficients $\bar{A}$, $\bar{B}$ and $\bar{C}$, which are functions of the geometry of the shell (and the elastic moduli), are to be determined: see fig. 11.22 for the sign conventions used.

Fig. 11.21. (a)–(c) Analysis of the problem of fig. 11.20 by the 'traditional' scheme in which the vessel is cut into its component parts. (d) A special case in which the branch has zero thickness.

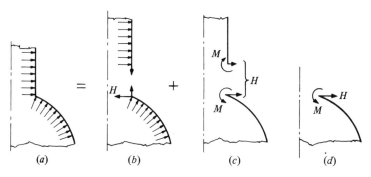

The determination of the coefficients is made much easier if we work within the context of *shallow-shell theory*. For the present problem this is tantamount to using the approximations

$$\sin\phi \approx \phi, \quad \cos\phi \approx 1$$

in the various equations. In particular the distinction between radii r and r_3 in figs 4.10 and 6.9 disappears. It turns out that the various coefficients may be determined analytically (see Calladine & Paskaran, 1974) and the results may be presented as follows:

$$\bar{A} = \frac{\alpha}{rEt}; \quad \bar{B} = \frac{2 \times 3^{\frac{1}{2}}\beta}{Eth}; \quad \bar{C} = \frac{4 \times 3^{\frac{3}{4}} A^{\frac{1}{2}} \gamma}{Eth^{\frac{3}{2}}}. \tag{11.57}$$

Here α, β, γ are dimensionless coefficients, which are plotted against the dimensionless parameter $r/(Ah)^{\frac{1}{2}}$ in fig. 11.23. Curves are given for two values of Poisson's ratio. Here r is the radius of the hole.

Before we proceed to use these coefficients it is pertinent to enquire in what conditions the shallow-shell equations are valid. We have already seen that, according to Geckeler's simplification, a spherical shell may be represented by a cylindrical one of the same radius. It is therefore of interest to investigate the form of the matrix equation corresponding to (11.56) for a *cylindrical* shell of radius A. This is given in (3.33); and we find that it agrees precisely with (11.56) provided we put

$$\alpha = 2 \times 3^{\frac{1}{4}} r/(Ah)^{\frac{1}{2}}, \quad \beta = \gamma = 1. \tag{11.58}$$

These coefficients have also been plotted in fig. 11.23. They agree with the

Fig. 11.22. Variables for use in the 'shallow-shell' analysis of a spherical shell in the vicinity of a hole.

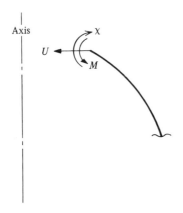

11.4 Some pressure-vessel junction problems

correct coefficients to within about 10% for $r/(Ah)^{\frac{1}{2}} > 3$, say. This is very fortunate, for it indicates that the coefficients derived from 'shallow-shell' calculations merge directly with those from 'deep-shell' calculations (via Geckeler's simplification) without the need for an intermediate 'bridging' calculation. (The transition is even more smooth if a special transformation of variables, due to Hetényi, is used. This gives an improved version of Geckeler's coefficients in the range $1 < r/(Ah)^{\frac{1}{2}} < 3$: see Calladine & Paskaran (1974).)

Consider the coefficients α, β, γ when $r/(Ah)^{\frac{1}{2}} \to 0$. Physically, this corresponds to the shell being so shallow that it becomes a flat plate for practical purposes. It is easy to show by the theory of plane stress that in this case $\alpha = 1 + \nu$ and $\beta = 0$. It is perhaps surprising that the value of coefficient γ should also be zero, since we expect the application of M to the edge of a hole in a flat sheet to produce some rotation. However, analysis by standard

Fig. 11.23. Coefficients α, β, γ in (11.57), as functions of $r/(Ah)^{\frac{1}{2}}$, for $\nu = 0$ and 0.5. The lines labelled G correspond to Geckeler's simplification of the equations (see (11.58)).

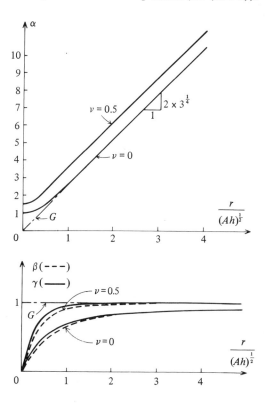

methods (e.g. Timoshenko & Woinowsky-Krieger, 1959 §17) shows that the application of a bending moment M_r to the edge of a hole of radius b in an infinite flat plate causes a rotation there of $Mb/(1-\nu)D$. This leads directly to $\gamma = (3^{\frac{1}{4}}/(1-\nu))(r/(Ah)^{\frac{1}{2}})$, which correctly gives the slope at the origin of the curve for γ in fig. 11.23.

The functions α, β, γ all diverge from their asymptotes in the range $r/(Ah)^{\frac{1}{2}} < 2$, say. There is a simple physical interpretation of this. For the cylindrical shell and the deep spherical shell there is a characteristic length of order $(Ah)^{\frac{1}{2}}$, as we have remarked on several occasions. For a flat sheet, on the other hand, the only possible characteristic length – apart from the thickness – is the radius of the hole. It is precisely the change-over from one characteristic length to another as the value of $r/(Ah)^{\frac{1}{2}}$ steadily increases which gives their form to the curves of fig. 11.23.

Returning to the problem of fig. 11.21, we are now in a position to solve the relevant simultaneous equations by reading off the values of α, β, γ from the curves of fig. 11.23 for the spherical shell, and using the corresponding analytical expressions for the cylindrical shell.

Let us investigate the simplest possible case. It is rather an artificial one which occurs when the thickness of the cylindrical branch tends to zero. This would correspond physically either to a branch with a very thin wall, or alternatively to a hole closed by a simple 'cover-plate' with frictionless edges. The simplicity comes because the H- or U-loading of fig. 11.21c is transferred directly to the spherical shell, and also $M = 0$; and moreover there is no matching of displacements to be done. From (11.12) we may evaluate ϵ_θ and χ at the edge of the hole, and use these to evaluate the peak stress. This occurs at the edge of the hole, on the outer surface: in shallow shells we must not ignore the change of curvature κ_θ (see (11.10)) as we do for deep shells. At this location the stress is essentially uniaxial in the circumferential direction, and we find (see problem 11.7) that

$$\Sigma = 1 - \nu + \alpha + [3(1-\nu^2)]^{\frac{1}{2}}\beta. \qquad (11.59)$$

This has been plotted in fig. 11.24.

It is straightforward to do more cumbersome calculations corresponding to different thicknesses of the cylindrical branch. The results shown in fig. 11.24 are taken from Leckie & Penny (1963) (cf. Leckie & Payne, 1966). These authors discovered that the peak stress (defined as in section 11.4.3) is almost always located in the spherical shell. Equation (11.59) is also plotted in fig. 11.24: given the sweeping assumptions on which it is based, it gives a satisfactory formula for sufficiently small values of $r/(Ah)^{\frac{1}{2}}$.

The curves of fig. 11.24 suggest that as the value of $r/(Ah)^{\frac{1}{2}}$ steadily rises,

11.5 Present scheme and the two-surface approach

so also does the value of Σ. This is certainly true in the range of geometry for which the shell is *shallow* in the vicinity of the branch. For wide-angle junctions it is important to realise that the line-loading at the junction (figs. 11.15c and 11.20c) is directly proportional to the areas shaded in figs. 11.15a and 11.20a; and in particular that it reaches a *maximum* when the opening subtends $\phi = 45°$, and is zero for openings with $\phi = 90°$ = i.e., $a = A$.

The methods of the present section can also be used to study the behaviour of the intersection under an axial tension or thrust: see problem 11.8.

11.5 Reconciliation of present scheme with the 'two-surface' approach

Let us now examine the relationship between the two different schemes, represented in figs. 8.1 and 11.3, respectively, which we have used in order to separate the 'bending' and 'stretching' effects within the shell. The key to the analysis of sections 8.2-8.4 is the notion that in some circumstances the tangential force-interaction between the S- and B-surfaces may be negligible; and this leads directly to the well-known 'shallow-shell' version of the governing equations. In the present chapter we made no comparable simplifications in setting up (11.5) - (11.7), whereas we did make subsequent

Fig. 11.24. Stress-concentration factors for the problem of fig. 11.20, showing results of present calculations for a zero-thickness junction, together with results of numerical calculations from Leckie & Penny (1963).

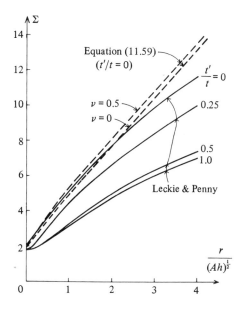

simplifications, after Geckeler, to the governing equations when we came to solve them. It would be interesting to know whether or not these two different sets of simplifications amount to roughly the same thing.

The most convenient way to do this is to set up the equations of the present problem in terms of the 'two-surface' idea, and then to examine the ratio of the tangential and normal intersurface forces.

Consider the equations of equilibrium for a symmetrically-loaded shell of revolution having a given arbitrary meridian, which is subjected to surface tractions of intensity p and q per unit area in directions normal to the surface and tangential to the meridian, respectively, as shown in fig. 11.3a. Also let subscripts S and B refer the portions of these fractions which are carried by the S- and B-surfaces, respectively; thus

$$p_S + p_B = p, \quad q_S + q_B = q. \tag{11.60}$$

Now according to the *membrane hypothesis* the loading p, q is carried by the stress resultants N_θ^*, N_s^*; and the corresponding equations of equilibrium for a shell having an arbitrary meridian are, from section 4.5.1;

$$\frac{N_s^*}{r_1} + \frac{N_\theta^*}{r_2} = p, \tag{11.61a}$$

$$-\frac{1}{r}\frac{d}{ds}(r N_s^*) + \frac{N_\theta^*}{r_3} = q. \tag{11.61b}$$

On the other hand, in the context of the two-surface analysis, the *actual* stress resultants N_θ, N_s are in membrane-hypothesis equilibrium with those portions of p and q which are carried by the S-surface. Thus we may also write

$$\frac{N_s}{r_1} + \frac{N_\theta}{r_2} = p_S, \tag{11.62a}$$

$$-\frac{1}{r}\frac{d}{ds}(r N_s) + \frac{N_\theta}{r_3} = q_S. \tag{11.62b}$$

Subtracting these equations in pairs, and expressing $(N_\theta - N_\theta^*)$ and $(N_s - N_s^*)$ in terms of U by means of (11.5) and (11.6), we obtain the following equations for p_B and q_B:

$$p_B = -\frac{1}{r_2}\frac{dU}{ds} - \frac{U}{r_1 r_3}, \tag{11.63a}$$

$$q_B = \frac{1}{r}\frac{d}{ds}\left(\frac{rU}{r_3}\right) - \frac{1}{r_3}\frac{dU}{ds}. \tag{11.63b}$$

The right-hand side of these expressions may be simplified by means of various identities (see section 6.9), and we thus obtain

$$p_B = -\frac{1}{r}\frac{d}{ds}\left(\frac{rU}{r_2}\right), \tag{11.64a}$$

$$q_B = -\frac{U}{r_1 r_2}. \tag{11.64b}$$

Hence, finally, we obtain a general expression for the ratio

$$\frac{p_B}{q_B} = \frac{r_1 r_2}{rU}\frac{d}{ds}\left(\frac{rU}{r_2}\right). \tag{11.65}$$

In a given problem, i.e. for a shell of given geometry under prescribed loading, it is possible to solve the governing equations for $U(s)$ and hence to determine the functions $p_B(s)$ and $q_B(s)$. Then, if the magnitude of q_B is everywhere small in comparison with that of p_B, we may conclude that the problem could have been analysed satisfactorily in principle in terms of shallow-shell theory. It would, of course, be much more convenient if we could decide on the circumstances in which shallow-shell theory is valid without first having to solve the full equations.

Consider some simple special cases. First, in a *cylindrical* shell $r_1 \to \infty$, so by (11.64b), $q_B = 0$; and thus the 'shallow-shell' equations hold. (Note that we are discussing here only the *axisymmetric* behaviour of cylindrical shells.) Second, in a *spherical* shell $r_1 = r_2 = a$, say, and we find that

$$\frac{p_B}{q_B} = \frac{a}{rU}\frac{d}{ds}(rU). \tag{11.66}$$

Thus, provided rU is a function which varies sufficiently rapidly with s, the shallow-shell scheme will be satisfactory. We know, of course, that the characteristic meridional length for the variation of U is of the order $(r_2 t)^{\frac{1}{2}}$.

Next consider the torispherical pressure-vessel. This provides an example in which we may expect the shallow-shell scheme *not* to be satisfactory; for in the region of a small-radius knuckle the value of r_1 may well be of the same order as that of $(r_2 t)^{\frac{1}{2}}$, and thus the ratio p_B/q_B will not necessarily be large. The effect is even more pronounced in the case of a vessel having an abrupt change of inclination of its meridian at a certain point. In terms of the S- and B-surface description we see clearly that in this case the concentrated junction loading (fig. 11.15c) is transferred directly to the B-surface, and since there is a substantial component of force in the direction tangential to the spherical cap, we cannot expect that q_B will be negligible.

On the basis of these cases we may conclude, tentatively, that the shallow-shell scheme is valid in general for symmetrically-loaded shells of revolution *except* in regions of high meridional curvature and in the vicinity of abrupt changes of slope in the meridian.

For present purposes it is adequate to note that our method of analysing symmetrically-loaded general shells of revolution by means of the shear variable U is not inconsistent with the shallow-shell approach except in certain well-defined situations. We shall use an explicitly shallow-shell approach to some axisymmetric problems in the context of plastic theory in section 18.5.1.

11.6 General discussion

Throughout this chapter we have made a variety of simplifying assumptions in order to expedite the analysis and keep the calculations simple. Thus we have considered shells of constant thickness, and have concentrated mainly on cases in which the Geckeler simplification is valid. In cases where our simplifying assumptions are not justified it is, of course, necessary to undertake detailed numerical studies of a kind which we have not considered. In an important sense, however, these are merely *computational* complications which alter the details without affecting the main features of behaviour.

For the sake of simplicity we have also considered only shells whose meridians can be described by a simple curve with no branches. The analysis of pressure-vessels supported on 'skirts', for example, requires more elaborate treatment by what we have described as the 'traditional' methods; nevertheless the general ideas of this chapter are still of some use in a qualitative sense. Another kind of complication which we have not considered occurs when there is a condition of statical indeterminacy in the axial direction; for example a vessel connected to a rigid foundation by means of *two* skirts, as in fig. 11.2c.

There is also another class of simplifications which we have made, which affects not so much the computational procedures but rather the basic conception of the analysis. Thus, throughout the chapter we have considered the original state of the structure as 'stress-free', and consequently the mechanical properties of the material have entered in the specially simple form of (11.12). But there are many important practical problems in which a 'stress-free' initial state is not appropriate; and it is therefore necessary at this point at least to make some comments on various possible practical situations which an engineer may encounter. Later on we shall mention some other limitations of the work of the present chapter.

Practical situations in which it is not reasonable to regard the initial state of the structure as stress-free may be listed under the headings of thermal stress, initial stress and prestress, respectively.

In *thermal stress* problems the engineer is concerned with questions about the distribution of stress on account of temperature differences through a

11.6 General discussion

structure. Thus at a localised hot-spot in a shell we must expect some compressive stress on account of the fact that thermal expansion is restrained by the presence of the surrounding material. Another example is a vertical cylindrical shell whose lower part contains a hot liquid while the upper part is relatively cool. In problems such as these the stresses are due not to externally applied forces but to differential thermal expansion. The equations of section 11.2.4, etc. are inadequate for problems of this sort. If we take as a base state a stress-free structure in isothermal conditions, then Hooke's law must be modified to take account of the fact that changes of strain occur not only as a result of changes of stress but also on account of thermal expansion. The necessary changes to the equations of uniform cylindrical shells under axisymmetric and doubly-periodic temperature variation have been examined in problems 3.23 and 8.9 respectively. Some kinds of thermal stress problem may be analysed by simple intuitive methods: see problem 11.9.

Many structures are in a state of *initial stress* as a consequence of manufacturing processes. This is true of any statically indeterminate structure assembled from pieces which do not fit precisely ('lack of fit' problems). For example, in welded pressure-vessels we must expect high longitudinal tensile stresses along the line of welds, in consequence of the cooling, shrinking and plastic deformation of the weld-metal from its molten state. The level of these stresses is normally high, and is indeed in the region of the room-temperature yield stress of the material. Various procedures of 'stress-relief' are available in metal vessels: by holding the vessel at a uniform elevated temperature over a sufficiently long period of time, followed by a slow cooling, we can reduce the locked-in stresses to an acceptable magnitude.

Another source of initial stresses is the traditional 'overpressure' test of a vessel, in which the structure is subjected to a pressure considerably in excess of the normal working pressure. At a feature such as a junction there will always be some concentration of stress in the elastic range; and so some relatively small zones of material may enter the *plastic* range in the course of such a test, and suffer irreversible deformation. When the pressure is removed, the material in and around these zones will not return to its former state of stress. Locked-in stresses of this sort may be beneficial for the future behaviour of the vessel under cyclic loading conditions, for it is possible that all subsequent changes of loading may be met by purely elastic changes. This important process is known as *shakedown*, and there is a substantial literature on it: see, e.g. Prager (1959 §2.6), Symonds (1951), Leckie (1965), Leckie & Payne (1966).

The third form of initial stress, *prestress*, is best illustrated by the example of a reinforced concrete shell or dome which is supported at its edge by abutments which are capable of exerting only vertical reactions: see fig. 11.25. If we imagine that this structure has somehow been erected in gravity-free conditions, and that gravity is now switched on, we can see immediately that there will be a concentrated zone of high hoop stress and bending stress near the supports; and this in turn will allow some vertical displacement of the bulk of the shell, accompanied by cracking of the concrete. It would obviously be prudent to have a concentration of hoop reinforcement in this zone. But even better would be to have a *prestressed tendon* around the circumference; for this tendon, being prestressed, could be arranged to exert the concentrated hoop of tension which is required according to the membrane hypothesis, *without the need for tensile hoop strain in the concrete.* Without prestress the necessary hoop stress can only be developed at the expense of hoop strain; but with a prestressing cable the hoop tension can be provided independently. For a large shell roof of this sort, in which the self-weight ordinarily provides the major part of the loading, the provision of prestress at the edge is a satisfactory feature. But for thin-walled pressure-vessels, on the other hand, the use of prestressing cables is not nearly so satisfactory, since the main loading comes from internal pressure, which varies with time, whereas a prestressing tension does not.

Several other remarks should be made on the question of the limitations of the methods of this chapter. The analysis has been based on the supposition that the displacements are so small that it is adequate to write the equations in terms of the original undistorted configuration rather than the current, distorted configuration of the shell. In some cases this can give misleading results. For example, in the case of a torispherical closure to a cylindrical vessel it may happen that the elastic distortion of the meridian, although small, can have an appreciable effect upon the stress resultants N_θ^*, N_s^* calculated according to the membrane hypothesis. In pressure-vessels which operate at a temperature so high that there is appreciable *creep* of the material under sufficiently high stress, the profile of a pressure-vessel head may change appreciably in the course of time (Calladine, 1972*b*). A

Fig. 11.25. A large shallow dome, supported vertically at the edges.

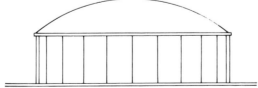

somewhat similar sequence of geometries is found if a torispherical pressure-vessel head is loaded into the plastic range by a steadily increasing interior pressure.

The *buckling* of shells also involves geometry-changes, as we shall see in chapters 14 and 15: and thus the methods of the present chapter explicitly ignore the possibility of buckling. In some circumstances the buckling of the toroidal region of a torispherical pressure-vessel head under internal pressure – on account of the compressive hoop stresses there – can be an important problem in design: see, e.g. Bushnell & Galletly (1977).

Lastly we should mention that in this chapter (and throughout the book) we have assumed that the shell may be modelled by, essentially, a *surface* endowed with certain mechanical properties. It is clear therefore that such important points as the concentration of stress in local regions around weld-fillets, etc., are beyond the scope of this work.

11.7 Conclusion

Bearing in mind the various limitations listed above, we may summarise the general conclusions of the present chapter as follows. In any given problem of the symmetric loading of an initially stress-free shell of revolution of uniform thickness, the first step is to compute the stress resultants N_θ^* and N_s^* according to the membrane hypothesis. Provided these change only gradually with meridional distance, they give a good approximation to the actual state of stress in the shell. But if there are any discontinuities or rapid changes in these variables, we must expect to find bending- and normal shear-stress resultants in the vicinity. By far the most severe effect occurs when, according to the membrane hypothesis, there is a concentrated hoop of tension (or compression) at a particular point on the meridian. In this case, unless special reinforcing rings, etc. are provided, high bending stresses will be developed.

11.8 Problems

11.1 Verify that when $P(s)$ as defined in (11.16) is zero, the elastic strains due to stress resultants N_s^*, N_θ^* in a shell of uniform thickness make the contents of the square brackets in (6.54) zero; and hence, by (6.53), $g = 0$.

Verify that $P(s) = 0$ for the following distribution of stress resultants (see problem 4.14):

$$N_s^* = -N_\theta^* = C/\sin^2\phi; \quad C = \text{constant}.$$

11.2 Stress resultants in a spherical dome due to a uniform snow-load were investigated in problem 4.17, according to the membrane hypothesis; and in problem 4.18 the corresponding calculation was done for a somewhat similar surface consisting of a paraboloid of revolution.

Use the following version of (11.29) to estimate the change of stress resultant N_θ^* when the meridional profile of a spherical shell under this kind of loading is changed by a small amount into a parabolic shape:

$$\Delta N_\theta^* \approx -r_2 N_s^* \Delta(1/r_1).$$

Hence obtain an approximation to N_θ^* in the paraboloidal shell, and compare it with the result of problem 4.18.

Work in terms of the variable r/a. Note that for the original spherical shell r_2 and N_s^* are both constant, and prove the following expression by rearrangement of various results in the two problems:

$$\Delta(1/r_1) = (1/a) \{1 - [(1 - (r/a)^2)^{\frac{3}{2}}/0.854]\}.$$

11.3 Verify expression (11.36) in the following way. Let N_θ^p represent the circumferential stress resultant in the original, 'perfect' shell, so that $\Delta N_\theta = N_\theta - N_\theta^p$ while $\Delta N_\theta^* = N_\theta^* - N_\theta^p$. Hence obtain

$$\Delta N_\theta = \Delta N_\theta^* + (N_\theta - N_\theta^*).$$

Use equilibrium equations (11.6), (11.7) (putting $r_3 \to \infty$ and $r_2 =$ constant) to write $(N_\theta - N_\theta^*)$ as $r_2 d^2 M_s/ds^2$; then use (11.34) (putting $N_s = N_s^* =$ constant) and (11.29) to obtain the required result.

11.4 Obtain fig. 11.13 from the results of section 3.4 in the following way. Observe that the introduction of the imperfection in the meridian of the cylindrical shell imparts a uniform additional normal pressure p^* (by 11.30) to the portion of the shell beyond the change-point. Consider the shell as two separate shells whose discontinuity of radial displacement is countered, just as in problem 3.22, by the imposition of self-balancing radial shear alone. Hence determine the distribution of M_s, and evaluate e_s by means of (11.34). In particular, show that the maximum $|e_s|$ is equal to $0.046 \, r_2 h \Delta(1/r_1)$.

11.5 A semi-infinite cylindrical shell of radius a has the following axisymmetric displacement imposed at an edge: $w = a\epsilon_\theta$, $\chi = 0$. Use (3.33) to express the necessary dimensionless stress resultants m, q in terms of the edge strain ϵ_θ. Hence show that the stress resultants required at the edge are

$$M_x = Eth\epsilon_\theta/2 \times 3^{\frac{1}{2}}, \quad Q_x = (-) Eth^{\frac{1}{2}} \epsilon_\theta / 3^{\frac{1}{4}} a^{\frac{1}{2}};$$

and in particular that M_x is independent of the radius a. Hence verify the solution of the problem in section 11.4.1 concerning the junction between two shells having the same thickness but different values of r_2.

11.8 Problems

11.6 By specialisation of results from section 4.5.2 show that the differential 'swept area' for a toroidal knuckle in which $r_1 \ll r_2$ is given by

$$dA \approx \tfrac{1}{2} r_2^2 \, d\phi;$$

and that the total swept area is given by

$$A = \tfrac{1}{2} a^2 \tan\beta,$$

where a is the radius of the cylindrical shell and β is the total angle through which the tangent to the meridian turns at the knuckle. Hence show that a *mean* value, R, of r_2 may be defined by

$$R^2 = a^2 \tan\beta/\beta,$$

and that the ratio of the *peak* value of N_θ^* to the *mean* value in the knuckle is equal to $\beta/(\sin\beta\cos\beta)$. Also show that the ratio of the peak value of N_θ^* to the peak value of N_θ^* for the 'triangular' distribution (fig. 11.18) is equal to $\beta/\sin2\beta$. Verify that the value of this ratio lies between approximately 0.8 and 1.2 when β lies between 45° and 60°.

11.7 A spherical pressure-vessel of radius A and thickness t is pierced by a hole of radius r_0 which is closed by a simple cover-plate. The vessel contains an interior pressure p. Show that the state of stress in the shell may be considered as the sum of two parts (as in fig. 11.21), in one of which there is (on the assumption that the shell may be treated as *shallow*) an edge-loading

$$U = \tfrac{1}{2} p A r_0.$$

The peak circumferential strain due to this loading is at the outer surface of the shell, at the edge of the hole; and it is equal to $\epsilon_\theta + \tfrac{1}{2} t \kappa_\theta$. By using (11.56) and (11.57) evaluate this for the U-loading; and hence, by adding a contribution from the other loading case show that

$$\epsilon_\theta)_{\max} = (pA/2Et) \{\alpha + [3(1 - \nu^2)]^{\tfrac{1}{2}} \beta + 1 - \nu\}.$$

Thus, by noting that the state of circumferential stress is uniaxial, and normalising with respect to the basic 'membrane' stress $\sigma = pA/2t$, establish (11.59).

11.8 The vessel of problem 11.7 is now subjected to a line load of intensity F at the edge of the hole, directed outwards from the vessel and parallel to the axis of rotational symmetry.

Show that if this loading is combined with an inward-directed edge-load $U = FA$, the stress resultants in the shell as calculated according to the membrane hypothesis satisfy $P = 0$ in (11.16), and are thus a solution of the full equations (cf. problem 11.1). Hence, by adding an outward-directed edge-load $U = FA$ and using (11.56), show that the peak circumferential stress σ at the edge of the hole is given by expression (11.59) but with 1 on the right-hand side replaced by −1; where for this problem Σ is defined by

$$\Sigma = \sigma r_0 t / FA.$$

11.9 At a cross-section remote from the ends of a long cylindrical pipe, a plane annular collar is firmly attached. When the temperature of the (unpressurised) fluid within the pipe suddenly changes from one steady value to another, the temperature of the pipe rapidly adopts the new temperature of the fluid. In contrast, the thermal response of the collar is relatively slow; and for practical purposes the collar may be regarded as showing no response until long after the pipe has reached its new steady temperature.

By use of the methods of chapter 3 (cf. also problem 11.5) determine the stress resultants N_θ and M_x in the pipe adjacent to the collar soon after a temperature rise ΔT. The pipe has radius a and thickness t. The material of the pipe has Young's modulus E, Poisson's ratio ν and coefficient of linear thermal expansion α.

12

Flexibility of axisymmetric bellows under axial loading

12.1 Introduction

This chapter is concerned with the analysis of simple bellows under axial loading. A bellows is a thin-walled shell of revolution with a corrugated meridian of the sort shown in fig. 12.1; and the purpose of the design is to achieve a high flexibility in the axial direction. Units of this sort have widespread application in many kinds of industrial plant. They are sometimes equipped with external constraints which limit the elongation and shearing distortion of the thin corrugated shell, and are often used to provide a simple hinge of low stiffness in a pipe: the high axial flexibility allows the bellows to bend easily.

In this chapter we shall be concerned primarily with the calculation of the overall flexibility of a bellows under axial loading. It is convenient to define a

Fig. 12.1. Meridional profiles for a bellows, consisting of circular arcs but subtending different angles α. In each case the mean radius of the bellows is a: the axis of rotational symmetry lies to the left of each diagram.

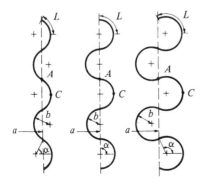

dimensionless *flexibility factor* F as the overall axial flexibility of the unit divided by the flexibility of a straight 'comparison tube' whose radius and length are equal to the mean radius and overall (unextended) length of the bellows, and whose thickness is the same as that of the bellows. It is possible to achieve, by suitable choice of dimensions, values of F in the region of several hundreds. One of our aims will be to present results in such a way that the influence of the choice of the geometry of the bellows upon the value of F may be seen readily.

We shall also investigate, in an approximate fashion, the *maximum strain* in the bellows in relation to that of the comparison tube when both are subjected to equal axial elongation. In general, of course, we hope that the maximum strain in the bellows will be only a small fraction of that in the plain 'comparison tube'.

Following other workers, we shall assume throughout the analysis that the material is linear-elastic, and that the displacements are sufficiently small for the assumptions of classical theory to be valid.

For the sake of brevity we shall consider only bellows of uniform thickness, whose meridians are composed of circular arcs, as shown in fig. 12.1. Nevertheless, the method of analysis which we shall develop may easily be applied to bellows whose meridians have other geometries.

Various workers have applied the classical theory of shells to the analysis of bellows. Notable among these are Turner & Ford (1957) and Clark (1950): see also Marcal & Turner (1963). A noteworthy feature of the work of Turner & Ford is the excellent agreement which they obtained between analytical and experimental work; and this furnishes a justification of the small-displacement method of analysis for bellows having a wide range of geometries.

The equations of the problem are those which have been developed in chapter 11. Unfortunately, Geckeler's simplification is not useful when the meridian is highly convoluted; and it therefore appears that we must on this occasion commit ourselves to a thorough and laborious solution of a differential equation having coefficients whose values fluctuate widely. Happily, this is not necessary. We can develop instead a relatively simple *strain-energy method* which not only produces results which are in excellent agreement with those of other workers (see Calladine, 1974*a*) but also affords a glimpse of those factors which are most important for design. The method shares a number of simplifying assumptions with the work of others. Thus, as already mentioned, we shall regard displacements of the shell as small, even though their cumulative effect may in fact be large: the experiments of Turner & Ford (1957) show that the overall force/extension characteristics of typical bellows have a substantial linear region. Also we shall regard the radius b of the convolutions as small in comparison with the overall radius a of the bellows, (see fig. 12.1), in order to simplify some of the calculations. We shall make a somewhat similar assumption in the analysis of curved tubes in chapter 13. The chapter is based on Calladine (1974a), but with some changes of notation.

12.2 Analysis of flexibility by an energy method

The starting point of our analysis is the observation that symmetric deformation of a bellows normally incurs mainly *bending* in the meridional direction and *stretching* in the circumferential direction. It is clear that there must be some bending of the meridian if the overall length of the shell is to change; and it is not difficult to show that the associated changes of curvature in the circumferential direction are normally relatively small in magnitude (see problem 12.1). Now, any flexure of the meridian will give, in general, a component of displacement normal to the axis of the bellows; and such displacements will incur some circumferential strain in 'hoops' of the shell. We may therefore think of the bellows as a collection of meridional slices which are coupled together by a circumferential stretching constraint; and this picture leads directly to the idea of a convoluted meridional 'beam' which is restrained by an elastic foundation which derives its stiffness from the circumferential stretching in the shell.

Our method of tackling this problem will be to use a straightforward strain-energy method based on a hypothetical mode of deformation: see appendix 1. For a given mode we shall evaluate the strain energy of meridional bending and circumferential stretching; and by equating the total strain energy (i.e. the sum of these two parts) to the work done by the external axial load, we shall obtain an upper bound on the stiffness of the shell – which is the same thing as a lower bound on the flexibility.

The hypothetical modeform which we shall use is determined by two arbitrary assumptions, which we shall attempt to justify later. First we shall assume that the meridian deforms *inextensionally*; and second we shall assume that the mode of deformation of the meridian is precisely the same as that which would occur in a uniform beam having initially the curved form of the meridian, but unrestrained in the circumferential direction. This idea is suggested by the behaviour of the simpler nearly-cylindrical shell having a *gently* undulating sinusoidal meridian: see section 11.3.2. The scheme leads to relatively simple calculations, as we shall see.

It is quite possible, of course, that this hypothetical mode is unrealistic, and therefore constitutes an unsatisfactory basis for an energy calculation. We shall investigate this point later. Briefly, we shall find that the mode is valid

416   *Flexibility of bellows under axial loading*

provided a certain geometrical ratio satisfies a certain relation. Fortunately, the range in which the hypothesis is satisfactory includes a design optimum.

Our first task is to analyse the geometry of deformation of the shell in the hypothetical mode. Once this has been done, the required strain-energy calculation is straightforward.

12.2.1 Geometry of deformation

We shall assume that the bellows has many convolutions, and that it is necessary to analyse only one typical *half convolution* whose meridian is shown as AC in fig. 12.2. It will be convenient to describe the form of the meridian of the unstressed shell by means of the Cartesian coordinate system XY, with origin located at A, and also to use an arc-length coordinate S, measured from A, as shown. It is also convenient to denote displacements of points on the meridian relative to A by means of the components u, v in the X- and Y-directions, respectively. For the present problem this is a more convenient scheme of components than (say) the tangential and normal components as used in chapter 11 (cf. fig. 6.9b), since we are directly concerned here with changes in length of the bellows in the X-direction and the circumferential strain ϵ_θ, which depends on the component of displacement in the Y-direction.

It will be convenient to normalise length variables with respect to the peripheral distance $AC = L$. Thus we define dimensionless variables x, y, s as follows:

$$x = X/L, \quad y = Y/L, \quad s = S/L. \tag{12.1}$$

Fig. 12.2. Notation used for a general half convolution AC (cf. fig. 12.1). A is the origin of an X, Y Cartesian coordinate system in the plane of the meridian, and u, v are the components of (small) displacement parallel to the coordinate directions.

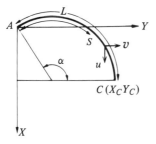

12.2 Analysis of flexibility by an energy method

Here the arc AC is circular, and it subtends angle α as shown. The initial configuration of the meridian is therefore given by

$$x = (1/\alpha)(\sin\alpha - \sin(1-s)\alpha), \tag{12.2a}$$

$$y = (1/\alpha)(\cos(1-s)\alpha - \cos\alpha). \tag{12.2b}$$

The key feature of the hypothetical mode is that the change of curvature of the meridian, κ_s, is directly proportional to $y(s)$. This follows directly from the behaviour under axial load of a uniform and unconstrained elastic beam in the form of the meridian, since the bending moment is everywhere proportional to distance from the 'line of thrust', which lies along the x-axis. The strain-energy calculation is ultimately indifferent to the amplitude of the mode, and we therefore put

$$\kappa_s = y/L \tag{12.3}$$

for the purposes of computation: the denominator L has been introduced here in order to preserve the correct dimensions.

Our next task is to determine the displacement components u, v corresponding to (12.3). We need to know v at all points on AC, as a step in the calculation of $\epsilon_\theta(s)$; but we are only interested in the value of u at the single point C, since the complete bellows contains, by assumption, an integral number of convolutions.

We could in principle use the strain-displacement relations developed in section 6.9 in order to do the necessary geometrical calculations. For the present problem, however, it is more straightforward and direct to use the principle of virtual work (see appendix 1). Let us first use this method to calculate u_C for a meridional strip which experiences a change in curvature according to (12.3), is inextensional, and has zero rotation at C for reasons of symmetry. The method of virtual work is advantageous in this case because it is very straightforward to analyse the statical equilibrium of a meridional 'wire' AC which is subjected to the ('dummy') external forces which are shown in fig. 12.3a. The bending moment is given by

$$m(s) = Ly;$$

and there are also tensile and shearing forces which we do not need to evaluate for present purposes. Taking this as a (dummy) statically admissible set of forces and stress resultants, together with a kinematically admissible set consisting of actual displacements and changes of curvature, we have, from the principle of virtual work:

$$u_C = \int_0^L \kappa_s m \, dS = L \int_0^1 y^2 \, ds = LI, \tag{12.4}$$

where

$$I = \int_0^1 y^2 \, ds. \tag{12.5}$$

Note that here the couple Y_C does not appear in the left-hand side, since the meridian at C does not rotate; and neither does the force at A, since $u = 0$ there, by definition. In this calculation we have taken the positive sense of m as giving tension on the outside of the arc, and correspondingly the positive sense of κ_s as tending to increase the original curvature.

A variant of the same method may be used to evaluate v at a typical point P. The appropriate dummy-load system for this calculation is shown in fig. 12.3b, and the corresponding statically determinate bending moment m is given by

$$m = Lx, \quad 0 \leq s \leq s_P; \tag{12.6a}$$

$$m = Lx_P, \quad s_P \leq s \leq 1. \tag{12.6b}$$

Here the subscript P denotes values of the variables at the point P. The equation of virtual work gives

$$-v_P = L \left(\int_0^{s_P} xy \, ds + x_P \int_{s_P}^1 y \, ds \right). \tag{12.7}$$

The circumferential strain at P is given by $\epsilon_\theta = v_P/(a + yL)$, where a is the mean radius of the bellows. We have already decided to regard $|yL|$ as small in comparison to a, and so we may write, approximately,

$$\epsilon_\theta = v/a \tag{12.8}$$

when we drop the subscript P.

Equations (12.3) and (12.8) enable us to calculate both the change of curvature in the meridional direction and also the circumferential strain for a bellows whose meridian has a given shape, and according to our hypothesis about the modeform of the deformation.

Fig. 12.4 displays the results which are obtained for different values of the angle α when (12.2) are substituted into (12.3) and (12.7). Note that for $\alpha \leq 90°$ the 'unwrapping' of the meridian gives negative values of v, and hence

Fig. 12.3. The meridional segment AC subjected to 'dummy' loads for the purposes of calculating (a) the axial extension of the portion AC, and (b) the component v of displacement of a typical point P.

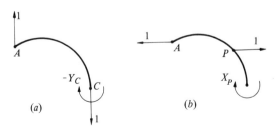

12.2 Analysis of flexibility by an energy method

of ϵ_θ, over the entire arc AC. On the other hand for $\alpha > 90°$ there is a section near A where the value of ϵ_θ is positive. This is directly attributable to the slope of the meridian at A: see problem 12.2. Note that, in contrast, the form of $\kappa_s(s)$ changes relatively little when the value of α is changed.

12.2.2 Expressions for strain energy and flexibility

Let us now evaluate the strain energy of bending, U_B, in the portion of shell formed by revolving the arc AC about the axis of the bellows. On the assumption that $\kappa_\theta^2 \ll \kappa_s^2$, we have (see section 2.3.3)

$$U_B = \pi D \int_0^L a \kappa_s^2 \, dS; \tag{12.9}$$

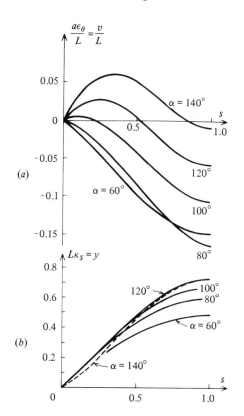

Fig. 12.4. (*a*) Hoop strain and (*b*) change of meridional curvature, plotted in dimensionless terms against contour-length *s*, for different values of the angle α.

where D is the flexural rigidity of the shell. Here, as before, we have neglected yL in comparison with a. This expression may be simplified to

$$U_B = \pi D a I/L \qquad (12.10)$$

by the use of (12.3), (12.4) and (12.5).

Similarly we find that the strain energy of stretching of the same piece of shell is given by

$$U_S = \pi E t L^3 J/(1-\nu^2)a, \qquad (12.11)$$

where

$$J = \int_0^1 (v/L)^2 \, ds. \qquad (12.12)$$

Here t is the thickness of the shell. The factor $(1-\nu^2)$ appears in the denominator of (12.11) since $\epsilon_s = 0$ in our hypothetical mode.

Note that the integrals I and J are functions only of the angle α: the integrands are the squares of the functions plotted in fig. 12.4. We shall discuss the values of I and J later.

Suppose now that the half convolution of the bellows is given an axial elongation u_C. According to our calculation the strain energy stored is equal to $U_B + U_S$; but this is in fact an upper bound on the actual energy stored, by virtue of the strain-energy theorem (see appendix 1). Next, suppose that an equal axial displacement $u_C = LI$ is applied to a plain 'comparison' tube of radius a, thickness t and length X_C. The strain energy per unit volume of this tube is equal to $\frac{1}{2}E(u_C/X_C)^2$, and so the total strain energy $\bar{U}$, is given by

$$\bar{U} = \pi a t E L^2 I^2/X_C. \qquad (12.13)$$

Now the strain energy of two different elastic structures when they are deformed by the same amount is inversely proportional to their flexibility. Thus our required flexibility factor is given by

$$F = \bar{U}/(U_B + U_S). \qquad (12.14)$$

Using (12.9), (12.11) and (12.13) we obtain directly the required expression for the flexibility factor:

$$F = 12(1-\nu^2)\frac{a}{t}\frac{I}{x_C} \bigg/ \left[\frac{at}{L^2} + 12\frac{J}{I}\frac{L^2}{at}\right]. \qquad (12.15)$$

Hence, when the geometry of the bellows has been fixed by the specification of the four defining parameters a, b, α, t (so $L = b\alpha$), it is a straightforward matter to evaluate F. Note that $x_C = \sin \alpha$. Clearly it is useful to evaluate first the parameters I, J and I/J for different values of α. Numerical integration of

12.2 Analysis of flexibility by an energy method

the data of fig. 12.4 gives the curves shown in fig. 12.5. The most remarkable feature, which may be visualised directly from fig. 12.4, is that J varies strongly with α, whereas I does not.

It is useful to investigate the way in which the factor F can be *optimised* by means of a suitable choice of the disposable geometric parameters. Suppose that we have decided upon values of a, t and α, and that we now wish to choose the value of b so as to maximise F. The only variable on the right-hand side of (12.15) is L; and we can see immediately that the denominator will be a minimum with respect to L (and hence F a maximum, say F_{max}) when

$$\frac{L^2}{at} = \left(\frac{I}{12J}\right)^{\frac{1}{2}} = \frac{L_{opt}^2}{at}. \tag{12.16}$$

This dimensionless group is shown as a function of α in fig. 12.6, together with the more useful group b_{opt}^2/at which is found directly from the relation $b = L/\alpha$. A remarkable feature is that the optimum value of b lies between $0.65(at)^{\frac{1}{2}}$ and $1.2(at)^{\frac{1}{2}}$ over a wide range of values of α. The corresponding values of F are also plotted against α by means of the parameter $F_{max}t/a$. It is clear that by changing the value of α from $60°$ to $120°$ – and using the optimum value of b for each value of α – we may increase the value of F by a factor of 20.

Fig. 12.5. Logarithmic plot of the integrals I, J and the ratio J/I against angle α. See (12.5) and (12.12).

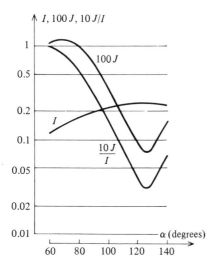

12.3 Comparison with previous work

The whole of the preceding analysis has been based on a specially simple hypothetical mode of deformation. Although we are assured by the strain-energy theorem that our calculated flexibility is a lower bound on the actual flexibility, it is useful to have an idea of the closeness of this bound to the correct value by comparing the results of our calculations with those which have been obtained previously by the use of conventional shell theory. Fig. 12.7 shows this comparison for two values of α which were considered by Turner & Ford (1957). It is clear from the diagram that our results agree very closely with those of Turner & Ford for values of b which are smaller than optimum, but that they begin to diverge for values of b which are larger than optimum. But the agreement is sufficiently close in the region of maximum flexibility for our formulae (12.15), (12.16) to be valid, at least as first approximations.

When $b \ll b_{opt}$ there is relatively little circumferential stretching; and thus it is not surprising that our analysis gives good results. When $b \approx b_{opt}$ our analysis gives good results for the flexibility factor F. This does not necessarily imply that our hypothetical mode is an accurate approximation to the correct mode in this region, since the value of F, being determined by inte-

Fig. 12.6. Logarithmic plot of maximum flexibility (in a dimensionless form) against angle α, together with the values of L and b for which the maximum flexibility is achieved. The broken curve shows the value of L for which the peak strain in the bellows is minimised for a given overall change in length.

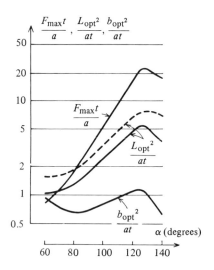

12.3 Comparison with previous work

gration, is insensitive to small contributions from other modes: see problem 12.3. Nevertheless it seems reasonable to suppose that our hypothetical mode is satisfactory as a first approximation in this region. On the other hand, when $b > 2b_{opt}$ (say), it seems clear that our hypothetical mode is not realistic. In this regime the behaviour of the shell tends towards that which is indicated by the membrane hypothesis, except in localised regions where the tangent to the meridian is almost normal to the axis of the bellows. In these zones the membrane hypothesis produces absurd results (see problem 12.4) and the shell actually carries the applied loading essentially by a bending action. This sort of behaviour has been studied by Clark (1950) for the case $\alpha = 90°$, and the corresponding stiffness factor is plotted in fig. 12.7.

In fig. 12.7 our results have been compared with those of conventional shell theory for two particular meridional geometries; but it seems reasonable to suppose that broadly similar results will be obtained for other geometries also.

Returning to our formula (12.15), we find that our analytical result may be expressed most conveniently as follows:

$$F = \frac{2F_{max}}{(b_{opt}/b)^2 + (b/b_{opt})^2}.$$

Here F_{max} and b_{opt} are the functions of α, a and t which are shown in fig. 12.6. The formula is reasonably accurate in the range

$$\frac{b}{b_{opt}} < 2.$$

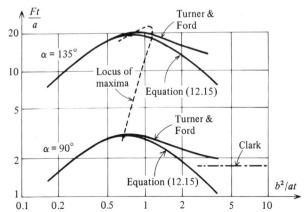

Fig. 12.7. Comparison of the present approximate calculation with results obtained numerically by Turner & Ford (1957). An asymptotic calculation by Clark (1950) is also shown.

12.4 Approximate analysis of strain in bellows

Bellows normally operate in cyclic loading conditions, and the possibility of failure by fatigue is therefore an important design consideration. So far we have discussed only the *flexibility* of the bellows in comparison with that of a comparison cylindrical tube; but now we see that a comparison of the peak strain in the bellows with that of the comparison tube will also be appropriate. This problem has been considered in detail by Marcal & Turner (1963) for specific meridional geometries. Our present approach leads to some useful approximate calculations provided we make some simplifying assumptions, as follows.

First, we assume that the appropriate measure of peak strain is the magnitude of the maximum principal strain. This is reasonable as a first approximation to a criterion of fatigue failure; and it is particularly simple to use in the present case since the critical condition will involve either circumferential stretching or meridional bending.

Second, we assume that the geometry of the bellows is such that our simple hypothetical mode is a reasonably good approximation to the actual deformation of the shell. We have argued already that an assumption of this sort should be satisfactory as a first approximation provided b is not significantly larger than b_{opt}.

Let us now define a maximum 'extreme fibre bending strain' ϵ_b, for purposes of comparison with the longitudinal strain in the 'comparison tube':

$$\epsilon_b^m = \tfrac{1}{2} t \kappa_s^m. \tag{12.17}$$

Here, and throughout this section, a superscript m denotes a maximum value.

The functions κ_s and ϵ_θ are plotted in a dimensionless form in fig. 12.4. For a given value of α we may therefore read off the corresponding maximum value of y and v/L and obtain

$$\epsilon_\theta^m = (L/a)(v/L)^m, \quad \epsilon_b^m = t y^m / 2L. \tag{12.18}$$

These expressions show that ϵ_θ^m increases, and ϵ_b^m decreases, as L increases. The optimum arrangement therefore occurs when the two values are equal, which requires that

$$\frac{L^2}{at} = \frac{y^m}{2(v/L)^m}. \tag{12.19}$$

When this condition is fulfilled we have

$$\epsilon^m = \epsilon_b^m = \epsilon_\theta^m = [(t/2a) y^m (v/L)^m]^{\frac{1}{2}}. \tag{12.20}$$

From the data of fig. 12.4 we find that (12.19) gives values of L somewhat larger than those for which the flexibility is maximum, by a factor of about 1.15 over the range $60° < \alpha < 140°$. Since this just lies within the range of

12.4 Approximate analysis of strain in bellows

variables for which our hypothetical mode of deformation is approximately correct, we may conclude tentatively that for bellows whose meridians consist of circular arcs, those geometries which maximise the flexibility also approximately minimise the peak strain in the bellows.

Let us denote by e the uniform axial strain in the comparison tube when a length X_C is subjected to an elongation u_C:

$$e = u_C/X_C.$$

From (12.20) we find

$$\epsilon^m/e = A(\alpha)\,(t/a)^{\frac{1}{2}}, \tag{12.21}$$

where

$$A^2 = y^m\,(v/L)^m\,(\sin^2 \alpha)/2l^2. \tag{12.22}$$

The function A is plotted against α in fig. 12.8. Its value is of order unity, and it is clear that the peak strain is a smaller fraction of the comparison strain e for the larger values of α.

If we restrict attention to designs for which $b \approx b_{\text{opt}}$, we find that the data of figs. 12.6 and 12.8 are consistent with a simple approximate formula

$$\epsilon^m/e \approx 1.8/F^{\frac{1}{2}}: \tag{12.23}$$

see problem 12.5.

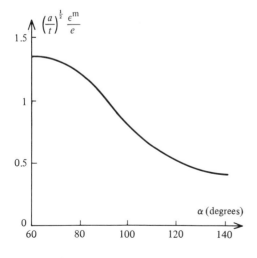

Fig. 12.8. Dimensionless plot of peak strain ϵ^m for a given overall extension e, in terms of the angle α: see (12.21), (12.22).

12.5 Discussion

In this chapter we have tackled in an approximate but direct way two problems in connection with the performance of simple bellows.

In relation to the book as a whole, the chapter illustrates the value of special methods for particular problems, established on the basis of various clues provided in other chapters. When these special methods concentrate directly on the most significant aspects of behaviour, they can reveal in a simple way such matters as conditions of optimality. But it must be acknowledged that a method of this sort can only be relied upon if there are independent means of checking the validity of the various hypotheses on which the method is based. In this respect, conventional solutions are indispensable.

In relation to the technology of bellows we have confronted only two of a large number of problems. A list of aspects which warrant further study would include the following topics: the effect of interior pressure on the axial stiffness of bellows, the behaviour of bellows in shear (see problems 12.6 and 12.7), and various buckling problems concerned with pressure-loading. Problems of *manufacture* are also far from trivial, as they involve the large-scale plastic straining which must occur when convolutions are imparted to an originally cylindrical tube.

12.6 Problems

12.1 Investigate the relative magnitude of the peak values of κ_s and κ_θ for a bellows having $\alpha = \frac{1}{2}\pi$, and deforming in such a way that $\kappa_s = \kappa_0 \sin(\pi S/2L)$ (see fig. 12.2). Show, in particular, that χ is maximum and r_3 is minimum at $S = 0$ (see fig. 6.9); and that

$$|\kappa_\theta|_{max} = (b/a)\kappa_0 \ll \kappa_0 \text{ for } b/a \ll 1.$$

Also show, by means of appropriate sketches, that the above result is approximately true for other values of α.

12.2 Consider the extension of a bellows of the type shown in fig. 12.1, and in which the meridian deforms inextensionally. Arguments from symmetry may be adduced to show that at point A of fig. 12.2 the component v of displacement is zero. Use physical reasoning to show that $v > 0$ in the part of the meridian immediately to the right of A when $\alpha > 90°$, and that $v < 0$ in this region when $\alpha < 90°$.

12.3 An approximate strain-energy analysis of the response of a uniform simply-supported elastic beam to a central transverse point-load W may be performed by considering a hypothetical mode of transverse displacement

$$w = w_1 \sin(\pi x/L),$$

12.6 Problems

evaluating the strain energy of elastic bending, and equating it to the work, $\frac{1}{2}Ww_1$, done by W: see appendix 1.

Show that the central deflection w_1 obtained in this way is given by

$$w_1 = \frac{2}{\pi^4} \frac{WL^3}{EI},$$

where EI is the flexural rigidity of the beam; and that this is approximately 1.4% less than the correct value.

Also show that the change of curvature at the centre of the beam is equal to $2WL/\pi^2 EI$ according to this method; and that this differs by about 19% from the correct value.

(This illustrates the remark in section 12.3 to the effect that a hypothetical mode which is not in itself particularly accurate in detail can nevertheless give a relatively accurate estimate of the total strain energy of the system (which is here equal to $\frac{1}{2}Ww_1$).)

12.4 A convoluted bellows under axial tension is analysed by means of the *membrane hypothesis*, as in section 4.5. Show that there is a singularity in N_s^* at any point on the meridian where the tangent to the meridian is perpendicular to the axis of the bellows; and that consequently the membrane hypothesis is certainly invalid in the immediate vicinity of this point.

12.5 By reading off values from figs. 12.6 and 12.8 show that formula (12.23) is accurate to within about 10% over the range $80° < \alpha < 140°$, but that the coefficient 1.8 in (12.23) should be lowered for smaller values of α, e.g. to about 1.2 for $\alpha \approx 60°$.

12.6 It may be argued that the flexibility factor F, as derived for the axial deformation of bellows, also applies to the *bending* flexibility of a bellows in relation to that of the comparison plain tube, provided $b/a \ll 1$.

A bellows of length L is connected to rigid plates at its ends. It is now subjected to a *shearing* deformation in which one of these plates is fixed while the other is moved a distance Δ in its own plane. On the assumption that the overall strain in the bellows corresponds broadly to that of the comparison tube, which moreover deforms according to the classical bending theory of beams, show that the overall tensile and compressive strains at the two ends of the unit (i.e. the proportional changes in length per convolution) are of magnitude $6\Delta a/L^2$. On the further supposition that this strain is limited to 0.2 – corresponding to contact between neighbouring convolutions – show that the overall shear deflection Δ is limited by

$$\Delta/a < (L/a)^2/30.$$

12.7 Our analysis of bellows has presupposed that the strains and displacements are relatively small; and that in particular adjacent convolutions do not come into contact during compression. A crude analysis of the circumstances in which convolutions clash may be made in the case of bellows whose meridional profile consists of circular arcs by assuming that the inextensional meridional profile of each convolution remains *circular* during all subsequent deformation of the bellows.

On this assumption let the angle of embrace of a convolution increase from its original value of α (fig. 12.1) to β in the course of compression of the bellows.

Show that if the thickness of the material is negligible, the condition for the meridional centre-lines to touch is $\beta = 150°$; and that the proportional shortening of axial length per convolution at this point is equal to

$$1 - (\alpha/300\sin\alpha),$$

where α is measured in degrees. Hence show that the capacity for shortening in such bellows decreases as α increases in the range $\alpha > 120°$.

13

Curved tubes and pipe-bends

13.1 Introduction

Piping systems are an indispensable feature of many industrial installations. In such systems straight tubes predominate; but problems of plant layout, etc., obviously make it necessary for pipes to turn corners. There are, broadly, four ways of getting the line of a pipe to turn a corner. First, fig. 13.1*a* shows a so-called *long-radius* bend in which the radius b of the centre-line of the curved portion is much larger than the radius a of the tube itself. A right-angle bend is illustrated, but it is obvious that the angle through which the line of the pipe turns is arbitrary, in general. On the domestic scale, bends of this sort may be made, *ad hoc*, in ductile metal pipes by the use of a pipe-bending machine; but the resulting cross-section of the curved portion is usually not circular: see later. Second, fig. 13.1*b* shows a so-called *short-radius* bend, in which the ratio b/a has a value of less than 4, say. The curved section is specially fabricated by casting, or welding together suitably curved panels; and the curved unit is connected to the straight pieces by bolted or welded joints. The types shown in fig. 13.1*a* and *b* are known as *smooth bends*. Third, fig. 13.1*c* shows a *single-mitre* bend, which is made by joining a pipe which has been 'mitred' by a plane oblique cut. A mitre joint may either be unreinforced (as shown) or reinforced by an elliptical ring or flange. Fourth, fig. 13.1*d* shows a multi-mitre (or 'lobster-back') bend.

There are yet other ways of making a corner, by means of a 'junction box', but we shall not consider them here.

The main structural feature of a curved pipe is that it is more flexible in bending than an equivalent straight pipe. This was first noticed in 1910 by Bantlin, and explained in 1911 by von Kármán (Clark & Reissner, 1951). The extra flexibility is associated with the ability of the tube to 'ovalise' or flatten when a bending moment is applied to it. The amount of additional flexibility depends on the proportions of the tube, and is determined by the value of a dimensionless group involving the radii a, b (see fig. 13.1) and the thickness t of the tube. In general a curved tube which has a high flexibility factor also has a relatively high *stress-concentration factor* in the sense that the peak local stresses are larger than the peak stresses in a corresponding straight tube under equal bending moment. On the other hand, the application of interior pressure to a curved tube produces stresses which are not appreciably higher than those in the corresponding straight tube. The high flexibility of pipe bends plays an important part in the overall behaviour of piping systems.

A single-mitre joint is also, in general, flexible for in-plane bending. The reasons are much the same as for smooth bends, but the flexibility, for given dimensions a, t, will usually be less than one-half of that of a smooth bend with, say, $b = 2a$. Flexible mitres also have high stress-concentration factors for pure bending and – in contrast with smooth bends – they have high stress-concentration factors for internal pressure-loading.

Fig. 13.1. Four ways of making a right-angle bend in a tube: (*a*) 'long-radius bend', (*b*) 'short-radius bend', (*c*) single-mitre bend and (*d*) multi-mitre ('lobsterback') bend.

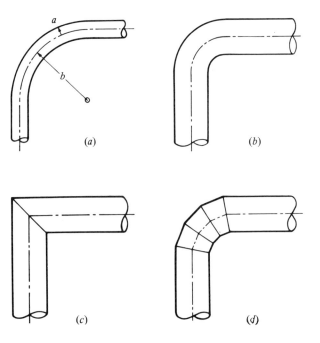

13.1 Introduction

A typical piping system in an industrial plant will follow a three-dimensional path, with straight members interconnected by bends and T-pieces of various kinds, together perhaps with flexible 'hinge' and 'expansion' units incorporating bellows, etc. (see chapter 12). The process of design of pipework systems is, in general, an extremely complex one which involves the consideration of constraints imposed by the geometry of spatial layouts, material properties, operating conditions, etc..

The main *structural* problems in piping systems arise from changes in loading (notably interior pressure) and from differential thermal expansion of the pipework itself and components such as pressure-vessels. The severity of these effects depends very largely on the details of the individual system, and the range of possibilities is enormous. In general, however, the structural problem may be treated in two parts.

(i) For given loading conditions, what are the main forces in the system? That is, what are the tensions, shearing forces, bending and twisting moments, etc. in the various component pipes and joints?

(ii) Given the answers to (i), what are the worst local effects (stress concentration, localised plastic strain, etc.) in the individual components?

The first of these sub-problems is usually tackled by replacing the actual pipes and junctions, etc., conceptually, by 'centre-line' members which are endowed with the appropriate bending and torsional rigidity, etc. This is essentially the same as the process which is used in the elastic analysis of building frameworks, in which the beams, columns and joints are replaced by idealised elastic members having appropriate elastic properties. However, while many three-dimensional building frames may reasonably be treated as loosely interconnected arrays of two-dimensional sub-frames, a three-dimensional piping system must almost always be treated as three-dimensional. Thus the idealised element which replaces, say, a smooth bend must reflect the response of that member to bending moments applied not only about an axis normal to the plane of the joint, but also about other axes.

It follows that the structural analysis of a component such as a smooth bend is necessary for two distinct purposes. First, we need to know the *overall* response of the component to applied bending moments etc. (i.e. its stiffness as a joint) as a prerequisite for calculation (i) of the piping system as a whole. Second, (part (ii), above) we need to know something of the *details* of its structural behaviour under the bending moments, etc., which, according to (i), it is required to sustain in practice.

The main purpose of the present chapter is to analyse various aspects of the behaviour of curved tubes, with a view to understanding both their overall behaviour and also something of the details of stress distributions, etc..

The analysis and design of piping systems is an enormous subject, on which much has been written: see the M.W. Kellog Company handbook (1956) for a guide.

Structural analysis of various components has been done by many investigators both theoretically and experimentally, in static and fatigue conditions and in elastic, plastic and creep ranges of behaviour of the material. The present chapter is not intended to give a review or even a summary of the available theoretical and experimental data. Its main aim is to present relatively simple and clear structural analyses of the curved tube, with particular emphasis on the identification of the main structural effects, and the presentation of results in a suitable dimensionless form.

The layout of the chapter is as follows. We begin by analysing the behaviour of a curved pipe under interior pressure according to the membrane hypothesis in section 13.2. Next, as a preliminary exercise, we consider the bending of a simple 'two-flange' curved beam, before turning to the analysis of the pure bending of a curved tube in section 13.4. Next, in section 13.5 we investigate the consequences of interior pressure on the bending behaviour. Finally, in section 13.6 we consider the restraining effects which occur when a portion of curved tube is connected to straight pipes, with or without the use of flanges. Most of the pieces of analysis are based on existing work, but there are some new extensions. Thus, the analysis of in-plane bending of a curved pipe in section 13.4.1 follows well-worn lines; but the discussions of the effects of interior pressure on the bending stiffness, and the stiffening which is due to the presence of nearby flanges, are novel and follow directly the work of chapter 9.

All of the work in this chapter presupposes linear-elastic behaviour, and assumes that the unloaded structure is stress-free. Inelastic behaviour must always be expected, of course, when elastic analysis reveals relatively high 'stress-concentration factors', as it does here on several occasions. The important topic of plastic behaviour of shells is the subject of chapter 18.

In most cases our analysis is 'approximate', and freely employs the well-known energy theorems of classical structural analysis (see appendix 1). The main reason for not seeking 'exact' solutions is that the geometry of curved pipes is more complicated than that of a straight pipe: consequently the prospect of using relatively simple calculations, which is afforded by the use of approximate methods, is attractive, even at the expense of loss of rigour. Since a main interest lies in the dimensionless groups which are relevant to particular problems, rather than in the precision of numerical constants, this approach is reasonable.

For reasons of space, mitred bends (fig. 13.1c and d) are not analysed in this chapter. Practical engineers will be quick to notice that the chapter does not include analyses of some loading cases which are practically important, such as twisting and 'out-of-plane' bending of curved tubes (see Jones, 1966). It is to be hoped that the loss in value arising from such omissions will be outweighed by the positive results which are presented here for various problems which have hitherto seemed intractable in general terms.

13.2   A curved tube subjected to internal pressure

Let us investigate first the stress resultants which are required in a curved tube of circular cross-section, according to the membrane hypothesis, when an internal gauge pressure p is applied to it. It is convenient to begin by considering a complete, toroidal shell and to work in terms of the θ, ϕ coordinates which we have used previously for shells of revolution. We cannot use the work of section 4.5 directly, however, because the diametral meridional cross-section of a toroid consists of two separate circles rather than a single closed curve as in fig. 11.2a: the topology of the surface makes an important difference to the calculation. Fig. 13.2 shows a typical meridional cross-section of the shell and defines radii a, b in addition to the meridional angular coordinate ϕ.

Fig. 13.2. Geometry of a toroidal surface having a circular cross-section: cf. fig. 6.9a.

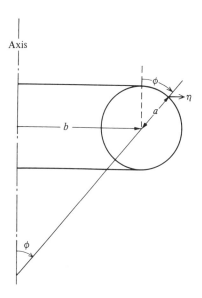

434 Curved tubes and pipe-bends

At a point on the shell defined by ϕ the principal radii are given by (see section 4.5)

$$r_1 = a, \quad r_2 = a + b/\sin\phi. \tag{13.1}$$

By symmetry, the only nonzero tangential-stress resultants are N_ϕ and N_θ in the meridional and circumferential directions, respectively: see fig. 13.3a. The equilibrium equation in the direction normal to the surface is thus

$$(N_\phi/a) + (N_\theta\sin\phi)/(b+a\sin\phi) = p. \tag{13.2}$$

It is straightforward to derive a second equation of equilibrium for an isolated element of the shell by resolving in the meridional direction the forces which act on it. It is more convenient, however, in the present case to derive a second equation of equilibrium by considering the equilibrium of a system consisting of the hoop-shaped portion of the shell shown in fig. 13.3b, together with an associated ring of fluid. Stress resultants $N_\phi)_0$ and N_ϕ have been applied externally at the two circumferential cuts, and the other forces acting on the system come from the pressure which is applied to the plane and cylindrical boundaries of the fluid.

The stress resultant $N_\phi)_0$ and the pressure acting on the cylindrical boundary of the fluid act normal to the axis. Consequently the equation of equilibrium in the axial direction is:

$$2\pi(b+a\sin\phi) N_\phi \sin\phi = 2\pi p(b+\tfrac{1}{2}a\sin\phi) a \sin\phi. \tag{13.3}$$

On the left-hand side we have the axial component of the total N_ϕ force, while

Fig. 13.3. (a) Perspective sketch of a small element of a toroidal shell and the stress resultants acting on it according to the membrane hypothesis. (b) An annular element, consisting of a portion of the shell and an associated volume of pressurised fluid, from the equilibrium of which N_ϕ may be determined directly.

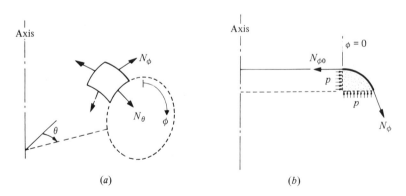

the right-hand side consists of the pressure multiplied by the area of the plane annular boundary of the fluid. Rearranging (13.3) we obtain

$$N_\phi = pa\,(b + \tfrac{1}{2}a\,\sin\phi)/(b + a\,\sin\phi), \tag{13.4}$$

and on substitution into (13.2) we find that

$$N_\theta = \tfrac{1}{2}pa. \tag{13.5a}$$

It is surprising that the 'longitudinal' stress resultant N_θ should be uniform around the circumference of a meridian. The expression is exactly the same as for a pressurised straight pipe with closed ends. The value of N_ϕ varies somewhat around the circumference according to (13.4), but when $b \gg a$ we obtain

$$N_\phi \approx pa. \tag{13.5b}$$

In general the pressure acting on an isolated small element of the shell is balanced by a combination of N_ϕ and N_θ: see (13.2). When $b \gg a$ the contribution from N_θ is negligible and we obtain the well-known result for a straight pipe. Now for $0 < \phi < \pi$ (i.e. $r > b$) the coefficient of N_θ in (13.2) is positive, and consequently the value of N_ϕ is somewhat less than the straight-pipe value pa; but in the region $r < b$, conversely, the contribution from N_θ is acting in the same sense as p, and consequently the value of N_ϕ is somewhat more than pa.

Although we have considered here a complete toroidal shell, we may also use the solution for a curved segment of pipe provided we make appropriate closure arrangements. The uniformity of N_θ (see (13.5a)) ensures that the solution may be joined directly (with no discontinuities of N_θ) to that for a straight connecting pipe with closed ends: alternatively (13.5) satisfies the equilibrium requirement for a simple plane end-closure to the curved segment itself.

The membrane hypothesis which we have used so far is only justified, of course, provided the surface strains (here $\epsilon_\theta, \epsilon_\phi$) which follow from Hooke's law are geometrically compatible without the necessity for high changes in curvature (here $\kappa_\theta, \kappa_\phi$). The most direct approach to this question is to use the compatibility equation (6.61) in order to compute the rotation χ of the tangent to the meridian. This may be rearranged as follows

$$\chi = -\frac{r}{a\,\sin\phi}\frac{d\epsilon_\theta}{d\phi} + (\epsilon_\phi - \epsilon_\theta)\cot\phi. \tag{13.6a}$$

When $\epsilon_\theta, \epsilon_\phi$ are computed from N_θ, N_ϕ by the application of Hooke's law, it is found that χ has a singularity at $\phi = 0$ (and π). This may be illustrated adequately by means of a case in which $\nu = 0$. Then, since $N_\theta = $ constant, the first term on the right-hand side of (13.6a) vanishes; but since $N_\phi \approx 2N_\theta$, $\epsilon_\phi \approx 2\epsilon_\theta$ and the second term gives a singularity in χ like $\cot \phi$, corresponding to a sharp outward-pointing crease (cf. Jordan, 1962, Fig. 3). Equation (6.61) is based, of course, on the assumption that all strains and rotations are small, so the above result is not strictly valid. Nevertheless, it provides clear evidence that the membrane hypothesis is inadequate for the discussion of a pressurised complete toroidal shell.

In fact, most interest has been shown in this problem in relation to shells which are so thin that they behave as flexible membranes: in this case the incompatibility of strain is not countered by the setting up of bending-stress resultants within the shell, but by the distortion of the surface so that the equilibrium equations themselves change a little by virtue of small changes in the meridional profile. This is, of course, a nonlinear problem of a kind which is beyond the scope of the methods of chapter 4, and indeed of this book. But see Jordan (1962) and Sanders & Liepins (1963).

The above remarks apply to *complete* toroidal shells. For incomplete toroids the situation is different. Imagine, for the sake of definiteness, a complete toroidal shell which has been cut around one meridional cross-section and the two ends sealed with separate circular discs. Suppose that on application of internal pressure the angle subtended by the toroid increases from 2π to $2\pi(1 + \eta)$, where $\eta \ll 1$: the cut toroid 'opens out' a little. In such a case (6.61) no longer applies. However, it can readily be adapted by inspection of fig. 6.10. The key to the situation is that the opening-up of the toroid turns a latitude circle of radius r into a circle of radius $r(1 + \eta)$; and all such (incomplete) circles have a common axis of symmetry. It follows that each point on the meridian undergoes a displacement $r\eta$ normal to the axis on account of the opening up, in addition to the displacement $r\epsilon_\theta$ on account of ϵ_θ. Thus we find that (13.6a) still holds provided ϵ_θ is replaced by $(\epsilon_\theta + \eta)$; and the equation thus becomes

$$\chi = -\frac{r}{a \sin\phi} \frac{d\epsilon_\theta}{d\phi} + \cot \phi \, (\epsilon_\phi - \epsilon_\theta - \eta). \tag{13.6b}$$

The singularity which we found earlier now disappears if we set

$$\eta = (\epsilon_\phi - \epsilon_\theta)_{\phi=0}.$$

The value of η may thus be computed readily; and clearly it is positive for positive internal pressure.

Further investigation shows that when the singularity in χ has been removed in this way, the remaining incompatibilities in strain are rather trivial. Thus we may conclude that the membrane hypothesis is satisfactory for toroidal pipes *provided* there is freedom to 'open out' a little. If there is no such freedom (e.g. as in a complete toroid), bending moments will be set up in the tube. The response of the tube to bending moments is the main subject of section 13.4.

The above analysis is concerned essentially with the geometry of distortion of a thin hoop of the shell in the region $\phi = 0$. As Flügge (1973, § 2.5.4) has remarked, this hoop is constrained to deform in its own plane. When $\epsilon_\phi = \epsilon_\theta$ the (cut) hoop enlarges isotropically without difficulty; but when $\epsilon_\phi \neq \epsilon_\theta$ it tends to 'open out' in much the same way as the curved beam of fig. 2.6 changes its curvature a little in response to 'through-thickness' strain.

The above analysis has been based explicitly on the assumption that the toroidal tube has a *circular* cross-section. If the circular cross-section of fig. 13.2 is replaced by, say, an ellipse having its major axis parallel to the axis of revolution and minor axis of length $2a$, it is possible to go through the same steps as before (at least, in principle), and thus to determine the distribution of N_θ and N_s around the circumference of the cross-section. In this case it turns out that N_θ is non-uniform, having larger values on the side $r > b$ of the cross-section than on the other side. Consequently the stress resultants N_θ provide an overall *bending moment* in the toroidal tube (about the axis of revolution) which is uniform and self-balancing for a complete toroid. This bending moment does not disappear if we let $b \to \infty$; which constitutes an embarrassment if we wish to analyse the response to internal pressure of a *straight* tube having the same cross-sectional shape. But on the other hand, it is not possible to analyse such a tube according to the membrane hypothesis anyway: so we should clearly expect paradoxical behaviour of the toroidal tube as $b \to \infty$. In fact we can only analyse a pressurised toroidal tube of non-circular cross-section satisfactorily according to the membrane hypothesis if the line connecting the locations in the cross-section at which $\phi = 0$ and $\phi = \pi$ (see fig. 13.2) is parallel to the axis of revolution: otherwise normal shearing-stress resultants are required at one of these locations (at least) in order to satisfy the conditions of equilibrium in the axial direction of either of the two portions of the toroidal surface which are isolated by cuts at these special latitude circles (see Flügge, 1973, p.30).

Curved tubes of non-circular cross-section have important practical applications in the 'Bourdon tube' pressure gauge. Throughout this chapter, however, we shall be concerned almost entirely with tubes which have a truly circular cross-section in their original, unloaded, configuration.

13.3 Pure bending of a curved two-flange beam

It is convenient to begin our analysis of bending of a curved tube by considering the bending of a simple 'two-flange' beam which is initially curved (see fig. 13.4), and is transversely compressible. The aim is to find the change of curvature of the beam in response to the application of a uniform bending moment. As we shall see later, this model constitutes a crude analogue of the bending of a curved tube; and a similar model will be useful as an introduction to the so-called 'Brazier' effect in chapter 16.

The beam is uniform. The two flanges are made from material having Young's modulus E, and each has cross-sectional area $\frac{1}{2}A$. The flanges act only in direct tension or compression: they are too thin to have any appreciable stiffness against bending separately. In the original, unstressed, configuration the centre-surfaces of the flanges are separated by H, and the flanges are circular arcs of radius $B + \frac{1}{2}H$ and $B - \frac{1}{2}H$ respectively: the centre-line of the beam lies in a plane and has radius B. The two flanges are separated by linear-elastic springs, and associated with unit length of centre-line is a spring of stiffness k.

When a pure bending moment M is applied to the beam, one flange is clearly put into tension and the other into compression; and since the flanges are curved there must also be some tension or compression in the transverse springs. The flanges and the transverse springs all distort elastically to some extent; and we must therefore investigate the effect of these distortions upon the configuration of the beam.

We shall assume that the strains in the flanges and the relative change in length of the transverse springs are all small, in order to simplify the analysis.

Consider a portion of the beam which subtends angle α at the centre of curvature in its original configuration. The lengths l_1, l_2 of the inner and outer flanges, respectively, are given by

$$l_1 = \alpha(B + \tfrac{1}{2}H), \quad l_2 = \alpha(B - \tfrac{1}{2}H): \tag{13.7}$$

see fig. 13.4b. Hence, on elimination of B from these expressions we find

$$\alpha = (l_1 - l_2)/H. \tag{13.8}$$

In the subsequent distortion let l_1, l_2 and H change by small amounts δl_1, δl_2 and δH, respectively. Differentiating (13.8) as a product, we obtain an expression change $\delta \alpha$:

$$\delta \alpha = \frac{\delta l_1 - \delta l_2}{H} - \frac{(l_1 - l_2)\delta H}{H^2}. \tag{13.9}$$

Now the small change, κ, in curvature of the beam may be defined as the change in angle α divided by the original length of centre line; and since this may be written as

$$\tfrac{1}{2}(l_1 + l_2) = B\alpha, \tag{13.10}$$

13.3 Pure bending of a curved two-flange beam

we obtain

$$\kappa = \frac{2(\delta l_1 - \delta l_2)}{H(l_1 + l_2)} - \frac{\delta H}{BH}. \tag{13.11}$$

We may calculate δl_1, δl_2 and δH by performing a statical analysis and applying Hooke's law. When a pure bending moment M is applied the flanges have tension $\pm M/H$, and on account of the (original) curvature of the flanges there is an inter-flange compressive force equal to M/BH per unit length of centre-line. Taking account of the cross-sectional area and the elastic coefficients we have

$$\delta l_1 = 2Ml_1/AEH, \quad \delta l_2 = -2Ml_2/AEH, \quad \delta H = -M/kBH. \tag{13.12}$$

Substituting these relations into (13.11) we thus obtain

$$\kappa = [(4/EAH^2) + (1/kB^2H^2)] M. \tag{13.13}$$

In this expression the first term in parentheses on the right-hand side represents the contribution to κ from the extension and compression of the flanges, while the second corresponds to the distortion of the transverse springs: these two effects are directly additive – at least for sufficiently small changes of curvature – as a consequence of the rule for differentiating a product.

It is easy to verify that the first term is exactly the same as it would have been if the beam had been *straight* originally. Therefore it is convenient to define a *flexibility factor f* as the ratio of κ as given by (13.13) to the corres-

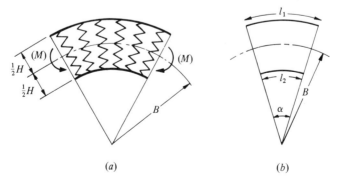

Fig. 13.4. (*a*) A curved beam whose two flanges are held apart by a series of uniform elastic springs. B is the mean radius of the beam in its original, unstressed, configuration. (*b*) A geometrical calculation; see text.

ponding value for a straight beam having the same cross-section. It follows directly that

$$f = 1 + EA/4kB^2. \qquad (13.14)$$

This expression suggests strongly the use of the dimensionless parameter EA/kB^2 to characterise the behaviour of the beam. When the value of this parameter is small the beam behaves almost as if it were straight; but when the value is large the beam is much more flexible. Note that the value of the parameter is always zero for a straight beam ($B \to \infty$) irrespective of the value of k. Note also that in this chapter we are concerned only with *small* changes from the initial configuration.

The flexural behaviour of a curved tube in pure bending is, in a broad sense, similar to that of our simple beam: the extra flexibility of the curved tube comes from the transverse compression or 'ovalisation' of the cross-section. The only major difference is that the curved tube has more than one degree of freedom with respect to changes of cross-sectional shape, as we shall see. Nevertheless, the 'flexibility factor' for a curved tube turns out to depend almost entirely on a single dimensionless group.

13.3.1 Alternative derivation by means of complementary energy

The analysis given above has used equilibrium equations, Hooke's law, and considerations of geometrical compatibility. It is of interest from several points of view to perform an alternative analysis of the flexibility of the curved beam by using the method of *complementary energy*. In general this will act as an introduction to the method which will be used in section 13.4 in order to compute the flexibility of a curved tube; but in particular it will reveal clearly the simple additive character of the flexibility which derives from the elasticity of the flanges and the transverse springs, respectively.

Suppose we have a uniform beam (in this case either curved or straight initially), which has the following linear relationship between bending moment, M, and corresponding small change of curvature, κ:

$$\kappa = \Phi M. \qquad (13.15)$$

The flexibility Φ is the reciprocal of the bending stiffness of the beam, and would thus be equal to $1/EI$ for a straight beam. For a piece of beam of length l which undergoes a small change of curvature κ, the relative rotation, say ψ, of the ends is given by

$$\psi = l\kappa \qquad (13.16)$$

Therefore a graph of M against the (corresponding) 'displacement' ψ has the form shown in fig. 13.5a: in particular the slope of the line is $(\Phi l)^{-1}$.

13.3 Pure bending of a curved two-flange beam

Now the *complementary energy*, $\overline{C}$, of the elastic system is defined, for the present problem, by

$$\overline{C} = \int_0^M \psi \, dM. \tag{13.17}$$

It is equal to the area shaded in fig. 13.5a. Note particularly that $\overline{C}$ is a function of M, and *not* of ψ. For the present problem we have, from (13.15) to (13.17),

$$\overline{C} = \tfrac{1}{2}\Phi l M^2. \tag{13.18}$$

Further, the complementary energy *per unit length of beam*, say C, is given by

$$C = \tfrac{1}{2}\Phi M^2. \tag{13.19}$$

Thus the 'specific' complementary energy is directly proportional to M^2 and also to the flexibility Φ of the beam. We may call C the *external* specific complementary energy, since it is calculated from the overall bending moment/curvature relationship.

Now when the beam carries a given bending moment M, it does so by setting up stresses within the beam which are in statical equilibrium with the applied loading. In the case of our curved beam there is longitudinal stress in the flanges and also compressive stress in the transverse springs. Associated with these internal stresses is what may be called 'internal' complementary energy. This quantity is summed from contributions from all the stressed ma-

Fig. 13.5. (a) Linear-elastic relationship between bending moment M and relative rotation ψ of the ends of a beam. The shaded area represents the total *complementary energy* of the beam. (b) Corresponding diagram for a bar in uniaxial tension which undergoes tensile strain ϵ due to stress σ: the shaded area represents the complementary energy per unit volume of material. Similarly, (c) relates to a spring under tension T which undergoes (overall) elongation e.

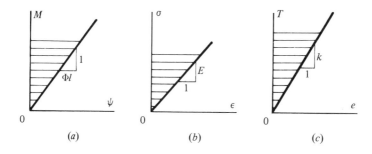

terial in the system. By analogy with (13.19) we may define the complementary energy per unit volume of an elastic material subject to uniaxial tensile stress σ as

$$\sigma^2/2E \qquad (13.20)$$

(see fig. 13.5b and section 2.3.3), while the complementary energy of a spring of stiffness k when subjected to a tension T is equal to

$$T^2/2k \qquad (13.21)$$

(see fig. 13.5c).

Now there is a basic theorem of linear elasticity (see appendix 1) which states that for a *statically determinate* system the 'external' complementary energy is equal to the total 'internal' complementary energy. (Later on, we shall use a more powerful complementary energy theorem for *statically indeterminate* structures.) Consequently we may use expressions (13.20) and (13.21) in the present case – with values of σ and T determined by considerations of statical equilibrium – to evaluate the complementary energy of the system and hence the flexibility factor Φ. In other words, we may use the complementary energy method to determine the flexibility of our curved beam. Note particularly that in this way we can determine the flexibility of the beam *without considering explicitly the geometry of strain and deformation*.

The details of the calculation are as follows. For the present example of a two-flange beam which is subjected to a pure bending moment M, the magnitude of the longitudinal stress in the flanges is found from considerations of statical equilibrium to be given by

$$\sigma = 2M/AH, \qquad (13.22)$$

as we have seen already. The volume of the flanges per unit length of beam is equal to A. The compressive force in the transverse springs occupying unit centre-line length of a curved beam of radius B is equal to M/HB, and the stiffness of the transverse spring per unit length is equal to k. It follows that the total ('internal') complementary energy per unit length of beam is given by

$$(2M/AH)^2 (A/2E) + (M/HB)^2 (1/2k).$$

Equating this expression to that for the 'external' complementary energy (see (13.18)) we obtain

$$\Phi = (4/EAH^2) + (1/kH^2B^2), \qquad (13.23)$$

which agrees exactly with the previous result (13.13).

Note particularly the way in which the *flexibility* (rather than its reciprocal, the *stiffness*) fits well with the complementary energy scheme, in that the final expression consists of a sum of two simple parts.

Two points should be made before we leave the subject of the two-flange beam. The first is that our analysis holds for arbitrary values of the geometrical curvature parameter B/H (see fig. 13.4*a*): the change of curvature on account of straining in the flanges depends on the (common) magnitude of the strain in the flanges multiplied by the *sum of the lengths of the two flanges*; and since this length is exactly equal to twice the centre-line length irrespective of the original curvature of the beam, the result is also valid for arbitrary initial curvature. The second is that one of the effects of the application of a pure bending moment M is to change, by a small amount, the centre-line length of the beam segment, since by (13.12)

$$\delta(\tfrac{1}{2}(l_1 + l_2)) = \tfrac{1}{2}(\delta l_1 + \delta l_2) \neq 0. \tag{13.24}$$

Thus the so-called 'neutral axis' of this two-flanged curved beam does not coincide with the geometrical centre of the beam. The straining of the geometrical centre-line is small in the present analysis. In particular, it does not adversely affect our definition of *change of curvature*.

13.4 Pure bending of a curved tube

In this section we shall begin an approximate analysis of the uniform bending of an initially curved tube by means of a complementary energy method. Our plan will be first to apply a given bending moment M to the tube; next to find a statically admissible variation around the circumference of longitudinal stress; then to find the distribution of bending moment in the circumferential direction which satisfies the equations of equilibrium for a hoop-like slice of the tube; and lastly to compute the total complementary energy per unit length of tube. When this quantity is equated to the 'external' complementary energy we shall obtain an estimate for the flexibility of the tube, by analogy with our previous calculation. There is, however, an important difference between the present computation and the previous one. Here, as we shall discover, the distribution of 'longitudinal' stress in equilibrium with the applied bending moment is not statically determinate, and therefore the magnitude of the complementary energy per unit length will depend to some extent upon our choice of longitudinal stress. In these circumstances the *theorem of minimum complementary energy* (see appendix 1) tells us that our calculated complementary energy (and hence our estimated flexibility) will always in general give an overestimate of the actual flexibility of the tube. Further, of all possible distributions of longitudinal stress which are statically admissible, the one which *minimises* the complementary energy (for a given applied bending moment) is also the one for which the computed flexibility factor is *correct*, and indeed for which the direct and bending strains (ϵ, κ) as computed by Hooke's law are geometrically compatible.

This kind of analysis was first performed by Beskin (1945), and we shall follow his calculations closely, apart from some minor changes of notation. But we shall begin by doing a particularly simple version of the calculation.

The tube and the applied loading are shown in fig. 13.6. Following Beskin and others, we shall assume that the radius of curvature b is large in comparison with a. And although it is not strictly proper, we shall use an 'axial' ('longitudinal') coordinate x instead of the circumferential coordinate θ in addition to the meridional angular coordinate ϕ, as shown. We shall argue in favour of these simplifications later. Note that the 'extrados' and 'intrados' are at $\phi = \frac{1}{2}\pi$ and $\phi = -\frac{1}{2}\pi$, respectively.

If the applied pure bending moment M were to act upon a *straight* tube of the same cross-sectional dimensions we should find, by the ordinary simple bending theory of beams, the following distribution of longitudinal stress – resultant N_x:

$$N_x = M(\sin\phi)/\pi a^2. \tag{13.25}$$

Later we shall represent the circumferential distribution of N_x by a Fourier series of which (13.25) is the first term, but for the present we shall use (13.25) as it stands: it certainly satisfies the overall equilibrium requirement, namely

$$\int_0^{2\pi} N_x \, a^2 \sin\phi \, d\phi = M. \tag{13.26}$$

Fig. 13.6. Perspective sketch of a portion of a curved tube of circular cross-section, subjected to a uniform bending moment M.

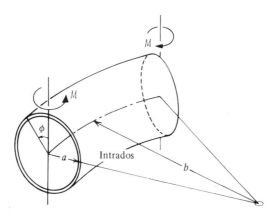

13.4 Pure bending of a curved tube

Fig. 13.7a shows a short length l of the tube cut by two planes normal to the curved central axis. These planes are inclined at a small angle l/b to each other, and so the stress resultants N_x at the two opposite ends of the segment are not exactly in line. Fig. 13.7b shows a small piece of this ring, cut out at angles ϕ and $\phi + \mathrm{d}\phi$. In order to maintain equilibrium an extra force must somehow be supplied, as shown. The magnitude of this force is equal to

$$N_x a(l/b)\mathrm{d}\phi, \tag{13.27}$$

and its line of action is in the central cross-sectional plane of the ring and is also parallel to the plane of the centre-line of the curved tube, as shown. The latter follows from the fact that the end-forces $N_x a \mathrm{d}\phi$ act tangentially to the latitude circle of radius $b + a\sin\phi$ which passes through the element: see fig. 13.6. Thus the ring as a whole must receive forces as indicated in fig. 13.7c if equilibrium is to be preserved. Note that these forces are all directed parallel to the line $\phi = \frac{1}{2}\pi$ in the cross-section.

So far we have been discussing the equilibrium requirements for a sort of '*S*-surface' of the shell. In the absence of an external agency, the parallel

Fig. 13.7. (a) A portion of the curved tube cut out by two nearby meridional planes. (b) Part of the 'ring' of (a), cut out by portions of two 'parallel' circles. The 'longitudinal' tensile stress resultants N_x are balanced statically by an outward-directed force. (c) Stress resultants N_x in the entire ring (a) require outward-directed forces for equilibrium, as shown. (d) The forces in (c) react against a 'bending ring' and set up circumferential bending moments M_ϕ.

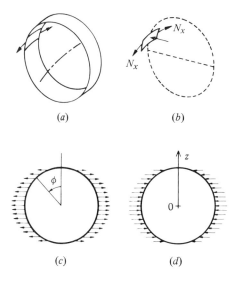

forces shown must be provided by the *bending* resistance of the ring-element acting as a kind of hoop. We could proceed as on previous occasions (see section 8.6) by thinking in terms of normal and tangential interactions between the S- and B-surfaces, but it is simpler in the present case to treat the B-surface directly as a ring loaded as in fig. 13.7d. This ring is in fact required to sustain some circumferential tension or compression; but since we are working in a regime in which the only significant straining of the S-surface is in the 'longitudinal' direction, this is of no consequence.

The analysis of a ring subjected to loads of this sort is particularly simple. Since all the loads are parallel we may treat the ring as a curved beam, using the equilibrium equation

$$\frac{d^2 M_\phi}{dz^2} = \text{Applied load per unit length in the } z\text{-direction,} \qquad (13.28)$$

where z is an auxiliary Cartesian axis as indicated in fig. 13.7d. Here M_ϕ has its usual meaning of bending moment per unit length, and the loading on the right-hand side is computed per unit length of tube. Since the element $d\phi$ occupies a length $a\sin\phi \, d\phi$ in the z-direction, the right-hand side of (13.28) is equal to

$$\frac{N_x}{b \sin\phi} = \frac{M}{\pi a^2 b}. \qquad (13.29)$$

Note that this is a *uniform* intensity of loading with respect to z. Also, $dz = a \sin\phi \, d\phi$; so we may rewrite (13.28) as

$$\frac{d}{d\phi}\left(\frac{1}{\sin\phi} \frac{dM_\phi}{d\phi}\right) = \frac{M \sin\phi}{\pi b}. \qquad (13.30)$$

On integration of this with respect to ϕ we have

$$\frac{1}{\sin\phi} \frac{dM_\phi}{d\phi} = -\frac{M \cos\phi}{\pi b} + A_1,$$

where A_1 is a constant of integration; and on rearrangement and a second integration we obtain

$$M_\phi = \frac{M \cos 2\phi}{4\pi b} - A_1 \cos\phi + A_2. \qquad (13.31)$$

The terms containing the two constants of integration represent distributions of M which are in equilibrium with zero external loading. It may be shown (see problem 13.1) that the complementary energy of circumferential bending is least when $A_1 = A_2 = 0$; so we shall use these values hereafter.

The variation of circumferential bending moment in the ring in proportion to $\cos 2\phi$ is what one might expect on intuitive grounds for the loading shown

13.4 Pure bending of a curved tube

in fig. 13.7c. Notice, however, that we have really been calculating the *total* bending moment in the ring, and that in expressing the result in terms of M_ϕ (a bending moment per unit length) we are implicitly assuming that the x-direction length of the ring varies relatively little around the circumference, i.e. that $b \gg a$.

The next step is to compute the complementary energy of circumferential bending. Per unit area, the complementary energy is $M_\phi^2/2D$, where D is the flexural rigidity of a plate element. (In this case the complementary energy does not involve M_x: cf. problem 2.9.) By integrating around the circumference, on the assumption that $b \ll a$, we obtain the following expression for the complementary energy of circumferential bending per unit length of tube:

$$C_B = (3/8\pi)(M^2 a/Eth^2 b^2). \tag{13.32}$$

Here we have expressed D in terms of E, t and h: see (2.18) and (3.25). This completes the first part of the calculation.

In computing the complementary energy which is associated with the tangential-stress resultants N_x and N_ϕ, we shall ignore N_ϕ in comparison with N_x. We have already mentioned the need for having nonzero N_ϕ in consequence of the loading of a cross-sectional ring, as shown in fig. 13.7d. It is obvious that at $\phi = 0$ and $\phi = \pi$ at least there must be nonzero compressive stress resultants N_ϕ to balance the applied transverse loading. Since this compressive load per unit length in the z-direction is constant (see 13.29) we find that at these two locations

$$N_\phi = -\frac{M}{\pi ab}.$$

On the other hand the stress resultant N_x has a peak value (see 13.25) of $M/\pi a^2$; consequently if $b \gg a$ it will be reasonable to neglect the contribution to the complementary energy from N_ϕ. Thus, per unit area of shell the complementary energy is given, approximately, by

$$N_x^2/2Et. \tag{13.33}$$

Using (13.25) and integrating around the circumference we find that the complementary energy of longitudinal stretching per unit length of tube is equal to

$$C_S = M^2/2\pi Eta^3. \tag{13.34}$$

Expressions (13.32) and (13.34) must now be added to give the total complementary energy; and then used in (13.19) to give an overestimate of the flexibility Φ of the curved tube. It is convenient to express this in terms of the flexibility of a straight 'comparison tube' of the same cross-sectional dimensions, multiplied by a flexibility factor f. For a straight tube $C_B = 0$ and C_S is given by (13.34) unchanged. Thus we find that

$$f = (C_S + C_B)/C_S. \tag{13.35}$$

Substituting from (13.32) and (13.34) we obtain

$$f = 1 + 3a^4/4b^2h^2. \tag{13.36}$$

It is convenient to define a dimensionless group, Γ, as follows:

$$\Gamma = a^2/bh. \tag{13.37}$$

Then (13.36) becomes, finally

$$f = 1 + \tfrac{3}{4}\Gamma^2. \tag{13.38}$$

Essentially the same dimensionless grouping appears in all writings on curved tubes. Some authors (e.g. Clark & Reissner, 1951; Crandall & Dahl, 1957) use (in the present notation a, b, h)

$$\mu = 12^{\frac{1}{2}}a^2/bh = 2 \times 3^{\frac{1}{2}}\Gamma,$$

while others (e.g. Pardue & Vigness, 1951; Jones, 1966; Smith & Ford, 1967; Blomfield & Turner, 1972) use

$$\Lambda \text{ or } \lambda = bt/a^2 = (1-\nu^2)^{\frac{1}{2}}/\Gamma$$

On the whole, it seems better to incorporate the factor $(1-\nu^2)$ into the group by using h, since this simplifies all other expressions. But on the other hand it is useful in design to have a group which, unlike μ, is free from numerical factors such as $2 \times 3^{\frac{1}{2}}$. Lastly, it is most convenient when plotting *flexibility* factor results to use Γ rather than its near-reciprocal λ.

Equation (13.38) represents our first attempt to find the functional form of f. The equation has a formal similarity with the corresponding expression (13.14) for the flexibility of a curved beam, in that the flexibility factor f differs from unity in each case by a single term which contains in its denominator the square of the radius of curvature of the centre-line of the beam or tube. From the point of view of complementary energy these terms come respectively from compressive stress in the transverse springs, which is required for equilibrium of the curved beam, and from circumferential bending-stress resultants which are required for equilibrium within the curved tube. It is clear that from the point of view of Hooke's law, these bending-stress resultants in the tube must be associated with a certain amount of *ovalisation* of the cross-section, corresponding broadly to a small change in length of the transverse springs in the beam. But whereas the compressive and tensile stresses in the flanges of the curved beam are statically determinate, the corresponding stress resultants N_x in the tube are not; and thus the simple

13.4 Pure bending of a curved tube

formula (13.25) for N_x is unlikely to be valid once the cross-sectional shape of the tube changes appreciably. Hence, in general we must expect to have to use more elaborate expressions for N_x if we are to achieve satisfactory results. But, as we shall see later, (13.38) is reasonably accurate over a limited range of Γ, say $\Gamma < 1$. Outside this range it overestimates f by significant amounts.

13.4.1 A more complete treatment

The weakness of the foregoing analysis in relation to larger values of Γ lies in the fact that we had no disposable parameters in the specification of the longitudinal-stress resultant N_x for a given applied bending moment M. Following Beskin we now try a more general expression:

$$N_x = (M/\pi a^2)(\sin\phi + 3c_3 \sin 3\phi + 5c_5 \sin 5\phi + \ldots). \qquad (13.39)$$

This satisfies the overall equilibrium equation (13.26) for arbitrary values of c_3, c_5, etc., because the functions $\sin 3\phi$, $\sin 5\phi$, etc. are all orthogonal to $\sin\phi$ (see appendix 4); but the coefficient of $\sin\phi$ is uniquely determined by (13.26). The infinite Fourier sine series contains only odd terms by virtue of the mirror symmetry of the problem about $\phi = \frac{1}{2}\pi$. There are no cosine terms of even argument by virtue of the skew-symmetry of the distribution of N_x about $\phi = 0$, which it is satisfactory to assume when $b \gg a$. The use of multiples 3, 5 etc. in the coefficients $3c_3, 5c_5$, etc., helps to make subsequent working easier.

Consider a typical term of the series

$$N_x = (M/\pi a^2) n c_n \sin n\phi \qquad (13.40)$$

in relation to the equilibrium condition for the circumferential bending moment M_ϕ in a hoop of the tube. Expression (13.27) still holds, and so indeed does the first expression in (13.29). The only difference between the previous case and the present one is the functional form of N_x. Hence, instead of (13.30) we now have, for this component of N_x,

$$\frac{d}{d\phi}\left(\frac{1}{\sin\phi}\frac{dM_\phi}{d\phi}\right) = \frac{M}{\pi b} n c_n \sin n\phi.$$

This may readily be integrated, and as before the two constants of integration are both zero. Hence we obtain for this term

$$M_\phi = \frac{M}{\pi b}\frac{c_n}{2}\left(\frac{\cos(n+1)\phi}{n+1} - \frac{\cos(n-1)\phi}{n-1}\right). \qquad (13.41)$$

This expression reduces to (13.31) (with $A_1 = A_2 = 0$) when $n = 1$. Thus we find, in general, that a single N_x-term of the form $\sin n\phi$ requires, for equilibrium, *two* cosine terms in M_ϕ with arguments $(n+1)\phi$ and $(n-1)\phi$, respectively. In particular, the presence of the sin 3ϕ-term in N_x can help to remove a portion of the $\cos 2\phi$-term in M_ϕ which is necessary (see the preceding analysis) for equilibrium against the fundamental $\sin\phi$ term in N_x; and so we see that the presence of a $\sin 3\phi$ term in the expression for N_x may help to reduce the total complementary energy even though, of course, it also tends to increase the energy on account of its contribution to N_x and the $\cos 4\phi$ term in M_ϕ.

The general expression for M_ϕ, corresponding to (13.39) is:

$$M_\phi = \frac{M}{2\pi b}\left[\left(\frac{1-c_3}{2}\right)\cos 2\phi + \left(\frac{c_3-c_5}{4}\right)\cos 4\phi + \left(\frac{c_5-c_7}{6}\right)\cos 6\phi + \ldots\right].$$

(13.42)

Since the components of the series in (13.39) and 13.42) are orthogonal, it is a simple matter to evaluate the two complementary energy quantities corresponding to the generalised distribution of N_x in (13.39): cf. (13.32) and (13.34).

$$C_S = \frac{M^2}{2\pi E t a^3}\left[1 + (3c_3)^2 + (5c_5)^2 + \ldots\right]$$

(13.43a)

$$C_B = \frac{3M^2 a}{2\pi E t h^2 b^2}\left[\tfrac{1}{4}(1-c_3)^2 + \tfrac{1}{16}(c_3-c_5)^2 + \tfrac{1}{36}(c_5-c_7)^2 + \ldots\right]$$

(13.43b)

The flexibility factor is given in general by

$$f = (C_S + C_B)/C_{S1},$$ (13.44)

where C_{S1} is the value of C_S corresponding to a straight tube (see (13.34)). Thus we find, in general,

$$f = 1 + (3c_3)^2 + (5c_5)^2 + \ldots$$
$$+ 3\Gamma^2\left[\tfrac{1}{4}(1-c_3)^2 + \tfrac{1}{16}(c_3-c_5)^2 + \ldots\right].$$ (13.45)

For a given curved tube, the value of the geometrical parameter Γ is fixed. For arbitrary values of c_3, c_5 etc. we obtain, as explained above, an *overestimate* of the flexibility factor f. What we should do, therefore, is to determine those values of c_3, c_5 etc. which *minimise* f for a given value of Γ. This is not difficult, at least in principle: we must simply solve the equations $\partial f/\partial c_3 = \partial f/\partial c_5 = \ldots = 0$. These equations are, by inspection,

$$k_3 c_3 - \tfrac{1}{16}c_5 \qquad\qquad = \tfrac{1}{4} \qquad (13.46a)$$
$$-\tfrac{1}{16}c_3 + k_5 c_5 - \tfrac{1}{36}c_7 \qquad = 0 \qquad (13.46b)$$
$$-\tfrac{1}{36}c_5 + k_7 c_7 - \tfrac{1}{64}c_9 \quad = 0 \qquad (13.46c)$$

. . . .

13.4 Pure bending of a curved tube

where

$$k_3 = \tfrac{1}{4} + (9/3\Gamma^2) + \tfrac{1}{16}, \quad k_5 = \tfrac{1}{16} + (25/3\Gamma^2) + \tfrac{1}{36}, \ldots;$$

$$k_n = \frac{1}{(n-1)^2} + \frac{n^2}{3\Gamma^2} + \frac{1}{(n+1)^2}. \qquad (13.46d)$$

Beskin solved these linear recurrence relations for various values of Γ. He found that in general (i.e. for all values of Γ) $c_3 > c_5 > c_7$ etc. His results are shown in fig. 13.8. Although they stop short at about $\Gamma = 30$, they suggest that the value of $c_3 \to 1$ as $\Gamma \to \infty$; and indeed that the values of the other coefficients also approach 1 for sufficiently large values of Γ. The diagram shows clearly that more nonzero coefficients are required for larger values of Γ. We shall comment later on the significance of the tendency for $c_n \to 1$ as $\Gamma \to \infty$.

It is a straightforward matter to evaluate f from (13.45) by reading off coefficients from fig. 13.8. This is shown in fig. 13.9, which indicates not only the total value of f but also the separate contributions from the various terms. Also shown is the first approximation, (13.38). It is clear that the process of minimisation works in such a way that any term which tends to grow too rapidly is countered by the growth of the next term in the series. Thus there is a sort of 'staircase' effect in the building up of the total.

A remarkable feature of fig. 13.9 is the fact that for $\Gamma > 1.5$, f is practically *linear* with Γ. This linearity persists for higher values of Γ, as may be seen from the logarithmic plot of fig. 13.10: the series (13.45) sums almost exactly to $3^{\frac{1}{2}}\Gamma$ for high values of Γ.

Even if we cannot understand completely why the sum has so simple a form, we can suspect that there are some aspects of the solution which may

Fig. 13.8. Coefficients in the Fourier series (13.39) for N_x, as computed by Beskin (1945) for a range of values of the geometrical parameter Γ.

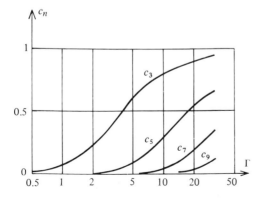

452 *Curved tubes and pipe-bends*

Fig. 13.9. Flexibility factor f for a curved tube, plotted against the geometrical parameter Γ. The broken curve corresponds to the preliminary analysis in section 13.4. The continuous curve shows Beskin's result, while the chain-dotted lines show the contributions from the various Fourier components.

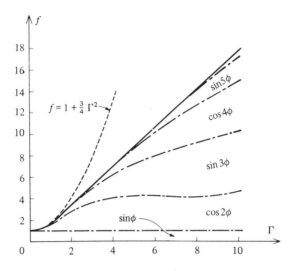

Fig. 13.10. Logarithmic plot of the flexibility factor and two stress-concentration factors against Γ. The broken lines are asymptotes.

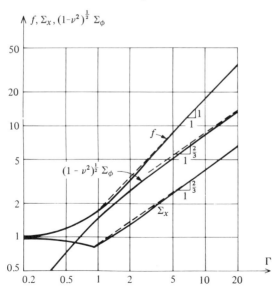

13.4 Pure bending of a curved tube

be more clear when viewed from a different standpoint. Now our prime intention was to use the complementary energy method in order merely to obtain *approximations* to the flexibility of a tube (e.g. (13.38)); and we must therefore be careful not to assign too much significance to the associated stress distributions N_x and M_ϕ, since these are to some extent the result of arbitrary choice of coefficients. But on the other hand, if we use the method, as above, in a *systematic* minimisation, we can argue that the stress resultants which are obtained by substituting the coefficients as determined by (13.46) are correct in the sense described in section 13.4. In fact (13.46) may be obtained from the compatibility condition relating ϵ_x and κ_ϕ, together with Hooke's law (see problems 13.2 and 13.3). Fig. 13.11a, b shows the distributions of N_x and M_ϕ which are found in this way for various values of Γ. An obvious feature is that when Γ increases, the stress resultants N_x and M_ϕ are crowded into a progressively narrower zone of the shell, near the line $\phi = 0$ (and also, of course, the line $\phi = \pi$); and there is a corresponding localisation of the deformation of the cross-section.

Clark & Reissner (1951) argued that in these circumstances it should be possible to set up an approximate differential equation, valid only for small values of ϕ, but which could be solved once and for all values of Γ. Their work, which we shall not describe in detail here, indicates that $\Gamma^{-\frac{2}{3}}N_x$ and $\Gamma^{-\frac{2}{3}}M_\phi$ are unique functions of $\Gamma^{\frac{1}{3}}\phi$ in this range, for a given value of M. Accordingly, the data of fig. 13.11a, b have been replotted in this way in fig. 13.11c and d; and it is clear that for $\Gamma > 4$, say, the distributions of N_x and M_ϕ are indeed standardised in this way for practical purposes. Clark & Reissner also show that in this range

$$f = 3^{\frac{1}{2}}\Gamma. \tag{13.47}$$

The main feature here is that for large values of Γ the 'boundary condition' at $\phi = \frac{1}{2}\pi$ is of little consequence, and the solution 'settles out' in a relatively simple way at small values of ϕ. It is interesting to note that in these circumstances the complementary energies of bending and stretching are equal.

13.4.2 Peak values of stress

It is useful to describe the distributions of N_x and M_ϕ around the circumference of the tube with reference to the *peak* values of these functions. It is convenient to define a longitudinal stress σ_x by

$$\sigma_x = N_x/t, \tag{13.48}$$

and an 'extreme fibre' hoop bending stress σ_ϕ by

$$\sigma_\phi = 6M_\phi/t^2. \tag{13.49}$$

454 Curved tubes and pipe-bends

The peak values of σ_x and σ_ϕ may then be normalised with respect to the peak longitudinal stress which is caused in the pure bending of a straight tube by a bending moment M, i.e. $M/\pi a^2 t$. In this way we may obtain expressions

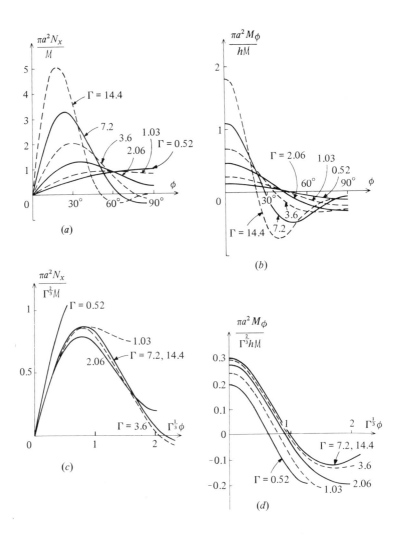

Fig. 13.11. (a), (b): Dimensionless plots of N_x and M_ϕ against the meridional angle ϕ, for different values of Γ. (c), (d): The same information replotted in a way (due to Clark, 1950) which gives a unique asymptote as $\Gamma \to \infty$ in each case.

13.4 Pure bending of a curved tube

for the longitudinal and circumferential 'stress-concentration factors' Σ_x and Σ_ϕ respectively. In particular we find

$$\Sigma_x \approx 0.865\, \Gamma^{\frac{2}{3}} \quad \text{for } \Gamma > 1, \tag{13.50a}$$

$$\Sigma_\phi \approx 1.8\Gamma^{\frac{2}{3}} \quad \text{for } \Gamma > 4, \text{ say.} \tag{13.50b}$$

Fig. 13.10 is a composite diagram showing the flexibility factor f and the two stress-concentration factors Σ_x, Σ_ϕ as functions of Γ. The discontinuous slope of the Σ_x curve corresponds to a change in location of $N_x^{\max}$: see fig. 13.11a.

It is important to realise that the 'peak stresses' defined by (13.48), (13.49) in terms of the peak values of N_x and M_ϕ do not do justice to the two-dimensional nature of the stress at a point on the surface of the shell: cf. section 11.4.3. Nevertheless, the factors Σ_x, Σ_ϕ should give at least a qualitative guide to the stress levels to be expected in the pure bending of a given curved pipe.

Kellog (1956, §3.1) discusses the relation between formulas (13.47), (13.50) and experimental observations from static and fatigue tests. It seems to be generally agreed that (13.47) is about correct but that (13.50) consistently overestimates experimentally observed values by a factor of about 2. The reason for this discrepancy is not clear.

13.4.3 Discussion

Throughout this section we have assumed that $b \gg a$ in order to avail ourselves of certain simplifications in the analysis. In particular, if $b \gg a$ a typical circumferential 'slice' has almost constant 'axial' length around the circumference, and the evaluation of the complementary energy is more straightforward than it would be otherwise.

Now it is clear from the principles of dimensional analysis that the behaviour of a structure whose geometry is specified by three variables (a, b, t) may be described completely in terms of *two* dimensionless groups. Hence we must conclude that f is, strictly, a function not only of Γ but also of a second group, say b/a. Therefore, it might be argued that in finding that f is a function of only one parameter, namely Γ, we have been led astray in consequence of our initial assumption that $b \gg a$.

Several authors, notably Symonds & Pardue (Appendix of Pardue & Vigness, 1951) have investigated the effect of this second dimensionless group on the flexibility factor for a curved tube and the distribution of stress within the tube. The general conclusion of these studies is that f depends on (b/a) to a negligible extent; but that the distributions of stress are affected somewhat more. This should not be a surprising result, particularly for $\Gamma > 7$, say: in this case the zones of interest are relatively narrow, in the regions of $\phi = 0$ and $\phi = \pi$, and there is virtually nothing left to be integrated in the regions $\phi = \frac{1}{2}\pi, \frac{3}{2}\pi$ in which the ratio b/a might be expected to have a significant effect: see fig. 13.11.

In this section we have presented results in terms of the flexibility of a curved pipe as a multiple, f, of the flexibility of a straight pipe having the same values of a and t. In the case of (say) a right-angle bend it is often more useful for the designer to think of the flexibility of the whole bend in terms of an *additional length of straight pipe*. If we assume that there are no end-effects at the junctions between the curved and straight pieces, we obtain the following simple expression for the additional 'equivalent length' L_b:

$$L_b = \tfrac{1}{2}(f-1)\,\pi b. \tag{13.51}$$

Now when $\Gamma > 1.5$ we may use $f = 3^{\frac{1}{2}}\Gamma$: and for $\Gamma > 4$, say, we may moreover use f as a reasonably close approximation to $f-1$. Hence we obtain, on expressing Γ in terms of a, b and t:

$$L_b/a \approx (\pi 3^{\frac{1}{2}}/2)(a/h) \approx 2.7\,(a/h) \quad \text{for } \Gamma > 4, \text{ say.} \tag{13.52}$$

The most remarkable feature of this result is that the 'equivalent length' is independent of the bend radius b, and is therefore a function only of a/t. For smaller values of Γ, of course, the value of b plays a part: see problem 13.4.

The fact that L_b is independent of b suggests that formula (13.52) may be approximately valid even for a single-mitre bend, although the differences between a smooth bend and a mitre joint are too great to warrant the adoption of the entire analysis without further investigation.

An unknown factor in the preceding calculation is the strength of any *end-effects* which may be present at junctions between the curved and straight pipes. This question is the subject of section 13.6.

13.5 The effect of internal pressure

We have seen in the preceding section that when a curved tube is subjected to pure bending, its originally circular cross-section is distorted into a slightly ovalised configuration, without an appreciable change in perimeter. Therefore there will in general be a small change in the enclosed cross-sectional area of the tube, since of all figures having a given perimeter the circle encloses the maximum area. Accordingly, the enclosed volume of the tube will be altered slightly. It follows that if the tube which is to be bent sustains a (constant) interior pressure, the overall flexural stiffness of the tube must be expected to be somewhat greater than in the absence of the pressure. This change in flexural stiffness on account of interior pressure is a separate

13.5 The effect of internal pressure

phenomenon from the change of curvature which the tube experiences when it is pressurised in the absence of bending moment (see section 13.2).

The most straightforward way of analysing the influence of internal pressure on the flexural stiffness of a curved tube is to observe that the 'ovalisation' of a typical cross-sectional ring is a response to 'pinching' forces imposed by the primary longitudinal bending stresses in the tube, as shown in fig. 13.7*d*. Now a short circular hoop which has been cut from a uniform thin cylindrical shell has a definite 'modal stiffness' against deflections of the form $w = w_n \cos n\phi$ (see fig. 13.12*a*), as we saw in section 8.6. If the ring is now subjected to an additional uniform constant internal pressure, as indicated in fig. 13.12*b*, the modal stiffness is increased by a simple factor which is a function of the pressure. This increase is due to a 'stretched string' effect in which the tension set up in a circular string as in fig. 13.12*b* is in equilibrium with additional loading of the type shown in fig. 13.12*a* when the string undergoes a small deflection of the form $w = w_n \cos n\phi$. This effect is directly analogous to the way in which the bending stiffness of a straight beam is apparently increased by the presence of a tension in the beam: see section 3.7.3. It is clear that this stiffening effect is directly proportional to p; and the factor by which the purely flexural stiffness of the hoop is increased may therefore be written

$$(1 + p/p^*), \qquad (13.53)$$

where p^* is a constant pressure, to be determined. The calculation, which combines results on strings and rings from sections 4.2 and 8.6, respectively (see problem 13.5) shows that for a circumferential mode-number n,

$$p_n^* = (n^2 - 1)Eth^2/12a^3. \qquad (13.54)$$

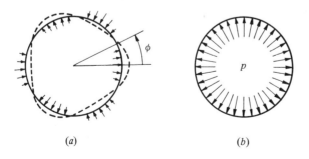

Fig. 13.12. (*a*) Modal response of a uniform circular ring to a simple periodic loading. (*b*) The ring is apparently stiffer in the same mode if it is pre-loaded uniformly as shown.

The factor (13.53) applies whether the value of p is positive or negative, i.e. whether the pressure is internal or external. In particular, the modal stiffness *vanishes* if $p = -p^*$; and in terms of classical buckling theory the hoop is then on the point of buckling under external pressure in the given mode. It may be verified that the magnitude of p_n^* in (13.54) agrees with an expression to be derived in the next chapter for the buckling of a long tube (see (14.77), $\Omega \to \infty$).

It follows from these considerations that the effect of interior pressure on the flexibility of a curved tube in pure bending may be studied by means of a relatively simple modification of the analysis which we developed in the preceding section. For a given interior pressure all we need do is to alter the various flexibility coefficients in our calculation of the complementary energy of circumferential bending. The only difficulty comes from the fact that there is a different factor for each of the circumferential bending modes having mode-numbers 2, 4, 6 etc.

Now from (13.54) we have

$p_2^* = Eth^2/4a^3$,

$p_4^* = 5p_2^*$, $\quad p_6^* = 11.7p_2^*$, $\quad p_8^* = 21p_2^*$, etc. $\hspace{2cm}$ (13.55)

Thus the modal stiffness factors on account of internal pressure are

$\quad (1+p/p_2^*) \hspace{2cm}$ for $n = 2$, $\hspace{4cm}$ (13.56a)

$\quad (1+0.2p/p_2^*) \hspace{1.5cm}$ for $n = 4$, $\hspace{4cm}$ (13.56b)

$\quad (1+0.085p/p_2^*) \hspace{1cm}$ for $n = 6$, etc. $\hspace{3.5cm}$ (13.56c)

It is clear from this that modes having lower wavenumbers are more strongly affected by interior pressure; and in particular that $n = 2$ is the mode which is most strongly affected.

As a first step, let us investigate the effect on the overall flexibility factor f of a change of the circumferential bending flexibility factor *for the mode $n = 2$ only*. This calculation is suggested by the above remarks. We shall be able to assess later the range of circumstances in which this simple calculation is adequate.

Suppose that, on account of interior pressure (or indeed, some other cause – see later), the *flexibility* of the 'cos 2ϕ' circumferential bending mode is altered by a factor ζ. The only consequent change in expression (13.43) is that the term $\frac{1}{4}(1-c_3)^2$ must now be premultiplied by the factor ζ. It follows that the recurrence relations (13.46) for the minimisation of f must be altered; but in fact only the *first* of these equations needs to be changed, thus:

$$c_3[\tfrac{1}{4}\zeta + (9/3\Gamma^2) + \tfrac{1}{16}] - \tfrac{1}{16}c_5 = \tfrac{1}{4}\zeta. \hspace{2cm} (13.57)$$

13.5 The effect of internal pressure

When $\zeta = 1$, this equation is exactly the same as before. For any given value of Γ let the values of c_3, etc, resulting from the solution of (13.46) be denoted by

$$c_3^0, c_5^0 \text{ etc..}$$

Since (13.46b), (13.46c)...) are not changed, they are *all* satisfied exactly by

$$c_n = \gamma c_n^0 \quad (n = 3, 5, 7 \text{ etc.}) \tag{13.58}$$

for an arbitrary single value of γ. In other words, the only consequence of the introduction of the factor ζ is that *all* of the coefficients c_3, c_5 etc. are altered by the *same* factor γ. We may find the value of γ as follows.

First we write (13.46a) for $\gamma = 1$, with the known values of c_3 and c_5:

$$c_3^0 \left[\tfrac{1}{4} + (9/3\Gamma^2) + \tfrac{1}{16}\right] - \tfrac{1}{16} c_5^0 = \tfrac{1}{4}. \tag{13.59}$$

Next we write (13.57) for a general value of ζ, using $c_3 = \gamma c_3^0$ etc:

$$\gamma c_3^0 \left[\tfrac{1}{4}\zeta + (9/3\Gamma^2) + \tfrac{1}{16}\right] - \tfrac{1}{16} \gamma c_5^0 = \tfrac{1}{4}\zeta. \tag{13.60}$$

Then we divide (13.60) throughout by γ, and subtract from (13.59) to obtain:

$$\tfrac{1}{4} c_3^0 (1-\zeta) = \tfrac{1}{4} [1 - (\zeta/\gamma)]. \tag{13.61}$$

From this we find

$$\gamma = \zeta/(1 - c_3^0 + \zeta c_3^0), \tag{13.62}$$

which enables us to evaluate γ for a tube having a given value of Γ (hence c_3^0 from fig. 13.8) and for an arbitrary value of the factor ζ. It is then a straightforward matter to re-evaluate f, since all of the terms $(3c_3)^2$ etc. and $\tfrac{1}{16}(c_3 - c_5)^2$ etc. in (13.45) have their previous values multiplied by γ^2. Only the term $\tfrac{1}{4}(1-c_3)^2$ is slightly awkward; but it is easy to show (see problem 13.6) that it must be multiplied by γ^2/ζ. The results of calculations along these lines are shown in fig. 13.13. By inspection of the diagram we see that for $f < 3$, say, the curves for different values of ζ may be obtained from each other by translation in the Γ-direction; whereas for large values of Γ and f this rule obviously does not apply.

We have already seen that for sufficiently small values of Γ the coefficients c_n are all approximately zero, and that consequently (13.45) reduces to (13.38):

$$f = 1 + 0.75\Gamma^2.$$

In this range the effect of the introduction of factor ζ is simply to change the formula to

$$f = 1 + 0.75\zeta\Gamma^2. \tag{13.63}$$

Thus the original curve (for $\zeta = 1$) may still be used if we put

$$\Gamma_{\text{eff}} = \Gamma \zeta^{\tfrac{1}{2}}. \tag{13.64}$$

460 Curved tubes and pipe-bends

This is valid for values of Γ and ζ for which the coefficients c_n are sufficiently small, i.e. for sufficiently small values of f.

Now the calculations leading to the above results are based explicitly on a change in the flexibility of a single circumferential bending mode, namely $\cos 2\phi$. To the extent that this is the dominant effect in the response of the tube when the internal pressure is raised, this figure may be read in terms of interior pressure by means of the relation

$$\zeta = (1 + p/p_2^*)^{-1}. \tag{13.65}$$

Crandall & Dahl (1957) made a thorough investigation of the effect of pressure on the pure (in-plane) bending of curved tubes (cf. Reissner, 1959). A detailed comparison of results obtained in the way described above with those of Crandall & Dahl shows that results obtained by our simple method are excellent for $\Gamma < 50$, but that they become progressively worse for larger values of Γ. The discrepancy is due, no doubt, to the increasing importance of the $\cos 4\phi$ and higher-order terms in circumferential bending as the value of Γ increases. We shall not pursue this matter further here, but we quote a result

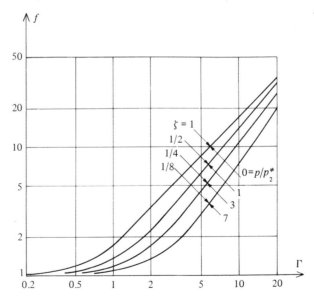

Fig. 13.13. Logarithmic plot of flexibility factor f against Γ for different values of the factor ζ by which the flexibility of the 'cos 2ϕ' mode of ring bending has been multiplied. Values of ζ correspond to different levels of interior pressure: high pressure gives small values of ζ.

13.6 End-effects in the bending of curved tubes

which may be obtained by inspection of some 'asymptotic' calculations done by Crandall & Dahl:

$$f = 3^{\frac{1}{2}}\Gamma \Big/ \left(1 + \frac{1.15}{\Gamma^{\frac{2}{3}}} \left(\frac{p}{p_{\frac{1}{2}}^*}\right)\right). \tag{13.66}$$

This asymptotic expression appears from the results of Crandall & Dahl to be reasonably accurate for $\Gamma > 20$. It gives curves which lie a little below those of fig. 13.13; but the discrepancy is not more than about 10% for $\Gamma > 10$ and $p/p_{\frac{1}{2}}^* < 7$. Formula (13.66) agrees closely with the experimental observations of Kafka & Dunn (1956) for tubes in the region of $\Gamma = 12$. Experimental results of Rodabaugh & George (1957) and numerical calculations of Blomfield & Turner (1972) also agree well with fig. 13.13.

13.6 End-effects in the bending of curved tubes

A segment of curved pipe may be connected to adjacent straight pipes by smooth welded joints, as shown schematically in fig. 13.14a, or alternatively it may be connected by flanged joints, as shown in fig. 13.14b.

Now an essential feature of the uniform in-plane bending of curved tubes, with which we have been concerned so far in this chapter, is that bending produces an 'ovalisation' in the cross-sectional shape, which in turn leads directly to the enhanced flexibility of the tube. It seems obvious, therefore, that the effect of flanges at the ends of a section of curved tube (fig. 13.14b) will be to inhibit to some extent the ovalisation of the tube, and thereby to reduce its flexibility. On the other hand it is not nearly so clear that the tendency for

Fig. 13.14. (a) Curved pipe joined smoothly to straight tubes. (b) As (a), but with substantial flange connections. (c) Curved tube with blanked-off flanges.

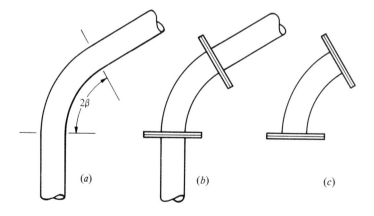

adjoining *straight tubes* (fig. 13.14a) *not* to ovalise will have an appreciable inhibiting effect in this respect.

The aim of this section is to give an approximate analysis of restraining effects of this sort, in order to give a qualitative understanding of the main factors which are involved in the stiffening effects of flanges. The results which we shall obtain agree reasonably well with the observations of a number of available experimental investigations, and thus seem appropriate for use in preliminary design studies.

When a tube like that which is shown in fig. 13.14a is subjected to a pure in-plane bending moment, there will be a stress resultant N_x which varies over the surface of the tube in both longitudinal and circumferential directions. We may express it as a Fourier series (13.39) in the circumferential direction, with each coefficient c_3, c_5 etc varying along the length of the tube. One thing is certain: the coefficient of the fundamental term ($\sin\phi$) does *not* vary along the length, since it is determined by overall equilibrium, and the bending moment M is constant. In section 13.4 we began our analysis by assuming, in effect, that all of the coefficients c_3, c_5, etc were zero. Then we investigated the 'pinching' effect produced by N_x in a curved tube, and determined the amount of distortion of the tube by considering the elastic deformation of a circumferential 'ring' of the tube. This particular calculation is tantamount, in fact, to an investigation of the ovalisation of a *straight* tube when it is subjected to the pinching effect due to N_x in a *curved* tube.

Consider now the tube shown in fig. 13.14a, also on the assumption that the coefficients c_3, c_5, etc. all have negligible magnitude. In the curved portion there is a 'pinching' effect, just as before; but in the straight parts there is not. So in terms of the effect on an equivalent straight tube we now have a tube (fig. 13.15) which suffers the pinching force over only a finite length equal to $2\beta b$, where 2β is the angle subtended by the bend (fig. 13.14a). How does a tube respond to a loading of this sort? As we have seen in chapter 9 the tube carries such a load primarily by a combination of circumferential bending and longitudinal stretching. In the preceding sections longitudinal stretching was not called on for this purpose, since the pinching effect was uniform over the entire length; but here it plays a crucial role. It is important to realise that we are here discussing longitudinal stress resultants N_x which vary as $\cos 2\phi$, which is quite distinct from the family of Fourier terms which we have considered previously (see (13.39)). In particular, it gives an additive longitudinal tensile stress in the region of $\phi = 0$ which appears to shift the 'neutral axis' of bending towards the extrados.

As we saw in chapter 9, the problem illustrated in fig. 13.15 boils down to the behaviour of a beam-on-elastic-foundation, with circumferential bending

13.6 End-effects in the bending of curved tubes

in the tube playing the part of the foundation, and longitudinal stretching playing the part of the beam: the pinching effect is analogous to uniform transverse loading on the beam.

It is straightforward to calculate the behaviour of such a beam. The deflection is shown schematically in fig. 13.16. The form of the deflection depends, of course, on the ratio of the length of the loaded part of the beam to the 'characteristic' length $\mu = (4B/k)^{\frac{1}{4}}$, where B and k are the elastic constants of the beam and foundation, respectively. We shall give quantitative results later, but for the present let us consider the consequences of this for the deflection of a curved tube. In section 13.3 we saw that the change of curvature of an initially curved beam consisted of two additive parts: (i) the change of curvature of an initially straight beam (ii) the change of curvature on account of change in separation of the two flanges. And in section 13.4 we found that the same additive feature occurs in the bending of curved tubes when the higher-order terms c_3, c_5 etc. are negligible. It follows that if the deflected form of fig. 13.16 is applied to the tube of fig. 13.14a, the extra bending of the joint on account of changes of cross-sectional shape is directly proportional to the area marked A_1. Note, in particular, that ovalising deflections

Fig. 13.15. Straight-pipe analogue of the tube of fig. 13.14a in pure bending, showing the 'pinching' effect over a limited section of the pipe.

Fig. 13.16. A beam-on-elastic-foundation analogue of the pipe shown in fig. 13.15, showing the deflected profile and the swept area A, which corresponds to change of curvature of the original curved tube.

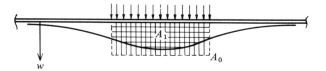

in the *straight* parts of the original tube do not contribute to additional curvature. Further, if A_0 is the area swept under the load for a uniformly loaded beam (where the deflection is determined by the elastic foundation constant alone), the ratio $A_1/A_0 = \zeta$, say, is equal to the 'extra' rotation of the bend of fig. 13.14a (on account of changes in cross-section) divided by the 'extra' rotation of an equivalent curved pipe without end restraint. In other words $\zeta = A_1/A_0$ is an overall flexibility factor against ovalisation, in the sense that the overall flexibility of the curved portion of the tube of fig. 13.14a is the same as that of a curved tube of the same dimensions under uniform bending moment (but with no end-effects, as in section 13.4.1) provided the flexibility against ovalisation of the cross-sectional rings is multiplied by the factor ζ. In fact the symbol ζ has precisely the same meaning as it did in section 13.5.

Before we can proceed further we must first investigate quantitatively the behaviour of a beam-on-elastic-foundation and determine the quantity $\zeta = A_1/A_0$ in terms of l/μ, where l is the length of the loaded region; and second we must express the characteristic length μ in terms of the dimensions of the shell, by using results from chapter 9.

Fig. 13.17, curve (a), shows the ratio $\zeta = A_1/A_0$ for a beam of infinite length. This corresponds to a tube like that of fig. 13.14a with flanges (if any) on the straight portions of tube sufficiently remote from the curved section. The curve was determined by the application of Hetényi's (1946) methods.

Fig. 13.17. The parameter ζ ($= A_1/A_0$, fig. 13.16) for the beam-on-elastic-foundation of fig. 13.16, for three different boundary conditions, shown inset.

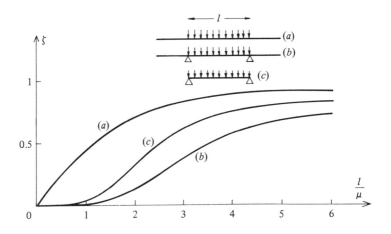

13.6 End-effects in the bending of curved tubes

Two other curves are also shown. Curve (b) is for the same arrangement but with additional unyielding simple supports at the ends of the loaded zone: this corresponds to the tube in fig. 13.14b with flanges which are rigid in their own plane but which do not inhibit axial displacements. Finally, curve (c) is for a short beam with simple supports, corresponding to an isolated curved segment which is blanked off by flanges or cover plates (see fig. 13.14c) which have the same properties as in curve (b) above. Real flanges are not, of course, perfectly rigid in their own plane, but it is obviously reasonable to treat them as such in a first calculation. Case (c) seems to correspond most closely with the experimental observations of Pardue & Vigness: see later.

Inspection of fig. 13.17 reveals clearly that for small values of l/μ the provision of simple supports (cases (b), (c)) severely reduces the deflection of the beam in comparison with case (a), whereas for sufficiently large values of l/μ there is not much difference in this respect between the three cases.

In order to express μ in terms of the dimensions of the shell we note that for the behaviour of a cylindrical shell in the 'long-wave' mode (which is dominated by longitudinal stretching and circumferential bending) the equivalent beam and elastic foundation stiffness are given by (see section 9.4)

$$B \leftrightarrow \frac{\pi E t a^3}{n^4}, \quad k \leftrightarrow \frac{(n^2-1)^2 \pi E t h^2}{12 a^3}, \tag{13.67}$$

where n is the circumferential wavenumber. Consequently we have, for $n=2$,

$$\mu = a^{\frac{3}{2}}/3^{\frac{1}{4}} h^{\frac{1}{2}}. \tag{13.68}$$

The parameter l/μ is closely related to the geometrical group Ω which was used in chapter 9.

Our scheme of calculation of the flexibility of bends such as those shown in fig. 13.14 may be summarised as follows.

(i) Calculate the value of Γ from the dimensions a, b and h, using (13.37).

(ii) Evaluate μ from (13.68), and then, taking $l = 2\beta b$, evaluate l/μ.

(iii) Using fig. 13.17 (curve (a), (b) or (c) as appropriate) read off the value of ζ for the known value of l/μ.

(iv) Using fig. 13.13 read off the flexibility factor for Γ (from (i)) and ζ (from (iii)).

Before we give examples of calculations of this sort we should emphasise some of the restrictive assumptions on which the analysis has been based. The major simplifying feature is the notion that the end-conditions on account of flanges, etc. apply most strongly to the $n = 2$ mode of ovalisation, and that a single equivalent flexibility factor ζ for this mode may be evaluated. The resistance of a curved tube to non-uniform ovalisation in the manner of a straight tube is straightforward to justify by inspection of expression (6.6) for change of Gaussian curvature. But what about the higher-order terms corresponding to the coefficients c_3, c_5 etc.? The main point here is that although the end-conditions will undoubtedly affect these higher modes, the corresponding values of μ are so small that the end-restraint will be much less significant than for the mode $n = 2$. Thus we may argue that the primary consequence of the end-fittings is to reduce the effective flexibility of the $n = 2$ bending mode; and the remaining terms in the energy expression follow suit, almost exactly in the manner of section 13.5.

13.6.1 Examples of joints with flanges

It is clear from the work of the preceding section that much depends on the value of l/μ in a given case: see fig. 13.17. The value of l depends, of course, on the value of β: thus, other things being equal, the value of l/μ for a 180° bend ($2\beta = 180°$) will be double that for a 90° bend ($2\beta = 90°$). It is therefore convenient first to work out the value of l/μ for a 90° bend, and then to scale it, if necessary, for other values of β. Thus we find from (13.37) and (13.68) that

$$l/\mu \approx 2(a/h)^{\frac{1}{2}}/\Gamma \quad \text{when } 2\beta = 90°. \tag{13.69}$$

Fig. 13.18 shows a logarithmic plot of b/a against a/h on which contours

Fig. 13.18. Double-logarithmic plot from which values of Γ and l/μ may be read for a given curved pipe *turning through* $2\beta = 90°$. For different values of 2β, l/μ is directly proportional to 2β. The points correspond to experiments performed by Pardue & Vigness (1951).

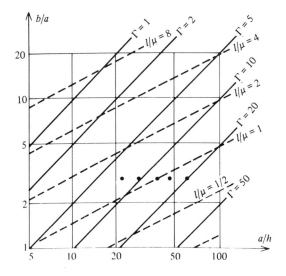

13.6 End-effects in the bending of curved tubes

of both Γ and l/μ have been marked. For any given design $(a/h, b/a)$ it is easy to interpolate approximate values of Γ and l/μ and then, with the aid of figs. 13.17 and 13.13 to find the corresponding flexibility factor. Some relatively simple general conclusions may be drawn by means of visual inspection of figs. 13.18 and 13.17. For example, we might decide, on the basis of fig. 13.13, that all arrangements having $\zeta > 0.5$ give relatively *small* reduction in the flexibility factor. Accordingly we find from fig. 13.17 that if there are no flanges (case (a)) there will be relatively little reduction in flexibility provided $l/\mu > 1$, which marks out a well defined zone of 'design space' in fig. 13.18. On the other hand, if there are flanges (cases (b), (c) in fig. 13.17) we must have $l/\mu > 3$ if we are to avoid significant reductions of flexibility, which restricts designs to rather large values of b/a: see fig. 13.18.

In the course of some experiments on pipelines, Smith & Ford (1967) measured changes in diameter in a smooth bend connected to straight tubes without flanges when an in-plane bending moment was applied. The value of l/μ was about 2.5, and the measurements correspond almost exactly to those which are indicated by the beam-on-elastic-foundation analysis.

Pardue & Vigness (1951) did some careful experiments on the flexibility of small-scale pipe-bends, and it is useful to discuss their results in the context of the present work. Their tubes were made from steel, and had $a \approx 2.5$ in. and $b \approx 7.5$ in. The specimens were made in five thicknesses, spaced uniformly in the range $23 \leq a/t \leq 69$: thus values of Γ were in the range $7.7 \leq \Gamma \leq 23$. The five 'design points' are marked on fig. 13.18. There were L- and U-bends ($2\beta = 90°$ and $180°$ respectively) for all thicknesses, and the bends were tested both without flanges and also as isolated bends with end-plates which will be referred to as 'flanges'. Careful measurements were made of the flexibility of the bends for a variety of in-plane and out-of-plane loadings. For in-plane bending the results may be summarised as follows.

(1) With no flanges the flexibility of both the U- and L-bends was close to the theoretical value for uniform bending.

(2) For the U-bends with two flanges the flexibility was less than that for no flanges. The factors by which the flexibilities were reduced on account of the flanges were in the region of 0.8, and are given for three specimens in table 13.1: the results for the intermediate specimens were consistent with these.

(3) For the L-bends with two flanges the results were of the same kind, but the reduction factors were more severe, being in the region of 0.4: see table 13.1.

The results (1) are consistent with our earlier remarks. For (2) and (3) we can use figs. 13.18, 13.17 and 13.13 in order to estimate the amount by which the flanges reduce the flexibility; but we must first decide what boundary conditions to adopt. Pardue & Vigness intended their end-plates to be so massive as to be equivalent to a flanged connection with neighbouring straight pipes. In terms of a beam-on-elastic-foundation this would correspond roughly to case (b) of fig. 13.17. It is possible, however, that in some cases at least the end-plates did not provide complete rigidity; and in such cases the values of ζ would lie somewhere between those corresponding to curves (b) and (c) of fig. 13.17. It turns out in fact that the values of ζ which may be deduced by plotting the experimental flexibility factors onto fig. 13.13 correspond closely to curve (c) for both sets of bends. In the following calculations we shall use curve (c) for the evaluation of ζ in a given case even though theoretical considerations suggest that curve (b) would be more appropriate. Consider first the U-bends. Either by application of (13.69) or by use of fig. 13.18 (doubling the labelled values of l/μ for a 180° bend) we obtain values of l/μ for the various tubes. Then from curve (c) of fig. 13.17 we read off the appropriate value of ζ. Finally, we use this on fig. 13.13 to interpolate the flexibility factor. We express this in table 13.1 as a number by which the flexibility factor f for uniform bending (curve $\zeta = 1$, fig. 13.3) must be multiplied in order to give the required flexibility factor. The corresponding

Table 13.1. Analysis of the experimental results of Pardue & Vigness (1951) on the flexibility of pipe-bends with flanges, in relation to the theory presented in section 13.6.1. In each case the value of ζ is taken from fig. 13.17, curve (c).

U-bends ($\beta = 90°$)	$\dfrac{b}{a}$	$\dfrac{a}{t}$	Γ	l/μ	ζ	Flexibility factor on account of flanges	
						calculated	measured
	3	69	23	1.4	0.13	0.6	0.72
	3	40	13.3	2.0	0.32	0.7	0.78
	3	23	7.7	2.4	0.47	0.8	0.90
L-bends ($\beta = 45°$)	$\dfrac{b}{a}$	$\dfrac{a}{t}$	Γ	l/μ	ζ	Flexibility factor on account of flanges	
						calculated	measured
	3	69	23	0.7	0.008	0.1	0.2
	3	40	13.3	1.0	0.03	0.2	0.47
	3	23	7.7	1.2	0.07	0.3	0.5

13.6 End-effects in the bending of curved tubes

numbers from the experiments of Pardue & Vigness are also tabulated; and the agreement is satisfactory.

Consider next the L-bends. Following the same procedure as before we obtain first values of l/μ and then values of ζ from fig. 13.17, curve (c). These are lower than those which are plotted in fig. 13.13; but it is not difficult to plot more curves, as required, and to deduce values of the flexibility number. These are given in the table, together with the corresponding experimental results. In general the agreement between our calculations and the experimental observations is not so good as for the U-bends. In each case the calculated flexibility reduction factor is about one-half the measured one: our analysis seems to be *overestimating* the restraining effect of the flanges, in spite of our use of curve (c) of fig. 13.17 in preference to curve (b). However, the trends in the results are of the correct sense, and we may regard our calculations as being satisfactory as a very crude first approximation.

A noteworthy feature of the experimental measurements on L-bends is that when there are flanges, the flexibility factor is insensitive to the value of Γ: the value of f lies between about 6 and 8 over the entire range of Γ, whereas the corresponding value of f for uniform bending varies between 13 and 40. The explanation of this seems to be that in the range $0 < l/\mu < 1$, $\zeta \propto (l/\mu)^4$, approximately: see problem 13.7. Consequently, since $l/\mu \propto t^{\frac{1}{2}}$ (when the values of a and b are fixed) the factor $\zeta \Gamma^2$ in (13.63) is independent of t. An analysis along these lines shows that the flexibility factor for a right-angle bend is given by

$$f \approx 1 + 0.46(b/a)^2 \tag{13.70}$$

(see problem 13.8). For the right-angle bends of Pardue & Vigness this gives $f = 5.1$, compared with the measured value of about 7.

It is clear from the work of this section that the presence of flanges at the ends of (or indeed *near*) a curved section of pipe may seriously reduce the flexibility of the bend, and that the scheme of calculation which we have devised gives a rough estimate of the effects of flanges. In particular, although some constants may not be quite right, the dimensionless groups which we have used should be the correct ones.

In terms of design our main conclusion must be that it is rather difficult to arrange for flanges on a 90° bend which do not seriously reduce the flexibility of the junction; but that for 180° bends the effect of flanges is not nearly so strong.

13.7 Problems

13.1 Show by direct integration of (13.31) that

$$\int_0^{2\pi} M_\phi^2 \, d\phi = M^2/16\pi b^2 + \pi A_1^2 + 2\pi A_2^2$$

(since the three parts of the right-hand side of (13.31) are mutually orthogonal: see appendix 4); and hence that the choice $A_1 = A_2 = 0$ minimises the complementary energy in the ring.

13.2 The cross-sectional profile of the toroid shown in fig. 13.2 undergoes a change of curvature κ_ϕ in the mode $\kappa_\phi = \kappa_m \cos m\phi$, while $\epsilon_\phi = 0$. By using

$$\kappa_\phi = -(d^2w/d\phi^2 + w)/a^2$$

(cf. (8.57)) and

$$dv/d\phi + w = 0$$

(cf. (6.9b)), obtain expressions for the normal (w) and tangential (v) components of displacement. Let $\eta = w\sin\phi + v\cos\phi$ be the component of displacement normal to the line $\phi = 0$. Show that (in ϕ, x coordinates)

$$\epsilon_x \approx \eta/b$$

provided $b \gg a$; and hence obtain the following compatibility relation between ϵ_x and κ_ϕ:

$$\epsilon_x = \frac{a^2 \kappa_m}{2bm} \left(\frac{\sin(m+1)\phi}{m+1} - \frac{\sin(m-1)\phi}{m-1} \right).$$

13.3 Take a single component

$$N_{xn} = \frac{n}{\pi a^2} c_n \sin n\phi \quad (n = 3, 5, \text{etc.})$$

of the Fourier series (13.39) for N_x, and use equilibrium equation (13.41) together with Hooke's law $\kappa_\phi = M_\phi/D$ to obtain an expression for κ_ϕ. By use of the result of problem 13.2 (twice; putting $m = n-1$ and $m = n+1$ respectively) obtain an expression for ϵ_x. Isolate from this the coefficient of $\sin n\phi$, and equate it to the strain ϵ_{xn} determined from the original expression by Hooke's law

$$\epsilon_{xn} = N_{xn}/Et.$$

Verify that equations (13.46) are obtained in this way.

13.4 Show that the formulae $f = 1 + 0.75\Gamma^2$ (13.38) and $f = 3^{\frac{1}{2}}\Gamma$ (13.47) touch, when plotted in Γ, f space, at the point $(2/3^{\frac{1}{2}}, 2)$.

13.7 Problems

Show that the use of (13.38) in (13.51) leads to the expression
$$L_b/a \approx 1.2a^3/bh^2 \quad (a^2/bh < 0.9)$$
while (13.47) gives
$$L_b/a \approx 2.7a/h - 1.6b/a \quad (a^2/bh > 0.9).$$

13.5 The elastic ring analysed in section 8.6 is now subjected to a uniform outward-directed load of intensity p_0 in addition to the load $p_n \sin n\theta$; and in general $p_n \ll p_0$. In consequence it is necessary to modify the radial equilibrium equation (8.56a) in order to allow for the effect of 'string tension' on the *deformed* element, whose curvature is, in general, $a^{-1} + \kappa_\theta$: thus

$$\frac{dQ}{d\theta} - T(1 + a\kappa_\theta) = -ap_0 - ap_n \sin n\theta.$$

Now put $T = ap_0 + \bar{T}$ (where $\bar{T}(\theta) \ll ap_0$); expand the second term on the left-hand side, neglecting the term $\kappa_\theta a\bar{T}$ in comparison with $a^2 p_0 \kappa_\theta$ and $\bar{T}$; and hence obtain

$$\frac{dQ}{d\theta} - \bar{T} - a^2 p_0 \kappa_\theta = ap_n \sin n\theta$$

as the modified form of (8.56a).

Using (8.56b) and (8.56c) and also Hooke's law $\kappa_\theta = M/EI$, express this equation in terms of M as follows:

$$\frac{d^2 M}{d\theta^2} + M(1 - a^3 p_0/EI) = -a^2 p_n \sin n\theta.$$

Hence show that M is now given by (8.61) but with the right-hand side divided by the factor $[1 + a^3 p_0/(n^2 - 1)EI]$; and that consequently the modal stiffness of the ring is increased by this factor.

13.6 When c_3 is changed to γc_3^0 (see (13.58)), the quadratic term in (13.45) corresponding to the circumferential bending mode $n = 2$ is changed from $\frac{1}{4}(1 - c_3^0)^2$ to $\frac{1}{4}\zeta(1 - c_3)^2$. Use (13.61) to show that

$$(1 - \gamma c_3^0) = (1 - c_3^0)(\gamma/\zeta);$$

and hence that the quadratic term must be multiplied by the factor γ^2/ζ.

13.7 In relation to the simply-supported beam on an elastic foundation (fig. 13.17, case c) an adaption of a result of Hetényi (1946, p. 61) shows that the fraction of the applied load which is transmitted to the foundation springs (i.e. not to the end-supports) is equal to $1 - (\mu/l)(S+s)/(C+c)$, where

$S = \text{Sh}(l/\mu)$, $s = \sin(l/\mu)$, $C = \text{Ch}(l/\mu)$, $c = \cos(l/\mu)$: $\mu = (4B/k)^{\frac{1}{4}}$. Verify that this fraction is precisely equal to ζ, as defined in fig. 13.17; and show that in the limit $(l/\mu) \to 0$, $\zeta = (l/\mu)^4/30$. Also verify, by inspection of fig. 13.17, that this expression is a good approximation provided $l/\mu < 1.2$, say.

13.8 Use the result of problem 13.7, together with $l = \frac{1}{2}\pi b$, (13.37), (13.63) and (13.68) to show that $f \approx 1 + (3\pi^4/640)(b/a)^2$ for a right-angle bend with flanged ends (fig. 13.14c), for which $(a/h) > 3(b/a)^2$, approximately.

14

Buckling of shells: classical analysis

14.1 Buckling of structures

Buckling is a word which is used to describe a wide range of phenomena in which structures under load cease to act in the primary fashion intended by their designers, but undergo instead an overall change in configuration. Thus a rod which was originally straight, but has bowed laterally under an end-to-end compressive load has *buckled*; and so has a cylindrical shell, which has crumpled up under the action of the loads applied to it.

The buckling of structures is an important branch of structural mechanics, because buckling often (but not always) leads to failure of structures. It is particularly important in shell structures because it often occurs without any obvious warning, and can have catastrophic effects.

The buckling of shells has been studied intensively for about four decades, and the information now available on the subject is enormous. The aim of this chapter and the two following is to give an introduction to the subject in the simplest possible terms.

The 'classical' theory to be described in this chapter is merely an extension into the field of shell structures of what is often described as the 'Euler' theory of buckling of simple struts which are initially straight. For some problems in shell buckling this kind of theory is adequate, and well attested by experiment; but for other problems it is inadequate and indeed can be positively misleading. Chapter 15 aims to present the main ideas of the *nonlinear* elastic buckling theories which have been developed since the mid-1940s in order to explain the inadequacies of classical theory for these problems. Chapter 16 investigates a problem which demands a different type of nonlinear analysis.

Most students are introduced to the analysis of buckling in structures through Euler's example of an initially straight uniform elastic rod under endwise compression. The basic arrangement is shown in fig. 14.1*a*. Euler investigated all of the plane configurations in which it is possible for the rod to exist in a state of statical equilibrium under the applied loading. For a given distorted configuration (see fig. 14.1b) there are three conditions which must be satisfied simultaneously at every point on the deformed centre-line of the rod.

(i) The *equilibrium* relation between the load P and the bending moment M in the rod.

(ii) The *elasticity* relationship (Hooke's law) between the bending moment and the curvature κ.

(iii) The *compatibility* relation between the profile of the displaced rod and the local curvature.

In his first analysis of this problem Euler explicitly used the exact expression for the local curvature κ of a plane curve $y(x)$ in Cartesian coordinates (see (2.7)):

$$\kappa = \frac{d^2 y}{dx^2} \Big/ \left[1 + \left(\frac{dy}{dx}\right)^2 \right]^{\frac{3}{2}}.$$

A small portion of Euler's results are indicated in fig. 14.1c, which is a plot of load P against central lateral deflection w of the rod. The line $w = 0$ represents a 'trivial' solution in which the rod remains perfectly straight: certainly

Fig. 14.1. Various aspects of the buckling of a uniform pin-ended elastic column subjected to a compressive load. (a) and (b) show the column in straight and buckled configurations, respectively. (c) Primary and secondary equilibrium paths in P, w space for a column which is initially straight. The broken line is the secondary path according to the simplified 'classical' calculation. (d) The corresponding diagram for an initially crooked column: there is no point of bifurcation.

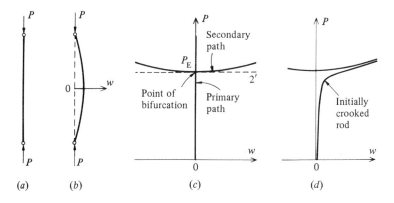

14.1 Buckling of structures

all of (i)–(iii) are satisfied in this case. This line is intersected by a curved path at

$$P = P_E = \pi^2 EI/l^2, \tag{14.1}$$

where EI is the flexural rigidity of the rod and l is the original length. Subscript E commemorates Euler. The portion of the curved path which is shown is rather shallow: the value of P rises only to $1.1 P_E$ when $w/l = 0.25$, which is certainly a 'large deflection' for most engineering purposes.

In modern terminology, fig. 14.1c shows two 'equilibrium paths' in P, w space corresponding to configurations for which all of (i)–(iii) above are satisfied. The secondary path intersects the primary path at a 'branching point' or 'point of bifurcation'.

In a later analysis Euler showed that the results of fig. 14.1c are only modified slightly for relatively small transverse displacements if (2.7) is replaced by the simplified form of (2.8):

$$\kappa = \frac{d^2 y}{dx^2}.$$

In this case the curved secondary path is replaced by the line $P = P_E$, which is labelled $2'$ in fig. 14.1c. In particular, however, the simplified treatment correctly gives the 'branching point' of the two equilibrium paths at $P = P_E$.

The 'small-deflection' analysis using the linearised expression (2.8) leads directly to an *eigenvalue* problem. It is this analysis which is usually presented first to students; and it is clear that for some buckling problems it is perfectly adequate. It certainly correctly emphasises the special value P_E of the load at which buckling of the initially straight rod begins.

Careful experiments made on slender elastic rods do not confirm Euler's results exactly: the experimental results invariably show a smooth path (see fig. 14.1d) which is asymptotic to Euler's two curves, and with a rounded corner which avoids the point of bifurcation of the curves. This discrepancy between theory and experiment is attributable to unavoidable lack-of-straightness of the rod and/or eccentricity of the loading. A small-deflection analysis of a rod which has a small initial bow (an analysis which was first done by Ayrton & Perry, 1886), indicates that as the load P increases the amplitude of the lateral bow is multiplied by a factor which is the following simple (but nonlinear) function of the applied load:

$$(1 - P/P_E)^{-1}. \tag{14.2}$$

We shall refer to this amplification factor for initial imperfections as 'Perry's rule' (cf. Shanley, 1967, §11.5; Timoshenko & Gere, 1961, p. 32). We have already come across related effects in sections 3.7.3 and 13.5. Not only does this relationship fit experimental results extremely well, but it also provides a way of extrapolating P to give P_E for the ideal or perfect structure, known as the Southwell Plot (e.g. Timoshenko & Gere, 1961, p. 190): see problem 14.1.

The success of the 'small-deflection' approach to the buckling of a simple elastic column has led to the development of the 'classical' theory of elastic buckling for a wide range of elastic structures. Attention is concentrated on the load condition required for bifurcation from a simple prebuckled configuration of the structure. Various approximate energy methods have been developed within the framework of classical theory for tackling awkward problems. A good account of this theory is given by Timoshenko & Gere (1961). Lorenz, Southwell, von Mises, Flügge, Schwerin and Donnell applied the classical theory of elastic buckling to cylindrical shells between 1911 and 1933: see Timoshenko & Gere (1961, Chapter 11); Brush & Almroth (1975, §5.1).

For many problems the classical theory predicted accurately the loading conditions under which an elastic shell would buckle under test. But it quickly became clear that there were certain exceptional cases, including thin cylindrical shells under uniform axial compression, for which the classical theory grossly overestimated the carrying capacity of actual shells. The classical theory to be presented in this chapter may be viewed as a linearisation of a strictly nonlinear problem. The paradoxical behaviour of some shells led to intensive work on the nonlinear theory of buckling of shells, and indeed of structures in general. There are various aspects of the nonlinear behaviour of shells, of which the Perry rule (14.2) is perhaps the simplest. Chapter 15 is devoted to a nonlinear study of the buckling of cylindrical shells under axial compression. In chapter 16 we study a different problem, concerning the buckling of a thin tube under pure bending. Here there is considerable nonlinear distortion of the structure before failure is precipitated by a bifurcation process.

The development of the classical theory in the present chapter is different from the treatment of most standard texts, mainly on account of our use of the 'two-surface' idea from chapter 8. The results, however, agree exactly with those of the more conventional treatment.

The structure of the chapter may be outlined as follows. First we introduce the 'two-surface' idea into the study of classical buckling in relation to simple columns and flat plates. Then we consider a cylindrical shell which buckles into a simple periodic waveform of the kind which we used in chapter 8; and indeed, our two-surface approach enables us to build on results which we obtained in that chapter. We consider buckling first under uniform axial load and second when this is combined with external pressure. In both cases we obtain relatively simple results when it can be assumed that the circumferential wavenumber is sufficiently large, so that the 'Donnell' or 'shallow-shell' approximations are valid; and we plot these in various graphs which are designed to show that the behaviour is characterised by the same two dimensionless groups which describe the geometry of the shell that we have encountered before. The next task is to extend the work to cases in which the circumferential wavenumber may be low. We do this by applying 'correction factors' to various terms of the equations, in the manner of section 8.6. Up to this point, only the simplest periodic buckling modes have been considered. The next section discusses the effects of various types of end-condition – clamped, free, etc. – on the buckling criterion. Torsional buckling of tubes is the subject of section 14.7: here we demonstrate the use of an energy method in connection with a simple approximate mode of buckling. The chapter closes with a brief discussion of experimental investigations, a design problem, and a relatively simple treatment of anisotropically reinforced cylindrical shells.

14.2 Eigenvalue calculations according to the 'two-surface' model

Although our aim is to analyse the buckling of shells, it will nevertheless be useful to introduce the proposed scheme of calculations with reference to two simpler cases concerning a column and a plate, respectively.

14.2.1 A simple pin-ended column

Most readers will no doubt be familiar with the usual derivation of the 'Euler load' for a simple pin-ended column. As we remarked earlier, there are several ways of tackling the problem, including a variety of energy methods. Here we shall use a different approach, based on a sort of 'two-surface' view of a column. This will act as an introduction to the method which we shall use subsequently for shells, and which will turn out to be much more expeditious than those which are normally given. Fig. 14.2 shows various diagrams illustrating a 'two-surface' rod. It is perhaps easiest to begin by thinking of the column being made in the form of a tube with a central core, as shown in fig. 14.2a, rather like a piece of bicycle brake cable. The idealised central core of the model is a member which can sustain tension or compression with little elongation, but offers no resistance to bending. Conversely, the outer tube, which is perfectly straight in the unloaded condition, offers elastic resistance to bending, but none to extension or compression. In other words, we have here a one-dimensional version of the 'two-surface' representation of a shell which was presented in section 8.2.

478   *Buckling of shells: classical analysis*

The diagrams in fig. 14.2b and c show what happens when the endwise compressive load reaches the value at which the rod is in equilibrium in a deformed state which involves a small lateral bow. The endwise load is applied of course to the central core – since the outer tube cannot withstand compression, by hypothesis – and for equilibrium to be preserved in the now curved configuration, lateral forces are necessary, as shown. For a given deflected profile these can be calculated by the methods of section 4.2. These lateral forces necessary for equilibrium of the core as a 'funicular curve' are provided by reactions from the tube, which can of course only supply them if it is bent elastically from its original straight configuration. In fig. 14.2 the deflected shape has been drawn as sinusoidal, for this is well known to be the eigenmode for the problem in question: but if we had not known this we could have begun with an arbitrary function, and obtained the eigenmode in the course of the calculations.

In general, the intensity of lateral load f, required for equilibrium of a slightly curved string in the absence of bending moments is given by

$$f_1 = P \frac{d^2 w}{dx^2}. \tag{14.3}$$

Fig. 14.2. A 'two-surface' scheme for investigating the buckling of a simple column. (a) The column is regarded as a central 'core' which is incapable of resisting bending moment, contained within a 'tube' which acts as a beam against transverse loading. (b) Equilibrium of the 'core' in the distorted configuration, as a 'funicular curve' requiring transverse forces of intensity f_1. (c) The tube is bent by forces imposed by the core.

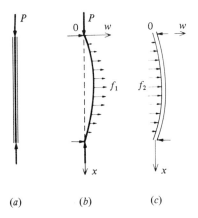

14.2 Eigenvalue calculations by the two-surface model

Thus for a rod of length l, sustaining an endwise compressive load P, and having a sinusoidal bow of small amplitude w_0, we have

$$f_1 = (-P\pi^2 w_0/l^2) \sin(\pi x/l). \tag{14.4}$$

Here the positive senses of w and f_1 are the same, mainly for the sake of consistency with more complicated problems later on. Equation (14.3) embodies the assumption that deflections are small: d^2w/dx^2 is the simplified form of the expression for curvature which was mentioned earlier.

Note that concentrated transverse forces are also required at the two ends of the rod in order to preserve overall equilibrium of the core.

The (small-deflection) governing equation of the tube as a simple beam (cf. (9.18)) is

$$f_2 = EI \frac{d^4 w}{dx^4}, \tag{14.5}$$

where EI is the flexural stiffness of the tube and indeed of the original rod. Consequently we have, for our chosen mode,

$$f_2 = (EI\pi^4/l^4) w_0 \sin(\pi x/l). \tag{14.6}$$

By hypothesis $f_1 + f_2 = 0$, since there are no externally applied transverse forces acting on the rod. Thus from (14.4) and (14.6) we obtain the corresponding eigenvalue of P:

$$P_E = \pi^2 EI/l^2. \tag{14.7}$$

Since $\sin(\pi x/l)$ cancels, this must have been a correct eigenfunction. Note also that w_0 cancels from the expression, which suggests that the equilibrium is neutral. This is, in fact, not quite true: as Koiter has pointed out, the linearisations which were implicit in the setting-up of (14.3) and (14.5), enable one to compute correctly the value of the bifurcation load P_E, but on the other hand they remove the possibility of making meaningful statements about the *stability* of equilibrium in the buckled configuration.

A trivial extention of the above calculation using $w = w_n \sin(n\pi x/l)$ as an alternative trial function, where n is a positive integer, gives a sequence of higher eigenvalues; see problem 14.2. Our choice of trial displacement functions is of course determined in part by the boundary conditions of the problem. Here, where the ends of the column are freely hinged, there is no bending moment in the outer tube, and our solution meets this condition.

14.2.2 Eigenvalue calculations for flat plates

Let us now extend the method of the previous section to plates and shells. Consider first a uniform rectangular plate of length L and breadth B under the action of uniform forces N_x and N_y per unit length applied normally

480 *Buckling of shells: classical analysis*

to the edges in the plane of the original configuration of the plate, as shown in fig. 14.3. For consistency with the notation of the preceding chapters we shall regard the positive sense of N_x and N_y as tensile. Consequently our eigenvalues will be predominantly negative, and we shall always write results as

$$(-N_x) = \ldots, \text{etc.}$$

The basis of our calculation is that in the course of *small* out-of-plane displacement of the S-surface of the plate, the distribution of in-plane stress resultants does not change significantly. We shall comment on this assumption later.

Since the S-surface was originally flat, it can only have curvature and twist on account of normal displacement w (cf. (6.9)):

$$\kappa_x = \frac{-\partial^2 w}{\partial x^2}, \quad \kappa_y = \frac{-\partial^2 w}{\partial y^2}, \quad \kappa_{xy} = \frac{-\partial^2 w}{\partial x \partial y}. \tag{14.8}$$

Consequently the equilibrium equation for a small element of the deformed S-surface in the direction normal to the original surface is

$$(-N_x) \frac{\partial^2 w}{\partial x^2} + (-N_y) \frac{\partial^2 w}{\partial y^2} = p_S, \tag{14.9}$$

where, as before, p_S is the surface pressure-loading on the S-surface – positive in the same sense as w.

The governing equation of the B-surface is exactly the same as that which was derived in section 8.3:

$$D\nabla^4 w = p_B. \tag{14.10}$$

Here D is the flexural rigidity of a plate element: see (2.18).

Fig. 14.3. Notation for the buckling of a simply-supported uniform rectangular plate.

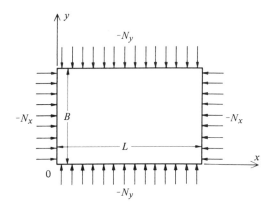

14.2 Eigenvalue calculations by the two-surface model

Lastly, we write down the condition that there is no surface loading on the plate, as a whole, i.e. on the S- and B- surfaces taken together:

$$p_S + p_B = 0. \tag{14.11}$$

Hence, substituting for p_S and p_B from (14.9) and (14.10) we obtain

$$(-N_x) \frac{\partial^2 w}{\partial x^2} + (-N_y) \frac{\partial^2 w}{\partial y^2} = -D\nabla^4 w \tag{14.12}$$

as the governing equation of the plate in its buckled condition. This is a well-known equation: cf. Timoshenko & Gere (1961, §9.1). Now suppose that the boundaries of the plate are simply supported. We may then use

$$w = w_{mn} \sin(m\pi x/L) \sin(n\pi y/B) \tag{14.13}$$

as a trial displacement function to be substituted into (14.12). Here m and n are integral wavenumbers. Making this substitution we find, after some obvious simplification,

$$(-N_x)(m\pi/L)^2 + (-N_y)(n\pi/B)^2 = D\left[(m\pi/L)^2 + (n\pi/B)^2\right]^2. \tag{14.14}$$

This equation gives those combinations of $(-N_x)$ and $(-N_y)$ at which the plate can be in equilibrium in a buckled configuration; that is, it represents the eigenvalues of the problem. For any given values m, n (14.14) gives a straight line in a Cartesian $(-N_x, -N_y)$ space; and for any given ratio N_x/N_y simple trial-and-error methods reveal which mode corresponds to the lowest possible buckling load: see problem 14.3.

Two remarks are appropriate here. First, if the plate were also subject to loading N_{xy} an extra term $-2N_{xy} \partial^2 w/\partial x \partial y$ would be required on the left-hand side of (14.9), and this in turn would give an extra term on the left-hand side of (14.12). Buckling of a rectangular plate (with sides parallel to the x- and y-axes) under edge-loading N_{xy} is more complicated to analyse, simply because the boundary conditions are necessarily more awkward than for N_x, N_y loading. There is a similar awkwardness in the torsional buckling of thin-walled cylindrical shells, as we shall see in section 14.7. The buckling of a uniform long rectangular plate under shear loading was first studied by Southwell & Skan (1924).

Second, we should examine critically our assumption that when the plate takes up its buckled form, the original uniform distribution of in-plane stress resultants is not changed. There is certainly in general a change of Gaussian curvature when the initially flat surface adopts a buckled configuration: consequently there must necessarily be some nonzero surface strains, which require, by Hooke's law, corresponding changes in N_x, N_y and N_{xy}. In principle we may do a check on the magnitude of these changes for any postulated modeform in comparison with the values of the stress resultants which are applied at the edges. Here the local change of Gaussian curvature is of the order of $\kappa_x \kappa_y$, i.e. $(\partial^2 w/\partial x^2)(\partial^2 w/\partial y^2)$ (see problem 5.4); so it is proportional to the square of the amplitude w_{mn}, and should certainly be negligible for sufficiently small displacements. Effects of this sort are well known in plate buckling as 'large-deflection' effects (see e.g. Timoshenko & Gere, 1961, §9.13), but although they are interesting and indeed important for the question of *stability* of equilibrium of the plate in its buckled form, they play no part in the classical eigenvalue calculation with which we have been concerned.

14.2.3 Eigenvalue calculations for cylindrical shells

It is now a relatively simple matter to extend our calculations so that they apply to a cylindrical shell. Throughout this chapter, as on previous occasions, we shall use an x, y coordinate system in the surface of the shell in its original configuration, with the x- and y-axes in the longitudinal and circumferential directions, respectively. Also, occasionally we shall use an equivalent angular coordinate $\theta = y/a$.

We shall suppose that the shell starts in its original cylindrical configuration with a uniform 'membrane' distribution of N_x, N_y and N_{xy}, as a result of external loading, and then undergoes a small displacement to a simple postulated periodic adjacent equilibrium configuration *during which N_x, N_y and N_{xy} remain constant*. Next we shall calculate the changes in pressure which are required to hold the 'membrane' in equilibrium in its new configuration; and we shall then argue that these pressures can only be provided by the forces which are required to distort the shell from its original configuration. This sort of calculation is analogous to that which we described in relation to the rod and the flat plate, but there are in fact two important new features in relation to the cylindrical shell.

First, the prebuckled state of a shell under uniform external loading will not normally be exactly as calculated according to the membrane hypothesis, on account of constraints imposed by end-fittings, etc.. Thus, if the radial expansion of an axially compressed shell on account of Poisson's ratio is prevented at the ends, there will occur some localised bending stresses, etc., as described in chapter 3. We shall ignore effects of this sort in the present chapter.

Second, whereas the rod and the plate could undergo small transverse displacements without developing appreciable in-plane stress resultants, this is no longer true in the case of a cylindrical shell. Recall that in section (8.4) we saw that when a shell is subjected to a pressure distribution

$$p = p_0 \sin(\pi x/l) \sin(\pi y/b) \tag{14.15}$$

14.2 Eigenvalue calculations by the two-surface model

its structural response involves a combination of the efforts of both the S- and B-surfaces in resisting an imposed displacement. This is in contrast with the behaviour of the rod and the plate, where, in the absence of endwise loading, (small) transverse displacements are resisted entirely by 'B-surface' effects. Thus we see that in the buckling of a cylindrical shell the S-surface plays a double role. First it acts as a 'funicular membrane' carrying the original stress resultants in a deformed configuration, and second it cooperates with the B-surface to carry the pressures imposed on the shell by the 'distorted membrane'. These two effects are, in a strict sense, mutually exclusive. Nevertheless, when we consider that the changes in S-surface stress resultants are directly proportional to the displacement w, we may argue that for sufficiently *small* amplitudes of any particular mode of displacement these changes are sufficiently small in comparison with the applied loads to warrant their neglect in consideration of the equilibrium of the 'membrane', even though they are sufficiently large to impart to the S-surface a significant stiffness. Our proposed method is therefore one which should be satisfactory for classical eigenvalue calculations; but by our neglect of various small quantities we rob ourselves of the ability to discuss the *stability* of equilibrium of the shell in its buckled configuration.

The dual role of the S-surface in this respect poses a problem of notation which was not present in the case of the plate. We shall, however, avoid this by doing the calculation in two separate parts. First, we shall enquire what distribution of normal pressure is required to hold the 'funicular membrane' in equilibrium in its deformed configuration; and second we shall ask what distribution of normal pressure would be required to deform combined S- and B-surfaces into the assumed configuration, in the manner of section 8.4. Indeed, this second calculation will in general be an example of precisely the type of problem which was considered in that section. Thus our overall scheme may be viewed as a method of employing the results of chapter 8, with the help of the device of a 'funicular membrane' to analyse buckling problems of a cylindrical shell, in exactly the same way that we used simple results on the bending of beams and plates in the two earlier examples. The present treatment of the S-surface in two different ways will be replaced by a unified, but more complicated and *nonlinear* treatment in chapter 15.

The only outstanding preliminary question concerns the local changes in pressure which are necessary to hold the funicular membrane in its distorted configuration. What will replace (14.9) when we change from a flat plate to a cylindrical shell?

It will be convenient to tackle this problem in two stages, following a scheme which we used in chapter 8. First, here, we shall consider the simpler case in which there is a high circumferential wavenumber, and later (in section 14.5) we shall consider what modifications to the resulting simple formulae are needed for low circumferential wavenumbers.

Consider a truly cylindrical membrane of radius a, which carries stress resultants N_x, N_y, N_{xy} which are uniform over the entire surface. Now, although uniform stress resultants N_x and N_{xy} can easily be provided by tension and torque applied to the ends of the cylinder, the provision of N_y demands a uniform internal pressure $p = N_y/a$. Here, as before, a positive pressure acts outwards. This requirement in the y-direction makes a contrast with the flat plate which we studied earlier. Suppose now that the membrane is distorted into an adjacent configuration, while the stress resultants N_x, N_y, N_{xy} are maintained unchanged at their former values. In general, equilibrium of a small element in the radial direction will no longer be satisfied by the original constant internal pressure p, but it is a simple matter to show (cf. section 4.3) that we need to supply locally a small additional pressure Δp_1 given by

$$-\Delta p_1 = (-N_x)\kappa_x - 2N_{xy}\kappa_{xy} + (-N_y)\kappa_y. \tag{14.16}$$

The subscript 1 refers here to a 'membrane' calculation. In this expression $\kappa_x, \kappa_{xy}, \kappa_y$ are changes in curvature and twist from the original configuration. They can be expressed in terms of the surface displacements by (6.9).

In accordance with our assumption that the circumferential wavenumber is high, we ignore the contribution of peripheral displacement v to κ_{xy} and also the second term in the expression for κ_y in (6.9). Thus we may write

$$\Delta p_1 = (-N_x)\frac{\partial^2 w}{\partial x^2} - 2N_{xy}\frac{\partial^2 w}{\partial x \partial y} + (-N_y)\frac{\partial^2 w}{\partial y^2}. \tag{14.17}$$

Apart from the use of Δp instead of p, and the inclusion of N_{xy}, this is precisely the equation which we obtained for the distortion of an originally *plane* membrane in section 14.2.2.

This equation, together with the results of chapter 8, enables us to solve directly classical buckling problems for cylindrical shells.

14.3 Cylindrical shell under uniform axial compression

As a first example, we shall consider a uniform circular cylindrical shell carrying uniform axial compression. The problem is to find the value of $-N_x$ at which a state of adjacent equilibrium is possible according to the classical theory.

Let us begin by assuming an eigenmode of the form

$$w = w_n \sin(\pi x/l) \sin(\pi y/b). \tag{14.18}$$

Here, as before, l and b are half wavelengths of the buckling mode. For the

14.3 Cylindrical shell under uniform axial compression

time being we shall regard l and b as continuously variable; but later on, when we take boundary conditions into account, we shall insist on integral numbers of half-waves in the axial length of the shell and in half of the circumference.

Using a result from chapter 8 (see (8.15)) we find that the distribution of pressure on the shell which is required to give the postulated deflection is

$$\Delta p_2 = w_n \left[\frac{Et}{a^2 l^4} \left(\frac{1}{l^2} + \frac{1}{b^2} \right)^{-2} + \pi^4 D \left(\frac{1}{l^2} + \frac{1}{b^2} \right)^2 \right] \sin(\pi x/l) \sin(\pi y/b). \quad (14.19)$$

Here the subscript 2 refers to the elastic shell, consisting of S- and B-surfaces. On the other hand the (outward) pressure required to hold the end-loaded membrane in its distorted configuration is (from (14.17))

$$\Delta p_1 = -w_n (-N_x) (\pi^2/l^2) \sin(\pi x/l) \sin(\pi y/b). \quad (14.20)$$

The sum of these two pressures is zero; and as the proposed displacement function and its amplitude are common to both expressions, we obtain our required eigenvalue:

$$(-N_x) = \frac{Et}{\pi^2 a^2 l^2} \left(\frac{1}{l^2} + \frac{1}{b^2} \right)^{-2} + \pi^2 l^2 D \left(\frac{1}{l^2} + \frac{1}{b^2} \right)^2. \quad (14.21)$$

This is a well-known result. Note that the second term on the right-hand side of this expression corresponds exactly to the buckling of the flat plate in (14.14); and the first term represents the restoring effect of the distorted S-surface.

In principle we can take this expression and substitute all possible values of l and b consistent with an assumed set of boundary conditions in order to find the mode which has the lowest eigenvalue $(-N_x)$. This can be an awkward problem if it is tackled haphazardly. Therefore we begin by studying the specially simple case in which the mode is axially symmetric; later we shall move on to the general case in which the modeform is periodic in both the axial and the circumferential directions.

14.3.1 Axisymmetric mode

This mode is obtained formally by shifting the origin of coordinates through $\frac{1}{2} b$ in the circumferential direction, so that the displacement is described by a cosine in the circumferential direction, and then by putting $b \to \infty$. Equation (14.21) still holds, and on substitution of $l/b = 0$ we obtain the required result

$$(-N_x) = \frac{Etl^2}{\pi^2 a^2} + \frac{\pi^2 Eth^2}{12 l^2}. \quad (14.22)$$

486 Buckling of shells: classical analysis

Here we have expressed the flexural rigidity D in terms of other factors (see (2.18) and (3.25)).

The first term on the right-hand side corresponds to circumferential stretching, and the second to longitudinal bending. The half wavelength l is the only variable parameter. The first term increases with l but the second decreases, and it is easy to show (see problem 14.4) that $-N_x$ has a minimum value

$$(-N_x)_{\min} = \frac{Eth}{3^{\frac{1}{2}}a} \approx 0.6 \frac{Et^2}{a} \quad \text{when} \quad \frac{l}{(ah)^{\frac{1}{2}}} = \frac{\pi}{2^{\frac{1}{2}} 3^{\frac{1}{4}}} \approx 1.69. \quad (14.23)$$

The half wavelength at which $(-N_x)$ is minimum is exactly the same as that for which the 'modal flexibility', in the sense of section 8.4, is maximum.

We must now take account of the boundary conditions of the shell. Let the overall length of the shell be L, and let the ends be simply-supported. In the prebuckled configuration the shell is not truly cylindrical, since the simple supports impede the radial displacement which is associated with uniaxial loading on account of the nonzero value of the Poisson's ratio of the material (see problem 3.20). Here, and indeed throughout this chapter, we shall ignore this effect. This is tantamount to an assumption that the state of stress in the prebuckled shell is represented adequately by the membrane hypothesis. The justification for this step is that detailed studies using the actual nonlinear prebuckled state of stress in a shell instead of the assumed membrane state usually gives buckling loads which differ by only about 15%: see Brush & Almroth (1975, §9.5b).

Our assumed mode of incipient deformation, i.e.

$w = w_0 \sin(\pi x/l)$,

therefore satisfies the boundary conditions provided there is an integral number, say m, of half-waves in the length L, i.e. provided

$l = L/m$.

A graph showing the consequent variation of $(-N_x)$ with L is shown in fig. 14.4a. There is a 'festoon' formed by the curves (14.22) for the different values of m, and we select, fro a given value of L, that curve which has the smallest ordinate. The same data is plotted on logarithmic scales in fig. 14.4b: again we have a festoon of curves, but in this logarithmic plot they are obtained from each other by translation.

It is clear from these diagrams that for, say,

$L/(ah)^{\frac{1}{2}} > 1.5$,

the lower limit of the festoon of curves is, for practical purposes, independent of L. In these circumstances we obtain a buckling formula which may be simplified to

$$-\sigma_x = 0.58\, Eh/a. \tag{14.24}$$

This is often quoted, approximately, as

$$-\sigma_x = 0.6 Et/a. \tag{14.25}$$

This formula should be memorised. The corresponding endwise strain in the shell at the inception of buckling is given by

$$-\epsilon_x = 0.58\, h/a \approx 0.6\, t/a. \tag{14.26}$$

We shall encounter festoons of curves of the same kind, with one curve for each mode-number, at various places in the remainder of this chapter. We shall usually be interested in the lower limit of the festoon, and although this is, strictly, a cusped curve as in fig. 14.4, it is often convenient to obtain the lower limit by pretending that the mode-number is continuously variable, and using standard formal mathematical techniques to find a bounding *envelope*.

For example, (14.22) may be written

$$(A/m^2) + Bm^2 - (-N_x) = 0, \tag{14.27}$$

where A and B are constants which may easily be determined. This equation has the form

$$f(m) = 0.$$

Fig. 14.4. The buckling of a uniform cylindrical shell under pure axial compression in a regular axisymmetric mode may be represented by a 'festoon' of curves in $-N_x, L$ space, in which each curve is associated with a wavenumber m in the longitudinal direction. The same information is shown (a) with linear scales and (b) with logarithmic scales.

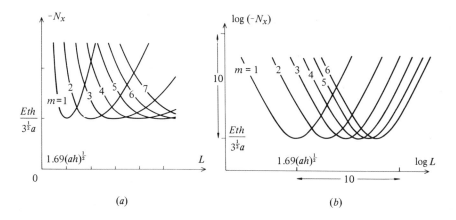

488   Buckling of shells: classical analysis

To find the equation of the envelope (e.g. Lamb, 1919, §139) we eliminate m between (14.27) and

$$\frac{\partial}{\partial m}(f(m)) = 0. \tag{14.28}$$

Here the latter equation gives

$$-(2A/m^3) + 2Bm = 0;$$

so

$$m^2 = (A/B)^{\frac{1}{2}} \tag{14.29}$$

and the equation of the required envelope is

$$-N_x = 2(AB)^{\frac{1}{2}}. \tag{14.30}$$

It is easy to check that this agrees with the previous result.

14.3.2 Doubly-periodic modes

When we widen the scope of our family of modes, and consider the possibility that the buckled form is not axisymmetric, we must regard l and b as variables; and it is then not nearly so straightforward to find the minimum value of $(-N_x)$ given by (14.21) for a shell of given dimensions.

The main problem is to grasp the nature of the function which constitutes the right-hand side of (14.21). The most convenient way of doing this is to introduce a change of variables from l, b to ξ, η, where

$$\xi = \lambda/l, \quad \eta = \lambda/b, \tag{14.31}$$

and λ is the constant length, related to a and h, which we used in chapter 8 (cf. 8.38):

$$\lambda = \frac{\pi}{3^{\frac{1}{4}}}(ah)^{\frac{1}{2}} \approx 2.39 \, (ah)^{\frac{1}{2}}. \tag{14.32}$$

ξ and η are thus dimensionless reciprocal half wavelengths in the longitudinal and circumferential directions, respectively. Equation (14.21) is then transformed to

$$(-N_x) = \frac{Eth}{3^{\frac{1}{2}}a} Z(\xi,\eta), \tag{14.33}$$

where

$$2Z = \frac{2\xi^2}{(\xi^2 + \eta^2)^2} + \frac{(\xi^2 + \eta^2)^2}{2\xi^2}. \tag{14.34}$$

We now investigate the form of the function $Z(\xi,\eta)$. Note that the two additive parts of $2Z$ are reciprocal. By inspection of (14.34) we find that whenever

$$\xi^2 + \eta^2 = c\xi, \tag{14.35}$$

where c is an arbitrary constant, the function Z has a fixed value. In other words (14.35) is the general equation for a *contour* of $Z(\xi,\eta)$ in the ξ, η plane. Equation (14.35) is in fact the description of a circle of diameter c which has its centre on the ξ-axis and passes through the origin; and the corresponding value of Z is $1/c^2 + \frac{1}{4}c^2$. Fig. 14.5 shows contours for various values of Z, , including the contour $c = 2^{\frac{1}{2}}$ for which Z has its minimum value of 1. This particular contour is known as the *Koiter circle* (Koiter, 1945, Fig. 8; 1963a, Fig. 7). One of the advantages of the use of variables ξ, η is that the contours of the function Z have this particularly simple circular form. The contour $Z = 1$ is particularly important, since it gives the combinations of ξ, η (and, correspondingly, of l, b) for which $(-N_x)$ is a minimum if l, b are regarded as continuously variable. It is also in fact precisely the condition that the mode-stiffness of the S- and B-surfaces, respectively, in the sense of section 8.4, are exactly equal: cf. (8.17) and problem 8.2. The axially symmetric mode considered in section 14.3.1 corresponds to the point A shown in fig. 14.5.

It is of course remarkable that there is a whole range of modeforms for which the eigenvalues $(-N_x)$ are exactly equal, at least if l and b are continuously variable. We shall consider later the restrictions which are imposed by the boundary conditions in relation to the discrete nature of allowable values of l, b; but first it is advantageous to examine some other features of the function $Z(\xi,\eta)$.

Fig. 14.5. Contours of the function Z (see (14.34)) are all circles in a Cartesian ξ, η space. Values of Z are labelled.

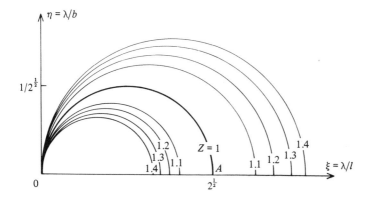

In some circumstances it is useful to express Z in terms of polar coordinates r, ϕ where

$$r^2 = \xi^2 + \eta^2, \tag{14.36a}$$

$$\tan\phi = \eta/\xi = l/b. \tag{14.36b}$$

Equation (14.34) may then be written

$$2Z = \frac{r^2}{2\cos^2\phi} + \frac{2\cos^2\phi}{r^2}, \tag{14.37}$$

and it is clear that along radial lines (ϕ = constant) Z has the simple form

$$Z = Ar^2 + B/r^2. \tag{14.38}$$

It is straightforward to show that the minimum value of Z along such a line is 1, and that it occurs when

$$r = r_K(\phi) = 2^{\frac{1}{2}} \cos\phi, \tag{14.39}$$

which is indeed the polar-coordinate representation of the Koiter circle: hence the subscript K. Thus we may rewrite (14.34) as

$$2Z = (r_K/r)^2 + (r/r_K)^2, \tag{14.40}$$

where $r_K(\phi)$ is given by (14.39).

The function $Z(\xi, \eta)$ is sketched in fig. 14.6 by means of contours and curves of constant ϕ. The function is a sort of curved valley whose floor is a

Fig. 14.6. Isometric view of the function $Z(\xi, \eta)$ for $Z \leqslant 1.4$.

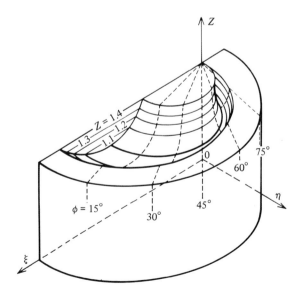

semicircle lying in a horizontal plane ξ, η. As one moves towards the origin along the valley floor the sides of the valley become progressively steeper. At the origin itself there is a singularity, and the sides are vertical.

In view of the work in section 8.4 which showed that for small values of l/b (i.e. small values of η/ξ) the dominant stretching and bending effects are in the circumferential and longitudinal directions, respectively, and vice versa for large values of l/b, it is clear that in various regimes of the space ξ, η different terms on the right-hand side of (14.33) will dominate. These dominant effects are marked in fig. 14.7. For modes corresponding to points in the ξ, η plane lying *inside* of the Koiter circle, *stretching* effects dominate while for points on the *outside*, *bending* effects dominate. There is a region (say)

$$\tfrac{1}{2} < l/b = \eta/\xi < 2 \qquad (14.41)$$

in which both the stretching and the bending effects receive contributions of comparable magnitude from the circumferential and longitudinal directions. But outside this zone the bending and stretching are each dominated by either circumferential or longitudinal effects: see the diagram for details.

14.3.3 Boundary conditions

We must now consider the boundary conditions of the problem; that is, the boundary conditions which must be satisfied by the mode of buckling deformation. These are, of course, distinct from the boundary conditions for the prebuckled state of the shell, which we treated summarily. In the first instance our basic strategy will be to find boundary conditions which can be

Fig. 14.7. Sketch of ξ, η space, showing regions in which the behaviour of the shell is dominated by longitudinal bending (LB), circumferential stretching (CS), etc.. In the region marked B the various effects cannot be separated in this way.

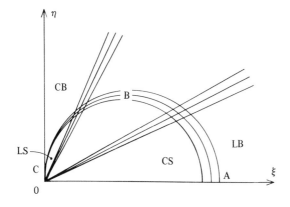

satisfied by our assumed mode of deflection; i.e. to 'tailor' boundary conditions to our proposed solution. Later on we shall consider some different boundary conditions which are important in practice.

If we declare that $w = 0$ at each end of the shell we shall imply 'simple' supports, which do not inhibit rotations $\partial w/\partial x$. But what about boundary conditions on the displacement components u and v? By using an equation which has a single variable w, we appear to be ignoring edge-conditions on u and v. In fact, as we saw in section 8.7, the assumed displacement mode does involve u and v, and we must concede that if we worked out the u- and v-functions we would be obliged to specify the boundary conditions sympathetically, in order to preserve the reasonableness of our chosen mode. In fact the doubly-periodic solution has $v = 0, N_{xy} \neq 0, u \neq 0, N_x =$ constant on the circumferential lines at which $w = 0$. The traditional approach to this problem, as explained by Batdorf (1947) is to argue that the above conditions will be met approximately by a thin diaphragm or deep stiffening rings which hold the shell circular (i.e. $w = 0$) and resist tangential displacement (i.e. $v = 0$), but do not prevent any *warping* displacements u which may wish to occur. We shall return to this question in section 14.6, where we shall investigate the consequences of providing edge-restraints which prevent warping, and also the effects of free edges. We shall also make some remarks on boundary conditions in section 14.11. But for the present we shall proceed with the simple, natural, 'conventional' boundary conditions

$$w = 0 \text{ at } x = 0 \text{ and } x = L. \tag{14.42}$$

This implies that there must be an integral number, m, of half-waves in the length of the shell. In the circumferential direction continuity demands an even number, $2n$, of half-waves: thus

$$l = L/m, \quad b = \pi a/n; \tag{14.43}$$

and consequently

$$\xi = (\lambda/L)m, \quad \eta = (\lambda/\pi a)n. \tag{14.44}$$

Now in a shell of given dimensions the values of λ/L and $\lambda/\pi a$ are fixed, and so the allowable values of ξ, η lie at the nodal points of a rectangular grid whose unit-cell dimensions are

$$\frac{\pi}{3^{\frac{1}{4}}} \frac{(ah)^{\frac{1}{2}}}{L} \approx 2.39 \frac{(ah)^{\frac{1}{2}}}{L}, \quad \frac{1}{3^{\frac{1}{4}}} \left(\frac{h}{a}\right)^{\frac{1}{2}} \approx 0.76 \left(\frac{h}{a}\right)^{\frac{1}{2}} \tag{14.45}$$

in the ξ and η directions, respectively. The ratio of the sides of the unit cell is $L/\pi a$, which is the aspect ratio of the cylinder in an imagined flattened condition achieved by making two longitudinal creases. This grid is shown in

14.3 Cylindrical shell under uniform axial compression

fig. 14.8. The uniformity of the grid is an advantageous consequence of the use of variables ξ, η. Some of the grid-points in the diagram are shown solid and some are open. The open ones are inadmissible for obvious reasons. Thus, it is impossible to have $m = 0$, since there must be at least one half-wave in the longitudinal direction; but on the other hand we may have $n = 0$, as this corresponds to the axially symmetric case. We must exclude $n = 1$ for the time being. The reason for this is that we have so far ignored the 'circumferential bending correction' which we found, in section 8.6, to be crucial for a proper discussion of the case $n = 1$. We shall discuss this factor again in section 14.5.

It is clear from (14.45) that for many thin-walled shells the grid mesh-size is much smaller than the diameter of the Koiter circle. Consequently a relatively large number of modes m, n must be considered as candidates for the lowest eigenvalue. However, there does not seem to be much point in seeking the lowest eigenvalue of all: the important feature is that there are likely to be *several* modes having eigenvalues within (say) 2% of the theoretical minimum. In any case our omission both of specific boundary conditions and of the 'circumferential bending correction' would not justify use of the present formulae for this purpose.

It is easy to make an estimate of the number of 'competing modes' within (say) 2% of the absolute minimum. The contours of $Z = 1.02$ are circles whose diameters are equal to that of the Koiter circle multiplied by 0.9 and 1.1, respectively, as shown in fig. 14.9. The area of the enclosed half-crescent is 0.1π. The number, N, of mesh points within this area is approximately equal

Fig. 14.8. Grid of 'solution points' in ξ, η space imposed by the boundary conditions, which require integral numbers of half-waves in the two directions.

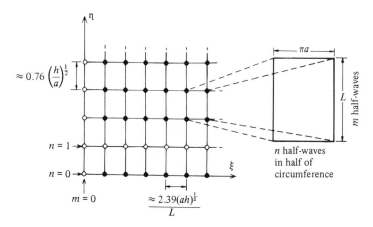

to the area divided by the area of the unit-mesh rectangle. Using (14.45) we find

$$N \approx 0.17 \, L/h. \qquad (14.46)$$

Clearly, for many practical problems this will be a large number.

It is instructive to examine the geometrical parameters of the modes which correspond to points lying on the Koiter circle, by replotting this curve in l/λ, b/λ space, as shown in fig. 14.10, with both linear and logarithmic scales. The points A, B, C in fig. 14.10a transform to A', B', C' in fig. 14.10b. A represents the axisymmetric mode, with $l/\lambda = 2^{-\frac{1}{2}}$, while B represents the mode with square panels $l = b$: note that the side of this square is just twice the half wavelength for the axisymmetric mode. Also observe that the 'square' mode is the one with the smallest width, b. Further note that for any given value of $b > b_{min}$ there are *two* possible values of l: it is easy to show that their geometric mean is equal to b, and that their aspect ratios are equal but reversed. Fig. 14.10c is also related to the plot of fig. 8.4a, but with the axes interchanged. On the left-hand edge of the Koiter circle, i.e. in the region $l/b \gg 1$, there are some specially simple relationships between l, b and n: see problem 14.5.

The remarkable result that a cylindrical shell may have a large number of eigenmodes each having practically the same eigenvalue depends crucially on having a loading condition which gives $N_y = 0$. When $N_y \neq 0$ the results are quite different, as we shall see in the next section. In chapter 15 we shall discover that the existence of 'competing' modes can have crucial consequences for the carrying capacity of cylindrical shells under axial loading.

Fig. 14.9. The region of ξ, η space for which $Z \leqslant 1.02$.

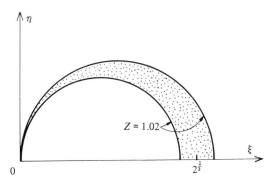

14.4 Axial compression and 'side' pressure combined

In many practical problems the axial loading of a cylindrical shell will be combined with an interior or exterior pressure. Thus, the tubular air-filled leg of a sea-based oil rig will carry a longitudinal compressive load together with exterior water pressure; and the body of a thin-walled rocket in flight will sustain axial compressive load and interior pressure. An isolated closed vessel which has been evacuated sustains both axial load and exterior pressure: the axial load is, of course, a result of pressure-loading on the ends, and in this case

$$N_x = \tfrac{1}{2} N_y. \qquad (14.47)$$

Also, in some applications, a closed vessel of this sort may support an additional, externally applied, axial load.

There is obviously scope for confusion of nomenclature when a cylindrical shell is loaded by pressure. Pressure is often referred to as 'side' pressure if it

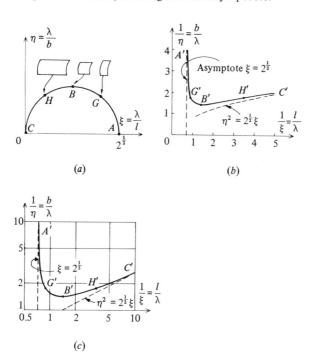

Fig. 14.10. (a) 'Panel' dimensions of the doubly-periodic modeform corresponding to different positions on the Koiter circle. (b) The same information replotted in $1/\xi$, $1/\eta$ space. (c) A replot of (b) on logarithmic scales, showing the two asymptotes.

does not act on closed ends, and 'hydrostatic' pressure if it does act on the closed ends of the vessel and thereby affects the axial loading. In all cases the quantities which matter in the determination of the eigenvalue are the stress resultants N_x and N_y; consequently we shall always specify loading cases in terms of these two variables: see problem 14.6.

It is clear from the way in which we have derived the buckling equations in this chapter that when we use an eigenmode as described by (14.18) in conjunction with loading N_x, N_y there will appear (see (14.17)) an extra term on the left-hand side of (14.21), but the right-hand side will remain unchanged. Hence, we find that under combined N_x, N_y loading the classical buckling equation is

$$(-N_x) + \left(\frac{l}{b}\right)^2 (-N_y) = \frac{Eth}{3^{\frac{1}{2}}a} Z(\xi,\eta). \tag{14.48}$$

There are various ways of dealing with this equation for a given loading ratio N_x/N_y. Perhaps the simplest approach, leading to a qualitative study, is to keep the right-hand side intact – since we now know its properties well – and to focus attention on the left-hand side.

This may be written

$$(-N_x) \left[1 + \frac{(N_y)}{(N_x)} \left(\frac{\eta}{\xi}\right)^2 \right]. \tag{14.49}$$

For a given value of N_y/N_x the expression in square brackets may be plotted as a function of ξ, η as shown in fig. 14.11. For a small (positive) value of the ratio N_y/N_x the function is almost constant for small values of ϕ (see (14.36b)), but it ascends like a spiral stair for larger values of ϕ, as shown in fig. 14.11a. For larger values of N_y/N_x the spiral part of the surface begins to rise at smaller values of ϕ; and of course when N_x is zero the entire surface is spiral. On the other hand, if N_y/N_x has a *negative* value (as it does for compressive axial load and internal pressure) the surface has a *descending* spiral part for large values of ϕ, as shown in fig. 14.11b.

Now for a given value of N_y/N_x we can imagine a sort of graphical procedure for determining N_x whereby we construct the surface of fig. 14.11a, for a small value of $-N_x$, underneath the surface of fig. 14.6. Then we steadily increase the value of $-N_x$, thereby raising the lower surface. The value of $-N_x$ at which the lower surface first *touches* the upper surface is the required eigenvalue; and the position of the point of first contact in the ξ, η plane defines the corresponding eigenmode.

In the special case $N_y/N_x = 0$ the lower surface is flat; therefore as the value of $-N_x$ is increased it first touches the upper surface along the entire Koiter circle. This, of course, describes exactly our previous result.

14.4 Axial compression and 'side' pressure combined

For the case $N_y/N_x > 0$ it is clear from fig. 14.11a that the lower surface will first encounter the upper surface, as $-N_x$ is increased, at a *high* value of ϕ, i.e. a high value of l/b. There will normally be just one eigenmode; and we shall find that in general this mode has $m = 1$ (one half-wave in the longitudinal direction) and lies on the side of the Koiter circle nearer the η-axis.

On the other hand, if the value of N_y/N_x is negative the critical mode is at *small* values of l/b; and it can be seen immediately that the axisymmetric mode always governs. Thus we may conclude that even small positive or negative values of N_y (in comparison to N_x) may have a decisive influence on the preferred buckling mode.

For the purposes of *quantitative* calculations it is more convenient to find critical values of $-N_y$, in terms of the loading ratio expressed as

$$\alpha = N_x/N_y. \tag{14.50}$$

For the sake of simplicity we shall consider here only three values of α, none of which is negative:

$$\alpha = 0, 0.5, 1.0.$$

The first case corresponds to 'pure side pressure', the second to 'hydrostatic' pressure and the third to equal bi-axial stress resultants. On this scheme pure axial loading, as in the preceding section, corresponds to $\alpha \to \infty$.

It is convenient to rearrange (14.48) thus:

$$\frac{(-N_y) 3^{\frac{1}{2}} a}{Eth} = \frac{Z}{\alpha + (\eta/\xi)^2}. \tag{14.51}$$

Fig. 14.11. Auxiliary functions for use with fig. 14.6 when the shell is subjected not only to axial compression but also (a) circumferential compression and (b) circumferential tension: see (14.49). The functions are shown for $|N_y/N_x| = 0.05$.

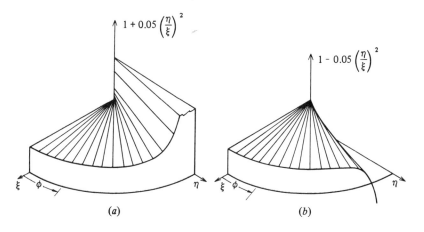

(a) (b)

498 *Buckling of shells: classical analysis*

We must find the condition for the right-hand side to be minimum. The right-hand side is a function of ξ and η, but subject to the condition that m and n are integral. It turns out, as we shall confirm later, that the minimum value of the right-hand side always occurs when $m = 1$, i.e. when there is just one half-wave between the simply-supported ends in the longitudinal direction. We are therefore concerned with values of the right-hand side of (14.51) at

$$\xi = \lambda/L = \xi_1.\tag{14.52}$$

It is simplest to begin by regarding the circumferential wavenumber n as continuously variable; thus we seek points at which

$$\frac{\partial}{\partial \eta}\left[\frac{Z}{\alpha + (\eta/\xi)^2}\right] = 0.\tag{14.53}$$

This calculation is much more awkward than the one in the previous section, although in principle it is straightforward. We give details of the calculation below, for the sake of completeness, but it is the results, given in figs. 14.12 to 14.14 which are of the most interest.

The most satisfactory way to proceed is to use polar coordinates in the ξ, η plane. On any given radial line defined by

$$\zeta = \tan\phi = \eta/\xi = \frac{l}{b},\tag{14.54}$$

Fig. 14.12. Preferred doubly-periodic modes are given by the Koiter circle (diameter $2^{\frac{1}{2}}$) for pure axial compression, and by other curves when there is a combination of axial load and external 'side pressure'. Shells having $\xi > 2$, say, are very short.

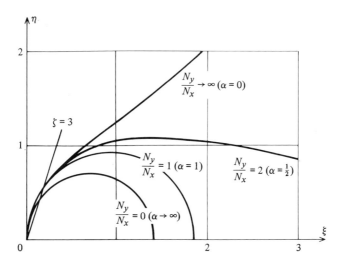

14.4 Axial compression and 'side' pressure combined

the minimum which we seek is at a point whose r-coordinate may be expressed by

$$r = \rho r_K. \tag{14.55}$$

Here, as before, r_K is the value of r at which the line ζ = constant intersects the Koiter circle, and ρ is a function of α and ζ, as yet undetermined. From (14.39) we may express r_K in terms of ζ as follows:

$$r_K = 2^{\frac{1}{2}} \cos\phi = [2/(1 + \zeta^2)]^{\frac{1}{2}}. \tag{14.56}$$

A detailed analysis shows that (14.53) is always satisfied when

$$\rho = \left[\frac{1 + 2\alpha + 3\zeta^2}{-1 + 2\alpha + \zeta^2} \right]^{\frac{1}{4}}. \tag{14.57}$$

By use of (14.57) we can find the loci of points in ξ, η space where the sought-for minima lie, for any given value of α. The results are shown in fig. 14.12. The corresponding curve when $N_y = 0$ is, as we already know, the Koiter circle $\rho = 1$: this result is recovered formally by putting $\alpha \to \infty$ in (14.57). Along each of these curves in the region $\rho > 1$ we find that $d(-N_y)/d\xi > 0$ (see problem 14.7); and hence in each case the critical mode is that for which the value of ξ is smallest, i.e. $m = 1$. It is now a simple matter to evaluate the minimum value of N_y for a shell of given length, by using fig. 14.12. First we evaluate ξ_1, by (14.52). Then, for a given value of α we read off the preferred value of η from the diagram: this then gives the value of $\zeta = l/b$, i.e. the preferred modal form, which we regard as continuously variable in the first instance. Also from the diagram we can scale off the value of ρ (see (14.55)) and obtain the corresponding value of Z by using (see (14.40))

$$2Z = \rho^2 + 1/\rho^2. \tag{14.58}$$

Lastly, we evaluate $(-N_y)$ by using (14.51).

Some results from calculations of this sort are shown in fig. 14.13a. The ordinate is a non-dimensional form of $(-N_y)$, cf. (14.51), and the abscissa is closely related to the reciprocal of ξ_1. Curves are shown for the three values of α. In fig. 14.13b curves are given for $(-N_x)$: for each value of α the corresponding curves in the two diagrams are the same except for translation, and in the second diagram result (14.33) for $N_y = 0$ has also been entered.

It is clear that for a given value of α the classical value of $(-N_y)$ and/or $(-N_x)$ is inversely proportional to L^2/ah for small values of $L/(ah)^{\frac{1}{2}}$, but is inversely proportional to $L/(ah)^{\frac{1}{2}}$ for larger values. Further, for *small* values of $L/(ah)^{\frac{1}{2}}$, N_x is the dominant effect for all values of α, while for *large* values of $L/(ah)^{\frac{1}{2}}$, N_y dominates (except, of course, when $\alpha \to \infty$). It is not difficult to see why this is so. For small values of $L/(ah)^{\frac{1}{2}}$ (i.e. large values of ξ_1) the form

500 *Buckling of shells: classical analysis*

of the Z-function is dominated by the bending terms (see fig. 14.7), and the critical values of $(-N_x)$, $(-N_y)$ and the corresponding preferred modes are essentially those which are obtained by regarding the shell as a flat plate: see problem 14.8. On the other hand, for large values of $L/(ah)^{\frac{1}{2}}$ there is genuine shell action, and it is clear, since ξ_1 is then small, that the dominant structural effects are longitudinal stretching and circumferential bending: see fig. 14.7. The fact that the curves in fig. 14.12 for the three finite values of α merge together in the region $\eta/\xi > 3$ is worth investigation; and it has some useful consequences. See problem 14.9.

The preferred modes have been indicated on fig. 14.13 by the values of l/b. It is often rather more instructive to interpret the preferred modes in terms of the circumferential wavenumber n. This can be worked out in a given case from the defining relation

$$n = \frac{\pi a}{b} = \left(\frac{l}{b}\right) \frac{\pi a}{l}, \qquad (14.59)$$

which can be evaluated in any given case by putting $l = L$.

Now consider what happens, in terms of fig. 14.12, as the value of $L/(ah)^{\frac{1}{2}}$ is steadily increased. The value of ξ_1 steadily decreases and the corresponding point on the diagram, for the given value of α, moves counterclockwise round

Fig. 14.13. Classical buckling criteria for different ratios N_y/N_x, plotted dimensionlessly in term of (a) N_y and (b) N_x against a dimensionless length $L/(ah)^{\frac{1}{2}}$: logarithmic scales. 'Panel' proportions of the preferred modeforms are indicated.

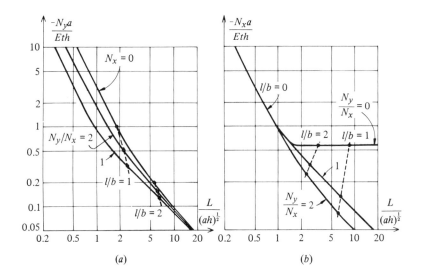

14.4 Axial compression and 'side' pressure combined

its curved path. In particular the ordinate η steadily decreases, and consequently the value of the preferred circumferential mode number n also decreases (see (14.59)). Eventually the point will be reached where $n = 2$, and further increases in the value of $L/(ah)^{\frac{1}{2}}$ will be dominated by the restriction of n to this value rather than the absolute minimum as indicated on fig. 14.12. Clearly we need to make a separate investigation of this end of the range of waveforms. This can be done in several ways. The simplest is to start with (14.21) but with the left-hand side augmented by the N_y term as in (14.48). It is clear from fig. 14.13a that in the range in question the value of N_x is practically irrelevant, so we shall drop that term. It is also clear from fig. 14.12 that the value of $\zeta = l/b = \eta/\xi$ will be large; so we shall neglect terms ξ^2 on the right-hand side of (14.48) in comparison with terms η^2. Lastly we put

$$l = L, \quad b = \pi a/n;$$

and we obtain, after some rearrangement,

$$(-N_y)a^2/Eth^2 = (\pi^4/n^6\Omega^4) + \tfrac{1}{12}n^2 \tag{14.60}$$

where, as before the dimensionless length parameter Ω is defined (see (9.9)) by

$$\Omega = Lh^{\frac{1}{2}}/a^{\frac{3}{2}}.$$

A formula of this sort (but in the more refined form (14.77)) was first given by Southwell (1913). (See Windenberg & Trilling (1934) for a collection of early design formulae.) The first term on the right-hand side of (14.60) corresponds to the effect of stretching in the longitudinal direction, while the second corresponds to the effect of bending in the circumferential direction. The appearance of the dimensionless group Ω in this expression should not be a surprise: we have encountered it on several previous occasions in connection with problems in which the dominant structural effects are longitudinal stretching and circumferential bending: see sections 9.2 and 9.5. Note also that the dimensionless group which includes $(-N_y)$ is different from that used in fig. 14.13. Equation (14.60) is plotted in fig. 14.14 as broken lines for various integral values of $n > 2$. The curve for each value of n touches an envelope, as shown, which may be determined by the method described above (cf. (14.27, 14.28)):

$$(-N_y)a^2/Eth^2 = 0.855/\Omega. \tag{14.61}$$

The value of n for which the curve (14.60) touches the envelope (14.61) is given by

$$n = 2.77/\Omega^{\frac{1}{2}}. \tag{14.62}$$

For a given value of n the second term on the right-hand side of (14.60) dominates when Ω exceeds about twice the value given by (14.62): in such cases the shell may be considered as being *infinitely long* for the purpose of classical buckling calculations.

Now the term $\frac{1}{12} n^2$ on the right-hand side of (14.60) represents the contribution from the circumferential bending of the tube. It is clear from previous remarks that this term ought to be re-examined in connection with the 'circumferential bending correction'; and indeed we shall do this in section 14.5.

The complete results of our calculations are shown in figs. 14.13 and 14.14. *Two* diagrams are needed because there is no single dimensionless group which covers the entire range of behaviour. In the 'intermediate' range given approximately by

$$L/(ah)^{\frac{1}{2}} > 1; \Omega = Lh^{\frac{1}{2}}/a^{\frac{3}{2}} < 2.5, \tag{14.63}$$

(cf. fig. 14.15), either graph gives essentially the same result (but with the preferred mode given in terms of l/b in one and n in the other); but outside this range we must use one of fig. 14.13 or fig 14.14, as appropriate.

Fig. 14.14. Logarithmic plot of buckling criterion for pure 'side pressure' against the dimensionless length Ω. The lower edge of the festoon is shown, for different integral values of n, together with the constituent curve for $n = 4$. Broken lines; Donnell equation (14.60): full lines; Donnell equation 'improved' (14.77) for low values of n. This diagram and fig. 14.13a overlap for shells of 'medium length'.

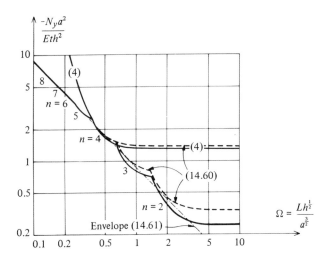

14.4 Axial compression and 'side' pressure combined

The information contained in figs. 14.13 and 14.14 is essentially the same as that given by Batdorf and others (e.g. Brush & Almroth, 1975, Figs. 5.14, 16 & 17). The main difference in presentation is that, in contrast to these workers, we have used dimensionless groups containing $-N_x$ and $-N_y$ which do not contain the length L of the shell: thus we can see at a glance the way in which the classical buckling pressure is affected by the length of the shell.

A condensed summary of the information contained in figs. 14.13 and 14.14 is given in fig. 14.15. This diagram is not intended to be particularly accurate, especially near the boundaries between the various regimes; but it does give a clear indication of the preferred modes which occur for cylindrical shells covering a wide range of geometrical parameters. It is useful for the purposes of preliminary design, as we shall see in section 14.9.

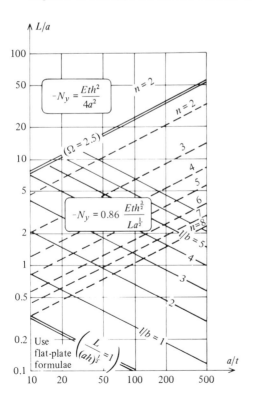

Fig. 14.15. Schematic chart for the buckling of a cylindrical shell under 'side pressure', when its ends are 'held circular'. Three distinct regions are shown, and preferred values of l/b and n are indicated. See comments at the end of section 14.6 in relation to the use of diagrams of this type for other boundary conditions.

14.5 Necessary corrections for small values of circumferential wavenumber n

The analysis which we have developed in the preceding section is based explicitly on the assumption that the circumferential wavenumber n has a sufficiently large value for various 'shallow-shell' approximations, leading to Donnell's equations, to be valid for the B-surface. We must now consider a relaxation of this restriction, since it is clear from fig. 14.14 that the value of n for the eigenmode is less than 3 when the value of Ω exceeds about unity. In cases of this kind we are concerned with the portion of the Koiter circle (or its equivalent for general values of N_x/N_y) which lies near the η axis in fig. 14.12. This corresponds (see fig. 14.7) to modeforms having high values of l/b, for which the dominant structural action involves longitudinal stretching and circumferential bending. We have already seen in section 8.6 that when the value of n is low, the circumferential bending stiffness according to the 'shallow-shell' theory should be multiplied by the 'correction factor' $(1 - n^{-2})^2$, whereas the shallow-shell S-surface behaviour needs no corresponding correction.

In the present problem there are two more 'correction factors' which must be introduced when the value of n is small. In section 8.6 the applied loading was prescribed independently. Here, however, the loading applied to the shell is the incremental pressure Δp_1, which is required for equilibrium of the 'funicular membrane' in its distorted configuration. Our previous formula (14.17) was derived, of course, within the context of 'shallow-shell' theory; and we must therefore re-examine it for our present purposes.

Consider first the contribution to Δp_1 corresponding to the stress resultant N_y. It is simplest to begin by investigating the analogous problem of an initially circular 'string' of radius a which carries a uniform tension T. Let it now be given a small radial displacement

$$w = w_n \sin n\theta \tag{14.64}$$

while the tension is maintained constant, and simultaneously let a normal loading of intensity Δp per unit length be applied in order to maintain equilibrium. The change of curvature, κ, in consequence of the displacement is given by

$$\kappa = -\frac{1}{a^2}\left(\frac{d^2 w}{d\theta^2} + w\right) = \left(\frac{n^2 - 1}{a^2}\right) w_n \sin n\theta. \tag{14.65}$$

Consequently, in calculating Δp_1 from

$$\Delta p_1 = T\kappa$$

14.5 Corrections for small values of wavenumber n

we find that Δp_1 is smaller, by the factor $(1 - n^{-2})$ than it was according to the earlier calculation in which we simply took

$$\kappa = -\frac{1}{a^2}\frac{d^2 w}{d\theta^2}.$$

Exactly the same argument applies to the N_y stress resultant acting in a 'ring' of the shell.

The second new effect concerns the stress resultant N_x. We have already seen that if a small radial displacement w is applied to the surface, an additional normal pressure Δp_1 of magnitude $N_x \partial^2 w/\partial x^2$ is required in order to preserve equilibrium of the momentless funicular membrane. Now we have already seen (e.g. section 6.5) that in inextensional displacements of circular rings, radial displacements of the form

$$w = w_n \sin n\theta$$

are accompanied by *tangential* displacements

$$v = (w_n/n) \cos n\theta. \tag{14.66}$$

It follows that when the shell undergoes a small deformation to an adjacent position of equilibrium, the stress resultant N_x demands not only a *normal* pressure as described above, bit also a *tangential* traction Δq_1, given by

$$\Delta q_1 = N_x \frac{\partial^2 v}{\partial x^2}. \tag{14.67}$$

From (14.17) we see that the amplitude of this traction is equal to $(1/n)$ times the amplitude of Δp_1 as originally calculated. Thus, for large values of n this effect is negligible; but clearly it must be taken into account in the present analysis. It is awkward in our two-surface theory to deal with tangential interactions: but we have seen in section 7.3.3 that they may be replaced by statically equivalent normal tractions with the help of small additional circumferential tensions which do not produce appreciable additional strains. Consequently, taking account of the signs of Δp and Δq we find that together they provide a pressure variation which is given by

$$\Delta p_1 = (1 + n^{-2}) N_x \frac{\partial^2 w}{\partial x^2}. \tag{14.68}$$

It is convenient at this stage to state explicitly the general buckling equation which we solved intuitively in the preceding sections for special cases involving a particularly simple modal form. Collecting various expressions together, we find that the following equation represents the buckling condition for the shell according to the 'quasi-shallow' approximation:

$$-(-N_x)\frac{\partial^2 w}{\partial x^2} + 2N_{xy}\frac{\partial^2 w}{\partial x \partial y} - (-N_y)\frac{\partial^2 w}{\partial y^2} = \frac{Et}{a^2}\nabla^{-4}\left(\frac{\partial^4 w}{\partial x^4}\right) + \tfrac{1}{12}Eth^2\nabla^4 w.$$
(14.69)

The left-hand side represents the (uncorrected) unbalanced pressure when a momentless membrane carrying N_x, N_y, N_{xy} is given a small displacement. On the right-hand side the second term represents the (uncorrected) pressure sustained by the bending of the B-surface, while the first term represents the pressure-carrying contribution from the S-surface. The first term on the right-hand side needs a little explanation. If we take the governing 'Donnell' equation (8.24), and simplify it by putting $a^* \to \infty$ (for a cylindrical shell) and omitting the 'bending' terms, we obtain the following governing equation for the S-surface:

$$\nabla^4 p = \frac{Et}{a^2}\frac{\partial^4 w}{\partial x^4}.$$

We have obtained the corresponding term in (14.69) by pre-multiplying both sides of this expression by the operator ∇^{-4}, which is defined by the relation

$$\nabla^{-4}\cdot\nabla^4(w) = w.$$
(14.70)

The operator ∇^{-4} implies a fourth-order integration, which of course cannot be done in general without a knowledge of some boundary conditions. However, in the present circumstances we are thinking only in terms of *periodic* eigenmodes applied to shells of effectively infinite length, or sections cut from such shells without disturbance: and consequently for our present purposes the integration does not present a real difficulty. All of our previous results correspond exactly to solutions of (14.69).

In our present investigation of the corrections which are necessary when the circumferential mode-number is small, we are interested only in circumstances where the ratio $l/b > 2$, say. Accordingly, in view of our previous work (cf. section 8.5.1) we may simplify the equation by retaining only one term (i.e. that related to changes in the y-direction) in each of the operators ∇^4 and ∇^{-4}. Furthermore, since we are not at present concerned with torsional buckling (but see section 14.7) we may omit the term containing N_{xy}. Consequently we may simplify equation (14.69), for present purposes, as follows:

$$-(-N_x)\frac{\partial^2 w}{\partial x^2} - (-N_y)\frac{\partial^2 w}{\partial y^2} = \frac{Et}{a^2}D^{-4}\left(\frac{\partial^4 w}{\partial x^4}\right) + \tfrac{1}{12}Eth^2\frac{\partial^4 w}{\partial y^4}.$$
(14.71)

Here, and in the remainder of this section, D stands for the differential operator d/dy: so

$$D^{-1}(\ldots) \equiv \int(\ldots)dy.$$
(14.72)

14.5 Corrections for small values of wavenumber n

At this point we may insert the correcting factors appropriate for small values of n, as discussed above, for any solution which is periodic in the circumferential direction. Hence we obtain a 'corrected' version of (14.71):

$$-(1+n^{-2})(-N_x)\frac{\partial^2 w}{\partial x^2} - (1-n^{-2})(-N_y)\frac{\partial^2 w}{\partial y^2} =$$
$$\frac{Et}{a^2}D^{-4}\left(\frac{\partial^4 w}{\partial x^4}\right) + (1-n^{-2})^2\frac{Eth^2}{12}\frac{\partial^4 w}{\partial y^4}. \quad (14.73)$$

The only term which is not 'corrected' is that which represents the effect of the S-surface: for a cylindrical surface the operator ∇^4 is correct as it stands, without any need of modification.

Let us now postulate an eigenfunction of the form

$$w = w_n(x)\sin n\theta, \quad (14.74)$$

and substitute into (14.73), using $y = a\theta$. Operator D^{-4} becomes the multiplier $(a/n)^4$, and we obtain

$$(-N_x)\left(\frac{n^2+1}{n^2}\right)w_n'' + (-N_y)\frac{(n^2-1)}{a^2}w_n = \frac{Eta^2}{n^4}w_n'''' +$$
$$\frac{(n^2-1)^2}{a^4}\frac{Eth^2}{12}w_n. \quad (14.75)$$

Here $'$ denotes differentiation with respect to x. We shall use this equation in section 14.6 to solve several buckling problems involving eigenfunctions which are not sinusoidal in the x-direction. At present, however, we are mainly concerned with the consequences of these changes on the results which we have already obtained according to 'quasi-shallow' cylindrical shell theory. Thus we now put

$$w_n(x) = w_0 \sin(\pi x/l). \quad (14.76)$$

We discovered earlier that in the range of solutions under discussion (at the left-hand end of the Koiter circle, fig. 14.5) the axial wavenumber m is equal to 1. Furthermore, the role of N_x is almost completely subordinate to that of N_y in this range; and it follows that an investigation of the case $N_x = 0$ will suffice for a wide range of cases. An exception to this rule occurs when $N_y = 0$: then, of course, N_x is dominant. For this reason we shall investigate two separate cases, as follows:

(a) $N_x = 0$,
(b) $N_y = 0$.

For the first of these problems, (a), we put $N_x = 0$ in (14.75) and obtain, after some manipulation,

$$(-N_y)a^2/Eth^2 = [\pi^4/n^4(n^2-1)](1/\Omega)^4 + \tfrac{1}{12}(n^2-1). \quad (14.77)$$

This should be compared with (14.60). The only change is that there are places where n^2 in (14.60) is replaced by $n^2 - 1$; and indeed this is not surprising. The second term on the right-hand side represents the effect of the tube in buckling like a ring, and the form of the expression is well known. But it is perhaps surprising that the contribution from the S-surface should also be altered: the effect enters not through the S-surface equations themselves, but through the pressure-load term for the equilibrium of the funicular membrane (see (14.65) etc.). It is a straightforward matter to alter fig. 14.14 in accordance with the new version of (14.60). For a given value of n each of the two asymptotes is altered in a simple way, and the corresponding 'transition curve' must be translated so that it fits the new asymptotes. The resulting alteration to the lower limit of the 'festoon' curve is shown by a full line. The lobe for each discrete value of n falls below the envelope (14.61) calculated earlier by a factor $(1 - n^{-2})^{\frac{1}{2}}$. This factor is smallest (= 0.866) at $n = 2$. The details of the lobes agree with the results of Flügge (1973, Fig. 8.15).

Note that in the special case $n = 1$, the formula (14.77) becomes meaningless.

For the second case, (b), we put $N_y = 0$ in order to obtain results for buckling under purely axial loading. Here we retain the axial half wavelength l, and we obtain

$$-N_x = \frac{\pi^2 E t a^2}{n^2(n^2 + 1)l^2} + \frac{n^2(n^2 - 1)^2}{(n^2 + 1)} \frac{Eth^2}{12} \frac{l^2}{\pi^2 a^4}. \tag{14.78}$$

This expression is equivalent to (14.21) when $b/l \to 0$. Bearing in mind that $bn = \pi a$, we find that the only difference between this equation and the simpler original version is that here we have terms $(n^2 - 1)$ and $(n^2 + 1)$ taking the place of n^2 in the original expression.

A simple way of investigating the impact of these modifications on our previous results is to investigate their effect on the function Z. Following (14.33) we write

$$-N_x = \frac{Eth}{3^{\frac{1}{2}}a} Z', \tag{14.79}$$

where Z' is the function Z modified in accordance with (14.78). We found earlier (see (14.40)) that on lines in the plot of fig. 14.5 for which the value of ζ $(= \eta/\xi = l/b)$ is constant, the function Z may be expressed in the following simple way:

$$2Z = (r_K/r)^2 + (r/r_K)^2.$$

Here, as before, r_K is the value of r where the line l/b = constant intersects the Koiter circle. In this expression the first and second terms on the right-

14.5 Corrections for small values of wavenumber n

hand side represent stretching and bending effects, respectively. For large values of l/b we may now arrange (14.78) to give

$$2Z' = \frac{n^2}{n^2 + 1}\left(\frac{r_K}{r}\right)^2 + \frac{(n^2 - 1)^2}{n^2(n^2 + 1)}\left(\frac{r}{r_K}\right)^2. \tag{14.80}$$

This function is plotted for various values of n in fig. 14.16; Z' has a minimum value of $(n^2 - 1)/(n^2 + 1)$ for $n \geqslant 2$. Reductions in the value of Z' below 1 have the effect of lowering the 'valley floor' near the η-axis (see figs. 14.5 and 14.6): formula (14.80) is only valid, of course, in this region of ξ, η space. These results agree with those of Flügge (1973, Fig. 8.10).

Our previous result, that the 'valley floor' is horizontal, was obtained by neglecting the various 'low-n' corrections. In view of our new result we must now be prepared to reject our previous conclusion that a thin cylindrical shell under axial compression has a large number of 'competing' modes, which all have practically equal eigenvalues $-N_x$. There is, however, an important qualification to be made in connection with these results. Throughout our analysis we have assumed a particularly simple mode of buckling (see (14.18)) and, in order to preserve the simplicity of our algebraic expressions we have postulated a set of rather artificial 'conventional' boundary conditions, in which the ends of the shell are held circular ($w = 0$) but are unre-

Fig. 14.16. Modification of the function Z for small values of the circumferential mode-number n, shown as normalised sections on planes η/ξ = constant (cf. (14.79) and fig. 14.6).

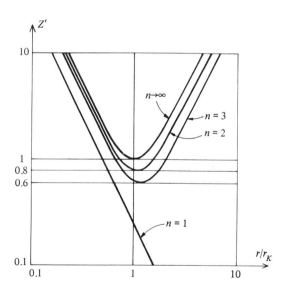

strained in the axial direction, i.e. are *free to warp*. Now in many practical problems the ends of a cylindrical shell are *not* free to warp, as we have already remarked; and this additional form of restraint must be expected to *raise* the eigenvalue load to some extent. This problem forms the subject of section 14.6; the effects are somewhat analogous to those which were discussed in section 9.5.1. We shall discover that the prevention of warping at the ends has a strong effect on modes having large values of l/b, but a relatively weak effect when $l/b < 1$, approximately. Consequently, the prevention of end-warping tends to raise the 'valley floor' in the left-hand part of the Koiter circle, but it has little effect on the right-hand part. Thus, when the effects of low circumferential wavenumbers and of prevention of warping are taken into account together, our earlier conclusion about a large number of competing modes survives virtually unscathed.

It is clear from fig. 14.16 that the case $n = 1$ is special in the sense that circumferential bending makes no contribution at all to the resistance of loading. In this case (14.78) reduces to

$$-N_x = \pi^2 E t a^2 / 2 l^2. \tag{14.81}$$

This is in fact precisely the 'Euler' result for a tube acting as a rod of length l: the value of I for a tube in pure bending is equal to $\pi a^3 t$, and (14.81) follows immediately from (14.1). Each cross-section of the tube remains plane, but rotates about a line normal to the axis. In particular, the ends of the tube rotate, just like the pinned ends of a rod. This is a special case of freedom to warp (i.e. to undergo nonzero u-displacements of the form $u = u_0 \sin\theta$), and we can see by analogy with an Euler column with ends restrained from rotation that prevention of rotation of ends of the tube would raise the line marked $n = 1$ in fig. 14.16 by a factor of 4.

This case of the buckling of a long tube in the manner of an 'Euler rod' demonstrates that the critical loading parameter is N_x. Indeed we should expect this, since a loading which produces N_y can have no effect for the mode $n = 1$. Thus we conclude that for buckling of a shell as an 'Euler rod' only the value of N_x is significant. Unfortunately, this result is not correct. It is well known that the application of exterior hydrostatic pressure to a *rod* can never produce elastic Euler buckling; and the same result holds when the rod is replaced by a circular tube. In particular, buckling in this mode $n = 1$ is not induced by a value of $-N_x (= -\frac{1}{2} N_y)$, however large. That this remarkable and undisputed result is not brought out by our analysis is traceable to our use of a *linearised* system of equations. We can, however, understand this result without doing any nonlinear analysis, by means of the following argument. Consider a curved tube of circular cross-section subjected to uniform external

14.5 Corrections for small values of wavenumber n   511

hydrostatic pressure, as shown in fig. 14.17a. Suppose we make a cross-sectional cut at an arbitrary point on the axis, and find what force is necessary to maintain either of the two separate pieces in equilibrium under the applied pressure. The answer is that the force required is equal to exactly pA, where p is the pressure and A the total cross-sectional area of the tube; and the line of action of this force passes through the *centre* of the cross-section. This follows immediately from a well-known theorem of hydrostatics. Note that the situation is quite different from that which obtains in a tube under the action of forces applied to the ends only: see fig. 14.17b. In that case the force which is necessary to keep one part in equilibrium does *not* pass through the centre of the section, and a bending moment is required; this in turn leads to the possibility of an initially straight rod buckling under the action of compressive force applied at the ends. The important point in relation to the tube of fig. 14.17a is that the theorem in hydrostatics which we have used applies quite generally for bodies of entirely arbitrary shape: it therefore applies, without correction, in circumstances of 'large deflections'. Fig. 14.17a represents a 'large-deflection' situation in this sense because the difference in length between the two sides of the tube – which is crucial for a correct hydrostatic analysis – is simply not accounted for in the equations which we have used. We may conclude from this argument that in applying formula (14.81) in the special case $n = 1$, we must not count towards N_x any axial force which is due to hydrostatic pressure-loading.

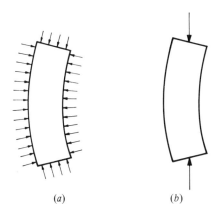

Fig. 14.17. Argument to show that external hydrostatic pressure (a) cannot induce overall 'column' buckling of a long, cylindrical pressure-vessel; in contrast to (b) purely axial compressive loading.

(a)   (b)

14.6 The effect of clamped and other boundary conditions

In this section we shall investigate the consequences of imposing the boundary condition of zero warping ($u = 0$) in addition to the 'held circular' condition ($w = 0$) which we have used throughout the previous section. We shall not give an exhaustive treatment of this problem. Instead, we shall restrict attention to those cases in which the predominant structural actions are longitudinal stretching and circumferential bending, for which, as we have seen, the equations are analogous to those for the relatively simple problem of a beam on an elastic foundation. This approach is not unreasonable; for we have already studied examples which show that there is relatively little tendency for warping to occur either when the predominant structural action involves circumferential stretching and longitudinal bending (see section 8.5) or in the 'quasi-flat-plate' region in which bending effects alone dominate. Thus it is clear that the prevention of warping has effects which are strongest in the region which we propose to investigate. In any given problem it is easy to ascertain which are the main structural actions by evaluating the panel proportions l/b: if $l/b \geqslant 3$, say, the shell will be well within the range for which circumferential bending and longitudinal stretching predominate: see fig. 8.4a.

Let us consider first the general question of a thin cylindrical shell loaded by side pressure only, i.e. having $N_y \neq 0$, $N_x = 0$. We shall consider a mode of the kind

$$w = w_n(x) \sin n\theta, \tag{14.82}$$

for which the general equation, including the 'n-small' effects may be written

$$-N_y \frac{\pi(n^2 - 1)}{a} w_n = \frac{\pi E t a^3}{n^4} w_n'''' + \frac{\pi(n^2 - 1)^2}{a^3} \frac{E t h^2}{12} w_n. \tag{14.83}$$

Here we have multiplied (14.75) throughout by πa, so that the two terms on the right-hand side may be associated directly with 'beam' and 'foundation' effects for the analogous beam on an elastic foundation, as before (cf. section 9.4):

$$B \leftrightarrow \frac{\pi E t a^3}{n^4}, \quad k \leftrightarrow \frac{\pi(n^2 - 1)^2}{12} \frac{E t h^2}{a^3}. \tag{14.84}$$

In contrast to the examples of chapter 9 the loading applied here to the analogous beam (i.e. the term on the left-hand side of (14.83)) is proportional to w. This follows from the fact that the loading term is derived from the out-of-balance pressure which occurs when the load-carrying membrane is distorted; and indeed it is precisely this effect which turns the problem into one involving an eigenvalue. It is convenient to define a term

$$\psi \leftrightarrow -N_y \pi(n^2 - 1)/a, \tag{14.85}$$

14.6 Clamped and other boundary conditions

so that the analogous beam equation can be written in the simple form

$$Bw'''' + (k - \psi)w = 0. \tag{14.86}$$

(ψw denotes the quantity called η in chapter 9, sections 9.3 and 9.4). Clearly this equation is satisfied, trivially, by $w = 0$ for an arbitrary value of ψ: correspondingly, the shell can be in equilibrium in an unbuckled configuration. But as the loading ($-N_y$) is steadily increased from zero, the constant ($k - \psi$) becomes negative and eventually reaches a critical value at which a state of equilibrium adjacent to the original undeformed state is possible. In fact ψ combines directly with k to produce what amounts to a kind of 'negative spring stiffness'; and this has a critical value at which a nontrivial solution is possible.

Mathematically the problem is to find the lowest value of (say)

$$\psi - k = \Psi$$

at which the equation

$$Bw'''' - \Psi w = 0 \tag{14.87}$$

has a nontrivial solution, subject to appropriate boundary conditions; then the required loading is found from

$$\psi = \Psi + k. \tag{14.88}$$

Now (14.87) is formally the same as the equation which must be solved in order to find the natural frequencies for transverse vibrations of a uniform elastic beam (where $\Psi =$ (mass/unit length) x (angular frequency)^2: see, e.g. Young, 1962). Solution techniques for this equation are well known; and indeed we shall be able to make use of results which have been tabulated.

We discovered in section 9.5.1 that when the edge of a cylindrical shell is held circular and warping is not prevented, the boundary condition for the corresponding beam is one of simple support; that a clamped edge of the shell corresponds to an encastré end of a beam; and that a free edge of the shell corresponds to a free end of a beam.

In our investigation of the effects of different boundary conditions on the buckling of a cylindrical shell due to 'side pressure' loading, let us examine first the case of simply-supported ends: this corresponds exactly to our previous boundary conditions, and we shall therefore be in a position to check the results. The boundary conditions at ends $x = 0, x = L$ are satisfied by a postulated eigenfunction

$$w = w_m \sin(m\pi x/L),$$

where m is an integer. This clearly satisfies (14.87) provided

$$\Psi = Bm^4\pi^4/L^4. \tag{14.89}$$

From this we find, by using (14.84) and (14.88)

$$\frac{(-N_y)\pi(n^2 - 1)}{a} = \frac{\pi E t a^3 m^4 \pi^4}{n^4 L^4} + \frac{\pi(n^2 - 1)^2}{12} \frac{E t h^2}{a^3}. \tag{14.90}$$

The right-hand side is smallest when $m = 1$; and with this substitution we recover precisely the equation (14.77) which was our previous solution of the same problem. It is clear from the form of this equation that the first and second terms on the right-hand side correspond to contributions from the S- and B-surfaces, respectively; and hence that the simplifying parameter Ψ represents, in fact, precisely the S-surface contribution. This feature emerges in the present problem because both the circumferential bending effect and the Δp_1 loading term are directly proportional to displacement w.

Consider next a case in which both ends of the shell are held circular, and also prevented from warping. The corresponding boundary conditions for the beam are

$$w = w' = 0 \text{ at } x = \pm\tfrac{1}{2}L: \tag{14.91}$$

in this case it is most convenient to put the origin in the middle. The complementary function of (14.87) is easily found:

$$w = A_1 \sin(2\beta x/L) + A_2 \cos(2\beta x/L) + A_3 \text{Sh}(2\beta x/L) + A_4 \text{Ch}(2\beta x/L), \tag{14.92a}$$

where

$$\beta = \tfrac{1}{2}L (\Psi/B)^{\frac{1}{4}}. \tag{14.92b}$$

The boundary conditions immediately give $A_1 = A_3 = 0$. Thus

$$w = A_2 \cos(2\beta x/L) + A_4 \text{Ch}(2\beta x/L), \tag{14.93}$$

and the two remaining boundary conditions become

$$0 = A_2 \cos\beta + A_4 \text{Ch}\beta$$

and

$$0 = -A_2 \sin\beta + A_4 \text{Sh}\beta.$$

On elimination of A_2/A_4 from these two equations we have

$$\tan\beta = -\text{Th}\beta \tag{14.94}$$

as the required condition. This equation has many roots; but since we are concerned with the lowest value of Ψ we seek only the lowest (nontrivial) one.

By trial-and-error we find the lowest root $\beta = 2.365$, from which we obtain $A_4/A_2 = 0.133$ (which gives the form of the eigenmode) and hence

$$\Psi = 16B\beta^4/L^4 = 500B/L^4. \tag{14.95}$$

This is larger than the previous critical value for simply-supported ends (cf.

14.6 Clamped and other boundary conditions

(14.89) with $m = 1$) by a factor $500/\pi^4 = 5.1$. We have already seen that Ψ is related directly to the part played by the S-surface in resisting applied loads by means of longitudinal stretching effects; and moreover that these effects are represented by the 'beam' part of the beam-on-elastic-foundation analogue. It is remarkable that this factor of 5.1 is so close to the factor 5.0 by which the central static deflection of a uniformly loaded uniform elastic beam is reduced when the end-conditions are changed from simply-supported to clamped.

Equation (14.95) gives in turn an equation similar to (14.90) but with the first term on the right-hand side multiplied by 5.1, and with $m = 1$. It is easy to use this equation to obtain a festoon of curves for different values of n, after the fashion of fig. 14.14. The lower limit of such a festoon is shown in fig. 14.18: the effect of the change in boundary conditions is to shift the entire curve in the Ω-direction by a factor $5.1^{\frac{1}{4}} \approx 1.5$ and thus to raise the critical value of N_y by the same factor over the entire 'middle range' of Ω. It is remarkable that although the prevention of warping at the ends enhances the performance of the S-surface by a factor of 5.1, the buckling pressure itself rises only by a factor of 1.5. This is a consequence of the interaction between stretching and bending effects which involves, *inter alia*, an increase in the circumferential wavenumber n: see problem 14.10.

Calculations relating to two other boundary conditions are also shown in fig. 14.18. (Further details concerning the solution of essentially the same equation are given in section 17.5.1.) In each case the results of calculations on simple beams have been adapted, in exactly the same way as in the preceding example. The equations of the envelopes (cf. fig. 14.14) and the corresponding formulae for n are given in the caption. These are virtually identical to formulae which appear in a new German Code of Practice (D.A.St, 1980, Fig. 2.4) and which are quoted by Malik, Morton & Ruiz (1980, p. 110).

The results displayed in fig. 14.18 agree well with the work of Malik *et al.* and Sobel (1964). Malik *et al.* used a simplified fourth-order equation in terms of the radial displacement w for a given value of the circumferential wavenumber, and found the corresponding buckling pressure for a variety of boundary conditions. Their method is almost exactly equivalent to the one which we have used. They found that all of their results agreed well with calculations performed by the comprehensive BOSOR (= buckling of shells of revolution) computer package (Bushnell, 1970; 1976). The BOSOR program uses a finite-difference representation of the shell equations in terms of three components, u, v, w of displacement; and in particular four boundary conditions must be specified at each end of the shell.

The simplified scheme of the present section, and of Malik *et al.*, works well in circumstances where the behaviour of the shell is dominated by an interaction between circumferential bending and longitudinal stretching. We have found previously, in problems concerning cylindrical shells with simply-supported edges, that a satisfactory test for the validity of this scheme is that the panel aspect ratio l/b should have a sufficiently large value: say $l/b > 3$ (cf. fig. 14.13). It is reasonable to use a similar test in the present circumstances, using the length L of the shell as a measure of l, irrespective of the boundary conditions; and it is therefore easy to make the check in any given case by using the formula for n in fig. 14.18, together with the relation

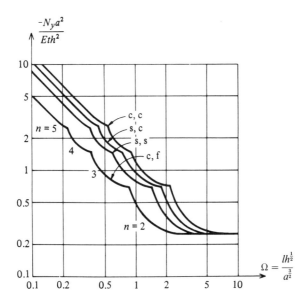

Fig. 14.18. The festoon of fig. 14.14 (labelled s, s) together with corresponding curves for different boundary conditions: s = simply supported (i.e. 'held circular'), c = clamped (i.e. held circular and plane), f = free. The (inclined) asymptotes and corresponding values of n are given by the formulae
$$-N_y a^2/Eth^2 = C_1/\Omega; \quad n = 3(C_1/\Omega)^{\frac{1}{2}}.$$
(These should be used with caution when $l/b < 3$: cf. fig. 14.13). Values of the constant C_1 are given below.

Boundary conditions	C_1
c, c	1.29
c, s	1.07
s, s	0.86
c, f	0.51

14.6 Clamped and other boundary conditions

$b = \pi a/n$. Malik *et al.* were concerned mainly with oil-storage tanks (see the following section), which usually satisfy this requirement with ease.

Sobel's (1964) calculations concerned the buckling under hydrostatic pressure of a shell having $a/t = 100$ and values of L/a ranging from 0.5 to 10. He used governing equations set up in terms of the three components u, v, w of displacement; and he investigated carefully the influence on the buckling pressure of various types of 'simply-supported' and 'clamped' boundary conditions – which were always of the same kind at both ends of a given shell. In all cases Sobel imposed the condition $w = 0$. He used alternative versions for each of the other three boundary conditions, in which *either* a displacement *or* its corresponding force were set to zero. The three conditions were thus:

either $\quad u = 0 \quad$ or $\quad N_x = 0$,

either $\quad \dfrac{\partial w}{\partial x} = 0 \quad$ or $\quad M_x = 0$,

either $\quad v = 0 \quad$ or $\quad N_{xy} = 0$.

The first of these represents either a complete prohibition of warping or a complete freedom to warp: here N_x refers to changes from the prescribed edge-loading. The second condition corresponds to either clamped or simple supports in relation to beam-like action of a longitudinal strip of the shell. The third condition represents either a prohibition of circumferential displacement (as at the circumferential panel boundaries of an endlessly repeating doubly-periodic mode) or a lack of such restraint. Altogether, therefore, Sobel investigated $2^3 = 8$ different sets of boundary conditions for each shell.

Sobel's results show some striking correlations. For the *longer* shells the critical buckling pressure is determined almost completely by the boundary condition on warping (either $u = 0$ or $N_x = 0$): the conditions on $\partial w/\partial x$ and v make practically no difference to the numerical results, which agree well with the corresponding curves in fig. 14.18 labelled c, c (for $u = 0$) and s, s (for $N_x = 0$). On the other hand, for the *shorter* shells the results are quite different. Here the buckling pressure depends primarily on the boundary condition on $\partial w/\partial x$: the other two conditions make relatively little difference.

This set of computations provides a striking demonstration of some general points which we have made in this chapter. The longer shells ($L/a \geqslant 1.5$) lie in the range for which Ω is the key geometrical parameter, and the prevention of warping at the ends raises the buckling pressure by a factor ≈ 1.5: $l/b \geqslant 3$ for all these cases. On the other hand Sobel's shortest shell ($L/a = 0.5$) is nearly short enough to lie in the range (cf. fig. 14.13) where the behaviour is determined almost entirely by 'flat-plate' effects: N_x is the key stress resultant, and the prevention of rotation ($\partial w/\partial x$) at the ends raises the buckling pressure by a factor ≈ 4. As we remarked earlier, there is practically no tendency for the edge of the shell to warp in this range, and the results are indeed indifferent to the boundary condition on warping. Some of Sobel's shells – particularly around $L/a \approx 0.75$ – lie in an intermediate range in which the buckling pressure is not influenced predominantly by either of the boundary conditions on u or $\partial w/\partial x$, respectively. In this region (for which $L/(ah)^{\frac{1}{2}} \approx 7.5$) both sets of conditions have a significant part to play in determining the buckling pressure, and a full investigation cannot be made by means of our simplified method. Nevertheless our present scheme helps us to interpret the numerical results in this regime as a 'crossover' from one relatively simple pattern of behaviour to another. This 'crossover' is directly analogous to the one which we have already investigated in the context of 'conventional' simply-supported edges: fig. 14.13 shows a crossover region corresponding to values of $L/(ah)^{\frac{1}{2}}$ lying between 1 and 10, with a central value ≈ 3. The fact that Sobel's crossover region is centred on a higher value of $L/(ah)^{\frac{1}{2}}$ is a consequence of the different boundary conditions.

In all of Sobel's examples the boundary condition on v or N_{xy} makes very little difference to the buckling pressure. In section 14.11, however, we shall mention a problem in which this particular aspect of the boundary conditions is crucial.

14.6.1 Buckling of oil-storage tanks

As a practical application of the preceding analysis we shall consider the buckling under wind-loading or internal vacuum of large vertical oil-storage tanks made from sheet steel.

The two cases shown in fig. 14.18 in which one end of the shell is clamped while the other is either 'free' or 'held circular' are relevant to the performance of oil-storage tanks. Typical dimensions of such tanks are: radius $a \approx 30$ m, height $L \approx 15$ m, thickness $h \approx 30$ mm; thus $\Omega \approx 0.02$. These tanks are usually built on concrete bases, and the vertical cylindrical walls are built up in the form of 'courses' of diminishing thickness. The roofing arrangements are of two types: (i) a shallow conical roof, connected to the top of the cylindrical shell; (ii) a flat roof which floats on the surface of the stored liquid, and rides up and down as the level changes. Tanks having roofs of type (ii) normally have the top (free) edge stiffened by the provision of a circumferential stiffening ring. There is a flexible seal between the roof and the shell wall. Both types of tank may be buckled by wind pressure-loading. Wind pressure varies around the circumference in the manner described in problem 4.10; but since

14.6 Clamped and other boundary conditions

the critical buckling mode for these tanks is usually in the region of $n \approx 10$, it is reasonable to consider the pressure as uniform for the purposes of calculation.

The boundary conditions for buckling of a tank of roof-type (i) will be c, s for practical purposes: the base is effectively clamped, while the roof provides a 'held circular' condition. The same boundary conditions also apply to tanks of type (ii) if the stiffening ring is sufficiently rigid in the critical circumferential mode, which is usually the case. However, a tank of either sort may be caught by the wind before the roof or stiffening girder has been completed; in which case the appropriate boundary conditions are c, f and the critical buckling pressure is lowered by a factor of about 2, according to fig. 14.18. Tanks with roof-type (i) may also buckle on account of a partial vacuum due to a sudden discharge of liquid from the tank if the air-vents are partially blocked.

In all of these cases fig. 14.18 may be used directly to predict the critical buckling pressure of tanks which have *uniform* wall thickness.

In practice, tanks almost always have variable wall thickness, since the main design criterion is the circumferential stress set up by the liquid in a full tank (British Standard 2654, 1973). In principle, the buckling of a given tank may be investigated by means of the analogous non-uniform elastic foundation: the flexural rigidity of the beam is proportional to the thickness t, while the stiffness of the foundation is proportional to t^3. Although such calculations are obviously much more simple than computations which do not recognise that the predominant structural actions are longitudinal tension and circumferential bending, they are by no means trivial. Calculations of essentially this type have been performed by Malik et al. (1980).

The results of fig. 14.18 may be pressed into service in the following way for an approximate analysis of the buckling of tanks having non-uniform wall thickness.

A computer study – by means of a finite-difference method – of the buckling of tanks having a *linear* variation of thickness with height from t_0 at the clamped base to t_1 at the top showed that for cases in which $1 < t_0/t_1 < 3$, approximately, fig. 14.18 correctly gives the buckling pressure p and critical mode-number n provided the following 'effective thicknesses' are used:

$t = 0.25\, t_0 + 0.75\, t_1$ for calculation of p ⎫
$t = t_1$ for calculation of n ⎬ in case c, f.
$t = 0.4\, t_0 + 0.6\, t_1$ for calculation of p ⎫
$t = t_1$ for calculation of n ⎬ in case c, s.

When the ratio t_0/t_1 exceeds 3, approximately, the above scheme ceases to be accurate. In these cases, however, it turns out that there is relatively little displacement in those parts of the shell for which the thickness exceeds $3t_1$; and indeed it is satisfactory to regard the shell as completely rigid in this region. In other words, when $t_0/t_1 > 3$ the shell may be regarded as being foreshortened, with an artificial clamped base provided at the level where $t = 3t_1$. Thus the rules given above may be used for *all* values of t_0/t_1 provided the lower end of the shell is taken as being clamped at its real or artificial base, as appropriate. (The subordination of the actual clamped boundary in cases $t_0/t_1 > 3$ is analogous to the ineffectiveness of the boundary condition at $\phi = \frac{1}{2}\pi$ in the bending of curved tubes for which $\Gamma > 5$, approximately: see fig. 13.11.) The apparent abruptness of the change of formulae in the above scheme at $t_0/t_1 = 3$ is not reflected in the computed results: see problem 14.11.

The above formulae give results for p and n within about 5% of those determined by the finite-difference procedure.

How can these results be applied to shells whose thickness changes stepwise with height? The following algorithm predicts p and n for such shells to within about 15% of those obtained by numerical computation (J. Morton, private communication).

(1) In cases where the thickness of any course exceeds 3 times the thickness of the top course, replace the course by a *rigid* course – thereby introducing an artificial clamped base to the shell.

(2) In the above formulae replace t_1 (for the calculation of n) by the thickness of the top course. Replace $0.25\, t_0 + 0.75\, t_1$ by the following weighted mean thickness:

$$\bar{t}_{cf} = \int_0^1 tz^2 \mathrm{d}z \Big/ \int_0^1 z^2 \mathrm{d}z, \qquad (14.96a)$$

where z is a dimensionless height co-ordinate so arranged that its values are 0 and 1 at the (real or artificial) base and top, respectively. Replace $0.4t_0 + 0.6\, t_1$ by the weighted mean thickness

$$\bar{t}_{cs} = \int_0^1 t(z^2 - z^3)\mathrm{d}z \Big/ \int_0^1 (z^2 - z^3)\mathrm{d}z. \qquad (14.96b)$$

In either case the weighted mean thickness gives precisely the previous formula for a linearly tapering shell: see problem 14.12.

Many practical shells have successive courses of equal height. In this case the above mean-thickness formulae may be replaced by tables of weighting-factors: see problem 14.13. An example concerning a multi-course tank is given in problem 14.14.

14.7 Buckling of cylindrical shells in torsion

So far in this chapter all of our examples have involved buckling under the action of stress resultants N_x and N_y; and all of the corresponding eigenmodes (see (14.18)) have been of the form

$$w = w_0 f(x) \sin(\pi y/b).$$

We turn now to the buckling of a thin cylindrical shell which is loaded in pure torsion, and therefore sustains stress resultant N_{xy} alone. As before, we shall use (14.69) as our basic equation. Also, as before, we shall take a 'relaxed' view of boundary conditions: just as in the previous examples we can investigate separately the effects of prevention of warping at the ends and corrections due to the 'circumferential' effects if we so wish. It was in connection with this problem that Donnell (1933; 1976 §7.4) first set up his celebrated shallow-shell equations. Our line of attack will be somewhat different from Donnell's, but we shall obtain results which are virtually identical to his. Donnell recognised that the full range of behaviour could only be described properly by means of two dimensionless groups containing the length of the shell. These groups are identical in effect to the two with which we are already familiar, namely $L/(ah)^{\frac{1}{2}}$ and $\Omega = Lh^{\frac{1}{2}}/a^{\frac{3}{2}}$. Donnell also performed some careful experiments covering the full range of variables, and found that his theoretical predictions were in satisfactory agreement with experimental observations, not only in relation to buckling loads but also in relation to critical modeforms (Donnell, 1976, §7.4).

As a first exercise let us investigate what happens if we try to use the mode of (14.18) in (14.69). Just as in the previous examples the right-hand side gives a function proportional to $\sin(\pi x/l) \sin(\pi y/b)$; but the left-hand side (when $N_x = N_y = 0$) involves instead a function proportional to $\cos(\pi x/l) \times \cos(\pi y/b)$. Clearly, then, this hypothetical mode is quite unsuitable for the study of buckling under the action of torsional loading N_{xy}.

14.7.1 Buckling of a 'long' cylindrical shell

As we shall see later, an eigenfunction which satisfies boundary conditions $w = 0$ at the ends $x = 0$ and $x = L$ of the shell is difficult to determine precisely. However, if we take instead a tube of infinite length – in order to avoid boundary effects – we find that the simple function

$$w = w_n \sin[n\theta - (nmx/a)] \qquad (14.97)$$

satisfies the governing equation. At any cross-section x = constant of the shell the radial displacement w in this mode is sinusoidal in the circumferential direction, with wavenumber n; but the phase varies with x. Nodal lines $w = 0$ have the form

$$a\theta - mx = \text{constant};\tag{14.98}$$

and these are helical lines drawn on the cylindrical surface, inclined at $\tan^{-1}m$ to the generators, as shown in fig. 14.19. Note that here the symbol m represents a *slope*, and not a wavenumber.

Now when w is given by (14.97) it follows that

$$\frac{\partial^2 w}{\partial x \partial y} = \frac{n^2 m}{a^2} w, \quad \frac{\partial^4 w}{\partial x^4} = \frac{n^4 m^4}{a^4} w, \quad \nabla^4 w = \frac{n^4}{a^4}(1+m^2)^2 w. \tag{14.99}$$

Also, since there are no problems over constants of integration in this case, we find

$$\nabla^{-4} w = \frac{a^4}{n^4(1+m^2)^2} w. \tag{14.100}$$

We now use these expressions in (14.69); and we find, after some rearrangement, the following formula for the eigenvalue:

$$N_{xy} = \frac{Eth^2}{24a^2} n^2 \frac{(1+m^2)^2}{m} + \frac{Et}{2} \frac{1}{n^2} \frac{m^3}{(1+m^2)^2}. \tag{14.101}$$

Here the first term on the right-hand side represents the effect of bending and the second the effect of stretching. The critical value of N_{xy} is positive because we chose to have the nodal lines making left-hand helices on the surface of the cylinder: in general the sign of N_{xy} is such that the direct stress component is tensile in the direction of the nodal lines, and compressive across the 'furrows' of the modal pattern.

The formula gives N_{xy} as a function of the two parameters n and m. Suppose for the moment that the value of n is fixed. We must now seek the value of m which minimises the value of the right-hand side. It turns out, as we shall

Fig. 14.19. The modeform (14.97) has helical nodal lines, shown here as parallel lines of slope m on the developed surface of the cylinder.

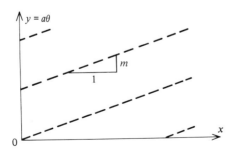

14.7 Buckling of cylindrical shells in torsion

see, that the expression is a minimum when the value of m is so small that $m^2 \ll 1$; therefore we may simplify (14.101) by replacing $(1 + m^2)$ by 1, and we obtain

$$N_{xy} = \frac{Eth^2}{24a^2} \frac{n^2}{m} + \frac{Et}{2} \frac{m^3}{n^2}. \tag{14.102}$$

This expression is a minimum with respect to m when

$$m = \frac{n}{6^{\frac{1}{2}}} \left(\frac{h}{a}\right)^{\frac{1}{2}}, \tag{14.103}$$

which confirms our hypothesis that $m^2 \ll 1$, unless n is large; see later. Using this value of m in (14.102) we find

$$N_{xy} = 0.136 n Et (h/a)^{\frac{3}{2}}. \tag{14.104}$$

The critical value of N_{xy} is thus directly proportional to n; and in practice, of course, it depends upon the requirement that n is an integer. When $n = 1$ the entire tube deforms into a helix, and the cross-sections remain circular. This is a possible mode in some circumstances, but our present formula is not at all reliable when $n = 1$, since the 'circumferential bending correction' is crucial in this case, as we have seen in section 14.5. If the ends of the tube are generally prevented from rotating about axes normal to the axis of the tube (but are nevertheless permitted to warp), this mode is excluded; then $n = 2$ is the lowest practical mode. Hence we obtain

$$N_{xy} = 0.27 \, Et(h/a)^{\frac{3}{2}} \tag{14.105}$$

as the buckling condition for an indefinitely long tube. Schwerin (see, e.g., Timoshenko & Gere, 1961, §11.11) was the first to obtain a buckling formula of this kind. The constant in (14.105) exceeds the corresponding value according to Timoshenko & Gere (1961, equation 11.27) by the factor $(4/3)^{\frac{1}{2}}$. This discrepancy is due to our neglect of the 'circumferential bending' correction for small values of n.

For a given long tube the critical value of N_{xy} is larger than the critical value of N_y (see (14.77)) by a factor whose order of magnitude is $(a/h)^{\frac{1}{2}}$. On the other hand it is less than the critical value of N_x for axial loading by a similar factor.

The value of m which minimises N_{xy} for a given value of n is, in general, small. If we define a circumferential half wavelength $b = \pi a/n$ and a longitudinal half wavelength $l = \pi a/mn$ (along a generator), we find that the ratio $b/l = m$ as given by (14.103) depends on n in almost exactly the same way as it did for the simpler buckling pattern (14.18) at the left-hand end of the Koiter circle: see problem 14.5. The two modes of deformation are of course quite different: nevertheless there are some important common features. In particular, the term

$$(1 + m^2)^2 \equiv 1 + 2m^2 + m^4$$

in (14.101) has three parts which represent, in the context of bending of the B-surface, circumferential, twisting and longitudinal effects, respectively: the circumferential bending term is by far the strongest. Similarly, in relation to the S-surface we find that the longitudinal stretching effect dominates. These are precisely the dominating features at the left-hand end of the Koiter circle, as we have observed before.

14.7.2 Torsional buckling of shells of finite length: an energy method

Let us now consider the torsional buckling of a tube of finite length L whose ends are supported in such a way that $w = 0$. Clearly the mode of (14.97) is not satisfactory in these circumstances; and so we modify it by including an extra multiplicative term:

$$w = w_n \sin[n\theta - (nmx/a)] \sin(\pi x/L). \qquad (14.106)$$

We can thus satisfy the boundary conditions rather easily. Unfortunately, however, this new function does not satisfy the basic equation (14.69), and in consequence relatively elaborate methods are required to give an accurate solution: see Donnell (1976, §7.4); Flügge (1973, §8.2.3.2); Timoshenko & Gere (1961, §11.11).

However, we shall proceed here to employ this simple mode in an *approximate* analysis, in which we reconcile the two (strictly, dissimilar) sides of (14.69) by means of an *energy method*. Energy methods have been used widely in continuum mechanics, ever since the time of Rayleigh, for providing relatively quick and simple approximate solutions to difficult problems. We have already employed them in chapters 12 and 13; and we shall also use them at various places in subsequent chapters. The basic scheme of the particular energy method which we shall use may be described as follows.

Consider first a simple shell structure which is currently buckling at constant load in the classical manner, and for which the exact eigenmode is known precisely. For any given (small) amplitude of this mode it is relatively easy to calculate the total elastic strain energy which is involved in the distortion of the shell, both by stretching and by bending, from its original to its current configuration. Throughout the process of growth in amplitude of this mode under constant load to its current value, energy must have been supplied by the loading agency, since at each stage the system is in a state of 'neutral' equilibrium according to the classical scheme. It follows that the

14.7 Buckling of cylindrical shells in torsion

current total strain energy must be equal to the total energy which has been supplied by the loading device during the process. Therefore, if we can calculate the 'energy input' corresponding to a load of arbitrary magnitude, we can use the equality of the two energies in order to determine the classical buckling load. So far in this description we have assumed that we know *a priori* the actual buckling mode; in which case the calculation gives correctly the actual classical buckling load. The same sort of calculation may obviously be performed in relation to a *hypothetical* mode which is not necessarily exact, but which nevertheless satisfies the boundary conditions on displacements for a given problem. Such a calculation will clearly not yield the correct buckling load in general; but by *Rayleigh's principle* it always gives a buckling load which is *greater than or equal to* the correct one. We shall also use a variant of this principle in the analysis of vibration in chapter 17: see appendix 3. This method, with the use of a hypothetical modeform, clearly constitutes a powerful tool for the approximate analysis of classical buckling in problems where, for a variety of reasons, the actual eigenmode cannot be expressed as a simple function.

In calculations of this kind it turns out that the 'energy input' is by far the most awkward term to calculate: see, eg. Timoshenko & Gere (1961, §9.2). However, the scheme which we have introduced in the present chapter enables us to perform this particular calculation with relative ease. The key to the situation is that the 'funicular membrane', which is in equilibrium under the applied loading and the 'interface' pressure Δp_1, itself absorbs no energy, since it may be regarded as inextensional without affecting the buckling calculation. This point is most easily made with respect to the example of a simple column in fig. 14.2: the resulting buckling load is just the same whether the central 'core' is regarded as inextensional or compressible. In relation to a cylindrical shell, of course, buckling deformations are not possible if the funicular membrane is absolutely inextensional; but this is a paradox whose resolution takes us into the *nonlinear* theory of buckling: see chapter 15. Thus, within the context of classical theory we may regard the 'funicular membrane' as an 'external' device for transferring the loads which are actually applied to the shell proper into another kind of loading, Δp_2, which is applied to the elastic system comprising the deformable S- and B-surfaces of the shell. It is, strictly, to this system that we shall be applying Rayleigh's principle. It follows that the work done by the external loads applied to the shell may be evaluated indirectly by means of the integral

$$\int_A \int_0^w \Delta p_2 \mathrm{d}w \mathrm{d}A; \tag{14.107}$$

which represents the energy input to the deformable system; and since Δp_2 is directly proportional to w we may write this as

$$\int_A \tfrac{1}{2} \Delta p_2 \, w \, dA. \tag{14.108}$$

Here A represents the total surface area of the shell. This is an easy calculation to do, once a modal form has been assumed.

In order to illustrate the use of this method let us consider again the case of a cylindrical shell loaded in compression by uniform stress resultants N_x, N_y. As before (see (14.18)), we use a mode of deflection

$$w = w_0 \sin(\pi x/l) \sin(\pi y/b).$$

For the sake of simplicity we shall assume here that $l \gg b$. This will enable us to compute the strain-energy expressions rather more easily than for the general case, since we need only be concerned with the circumferential curvature κ_y and the longitudinal strain ϵ_x.

From (14.8) and (6.10) (putting $\epsilon_y = \gamma_{xy} = 0$) we obtain

$$\kappa_y = (\pi^2/b^2)w; \quad \epsilon_x = (b^2/l^2)(w/a). \tag{14.109}$$

The strain energy of bending is given by $\tfrac{1}{2} D \kappa_y^2$ per unit area, while the strain energy of stretching may be taken as $\tfrac{1}{2} Et \epsilon_x^2$: see section 2.3.3. Now both of these strain-energy functions have the form $\sin^2(\pi x/l) \sin^2(\pi y/b)$, and this function has a mean value of $\tfrac{1}{4}$ over the entire panel, as may be shown easily. Hence the *mean* value of strain energy per unit area is given by

$$\left(\frac{\pi^4 D}{8b^4} + \frac{Etb^4}{8l^4 a^2} \right) w_0^2. \tag{14.110}$$

Next, we seek an expression for Δp_2. Since this is the negative of Δp_1 we may use (14.17) for arbitrary values of N_x, N_y due to the external loads:

$$\Delta p_2 = -\pi^2 \left(\frac{N_x}{l^2} + \frac{N_y}{b^2} \right) w, \tag{14.111}$$

and it follows that the mean value of the energy supplied per unit area of the shell is equal to

$$\frac{\pi^2}{8} \left(\frac{(-N_x)}{l^2} + \frac{(-N_y)}{b^2} \right) w_0^2. \tag{14.112}$$

The classical buckling criterion is simply that expressions (14.112) and (14.110) are equal. Hence

$$\frac{(-N_x)}{l^2} + \frac{(-N_y)}{b^2} = \frac{Etb^4}{\pi^2 a^2 l^4} + \frac{\pi^2 D}{b^4}. \tag{14.113}$$

It is easy to verify that this agrees exactly with previous results (cf. (14.21),

14.7 Buckling of cylindrical shells in torsion

(14.33) and (14.48)) when $l \gg b$. Note particularly that the calculation of the energy supplied by the external loads by means of (14.108) and (14.111) poses no difficulties.

Let us now apply the same method to the torsional buckling problem, but now using the hypothetical modal form (14.106) and again assuming (as indeed may be justified later) that the dominant effects are longitudinal stretching and circumferential bending. It is convenient to introduce the following notation for use in the present problem:

$$S = \sin[n\theta - (nmx/a)] \sin(\pi x/L),$$
$$C = \cos[n\theta - (nmx/a)] \cos(\pi x/L). \tag{14.114}$$

Then our modeform may be written

$$w = w_0 S, \tag{14.115}$$

and it is easy to show that the various required functions are:

$$\frac{\partial^2 w}{\partial x \partial y} = \frac{mn^2}{a^2} w_0 S + \frac{\pi n}{aL} w_0 C \tag{14.116a}$$

$$\frac{\partial^2 w}{\partial y^2} = -\frac{n^2}{a^2} w_0 S \tag{14.116b}$$

$$\frac{\partial^2 w}{\partial x^2} = -\left(\frac{n^2 m^2}{a^2} + \frac{\pi^2}{L^2}\right) w_0 S - \frac{2nm\pi}{aL} w_0 C. \tag{14.116c}$$

Hence, by use of (6.10) as before,

$$\epsilon_x = \left(\frac{m^2}{a} + \frac{a\pi^2}{n^2 L^2}\right) w_0 S + \frac{2m\pi}{nL} w_0 C; \tag{14.117}$$

and from (14.17)

$$\Delta p_2 = 2N_{xy} \left(\frac{mn^2 w_0}{a^2} S + \frac{\pi n}{aL} w_0 C\right). \tag{14.118}$$

It is easy to show that the mean value of both S^2 and C^2 over the surface is $\frac{1}{4}$, and that the mean value of SC is zero: these functions S and C are orthogonal. Hence we may work out the mean values of various quantities, as follows:

$$\tfrac{1}{2}D\kappa_y^2 \ldots \tfrac{1}{8}Dn^4 w_0^2/a^4, \tag{14.119a}$$

$$\tfrac{1}{2}Et\epsilon_x^2 \ldots Et\left\{\tfrac{1}{8}\left(\frac{m^2}{a} + \frac{a\pi^2}{n^2 L^2}\right) + \frac{m^2 \pi^2}{2n^2 L^2}\right\} w_0^2 \tag{14.119b}$$

$$\tfrac{1}{2}\Delta p_2 w \ldots \tfrac{1}{4}N_{xy} mn^2 w_0^2/a^2. \tag{14.119c}$$

Note in particular that although Δp and w are not 'in phase', the calculation of the energy input is not difficult by virtue of the orthogonality of the functions S and C.

By equating the third of these quantities to the sum of the first two, and dividing throughout by $\frac{1}{8}w_0^2$, we obtain the following buckling equation:

$$\frac{2N_{xy}mn^2}{a^2} = \frac{n^4 E t h^2}{12 a^4} + \left[\frac{m^4}{a^2} + \frac{6m^2\pi^2}{n^2 L^2} + \frac{\pi^4 a^2}{n^4 L^4}\right] Et. \tag{14.120}$$

When we put $L \to \infty$ in this expression, the last two terms enclosed within the brackets disappear, and we recover our previous expression (14.102) with the exception that terms $(1 + m^2)$ are replaced by 1: this small difference is, of course, a direct consequence of our simplifying assumption about the mode of action of the shell.

From (14.120) we can obtain an expression for N_{xy}, and in principle this can be minimised with respect both to n and to m. This calculation is easier if we do some preliminary rearrangement of (14.120). It is convenient to define a function ϕ by

$$\phi = mnL/\pi a, \tag{14.121}$$

and a function Φ by

$$\Phi = \phi^3 + 6\phi + \phi^{-1}. \tag{14.122}$$

Then (14.120) may be written

$$\frac{N_{xy} a^{\frac{3}{2}}}{E t h^{\frac{3}{2}}} = \frac{n^3 \Omega}{24\pi\phi} + \frac{\pi^3 \Phi}{2n^5 \Omega^3}. \tag{14.123}$$

Here, as before (see (9.9))

$$\Omega = L h^{\frac{1}{2}}/a^{\frac{3}{2}}.$$

The form of the left-hand side is dictated by our previous result, which we hope to recover as a special case in due course.

The variable ϕ has a very simple physical meaning. It is equal to the circumferential displacement of a typical nodal line of the proposed hypothetical mode over the entire length of the shell, and expressed as a fraction of the circumferential half wavelength: see fig. 14.20.

It is easiest to begin the process of minimisation by holding ϕ (and hence Φ) constant, and minimising the expression on the right-hand side with respect to n. We find that the minimum occurs when

$$n = (\pi/\Omega)^{\frac{1}{2}} (20\Phi\phi)^{\frac{1}{8}}, \tag{14.124}$$

and the corresponding value of the right-hand side of (14.123) is

$$\frac{(20)^{\frac{3}{8}}}{15} \left(\frac{\pi}{\Omega}\right)^{\frac{1}{2}} \Phi^{\frac{3}{8}} \Big/ \phi^{\frac{5}{8}}. \tag{14.125}$$

14.7 Buckling of cylindrical shells in torsion

We complete the process by finding the minimum value of $\Phi/\phi^{\frac{5}{3}}$. From (14.122) we obtain the following condition by differentiation:

$$\phi^4 - 3\phi^2 - 2 = 0. \tag{14.126}$$

This is a quadratic in ϕ^2 which has a single positive root $\phi^2 = 3.56$. Therefore

$$\phi = 1.88 \tag{14.127}$$

is the condition which we seek: the negative root corresponds, of course, to buckling under a torque of opposite sign. It is remarkable that the best value of ϕ is independent of the length of the shell, at least when n is continuously variable.

Hence we obtain by the method of section 14.3.1 our required formula for the envelope of the festoon of curves for different values of n:

$$\frac{N_{xy}a^{\frac{3}{2}}}{Eth^{\frac{3}{2}}} = \frac{0.723}{\Omega^{\frac{1}{2}}}. \tag{14.128}$$

Also, the corresponding value of n is given by

$$n = 4.0/\Omega^{\frac{1}{2}}. \tag{14.129}$$

Equation (14.128) is plotted on fig. 14.21a, and some integral values of n are marked. These results agree closely with Donnell's (1976, §7.4).

The lowest curve of the festoon is for $n = 2$. This has also been plotted on the diagram. The calculation of this curve is not nearly so easy as the calculation of the curves in figs. 14.14 and 14.18. When $n = 2$ the right-hand side of (14.123) becomes

$$\frac{\Omega}{3\pi\phi} + \frac{\pi^3}{64}\frac{\Phi}{\Omega^3}; \tag{14.130}$$

and for any given value of Ω we must find the value of ϕ which minimises the

Fig. 14.20. 'Panel' of the hypothetical modeform (14.106) for torsional buckling of a shell having a finite length L.

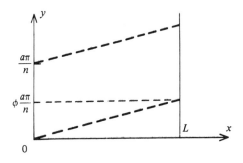

expression. Some values of ϕ are shown against the curve labelled $n = 2$ in fig. 14.21a. For large values of Ω, the optimum value of ϕ is also large: so (14.122) becomes $\Phi \approx \phi^3$ and we find that the expression is minimised by $\phi = 0.52\Omega$. This corresponds exactly to (14.103) for $n = 2$, and the buckling condition reverts to (14.105), giving a constant ordinate in the diagram.

All of our calculations on torsional buckling have been based on the hypothesis that the dominant structural effects are circumferential bending and longitudinal stretching. For very short tubes, when the value of $L/(ah)^{\frac{1}{2}}$ becomes sufficiently small, the stretching effects become relatively insignificant; and then the structural behaviour involves an interaction between bending effects in different directions. In fact, the problem becomes practically equivalent to that of the elastic buckling of a long flat plate which is loaded in pure shear. This problem was studied by Southwell & Skan (1924), who obtained the following formula for classical buckling when the edges are simply-supported:

$$N_{xy} = 4.4 \frac{Et^3}{L^2(1-\nu^2)}.$$

This relationship has been plotted in fig. 14.21b as

$$\frac{N_{xy}a}{Eth} = 4.4 \left(\frac{L}{(ah)^{\frac{1}{2}}}\right)^{-2}, \qquad (14.131)$$

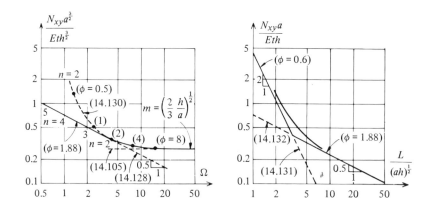

Fig. 14.21. (a) Lower envelope of the festoon of curves for torsional buckling of long and medium-length shells. (b) Corresponding diagram for medium-length and short shells; schematic.

14.7 Buckling of cylindrical shells in torsion

together with our envelope formula (14.128) which may be rearranged similarly to give

$$\frac{N_{xy}a}{Eth} = 0.723 \left(\frac{L}{(ah)^{\frac{1}{2}}}\right)^{-\frac{1}{2}}. \tag{14.132}$$

A transition curve between these asymptotes is also sketched: it is taken from Donnell's work, and it deviates from the upper of the two asymptotic lines by a factor of 1.5 at most. The mode which is obtained in the quasi-flat-plate region corresponds roughly to $\phi = 0.6$; therefore the transition region involves a considerable change in modal pattern.

Figs. 14.21a and b together cover the complete range of variables. Either can be used in the intermediate range in which, approximately,

$$3 < L/(ah)^{\frac{1}{2}}, \quad \Omega < 7 \tag{14.133}$$

(cf. figs. 14.13 and 14.14). If the dimensions of the shell put it outside this range, only one of the two graphs is relevant. The data presented here agree closely with those given by several authors, including Batdorf (1947, Fig. 7), Timoshenko & Gere (1961, §11.11), Brush & Almroth (1975, §5.5c) and Flügge (1973, Fig. 8.21); however, only Donnell and Flügge recognise explicitly the necessity for a parameter equivalent to Ω. (Flügge also gives results for combined torsion and axial compression.)

Several investigators, notably Donnell, have done a much more thorough study than the one we have described here. For instance, they have used much more elaborate expressions for the mode of deflection than our one-term expression (14.106). Nevertheless, our result (14.123) (or (14.128)) gives buckling loads only about 3% higher than those found by more thorough methods.

In general, the aim of the present work has been to pick out simple relationships which can be used for the three almost distinct ranges of geometrical parameters.

Various refinements can be made to fig. 14.21. For example, the effect of *clamping* the boundaries of the shell raises the critical value of N_{xy} by a factor of about 1.7 in the quasi-flat-plate region: this is a consequence of prevention of rotation of the wall of the shell at the edges. In the intermediate-length region the effect of clamping enters mainly in terms of a suppression of warping, and has effects similar to those discovered in section 7.3.4.

It is interesting to compare fig. 14.21 with the corresponding diagrams (figs. 14.13 and 14.14) for buckling under the action of N_y alone. The 'intermediate' region, defined by (14.133), extends over a slightly wider range for N_{xy} than for N_y; and the difference in slope of the corresponding curves in the two cases ($-\frac{1}{2}$ and -1, respectively) may thus be linked with the different magnitudes of N_y and N_{xy} for the 'very long' shells, and the comparable magnitudes for the 'very short' shells. Apart from the 'skew' effect shown in fig. 14.19, the modal patterns for buckling under N_y and N_{xy} are similar in the intermediate range: cf. (14.62) and (14.129).

14.8 Experimental observations

Many experiments on the buckling of uniform thin-walled cylindrical shells have been performed by numerous investigators using a wide range of materials and geometric parameters: for collections of experimental results see Gerard & Becker (1957), Fung & Sechler (1960), Brush & Almroth (1975). The general conclusion from all of these studies is that under external pressure (both lateral and hydrostatic) and torsion, experimentally observed buckling loads agree closely with classical theoretical results when due allowance is made for experimental errors, etc.; only in a few hydrostatic pressure tests of Windenberg & Trilling (1934) on shells in the region $3 < l/(ah)^{\frac{1}{2}} < 9$ was the buckling load less than about 80% of the theoretical prediction. (Many buckling tests have also been made on thicker tubes under external pressure, in connection with problems of collapse of boiler-tubes, etc. Tubes of this sort buckle in the plastic range, and they thus lie outside the scope of the present work.) On the other hand, the buckling loads which have been measured for purely axial loading – both uniform and non-uniform – have fallen far short of the theoretical results which we have given: typically they lie in a range between 0.2 and 0.5 of the theoretical values. As we mentioned earlier, this paradox was discovered as soon as experimental work was begun in the 1930s. A resolution of the paradox forms the subject of chapter 15.

14.9 A simple design problem

In the preceding sections we have solved a variety of classical buckling problems involving thin cylindrical elastic shells, and we have made compact plots of the results by the use of suitable dimensionless groups and logarithmic scales. In principle, these results can be used by engineers for the purposes of design, as soon as suitably factored loads have been agreed and decisions made about prudent corrosion allowances, etc. In practice, however, the form of our graphs is not necessarily the most convenient for design purposes, and it may therefore be desirable to rearrange the data in order to construct a 'design chart' which is more immediately useful for a particular class of problem. This process of rearranging the data is usually not difficult, and a single example should suffice to illustrate the lines of attack which may be taken in general.

14.9 A simple design problem

Let us consider the following problem. A long circular tube which forms part of a submarine structure is to be designed. It is known that exterior pressure may cause buckling. One possible solution is to use a plain tube of uniform wall thickness. An alternative possibility is to have a tube whose wall thickness is smaller, but which is strengthened by a series of plane bulkheads or diaphragms, regularly spaced. What is the most convenient way of presenting the relation between the spacing of the bulkheads and the required thickness of the shell, in relation to elastic buckling?

Let the radius and effective thickness of the shell be a, h respectively, and let the spacing of the bulkheads be L. If the bulkheads are thin, their primary effect is to hold the shell in a circular configuration. In particular, they are not likely to have an appreciable effect in preventing warping of cross-sections of the tube. Consequently, we may consider each length L of the tube as if it were simply-supported. The corresponding curve on fig. 14.14 is given approximately by the following equation, provided (see fig. 14.15) $L/(ah)^{\frac{1}{2}} > 1, \Omega = Lh^{\frac{1}{2}}/a^{\frac{3}{2}} < 2.5$:

$$(-N_y)a^2/Eth^2 = 0.86\, a^{\frac{3}{2}}/Lh^{\frac{1}{2}}.$$

Now for an external pressure p we have

$$-N_y = pa.$$

Making this substitution we may rearrange (14.61) as follows:

$$\frac{p}{E} = 0.86\, (a/L)(t/a)(h/a)^{\frac{3}{2}}. \tag{14.134}$$

This is one version of our required design formula. It has been plotted in fig. 14.22 as a graph of h/a against p/E for various values of the variable L/a, which is perhaps the most convenient dimensionless measure of L at this stage: for this purpose the distinction between t and h has been suppressed. Also plotted on the graph is the 'very long tube' formula

$$(-N_y)a^2/Eth^2 = 0.25$$

which may be rearranged as

$$\frac{p}{E} = 0.25\, (t/a)(h/a)^2. \tag{14.135}$$

The third line which is plotted is intended to give a rough guide to the circumstances under which buckling is likely to involve plasticity rather than elasticity. The compressive hoop stress σ is given by

$$\sigma = pa/t,$$

and if this is to be below the yield stress σ_y of the material we must have

$$p/E < (\sigma_y/E)(t/a). \tag{14.136}$$

For typical structural steels the value of σ_y/E is in the region of 10^{-3}. Allowing a margin of, say, 3 to allow crudely for the effect of imperfections, we find that elastic buckling will predominate provided

$$p/E < 3 \times 10^{-4} \, t/a. \tag{14.137}$$

It is clear from an inspection of fig. 14.22 that if the value of L/a is in the region of 2 to 5, the thickness of the shell with bulkheads can be reduced to between $\frac{1}{2}$ and $\frac{1}{3}$ of that of a plain tube. The precise ratio depends, of course, on the value of p/E, but is rather insensitive to it because the slopes of the lines corresponding to (14.134) and (14.135) are not very different. It seems clear that the provision of bulkheads can lead to considerable savings in material, provided the bulkheads themselves do not need to be too thick: see problem 14.15.

A disadvantage of the bulkhead scheme, which is clear from the diagram, is that for a given value of p/E the use of thinner material for the shell may bring the design near to the plastic range. A depth of 1 m of water gives a hydrostatic pressure of about 10^4 N/m². Taking E for steel as 210×10^9 N/m² we find that a depth of 1 m of water corresponds to $p/E = 5 \times 10^{-8}$. A scale of depth of water is also provided in fig. 14.22. In design, of course, the anticipated pressure would be increased by a suitable margin of safety.

Fig. 14.22. Design chart for buckling of long tube with evenly spaced diaphragms, under hydrostatic pressure; together with plastic-range cutoff for typical low-carbon steel.

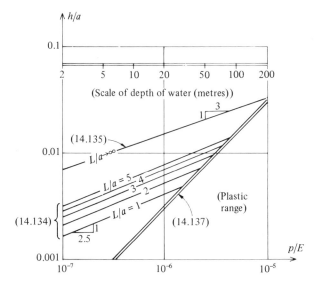

14.10 Unidirectionally reinforced shells

Many thin-walled shell structures, particularly in aerospace engineering, are reinforced against buckling by the provision of closely spaced stiffening ribs. Shells with circumferential ribs are said to be *ring-stiffened*, and shells with straight longitudinal ribs *stringer-stiffened*.

The addition of ribs to a sheet or plate obviously alters its mechanical properties. When the ribs are sufficiently close together, it is reasonable to treat them for purposes of analysis as if they were 'smeared out'; and then to perform calculations on a modified *uniform* but *anisotropic* plate.

Consider the reinforced plate whose cross-section is shown in fig. 14.23a. For the sake of definiteness let the volume of material in the ribs be equal to the volume of the original plate. It is obvious that the addition of these ribs increase the uniaxial *stretching* stiffness of the plate by a factor of 2 in the direction parallel to the ribs; but that these ribs have practically no effect either on the transverse stretching stiffness or on the in-plane shear stiffness with respect to the longitudinal and transverse directions. In relation to pure *bending* in the longitudinal direction, the ribs obviously increase the flexural stiffness of the plate. The stiffening factor would have been $2^3 = 8$ if the same amount of material had been added in the form of uniform extra thickness; but it will obviously be much greater if the ribs are proportioned as in fig. 14.23a. The additional flexural stiffness provided by the ribs depends on the precise form of the cross-section: typically, however, flexural stiffening factors are in the region of 50 in the direction of the ribs. In contrast, bending in the direction transverse to the ribs is practically unaffected by the provision of ribs. The *torsional* stiffness for twisting with respect to the longitudinal and transverse directions of the plate is also increased moderately by the provision of ribs.

Fig. 14.23. (a) Schematic cross-section of thin plate with reinforcing ribs extending on both sides: the additional cross-sectional area is equal to that of the original plate (i.e. $\gamma = 2$). For specific proportions see problem 14.16. (b) As (a), except that the ribs lie on one side of the plate. The broken line shows the centroidal axis of the combined cross-section.

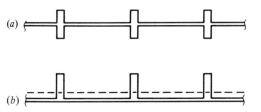

Most practical stiffeners are attached on one side of the sheet, as shown in fig. 14.23b. In this case the various bending and stretching effects are *coupled*, and the computation of the appropriate stiffness matrix for a plate element is rather complicated. The details of the appropriate stiffness matrix were first worked out by Baruch & Singer (1963), and the matrix is displayed by Brush & Almroth (1975, §5.6). When the full range of stiffness factors for such a shell is put into the governing equations, these become very complicated; and it is usual to resort to numerical methods of solution (e.g. Singer, Baruch & Harari, 1967).

In this section we shall attempt to give a brief, basic treatment of the classical buckling of simple anisotropic cylindrical shells by exploiting some of the ideas which we have developed in the preceding sections.

Consider first a shell which is reinforced in the manner of fig. 14.23a. We may characterise its anisotropic features in relation to the behaviour of the original unreinforced shell primarily in terms of a *stretching stiffness factor* α and a *bending stiffness factor* β in the direction of the ribs, where $(\alpha, \beta) \approx$ (2, 50), say. Now in much of our discussion of cylindrical shells we have concentrated on the description of almost-separate 'short-wave' and 'long-wave' regimes of behaviour, in which the dominant structural effects are longitudinal bending and circumferential stretching on the one hand, and circumferential bending and longitudinal stretching on the other. Thus it seems that in either of these two regimes only *one* structural effect can possibly be altered by the addition of one-way reinforcement, whether this is circumferential or longitudinal; and hence it should not be difficult to give a useful qualitative description of the consequences of adding reinforcement in either direction to a cylindrical shell which is subjected to axial compressive loading or external pressure, or indeed a combination of the two. The general description which can be obtained in this way does agree remarkably well with the results of detailed calculations by traditional methods on shells having 'balanced' reinforcement (fig. 14.23a) which are reported in the literature.

In relation to the more practical reinforcement of the kind shown in fig. 14.23b, the situation is obviously more complicated. There are two main consequences of a change of design from fig. 14.23a to b. The first is that the flexural stiffness of the cross-section increases. The flexural stiffness is, of course, directly proportional to the second moment of area of the cross-section about its centroidal axis; and it is straightforward to show (see problem 14.16) that this increases considerably – typically by a factor of about 2 – when the centroid of the ribs is moved away from the centre of the plate. The consequences of an increase in the factor β on account of changes of this sort can be seen clearly in fig. 6 of Singer *et al.* (1967).

14.10 Unidirectionally reinforced shells

The effect just described is independent of whether the reinforcing ribs are attached to the outside or the inside of the shell. The second consequence of a change of design from fig. 14.23a to b is more subtle, and is not the same for external and internal stiffeners. It was first pointed out by van der Neut (1947) that according to classical buckling theory stringers mounted on the *outside* of a cylindrical shell are more effective against buckling than stringers mounted on the *inside* for shells whose proportions lie in a certain range; and the detailed calculations and experiments of Singer *et al.* (1967) have confirmed that this is so.

Although this effect is somewhat surprising, it is not difficult to understand in principle. Consider an initially uniform shell which is deforming in the simple mode $w = w_0 \sin(\pi x/l) \sin(\pi y/b)$. In a typical 'panel' $l \times b$ which is deflecting outwards, the longitudinal strain ϵ_x on account of this displacement is positive (cf. section 7.3). Now a longitudinal stiffening rib attached to such a panel will experience a change of curvature in the sense which produces positive and negative longitudinal strains on its outside and inside edges, respectively. If such a rib is now attached to the *outside* of the panel, so that the longitudinal strain on its inside edge must match that of the panel itself, the strain at the centre-line of the rib will *exceed* that of the panel. On the other hand the centre-line of a rib attached on the *inside* of the panel would experience *less* longitudinal strain than the panel itself. Thus, in terms of a strain-energy assessment, we can see that ribs which are attached externally to the panel will produce a stretching-stiffness factor α in the direction of the ribs which exceeds that which would be found for a flat plate on the simple basis of added cross-sectional area; while, conversely, ribs attached internally will be less effective than they would be in the plane stretching of a plate. Exactly the same conclusion is reached for inward-deflecting panels.

This type of argument is open to criticism in several details; and in particular it cannot be expected to give direct quantitative results. Nevertheless the general conclusion is correct. In fact this stringer-eccentricity effect is a relatively minor one in comparison with the first effect discussed above, and we shall temporarily disregard it in performing the following piece of analysis. In some circumstances, however, it can be important; and we shall consider these separately later on.

The main task of the present section is therefore to discuss the consequences for classical buckling of the application of reinforcement which increases the stretching and bending stiffness of the shell in the direction of the reinforcement by the factors α and β respectively. For the sake of simplicity we shall consider mainly the case of buckling under a compressive axial load, for we shall then be directly concerned with the changes to the function Z (fig. 14.6) on account of the reinforcement; but it will then be a simple matter to extend the analysis to combinations of loading N_x, N_y in the manner of section 14.4. It will be interesting to compare the changes to Z on account of rib reinforcement with those which would occur if the additional material were used instead simply to increase the uniform thickness of the shell by a factor α; and we shall find that in some circumstances the latter arrangement is more satisfactory than the addition of ribs, at least in relation to the classical theory of buckling.

Consider first a case where *longitudinal* stiffeners increase the longitudinal stretching and bending stiffness by factors α and β, respectively. It is relatively straightforward to evaluate the corresponding changes to the 'panel stiffness' of the S- and B-surfaces, and thence to recalculate the function Z. We thus obtain (cf. (14.34))

$$2Z = \frac{2\xi^2}{(\xi^4 + 2\xi^2\eta^2 + \eta^4/\alpha)} + \frac{(\beta\xi^4 + 2\xi^2\eta^2 + \eta^4)}{2\xi^2}. \tag{14.138}$$

In this equation the factor β multiplies the term corresponding to longitudinal beam-action in the panel, while the factor α divides the term corresponding to the deflection due to stretching in the longitudinal direction: cf. section 7.3.1. The above expression ignores any contribution which the ribs make towards the torsional stiffness of the assembly.

At the risk of introducing spurious effects – which we shall discuss later – we shall now simplify the form of (14.138) by adjusting the coefficient of $\xi^2\eta^2$ in each of the two expressions, so as to make them perfect squares. Thus we now write

$$2Z = \frac{2\xi^2}{(\xi^2 + \eta^2/\alpha^{\frac{1}{2}})^2} + \frac{(\beta^{\frac{1}{2}}\xi^2 + \eta^2)^2}{2\xi^2} \tag{14.139}$$

as our modification to (14.34) on account of the longitudinal stiffeners. Similarly, for the case of circumferential stiffeners we write

$$2Z = \frac{2\xi^2}{(\xi^2/\alpha^{\frac{1}{2}} + \eta^2)^2} + \frac{(\xi^2 + \beta^{\frac{1}{2}}\eta^2)^2}{2\xi^2}. \tag{14.140}$$

In relation to the problem of buckling of a simply-supported shell under axial load, our main task in principle is to evaluate $Z(\xi, \eta)$ and then to find its minimum value subject to the constraints which are imposed by the boundary conditions in any given case. We saw in section 14.3.2 that this task is specially simple for the isotropic shell because the contours of Z are a family of similar circles (fig. 14.5). This picture will obviously be distorted by the introduction of factors α and β; and indeed our present aim is to see precisely what form this distortion takes. Fig. 14.24*a* is a graphical representation

14.10 Unidirectionally reinforced shells

of expression (14.139) when $\alpha, \beta = 2, 100$. This diagram shows the ξ, η trajectories on which $\partial Z/\partial \eta = 0$; and some Z values are marked against those trajectories for which $\partial^2 Z/\partial \eta^2 > 0$, i.e. the trajectories for which Z is *minimum* for a given value of ξ. The motivation behind this scheme of presentation is that it often turns out that the lowest value of Z is achieved when there is one half wavelength in the longitudinal direction: thus for a given shell the value of ξ is fixed and we need only determine the minimum of Z with respect to η, i.e. with the circumferential mode-number. It is clear from fig. 14.24a that in the 'short-wave' region $\xi \gg \eta$ the trajectory $\partial Z/\partial \eta = 0$ is shifted closer to the origin, relative to the original Koiter circle, on account of the reinforcement (see problem 14.17), while the corresponding value of Z is increased considerably. In this region, of course, bending in the longitudinal

Fig. 14.24. (a) ξ, η plot according to (14.139) for a cylindrical shell having *longitudinal* reinforcing ribs with $\alpha, \beta = 2, 100$, and showing the locus on which $\partial Z/\partial \eta = 0$ (and $\partial^2 Z/\partial \eta^2 > 0$), together with some spot values of Z. The Koiter circle for an unreinforced shell is also shown. (b) As (a), except that the reinforcement is now *circumferential*, and (14.140) is used.

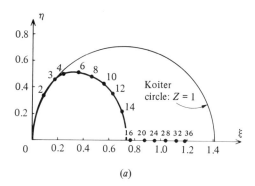

(a)

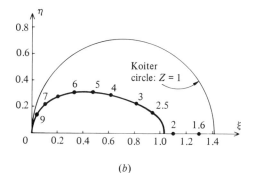

(b)

direction is one of the two interacting structural effects, and the increase in Z is by a factor in the region of 10 (or $\beta^{\frac{1}{2}}$ in general). At the left-hand or 'long-wave' end of the diagram, in which $\xi \ll \eta$, longitudinal stretching is one of the two interacting effects. The trajectory is pushed a little outside the original Koiter circle, while the value of Z is raised by a factor ≈ 1.4 (or $\alpha^{\frac{1}{2}}$ in general): see problem 14.18. In the intermediate region, in which ξ and η are of the same order of magnitude, the trajectory $\partial Z/\partial \eta = 0$ forms a roughly elliptical curve joining the two portions already discussed. It is only in this region that the simplification of (14.138) to (14.139) can alter the trajectory significantly.

Although the diagram has been drawn for a particular pair of values α, β, it is not difficult to appreciate in a qualitative way the consequences of adding arbitrary longitudinal reinforcement: the Koiter circle is shrunk to a smaller and roughly elliptical shape, while the values of Z are increased substantially at the right-hand end but only moderately at the left-hand end.

It is clear from the form of those results that if a shell of given design is made progressively longer, so that the value of ξ ($\approx 2.39(ah)^{\frac{1}{2}}/l$) steadily decreases, the value of Z – and hence the classical buckling load – becomes progressively smaller. This is in sharp contrast to the behaviour of a *uniform* shell, for which the value of Z does not vary around the Koiter circle, and for which therefore the classical buckling load is independent of length (provided this exceeds the short-wave half wavelength).

At the left-hand end of the diagram, $Z \approx \alpha^{\frac{1}{2}} \approx 1.4$, as we have seen. In this region the reinforcement is effective only through the longitudinal stretching stiffness factor α, which of course is generally much smaller than the flexural stiffness factor β. And indeed in this region the performance of the shell is particularly disappointing, since the value of Z would have been α^2 if the additional material had been deployed as uniform extra thickness. This factor can be deduced directly from the form of (14.23); or alternatively we can regard $\alpha^{\frac{1}{2}}$ of it as coming from an increase of stretching stiffness, while $\alpha^{\frac{3}{2}}$ comes from an increase of flexural stiffness. The absence of the latter effect in the unidirectionally reinforced shell is responsible for the relatively poor performance.

Fig. 14.24*b* shows the corresponding diagram for a shell which is reinforced by *circumferential* stiffeners also having $(\alpha, \beta) = (2,100)$. In this case the Koiter circle is also deformed into a roughly elliptical shape: the main change is a flattening in the η-direction, corresponding to an increase in the preferred circumferential half wavelength for doubly-periodic modes. The value of Z decreases from 10 ($= \beta^{\frac{1}{2}}$) at the left-hand end of the curved trajectory to about 2.1 at the right-hand end. The value of Z continues to fall along the ξ-axis beyond the end of the curve, reaching a minimum value of

14.10 Unidirectionally reinforced shells

$2^{\frac{1}{2}}$ ($= \alpha^{\frac{1}{2}}$) at $\xi = 1.68$ ($= 2^{\frac{1}{2}}\alpha^{\frac{1}{4}}$). Thus, in contrast to the previous diagrams, Z *increases* as ξ decreases in the region $\xi < 1.68$. This does not mean, however, that the axial buckling load increases as the length of the shell increases: in this case the shell will prefer to adopt a mode in which there are several axisymmetric half-waves in the length of the shell, each with a half wavelength in the region of $l/(ah)^{\frac{1}{2}} \approx 1.4$. In other words, the shell will buckle in a *short-wave mode* in which the main effects are (enhanced) circumferential stretching and (unaffected) longitudinal bending: the corresponding axial half wavelengths will be shorter than those corresponding to the uniform shell.

We might conclude from this example that circumferential reinforcement is of relatively poor value, since its performance is never better than that of longitudinal reinforcement; and indeed always worse than that of a uniformly thickened shell containing the same total amount of material. Such a conclusion only applies, of course, for simple uniaxial compressive loading: in the presence of exterior pressure it will often be advantageous to use circumferential stiffening: see fig. 14.11a and problem 14.19.

It is obvious from the above examples that if the preferred mode of buckling of a given shell under given loading makes use only of the enhanced unidirectional *stretching* stiffness of the reinforcement, it would actually be preferable to use instead a uniform thickening of the shell. In both fig. 14.24a and b the value of Z along the trajectory $\partial Z/\partial \eta = 0$ falls below α^2 – the constant value corresponding to uniform thickening – in the region of ξ, η space where the structural action involves genuinely two-directional action in both bending and stretching.

Let us now consider again the *van der Neut* effect whereby external stiffeners are more effective than internal ones. In the previous discussion we argued in effect that the stretching stiffness factor would exceed the nominal value of α for external stiffeners, but fall below it for internal ones. The argument involved a *cooperation* between longitudinal stretching and longitudinal bending effects. Now we have seen previously, in relation to uniform shells, that in the 'long-wave' region where longitudinal stretching is dominant, longitudinal bending is relatively unimportant (see fig. 14.7); and conversely that in the 'short-wave' region where longitudinal bending is dominant, longitudinal stretching is relatively unimportant. Hence the only region in which we can expect to find a strong van der Neut effect is that which corresponds to the central or upper part of the Koiter circle where longitudinal and circumferential effects cooperate on roughly equal terms. We may therefore conclude that situations in which the van der Neut effect is strongest are in general the same as those in which unidirectional reinforcement has only a marginal advantage, if any, over uniform thickening. These simple considerations, of course, give no indication of the maximum strength of the van der Neut effect. According to the extensive numerical study by Singer *et al.* (1967, Fig. 6), the transfer of reinforcing ribs from the inside to the outside of a cylindrical shell can at most double the classical buckling load; but neither of these buckling loads differs by very much from the classical value corresponding to a *uniform* thickening of the shell which uses the same total amount of material.

Two further points should be mentioned here. First, the entire analysis has been based upon the assumption that it is legitimate to consider an anisotropic shell in which the effect of the discrete ribs is *smeared out*. It was pointed out that this requires the ribs to be sufficiently close to each other. Here the appropriate measure of closeness is the wavelength of the buckling pattern in the direction transverse to the ribs. Thus, if we find in a particular case of longitudinal reinforcement that the critical mode is of the long-wave type with a circumferential mode-number $n = 7$ (say), then it will be necessary to have at least 45 equally spaced longitudinal ribs, so that at least 3 (say) are to be found in any circumferential half-wavelength. Similar remarks apply to circumferential reinforcement. The ribs would need to be very closely spaced to be effective in short-wave axisymmetric mode described above. For long-wave modes the analysis of problem 3.18 is relevant. In cases where substantial ribs are widely spaced it is unreasonable to consider a smeared-out version; and it is necessary to adopt different schemes of analysis. This subject is beyond the scope of this book; but see Bushnell (1973).

Second, we should include a word of caution in relation to the credibility of the *classical* theory of buckling, particularly in relation to comparisons with a shell which is 'reinforced' simply by being thickened uniformly. We have already remarked that uniform thin cylindrical shells, particularly under axial loading, sometimes buckle at loads far short of the classical buckling load. Singer *et al* (1967) and Singer (1969) have argued on the basis of extensive experimental work that *stiffened* shells are less prone to non-classical effects than are shells having uniform thickness.

14.11 A special boundary condition

It is appropriate to finish this chapter by mentioning a particular set of boundary conditions which can give a specially low classical buckling load for a thin cylindrical shell under uniaxial compression. The problem is of little practical significance; but it illustrates a number of important points.

Consider a long cylindrical shell which is loaded in uniform axial compression. The ends of the shell are held in such a way that they stay circular, but radial displacement w is not inhibited. Moreover, the entire mode of

14.11 A special boundary condition

displacement is constrained to be axially symmetric: $w = w_0(x)$. This problem is directly analogous to the buckling of a long uniform beam on an elastic foundation with free ends: the beam corresponds to flexure of longitudinal strips of the shell, while the foundation corresponds to symmetric hoopwise stretching. It is well known (Hetényi, 1946, §§40, 41) that when the end-conditions of such a beam are changed from simply-supported (i.e. $w = w'' = 0$) to free (i.e. $w'' = w''' = 0$) the axial buckling load is reduced by a factor of 2. Now the end-conditions implicit in section 14.3.1 are practically equivalent to those of simple support in the analogous beam. Hence we see that the present condition of 'free' boundaries leads to a halving of the previously determined classical buckling load.

Suppose next that the freedom of radial displacement at the ends is retained, but that the mode is now constrained to be periodic in the circumferential direction, i.e. to have the form $w = w_0(x) \sin n\theta$, where $n = 2$ or 3, say. In these circumstances the modeform $w_0(x)$ and the corresponding buckling load will be practically the same as in the axisymmetric case, since the axial wavelength of the mode (of order $(ah)^{\frac{1}{2}}$) will be but a small fraction of the circumferential wavelength for a thin shell. Furthermore, there will be virtually no circumferential component of displacement in this 'short-wave' mode, by the arguments of section 8.7.

Lastly, suppose that an *inextensional* mode $w = -w_0(0) \sin n\theta$ is superposed onto the preceding mode. This will restore the condition $w = 0$ at the edge $x = 0$. Since this inextensional mode is highly flexible it will require little strain energy, and it will raise the buckling load very little.

In this way we have succeeded in getting the shell to buckle with simply-supported edges ($w = 0$) at approximately *half* of the load determined in section 14.3. Notice, however, that in the superposed mode of inextensional deformation, periodic *tangential* displacements of amplitude $w_0(x)/n$ occur throughout the shell, while the stress resultants N_{xy} are practically equal to zero. If this freedom of circumferential displacement is not allowed at the edges, the low buckling mode will not occur. This strong influence of the tangential boundary condition on the buckling load is in sharp contrast to the cases described in section 14.6.

The mode which we have just described fits well the essential features of an analysis by Hoff (1965) on the influence of boundary conditions on the buckling of a cylindrical shell under axial load. Hoff was motivated by a desire to explain the anomalous poor buckling performance of axially compressed cylindrical shells which forms the subject of chapter 15. However, the special boundary conditions of simultaneous radial restraint and tangential freedom ($w = 0, N_{xy} = 0$) are difficult to achieve in practice; and they cer-

14.12 Problems

14.1 A simple pin-ended elastic column has an initial sinusoidal bow $w = w_0 \sin(\pi x/l)$, where l is the length of the column and w is measured, as in fig. 14.2b, from the line joining the two ends. A compressive load P is applied to the column, and the transverse displacement increases to $w = [w_0 \sin(\pi x/l)]/(1 - P/P_E)$. In an experimental study the transverse displacement u at the mid-point of the column is measured relative to the initial configuration; i.e.

$$u = (w - w_0)_{x=\frac{1}{2}l}.$$

By manipulation of these expressions show that

$$u = P_E(u/P) - w_0;$$

and hence that in a Cartesian plot of u against u/P (the 'Southwell' plot) the experimental points lie on a straight line of slope P_E.

(This result presupposes a particular simple modeform for the initial deflection of the column. If an experimental column has an initial bow which does not conform to this special mode – which is usually the case – it turns out that the experimental points in the $u, u/P$ space lie on a *curve* which becomes asymptotic to a line of slope P_E. In practice the points lie close to this asymptote for $P > 0.3 P_E$, approximately. This more complicated pattern of behaviour may be analysed by means of Fourier series; and the term corresponding to the buckling mode tends to dominate except in the initial stages.)

14.2 In relation to the problem shown in fig. 14.2, show that all of the conditions are met by a function $w = w_n \sin(n\pi x/l)$, where n is a positive integer; and that the corresponding load is given by $P = n^2 P_E$.

14.3 Plot lines in $-N_x, -N_y$ space corresponding to the buckling of a square plate ($B = L$) according to (14.14) for the following values (m, n): (1, 1), (2, 1), (1, 2), (2, 2), (3, 1) and (1, 3). In what circumstances are modes (2, 1) and (1, 2) preferred to (1, 1)?

14.4 Show that the minimum of $f(x) = Ax^p + Bx^{-q}$ with respect to x occurs when $pAx^p = qBx^{-q}$: A and B are constants, and both $p > 0$ and $q > 0$.

14.5 The Koiter circle (fig. 14.5, curve $Z = 1$) is described by (14.35), with $c = 2^{\frac{1}{2}}$. Show that the left-hand edge of the curve, where $l/b = \eta/\xi \gg 1$, may be represented approximately by the equation $\eta^2 = 2^{\frac{1}{2}}\xi$. Hence show that, in this region,

$$n^2 \approx 12^{\frac{1}{4}}\pi a^{\frac{3}{2}}/lh^{\frac{1}{2}} \approx 5.9\, a^{\frac{3}{2}}/lh^{\frac{1}{2}};$$

and

$$b/l \approx n(h/a)^{\frac{1}{2}}/12^{\frac{1}{4}} \approx 0.54\, n(h/a)^{\frac{1}{2}}.$$

(cf. problem 8.1).

14.6 A cylindrical shell with closed ends is subjected simultaneously to an interior gauge pressure p and an externally applied axial compressive force P. Show that

$$N_x = -P/2\pi a + \tfrac{1}{2}pa, \quad N_y = pa,$$

according to the membrane hypothesis, where a is the radius of the shell.

14.7 Let $Q = Z/[\alpha + (\eta/\xi)^2]$ (cf. (14.53)). Regard Q as a function of ζ (cf. 14.54) and ξ. Show that along any path $\alpha = \text{const}$ in fig. 14.12,

$$\frac{dQ}{d\xi} = \left(\frac{\partial Q}{\partial \zeta}\right)_\xi \frac{d\zeta}{d\xi} + \left(\frac{\partial Q}{\partial \xi}\right)_\zeta,$$

where $d\zeta/d\xi$ is evaluated along the path. Show that since Q is minimum with respect to ζ on $\xi = \text{constant}$ at every point on the path, the first term on the right-hand side vanishes. Also show that $\partial Q/\partial \xi > 0$ for $\rho > 1$ on any radial path $\zeta = \text{constant}$. Hence show that $dQ/d\xi > 0$ along each path in fig. 14.12 except the one for $\alpha \to \infty$.

14.8 Show that for points lying well outside the Koiter circle (fig. 14.5) the first term on the right-hand side of (14.34) is negligible. In these circumstances show that (14.48) reduces to the flat-plate expression (14.14).

14.9 Show that for sufficiently large values of ζ and finite values of α, (14.57) gives $\rho \approx 3^{\frac{1}{4}}$ and hence $Z = 2/3^{\frac{1}{2}} \approx 1.15$. Show also that in this range the curve approximates to the parabola $\eta^2 = (12)^{\frac{1}{4}}\xi$; and that hence $(l/b)^2 \approx (6^{\frac{1}{2}}/\pi) \times (l/(ah)^{\frac{1}{2}})$ provided $l/b > 3$. Use this information together with (14.48) to obtain a simple buckling formula in the range $L/(ah)^{\frac{1}{2}} > 12$.

14.10 Take the simplified form (14.60) of (14.77), to which (14.90) reduces when $m = 1$ and $n \geqslant 3$, approximately. Now multiply the first term on the right-hand side of (14.60) by a factor F, thus representing an increase in the stiffness of the S-surface in the manner described at the end of section 14.6. Show that the corresponding forms of (14.61) and (14.62) become

$$-N_y a^2/Eth^2 = 0.855 \, F^{\frac{1}{4}}/\Omega; \quad n = 2.77 \, F^{\frac{1}{8}}/\Omega^{\frac{1}{2}}.$$

14.11 According to fig. 14.18 the buckling pressure p of a cylindrical shell may be expressed as follows in terms of the uniform thickness t and length L, when the radius a and the material properties E, ν, are fixed, and provided the value of Ω is not too large:

$$p = Bt^{2.5}/L. \tag{i}$$

Here B is a constant (but not dimensionless) whose value depends on a, E, ν and the boundary conditions of the shell.

Consider an oil-storage tank whose thickness increases linearly with distance x measured vertically downwards from the top, as in section 14.6.1:

$$t = t_1 + kx. \tag{ii}$$

The thickness at the top and at the clamped base are t_1 and t_0, respectively; hence

$$kL = t_0 - t_1. \tag{iii}$$

Imagine that the height L of the tank is steadily increased by adding new material at the base in such a way that equation (i) still holds, with the same constant values of k and t_1. Examine the way in which the buckling pressure varies as L increases, according to (i) but with an 'effective' thickness $\bar{t}$ given by

$$\bar{t} = 0.25 t_0 + 0.75 t_1. \tag{iv}$$

Plot a graph of $p/Bkt_1^{1.5}$ against kL/t_1: these are convenient dimensionless forms of p and L, respectively. (Suitable values of kL/t_1 are 0.5, 1, 1.5,...,3.)

Show in particular that $p/Bkt_1^{1.5}$ reaches a *minimum* value at $kL/t_1 = 2.67$; but that this differs by little from the value of $p/Bkt_1^{1.5}$ at $kL/t_1 = 2$, which is recommended in the text as a good approximation in all cases where $t_0/t_1 > 3$.

Repeat the calculation for

$$\bar{t} = 0.4 \, t_0 + 0.6 \, t_1. \tag{v}$$

(It may be argued independently that the (classical) buckling pressure cannot increase as L increases, and therefore that formulae (iv) and (v) are certainly invalid when the length is greater than that for which the formula suggests that p is minimum. Note that a change of formula at the height for which $t_0 = 3t_1$ is satisfactory as a first approximation for both cases.)

14.12 Substitute $t = t_0 - (t_0 - t_1)z$ into (14.96a) and (14.96b), and evaluate $\bar{t}$ in each case. Hence check that (14.96a), (14.96b) are consistent with (iv) and

(v) of problem 14.11, respectively, in the case of a shell whose thickness varies linearly with height.

14.13 When an oil-storage tank consists of successive courses of uniform height, the evaluation of $\bar{t}_{cf}$ according to (14.96a) may be reduced to a simple summation, as follows:

$$\bar{t}_{cf} = \sum_{M=1}^{N} \alpha_{NM} \, t_M. \qquad \text{(i)}$$

Here N is the number of equal-height courses, t_M is the thickness of the Mth course (counting from bottom to top) and α_{NM} is a *weighting factor* computed according to

$$\alpha_{NM} = \int_{(M-1)/N}^{M/N} z^2 \, dz \bigg/ \int_0^1 z^2 \, dz. \qquad \text{(ii)}$$

The calculation $\bar{t}_{cs}$ according to (14.96b) may be reduced similarly to a formula like (i) but having weighting factors β_{NM} calculated according to (ii), and with integrand z^2 replaced by $(1-z)z^2$.

Verify the table of weighting factors given below (to 2 significant figures).

Values of α_{NM}

		\multicolumn{6}{c}{$N =$}
		6	5	4	3	2	1
$M =$	6	0.42					
	5	0.28	0.49				
	4	0.17	0.30	0.58			
	3	0.09	0.15	0.30	0.70		
	2	0.03	0.05	0.11	0.26	0.88	
	1	0.01	0.01	0.01	0.04	0.12	1.0

Values of β_{NM}

		\multicolumn{6}{c}{$N =$}
		6	5	4	3	2	1
$M =$	6	0.13					
	5	0.28	0.18				
	4	0.28	0.34	0.26			
	3	0.20	0.30	0.43	0.41		
	2	0.10	0.15	0.26	0.48	0.69	
	1	0.01	0.03	0.05	0.11	0.31	1.0

14.14 A cylindrical oil-storage tank with a clamped base has 6 courses of equal height. The thicknesses of courses, starting from the bottom, are 3.0, 2.5, 2.0, 1.5, 1 and 1 unit, respectively. Using the method of section 14.6.1, with coefficients from tables given in problem 14.13, show that the external buckling pressure at the completion of each course is proportional to the following numbers:

15.6, 5.2, 2.3, 1.1, 0.46, 0.29, 0.88.

The top edge is to be regarded as free at the end of the first 6 stages of construction. The seventh stage involves the addition of a stiffening ring at the top which then holds the top circular.

(It is obvious in this example that the tank is most susceptible to buckling immediately before the stiffening ring is attached, Note that in this case the provision of a stiffening ring raises the buckling pressure by a factor of ≈ 3, in comparison with the factor ≈ 2 for a tank having uniform wall thickness.)

14.15 A submarine tube is to be designed so that it will be on the point of buckling elastically at an external pressure head of 40 m of water. If bulkheads are used, their thickness is to be the same as that of the tube. Use fig. 14.22 to show that for bulkheads spaced at $L = 5a$, the total volume of material required is approximately 60% of the volume required for a design having no bulkheads; and that when $L = 2a$ the corresponding figure is approximately 47%.

14.16 Evaluate the second moment of area about the centroidal axis for one bay of the cross-section shown in fig. 14.23a. Take the thickness of the plate as t, the spacing between the centres of the ribs as $30t$, the thickness of the ribs as $3t$, and the overall height of the ribs as $11t$.

Now repeat the calculation for the arrangement shown in fig. 14.23b; and hence show that $I_b/I_a \approx 2.1$.

14.17 Using (14.139) determine the condition for $\partial Z/\partial \eta = 0$, other than $\eta = 0$, for the case of longitudinal reinforcement. By putting $\eta = 0$ show that the right-hand end of the curved trajectory of fig. 14.24a is at $\xi = \xi^* = 2^{\frac{1}{2}}(\beta\alpha)^{-\frac{1}{8}}$.

Similarly show that in case of circumferential reinforcement the corresponding point is at $\xi = \alpha^{\frac{1}{2}}\xi^*$.

14.18 Show that in the region $\eta/\xi \gg 1$ (14.139) is given by

$$Z \approx \alpha\xi^2/\eta^4 + \eta^4/4\xi^2 \, ;$$

that the locus $\partial Z/\partial \eta = 0$ is the parabola

$$\xi \approx \eta^2/2^{\frac{1}{2}} \alpha^{\frac{1}{4}};$$

and that the value of Z along this locus is

$$Z \approx \alpha^{\frac{1}{2}}.$$

These results are for a shell with longitudinal reinforcement. Also show that for a shell with circumferential reinforcement the corresponding formulae in the same region are

$$\xi = \beta^{\frac{1}{4}} \eta^2/2^{\frac{1}{2}}; \quad Z \approx \beta^{\frac{1}{2}}.$$

14.19 A cylindrical shell is subjected to external hydrostatic pressure (so $N_y = 2N_x$). The shell has length L, radius a and thickness t, and its ends are 'held circular'. It is to be reinforced by closely-spaced circumferential rings which increase the circumferential stretching and bending stiffness by factors α and β, respectively.

First consider the case $\alpha, \beta = 2, 100$. Show that the classical buckling pressure is proportional to $Z/[0.5 + (\eta/\xi)^2]$. Evaluate this at each point on the curve in fig. 14.24b at which a value of Z is given. Hence show that in this case a long-wave mode (at the left of the diagram) will give a lower buckling pressure than the axisymmetric mode if the shell is so long that $\xi < 0.3$, approximately; i.e. $L/(ah)^{\frac{1}{2}} > 8$.

For cases in which $\xi < 0.1$ (i.e. $L/(ah)^{\frac{1}{2}} > 24$) show that a modified form of (14.60) may be used for general values of β (with the second term on the right-hand side multiplied by β); and that this gives buckling formulae as in fig. 14.18 except that the right-hand side of the formula for $-N_y a^2/Eth^2$ must be multiplied by a factor $\beta^{\frac{3}{4}}$, and the right-hand side of the formula for n must be divided by $\beta^{\frac{1}{8}}$.

15

Buckling of shells: non-classical analysis

15.1 Introduction

The 'classical' analysis of the buckling of thin elastic cylindrical shells was presented in the preceding chapter. There it was pointed out that in some circumstances the results of the classical analysis are reliable, and can form the basis of rational design procedures; but also that there are other circumstances in which the results of the classical analysis can be grossly unsafe. The task of the present chapter is to consider this second class of problem. It was pointed out in sections 14.1 and 14.2.3 that the key to the situation is the use of *nonlinear* theories of elastic buckling. This is a very large subject indeed, which has attracted much attention from many workers since the early days of the use of thin-sheet metal for the construction of aircraft. The aim of this chapter is to investigate the particular problem of a thin-walled cylindrical shell under axial compression, and to attempt to describe, by the means at our disposal, some important features of the behaviour. In chapter 16 we shall tackle a different problem which also demands nonlinear analysis; and indeed we shall find some common aspects with chapter 15. At the end of the two chapters the reader should be in a position to appreciate the literature of the field (e.g. Hutchinson & Koiter, 1970; Brush & Almroth, 1975; Bushnell, 1981): this covers not only the general theory of elastic post-buckling behaviour, but also specific applications to cylindrical and non-cylindrical shells and panels, with and without reinforcing ribs, etc.

The most satisfactory way of introducing the main ideas contained in this chapter is to give a brief historical survey of the assaults which have been made on the problem of the buckling of a cylindrical shell under uniform axial loading.

During the 1930s many experimental studies were made of the buckling of thin circular cylindrical shells under uniform axial loading. The results showed

15.1 Introduction

clearly that the maximum load which could be sustained was usually much lower than the 'classical' buckling load, derived along the lines of chapter 14, except for short shells: see fig. 15.1. Furthermore, these experiments showed that the buckling was of a different *character* from that of an ordinary elastic column; for whereas an elastic column continues to support the buckling load as deflections increase (see fig. 14.1), a shell of this kind appears to lose its load-carrying capacity as soon as buckling commences; and moreover it collapses in a dynamic and catastrophic fashion. Lastly, there is much more 'scatter' in the experimentally obtained maximum loads (see fig. 15.1) than in tests on structures which buckle according to the classical paradigm.

Various attempts were made to explain this paradoxical behaviour. Inelastic deformation of the material, and various aspects of boundary conditions were invoked as crucial factors; but none of these hypotheses could be sustained against the experimental evidence.

A most important contribution to the elucidation of this problem was made by von Kármán, Dunn & Tsien (1940). They argued that under axial compression a shell could be regarded as a sort of longitudinal beam supported transversely by a 'foundation' of elastic springs. Now if these supporting springs were *linear* elastic (as they are in effect reckoned to be in the classical theory), the behaviour of the whole system would be in accordance

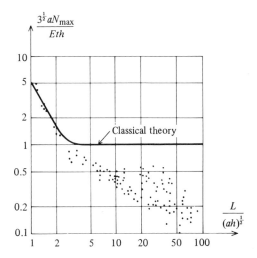

Fig. 15.1. Experimental observations on the buckling of thin cylindrical shells under axial compression, adapted from Brush & Almroth (1975, Fig. 5.14). The ends of the shells were clamped. The curve corresponds to the classical theory as in fig. 14.13b, but takes account of the different end-conditions.

with the Eulerian paradigm. Therefore, since the actual behaviour of some shells is markedly different from this pattern, the supporting springs must be *nonlinear*. This remarkable argument led to both theoretical and experimental studies of simple columns supported laterally by nonlinear elastic springs (see fig. 15.2); and it was discovered that a relatively *small* amount of nonlinearity in the spring's tension/extension characteristic can make a dramatic difference to the buckling behaviour of the assembly. In particular, the presence of a small initial crookedness or misalignment of the column can lead to a disproportionately large reduction of load-carrying capacity below that which occurs when the nonlinear factor is excised from the analysis; and moreover the equilibrium of the system becomes very unstable after the maximum load has been reached. The model thus displays all three experimentally observed features of the buckling of cylindrical shells under axial compression: reduced load-carrying capacity, 'scatter' of results on account of variable imperfections, and catastrophic failure. What the model does *not* do, however, is to make a direct connection between the nonlinearity of the supporting spring on the one hand and the equations of the cylindrical shell on the other. It was perhaps for this reason that this paper, and also one by Cox (1940) containing an appendix on broadly similar lines, failed to make the impact which was deserved. Later in the present chapter we shall attempt to rehabilitate this work by providing explicitly the quantitative missing link between the shell and the simple conceptual model.

The next important step in the analysis of buckling of thin cylindrical shells under axial load was taken by von Kármán & Tsien (1941). They did some complicated and painstaking work on the nonlinear version of the cylindrical shell equations which had been proposed earlier by Donnell. They argued that in view of the evident complexity of the equations, the best line of attack was an *energy* calculation, since this would be insensitive to certain assumptions about the modeform of the deformation. One aspect of their study is shown in fig. 15.3a. For a given modeform the relation between compressive load and endwise shortening turned out to be of the kind shown; in particular they discovered equilibrium states at loads far less than the classical

Fig. 15.2. The conceptual model of a pin-ended column supported by a nonlinear elastic spring, as used by von Kármán *et al.* (1940) in their explanation of the buckling of shells.

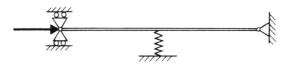

15.1 Introduction

critical load. With a 'stiff' testing machine the structure could 'jump' from the classical load level to a much lower load corresponding to a post-buckled configuration; and they conjectured that the presence of a small geometrical imperfection could give a significant and disproportionately large reduction in the maximum axial load which the shell could support.

Donnell & Wan (1950) did a modified version of this energy calculation into which they introduced a small initial geometrical imperfection. Their work has been criticised on the grounds that it made certain untenable assumptions about the magnitudes of certain components of the modeforms; but these workers succeeded nevertheless in showing that a small imperfection produces a local maximum in the axial load/endwise shortening curve considerably lower than the classical load level: see fig. 15.3b.

Meanwhile, in wartime Holland, W. T. Koiter (1945) perceived that it was possible to set up a new *general* theory of the stability of quasi-static elastic equilibrium. He focussed attention on the immediate vicinity of the *branching point* which is indicated by the classical theory of buckling. He conceived the idea of a multi-dimensional space with axes for load and each of various possible buckling modes, and showed not only how to find the form of the *equilibrium paths* in this space by taking the first derivatives of the total potential energy function, but also how to determine the *stability* of the states of equilibrium along these paths by taking second derivatives. Some powerful general results emerge quickly from a study of this kind: for instance, there is always

Fig. 15.3. (*a*) Load, displacement path for the buckling of an initially perfect cylindrical shell under axial load, according to von Kármán & Tsien (1941). (*b*) Calculations of Donnell & Wan (1950) on an initially imperfect cylindrical shell.

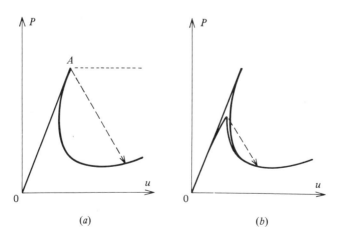

at least one path of *unstable* equilibrium emerging from a point of bifurcation.

In relation to the buckling of cylindrical shells under axial compression Koiter discovered that several different modes which according to classical theory occur at the same critical load can interact destructively in a nonlinear fashion to produce extremely unstable equilibrium paths of the kind which had been revealed by von Kármán & Tsien (1941): see fig. 15.4. A strength of Koiter's approach was that it allowed him to consider the consequences of introducing small geometrical imperfections by means of a simple scheme of perturbation. In this way he was able to give explicit formulae for the reduction in load-carrying capacity on account of imperfections having various modal forms. In particular he showed that geometrical imperfections of certain kinds reduce the load-carrying capacity of a shell approximately in proportion to the *square root* of the amplitude of the imperfection. Thus the buckling is 'imperfection-sensitive' in the sense that a small imperfection can have a disproportionately large effect upon the load-carrying capacity of the shell. Specifically, Koiter established the formula

$$(1 - \Lambda)^2 = \tfrac{3}{2} (3)^{\tfrac{1}{2}} \Lambda w_0/h, \qquad (15.1)$$

where Λ is a dimensionless buckling load whose 'classical' value is 1, and w_0 is the amplitude of the initial imperfection. We shall refer to (15.1) as 'Koiter's first formula'.

Koiter's work did not become available to the English-speaking world in translation until 1967. Since then it has had a profound influence on the study of elastic buckling: see, eg., Croll & Walker (1972); Thompson & Hunt (1973); and Budiansky (1974).

Fig. 15.4. According to Koiter (1945), nonlinear interaction may occur between three classical buckling modes, represented by three points on the 'Koiter circle': cf. fig. 14.10.

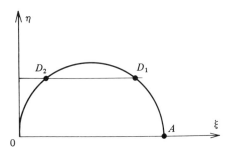

15.1 Introduction

One of the strengths of Koiter's work is that since it deals with conditions in the immediate vicinity of a bifurcation point it is relatively *compact* in comparison with, say, the calculations of Donnell & Wan already mentioned. But the scheme also contains a weakness, since it is only capable of dealing convincingly with very small initial imperfections. As a counter to this point, Koiter (1963b) presented a second analysis of the buckling of a thin cylindrical shell under axial load, in which he assumed a finite initial imperfection in the form of the axisymmetric classical buckling mode. The nonlinear growth of such an imperfection can be followed without too much difficulty (the problem is analogous to Perry's: see section 14.1), and Koiter showed that at a certain point there is a bifurcation from the nonlinear prebuckled state into an asymmetric (doubly-periodic) mode in the form of another of the classical modes. The calculations showed that the relationship between this bifurcation load and the amplitude of the initial imperfection was fairly close to that which was indicated by his first formula, based on an asymptotic analysis, even for moderately large initial imperfections.

Much experimental and theoretical work has been done on the subject of nonlinear buckling in the last 15 years. This is not the place for a sustained description of experimental work; but we should nevertheless mention the simple work of Thompson (1962) on the buckling of complete spherical shells under external pressure; the careful assays of Singer (1969) and his colleagues on stiffened shells of various sorts; the work of Arbocz & Babcock (1969) (also Arbocz, 1974) on the correlation of measured initial geometrical imperfections with the load-carrying capacity of cylindrical shells; the work of Tennyson (1969, 1980) on the static and dynamic buckling of carefully made epoxy-resin shells; and the thorough work of Yamaki (1976) and his colleagues on the post-buckling of cylindrical shells in torsion.

For our purposes an important contribution was made by Hutchinson (1965). He reworked Koiter's problem of 1963, but with a *two-mode* initial imperfection. He solved approximately the nonlinear equations in a way which allowed the growth of both modes of imperfection to be followed simultaneously. In particular he applied this method to the buckling of cylindrical shells under a combination of axial load and interior pressure. As we have seen in chapter 14, loading of this sort produces a single classical buckling mode in place of the large number of competing modes which occur under purely axial loading. Hutchinson found that the addition of interior pressure did not eliminate the imperfection-sensitive behaviour of the shell; and he thereby demonstrated that coincident modes in the classical analysis are not a necessary concomitant of imperfection-sensitive behaviour.

The layout of the present chapter is as follows. First we investigate the buckling behaviour of a simple system consisting of a column which is restrained by a nonlinear spring. The point of this is threefold. First it furnishes a simple illustration of Koiter's scheme of stable and unstable equilibrium paths; second it gives a simple representation of the model which von Kármán, Dunn and Tsien used to describe the buckling of a thin cylindrical shell under axial loading; and third it provides the basis for a quantitative comparison between the behaviour of this simple system and an actual shell.

In section 15.3 we re-examine the classical analysis of chapter 14 in order to show clearly the need for a nonlinear treatment of the present problem. In section 15.4 we derive the equations of a simplified version of Hutchinson's problem concerning the simultaneous growth of two modes of initial imperfection, but we tackle the analysis by using the two-surface idea. This enables us to present as a special case Koiter's analysis of 1963, and to study the consequences of an imperfection in the form of the second mode.

One of the main aims of the present chapter is to devise 'physical' interpretations of nonlinear buckling phenomena. To this end we give, in section 15.5, an approximate interpretation of Koiter's 1963 buckling criterion in terms of a hypothetical 'locally classical' condition. The same idea is also successful in relation to Hutchinson's work of 1965. There is a formal similarity here with Perry's well-known analysis of 'first yielding' in an initially bowed column made of elastic-plastic material: see Calladine (1973b).

Next, in section 15.6, we return to the simple nonlinear spring model of section 15.2. We show that the two-surface idea may be used to associate a definite nonlinear spring with the 'S-surface' of the shell, and we find that this correspondence enables us to establish satisfactorily an approximate buckling equation for an axially loaded cylindrical shell. Furthermore, we see from this analysis why other forms of loading applied to a cylindrical shell – such as external pressure – do *not*, in general, lead to imperfection-sensitive buckling.

The two-surface scheme thus leads to several simple ways of detecting situations in which the buckling is likely to be imperfection-sensitive. We then engage in a discussion of various practical points which have been raised in the chapter, and finish in section 15.8 by applying successfully the simple 'local buckling' hypothesis to Hutchinson's problem of the buckling of a pressurised cylindrical shell under axial compression.

15.2 A simple model for the study of buckling

Fig. 15.5a shows a useful conceptual model which is essentially a simplified version of the system proposed by von Kármán *et al.* (1940). It is, of course, a single-degree-of-freedom system and as such it cannot be expected

15.2 A simple model for the study of buckling

to model a multi-mode problem: in this respect it suffers from the same restrictions as all simple conceptual models. The system consists of a rigid lightweight rod of length l, freely pivoted to a support at the base and restrained by a spring which exerts a restoring couple

$$C = \eta\theta - \nu\theta^2 \qquad (15.2)$$

(see fig. 15.5b), where θ is the angle of rotation of the rod from its original, vertical position. The upper end of the rod carries a vertical force P, applied eccentrically by a small distance μl. At this point we are not concerned with the physical origin of the nonlinearity of the spring (represented by the term $\nu\theta^2$ in 15.2), but merely with the consequences of its presence. But it is easy to construct such devices in the manner of fig. 15.5c and other ways: see von Kármán et al. (1940) and problem 15.1.

By taking moments about the pivot we obtain the equation of statical equilibrium of the system:

$$Pl \sin(\theta + \mu) = \eta\theta - \nu\theta^2. \qquad (15.3)$$

We now introduce the small-angle linearisation of $\sin(\theta + \mu)$, since the neglected terms are all of a higher order than θ^2; hence we obtain the adequately accurate equation

$$Pl\,(\theta + \mu) = \eta\theta - \nu\theta^2. \qquad (15.4)$$

Fig. 15.5. (a) Simple model consisting of a rigid rod restrained by a nonlinear elastic spring, for the study of a class of buckling problems. (b) Plot of couple C against rotation (θ) for the nonlinear spring. (c) A physical realisation of such a spring.

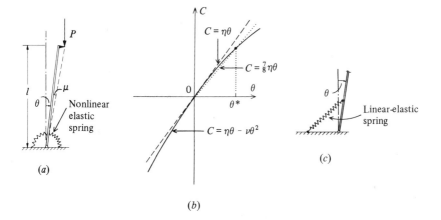

Consider first the 'perfect' case $\mu = 0$. Equation 15.4 is clearly satisfied both by

$$\theta = 0, \quad (15.5a)$$

and also by

$$Pl/\eta = 1 - (\nu\theta/\eta). \quad (15.5b)$$

These two lines are shown in the Cartesian P, θ space of fig. 15.6. Just as in fig. 14.1, there is a branching point where the secondary and primary equilibrium paths intersect. But in contrast to fig. 14.1c the secondary path leaves the branching point with a definite *nonzero slope*. This feature is clearly attributable to the nonlinear term $\nu\theta^2$ in the spring characteristic; and if we put $\nu = 0$ we recover immediately a curve of the same kind as the simplified secondary path of fig. 14.1c.

15.2.1 The total potential energy function

The nonzero slope of the secondary path indicates in fact a profound difference in the behaviour of the system. One way of seeing this difference is to enquire about the *stability* of the state of equilibrium along the various equilibrium paths. Now in reversible thermodynamic systems the concepts of equilibrium and stability are associated with stationary values and minima, respectively, of the global 'free energy' of the system with respect to the state variables. These ideas may be applied to a conservative mechanical system by virtue of the remark that the free energy is equal to the total potential energy V, being the sum of the potential energy of the loads and the elastic strain

Fig. 15.6. Equilibrium paths for the system of fig. 15.5a, with $\mu = 0$ ('perfect').

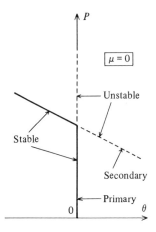

15.2 A simple model for the study of buckling

energy U of distortion of the deformable parts of the system. Thus, in the present problem we may write

$$V = V(P, \theta) \tag{15.6}$$

and since we suppose that the value of P is given, θ is the sole state variable of the problem. Hence we expect that configurations of statical equilibrium will correspond to the condition

$$\frac{dV}{d\theta} = 0. \tag{15.7}$$

In the present problem it is convenient to take the height of the pivot as the datum for the gravitational potential energy of the load. Hence we obtain, when $\mu = 0$,

$$V = Pl\cos\theta + U, \tag{15.8a}$$

where

$$U = \int_0^\theta C d\theta = \tfrac{1}{2}\eta\theta^2 - \tfrac{1}{3}\nu\theta^3. \tag{15.8b}$$

Replacing $\cos\theta$ by the small-angle approximation $1 - \tfrac{1}{2}\theta^2$ we thus obtain

$$V = Pl - \tfrac{1}{2}Pl\theta^2 + \tfrac{1}{2}\eta\theta^2 - \tfrac{1}{3}\nu\theta^3. \tag{15.9}$$

It is easy to verify that (15.9) corresponds precisely to (15.4) when $\mu = 0$.

The total potential energy function V is plotted against P and θ in fig. 15.7. Curves of constant P are shown, and the projection of the loci (15.5) are marked in the P, θ plane; these curves correspond precisely to those of fig. 15.6. For present purposes the main advantage of this approach is that we may evaluate $d^2 V/d\theta^2$ along the equilibrium paths in order to investigate the stability of equilibrium: for those paths along which this quantity is positive, the equilibrium is stable, whereas if it is negative the equilibrium is unstable. By inspection of fig. 15.7 we see that the primary path from the origin to the point of bifurcation is *stable* (full line) while the path which descends from the bifurcation point is *unstable* (broken line). The other segments of the equilibrium paths in P, θ space are labelled appropriately.

In the present example we see that on both the primary and the secondary path there is a *change* of stability at the bifurcation point. This illustrates a general theorem, due to Koiter (1945), that there is always a change of stability at a point of bifurcation, and that at least one unstable path emerges from a point of bifurcation. The proof of this theorem depends essentially upon the observation that if a smooth curve passes through a sequence of maxima and minima, each maximum is followed by a minimum and vice versa. Thus at any given value of the loading parameter P, adjacent paths have opposite stability.

15.2.2 The introduction of an imperfection

Consider next the behaviour of the system when $\mu \neq 0$, i.e. when there is an initial imperfection. The curve described by (15.4) now has two separate branches, but the form of these curves depends on the sign of μ, as shown in fig. 15.8a and b. In these diagrams the ordinate is the dimensionless load Λ. Thus, when $\mu > 0$ (so that the bar is tilted in the 'weakening' direction of the spring) the curve starting from the origin reaches a *maximum* load, whereas when $\mu < 0$ the system carries an ever-increasing load as the inclination of the bar increases. In these diagrams the various curves are labelled 'stable' or 'unstable' as appropriate (see problem 15.2).

We have already seen that a change of stability can occur at a point of bifurcation. In fig. 15.8a we have an example of the fact that changes in stability can also occur along (unbranched) curves at points where the load passes through either a maximum or minimum. Such changes are called *limit points*, since the load reaches a limiting (i.e. maximum or minimum) value. Once the point of maximum load has been reached, the equilibrium becomes

Fig. 15.7. The total potential energy (V) as a function of P and θ for the system of fig. 15.5a with $\mu = 0$. The equilibrium paths of fig. 15.6 correspond to the condition $\partial V/\partial \theta = 0$. The portion shown has arbitrary boundaries.

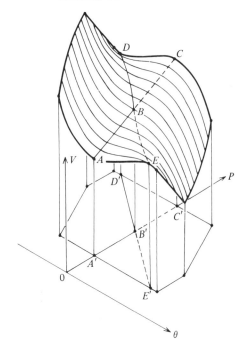

15.2 A simple model for the study of buckling

unstable, and catastrophic behaviour ensues. The situation is analogous to that shown in fig. 15.3b, where buckling also occurs by the reaching of a limit point. Clearly the point of maximum load is of great interest for our present purposes, so we shall now investigate it in detail. Throughout the subsequent analysis we shall consider only the case $\mu > 0$.

For a given value of P, (15.4) may be treated as a quadratic in θ, and its roots may be obtained in the standard way. For small values of P there are two real roots. The maximum value of P corresponds to the point where these two roots are coincident. Rewriting (15.4) as a quadratic in θ, and introducing a dimensionless load

$$\Lambda = Pl/\eta, \tag{15.10}$$

we have

$$\nu\theta^2 - \eta(1 - \Lambda)\theta + \mu\eta\Lambda = 0. \tag{15.11}$$

Thus the condition we seek is

$$(1 - \Lambda)^2 = 4\Lambda(\nu/\eta)\mu. \tag{15.12}$$

This relation is plotted in fig. 15.8c. The behaviour is clearly *imperfection-sensitive* in the sense that a small value of μ produces a disproportionately large reduction of load-carrying capacity. Indeed, (15.12) is formally identical to Koiter's first formula (15.1).

The form of (15.12) suggests clearly that the strength of the imperfection-sensitivity depends upon the relative values of ν and η. It is sufficiently general

Fig. 15.8. (a), (b) Equilibrium paths corresponding to fig. 15.6 when $\mu > 0$ and $\mu < 0$, respectively. Stable and unstable paths are marked s, u, respectively. (c) Plot of maximum and minimum values of Λ, as in (a), against a measure of initial imperfection.

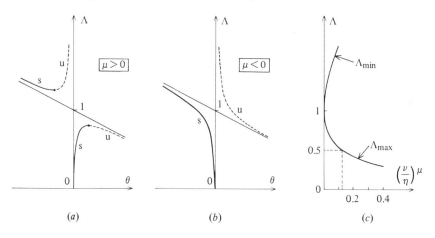

to investigate the circumstances in which the value of Λ_{max} is equal to (say) 0.5, i.e. to find the size of imperfection required to halve the load-carrying capacity. Putting $\Lambda = 0.5$ in (15.12) we have

$$\mu = \tfrac{1}{8} \eta/\nu. \tag{15.13}$$

Now let θ^* be the value of θ at which $C = \tfrac{7}{8}\eta\theta$, i.e. the value at which the restoring couple in (15.2) is $\tfrac{7}{8}$ of the value it would have in the absence of the second term. From (15.2) and (15.8) we have

$$\theta^* = \tfrac{1}{8} \eta/\nu. \tag{15.14}$$

Thus we see from (15.13) and (15.14) that θ^* is in fact the required value of μ. It follows that as soon as we know the details of the nonlinear response of the spring, we can immediately deduce the value of μ required to halve the load-carrying capacity of the system. The required value of μ may be small (if θ^* is small) or large (if θ^* is large): in the latter case the imperfection-sensitivity will perhaps hardly be significant in practical terms.

In relation to the buckling of cylindrical shells under axial compression we shall find later that in some circumstances the equivalent value of θ^* for the shell corresponds to normal displacements of the shell equal to only a fraction of the thickness. In these circumstances we must expect the buckling of the shell to be strongly imperfection-sensitive.

It is useful to make a final remark in relation to the model of fig. 15.5a in the special case $\nu = 0$, in which the behaviour of the spring is *linear*. Equation (15.4) then gives

$$(\theta + \mu) = \mu/(1 - \Lambda).$$

This equation may be interpreted in terms of the original imperfection μ being 'amplified' by the factor $1/(1 - \Lambda)$ under the action of the load P: for this purpose we are considering the inclination to the vertical of a line from the pivot through the point of application of P, both in its original and also its current configuration. In section 14.1 we referred to the same amplification factor in relation to a uniform pin-ended column as 'Perry's rule'. Essentially the same formula will emerge at the end of section 15.4 in relation to the amplification of initial imperfections in a cylindrical shell when we neglect certain nonlinear terms, just as it appears in the present example when we put $\nu = 0$.

15.3 A re-examination of the 'classical' calculation

Before we begin our nonlinear analysis of the buckling of an imperfect cylindrical shell, it is useful to examine again, more critically, the kind of

15.3 A re-examination of the classical calculation

calculation which we performed in chapter 14. There we imagined that the applied loading was carried by a sort of funicular membrane; and if the shape of this deviated from a true cylindrical form, the lateral pressures required for the equilibrium of the membrane were supplied by the shell proper, acting as a combination of a B-surface and an S-surface. Note particularly that the shell was considered in terms of the interaction of *three* separate surfaces:

(i) A 'funicular membrane' which carried the applied loading,

(ii) a 'stretching' surface which responded to normal displacements by developing in-plane stress resultants, and

(iii) a 'bending' surface which responded to normal displacements by developing bending-stress resultants.

Why was the extra 'membrane' surface necessary for this problem? Are not (i) and (ii) really performing the same function? The answer to these questions lies in the fact that in the *classical* analysis of buckling the membrane surface (i) is endowed with the special property that *however it deforms*, the stress resultants N_x, etc. remain unchanged.

Now it is clear that the actual S-surface does not behave in this way, except in special circumstances (e.g. a plate, with free lateral edges, behaving as an Euler column). The fictitious 'membrane' was introduced *in order to simplify the analysis, and to yield a linear eigenvalue problem*. Obviously it would be more realistic in general to make the S-surface play both roles (i) and (ii); but then of course the distribution of N_x etc., which was used in the normal equilibrium equation of the S-surface for the calculation of the interface pressure p_S, would depend on the distortion of the S-surface.

As an example of this effect, consider an initially perfect cylindrical S-surface which, under the action of axial compressive loading, develops a 'square' mode of buckling:

$$w = w_0 \cos(\pi x/l) \cos(\pi y/l). \tag{15.15a}$$

It is a simple exercise in the methods of section 7.4 to find the corresponding change of Gaussian curvature; to solve equation (7.43) for the stress function ϕ; and then to evaluate the change in stress resultants N_x etc. by means of (7.36). If we do this we find that

$$N_x = N_y = (w_0 \, Et/4a) \cos(\pi x/l) \cos(\pi y/l); \tag{15.15b}$$

and that in particular there is longitudinal tension in the panels which deflect outwards, and a compression in those which deflect inwards. These stress resultants are directly proportional to the amplitude of displacement, and they must clearly be added to any compressive load which is applied externally to the shell. Consider the consequences of this for an initially perfect cylindrical shell which is supporting the classical buckling load, i.e.

$$N_x = -Eth/3^{\frac{1}{2}} a. \tag{15.16}$$

Suppose that buckling occurs according to a classical mode described by (15.15a) while the compressive load remains constant. The periodic addition (15.15b) to (15.16) grows with w, and we find from these that when

$$w_0 \approx 2.4t \tag{15.17}$$

(which is admittedly rather a large displacement) the local value of N_x varies over the surface of the shell between zero and twice the classical buckling value.

It is unlikely, of course, that this would happen without inducing prior failure of the shell. But we are driven to the conclusion that the classical analysis, which explicitly disregards this kind of effect, is of somewhat doubtful value. Our next task, therefore, is to bring the effect of changes in N_x, etc., on account of the growth of buckles, within the scope of our investigation. We shall return to some more simple calculations in section 15.5.

15.4 A nonlinear analysis of buckling

The method which we shall adopt in this section is an extension of our previous 'two-surface' treatment. It forms a chain of relatively simple steps which differ in some respects at specific points from those of the 'classical' analysis. Since the calculation turns out to be both lengthy and also complicated in detail, it is useful first to summarise the scheme as follows.

First we assume a normal displacement function which is the sum of two 'classical' modes, as mentioned above. We then calculate accurately (i.e. more accurately than by (6.6)) the *change* in Gaussian curvature, g, which has occurred on account of changes in the configuration of the shell from the original state. Next we solve the S-surface equation

$$\nabla^4 \phi = -Etg \tag{15.18}$$

for the stress function ϕ, and use the result to obtain expressions for the stress resultants N_x, N_y, N_{xy} which are present, in addition to the applied axial loading, on account of changes in configuration.

We are then in a position to compute the outward-directed pressure p_S which is required for equilibrium of the S-surface when the displacements and tangential-stress resultants are as calculated above (15.18). Considering

15.4 A nonlinear analysis of buckling

the equilibrium of a small element in its current *distorted* configuration we have:

$$-p_S = N_x \frac{\partial^2 w}{\partial x^2} + N_y \left(-\frac{1}{a} + \frac{\partial^2 w}{\partial y^2}\right) + 2 N_{xy} \frac{\partial^2 w}{\partial x \partial y}. \tag{15.19}$$

Here N_x, etc. are the total tangential-stress resultants, and the multiplying factors are the total curvatures and twist with respect to the x, y coordinate system. This equation is a more accurate version of (14.9), since it takes into account the distorted configuration of the shell. Note that the term enclosed in brackets in (15.19), which is an expression for the current circumferential curvature, omits a term w/a^2: cf. (6.9e). This is justified in the present problem because we shall be concerned almost entirely with an axisymmetric mode and with a doubly-periodic mode which in practice has a high circumferential mode-number.

Next we consider the deformation of the B-surface. There are no complications on account of nonlinearity here, and we may use unaltered the governing equation which we have developed earlier (see (8.9)):

$$p_B = \tfrac{1}{12} Eth^2 \, \nabla^4 w. \tag{15.20}$$

It is easy to solve this equation to obtain the pressure p_B required by the B-surface to sustain the imposed displacement: clearly p_B will consist of two terms, of the same form as the displacement function.

We are now ready to use the overall equilibrium equation

$$p_S + p_B = 0.$$

This will give a separate equation with respect to each of the two constituent modes. The final step is to solve these (nonlinear) equations in order to determine the relationship between the two modal amplitudes and the applied load, and to find, in particular, the maximum load which can be sustained by the shell once the amplitude of the initial geometrical imperfections has been specified.

The main differences between this scheme and the one which we used previously are that here we are much more careful about the calculation of g, and the evaluation of p_S in (15.19).

15.4.1 *The two-mode calculation*

It is clear that the required calculations are likely to involve somewhat more cumbersome manipulations than previous problems which we have tackled. Therefore we shall define the main variables in a way which simplifies the subsequent working, following Koiter (1963b).

Let the current distorted two-mode configuration be given by

$$w = -(h\Delta/3^{\frac{1}{2}})\cos 2X + (h\Gamma/3^{\frac{1}{2}})\cos X \cos Y, \qquad (15.21a)$$

where

$$X = x(3^{\frac{1}{2}}/2ah)^{\frac{1}{2}} \qquad (15.21b)$$

$$Y = y(3^{\frac{1}{2}}/2ah)^{\frac{1}{2}}. \qquad (15.21c)$$

In these expressions the variables have been defined so that the subsequent lengthy algebra can be written compactly. Thus the dimensionless coordinates X, Y have been chosen so that $\cos 2X$ and $\cos X \cos Y$ represent the classical axisymmetric and 'square' modes, respectively. Also the use of Δ and Γ as dimensionless mode-amplitudes have been chosen for the same purpose: when we have established the required results we shall revert to the use of the actual displacement w to describe mode-amplitudes. It is convenient to use cosine functions for this problem because the interaction between the modes occurs when the circumferential lines of the axially symmetric mode where the outward-directed deflection is *maximum* coincide with the circumferential *nodal* lines of the 'square' mode. And since the inward-displaced parts of the axially symmetric mode pass through the middle of the square 'panels', the negative sign in (15.21) is appropriate. A more complete analysis would consider the relative phase of the two modes: this has been omitted here for the sake of clarity.

Equation (15.21) describes the current configuration of the shell with respect to a truly cylindrical reference surface. Γ and Δ define the *current* mode-amplitudes. Initially the shell is supposed to contain a geometric imperfection described by Γ_0, Δ_0; and in this configuration it is reckoned to be completely stress-free.

Equation (15.21) contains no term to represent the uniform radial displacement which is associated with uniform axial loading on account of Poisson's ratio. Although such a term is strictly required, it plays no part in the subsequent analysis and has therefore been omitted.

As we have stated already, our first task is to determine N_x, N_y and N_{xy} for the S-surface in its deformed configuration. As in chapter 8 we shall approach the calculation via the change in Gaussian curvature. In order to avoid the possibility of discarding terms which appear to be negligible but which in fact are important, we shall compute the change of Gaussian curvature by first evaluating the *absolute* Gaussian curvature in the current deformed configuration and then determining formally the *change* in this quan-

15.4 A nonlinear analysis of buckling

tity when the mode-amplitudes change from their initial values of Γ_0, Δ_0 to their current values Γ, Δ.

Using, as before (see (6.9)) the simple relations which are appropriate for small displacements w from a truly cylindrical reference surface, we have the following expressions for current *absolute* curvature, and twist (cf. (15.19)):

$$\kappa_x = -\frac{\partial^2 w}{\partial x^2}; \quad \kappa_{xy} = -\frac{\partial^2 w}{\partial x \partial y}. \tag{15.22}$$

In the circumferential direction the *total* curvature is given by

$$\left(\frac{1}{a} + \kappa_y\right) = \frac{1}{a} - \frac{\partial^2 w}{\partial y^2}. \tag{15.23}$$

Using equations (15.21) we thus find:

$$\left. \begin{array}{ll} \kappa_x &= -(2\Delta/a)\cos 2X + (\Gamma/2a)\cos X \cos Y, \\ 1/a + \kappa_y &= 1/a \quad\quad\quad\quad\quad + (\Gamma/2a)\cos X \cos Y, \\ \kappa_{xy} &= \quad\quad\quad\quad\quad\quad -(\Gamma/2a)\sin X \sin Y. \end{array} \right\} \tag{15.24}$$

In general, the local Gaussian curvature K of this surface is given by

$$K = \kappa_x(1/a + \kappa_y) - \kappa^2_{xy}: \tag{15.25}$$

see problems 5.4 and 15.3. We thus obtain, after some manipulation,

$$\begin{aligned} Ka^2 &= (-2\Delta + \tfrac{1}{8}\Gamma^2)\cos 2X + (\tfrac{1}{2}\Gamma - \tfrac{1}{2}\Gamma\Delta)\cos X \cos Y \\ &\quad + (\tfrac{1}{8}\Gamma^2)\cos 2Y \quad + (-\tfrac{1}{2}\Gamma\Delta)\cos 3X \cos Y. \end{aligned} \tag{15.26}$$

Here we have expressed the result in terms of the original and other modes by use of the following trigonometrical identities:

$$\cos 2X \cos X \cos Y = 0.5 \cos X \cos Y + 0.5 \cos 3X \cos Y, \tag{15.27a}$$

$$\cos^2 X \cos^2 Y - \sin^2 X \sin^2 Y = 0.5 \cos 2X + 0.5 \cos 2Y. \tag{15.27b}$$

It is interesting to compare the expression (15.26) with that which would have been obtained by our usual but less accurate expression (see (6.6)): this formula gives correctly the first-order terms in Γ and Δ, but it omits the various second-order terms Γ^2 and $\Gamma\Delta$. Note in particular that the 'new' modes represented in (15.26), i.e. $\cos 2Y$ and $\cos 3X \cos Y$, have coefficients which are entirely second order. At this stage we shall retain all of these terms.

In this connection it should be emphasised that the nonlinearities of the problem are quite distinct from those which we mentioned in section 14.1 in relation to the simple 'Euler' column. The present nonlinearities occur well within the range for which the 'small-displacement' expressions for change of curvature are valid.

Equation (15.26) has been set up for the *current* mode amplitudes Δ, Γ. Originally, in the unstressed state of the shell, let the mode amplitudes be Δ_0, Γ_0, respectively. Hence we may write the corresponding expression for the original Gaussian curvature, and thus by subtraction obtain the following expression for the *change* of Gaussian curvature:

$$a^2 g = C_1 \cos 2X + C_2 \cos X \cos Y + C_3 \cos 2Y + C_4 \cos 3X \cos Y, \quad (15.28a)$$

where

$$C_1 = -2\Delta + 2\Delta_0 + \tfrac{1}{8}\Gamma^2 - \tfrac{1}{8}\Gamma_0^2, \qquad (15.28b)$$

$$C_2 = \tfrac{1}{2}\Gamma - \tfrac{1}{2}\Gamma_0 - \tfrac{1}{2}\Gamma\Delta + \tfrac{1}{2}\Gamma_0\Delta_0, \qquad (15.28c)$$

$$C_3 = \tfrac{1}{8}\Gamma^2 - \tfrac{1}{8}\Gamma_0^2, \qquad (15.28d)$$

$$C_4 = -\tfrac{1}{2}\Gamma\Delta + \tfrac{1}{2}\Gamma_0\Delta_0. \qquad (15.28e)$$

We now use this expression in order to compute the stress resultants in the S-surface. Equation (15.18) may be solved separately for each of the four terms on the right-hand side of (15.28a), and the solutions superposed: there are no problems over constants of integration since we are assuming 'natural' boundary conditions: see section 14.3.3 The evaluation of stress resultants N_x, etc. by means of (7.36) is also straightforward. Hence we obtain the following expressions, which have been written to include also the component

$$N_x = N_0$$

of axial loading which is applied externally. Here, as before, the sign of N_0 is positive when tensile; so we shall eventually be concerned with the critical value of $(-N_0)$.

$$N_x = N_0 + (Eth/3^{\frac{1}{2}}a)(\tfrac{1}{2}C_2 \cos X \cos Y$$
$$+ \tfrac{1}{2}C_3 \cos 2Y + \tfrac{9}{50} C_4 \cos 3X \cos Y) \qquad (15.29a)$$

$$N_y = (Eth/3^{\frac{1}{2}}a)(\tfrac{1}{2}C_1 \cos 2X + \tfrac{1}{2}C_2 \cos X \cos Y$$
$$+ \tfrac{1}{50} C_4 \cos 3X \cos Y) \qquad (15.29b)$$

$$N_{xy} = (Eth/3^{\frac{1}{2}}a)(\tfrac{1}{2}C_2 \sin X \sin Y$$
$$+ \tfrac{3}{50} C_4 \sin 3X \sin Y) \qquad (15.29c)$$

If we discount for the moment the purely second-order terms C_3 and C_4 we can see clearly the main effects of the two modes upon the stress distribution in the S-surface. Thus the axisymmetric mode enters through C_1 and

15.4 A nonlinear analysis of buckling

involves only a hoop stress N_y, while the square-panel mode gives contributions to N_x, N_y and N_{xy}, just as we expect.

The next task is to evaluate p_S by means of (15.19). Thus we must consider the expression

$$-ap_S = N_x (2\Delta \cos 2X - 0.5 \Gamma \cos X \cos Y)$$
$$+ N_y (-1 - 0.5 \Gamma \cos X \cos Y) + N_{xy} (\Gamma \sin X \sin Y). \quad (15.30)$$

It is clear that when N_x, etc. are substituted from (15.29) there will be coefficients of order 1, 2 and 3 in Δ and Γ. At this point we shall neglect the third-order terms, while retaining all others. It is also clear that the trigonometric expressions will involve terms such as

$$\cos X \cos 3X \cos^2 Y. \quad (15.31)$$

Now our main interest is in terms which make contributions to the two modes $\cos 2X$ and $\cos X \cos Y$. Therefore it is appropriate to rewrite (15.31), by means of trigonometrical identities, as follows:

$$\cos X \cos 3X \cos^2 Y \equiv \tfrac{1}{4} (\cos 2X + \cos 4X + \cos 2X \cos 2Y +$$
$$+ \cos 4X \cos 2Y). \quad (15.32)$$

Other relevant identities, in addition to this and (15.27) are:

$$\cos 2X \cos 3X \cos Y \equiv \tfrac{1}{2} (\cos X \cos Y + \cos 5X \cos Y),$$
$$\cos X \cos Y \cos 2Y \equiv \tfrac{1}{2} (\cos X \cos Y + \cos X \cos 3Y),$$
$$\cos^2 2X \equiv \tfrac{1}{2} (1 + \cos 4X), \quad (15.33)$$
$$\cos^2 X \cos^2 Y \equiv \tfrac{1}{4} (\cos 2X + 1 + \cos 2Y + \cos 2X \cos 2Y),$$
$$\sin^2 X \sin^2 Y \equiv \tfrac{1}{4} (-\cos 2X + 1 - \cos 2Y + \cos 2X \cos 2Y),$$
$$\sin X \sin 3X \sin^2 Y \equiv \tfrac{1}{4} (\cos 2X - \cos 4X - \cos 2X \cos 2Y + \cos 4X \cos 2Y).$$

In expressions of this sort we can easily pick out the contribution from a given term towards the two main modes.

Our process of deriving an expression for p_S thus also generates many other terms, as we can see, corresponding to modes other than the two original modes. In the following work we shall ignore all of these 'extra' modes, on the basis of the following argument. The simplest way of assessing the significance of the 'extra' modes is to plot them onto the 'Koiter circle' diagram which we investigated in chapter 14. The plot is shown in fig. 15.9. The mode corresponding to the term $\cos 2X$ is marked as point A: the abscissa is the coefficient of X and the ordinate is the coefficient of Y. Similarly, point B represents the 'square' mode. These two points lie on the Koiter circle, as we expect. The other points (not labelled) correspond to those terms which appear on the right-hand side of the various relevant identities, and which therefore appear in our final expression for p_S; and there is obviously a simple pattern in the arrangement of these points. Now except for the point at the origin (corresponding to a constant term) and the two points A and B, *all of the points lie well outside the Koiter circle*. The significance of this remark is that if we drop a particular term, say $\cos 3X \cos Y$, from the expression for p_S we shall be left with an 'unbalanced' pressure of this form at the end of the analysis. Now the application of a pressure distribution of this form to the shell in ordinary circumstances (cf. fig. 8.4) would elicit a response from the shell involving predominantly *bending* effects: and indeed this is true for all modes which map onto points lying sufficiently far outside the Koiter circle. Therefore the neglect of a term like this will not introduce an appreciable change in N_x, etc. We may therefore conclude that by neglecting the higher-mode terms in our final expression for p_S we are not introducing any serious error (The entire pattern of points in fig. 15.9 has been transferred to fig. 8.4.)

The one remaining point, at the origin, represents a uniform state of stress; and this suggests that the form of deflection (15.21) with which we started ought to have included a constant term also. It turns out, however, when the various terms are evaluated, that the constant term in the expression for p_S has a magnitude of identically zero: consequently the original expression (15.21) is satisfactory.

We are now in a position to evaluate p_S by the use of (15.29) and (15.30).

Fig. 15.9. Map of modes occurring in the calculation of section 15.4.1. The two 'input' modes A, B lie on Koiter's circle, while the modes which are generated in the nonlinear calculations lie well outside it. Subscript B (in l_B) refers to mode B. The broken curve corresponds to modes for which the stiffness of the B-surface is ten times that of the S-surface. Corresponding points are also marked on fig. 8.4a.

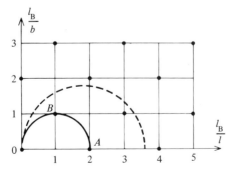

15.4 A nonlinear analysis of buckling

Each trigonometric term can be 'looked up' in (15.27), (15.32) or (15.33), and its contribution to the terms $\cos 2X$ and $\cos X \cos Y$ can be noted. In this way we obtain the following expression, from which all of the non-relevant terms have been omitted:

$$-p_S = \cos 2X \left[\frac{2N_0 \Delta}{a} - \frac{Eth}{3^{\frac{1}{2}}a^2} \left\{ \frac{C_1}{2} + \frac{\Gamma C_2}{4} + \frac{\Gamma C_4}{100} \right\} \right] +$$

$$+ \cos X \cos Y \left[-\frac{N_0 \Gamma}{2a} - \frac{Eth}{3^{\frac{1}{2}}a^2} \left\{ \frac{C_2}{2} + \frac{\Gamma C_1}{8} - \frac{\Delta C_2}{2} + \frac{\Gamma C_3}{8} - \frac{9}{50} \Delta C_4 \right\} \right]. \tag{15.34}$$

Now since there is zero external pressure-loading on the shell, we have

$$-p_S = p_B$$

thus (15.34) may be regarded as an expression for p_B. The normal deflection of the B-surface which corresponds to p_B is easily computed, since we know from previous work (see problem 8.3) that the mode-stiffness of the B-surface in the axisymmetric and 'square' modes of classical buckling are Et/a^2 and $Et/4a^2$, respectively. Hence the coefficient of $\cos 2X$ in (15.34), when multiplied by a^2/Et, gives the amplitude of the corresponding component of displacement of the B-surface which is (from (15.21a)) equal to $-(h/3^{\frac{1}{2}})(\Delta - \Delta_0)$. Thus we obtain the following equation in relation to the mode $\cos 2X$:

$$\frac{2 \times 3^{\frac{1}{2}} N_0 a \Delta}{Eth} \left\{ \frac{C_1}{2} + \frac{\Gamma C_2}{4} + \frac{\Gamma C_4}{100} \right\} = -(\Delta - \Delta_0). \tag{15.35}$$

Now the classical buckling value of N_0 has magnitude $Eth/3^{\frac{1}{2}}a$ (see section 14.3). Let us therefore define a dimensionless *compressive* load Λ by

$$\Lambda = -3^{\frac{1}{2}} N_0 a / Eth. \tag{15.36}$$

Rearranging (15.35) and substituting for C_1 etc from (15.28) we obtain, finally,

$$\Delta(1 - \Lambda) = \Delta_0 + \tfrac{1}{32}(3\Gamma^2 - 2\Gamma\Gamma_0 - \Gamma_0^2 + \ldots) \tag{15.37}$$

Here we have neglected all terms of order Γ^3, $\Gamma^2 \Delta$, etc. In particular, terms C_3 and C_4 in (15.28) make contributions to (15.37) which are entirely negligible.

If for the moment we further neglect terms of order Γ^2, etc., we recover the simple expression

$$\Delta = \Delta_0 / (1 - \Lambda), \tag{15.38}$$

which is the well-known 'Perry' formula for the amplification of a single mode by the action of the compressive load: see section 15.2. It is the *quadratic* terms in (15.37) which represent the nonlinear coupling from the other mode, which we have been seeking in this section.

In a precisely similar way the term $\cos X \cos Y$ in the expression for $-p_S$ yields the equation

$$\Gamma(1 - \Lambda) = \Gamma_0 + \tfrac{1}{2} (3\Gamma\Delta - \Gamma\Delta_0 - \Gamma_0\Delta - \Gamma_0\Delta_0 + \ldots) \qquad (15.39)$$

This is a generally similar expression to (15.37), except that the quadratic coupling term in this case involves both Γ and Δ. Equations (15.37) and (15.39) are the two expressions which we have been seeking.

15.4.2 Equilibrium paths for imperfect shells

Our next task is to examine the loading paths in Γ, Δ, Λ space which are indicated by (15.37) and (15.39) for imperfect cylindrical shells. It is convenient to begin with the simplest case, in which there is only an axially symmetric imperfection, i.e. $\Gamma_0 = 0$. This is the case first studied by Koiter in 1963. It is convenient to define

$$\alpha = \Delta/\Delta_0, \quad \beta = \Gamma/\Delta_0. \qquad (15.40)$$

Thus α, β are measures of the amplitudes of the axisymmetric and square modes, respectively, relative to the initial axially symmetric form. Equations (15.37) and (15.39) become, in these circumstances,

$$\alpha(1 - \Lambda) = 1 + \tfrac{3}{32} \Delta_0 \beta^2, \qquad (15.41)$$

$$\beta(1 - \Lambda) = \tfrac{1}{2} \beta \Delta_0 (3\alpha - 1). \qquad (15.42)$$

What is the solution of these simultaneous equations in α, β? First we notice that in (15.42) the factor β is common to both sides: consequently this equation degenerates into two separate equations

$$\beta = 0, \qquad (15.43a)$$

$$1 - \Lambda = \tfrac{1}{2} \Delta_0 (3\alpha - 1). \qquad (15.43b)$$

These represent two planes in a Cartesian β, α, Λ space as shown in fig. 15.10a. Equation (15.41), on the other hand, defines a curved surface, as shown in fig. 15.10b. The required solution consists of the curves of intersection of these three surfaces. Equations (15.43a) and (15.41) intersect in the simple plane curve

$$\alpha = 1/(1 - \Lambda), \qquad (15.44)$$

which represents the usual 'Perry' rule for growth of an imperfection. Indeed, the equilibrium path follows this curve initially, since it must lie both on the plane $\beta = 0$ and on the surface (15.41). Sooner or later, however, the path will rise as high as the second plane (15.43b). Then there will be a branching point, as shown in fig. 15.10c. This is an example of a branching point on a nonlinear primary equilibrium path, as distinct from the examples given in figs. 14.1 and 15.6, which both show primary paths corresponding to trivial, 'membrane' states validly computed in relation to the original undeformed configuration of the structures concerned. The paths emerging in the $\pm\beta$ direction from the branching point in fig. 15.10c have their highest point at the point of bifurcation itself: thus the bifurcation load represents the largest load which can be carried when $\beta \neq 0$: consequently the buckling on this secondary path is unstable. The bifurcation load is found by solving (15.41), with $\beta = 0$, together with (15.43b). Thus

$$\alpha(1 - \Lambda) = 1, \quad (1 - \Lambda) = \tfrac{1}{2}\Delta_0 (3\alpha - 1).$$

Eliminating α from these equations, and expressing Δ_0 in terms of the initial displacement w_0, we find

$$(1 - \Lambda)^2 = (3^{\tfrac{1}{2}} w_0 / h)(1 + \tfrac{1}{2}\Lambda). \tag{15.45}$$

Apart from a term corresponding to higher-order terms which we neglected in obtaining (15.37), expression (15.45) is identical to formulae derived by Koiter (1963b) and Hutchinson (1965). Koiter's method involved a linearised bifurcation analysis on a nonlinear prebuckled form. His result is strictly an

Fig. 15.10. Equilibrium paths for an axially compressed cylindrical shell containing an axisymmetric imperfection in a space which represents the dimensionless load Λ and dimensionless amplitudes α, β of the classical axisymmetric (A) and 'square' (B) modes respectively. (a) Two planes representing the solution of (15.42). The inclined plane intersects the Λ-axis a little above $\Lambda = 1$. (b) A curved surface representing (15.41). It intersects the α-axis at $\alpha = 1$. (c) Equilibrium paths consisting of the intersection of (a) and (b).

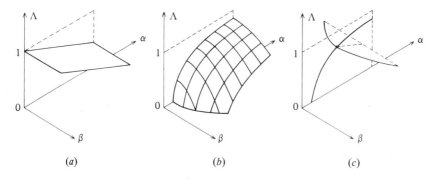

upper bound, whereas our result can only be classified as 'approximate'. Koiter also investigated bifurcation into a mode of the form $\cos X \cos q Y$ where q is a variable; and he found that for larger imperfections the lowest bifurcation point corresponded to $q < 1$. It is clear that our present method correctly reproduces the important features of Koiter's calculation.

For any postulated value of Λ at buckling, the required initial imperfection may be evaluated from (15.45); and the resulting curve is shown in fig. 15.11. Also shown is a curve corresponding to Koiter's first formula (15.1). This formula agrees well with (15.45) in the region $\Lambda \approx 1$, since the factor $(1 + \frac{1}{2}\Lambda)$ in (15.45) is then numerically close to the 1.5 in (15.1).

Since these curves are both smooth and tangential to the Λ-axis at $\Lambda=1$, the value of Λ is obviously very sensitive to the amplitude of imperfection for small imperfections: note that an imperfection of only 0.115 of the wall thickness is needed to reduce Λ to a value of 0.5.

Consider next an example in which there is originally only a 'square' waveform, i.e. $\Delta_0 = 0$. Let us define for this problem

$$\alpha = \Delta/\Gamma_0, \quad \beta = \Gamma/\Gamma_0. \tag{15.46}$$

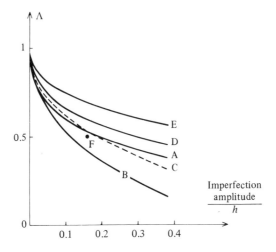

Fig. 15.11. Dimensionless load plotted against amplitude of initial imperfection according to various analyses. Curve A: Koiter's (1945) 'first formula' (15.1). B: Koiter's (1963b) and Hutchinson's (1965) calculations (15.45), both for axisymmetric mode imperfection. C: present work, 'square' mode imperfection. D: 'local buckling hypothesis' (15.60) for axisymmetric mode. E: Ditto, (15.61) for 'square' mode. F: Point calculated by 'nonlinear spring' model for 'square' mode: the entire curve has the same form as A.

15.4 A nonlinear analysis of buckling

Equations (15.37) and (15.39) now become

$$\alpha(1 - \Lambda) = \tfrac{1}{32}\Gamma_0 (3\beta + 1)(\beta - 1) \tag{15.47}$$

$$\beta(1 - \Lambda) = 1 + \tfrac{1}{2}\Gamma_0 \alpha (3\beta - 1). \tag{15.48}$$

In the previous case we found that (15.37) was the main equation describing the growth of the imperfection, while the second equation could always be satisfied by $\beta = 0$, i.e. if the second imperfection did not grow. In the present problem the roles of the variables are reversed, and (15.48) describes the growth of the imperfection. But in the present case the second (α) imperfection must also begin to grow immediately, because the equations cannot be satisfied with $\alpha = 0$. The most straightforward way of dealing with these equations is to eliminate α between (15.47) and (15.48), which gives the following equation in β and Λ:

$$\beta(1 - \Lambda)^2 - (1 - \Lambda) = \tfrac{1}{64}\Gamma_0^2 (9\beta^2 - 1)(\beta - 1). \tag{15.49}$$

In principle, for a given value of Γ_0 this can be solved for $(1- \Lambda)$ for an arbitrary value of β; hence the maximum value of Λ can be determined. It is more convenient to work in terms of a graphical construction, shown in fig. 15.12, in which the left-hand and right-hand sides of (15.49) are plotted separately as functions of β; we shall call these functions $L(\beta)$ and $R(\beta)$ respectively. $R(\beta)$ has been sketched for a given value of Γ_0, while $L(\beta)$ has been drawn for various values of $(1 - \Lambda)$. For a given value of $(1 - \Lambda)$ the value of β is

Fig. 15.12. Graphical construction for finding the maximum value of β which satisfies (15.49) for a given value of Γ_0 (here = 0.2).

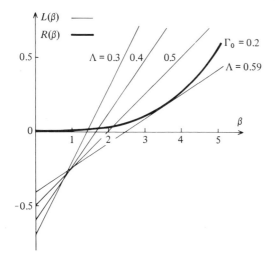

determined by the intersection of the corresponding line with the curve. The slope of the line is proportional to $(1 - \Lambda)^2$; thus the slope decreases as the value of Λ increases. The limiting value of Λ is therefore determined by the condition that the line touches the curve. It is easy to find the equation of the tangent to the curve at an arbitrary value of β; and since this line intersects the β-axis at $1/(1 - \Lambda)$ and the $L(\beta)$-axis at $-(1 - \Lambda)$ it is a simple matter to work out corresponding values of $(1 - \Lambda)$ and Γ_0. The results are:

$$\Gamma_0 = 8A^{\frac{1}{2}}/B; \quad 1 - \Lambda = A/B, \qquad (15.50a)$$

where

$$A = 27\beta^2 - 18\beta - 1, \quad B = 18\beta^3 - 9\beta^2 - 1. \qquad (15.50b)$$

These expressions have been evaluated for several values of β, and are plotted in fig. 15.11. For large values of β (say $\beta > 5$) the leading terms dominate in the expressions for A and B, and the relation

$$(1 - \Lambda)^2 = (9 \times 3^{\frac{1}{2}}/16)\Gamma_0 \approx 0.97\Gamma_0 \qquad (15.51)$$

emerges: it provides a good approximation for $\Lambda > 0.7$, say.

Equation (15.51) is of broadly the same form as the corresponding relation for axisymmetric imperfections, although the mechanics of the two situations are quite different. In the case of a purely axisymmetric imperfection the 'square-panel' mode is entirely absent until the bifurcation point is reached, as we have seen. On the other hand, when the initial imperfection is purely of the 'square-panel' mode, the axisymmetric mode begins to grow as soon as a small load is applied; and it is easy to show that at the point when Λ reaches its maximum value, $\alpha \approx \frac{1}{7}\beta$ when the value of Γ_0 is sufficiently small: see problem 15.4. Thus, although the axisymmetric mode necessarily grows, its amplitude is never more than a small fraction of that of the 'square-panel' mode.

It is not difficult to extend the foregoing analysis to cover the situation in which the initial imperfection involves *both* modes, at least when the amplitudes of the initial imperfections are so small that only the leading term in the quadratic expressions on the right-hand side of (15.37) and (15.39) need be retained. Let us write, for this case,

$$\alpha = \Delta/\Delta_0, \quad \beta = \Gamma/\Gamma_0. \qquad (15.52)$$

Then the equations become

$$(1 - \Lambda) = (1/\alpha) \left[1 + \tfrac{3}{32} \Gamma_0^2 \beta^2/\Delta_0\right]$$

$$(1 - \Lambda) = (1/\beta) + \tfrac{3}{2} \Delta_0 \alpha.$$

15.4 A nonlinear analysis of buckling

It is easy to eliminate α between these two expressions; and we thus obtain

$$\beta(1 - \Lambda)^2 - (1 - \Lambda) = \tfrac{3}{2}\Delta_0 \beta + \tfrac{9}{64} \Gamma_0^2 \beta^3. \tag{15.53}$$

This relation is similar to (15.49) except that the right-hand side now includes a term in Δ_0, and some high-order terms in β have been omitted. The previous method can still be used, and after some manipulation the following result emerges:

$$9 \times 3^{\tfrac{1}{2}}\Gamma_0/16(1 - \Lambda)^2 = \{1 - [\tfrac{3}{2}\Delta_0/(1 - \Lambda)^2]\}^{\tfrac{3}{2}}. \tag{15.54}$$

In the special cases $\Delta_0 = 0$ and $\Gamma_0 = 0$, this reduces to (15.51) and a simplified version of (15.45), respectively. The 'interaction diagram' of fig. 15.13 shows those combinations of Δ_0 and Γ_0 which together give maximum load-carrying capacity equal to that which would be obtained with a purely axisymmetric imperfection of amplitude Δ^*, according to (15.54). The diagram has been plotted only in the first quadrant of Δ_0/Δ^*, Γ_0/Δ^* space. It has not been extended further in order to draw attention to the fact that we have only studied the interaction between two specific modes, and there may well be other combinations of modes for which the buckling is more sensitive. Nevertheless it is of interest to observe that to a very crude first approximation the load-carrying capacity depends on the value of $\Delta_0 + \Gamma_0$: see fig. 15.13.

Fig. 15.13. Combinations of initial axisymmetric (Δ_0) and 'square' (Γ_0) mode amplitudes for which Λ is constant.

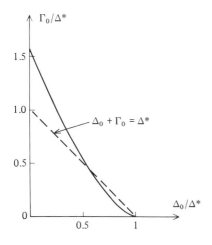

578   *Buckling of shells: non-classical analysis*

15.5   Distribution of tangential-stress resultants during buckling

A feature which emerges clearly from the present analysis, but is rather obscure in the usual potential energy formulations of the nonlinear problem, is that when imperfections grow under the application of axial loading, there are corresponding changes in the tangential-stress resultants, which can be of significant magnitude in comparison with the applied loading. We touched on this point earlier, in section 15.3, but now we are in a position to do a more systematic treatment (Calladine & Robinson, 1980).

Consider first a thin cylindrical shell having a small axisymmetric imperfection, which under axial loading has just reached the point of bifurcation at a load Λ according to Koiter's first formula (15.1). The amplitude Δ of the imperfection at this point is given by (15.44):

$$\Delta = \Delta_0/(1 - \Lambda)$$

and thus, by (15.1),

$$\Delta = \tfrac{2}{3}(1 - \Lambda)/\Lambda;$$

the amplitude of the radial displacement w is equal to $h\Delta/3^{\frac{1}{2}}$ by the definition (15.21a). In particular, at those places where the inward-directed displacement is maximum there is a compressive circumferential strain equal in magnitude to $(\Delta - \Delta_0)h/3^{\frac{1}{2}}a$. Consequently we may calculate the peak value of the compressive circumferential stress resultant:

$$N_y \approx -(\tfrac{2}{3}/3^{\frac{1}{2}})(1 - \Lambda)(Eth/a). \tag{15.55}$$

Stress resultants N_x and N_{xy}, of course, are unaffected by the w-displacements of the axisymmetric mode.

Next we recall that the classical buckling value of N_x is equal in magnitude to $Eth/3^{\frac{1}{2}}a$, and that at present the applied compressive load is a fraction Λ of the classical load. Hence from (15.55) we conclude that when the point of bifurcation is reached there are small regions of the surface where

$$N_x + \tfrac{3}{2}N_y \approx N_{cl}; \tag{15.56}$$

here the subscript cl stands for 'classical'. Furthermore, since the half-wavelength of the axisymmetric mode is exactly one-half that for the 'square' mode, condition (15.56) is reached in particular at the centre of every panel of a prospective 'square' mode of buckling. (Indeed, the phase of the two modes in (15.21) was fixed in precisely this way.)

Now according to our analysis of chapter 14, the classical buckling condition for a thin cylindrical shell under simultaneous loading N_x and N_y is given by (14.48); and in particular we may write, for the 'square-panel' mode

$$N_x + N_y = N_{cl}. \tag{15.57}$$

15.5 Distribution of stress resultants during buckling

Thus, by comparing (15.56) and (15.57) we see that except for the factor $\frac{3}{2}$ (which is nevertheless of order unity) the bifurcation condition (15.56) is almost equivalent to a buckling condition of a classical kind being met *locally* by the *current* distribution of stress. Here, of course, we have considered the 'worst' locations from the point of view of a hypothetical 'local' buckling condition, and clearly there will be other regions in which N_y is tensile, where the 'local' buckling condition will obviously not be met. Nevertheless, it seems reasonable to regard the localised compressive circumferential stress resultants N_y, which are developed by the straining of the S-surface in the course of enlargement of the original imperfection, as an important destabilising factor; this point was first made by Koiter.

This idea may be expressed in a more striking form as follows. We shall assume that an imperfection is present in the form of the classical axially symmetric mode, with amplitude w_0. As the axial compressive load is increased, we may calculate the growth in amplitude by means of the 'Perry' rule, and consequently the peak compressive hoop stress in the inward-deflecting parts. Then we make the *hypothesis* that buckling will occur when the worst *local* combination of axial and circumferential stress resultants reach the *classical* condition for buckling in the 'square' mode, and see what result emerges.

Defining a dimensionless compressive load Λ as before, we have the current amplitude of imperfection, w as follows, according to the 'Perry' formula:

$$w = w_0/(1 - \Lambda).$$

This is superposed onto a uniform radial displacement on account of the Poisson ratio effect. At points of maximum inward deflection the periodic part of the circumferential strain is equal to $-(w - w_0)/a$; and hence there is a circumferential stress resultant of

$$N_y = -\left(\frac{\Lambda}{1 - \Lambda}\right) \frac{w_0 E t}{a}. \tag{15.58}$$

The local buckling condition is thus

$$N_0 - \left(\frac{\Lambda}{1 - \Lambda}\right) \frac{w_0 E t}{a} = -\frac{E t h}{3^{\frac{1}{2}} a}.$$

Dividing throughout by the right-hand side, we obtain

$$\Lambda + \left(\frac{\Lambda}{1 - \Lambda}\right) \frac{3^{\frac{1}{2}} w_0}{h} = 1. \tag{15.59}$$

In this equation the first term on the left-hand side represents the applied axial load; the second term represents the local compressive circumferential stress on account of the grown imperfection; and the right-hand side represents the classical buckling condition. Rearranging (15.59) we have, finally

$$(1 - \Lambda)^2 = 3^{\frac{1}{2}} \Lambda w_0 / h. \tag{15.60}$$

This has precisely the same form as Koiter's first formula (15.1); and would indeed be identical to it if the right-hand side of (15.60) were multiplied by 1.5.

In view of the success of this simple idea, consider next the behaviour of a shell which contains an imperfection in the form of the classical 'square' mode, again of amplitude w_0. According to the linearised theory this also grows according to the 'Perry' rule since the classical buckling load Λ is the same for each of these modes. (Note, however, that this represents a simplification, since, according to our previous analysis, the 'square' mode cannot grow entirely in isolation from the axisymmetric mode.) We saw in section 15.3 that when an S-surface undergoes a displacement

$$w = w^* \cos(\pi x/l) \cos(\pi y/l),$$

there are induced, in particular, stress resultants

$$N_x = N_y = (w^* E t / 4a) \cos(\pi x/l) \cos(\pi y/l).$$

Consequently, at the centre of an inward-deflecting panel there are induced, on account of the growth of the imperfection,

$$N_x = N_y = -\left(\frac{\Lambda}{1 - \Lambda}\right) \frac{w_0 E t}{4a}.$$

Note that each of N_x and N_y is one-quarter of the corresponding change in the case of an axisymmetric imperfection. Thus, according to the 'local buckling hypothesis' the criterion of buckling (15.57) is given by:

$$N_0 - \left(\frac{\Lambda}{1 - \Lambda}\right) \frac{w_0 E t}{2a} = -\frac{E t h}{3^{\frac{1}{2}} a},$$

which in turn leads to

$$(1 - \Lambda)^2 = \tfrac{1}{2} \times 3^{\frac{1}{2}} \Lambda w_0 / h. \tag{15.61}$$

Again, this expression is of the same form as Koiter's first formula, but the right-hand side is about one-third of the usual value.

Equations (15.60) and (15.61) are plotted in fig. 15.11, together with results from section 15.4.1. The most remarkable feature, of course, is that this crude and simple analysis produces precisely the form of result which we obtained from the full nonlinear analysis. Thus we may conclude that the overall results of the nonlinear analysis are broadly consistent with a view of buckling in terms of the achievement *locally* of a 'classical' buckling con-

15.6 Nonlinear behaviour of the S-surface

dition. In this view the applied loading is 'assisted' by tangential stress resultants in the inward-displacing 'panels'; and these are simply a feature of the way in which the S-surface behaves when imperfections grow.

15.6 Nonlinear behaviour of the S-surface

In the preceding section we focussed attention on the surprisingly large changes of stress in the S-surface which occur when imperfections grow on account of axial loading. In the present section we shall investigate in an approximate fashion another aspect of nonlinear buckling phenomena.

One of the foundations of our treatment of classical buckling of a cylindrical shell was the response of the S-surface to a periodically distributed normal pressure-loading; in section 7.3 we found a linear relationship between this pressure and the corresponding modal displacement. The linearity was partly a consequence of the assumption that displacements were small. In particular the equilibrium equation of the S-surface in the normal direction was

$$p_S = N_y/a. \tag{15.62}$$

On the other hand, in writing down the equilibrium equation (15.19) in our more detailed study we included three extra terms which were implicitly neglected in the linearised analysis. Thus, comparing (15.19) and (15.62) we may write

$$\Delta N_x \frac{\partial^2 w}{\partial x^2} + \Delta N_y \frac{\partial^2 w}{\partial y^2} + 2\Delta N_{xy} \frac{\partial^2 w}{\partial x \partial y} = -\Delta p_S: \tag{15.63}$$

here ΔN represents a tangential-stress resultant in the S-surface on account of changes of Gaussian curvature, and Δp_S is the additional pressure which is included in (15.19) but not in the simpler expression.

For a doubly-periodic waveform of the usual kind it is straightforward to evaluate Δp_S from (15.63) with the help of the expressions of chapter 7. Thus, whereas previously we found that the displacement

$$w = w_0 \sin(\pi x/l) \sin(\pi y/b)$$

required

$$p_S = \frac{Etw_0}{a^2} \left(1 + \frac{l^2}{b^2}\right)^{-2} \sin(\pi x/l) \sin(\pi y/b), \tag{15.64}$$

we now find that an additional term is required, as follows:

$$\Delta p_S = \frac{2Et\pi^2 w_0^2}{ab^2} \left(1 + \frac{l^2}{b^2}\right)^{-2} \left[\sin^2(\pi x/l) \sin^2(\pi y/b) - \cos^2(\pi x/l) \cos^2(\pi y/b)\right]. \tag{15.65}$$

Note in particular that Δp_S is positive in the centre of each panel, i.e. at $x, y = \frac{1}{2}l, \frac{1}{2}b$, and is proportional to w_0^2. Thus in the centre of the outward-deflecting panels the pressure has a larger absolute value than was calculated previously, while in the centre of the inward-deflecting panels the pressure has a smaller absolute value than before. In fact the sense of these changes can be deduced correctly from the term

$$(\Delta) N_y \left(-\frac{1}{a} + \frac{\partial^2 w}{\partial y^2} \right)$$

in (15.19): the key point is simply that the circumferential curvature is slightly larger than before in the outward-deflecting panels and slightly smaller than before in the inward-deflecting panels, so that in the outward-deflecting panels a larger positive pressure is required, while in the inward-deflecting panels the absolute value of the negative pressure is smaller. We may express this result by stating that the *stiffness* of the inward-deflecting panels of the S-surface *decreases* with increasing amplitude, while that of the outward-deflecting panels *increases*.

This feature is in contrast with our earlier view (based on equation (15.62), which ignores changes of curvature) that these two stiffnesses were equal and constant. Furthermore we can see that the nonlinear effects are of precisely the same form as those with which the spring in section 15.2 was endowed, since we have (at least at the *centre* of each panel)

$$p_S = C_1 w_0 + C_2 w_0^2, \qquad (15.66)$$

where C_1, C_2 are constants. Note than an *inward*-directed displacement of the shell corresponds to a *positive* rotation θ of the simple bar/spring system.

In this way we may establish a direct link between the shell on the one hand and the simple system of fig. 15.5 on the other: the key factor is a 'softening' of the inward-deflecting parts of the S-surface.

It is true, of course, that we have cut some corners in order to establish this result. In effect we have isolated certain aspects of the detailed analysis of section 15.4.1; and in particular we have not considered what happens to the various higher harmonics of the problem; and we have also considered only the central point of each panel. But since the effect which we are considering is clearly defined, we shall pursue it further.

We saw earlier that there is a close correspondence between the simple system of section 15.2 and Koiter's formula (15.1) in relation to the reduction in load-carrying capacity on account of the presence of imperfections. All that remains, therefore, is to make an estimate of the amplitude of the geometric imperfection in the shell which is required in order to reduce the load-carrying capacity to, say, one-half of the classical value: thus we need to determine the deflection at which the second-order stiffness of the S-surface contributes an amount equal to one-eighth of the first-order stiffness (cf. (15.4)).

At this point we recall that in the classical analysis according to the 'three-surface' idea, the normal pressure required for the equilibrium of the compressed 'funicular' membrane surface is provided partly by the S-surface and partly by the B-surface. So far we have confined attention to the S-surface because it alone suffers from strong nonlinear effects. But we recall from section 14.3.2 that for every one of the numerous classical modes, each of the B- and S-surfaces had precisely equal stiffness. It follows that in relation to *any* of the classical buckling modes we need only work out the deflection for which the second-order term from the S-surface contributes one-quarter of the first-order pressure of the S-surface at the centre of a typical panel; and this will be equal to the required amplitude of initial imperfection. From (15.64) and (15.65) we have

$$2Et\pi^2 w_0^2/ab^2 = \tfrac{1}{4} Etw_0/a^2 ,$$

and hence

$$w_0 = b^2/8\pi^2 a. \tag{15.67}$$

It follows directly from this equation that out of all of the classical modes, *the one with the smallest value of b is most imperfection-sensitive.* From section 14.3.3 we find that the 'square' mode (having $l = b$) has the smallest value of b, with

$$b = (2^{\tfrac{1}{2}}\pi/3^{\tfrac{1}{4}})(ah)^{\tfrac{1}{2}}.$$

Putting this into (15.67) we obtain, finally,

$$w_0/h = \tfrac{1}{4}/3^{\tfrac{1}{2}} \approx 0.14, \tag{15.68}$$

which lies close to curve A of fig. 15.11 when $\Lambda = 0.5$.

Thus, in spite of obvious points of possible criticism, this analysis produces a result which is in remarkably good agreement with Koiter's first formula.

15.7 Discussion

In section 15.4 we presented a nonlinear analysis of the behaviour of the shell in the presence of imperfections in the form of two of the classical modes, and studied in detail the interaction between them and the resulting imperfection-sensitivity of the structure. Then in sections 15.5 and 15.6 respectively we investigated two different simplified views of the analysis, each of which produced independently Koiter's first formula, albeit with somewhat different values of the constant. The main point about these two simplified analyses is that they enable us to concentrate on one particular effect, to ignore a variety of others, yet nevertheless to obtain a result of the appropriate form: thus they provide useful ways of thinking 'physically' about a difficult problem.

Thus, in the first method we assume that imperfections grow with applied load according to the Perry rule (which is certainly an oversimplification in the case of an asymmetric mode) but we concentrate on the changes of *tangential-stress resultant* when the S-surface deforms. The hypothesis that buckling occurs when the (total) tangential-stress resultant *locally* reaches the classical buckling value is, in general, successful.

In the second method we also acknowledge the changes in tangential-stress resultants on account of deformation of the S-surface; but on this occasion we interpret them differently, in terms of a nonlinearity in the response of the S-surface to doubly-periodic normal pressure-loading. By ignoring the fact (discovered in section 15.4.1) that the growth of one mode induces effects in a variety of other modes, and treating the shell as a simple system with one degree of freedom, we can immediately make use of some simple results which are easily obtained for a bar supported by a nonlinear spring (section 15.2).

A fundamental feature of scientific method is the isolation of 'important' effects by a conscious overlooking of less important effects. An interesting aspect of the present work is that our two distinct schemes of simplification lead to practically the same result.

Now it may be argued that if we can solve a problem 'properly' (i.e. as in section 15.4) then it is rather pointless to produce simplified views of the same analyses, since they will necessarily be less complete than the original. In this connection it is important to realise that the 'complete' analysis of section 15.4 itself contains at least two highly questionable features. Firstly, no attempt is made to satisfy 'practical' boundary conditions (i.e. those which actually occur in real shells); and secondly the shell under examination is initially furnished with uniform regular, periodic imperfections of rather special wavelength and phase, which are most unlikely to be imparted in fact to any real shell by any known manufacturing process. The conventional defence of these two features is, of course, well known. It is that since the key result concerns the sensitivity of the structure to the presence of imperfections, it is of course necessary to build some sort of postulated pattern of imperfections into the analysis; and the two simple periodic forms are in fact precisely those for which the imperfection-sensitivity is greatest. Moreover, since imperfections of even the smallest amplitudes can have a significant ef-

15.7 Discussion

fect, it is reasonable to suppose that in any actual pattern of unavoidable imperfections in a real shell there will be a nonzero Fourier coefficient corresponding to one or other of these specially critical modal forms. As for the objection about boundary conditions; it is well known that attempts to explain the buckling behaviour of thin cylindrical shells in compression with reference to real boundary conditions but *without* imperfections have not reproduced the three distinctive experimental features listed in section 15.1. Nevertheless, since particularly short shells ($L \approx (ah)^{\frac{1}{2}}$) do not show these three characteristic features (see fig. 15.1) it has to be acknowledged that the non-linear theory is not satisfactory in this range. But this is hardly surprising, since the fixed ends of a shell of this sort would clearly inhibit the growth of imperfections having the special mode forms.

In the area of this general debate, our two distinct simple approaches make some interesting contributions. Specific points in relation to the first method (section 15.5) may be listed as follows.

(i) Imperfection-sensitivity is sometimes a consequence of mode-interaction effects. The term $(1 - \Lambda)^2$ in (15.61) receives one factor $(1 - \Lambda)$ from the denominator of the Perry 'imperfection amplification factor', and another from the fact that the shortfall $1 - \Lambda$ of the applied load below the classical buckling load has to be made up by distortion of the S-surface. The coincidence of classical buckling loads for two distinct modes gives the two identical factors of $(1 - \Lambda)$. Historically the first discussion of imperfection-sensitive buckling was given by Ayrton & Perry (1886) in the context of *plastic* buckling of initially curved uniform pin-ended struts. Their idea was that in an initially stress-free strut, the buckling load would not exceed significantly the load at which 'first yield' occurred in the 'extreme fibre' of the central cross-section. In particular they discovered that the buckling load is most sensitive to the presence of imperfections when the Euler load for the column is equal to the product of yield stress and cross-sectional area; and in these circumstances a formula directly analogous to (15.1) emerges, as may be demonstrated easily. The imperfection-sensitivity is due here to an interaction between Euler buckling and plastic squashing; and it is particularly sensitive when the corresponding loads are equal.

(ii) Although the Perry approach to buckling of columns in the plastic range is often disregarded in favour of the 'tangent-modulus' approach (which seeks bifurcation conditions in initially perfect structures: see Shanley (1947); Sewell (1972)), it clearly has important advantages in relation to the buckling of *imperfect* structures, particularly in conditions when such buckling is imperfection-sensitive: see Hutchinson (1974). Our simplified scheme thus affords the possibility of performing relatively simple analyses of buckling of shells in the plastic range; and it leads us to expect strong imperfection-sensitivity in some circumstances.

(iii) It is the *inward-deflecting* parts of the shell which first reach the classical buckling condition locally, by virtue of features of the behaviour of the S-surface pointed out in section 15.5. This is true for both axisymmetric and asymmetric modes.

(iv) In all of our studies so far we have assumed that the *imperfections* have been of a purely *geometric* kind, and in particular that the imperfect structure is initially stress-free. Imperfections produced by manufacturing procedures or accidental damage are unlikely to leave a shell in a perfectly stress-free condition. For example, a localised inward-directed blow which gives some plastic deformation and leaves behind a small dent will in general leave compressive tangential stresses at the centre of the dent, whose value is perhaps about one-fifth of the stress level for classical buckling. The present method suggests that such initial stresses will contribute to the reaching of a local buckling condition, and possibly to premature failure of the structure.

Next we give a list of features of the buckling of shells which are emphasised by the second simple method (section 15.6).

(i) Again the significance of *inward* deflection emerges; this time in relation to the stiffness of the S-surface, which decreases for inward deflection. This point is closely related to (iii) above, as we remarked earlier.

(ii) Imperfection-sensitivity does not necessarily involve strong mode-interaction. This may appear to be rather a bold claim on the basis of an analysis which idealises the system at the outset as having a single degree of freedom. But recall that in the first method we analysed buckling in the square mode more-or-less successfully by calculating the condition for local buckling at the centre of the inward-deflecting panel in the same mode. In other words we found that the term $(1 - \Lambda)^2$ occurred through a sort of interaction of a mode *with itself*. We did not mention this point in (i) of the previous list, because it seems on balance more satisfactory to think of this aspect in relation to a system having a nonlinear spring. It is important to realise that imperfection-sensitivity can occur in shells which do not have competing classical modes at exactly the same buckling load. Thus, Budiansky & Amazigo (1968) and Budiansky (1969) have shown by application of Koiter's general theory that in some circumstances the initial post-buckling of cylindrical shells under external pressure and torsion, respectively can be strongly unstable; but in neither case does the classical theory (sections 14.4; 14.7) indicate competing modes. We shall investigate another example of this in the following section.

(iii) The deleterious 'nonlinear spring' effect in the S-surfaces is most strong

15.7 Discussion

for modes having the smallest half wavelength b in the circumferential direction. This remark suggests that imperfection-sensitive buckling in cylindrical shells should be predictable from a study of the circumferential half wavelength of the preferred *classical* mode for any given type of loading. And indeed this correlation is remarkably good. We have already dealt at length with the case of pure axial compression. In the cases both of external pressure-loading and torsion, the classical analysis of chapter 14 shows clearly that the critical mode involves a single half wavelength in the axial direction (at least, for 'conventional' end-conditions), and that as the length parameter $\Omega \; (= L h^{\frac{1}{2}}/a^{\frac{3}{2}})$ is decreased, the circumferential wavenumber n increases as $\Omega^{-\frac{1}{2}}$: see figs. 14.14 and 14.21a. The circumferential half wavelength continues to fall until the 'flat-plate' regime is reached in the region of $L/(ah)^{\frac{1}{2}} \approx 1$ or 2; and in this regime the S-surface plays a relatively minor part. We thus expect that the strongest imperfection-sensitivity will occur in these cases in the region of the 'crossover' from the 'flat-plate' to the full shell behaviour. This is precisely what is found in the results of Budiansky cited above; and moreover the way in which the imperfection-sensitivity decreases as the length of the shell moves away from this crossover region follows qualitatively the increase of the circumferential half wavelength b with the length of the shell. The specimens for which Windenberg & Trilling (1934) found particularly low experimental hydrostatic buckling pressures (see section 14.8) lay precisely in the range of $L/(ah)^{\frac{1}{2}}$ for which the value of b is minimum: $3 < L/(ah)^{\frac{1}{2}} < 9$. This may be seen clearly in fig. 14.12, curve $N_y/N_x = 2$: recall that $\xi \approx 2.4(ah)^{\frac{1}{2}}/l$ and $\eta \approx 2.4(ah)^{\frac{1}{2}}/b$. The minimum value of b in this case is actually smaller than that for the 'square' mode under pure axial loading, which corresponds to the highest point of the Koiter circle in fig. 14.12; but the corresponding extra strength of the nonlinearity of the S-surface is more than counterbalanced by the fact that in the region lying outside the Koiter circle the basic S-surface stiffness is less than that of the B-surface. Note also that in unidirectionally reinforced shells with either longitudinal or circumferential stiffeners (see fig. 14.24) the minimum values of circumferential half wavelength b are larger than for unreinforced shells. This may well provide at least a partial explanation of Singer's (1969) claim that stiffened shells are less prone to nonlinear buckling than unstiffened ones.

We shall consider the experimental behaviour of thin cylindrical shells under torsional loading in section 15.9.

In this connection it is also worth noting that the highly unstable buckling of complete spherical shells under external pressure (Thompson, 1962; Hutchinson, 1967; Koiter, 1969b) correlates with modes of classical buckling in which the S-surface behaviour is highly nonlinear on account of small half wavelengths.

(iv) It seems likely that the presence of a small inward-directed *localised* imperfection will be almost as severe as that of a *periodic* imperfection. Simple studies suggest that the geometric imperfection in a shell after a localised inward-directed blow has been applied, causing localised plastic deformation, will have superficial dimensions of order $(ah)^{\frac{1}{2}}$. Koiter (1978) has recently applied his original method to 'more-or-less localised' periodic imperfections, and found an imperfection-sensitivity of the same order as that which occurs in the presence of periodic imperfections.

15.8 Axial buckling of pressurised cylindrical shells

So far in this chapter we have been concerned almost exclusively with the buckling of uniform cylindrical shells under purely axial compressive loading. In this section we shall extend the analysis to a shell which also contains an internal pressure. This is an important technological problem in the operation of large rockets: interior pressure can improve substantially the structural performance of thin shells. In section 14.4 we studied this problem in the context of classical theory; but now our aim is to investigate the imperfection-sensitivity of the system.

Hutchinson (1965) investigated this problem in relation to geometric imperfections in the form of the classical axisymmetric and 'square' modes, both separately and in combination. Here we shall adopt the same two modes, but we shall tackle the problem by means of the 'local buckling hypothesis' of section 15.5. By comparing our results with those of Hutchinson, we shall be able to test the 'local buckling hypothesis' as a predictive tool more stringently than in the previous examples.

The general arrangement of the shell and its loading are shown schematically in fig. 15.14. The shell is compressed axially by external forces of

Fig. 15.14. Axial compressive loading of a pressurised thin cylindrical shell.

15.8 Axial buckling of pressurised cylindrical shells

intensity F per unit length distributed uniformly around the circumference, and there is also an interior pressure P. The classical analysis shows that the buckling condition is most conveniently expressed in terms of the stress resultants N_x, N_y resulting from the combined loading. Thus if we find in a given case that buckling occurs at $N_x = N_0$, we have the corresponding value of F:

$$F = (-N_0) + \tfrac{1}{2} Pa. \tag{15.69}$$

We expect, of course, the value of N_0 to be negative; so the presence of interior pressure gives a direct boost to the buckling load, other things being equal. The present problem thus concerns the imperfection-sensitivity of the quantity $(-N_0)$ in the presence of interior pressure.

It is convenient to define a dimensionless pressure p as follows:

$$p = 3^{\frac{1}{2}} Pa^2 / Eth. \tag{15.70}$$

Also defining Λ as before (see (15.36)) we find from section 14.4 the following results for classical buckling:

$$\left. \begin{array}{l} \text{axisymmetric mode: } \Lambda = 1 \\ \text{'square' mode: } \quad \Lambda = 1 + p. \end{array} \right\} \tag{15.71}$$

The circumferential tension on account of the interior pressure thus raises the classical buckling load for the case of the 'square' mode; but it has no effect upon the buckling load for the axisymmetric mode.

Consider first the growth of an axisymmetric imperfection. As before, the formula is

$$w = w_0/(1 - \Lambda):$$

the denominator is such that $w \to \infty$ when Λ approaches the classical value. Following the steps of section 15.5, we find that the condition for local buckling in the square mode is now

$$\Lambda + \left(\frac{\Lambda}{1 - \Lambda} \right) \left(\frac{3^{\frac{1}{2}} w_0}{h} \right) = 1 + p.$$

This is the same expression as before, except for the enhanced classical buckling condition for the 'square' mode. Thus we obtain finally

$$(1 + p - \Lambda)(1 - \Lambda) = 3^{\frac{1}{2}} \Lambda w_0/h \tag{15.72}$$

in place of the previous expression. (This expression is formally analogous to the 'Perry' formula for plastic buckling of a column in the general case in which the 'Euler' and 'plastic squash' loads are unequal.)

Next consider the growth of an imperfection in the form of the 'square' mode. In this case we have

$$w = \frac{w_0}{1 + p - \Lambda},\tag{15.73}$$

so that when $p = \Lambda = 0$, $w = w_0$; and $w \to \infty$ when the classical buckling condition is reached. Following the same steps as before, we obtain in this case

$$(1 + p - \Lambda)^2 = 3^{\frac{1}{2}} \Lambda w_0/2h.\tag{15.74}$$

Here the expression $(1 + p - \Lambda)$ occurs twice, since it appears both in the imperfection-growth formula and also in the classical buckling target.

In order to make a comparison with Hutchinson's results, we shall consider in each case an imperfection whose amplitude is such that *in the absence of pressure* the shell buckles at $\Lambda = 0.5$. Hence (15.72) and (15.74) become:

$$(1 + p - \Lambda)(1 - \Lambda) = \tfrac{1}{2}\Lambda,\tag{15.75}$$

$$(1 + p - \Lambda)^2 = \tfrac{1}{2}\Lambda.\tag{15.76}$$

These relationships are plotted in fig. 15.15: it is easy to evaluate p in each case for a given value of Λ. A striking feature is that pressure reduces the imperfection-sensitivity for the 'square' mode much more effectively than for the axisymmetric mode. Fig. 15.15 also shows that the corresponding curves

Fig. 15.15. Comparison of results of the 'local buckling' hypothesis with Hutchinsons's (1965) calculations for the shell of fig. 15.14. The amplitude of each mode of imperfection is chosen so that when $p = 0$, $\Lambda = 0.5$.

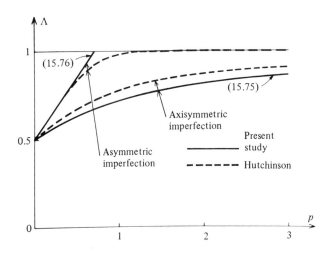

from Hutchinson's paper; the agreement between formulas (15.75) and (15.76) and Hutchinson's calculations is remarkably good, apart from the fact that our formulas do not indicate quite the correct amplitude of the imperfection necessary to give $\Lambda = 0.5$ for $p = 0$, on account of the discrepancies in the values of the constants which are produced by the 'local' buckling' hypothesis.

The physical interpretation of these effects, as pointed out by Hutchinson, is that the interior pressure tends to 'iron out' the 'square wave' imperfections, but to leave the symmetric imperfections untouched: when $\Lambda = 0$ (15.73) indicates that *p reduces* the amplitude of the imperfection below its initial value, with a consequent reduction in the degree of imperfection-sensitivity.

Hutchinson also studied the behaviour of the shell when the two kinds of imperfection are present simultaneously; and he found that for a given value of Λ in the absence of interior pressure, the curve corresponding to a two-mode imperfection lay between the two curves of fig. 15.15. Our present simplified scheme does not afford the possibility of a comparative study in this case, since our expressions for growth of the two modes are independent: by virtue of one of our simplifying assumptions we have ruled out interaction between the modes when the modal amplitudes are being enlarged by the axial load.

15.9 Nonlinear effects in the buckling of cylindrical shells under pure torsion

The nonlinear theory of elastic buckling was developed largely in order to explain the fact that thin cylindrical shells often buckle under axial loads which are much smaller than those which are predicted by the classical theory. Now many experimental studies have been made of the buckling of cylindrical shells under torsional loading; and these have all shown that buckling occurs within about 10%-20% of the loads predicted by classical theory. On the face of it, therefore, nonlinear theory should not be required for this particular loading case. On the other hand, however, Yamaki (1976) did find that the post-buckling behaviour under torsion of some of his specimens was weakly unstable. And, moreover, Budiansky (1969) has shown that for cylindrical shells having certain proportions, the initial post-buckling under torsional loading is highly unstable according to precisely the same kind of theory which was so successful in the elucidation of buckling in the case of axial loading. The paradox thus posed has recently been resolved by Yamaki & Kodama (1980). Their key point is that the simplest form of Koiter's nonlinear analysis involves a severe truncation of the power-series expansion of the total potential energy of the system. The fact that this theory, which is strictly valid asymptotically for small deflections, nevertheless agrees well with a less restrictive version (Koiter, 1963b) for the cylindrical shell under axial loading (as we have seen) does not necessarily imply that there will be equally good agreement for other loading conditions, or indeed for other types of structure (such as cylindrical shells having oval cross-sections (Hutchinson, 1968)). Yamaki & Kodama have shown clearly that, in the case of *torsional* loading the retention of higher-order terms in the power series produces qualitatively different results. In some cases, for example, it turns out that although the *initial* post-buckling of the perfect shell is unstable, the subsequent post-buckling is stable; and hence they conclude that an analysis only of the *initial* post-buckling behaviour can be misleading. (We shall discover an analogous situation in chapter 18 in relation to a perfectly plastic shell which is loaded radially through a rigid boss: see fig. 18.33c.) Hence we can perceive that any *initial* post-buckling analysis – including the work of the present chapter – could, in principle, fail to detect a practically important subsequent stabilising effect. The important lesson which emerges from this example is that while the relatively simple asymptotic nonlinear theory of buckling resolves an important class of problems, it cannot claim to provide a *universal* tool for the satisfactory analysis of all shell-buckling problems. It should always be viewed in the light of relevant experimental evidence.

15.10 Problems

15.1 The spring in fig. 15.5c has unstretched length $2^{\frac{1}{2}}b$ and stiffness k. Its ends are attached to the base and the rigid rod at points distant b from the pivot. Show that the couple C required to produce a rotation θ of the rod is given by

$$C = kb^2 \, (\sin\tfrac{1}{2}\theta - 2\sin\tfrac{1}{4}\theta \sin\tfrac{3}{4}\theta);$$

and hence that for sufficiently small values of θ

$$C \approx \tfrac{1}{2}kb^2 \, (\theta - \tfrac{3}{4}\theta^2).$$

Verify that the magnitude of the term $\tfrac{3}{4}\theta^2$ is equal to $\tfrac{1}{8}\theta$ when $\theta \, (=\theta^*) = \tfrac{1}{6}$; and that in the exact expression the corresponding condition is met when $\theta \approx \tfrac{1}{6} \times 1.002$.

15.2 Consider the stability of equilibrium along the paths which are shown in fig. 15.8a and b. First show that when $\mu \neq 0$ the total potential energy function is given by

$$V = Pl\cos(\theta + \mu) + U \quad (\text{cf. (15.8)});$$

and hence rewrite (15.9) with appropriate additional terms. Verify that for P = constant the condition $dV/d\theta = 0$ gives (15.4) precisely; and that the condition $d^2 V/d\theta^2 > 0$ for stable equilibrium gives

$$1 - \Lambda - 2(\nu/\eta)\theta > 0.$$

Confirm that the condition for Λ to be maximum or minimum with respect to θ (which corresponds to the condition for the quadratic equation (15.11) to have coincident roots) is satisfied by

$$1 - \Lambda - 2(\nu/\eta)\theta = 0;$$

and hence that as θ increases along an equilibrium path through a point of maximum or minimum Λ there is a change from stable to unstable equilibrium. Use this result together with spot-checks to establish the status of the equilibrium along the various paths.

15.3 Donnell (1933) (cf. von Kármán et al., 1940) showed that the (nonlinear) compatibility equation for an initially perfect uniform elastic cylindrical shell may be written as follows using our notation in terms of an Airy stress function ϕ and the radial component of displacement w:

$$\nabla^4 \phi = - Et \left(\frac{1}{a} \frac{\partial^2 w}{\partial x^2} + \frac{\partial^2 w}{\partial x^2} \frac{\partial^2 w}{\partial y^2} - \left(\frac{\partial^2 w}{\partial x \partial y} \right)^2 \right). \tag{i}$$

Show that this corresponds precisely to a combination of (15.18) and (15.25): note that $g = K$ for an initially perfect cylindrical surface.

Equation (i) was originally set up by means of an elimination of displacement components u and v from the following nonlinear strain–displacement relations (which are more accurate than (6.9)), together with Hooke's law (2.5) and the definition of ϕ (7.36):

$$\left.\begin{aligned} \epsilon_x &= \frac{\partial u}{\partial x} + \tfrac{1}{2}\left(\frac{\partial w}{\partial x}\right)^2 \\ \epsilon_y &= \frac{\partial v}{\partial y} + \tfrac{1}{2}\left(\frac{\partial w}{\partial y}\right)^2 + \frac{w}{a} \\ \gamma_{xy} &= \frac{\partial u}{\partial y} + \frac{\partial v}{\partial x} + \frac{\partial w}{\partial x}\frac{\partial w}{\partial y}. \end{aligned}\right\} \tag{ii}$$

Verify the derivation. Note the advantage of the method used in section 15.4, in which the last two terms in (i) are determined by means of a more accurate evaluation of the Gaussian curvature of the surface, and without the need for consideration of the nonlinear strain–displacement equations (ii).

15.4 An approximate analysis of the conditions at maximum compressive load of a cylindrical shell containing an imperfection in the 'square' waveform may be performed by putting $\beta \gg 1$, so that the right-hand side of (15.47) and

(15.48) become respectively, $3\beta^2 \Gamma_0/32$ and $3\Gamma_0 \alpha\beta/2$. Making corresponding changes to (15.49), show that (15.50) still applies, but with $A = 27\beta^2$ and $B = 18\beta^3$; and that at the point of maximum load

$$\beta = 2/(3^{\frac{1}{4}}\Gamma_0^{\frac{1}{2}}) = 1.5/(1 - \Lambda)$$

and

$$\alpha = \beta/(4 \times 3^{\frac{1}{2}}) \approx \beta/7.$$

16

The Brazier effect in the buckling of bent tubes

16.1 Introduction

When a drinking-straw is bent between the fingers into a uniformly curved arc of steadily increasing curvature, there comes a point when the tube suddenly collapses locally and forms a kink. If the experiment is repeated with a fresh straw, and the specimen is observed more carefully, it is found that the cross-section of the entire tube becomes progressively more oval as the curvature increases: and the kink or crease which suddenly forms involves a complete local flattening of the cross-section, which then offers virtually no resistance to bending.

These observations suggest that the buckling of a thin-walled cylindrical shell which is subjected to pure bending involves behaviour which is of a different kind from that which we have encountered in chapters 14 and 15; for we have previously not come across major changes in geometry, spread over the entire shell, before buckling occurs.

The first investigation of this effect was made by Brazier (1927). He showed that when an initially straight tube is bent uniformly, the longitudinal tension and compression which resist the applied bending moment also tend to flatten or ovalise the cross-section. This in turn reduces the flexural stiffness EI of the member as the curvature increases; and Brazier showed that under steadily increasing curvature the bending moment – being the product of curvature and EI – reaches a maximum value. Clearly the structure becomes unstable after the point of maximum bending moment has been passed; and it is therefore not surprising to find that in experiments the tubes 'jump' to a different kind of configuration, which includes a 'kink'. The buckling in this case occurs by the reaching of a maximum load or limit-point. Here, in contrast to the example of fig. 15.8a the limit-point is not associated with an initial imperfection, but with the fact that pure bending of an initially perfect tube is essentially nonlinear in character. Experiments performed carefully by Brazier showed that the bending moments which caused failure of the physical specimens agreed fairly well with the maximum bending moments predicted by his theory.

Brazier was concerned primarily with 'long' tubes, which were 'free' to ovalise. If such ovalisation is prevented – for example by the presence of bulkheads or diaphragms – an elastic tube can still buckle under pure bending moment, but in a different fashion. When ovalisation is thus prevented, the methods of chapters 14 and 15 can be applied. In the context of classical bifurcation buckling theory, Seide & Weingarten (1961) showed that under pure bending – or indeed bending combined with axial loading – buckling occurs at almost exactly the point when the 'extreme fibre' compressive stress, calculated according to the simple theory of bending of elastic beams, reaches the classical buckling stress of the shell under pure axial compression. This extremely simple approximate result is due largely to the fact that classical buckling of a long thin-walled cylindrical shell under pure axial loading (cf. section 14.3) involves doubly-periodic modes whose rectangular 'panels' have circumferential half wavelengths which are small in comparison with the radius of the shell; and so classical buckling can take place even if only a relatively small portion of the shell (in the vicinity of the 'extreme fibre') sustains a longitudinal stress whose level is the region of the classical buckling stress (see (14.23)). We shall refer to this type of buckling by bifurcation as 'local' buckling.

The critical pure bending moment calculated by Seide & Weingarten is almost double Brazier's maximum bending moment, irrespective of the radius/thickness ratio of the shell. This seems to suggest that there are two distinct kinds of buckling of elastic tubes in pure bending associated with 'long' and 'short' tubes, respectively; and thus that perhaps there is an intermediate length at which a crossover takes place. This suggestion is, in fact, wrong; for it turns out on closer examination that *local* buckling plays a part in the buckling of both short and long tubes, and indeed in the buckling of tubes of unlimited length (except possibly in some cases where there is also exterior pressure-loading).

The key to the situation is that the classical local bifurcation-buckling stress of a cylindrical shell in axial compression is given by formula (14.23); and that furthermore, as shown by Hutchinson (1968) this formula is still approximately correct for tubes having smooth non-circular cross-sections provided the *local* radius of curvature is used in place of the radius a of the circular tube. Therefore, when a long tube ovalises progressively under the action of a bending moment, the compressive stress level required for local

16.1 Introduction

buckling steadily falls. As we shall see, detailed calculations show that for long tubes the local buckling condition is reached just before Brazier's maximum bending moment is reached.

There is ample experimental confirmation of this general description. Tests on long elastic tubes performed by Brazier and others show that buckling occurs before the moment-curvature relationship reaches a maximum: see fig. 16.1.

The general approach of the present chapter will be to study the progressive ovalisation of tubes under steadily increasing bending moment, and then to do a check on the point at which the local buckling condition, as described above, is reached. We shall consider unpressurised tubes of various overall lengths, and also long tubes under internal and external pressure. The method relies on the idea that the local buckling condition can be treated as a separate calculation. Several authors, notably Stephens, Starnes & Almroth (1975) and Fabian (1977) have done much more thorough calculations in which the

Fig. 16.1. Bending-moment/curvature relationship for a silicone-rubber tube, determined experimentally by R. J. Anderson (see Reddy, 1977), together with theoretical curve of Brazier (1927). Curvature was determined experimentally from the relative rotation of rigid end-plates. The compliance of the attachment of the tube to the end-plates was about 0.18 of the total compliance in the early stages; and the Brazier curve has accordingly been offset from the broken line A. According to the theory presented in section 16.4, local buckling occurs at the point indicated by the open circle.

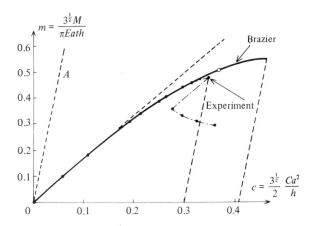

buckling conditions of the deformed tube have been evaluated by means of the general methods of elastic buckling theory, without recourse to a local buckling 'rule' of the kind described above. By comparison with the work of the present chapter, their calculations indicate that the local buckling rule is very satisfactory.

Almost all of the calculations in the present chapter will be based on Brazier's original analysis, and in particular on his rather drastic truncation of a power-series expression for the total strain energy of the system. One point which emerges clearly from Fabian's work – which was done in the context of 'large deflections' – is that Brazier's calculations agree with the much more accurate analysis surprisingly well, and certainly much better than might be expected in view of the crude approximations which Brazier used. This suggests that work along Brazier's lines will be satisfactory for most purposes: but at the end of the chapter we shall investigate Brazier's simplification in some detail and explain why it happens to work so well in practice.

The layout of the chapter is as follows. First we explore several aspects of the bending of simple two-flange beams, in order to demonstrate various mechanical and computational features free from complications connected with shell theory. Then, in section 16.3, we repeat Brazier's original calculation, and extend it by adding the 'local buckling' condition. In sections 16.5 and 16.6 we examine the consequences of having interior or exterior pressure in combination with bending moment, and the effects of finite length. Lastly, as already mentioned, we investigate some points concerning the computational accuracy of Brazier's calculation.

The chapter is concerned entirely with behaviour which occurs within the elastic range of the material. The ovalisation and local buckling of tubes bent into the plastic range are important aspects of the behaviour of submarine pipelines; however they are beyond the scope of this book. But see Reddy (1979) and Gellin (1980).

16.2   The Brazier effect in a simple beam

When an initially straight tube of circular cross-section is bent into a circular arc, there is a tendency for the cross-section to 'ovalise' or 'flatten'. The easiest way to grasp the factors which are involved in this process is to study the behaviour of the simple special beam shown in fig. 16.2*a* when it is subjected to a pure bending moment. The beam consists of two discrete flanges, which are separated by a series of closely-spaced springs. Initially the beam is straight, as shown in fig. 16.2*a*, and the entire arrangement is stress-free. When a uniform bending moment is applied, as shown in fig. 16.2*b*, one flange is put into tension and the other into compression. The flanges deform

16.2 The Brazier effect in a simple beam

according to Hooke's law, and since we shall suppose that originally plane sections of the beam remain plane and perpendicular to the curved axis – which follows, of course, by symmetry for uniform stable bending – the differential tensile strain in the flanges gives a uniform curvature to the beam. The two flanges are thus curved members in states of tension and compression, respectively; and elementary considerations of equilibrium demand that the transverse springs are in a state of compression. Now if these springs happen to be infinitely stiff they do not change in length: there is then no change in the cross-sectional dimensions of the beam, and we recover in fact the classical formulae for the bending of an elastic beam. But on the other hand, if these springs are relatively 'soft', the distance between the flanges decreases, and this in turn gives an increase in curvature to the beam for a given applied bending moment. This beam is, of course, somewhat similar to that which we investigated in section 13.3: but there is an important difference in that the earlier beam was initially *curved*, whereas the present beam is initially *straight*, and only becomes curved on account of the applied bending moment.

Let us do a general analysis of this relatively simple problem by considering the equations governing the behaviour of the arrangement in the deformed configuration.

In the original configuration (fig. 16.3a), each flange has cross-sectional area $\frac{1}{2}A$ and is made of material having Young's modulus E. The individual flanges as such offer no resistance to bending, but local buckling of the flanges is not permitted. The original separation of the flanges is H, and we shall denote the current separation by

$$(1-\zeta)H, \tag{16.1}$$

where ζ is a dimensionless measure of the change in distance between the flanges: see fig. 16.3b. The transverse springs occupying unit length in the original configuration have stiffness k, and we shall regard these as being uniformly distributed along the length of the beam.

Fig. 16.2. Two-flange beam used for initial study of the Brazier effect. (a) Under zero bending moment. (b) Under moment M, showing curvature and compression of transverse springs.

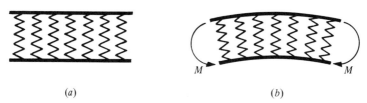

(a)   (b)

Let the tensile stress and strain in the outer flange be σ and ϵ, respectively. Since there is no net tension in the member as a whole, the tensile stress in the inner flange has the same magnitude, but opposite sign. Here we are assuming that the stresses and strains are sufficiently small for the usual small-strain assumptions to be valid. It will be sufficient to consider the equations in relation to only one flange, since the conditions in the two flanges are similar apart from differences in sign. Note that the plane midway between the flanges does not change in length: the 'neutral axis' remains central in the deformed configuration.

Finally, let the applied bending moment be M, and let the resulting curvature of the centre-line of the beam be C.

There are two equilibrium equations. The first expresses the statical equivalence between the tensile and compressive stresses in the respective flanges of the deformed cross-section, and the applied bending moment:

$$M = \tfrac{1}{2}\sigma A H (1 - \zeta). \tag{16.2}$$

The second relates the compressive force f in the transverse springs per unit length of centre-line to the tensile stress in a flange, just as for a curved string:

$$f = \tfrac{1}{2}\sigma A C. \tag{16.3}$$

An equation of geometric compatibility connects the curvature C and strain ϵ as follows (cf. section 2.3.2):

$$\epsilon = \tfrac{1}{2}CH(1 - \zeta). \tag{16.4}$$

Lastly, we write Hooke's law for the flanges and transverse springs, respectively:

$$\sigma = E\epsilon \tag{16.5}$$

$$f = kH\zeta. \tag{16.6}$$

Note that in deriving (16.2) and (16.4) we have used the *current* separation of the flanges.

The five equations (16.2)-(16.6) together describe the mechanics of the

Fig. 16.3. Cross-section of the two flange beam in (*a*) original and (*b*) distorted configurations.

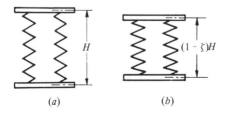

system. Our aim is to find a relation between M and C, by eliminating f, σ, ϵ and ζ from the equations. First, by eliminating f from (16.3) and (16.6), and ϵ from (16.4) and (16.5) we obtain

$$\sigma C/\zeta = 2kH/A, \qquad (16.7)$$

$$C(1 - \zeta)/\sigma = 2/EH. \qquad (16.8)$$

Also, from (16.2) we have

$$\sigma(1 - \zeta)/M = 2/AH. \qquad (16.9)$$

In these equations all of the variables are on the left-hand side. Eliminating σ between (16.8) and (16.9) we find

$$M = \tfrac{1}{4}CEAH^2 (1 - \zeta)^2. \qquad (16.10)$$

Also, by eliminating σ between (16.7) and (16.8) we have

$$1/(1 - \zeta) = 1 + (AEC^2/4k). \qquad (16.11)$$

Hence, by substituting for $(1 - \zeta)$ from (16.11) in (16.10) we may express M in terms of C. At this point it is convenient to define a dimensionless curvature c as follows:

$$c = \tfrac{1}{2}C(AE/k)^{\tfrac{1}{2}}. \qquad (16.12)$$

In terms of this measure of curvature we thus obtain, finally,

$$M = [\tfrac{1}{2}H^2 (kEA)^{\tfrac{1}{2}}] c/(1 + c^2)^2. \qquad (16.13)$$

This is plotted in fig. 16.4 (broken curve). For small values of c the term $(1 + c^2)$

Fig. 16.4. Bending moment and 'flattening' (ζ) plotted against curvature for the two-flange beam, according to a complete analysis and an approximate energy analysis using a truncated energy function, respectively.

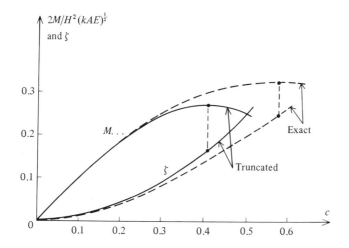

differs from 1 by a negligible amount, and so the relation between M and c is almost linear. If the value of k is sufficiently large in relation to AE, the elastic limit (or else a yield or fracture criterion) in the flanges may well be reached before there has been an appreciable change in thickness of the beam: and in these circumstances the classical analysis, which neglects changes of thickness, is adequate. However, when the value of k is sufficiently small, the relationship between bending moment and curvature is nonlinear, and it is easy to show that the value of M reaches a *maximum* when $c = 3^{-\frac{1}{2}} \approx 0.577$. Fig. 16.4 shows also the relationship

$$\zeta = c^2/(1 + c^2) \tag{16.14}$$

which may be obtained by rearrangement of (16.11). It follows that the point where M is maximum is characterised by

$$\zeta = \tfrac{1}{4}. \tag{16.15}$$

In other words, the maximum value of M is reached when the thickness of the beam is equal to $\tfrac{3}{4}$ of its original value, *whatever the values of H, A, E and k*. The maximum value of $c/(1 + c^2)^2$ is equal to $3^{\frac{3}{2}}/16$, and it follows that

$$M_{\max} = 0.162 \, (kEA)^{\frac{1}{2}} H^2. \tag{16.16}$$

Thus $M_{\max}$ is proportional to the geometric mean of the two stiffnesses E and k, other things being equal.

The above analysis brings out clearly the factors which work to compress the cross-section, thereby altering the geometry of the cross-section in relation to its resistance to bending moment, and eventually giving a distinct *maximum* to the bending moment which can be sustained. Although the material is all linear-elastic, the analysis as a whole is essentially nonlinear: all of the equations are written in terms of the *current* configuration, which is (particularly in the later stages) significantly different from the original configuration.

As we shall see later, the behaviour of a thin-walled tube of circular cross-section is rather similar to that of the simple two-flange beam in conditions of pure bending. The tension and compression on the opposite sides of the tube tend to compress the cross-section; but the elastic resistance to this compression is provided not by a transverse spring, as in our simple model, but by the resistance of cross-sectional 'rings' of the tube to distortion into an oval configuration.

There is, however, one important respect in which the simple beam is not an adequate model for the behaviour of a tube. As we have mentioned already, the strength in bending of a real tube is not usually governed by a *maximum* in the bending moment, as in fig. 16.4, but by the reaching of a *local* bifurcation condition on the compressive side of the tube. There is nothing equiv-

16.2 The Brazier effect in a simple beam

alent to this in our simple model. Mainly on account of this feature we have not defined a dimensionless bending-moment variable for our model beam.

Before we proceed to an analysis of the bending of a tube, we shall consider an 'energy' analysis of our beam. This will suggest useful lines of attack on the more complicated problem of the bending of a tube in later sections.

16.2.1 An energy method

The classical formula for linear-elastic bending of an elastic beam is

$$M = CEI, \tag{16.17}$$

where I is the second moment of area of the cross-section about the relevant centroidal axis, and the symbols M, C and E have the same meanings as before. If we compare this with (16.10) we find that

$$I = \tfrac{1}{4}AH^2 (1 - \zeta)^2, \tag{16.18}$$

and the expression on the right-hand side is indeed the second moment of area of the cross-section about the relevant centroidal axis *in its current (deformed) configuration*. The moment-curvature relation can thus be determined easily provided we can find an expression for the amount of transverse contraction, ζ. Let us try to do this by use of the *strain-energy theorem*: see appendix 1.

Consider a unit length of the beam, which is constrained to adopt a curvature C. The strain energy of stretching of the two flanges is equal to $\tfrac{1}{2}C^2EI(\zeta)$, where $I(\zeta)$ is the current second moment of area, expressed by (16.18) as a function of ζ: as far as the strain energy of stretching of the flanges is concerned, the situation is exactly that of a simple beam having the correct current separation of the flanges. But there is also, in addition, some strain energy of compression of the transverse springs, which is equal to $\tfrac{1}{2}kH^2\zeta^2$ per unit length of beam. Thus the total strain energy, U, of the section is given by

$$U = \tfrac{1}{8}C^2E\,AH^2 (1 - \zeta)^2 + \tfrac{1}{2}kH^2 \zeta^2. \tag{16.19}$$

This expression is a function of the curvature C and the transverse contraction ζ. By the theorem of minimum strain energy the value of ζ will be determined by the condition

$$\frac{\partial U}{\partial \zeta} = 0. \tag{16.20}$$

This gives

$$C^2EA (1 - \zeta) = 4k\zeta, \tag{16.21}$$

which agrees precisely with the previous result (16.11). Equation (16.21) may be solved for ζ, and the resulting expression substituted into (16.19) to give an expression for U in terms of the single variable C. We can then obtain an expression for M by means of the relation

$$M = \frac{dU}{dC}, \tag{16.22}$$

and it is easy to verify that the previous expression (16.13) is recovered exactly.

In his analysis of the bending of a *tube*, Brazier used an analogous strain-energy approach. However, he simplified the expression for strain energy of longitudinal stretching, which is directly analogous to $\frac{1}{2}C^2 EI(\zeta)$ (as we shall see later) by omitting a term analogous to ζ^2 in the first term on the right-hand side of (16.19). Thus, in relation to the present problem, instead of using (16.19) he used, in effect:

$$U = \tfrac{1}{8} C^2 EAH^2 (1 - 2\zeta) + \tfrac{1}{2} kH^2 \zeta^2. \tag{16.23}$$

This, of course, gives a different equation for ζ, and it is easy to verify that the final equation, corresponding to (16.13) is

$$M = \left[\tfrac{1}{2} H^2 (kAE)^{\frac{1}{2}}\right] c(1 - 2c^2). \tag{16.24}$$

This expression is plotted as a full curve in fig. 16.4. Although it represents correctly the first three terms of a Taylor power-series expansion of (16.13), and thus agrees well with it for sufficiently small values of c, the location of the point of maximum bending moment is significantly different, as may be observed: see problem 16.1. Fig. 16.4 also shows the relationship

$$\zeta = c^2, \tag{16.25}$$

which replaces (16.14) when the same simplification is made.

It is clear that in this example the consequences of a truncation of the strain-energy expression are to alter the *details* of the results while preserving the general character of the solution. We shall return to this point in section 16.7.

An expression for bending moment as a function of curvature may be obtained, as above, by using (16.20) in order to rewrite the general expression for U as a function of C alone. Alternatively, we may proceed by using

$$M = \frac{\partial U}{\partial C}$$

in relation to the original expression for U as a function of two variables, and then using $\partial U/\partial \zeta = 0$ as a way of eliminating ζ. The result is precisely the same by either route. An advantage of the above expression over (16.22) is that we recover directly the moment–curvature relationship

$$M = CEI(\zeta).$$

16.3 The Brazier effect in a tube of circular cross-section

When a thin-walled tube of circular cross-section, and originally straight, is subjected to a pure bending moment M, the tensile and compressive longitudinal stresses on opposite sides of the neutral plane combine with the curvature of the axis of the tube to flatten the cross-section into an oval shape. It seems reasonable to assume, as Brazier did, that a typical cross-section deforms according to

$$w = a\zeta \cos 2\theta, \qquad (16.26)$$

where w is the radial component of displacement, a is the radius of the tube and θ is an angular coordinate measured from the neutral plane in the original configuration: see fig. 16.5. The parameter ζ is a dimensionless measure of the flattening of the cross-section across the 'extreme fibres' in bending. Following Brazier further, we shall assume that the cross-section deforms inextensionally in its own plane; thus the circumferential component, v, of displacement is given by

$$\frac{dv}{d\theta} + w = 0$$

(see (6.9b)). Hence we obtain

$$v = -\tfrac{1}{2} a\zeta \sin 2\theta. \qquad (16.27)$$

Here we have set a constant of integration to zero, since we shall suppose that there is no net rotation of the section about the axis of the tube. Note that we are treating the problem as a two-dimensional one, in the plane of a typical cross-section.

The methods of chapter 6 may be used to determine the longitudinal strain as a function of θ for the assumed displacement (16.26), (16.27), together with an overall curvature of the tube; and then the strain energy of longitudinal stretching may be evaluated. However, there is a simpler and more direct way of tackling the problem, which is based on the following physical argument. The strain energy associated with the longitudinal stretching of the tube is a function of the *current configuration* of the shell; and in particular it does not depend upon the *path* by which the current configuration is reached from the initial configuration. Thus, although the ovalisation of the tube actually increases steadily as the curvature increases, we may perform the strain-energy calculation correctly by imagining a different process, involving two distinct stages, which arrives finally at the current configuration. In this alternative process, stage 1 is a uniform ovalisation of the straight tube, and stage 2 is the uniform bending of this tube at a constant degree of ovality. In stage 1, of course, no strain energy of stretching is incurred, since the deformation is *inextensional*. And in stage 2 the calculation is precisely that of the pure bending of a beam having a given oval cross-section. This is perfectly straightforward once we have computed the second moment of area, I, of the cross-section in the deformed configuration. This constitutes our next task; but it is appropriate first to make some remarks.

The fact that the 'stage 1' ovalisation incurs no strain energy of stretching obviously simplifies the calculations. This is true only for *uniform* ovalisation of a *straight* tube. Thus, if the ovalisation were not uniform along the length of the tube there would in general be changes of Gaussian curvature, and some stretching would necessarily occur. We shall encounter this case in section 16.6. Also, the pure ovalisation of an initially *curved* tube (like that of chapter 13) would also require some longitudinal stretching: the present 'short-cut' is not useful for problems of that sort.

Since the evaluation of the second moment of area, I, of the deformed cross-section involves an integration of the square of the perpendicular distance of points on the circumference from the neutral axis, we shall need to calculate η (see fig. 16.5), the component of displacement of a typical point on the cross-section in the direction perpendicular to the axis $\theta = 0$. The calculation is obviously simplified if we make the assumption that the displacements from the original circular configuration are relatively *small*. This we shall do throughout the section, even though it is clear from Brazier's work that radial displacements of the order of $0.2a$ may be indicated by the subsequent calculations. We shall investigate this point in detail in section 16.7.

On the basis of this assumption we may clearly write

$$\eta = w \sin\theta + v \cos\theta; \tag{16.28}$$

so we obtain, from (16.26) and (16.27)

$$\eta = \tfrac{1}{4}a\zeta(-3\sin\theta + \sin 3\theta) = -a\zeta\sin^3\theta. \tag{16.29}$$

Fig. 16.5. Displacement variables for the analysis of (small) distortion of the cross-section of a tube. $\theta = 0$ is the 'neutral axis'.

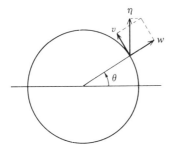

16.3 The Brazier effect in a round tube

(Note that at $\theta = \tfrac{1}{2}\pi$, $\eta = -a\zeta$, as expected.) Now by definition

$$I = \int_0^{2\pi}(a\sin\theta + \eta)^2 \, at \, d\theta, \tag{16.30}$$

where t is the thickness of the wall of the tube. Thus we find, on performing the integration,

$$I = I_0\left(1 - \tfrac{3}{2}\zeta + \tfrac{5}{8}\zeta^2\right), \tag{16.31}$$

where

$$I_0 = \pi a^3 t: \tag{16.32}$$

I_0 is the value of I for the cross-section in its *original* configuration.

Our next task is to evaluate the strain energy of bending of unit length of the tube, as a ring, into its deformed cross-sectioned shape. From (6.9e) we have the following expression for the change of curvature, κ, in the circumferential direction:

$$\kappa = -\left(\frac{1}{a^2}\right)\left(w + \frac{d^2 w}{d\theta^2}\right) = \left(\frac{3\zeta}{a}\right)\cos 2\theta. \tag{16.33}$$

On the assumption that there is, by comparison, a negligible change of curvature in the longitudinal direction, we find by integration around the cross-section that the strain energy of bending of a unit length of tube is equal to

$$\tfrac{3}{8}\pi E t h^2 \, \zeta^2/a. \tag{16.34}$$

Here we have expressed the flexural rigidity D of an element in terms of t and h, as usual.

Thus we may write down the following expression for the total strain energy per unit length, U, of distortion of a tube whose axis has been deformed into an arc of curvature C:

$$U = \tfrac{1}{2}C^2 E\pi a^3 t\left(1 - \tfrac{3}{2}\zeta + \left[\tfrac{5}{8}\zeta^2\right]\right) + \tfrac{3}{8}\pi E t h^2 \zeta^2/a. \tag{16.35}$$

In this expression the first term on the right-hand side represents the strain energy of longitudinal stretching, calculated via the second moment of area of the deformed cross-section, and the second term represents the strain energy of circumferential bending.

We now proceed, exactly as in the problem of the beam, to find the optimum value of ζ for a given value of C from the condition $\partial U/\partial \zeta = 0$, and then to obtain an expression for M from $M = dU/dC$.

As we have said, we shall also use Brazier's truncation of the expression, by omitting the ζ^2 term enclosed in square brackets in (16.35). Putting $\partial U/\partial \zeta = 0$ we obtain

$$\zeta = \tfrac{4}{3}c^2, \tag{16.36}$$

where

$$c = \tfrac{1}{2} \times 3^{\tfrac{1}{2}} Ca^2/h. \tag{16.37}$$

The dimensionless curvature c of the tube has here been defined in a way which will be most convenient for later purposes. We may now use (16.36) and (16.37) in order to express U as a function of c alone:

$$U = \tfrac{2}{3}\pi(Eth^2/a)(c^2 - c^4). \tag{16.38}$$

Hence we have

$$M = \frac{dU}{dC} = \frac{dU}{dc}\frac{c}{C} = \frac{2\pi}{3^{\frac{1}{2}}}Eath(c - 2c^3). \tag{16.39}$$

Clearly M reaches a maximum value when $(c - 2c^3)$ is maximum, which occurs when $c = 6^{-\frac{1}{2}} = 0.408$; thus

$$M_{\max} = 2^{\frac{3}{2}}\pi Eath/9 = 0.987\ Eath. \tag{16.40}$$

Also, with this value of c,

$$\zeta = \tfrac{2}{9}. \tag{16.41}$$

Expressions (16.36) and (16.39) are exactly the same as those found by Brazier (1927). It is of course remarkable that the amount of flattening ζ corresponding to the maximum value of M is independent of any of the parameters defining the tube; but this should not be too surprising in view of the corresponding result for the simple beam: cf. (16.15)

It is obviously desirable to define a dimensionless bending moment m, say. It turns out to be convenient to normalise M with respect to a 'classical' critical bending moment M_{cr} for buckling of the tube, which is obtained by assuming that the cross-section remains circular and then setting the compressive stress at the 'extreme fibre' equal to the stress level corresponding to classical buckling of the tube in uniform axial compression. From (14.23) we have

$$\sigma_{cr}\ (= N_x/t) = -Eh/3^{\frac{1}{2}}a \tag{16.42}$$

at the extreme fibre, and thus by elementary beam theory we obtain

$$M_{cr} = \left(\pi/3^{\frac{1}{2}}\right)Eath: \tag{16.43}$$

hence

$$m = M/M_{cr} = 2c(1 - 2c^2). \tag{16.44}$$

In particular,

$$m_{\max} = \left(\tfrac{2}{3}\right)^{\frac{3}{2}} = 0.544. \tag{16.45}$$

Equation (16.44) is plotted as a solid curve in fig. 16.6. According to Brazier therefore, the maximum bending moment is a little more than one-half of the 'classical' bending moment, irrespective of the properties of the tube. It is not obvious at the outset that the 'classical' moment is suitable as a normalising factor; and in making this step we have anticipated a later result.

16.4 Local buckling

In this section we shall supplement our previous calculations by incorporating a simple condition of local bifurcation buckling, as mentioned in section 16.1. The work of Seide & Weingarten (1961) (cf. Reddy & Calladine, 1978), together with that of Hutchinson (1968) suggests that we should use as a criterion of local buckling a modified form of (16.42), namely,

$$\sigma_{cr} = -Eh/3^{\frac{1}{2}}\rho, \tag{16.46}$$

where ρ is the radius of curvature of the deformed cross-section at the point where the compressive stress is greatest. It is convenient to define a dimensionless stress s by normalising σ with respect to the classical buckling stress σ_{cr} for a circular tube (16.42):

$$s = \sigma/\sigma_{cr}. \tag{16.47}$$

Thus the critical value of s is given, in general, by

$$s_{cr} = a/\rho. \tag{16.48}$$

Fig. 16.6. Dimensionless plot of bending moment (m), flattening (ζ), 'extreme fibre' stress (s) and critical stress for local buckling (s_{cr}) against curvature (c) for a tube. Full curves correspond to an approximate analysis after Brazier, while broken curves show the results of the more accurate analysis of section 16.7. In either case the values of m and ζ for local buckling are obtained by projection from the point $s = s_{cr}$.

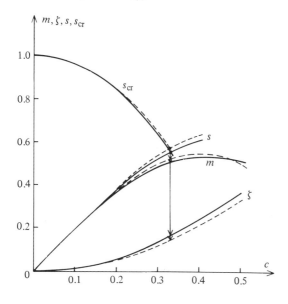

(Note that the signs have been arranged here so that s is positive on the compression side of the tube.) Now the original curvature of the cross-section was $1/a$, and the change of curvature at the location $\theta = \frac{1}{2}\pi$ is, by (16.33), $-3\zeta/a$. For sufficiently small values of ζ it follows that

$$\rho = a/(1 - 3\zeta), \qquad (16.49)$$

and thus that

$$s_{cr} = 1 - 3\zeta. \qquad (16.50)$$

Hence, by (16.36)

$$s_{cr} = 1 - 4c^2. \qquad (16.51)$$

Next we must obtain an expression for the 'extreme fibre' compressive stress in the tube. The simplest approach is to note that the corresponding strain is equal to

$$Ca(1 - \zeta), \qquad (16.52)$$

by elementary beam theory. Multiplying this strain by E, and taking account of the various dimensionless quantities, we obtain the following general expression for the dimensionless 'extreme fibre' stress:

$$s = 2c(1 - \zeta). \qquad (16.53)$$

Substituting for ζ from (16.36) we obtain

$$s = 2c(1 - \tfrac{4}{3}c^2). \qquad (16.54)$$

Fig. 16.6 shows graphs of m, s and s_{cr} against c, using expressions (16.44), (16.54) and (16.51). Local buckling occurs when $s = s_{cr}$, which corresponds to the intersection of the two full curves. From this diagram we see that although the local buckling condition for the tube in its original circular cross-section is reached at a bending moment of almost *twice* the maximum value according to Brazier's analysis, nevertheless the ovalisation of the tube in the course of bending increases the local radius of curvature on the compressive side to such an extent that the local buckling condition is reached *before* Brazier's maximum bending moment.

The graphical procedure of fig. 16.6 for finding the local buckling condition is straightforward. But for subsequent purposes it will be more convenient to use a different diagram in which the three qualities s, s_{cr} and m are each plotted as functions of ζ. It will also be useful to work throughout in dimensionless terms. Thus, we define a dimensionless strain energy u by

$$u = \tfrac{3}{2}aU/\pi Eth^2, \qquad (16.55)$$

so that the dimensionless form of (16.22) is

$$m = \frac{du}{dc}.$$

16.4 Local buckling

Thus, in place of the truncated form of (16.35) we have

$$u = c^2\left(1 - \tfrac{3}{2}\zeta\right) + \tfrac{9}{16}\zeta^2. \tag{16.56}$$

The condition $\partial u/\partial \zeta = 0$ gives

$$c = 3^{\tfrac{1}{2}}\zeta^{\tfrac{1}{2}}/2. \tag{16.57}$$

On this occasion we do not express u as a function of a single variable, but use instead

$$m = \frac{\partial u}{\partial c} = 2c\left(1 - \tfrac{3}{2}\zeta\right). \tag{16.58}$$

Also, as before (see (16.53) and (16.50)),

$$s = 2c(1 - \zeta)$$

$$s_{cr} = 1 - 3\zeta.$$

By means of these expressions we may evaluate c, m, s and s_{cr} as functions of ζ. This has been done in fig. 16.7a. The intersection of the curves s and s_{cr} gives the value of ζ at which local buckling occurs; and we may then read off the corresponding values of m and c. A graph of m against c may be obtained by plotting pairs of values for a sequence of values of ζ.

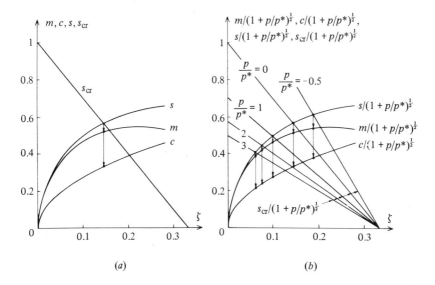

Fig. 16.7. (*a*) Replot of fig. 16.6 showing m, c, s and s_{cr} as functions of ζ, according to Brazier's scheme of approximation. (*b*) An adaptation of (*a*) which takes account of internal pressure.

16.5 The effect of interior pressure

Let us now investigate what changes are made to the preceding analysis when the tube is subjected to interior (or exterior) pressure. It seems intuitively obvious that the presence of an interior pressure will raise both the maximum bending moment, since the pressure will provide extra resistance of the cross-section to ovalisation, and also the local buckling stress, since it will provide an additional axial tension to be overcome before buckling takes place.

Consider first the effect of interior pressure on the Brazier analysis. To the energy expression (16.56) should be added a term equal to the interior pressure multiplied by the change in cross-sectional area of the tube on account of the ovalisation. This latter quantity is proportional to ζ^2, to the present level of approximation, since the circle is the figure which maximises the area of a curve of given perimeter. Fortunately it is not necessary to work out these details geometrically. We have seen already in section 13.5 that the presence of interior pressure effectively increases the circumferential bending stiffness by the factor $1 + p/p_2^*$, where

$$p_2^* = Eth^2/4a^3 \qquad (16.59)$$

for displacements in the mode $n = 2$.

Thus we may account for this effect by increasing the term $\frac{9}{16}\zeta^2$ in (16.56) by this same factor. On following through the same steps as before, we obtain in place of (16.57) and (16.58) (which is actually unchanged)

$$c(1 + p/p^*)^{-\frac{1}{2}} = 3^{\frac{1}{2}}\zeta^{\frac{1}{2}}/2, \qquad (16.60)$$

$$m(1 + p/p^*)^{-\frac{1}{2}} = 2\left[c(1 + p/p^*)^{-\frac{1}{2}}\right](1 - \tfrac{3}{2}\zeta). \qquad (16.61)$$

It follows immediately that the curves for m and c in fig. 16.7a may be re-used for the variables $m(1 + p/p^*)^{-\frac{1}{2}}$ and $c(1 + p/p^*)^{-\frac{1}{2}}$, as shown in fig. 16.7b. Thus, both the maximum bending moment and also the critical curvature at which it occurs are increased by the same factor $(1 + p/p^*)^{\frac{1}{2}}$. In particular, the initial slope of the moment/curvature graph is not changed, as we expect.

The same results hold for negative values of p, corresponding to exterior pressure, provided $p > -p^*$. This limit corresponds to a straight tube being on the point of buckling under exterior pressure alone, and therefore being incapable of sustaining any applied bending moment.

Consider next the local buckling condition. An interior pressure p causes a longitudinal tensile stress of $0.5\, pa/t$, which may be put in terms of our dimensionless stress s by means of (16.47). Thus we find a dimensionless tensile stress which may be written as

$$\frac{3^{\frac{1}{2}}}{8} \frac{h}{a} \frac{p}{p^*};$$

16.5 The effect of interior pressure

and since this will ordinarily be small in comparison with 1, we shall ignore it. The smallness of this contribution is a consequence of the fact that the circumferential stress resultant N_y for the buckling of a long thin-walled tube in the mode $n = 2$ is of a much smaller magnitude than the stress resultant N_x which is required for buckling in an axisymmetric mode: see section 14.7.1.

The local buckling condition may therefore still be expressed approximately by (see (16.50))

$$s_{cr} = 1 - 3\zeta$$

or, in view of the change of variables from fig. 16.7a to b, by

$$s_{cr}(1 + p/p^*)^{-\frac{1}{2}} = (1 - 3\zeta)(1 + p/p^*)^{-\frac{1}{2}}. \tag{16.62}$$

Lines corresponding to the right-hand side have been plotted in fig. 16.7b for several values of p/p^*; and the corresponding points at which local buckling occurs have been marked on the moment/curvature graph of fig. 16.8a. The

Fig. 16.8. (a) Unified dimensionless moment/curvature plot for a tube in the presence of interior pressure, showing points at which local buckling occurs, according to fig. 16.7b. (b) The data of (a) replotted as curves of bending moment against pressure. Theoretical results of Fabian (1977), using accurate theory, are shown as points: ● for maximum moment and ○ for local buckling. (c) The data of (a) replotted as curves of curvature against pressure: c_{max} is the value of c corresponding to $m = m_{max}$. Fabian's results for local buckling are shown as points ○. R. J. Anderson's experimental critical curvatures are shown as points ■.

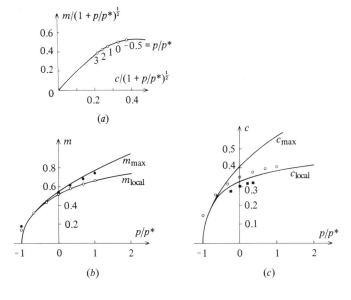

same data have been replotted in fig. 16.8*b* and *c* to show directly how *m* and *c* vary with pressure, both for the maximum bending moment and the local buckling conditions. Note that for external pressure of magnitude greater than approximately $0.6p^*$, the maximum bending moment is reached *before* local buckling, according to the present theory. Some results of Fabian (1977) are also shown on these diagrams. He did a much more comprehensive analysis by means of equations which correctly took account of large-displacement effects; but his results agree closely with those of the present analysis except in the external-pressure regime $p/p^* < -0.5$. Fig. 16.8*c* also shows dimensionless critical curvatures deduced by Reddy (1977) from the experiments of Anderson of silicone-rubber tubes. He found that the experimental moment-curvature relationship for various internal and external pressures lay very close to the single theoretical curve shown in fig. 16.8*a*; see fig. 16.1 for details of a typical test. Anderson measured p^* for his tube by means of Southwell plot. In these experiments the critical condition was reached in all cases at bending moments and curvatures a little less than those indicated by Fabian and by the present theory.

16.6 The effect of finite length

Fig. 16.9*a* shows schematically the curved profile of a portion of a long tube which has been subjected to a steadily increasing pure bending moment until $\zeta = 0.2$. Suppose now that we take a second tube, having a

Fig. 16.9. Diagram to show end-restraint effects. (*a*) Part of long tube. (*b*) Localised zones of restraint. (*c*) Short tube bent so that one generator is straight. (*d*) Simple hypothetical mode (see (16.64)) for approximate analysis.

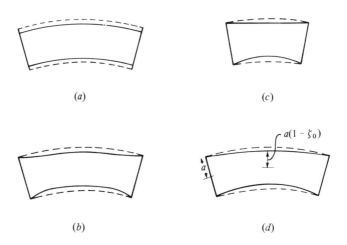

16.6 The effect of finite length

finite length, and bend it to the same extent. If the end-fittings of this tube preserve the original circular shape, there are likely to be 'transition zones' at the two ends, in which the value of ζ increases from zero (at the ends) to 0.2 over a central region: such zones have been sketched in fig. 16.9*b*. It is obvious that for tubes which are shorter than the one which is depicted, the central region in which ζ remains effectively constant will be shorter; and indeed for even shorter tubes the 'end-effects' may well dominate the behaviour over the entire length of the tube. In fig. 16.9*c* a tube has been sketched which is so short that the generator on the outer side remains approximately straight. This is, of course, only a schematic picture, but it does suggest that a tube having this geometry will be in a range in which the end-effects are dominant.

Let the length of the tube be L. Then the curvature, C, of the centre-line for which the 'sagitta' is equal to $0.2a$ is given by $C \approx 1.6a/L^2$: see appendix 8. But for a long tube, when $\zeta = 0.2$, $c \approx 0.4$ (see (16.36)). Hence from (16.37),

$$C \approx 0.45\, h/a^2.$$

By equating these two expressions we find a condition on L:

$$L^2 h/a^3 \approx 3.5$$

The dimensionless group on the left-hand side is familiar: it is precisely equal to Ω^2. Hence we find

$$\Omega \approx 1.9. \tag{16.63}$$

This argument suggest that end-effects are not important if the value of Ω is sufficiently large, but that they become important for sufficiently small values of Ω. It is not surprising that the dimensionless length Ω should be involved in this problem, in view of the work of chapter 9 and section 14.6.

Let us attempt to analyse the behaviour in bending of a tube having an arbitrary length L between the ends, which are 'held circular', as in chapter 9. We might expect that for small values of Ω the analysis of Seide & Weingarten (1961) is appropriate, while for large values of Ω the analysis of the previous section will be valid. This problem was first studied by Akselrad (1965), and the following approximate analysis is a somewhat simpler treatment of Akselrad's ideas.

It will be convenient to use an elaboration of Brazier's energy method, and to work as before with a truncated expression for the strain energy of longitudinal stretching. It will also be convenient to consider the deformation of the tube as a two-stage process. In the first stage the tube remains straight and ovalises in a way which is compatible with the 'held circular' end-conditions; and in the second stage the tube bends as a non-uniform beam subjected to constant bending moment.

For the sake of simplicity we shall assume that the variation of ζ along the length of the tube is given by

$$\zeta = \zeta_0 \sin(\pi x/L), \qquad (16.64)$$

where x is an axial coordinate whose origin is at one end, and ζ_0 is the value of ζ at the central cross-section: see fig. 16.9d. We shall justify the choice of this simple function later. Thus the normal component of small deflection has the form

$$w = a\zeta_0 \sin(\pi x/L) \cos 2\theta, \qquad (16.65)$$

where θ is the angular coordinate used earlier. Now in our previous calculation of *uniform* bending the only strain energy involved in the first stage of deformation was that of circumferential bending. In the present problem, however, some surface stretching is also involved, since the mode of deformation (16.65) involves a change of Gaussian curvature. We have investigated problems of this sort in chapters 8 and 9. Here we shall assume that

$$L/a > 3, \qquad (16.66)$$

so that the 'panel aspect ratio' of the mode exceeds 2; thus we shall be concerned only with circumferential bending and longitudinal stretching.

The strain energy of circumferential bending is easy to compute:

$$U_B = \tfrac{3}{16} \pi E t h^2 \zeta_0^2/a. \qquad (16.67)$$

This expression may be derived from the last term in (16.35), together with the observation that the mean value of ζ^2 along the length is $\tfrac{1}{2}\zeta_0^2$.

It is also a straightforward matter to compute the stress resultants N_x, etc. for an S-surface distorted according to (16.65), and then to evaluate the total strain energy. In this way we find that

$$N_x = \tfrac{1}{4}\pi^2 \, (Eta^2/L^2)\zeta_0 \sin(\pi x/L) \cos 2\theta. \qquad (16.68)$$

In particular, these stress resultants are negative, i.e. compressive, at the ends of the minor diameter of the distorted cross-sections (and positive at the ends of the major diameter). They will therefore contribute towards the reaching of the local buckling condition on the compressive side of the bent tube. We shall return to this point later.

Evaluating the total strain energy which is associated with this aspect of the distortion we find, on rearrangement,

$$U_S = \tfrac{1}{12}\pi^4 U_B/\Omega^4. \qquad (16.69)$$

Since the ratio U_S/U_B is essentially the same as the ratio of the stiffnesses of the S- and B-surfaces for this particular mode, we could have obtained the same result by specialisation of expression (8.68). And moreover, since U_S and U_B are both proportional to ζ_0^2, we may combine the two terms to give

16.6 The effect of finite length

a simple expression for the total strain energy of distortion of the tube in the first stage:

$$U_1 = U_S + U_B = U_B (1 + 8.12/\Omega^4). \tag{16.70}$$

The simplicity of this expression is due primarily to our assumption that longitudinal stretching and circumferential bending are the only structural actions which need be considered.

Next, consider the strain energy which is involved in the second stage of deformation. Suppose that a *mean* curvature $\bar{c}$ is applied to the tube; that is, that the ends are given a relative rotation of $L\bar{c}$. The bending moment is constant throughout the length of the tube, and therefore elementary beam theory indicates that the curvature c will vary along the length in inverse proportion to the second moment of area of the cross-section. Now according to Brazier's truncation, $I \propto 1 - 1.5\zeta$; and so on using the binominal theorem and truncating the resulting power-series, we find that the local curvature varies thus:

$$c = \text{constant} \times [1 + 1.5\zeta_0 \sin(\pi x/L)]. \tag{16.71}$$

On integrating this we obtain an expression for the mean curvature $\bar{c}$; and hence we find

$$c = \bar{c} \, [1 + 1.5\zeta_0 \sin(\pi x/L)]/(1 + 3\zeta_0/\pi). \tag{16.72}$$

In particular we determine from this an expression for c_0, the curvature at the central section:

$$c_0 = \bar{c} \, [1 + 1.5(1 - 2/\pi)\zeta_0] = \bar{c}(1 + 0.55\zeta_0). \tag{16.73}$$

Here again we have used the binominal theorem and truncated the resulting power-series.

The total strain energy of bending of this non-uniform beam is thus given by

$$U_2 = \int_0^L \tfrac{1}{2} c^2 EI_2 \, [1 - 1.5\zeta_0 \sin(\pi x/L)] \, dx, \tag{16.74}$$

where I_2 is (here) the second moment of area of the tube in its original circular configuration. Using (16.72) we find the truncated result

$$U_2 = \tfrac{1}{2}\bar{c}^2 LEI_2 (1 - 3\zeta_0/\pi) = \tfrac{1}{2}\bar{c}^2 LEI_2 (1 - 0.955\zeta_0). \tag{16.75}$$

This expression is in fact the same as one which we would obtain by considering the imposition of a *uniform* curvature $\bar{c}$ onto the non-uniform beam: see problem 16.2. According to the theorem of minimum strain energy (appendix 1) we should expect to find a somewhat smaller amount of energy as a result of our calculation using a varying curvature: but in fact the difference appears only in the terms of order ζ_0^2, which we have systematically truncated.

Expression (16.75) involves, of course, some longitudinal stretching of the material in the course of bending the beam tube. The quantity U_S, evaluated above, also involves longitudinal stretching. But the calculations may properly be regarded as independent, since the respective circumferential functions $\sin\theta$ and $\cos 2\theta$ are orthogonal. Note in particular that the stress-resultants N_x involved in the calculation of U_S do not sustain any overall bending moment in the tube.

When we compare our various strain-energy expressions above with those which we obtained for Brazier's problem, we find that for the present case we may write down a modified dimensionless energy expression in place of (16.56) as follows:

$$u = \bar{c}^2(1 - 0.955\zeta_0) + \tfrac{9}{32}(1 + 8.12/\Omega^4)\zeta_0^2. \tag{16.76}$$

This has precisely the same form as the earlier expression, and differs only in the values of the constants. Proceeding as before, we first put $\partial u/\partial \zeta_0 = 0$, and thus obtain

$$\bar{c}(1 + 8.12/\Omega^4)^{-\tfrac{1}{2}} = 0.767\zeta_0^{\tfrac{1}{2}}. \tag{16.77}$$

Then we put $m = \partial u/\partial \bar{c}$ (since $L\bar{c}$ is the displacement variable which corresponds to m) and find

$$m = 2\bar{c}(1 - 0.955\zeta_0). \tag{16.78}$$

These expressions are of exactly the same kind as before, but with different constants. They are plotted in fig. 16.10a. They are sufficient to determine the *maximum* bending moment as the mean curvature increases, but they must be supplemented by formulae for the 'extreme fibre stress' if we are to introduce the local buckling condition. In the first stage of our imaginary two-stage process we find from (16.68), (16.47) and (16.42) that

$$s_1 = \tfrac{1}{4}\pi^2 3^{\tfrac{1}{2}}\zeta_0/\Omega^2 = 4.27\zeta_0/\Omega^2 \tag{16.79}$$

while for stage 2 we have, from the simple bending theory of beams,

$$s_2 = 2c_0(1 - \zeta_0) = 2\bar{c}(1 - 0.45\zeta_0). \tag{16.80}$$

This relationship is also plotted in fig. 16.10a. The local buckling criterion is thus given by

$$s_2 + s_1 = 2\bar{c}(1 - 0.45\zeta_0) + 4.27\zeta_0/\Omega^2 = 1 - 3\zeta_0.$$

If we write this as

$$s_2(1 + 8.12/\Omega^4)^{-\tfrac{1}{2}} = 2\bar{c}(1 - 0.45\zeta_0)(1 + 8.12/\Omega^4)^{-\tfrac{1}{2}} =$$
$$(1 - (3 + 4.27/\Omega^2)\zeta_0)(1 + 8.12/\Omega^4)^{-\tfrac{1}{2}} \tag{16.81}$$

we find that the right-hand side is represented by a line on the graph of fig. 16.10a. Thus we discover a graphical procedure for determining the local

16.6 The effect of finite length

buckling condition, which is analogous to that which we used in fig. 16.7b for buckling of a long tube with internal pressure; and lines corresponding to several values of Ω are shown in the diagram. The intersection of the curve for $s_2(1 + 8.12/\Omega^4)^{-\frac{1}{2}}$ with this straight line for a given value of Ω indicates the value of ζ_0 for local buckling; and then the corresponding value of $m(1 + 8.12/\Omega^2)^{-\frac{1}{2}}$ may be read off, and m computed.

This procedure yields the results shown in fig. 16.10b. Values of ζ_0 at which the local buckling condition is reached are also marked on the curve. In the range $\Omega < 0.5$ there is relatively little ovalisation of the tube as the bending moment increases, and local buckling occurs essentially in the manner described by Seide & Weingarten. Over the range $0.5 < \Omega < 2$, however, there is a marked change of behaviour: as Ω increases so does the amount of ovalisation at the point of local buckling, and the value of m falls steadily towards a value approximately equal to that corresponding to an infinitely long tube.

In this calculation the contribution of s_1 towards local buckling is significant; that is, the longitudinal stress on account of ovalisation of the tube with constrained ends in the first stage is appreciable. In order to assess the magni-

Fig. 16.10. (a) Dimensionless plot (cf. fig. 16.7a) for investigating the effect of end-restraint. (b) Dimensionless bending moment m for local buckling as a function of the dimensionless length Ω $(= Lh^{\frac{1}{2}}/a^{\frac{3}{2}})$. The broken curve shows the consequence of ignoring the longitudinal stress s_1, due to non-uniform ovalisation of the tube.

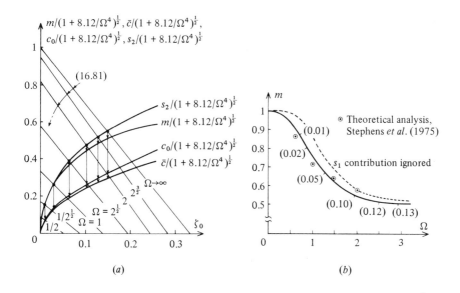

tude of this contribution we may note that if we were to omit the term $4.27\zeta_0/\Omega^2$ from (16.81), the straight lines in fig. 16.10a would all pass through the point $\zeta_0 = \frac{1}{3}$ on the ζ_0 axis, just as they do in fig. 16.7b. If the calculation is reworked in this way, we obtain the results shown by the broken curve in fig. 16.10b. The ratio s_1/s_2 reaches its largest value of about 0.2 in the region of $\Omega \approx 1.5$.

The results of four calculations by Stephens et al. (1975) are also shown on the graph. They used the STAGS computer package for thin-walled tubes having four specific geometries. Their distribution of longitudinal stress around the central cross-section (Figure 8 in Stephens et al. (1975)) shows clearly the $\cos 2\theta$ component (s_1) which features in our analysis. They describe this in terms of an apparent shift of 'neutral axis': see problem 16.3. Their method involved the deliberate insertion of a very small short-wave ripple imperfection in the form of the tube, and they found in all cases that local buckling occurred before the maximum bending moment was reached. On the whole, our results agree well with those of Stephens et al. The discrepancies (of the order of 0.04 in m) may be attributed to two factors, as follows. (i) Stephens et al. performed a nonlinear calculation in the presence of an initial imperfection. This must be expected to give lower critical values of m in comparison with those computed by our method, which uses a simple 'local-classical' criterion of buckling. (ii) Stephens et al. investigated a shell whose ends were held circular by two closely-spaced circumferential stiffening rings which were very stiff in flexure. Acting together, these tended to inhibit warping of the ends of the shell. Order of magnitude calculations suggest, however, that the restraint of warping was much stronger for the longer shells than for the short ones, mainly because the separation of the two rings was a fixed fraction of the total length of each shell. Consequently, the end-supports for the shorter shells were broadly similar to the conventional 'held circular' conditions of our analysis, whereas those for the longer shells were closer to fully-clamped conditions. This effect tends to increase the critical value of m computed by Stephens et al. for their longer shells.

Another feature of our hypothetical mode (16.65) is that it does not reflect the point made in fig. 16.9b concerning the concentration of the non-uniform deformation near the two ends in the case of a long tube. A direct consequence of our assumed mode is a prediction (see fig. 16.10b) that $m = 0.505$ for local buckling of an infinitely long tube, in contrast to our previous value of 0.515. The important point here, however, is that the two results differ by only 2%, in consequence of the overall similarity between the curves in fig. 16.10a and fig. 16.7a. It seems clear that the assumed modeform (16.65) is satisfactory for an approximate investigation of end-effects.

Akselrad's results are in broad general agreement with those shown in fig. 16.10b, except that that his asymptotic value of m for large values of Ω is somewhat smaller. His geometrical parameter λ is almost exactly proprotional to Ω: the slight difference in form seems to be a result of our assumption that the predominant effects in the resistance to ovalisation are circumferential bending and longitudinal stretching and that other terms may be ignored. This enables us to use a simplified expression (16.70) which is amply justified provided $L/a > 3$: consequently our parameter Ω is only strictly relevant if this condition is fulfilled.

16.7 An improvement on Brazier's analysis

All of the calculations performed so far in this chapter have been based on a simplification first introduced by Brazier, which involves an early truncation of the power-series expression for the strain energy of longitudinal stretching of the tube for a given postulated mode. There are strong *prima facie* reasons for supposing that this truncated expression will be inadequate when the ovalisation is as large as, say, $\zeta = 0.15$ or 0.2; but we have nevertheless employed it on the grounds that more exact and complete studies, notably by Fabian (1977) have produced results for maximum bending moment and for the local buckling condition which differ by only a few per cent from those of the present analysis. By far the largest discrepancy, in fact, comes in the magnitude of the *ovalisation* of the cross-section: at the point of maximum bending moment Fabian's analysis indicates a value of ζ some 15% larger than that given by Brazier.

The aim of the present section is to investigate the discrepancies between the results of Brazier and Fabian. In particular we shall try to discover why the discrepancies are so *small*, in view of the fact that our previous experience of a truncated expression, in section 16.2.1, led to rather *large* changes in the various quantities: see fig. 16.4.

It is clear at the outset that we must do a more accurate analysis of the geometry of distortion of the cross-section than the one which is described by (16.26) to (16.32). It will be adequate for present purposes to consider a cross-section which is originally circular but which deforms inextensionally by suffering a change of curvature κ according to

$$\kappa = (3\Gamma/a)\cos(2s/a), \quad (16.82)$$

where s is here the arc-length coordinate measured from the axis $\theta = 0$ of mirror symmetry: see fig. 16.11. The choice of the constant as $3\Gamma/a$ will become clear later. It will actually be more convenient to use the coordinate

$$\theta = s/a \quad (16.83)$$

instead of s, and thus to write

$$\kappa = (3\Gamma/a)\cos 2\theta; \tag{16.84}$$

but it must be emphasised that θ is here essentially an *arc-length* coordinate.

An immediate advantage of this prescription of the deformed configuration is that it is a straightforward matter to compute the strain energy of circumferential bending of the cross-section, since κ is directly proportional to $\cos(2s/a)$. Clearly, this part of the total strain energy is directly proportional to Γ^2, with no higher-order terms.

As we saw earlier, the calculation of strain energy of longitudinal stretching involves mainly the calculation of I, the second moment of area of the cross-section in its distorted condition. As a preliminary step towards this, we must first determine the coordinate Y of the deformed profile shown in fig. 16.11 as a function of θ and Γ.

In general, if the tangent to the curve makes angle ψ with the tangent at $\theta = 0$, the curvature of the arc is equal to $d\psi/ds$. Since the original curvature was $1/a$, we have in general

$$\frac{d\psi}{ds} = \frac{1}{a} + \frac{3\Gamma}{a}\cos\frac{2s}{a},$$

or

$$\frac{d\psi}{d\theta} = 1 + 3\Gamma\cos 2\theta. \tag{16.85}$$

By integration, therefore,

$$\psi = \theta + 1.5\Gamma\sin 2\theta. \tag{16.86}$$

Now by elementary trigonometry

$$dY = \cos\psi \, ds; \tag{16.87}$$

Fig. 16.11. Coordinate system used for analysis of moderately large deformations of the cross-section of a tube: see section 16.7.

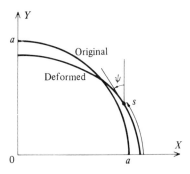

16.7 An improvement on Brazier's analysis

and thus, by defining a dimensionless variable

$$y = Y/a \tag{16.88}$$

we have

$$dy = \cos\psi \, d\theta. \tag{16.89}$$

It follows that

$$y(\theta) = \int_0^\theta \cos\psi \, d\theta$$

$$= \int_0^\theta \cos(\theta + 1.5\Gamma\sin 2\theta) \, d\theta. \tag{16.90}$$

This integration is potentially awkward; but since we are interested in relatively small values of Γ we can express the integral as a power-series in Γ by using the Taylor series expansion

$$\cos(\theta + h) = \cos\theta - h\sin\theta - \tfrac{1}{2}h^2\cos\theta + \tfrac{1}{6}h^3\sin\theta + \ldots \tag{16.91}$$

Putting $h = 1.5\Gamma\sin 2\theta$ and performing the necessary integration, we find the following expression for y as a power-series in Γ:

$$y = S - \Gamma S^3 - \Gamma^2\left(\tfrac{3}{2}S^3 - \tfrac{9}{10}S^5\right) + \Gamma^3\left(\tfrac{9}{10}S^5 - \tfrac{9}{14}S^7\right) + \ldots, \tag{16.92}$$

where

$$S = \sin\theta. \tag{16.93}$$

Consider first the point at the apex, $S = 1$. Since at this point $y = 1 - \zeta$, where ζ is our parameter of ovalisation, now defined at a single point, we find

$$\zeta = \Gamma + \frac{3\Gamma^2}{5} - \frac{9\Gamma^3}{35} + \ldots. \tag{16.94}$$

The first term on the right-hand side corresponds exactly to our previous, linearised, calculation; and we thus see that the present deformation parameter Γ is equal to ζ as a first approximation. Indeed, this was the reason behind the choice of constant in (16.84).

Our next task is to determine I in terms of Γ. This involves primarily the evaluation of $\int_0^{\frac{1}{2}\pi} y^2 \, d\theta$. Our policy here will be to square the expression on the right-hand side of (16.92), but to discard terms of order Γ^4 and above. The resulting integration is simplified if we note that we may write

$$\int S^n \, d\theta = \tfrac{1}{4}c_n\pi, \tag{16.95a}$$

where

$$c_2 = 1, c_4 = \tfrac{3}{4}, c_6 = \tfrac{5}{8}, c_8 = \tfrac{35}{64}. \tag{16.95b}$$

Thus we find

$$\int_0^{\frac{1}{2}\pi} y^2 \, d\theta = \tfrac{1}{4}\pi\left\{1 - \frac{3\Gamma}{2} - \frac{\Gamma^2}{2} + \frac{21\Gamma^3}{16} + \ldots\right\}. \tag{16.96}$$

The term in brackets is analogous to the corresponding term in (16.31) of our previous analysis. We thus find that the first two terms of our previous expression (16.31) were correct – as expected, of course – but that the term in ζ^2 was not correct. The discrepancy is due to the fact that in our previous work we included in effect only the first two terms on the right-hand side of (16.92) before squaring and integrating; and this procedure omitted some important contributions to the Γ^2 term.

We are now in a position to perform a more accurate version of Brazier's calculation. Taking Γ as the basic variable, we now have, in place of (16.56):

$$u = c^2 \left(1 - \tfrac{3}{2}\Gamma - \tfrac{1}{2}\Gamma^2 + \tfrac{21}{16}\Gamma^3\right) + \tfrac{9}{16}\Gamma^2. \tag{16.97}$$

For a given value of c we find Γ from $\partial u/\partial \Gamma = 0$; and then we have $m = \partial u/\partial c$. This enables us to express both m and c in terms of Γ, and we can also determine ζ from (16.94). In this way we obtain

$$\begin{aligned}
c &= 0.5 \times 3^{\tfrac{1}{2}} \left((1/\Gamma) + [\tfrac{2}{3} - \tfrac{63}{24}\Gamma]\right)^{-\tfrac{1}{2}}, \\
m &= 2c\left(1 - \tfrac{3}{2}\Gamma - [\tfrac{1}{2}\Gamma^2 - \tfrac{21}{16}\Gamma^3]\right), \\
\zeta &= \Gamma + [\tfrac{3}{5}\Gamma^2 - \tfrac{9}{35}\Gamma^3].
\end{aligned} \tag{16.98}$$

Each of these expressions is, of course, a truncation of an infinite series. If the expressions are truncated further by the omission of terms enclosed [ ], we obtain precisely the previous results.

Let us now examine the local buckling condition. For this purpose we need expressions for s and s_{cr}.

These are:

$$s = 2c(1 - \zeta) \tag{16.99a}$$

$$s_{cr} = 1 - 3\Gamma. \tag{16.99b}$$

The first of these comes direct from simple beam theory, while the latter is the exact equivalent of (16.50).

All of these expression have been evaluated, and plotted in fig. 16.6 as broken curves. In general, the extra terms make very little difference up to the point of maximum bending moment. The explanation of this is that the terms enclosed [ ] in the expressions for c and m are roughly self-cancelling in the region $\Gamma \approx 0.2$; and thus they make relatively little difference to the m, c plot. On the other hand, the additional terms in the expression for ζ are not nearly as well balanced, and make a more appreciable difference to the results.

These conclusions would have been markedly different if we had retained only *one* extra term in each of the series, instead of two; and we might therefore expect that the conclusions would again be changed if we were to use three or more extra terms instead of two. Further analysis shows that this is not so: any further terms make relatively little difference to the conclusions, at least up to the point of maximum bending moment.

It is clear that Brazier was lucky in being able to obtain good results for the maximum bending moment by using a power-series which was so crudely truncated. But if we are mainly interested in the conditions under which *local* buckling takes place, say in the region $\zeta \approx 0.1$, the terms which Brazier discarded are rather small, anyway. This constitutes the justification for using the Brazier truncation generally in the present chapter.

16.8 Problems

16.1 Show that the maximum value of M according to (16.24) is 0.84 (= 16 × $2^{\frac{1}{2}}/27$) of the maximum value according to (16.13), and that the ratio of the corresponding values of c is $2^{-\frac{1}{2}}$: cf. fig. 16.4.

16.2 A beam of length L, made of uniform elastic material, has a non-uniform cross-section, and its second moment of area I is given by
$$I = I_2 \left[1 - 1.5\zeta_0 \sin(\pi x/L) \right].$$
The beam is now constrained to deform into an arc of uniform curvature $\bar{c}$. Show that the total strain energy of bending is equal to $\frac{1}{2}\bar{c}^2 LEI_m$, where I_m is the mean value of I along the beam; and that this agrees with (16.75).

16.3 An initially straight tube of circular cross-section has been bent into a curved arc. At the central cross-section the distribution of longitudinal strain around the circumference is given by
$$\epsilon_x = \alpha \sin\theta + \beta \sin 3\theta + \gamma \cos 2\theta,$$
where θ is a circumferential angular coordinate. The first two terms on the right-hand side correspond to the uniform bending of a tube of oval cross-section, while the third term is a consequence of non-uniform ovalisation along the length of the tube. Show that the apparent neutral axis – i.e. the line through points for which $\epsilon_x = 0$ – is coincident with the axis $\theta = 0$ only when $\gamma = 0$; and that it moves away from this diametral axis by a fraction $\gamma/(\alpha + 3\beta)$ of the radius, approximately, provided $\gamma \ll \alpha + 3\beta$.

17

Vibration of cylindrical shells

17.1 Introduction

Most of this book is concerned with the performance of shells under static loading. In contrast, the present chapter is concerned with an aspect of the response of shells to dynamic loading. The response of structures to dynamic loads is an important part of design in many branches of engineering: examples are the impact loading of vehicles, the aeroelastic flutter of aircraft, and wave-loading on large marine structures.

In this chapter we shall be concerned with the *vibration* of cylindrical shells, and in particular with the calculation of undamped natural frequencies. Calculations of this kind sometimes give the designer a clear indication that trouble lies ahead for a proposed structure; but if the design can be altered so that the natural frequencies of vibration of the structure are sufficiently different from the frequencies of the exciting agency, the occurrence of vibration can often be avoided.

For reasons of brevity, this chapter is restricted to cylindrical shells. The methods of the chapter may be adapted to the study of other sorts of shell, e.g. hyperboloidal shells used for large natural-draught water-cooling towers: see Calladine (1982).

Two very early papers on the subject of shell structures, by Rayleigh and Love, respectively, were on the subject of vibration, and the present chapters represent in fact only a relatively small advance on their work. Rayleigh (1881) was concerned with the estimation of the natural frequencies and modes of vibration of bells. He was interested in obtaining definite numerical answers, and to this end he tackled the problem in relation to a bell idealised as a uniform thin hemispherical bowl. He began by making certain assumptions, among which are two which we shall also adopt: he considered only small displacements from the original configuration of the shell; and he treated the

17.1 Introduction

mass of the material as if it were concentrated at the central surface of the shell (see section 17.2.3). He also invented an 'energy' method – since known as 'Rayleigh's principle' – for estimating natural frequencies. We shall use this method in some parts of the chapter, in cases where it turns out to be more expeditious than setting up the equations of motion for the system.

Rayleigh's work has often been criticised on the grounds that he expressed most interest in *inextensional* modes of deformation, and only did brief calculations for extensional modes. Rayleigh's inextensional modes were indeed appropriate for the calculation of the 'gravest' tones of a thin hemispherical bowl, and Rayleigh must be given credit for finding a simple way of tackling a difficult problem. In the present chapter we shall of course include the possibility of having extensional modes: and in the light of chapters 8 and 9 it is not difficult to incorporate these into a 'Rayleigh method', as we shall see.

Flexural vibrations of shells were investigated in a general way by Love (1888). In the case of cylindrical shells he showed (Love, 1927, §334) how to set up an equation whose roots would give the entire range of natural frequencies. Flügge (1962) was the first to solve, in 1934, such an equation for a particular set of boundary conditions. We shall comment on this approach to the vibration of shells in section 17.7.

An important contribution to the subject of vibration of cylindrical shells was made by Arnold & Warburton (1953). They solved the frequency equations for simple boundary conditions and used a Rayleigh method for other boundary conditions. They also did a large number of experimental investigations, and found that in all cases there was excellent agreement between experiment and theory. A starting point for their work was the puzzling observation that for cylindrical shells of certain proportions and with certain boundary conditions, the fundamental mode is one with a relatively high circumferential wavenumber. They successfully disentangled the distinct contributions towards the strain energy of the shell from bending and stretching effects. Their work provides a major confirmation of the use of the classical small-deflection equations for the calculation of natural frequencies.

We shall assume throughout this chapter that damping has a negligible effect upon natural frequency, and that it may therefore be omitted entirely from the calculations. This is a well-known feature of the behaviour of vibrating systems. We do not wish to argue, of course, that effective damping is never present: the provision of damping either by means of special energy-dissipating devices or by the use of materials with high 'natural' damping (such as twisted wire cables) is a common device in the solution of vibration problems: see Warburton (1976).

The various aspects of vibration of shells are by now so well established that much of the analysis required is very straightforward. Sometimes it is an almost trivial adaptation, by means of d'Alembert's principle, of results which we have already established in earlier chapters in relation to the static behaviour of shells. In particular, we shall make use of the modal patterns developed in chapter 8 and elsewhere, in which the cylindrical surface is divided by longitudinal and circumferential nodal lines into 'panels' of length l and breadth b: on these lines the normal component w of displacement is zero, and all of the panels oscillate in phase, giving a 'chessboard' pattern of normal displacement at any given instant of time. Just as in other chapters, much of our effort will be devoted to presenting solutions to problems in the most economical way. It is not obvious from most of the available literature on the subject (see Leissa, 1973) that such data can be presented compactly: Forsberg (1964) is one of the few workers in this field to emphasise the usefulness of plotting results in a suitable dimensionless form. After all, even a uniform cylindrical shell of radius a, thickness t and length l, which is vibrating in a mode with circumferential and longitudinal wavenumbers n and m respectively, can only be completely specified by four dimensionless groups: thus it is not easy to envisage a straightforward plot of the results. However, if we are prepared to restrict attention to the *lower end* of the frequency spectrum, and in particular to the 'gravest' or *fundamental* mode – which seems in many cases to be the question of most practical interest – we can make considerable simplifications in the analysis, as we shall see, and then present the results in a compact and immediately useful form.

The general strategy of the chapter is to proceed from simple problems to more complex ones in relatively easy stages. We shall make some reflections on this as a general policy in section 17.7. The succession of stages is as follows. First, in section 17.2, we shall consider some simple problems in connection with the vibration of a circular *ring* in its own plane. This will enable us to introduce the two main computational schemes which we shall use throughout the chapter, and to investigate the consequences of making various approximations. Then in section 17.3 we shall consider the vibration of a uniform cylindrical shell by means of a 'shallow-shell' theory: the entire surface will be covered by rectangular 'panels' $l \times b$; w will be the only significant component of displacement; and 'conventional' boundary conditions will be assumed. This will lead, in section 17.4, to an analogy with the vibration of a beam-on-elastic-foundation for the low-frequency end of the spectrum; and this has important practical consequences, particularly in relation to boundary conditions. We shall find in section 17.5 that the impact of various sorts of practical boundary conditions is closely analogous with the work of chapter

17.2 Vibrations of a simple ring

equations are valid only for sufficiently high values of the circumferential wavenumber, as we have seen (cf. section 8.6); but Rayleigh's energy method provides a straightforward way of extending the results to low wavenumbers.

In this chapter we shall be concerned almost entirely with cases for which it is possible to obtain relatively simple analytical solutions. Many practical problems are more complicated in various ways, and yield only to attack by numerical means. Nevertheless, the results of the present chapter are of use to the engineer in various ways.

In some circumstances the natural frequencies of vibration of a structure may be affected by the state of stress which exists in the structure before vibration begins. The simplest way of appreciating this point is to consider the natural frequencies of a uniform simply-supported elastic beam. Suppose that the beam is subjected to an axial tension. This has the effect, broadly, of increasing the apparent flexural stiffness of the beam and thereby increasing the natural frequencies. Conversely, axial compression lowers the natural frequencies; and indeed we can easily envisage the interaction between vibration and buckling which has sometimes been used to predict the critical classical buckling load of a structure by the measurement of its natural frequencies at several sub-critical loads. We do not take account of effects of this sort in this chapter, although in some circumstances they can be important. Thus, it should be remarked that zones of residual tensile and compressive stress in flat-plate structures (a consequence of fabrication techniques) can have a significant effect on natural frequencies.

The calculations in this chapter are all made on the assumption that strains and displacements are small. For an introduction to nonlinear vibration of shells see Evensen (1974).

17.2 Vibrations of a simple ring

Consider the natural modes and frequencies of a uniform thin circular elastic ring vibrating in its own plane, as shown in fig. 17.1. The radius of the ring is a and the area of its (symmetrical) cross-section is A. The Young's modulus is E, and the flexural rigidity for in-plane bending is EI. The density of the material is ρ. We shall use a circumferential angular coordinate θ, and denote by w and v the radial and peripheral components, respectively, of displacement of a typical point on the centre-line of the ring: see fig. 17.1a for sign conventions. Finally we shall denote by ω (omega: not to be confused with w) the angular natural frequency of a given mode. Thus the natural *period* of vibration is equal to $2\pi/\omega$ and the frequency in cycles per unit time (hertz) is equal to $\omega/2\pi$.

We shall not give an exhaustive treatment here: for various other aspects of

630   *Vibration of cylindrical shells*

the problem see Love (1927, §293) or the summary of Young (1962, §61.6).

First we shall investigate *extensional* vibration and, second, *inextensional* or *flexural* vibration of the ring.

17.2.1 Extensional vibrations

The simplest possible mode is one in which the ring pulsates radially, always remaining circular and concentric with its original configuration.

At a particular time let its radial displacement be w: see fig. 17.1b. Then there is a uniform circumferential strain $\epsilon = w/a$, and consequently, by Hooke's law, a uniform tension T given by

$$T = EAw/a. \tag{17.1}$$

The effect of this tension on a small element subtending angle $d\theta$ is to give a resultant inward-directed radial force equal to $Td\theta$: see fig. 17.1c. This force is unbalanced, and it causes an inward-directed acceleration. Since the mass of the element is equal to $\rho Aad\theta$, we have the following *equation of motion*:

$$EAwd\theta/a = -\rho Aad\theta \ddot{w}, \tag{17.2}$$

which may be simplified to

$$\ddot{w} + (E/\rho a^2)w = 0. \tag{17.3}$$

Here a superior dot denotes differentiation with respect to time. The negative sign in (17.2) is due to the fact that the positive sense of $\ddot{w}$ is, like that of w, directed outwards.

Fig. 17.1. In-plane vibration of a uniform circular ring. (a) Notation. (b) Simple 'pulsating' mode. (c) Forces on an element in mode (b). (d) Flexural mode (here with $n = 3$).

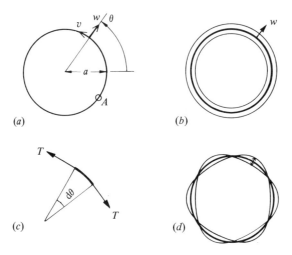

17.2 Vibrations of a simple ring

The general solution of (17.3) is

$$w = C \sin(\omega t + \phi), \qquad (17.4a)$$

where

$$\omega = (E/\rho a^2)^{\frac{1}{2}}. \qquad (17.4b)$$

Here C and ϕ are arbitrary constants, and t in (17.4a) represents time. Note that the circular frequency ω is independent of A: any uniform circumferential 'hoop' of material within the ring provides the force necessary to accelerate its own mass. Note also another feature of classical vibrations, namely that the amplitude C and phase ϕ are indeterminate. A more detailed study (Love, 1927, §293) shows that there are some other modes in which the main structural action is circumferential tension, and the vibration is analogous to the longitudinal vibration of a bar: the circumferential displacements are of the form $v = C_1 \cos n\theta$. All of these modes have higher frequency than the one which we have calculated. These modes are analogous to what Lamb described (Rayleigh, 1894, p.418) as a 'first class' of modes of vibration of shells in which there is zero normal component of displacement, as distinct from a 'second class' in which $w \neq 0$. Throughout this chapter we shall be concerned exclusively with modes in this 'second class'.

In the above analysis we found an equation of motion (17.3) which we solved formally to give (17.4). We must always expect, for a small-displacement analysis, a sinusoidal variation of displacements with time; and we might well have taken a valid short-cut to the answer by postulating a solution of the form

$$w = C \sin \omega t \qquad (17.5)$$

and then using d'Alembert's principle. For this purpose omission of the phase angle ϕ is of no consequence. We shall take short-cuts of this sort in subsequent sections.

17.2.2 Inextensional (bending) vibrations

Here the ring does not remain circular (see fig. 17.1d), and we therefore assume a radial displacement function of the form

$$w = C \cos n\theta \sin \omega t, \qquad (17.6)$$

where n, the circumferential wavenumber, is an arbitrary positive integer for a complete ring. The direction $\theta = 0$ has been chosen, by implication and without loss of generality, to suit (17.6). Throughout the motion, the ring remains in its original plane.

The deflection is thus assumed to be the product of a function of space and a function of time. The spatial function, here

$$w = C \cos n\theta, \qquad (17.7)$$

is known as the *mode* of vibration. We shall assume throughout that the magnitude of C is so small that the standard small-displacement relations are valid.

Now, since the ring is inextensional, by supposition, we may use the relation (cf. (6.9b))

$$a\epsilon_\theta = \frac{\partial v}{\partial \theta} + w, \qquad (17.8)$$

together with $\epsilon_\theta = 0$ in order to obtain an expression for v; and we obtain, in general,

$$v = -(C/n) \sin n\theta \sin \omega t + f(t). \qquad (17.9)$$

Now we shall suppose that the ring has zero mean angular momentum about its centre, and therefore we may put $f(t) = 0$. Note that we used partial derivative $\partial/\partial \theta$ in (17.8), since there are two independent variables θ and t; and that this produced in (17.9) an arbitrary function of time. But in practice we shall not go wrong if we use the ordinary derivative $d/d\theta$ instead, regarding the functions of time to be constrained to the form $\sin \omega t$, and excluding all rigid-body motions.

From section 6.4 we have a simple relation between w and the change of curvature, κ, in the circumferential direction:

$$\kappa = -\frac{1}{a^2} \frac{\partial^2 w}{\partial \theta^2} - \frac{w}{a^2}. \qquad (17.10)$$

This gives, on substitution of (17.6):

$$\kappa = \left(\frac{n^2 - 1}{a^2}\right) C \cos n\theta \sin \omega t. \qquad (17.11)$$

Here, as previously, κ has positive sense if the change of curvature involves stretching on the outside of the ring.

Let us proceed on this occasion by using Rayleigh's principle. The principle may be stated as follows. We wish to estimate the natural frequency of one of the natural modes of small vibration of a given elastic system. To do this we use the following algorithm. We begin with a hypothetical mode, which satisfies the kinematic boundary conditions of the system and is multiplied by the time function $\sin \omega t$ where ω is the frequency to be determined. For this mode, we calculate, by means of the kinematic relations of the system, (a) the maximum strain energy with respect to time, say U_m, and (b) the maximum kinetic energy T_m. Then we put

$$U_m = T_m; \qquad (17.12)$$

and since T_m is proportional to ω^2, this equation yields a value of ω. Ray-

17.2 Vibrations of a simple ring

leigh's principle states that if the hypothetical mode is a *fair* approximation to one of the natural modes of vibration of the system, then the value of ω obtained by the algorithm will be a *good* approximation to the corresponding natural frequency. Further, if the assumed mode is an approximation to the *fundamental* mode, the calculated frequency will be an *upper bound* on the actual fundamental frequency. The status of the terms 'fair' and 'good' in the above statement will be clear from an inspection of the proof given in appendix 3.

We shall use this method at various points in the present chapter as a means of estimating natural frequencies. As Rayleigh (1894, §89) remarks, an element of judgement is required in the choice of a hypothetical mode, the aim being 'to approach the truth as nearly as can be done without too great a sacrifice of simplicity'. In most of our applications it will be clear that the hypothetical mode is close to an actual mode, and the corresponding values of ω should then be almost exact.

Returning to our problem, we first calculate the total strain energy U by means of the well-known expression for flexure (see section 2.3.3)

$$U = \int \tfrac{1}{2} \kappa^2 EI \, ds, \qquad (17.13)$$

where s is arc-length. Thus

$$\begin{aligned}
U &= \tfrac{1}{2} aEI \int_0^{2\pi} \kappa^2 \, d\theta \\
&= (EI/2a^3)(n^2 - 1)^2 C^2 \sin^2 \omega t \int_0^{2\pi} \cos^2 n\theta \, d\theta \\
&= (\pi EI/2a^3)(n^2 - 1)^2 C^2 \sin^2 \omega t;
\end{aligned}$$

so

$$U_m = (\pi EI/2a^3)(n^2 - 1)^2 C^2. \qquad (17.14)$$

Next we compute the total kinetic energy T. Essentially we must evaluate

$$\int \tfrac{1}{2} (\text{velocity})^2 \, dm$$

over the volume of the ring, where dm is the mass of an elementary volume of material. Differentiation of expressions (17.6) and (17.9) with respect to time gives formulae for $\dot{w}$ and $\dot{v}$, the radial and tangential components of velocity:

$$\dot{w} = C\omega \cos n\theta \cos \omega t, \qquad (17.15a)$$

$$\dot{v} = (-C\omega/n) \sin n\theta \cos \omega t. \qquad (17.15b)$$

Now the mass of an elementary volume of material subtending angle $d\theta$ at the centre is equal to $\rho a A \, d\theta$; hence we obtain the following expression for the total kinetic energy:

$$\begin{aligned}
T &= \tfrac{1}{2} \rho a A C^2 \omega^2 \cos^2 \omega t \int_0^{2\pi} (\cos^2 n\theta + (1/n^2) \sin^2 n\theta) \, d\theta \\
&= \tfrac{1}{2} \pi \rho a A C^2 \omega^2 \cos^2 \omega t [(n^2 + 1)/n^2].
\end{aligned}$$

Thus
$$T_m = \tfrac{1}{2}\pi\rho a A \omega^2 [(n^2 + 1)/n^2] C^2 \tag{17.16}$$

The equation $U_m = T_m$ then gives our required expression for ω:

$$\omega = \frac{n^2(1 - n^{-2})}{(1 + n^{-2})^{\frac{1}{2}}} \frac{1}{a^2} \left(\frac{EI}{\rho A}\right)^{\frac{1}{2}}. \tag{17.17}$$

Observe that when $n = 1$, $\omega = 0$. This reflects the fact that the mode $n = 1$ involves simply a rigid-body translation of the ring, for which there is no strain energy. The formula does not reduce to (17.4b) when $n = 0$, because we have excluded extensional deformations. The lowest practical value of n is 2. For large values of n formula (17.17) is practically equivalent to the simpler expression.

$$\omega = \frac{n^2}{a^2} \left(\frac{EI}{\rho A}\right)^{\frac{1}{2}}, \tag{17.18}$$

and the accuracy of this simplified expression can be gauged from fig. 17.2. The factors $(1 \pm n^{-2})$ in (17.17) may be regarded as 'correction factors' to expression (17.18) for low values of n. The physical basis of this is that for sufficiently large values of n the relationship between displacement and change of curvature approaches that corresponding to a straight beam (see section 6.4), and also (see (17.6) and (17.9)) the effect of the peripheral component of

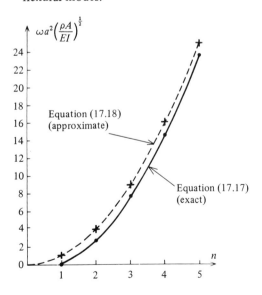

Fig. 17.2. Comparison of two formulae for natural frequency in flexural modes.

17.2 Vibrations of a simple ring

velocity becomes negligible. Thus, expression (17.18) corresponds precisely to the vibration of a straight beam in a mode whose half wavelength is equal to $\pi a/n$, as we shall see later.

We may use expressions (17.17) and (17.18) in relation to the vibration of a thin cylindrical shell in a mode (17.6) in which the generators are displaced, monolithically, parallel to themselves. For a 'hoop' of such a shell, having axial length b, we may put

$$EI = Db = \tfrac{1}{12} Eth^2 b, \quad A = bt \tag{17.19}$$

(where t and h are the thickness and the effective thickness of the shell, respectively: see section 3.3.1), and thus obtain

$$\omega = \frac{n^2(1 - n^{-2})}{(1 + n^{-2})^{\frac{1}{2}}} \frac{h}{a^2} \left(\frac{E}{12\rho}\right)^{\frac{1}{2}}. \tag{17.20}$$

This result is valid for all but very short cylindrical shells, for which a marginally smaller value of ω may be obtained: see problem 17.1.

It is of interest to compare the frequencies of the bending modes with the frequency of the stretching mode, as determined above. For the bending mode $n = 2$ we find

$$\frac{\omega_B}{\omega_S} = \left(\frac{3}{5}\right)^{\frac{1}{2}} \frac{h}{a} \approx 0.8 \frac{h}{a} \tag{17.21}$$

which, being of order t/a, is small. The higher modes of bending vibration have higher frequency, and we find that the frequency for a bending mode is equal to that for the stretching mode when

$$n \approx 1.8(a/t)^{\frac{1}{2}}. \tag{17.22}$$

For this value of n the mode has a circumferential half wavelength equal approximately to

$$1.7(at)^{\frac{1}{2}}. \tag{17.23}$$

17.2.3 A finite-thickness effect

We derived (17.17) by the use of Rayleigh's principle. We could have obtained exactly the same relation by using d'Alembert's principle and thereby turning the problem into one involving the elastic distortion of a ring under normal and tangential loading, on the same lines as in section 8.6.

In one respect, however, Rayleigh's principle is specially advantageous. So far we have considered the mass of the ring as if it were concentrated at the centre-line, instead of being distributed over the cross-sectional area, as it actually is. Following Rayleigh, we can make a simple assessment of the error which is introduced by this particular idealisation.

Consider a straight beam which is vibrating with a mode

$$w = C \sin(\pi x/l) \sin\omega t. \tag{17.24}$$

Let the cross-section be rectangular, $b \times t$, and assume – in accordance with the classical theory of bending of beams – that cross-sections of the beam in its undistorted configuration remain plane and perpendicular to the deformed centre-line: see fig. 17.3. The mass of the beam is $\rho b t$ per unit length, and if we consider it to be concentrated at the centre line, the kinetic energy associated with length dx of the beam is given by $\rho b t \dot{w}^2 dx$. If, on the other hand we consider the mass to be distributed through the thickness, we must envisage for each thin cross-sectional slice of the beam the motion of a lamina of dimensions $b \times t$, which of course has not only mass but also rotatory inertia. The moment of inertia of the lamina is $\frac{1}{12}\rho b t^3 dx$, where dx is the axial length of the thin slice; and the appropriate angular velocity is equal to

$$\frac{\partial^2 w}{\partial t \partial x} = \frac{\pi}{l} C\omega \cos\left(\frac{\pi x}{l}\right) \cos\omega t.$$

When we perform the relevant integrations over a half wavelength of the beam we find that the kinetic energy is larger by the factor

$$\left(1 + \frac{\pi^2}{12} \frac{t^2}{l^2}\right) \tag{17.25}$$

than it was when the mass was regarded as being concentrated on the centre-line. Now since the calculation of the total strain energy of the beam is unaffected by these considerations, we must conclude that this effect is one of significant magnitude only for modes which have half wavelengths of order t. In general, of course, we are not interested in modes having such short wavelengths (see (17.23)); and in any case it would be necessary to take into account shear deformations for such modes. For the remainder of the chapter, therefore, we shall treat the mass as if it were concentrated at the central surface of the shell.

Fig. 17.3. A half wavelength of a 'thick' beam, showing that a typical mass-element is actually a lamina rather than a point mass on the centre-line: see section 17.2.3.

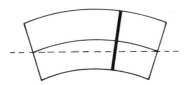

17.2 Vibrations of a simple ring

17.2.4 Standing and travelling waves

The type of vibration of a ring which may be characterised by the displacement function (see (17.6))

$$w = C \cos n\theta \sin \omega t$$

is known as a *standing wave*, since the $2n$ nodal points at which w vanishes are stationary. A second kind of vibration is also possible, in which the nodal points do not remain stationary but move steadily round the circumference. This kind of vibration is known as a *travelling wave*. The simplest way of analysing a travelling wave is to superpose onto (17.6) a second standing wave of the form

$$w = C \sin n\theta \cos \omega t. \tag{17.26}$$

This is of exactly the same type as (17.6) but it is out-of-phase with (17.6) in both space and time. The principle of superposition for small displacements of an elastic system is valid for steady undamped vibration by virtue of d'Alembert's principle, and we thus have, for the simultaneous modes,

$$w = C \cos n\theta \sin \omega t + C \sin n\theta \cos \omega t$$
$$\equiv C \sin(n\theta + \omega t). \tag{17.27}$$

Here the nodal points $w = 0$ are determined by the condition

$$\sin(n\theta + \omega t) = 0,$$

i.e.

$$n\theta + \omega t = r\pi,$$

where r is zero or a positive integer: consequently the locations of the nodes varies with the time according to

$$\theta = (r\pi/n) - (\omega t/n). \tag{17.28}$$

The entire system of nodes thus rotates steadily around the ring with a definite angular speed equal to $-\omega/n$, i.e. with a peripheral phase velocity equal to $-a\omega/n$. The frequency ω is a function of n (cf. (17.17)), so there is a distinct phase velocity associated with each circumferential mode-number. The negative sign in the above expressions has no particular significance: the phase velocity would have been in the opposite sense if the constant of the second standing wave had been $-C$ rather than $+C$. Indeed, if the two superposed vibrations (17.6) and (17.26) were to have arbitrary distinct constants, the resulting vibration could be described generally as the sum of two travelling waves moving with the same absolute phase velocity in opposite directions around the ring. In any of these cases the radial displacement of any given point on the ring is of the form

$$w = C\sin(\omega t + \phi), \qquad (17.29)$$

where ϕ is a constant phase angle.

It is obvious that vibrations in the form of a steady travelling wave are not possible in an ordinary *straight* beam of finite length, since a wave of the form

$$w = C\sin[(\pi x/\lambda) + \omega t] \qquad (17.30)$$

is incompatible with any ordinary boundary conditions. Such restrictions do not apply, of course, to a free circular ring, and we should therefore be open to the possibility that a particular mode of vibration of a ring, or indeed of a shell of revolution, which we investigate as a *standing* wave may appear in practice in the form of a circumferentially *travelling* wave. A travelling wave of this sort was observed by an eyewitness shortly before the collapse of one of the Ferrybridge cooling towers (C.E.G.B., 1965, §13).

17.3 Vibration of a cylindrical shell in 'shallow' modes

The aim of this section is to investigate the modes and natural frequencies of vibration of a thin cylindrical shell when the circumferential mode-number n is so high that the 'shallow-shell' version of the appropriate equations becomes adequate. We shall use d'Alembert's principle to turn the problem into an analogous statical one, and we shall then be able to determine the natural frequency associated with a given mode by making use of results already obtained in chapter 8. In particular we note that the equations developed in chapter 8 were only suitable for *normal* surface loading; and are therefore appropriate to the present case if the components of acceleration in the circumferential direction are generally smaller than those in the radial direction. This is, in fact, a feature of deformations for high values of n: see section 17.2.2.

A relaxation of the requirement that n be large will be investigated in the next section.

Consider a radial displacement field of the form

$$w = C\sin(\pi x/l)\sin(\pi y/b)\sin\omega t. \qquad (17.31)$$

Here x, y is a Cartesian coordinate system in the surface, and l, b are the longitudinal and circumferential dimensions of the 'panels' of the modeform, just as in section 8.4. C is an arbitrary constant amplitude, but of such a magnitude that the displacement may be regarded as 'small'. This is, of course, an extremely simple mode of vibration, and we shall suppose for the moment that the boundary conditions of the shell are so arranged that the mode is compatible with them.

The (outward-directed) acceleration of a typical point is given by

$$\ddot{w} = -C\omega^2 \sin(\pi x/l)\sin(\pi y/b)\sin\omega t, \qquad (17.32)$$

17.3 Vibration of cylindrical shell: 'shallow' modes

and since the mass per unit area of surface is equal to ρt, we have by d'Alembert's principle an equivalent outward-directed pressure

$$p = -\rho t \ddot{w} = C\omega^2 \rho t \, \sin(\pi x/l) \sin(\pi y/b) \sin\omega t. \quad (17.33)$$

We can now determine the frequency ω by stating that the pressure p in this expression must be precisely the same as that which is required according to the theory of chapter 8, to produce the given displacement (17.31). For a mode

$$w = C \sin(\pi x/l) \sin(\pi y/b) \sin\omega t,$$

(8.15) indicates a required pressure

$$p = C \left\{ \frac{Et}{a^2 l^4 (l^{-2} + b^{-2})^2} + \frac{\pi^4 E t h^2}{12} (l^{-2} + b^{-2})^2 \right\} \sin\left(\frac{\pi x}{l}\right) \sin\left(\frac{\pi y}{b}\right) \sin\omega t. \quad (17.34)$$

Here the term $\sin\omega t$ is simply a factor by which both the displacement and pressure from section 8.4 are multiplied. On substitution for p from (17.33) we obtain the required expression for ω^2. It is convenient to rearrange this by writing

$$b = \pi a/n, \quad (17.35)$$

where n is the circumferential wavenumber. Thus we obtain

$$\rho t \omega^2 = \frac{\pi^4 E t a^2}{n^4 l^4 (1 + (b/l)^2)^2} + \frac{n^4 E t h^2}{12 a^4} (1 + (b/l)^2)^2. \quad (17.36)$$

This formula may be rearranged in various ways. We obtain perhaps the most convenient form by defining a dimensionless circular frequency Y by the relation

$$Y = (\omega a^2/h)(\rho/E)^{\frac{1}{2}}, \quad (17.37)$$

and by expressing l in terms of a dimensionless group Ω, as in chapter 9, etc.:

$$\Omega = lh^{\frac{1}{2}}/a^{\frac{3}{2}}. \quad (17.38)$$

The frequency parameter Y is a function only of n for a circumferential bending mode in which the generators remain straight: cf. (17.20).

We shall use parameters Y and Ω extensively in the remainder of the chapter. Y will always be defined exactly as in (17.37), but the definition of Ω will sometimes incorporate a half wavelength l, as here, but more usually the overall length L of the shell.

$$Y = \left\{ \left[\frac{\Omega^2 n^2}{\pi^2} + \frac{h}{a} \right]^{-2} + \frac{1}{12} \left[n^2 + \frac{\pi^2}{\Omega^2} \frac{h}{a} \right]^2 \right\}^{\frac{1}{2}}. \quad (17.39)$$

$\quad\quad\quad\quad$ (LS) $\quad$ (CS) $\quad\quad\quad$ (CB) $\quad$ (LB)

The four components of the right-hand side may be identified (see section 8.4) with longitudinal (L) or circumferential (C) stretching (S) or bending (B), respectively, as indicated. Thus, if circumferential stretching dominates all three other effects, (17.39) reduces to

$$Y = a/h, \qquad (17.40)$$

which agrees exactly with (17.4b). On the other hand, if circumferential bending alone dominates, we recover $n^2/(12)^{\frac{1}{2}}$ on the right-hand side in agreement with (17.20) for sufficiently large values of n. We do not, of course, recover the exact equation (17.20) since the 'shallow-shell' approximation involves a simplified view both of circumferential bending and also of inertia effects, as we have remarked. But in practical cases it should be possible to make a correction for low values of n in accordance with the results of section 17.2.2.

The right-hand side of (17.39) is a function of the three variables Ω, n and a/h. It is easy to evaluate the right-hand side for any particular case and hence

Fig. 17.4. Logarithmic plot of dimensionless frequency Y ($= (\omega a^2/h) \times (\rho/E)^{\frac{1}{2}}$) against dimensionless panel length Ω ($= lh^{\frac{1}{2}}/a^{\frac{3}{2}}$) for doubly-periodic modes of vibration of a uniform cylindrical shell: n = circumferential mode-number. Curves are shown for three values of a/h.

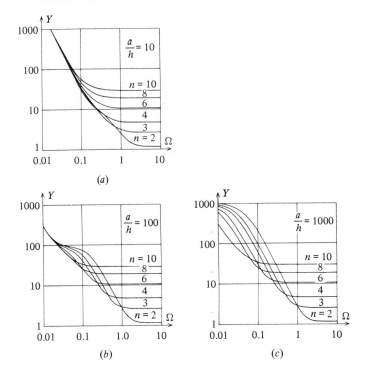

17.3 Vibration of cylindrical shell: 'shallow' modes

to determine the corresponding value of ω. But it is more instructive to display the equation graphically. This has been done in fig. 17.4 for three values of the parameter a/h. In each case the frequency parameter Y has been plotted against Ω for a range of values of the circumferential wavenumber $n \geq 2$.

At first sight the three plots look very different; but closer inspection shows some definite similarities. The most important is that the lower parts of the curves in the three diagrams for a given value of n are practically the same; that is, the relationship between Y and Ω is practically independent of the parameter a/h. The explanation of this may be seen in the schematic plot of fig. 17.5. This shows, with the same axes as in fig. 17.4, the four components of the right-hand side of (17.39) both separately, and in combination. Four significant 'crossing points' are marked A, B, C and D. Point B is where the longitudinal and circumferential stretching effects coincide, and similarly point D is where the two bending effects coincide. By virtue of the form of (17.39) (or (17.34)) the abscissae of B and D are always equal, irrespective of the values of a/h and n. In general, the resultant curve follows the path $EABCF$, but the corners are rounded and the precise final shape depends upon the dimension BD, which represents a factor of magnitude

$$\frac{2 \times 3^{\frac{1}{2}}}{n^2} \left(\frac{a}{h}\right). \tag{17.41}$$

Fig. 17.5. Schematic plot of a general constituent curve of fig. 17.4, showing the contributions from bending and stretching in both longitudinal and circumferential directions: cf. (17.39).

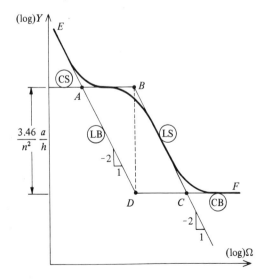

642   *Vibration of cylindrical shells*

Now the lower part of the combined curve is dominated by longitudinal stretching and circumferential bending; and since the coordinates of point C are independent of the parameter a/h, but depend only on the value of n, this part of the curve is common to all three diagrams in fig. 17.4 for a given value of n, for all practical purposes.

Another interesting feature is that A and C lie on a line of slope -1 in all cases, since the abscissae of B and D are the same and the slopes of the lines are either 0 or -2. Indeed, it is easy to show that points A and C lie on the same line in every case, irrespective of the values of a/h and n (see problem 7.2). Hence there is a line of slope -1 below which none of the curves goes: its equation is

$$Y\Omega \approx 2.4. \tag{17.42}$$

All of the above remarks apply specifically to a plot in which the variables are Y and Ω respectively; and indeed these particular variables have been chosen with a view to giving a simple picture in the low-frequency range of the diagram. It would be possible, of course, to define other frequency and wavelength variables in such a way that the curves were unified in the high-frequency range: see problem 17.3.

Formula (17.39) and the curves of fig. 17.4 agree exactly with the results of Arnold & Warburton (1953), except for low values of n (but see the next section). Apart from this, the only difference lies in the choice of dimensionless groups and our decision to use a logarithmic plot. As we have mentioned already, Arnold & Warburton found excellent agreement between their theoretical and experimental work.

17.4   Low-frequency approximations

From now on we shall proceed on the assumption that the main interest in the vibration of shells from an engineering point of view is in the low-frequency range of the relevant variables, and indeed that we shall often only be interested in the 'fundamental' mode, i.e. the one with the lowest frequency of all. In the light of the analysis of the preceding section we shall therefore simplify expression (17.39) by retaining only those terms which are derived from longitudinal stretching and circumferential bending, respectively. Hence from (17.39) we obtain the approximate formula

$$Y = [(\pi^4/n^4\Omega^4) + \tfrac{1}{12}n^4]^{\frac{1}{2}}. \tag{17.43}$$

Curves of Y against Ω according to this relation are plotted in fig. 17.6a for a range of values of $n \geq 2$. The equation of the envelope to which we have referred already has been drawn in, and its equation may be found by the standard process, as follows. Equation (17.43) may be rewritten

17.4 Low-frequency approximations

$$f(n) = (\pi^4/n^4\Omega^4) + \tfrac{1}{12}n^4 - Y^2 = 0. \tag{17.44}$$

The equation of the envelope is found by eliminating n between $f(n) = 0$ and $f'(n) = 0$ (cf. section 14.3.1). Here the second equation may be written down by inspection, and rearranged thus:

$$(\pi^4/n^4\Omega^4) - \tfrac{1}{12}n^4 = 0. \tag{17.45}$$

Hence

$$n = \pi^{\frac{1}{2}}12^{\frac{1}{8}}/\Omega^{\frac{1}{2}} = 2.42/\Omega^{\frac{1}{2}} \tag{17.46}$$

and, from (17.44)

$$Y = \pi/3^{\frac{1}{4}}\Omega = 2.39/\Omega. \tag{17.47}$$

This is the required equation of the envelope; and (17.46) gives the value of n of the curve which touches the envelope at a given value of Ω. For the purposes of finding the envelope we have regarded n as continuously variable, although physically n must be an integer. In practice the lower edge of the festoon of curves (17.43) for discrete values of n differs by little from the envelope (17.47); except, of course, that the curve for $n = 2$ provides an ultimate *base-line* for the diagram.

The curves of fig. 17.6, and their envelope, are formally similar to the broken curves of fig. 14.14, in connection with the classical buckling of a cylindrical shell under external pressure. In both of these problems the pre-

Fig. 17.6. The same plot as fig. 17.4, but omitting longitudinal bending and circumferential stretching terms (cf. fig. 17.5) and thereby making the curves independent of the value of a/h. (a) Shallow-shell equations. (b) 'Festoons' for shallow-shell and 'corrected' equations.

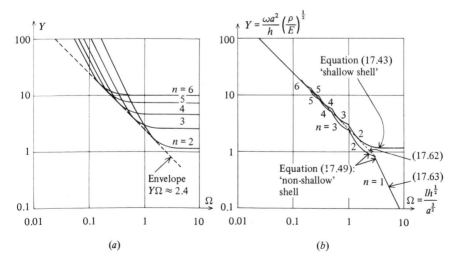

dominant structural effects are stretching in the longitudinal direction and bending in the circumferential direction, and the dimensionless length Ω is appropriate in both cases.

So far we have used the shallow-shell equation, mainly because it is so simple. We have already seen, in section 17.2.2, that the use of 'shallow-shell' equations can lead to significant discrepancies in the estimation of natural frequencies for low values of n. Therefore our next task must be to improve formula (17.43) so that it becomes more accurate in this range. Having done this, we shall be in a position to plot an 'improved' version of fig. 17.6a; and we shall find that this involves not only a minor repositioning of the curves for $n = 2, 3, 4 \ldots$, but also the addition of a curve of a different character for $n = 1$.

The most straightforward way of 'correcting' expression (17.43) for low values of n is to invoke an energy argument. Imagine that we are applying Rayleigh's method to the displacement function (17.31). In particular we must calculate (a) the strain energy of stretching (b) the strain energy of bending, and (c) the kinetic energy. Let us examine these three quantities in turn. We saw in chapters 6 and 8 that the relation between strain and displacement of the S-surface of a cylindrical shell involves straightforward operations in the developed plane of the surface, and the basic expressions are valid for all values of n, including low ones. Therefore there are no corrections to be made to expression (a) on this account. In the present problem we are reckoning that the value of l/b is so large that the only significant strain energy of bending comes from flexure in the circumferential direction. Now we saw in section 17.2.2 that the expression for strain energy of circumferential bending which we obtained for high values of n – equivalent to a 'shallow-shell' analysis – ought strictly to be multiplied by the factor $(1 - n^{-2})^2$, which is, of course, significantly different from 1 for small values of n. Therefore expression (b) should be multiplied by this factor. Lastly, in calculating the kinetic energy of the shell we have hitherto neglected the contribution from the peripheral component of velocity, which is justified for large values of n. Now the contribution from this term is easy to calculate if we assume that $\epsilon_\theta = 0$, i.e. that the mode is inextensional in the circumferential direction; and the previous expression for kinetic energy must be multiplied by the factor $(1 + n^{-2})$ on this account, as we saw in section (17.2.2). In the present case it is not absolutely true that $\epsilon_\theta = 0$; but it is a straightforward matter to show that the discrepancy introduced by this approximation is no larger than the terms which we have already discarded in setting up the simplified equation (17.43): cf. section 7.3.

In general the Rayleigh method leads to a calculation of the form

$$\omega^2 \propto ((a) + (b))/(c), \qquad (17.48)$$

17.4 Low-frequency approximations

where $(a), (b), (c)$ refer to the three total energies described above. It is therefore a straightforward matter, firstly to interpret equation (17.43) in terms of an energy calculation appropriate to a high value of n, then to introduce to (b) and (c) the correction factors as described above, and hence to determine a 'corrected' version of (17.43) which is valid for low values of n. In this way we obtain

$$Y = \left(\frac{\pi^4}{n^2(n^2+1)\Omega^4} + \frac{n^2(n^2-1)^2}{12(n^2+1)} \right)^{\frac{1}{2}}. \tag{17.49}$$

The lower edge of the festoon of curves obtained by plotting (17.49) for various values of n is shown in fig. 17.6b. For $n \geqslant 2$ the curves are of the same sort as those obtained by use of the simpler formula (17.43), but are displaced somewhat in the Y- and Ω-directions of the logarithmic plot. On the other hand when $n = 1$ (17.49) gives a different kind of relationship, namely,

$$Y = \pi^2/2^{\frac{1}{2}}\Omega^2, \tag{17.50}$$

which plots as a single straight line on the diagram. This corresponds physically to the vibration of the shell as a simply-supported beam with circular cross-sections: the case $n = 1$ corresponds to translation of the cross-sections as inextensional rings, without distortion, so there is no strain energy of bending, and therefore no second term on the right-hand side. In this connection it is important to realise that the ends of the shell are supported in a way which holds them circular ($w = 0$) but in particular does not restrict u-displacements and thus allows the end cross-sections to rotate just like the end-planes of a simply-supported beam.

We conclude (see fig. 17.6) that the lower edge of the festoon formed by the curves for $n = 2, 3, 4 \ldots$ lies close to the equation (17.47) developed earlier with $Y \propto \Omega^{-1}$; but that for $\Omega > 3$, approximately, there is a single line ($n = 1$) for which $Y \propto \Omega^{-2}$. Note that in fig. 17.6a we have not plotted (17.43) for $n = 1$, since the form of the expression is wrong in this case.

An important point made by Arnold & Warburton is brought out clearly by fig. 17.6. The circumferential mode-number n corresponding to the fundamental mode depends strongly on the proportions of the shell, i.e. on the value of the dimensionless group Ω.

In simplifying the curves of fig. 17.4 into the form of fig. 17.6 we should remember that the curves of fig. 17.6 become inaccurate in the high-frequency range. The simplest way of identifying this range is by means of the position of point A of fig. 17.5 in relation to our new diagram. The horizontal line AB in fig. 17.5 corresponds exactly to vibration in a mode involving simple symmetrical radial pulsations of the shell. Consequently we can state that the 'high-frequency' range begins, in a given case, roughly at the frequency corresponding to the lower, horizontal, part of the curve (17.43) for a particular value of n given by

$$n \approx 1.8 \, (a/t)^{\frac{1}{2}}. \tag{17.51}$$

This is, of course, a parameter which is very easy to determine.

17.5 Boundary conditions

So far we have been concerned only with modes in which w varies sinusoidally in both longitudinal and circumferential directions; and therefore our analysis has been restricted implicitly to problems having rather special boundary conditions. From the work of chapter 9 we see that in order to comply with this assumed mode of displacement we need supports at the ends of the shell which maintain a circular cross-section yet do not inhibit either longitudinal or circumferential displacement. An approximation to boundary conditions of this sort sometimes occurs in practice, but we are perhaps more likely to be concerned with shells which are either completely fixed or completely free at the ends. Thus a thin-walled steel grain-storage silo or oil-storage tank during the course of construction can be fixed at the base and free at the top; but when the roof has been fixed the top may in effect be 'held circular'. Now in chapter 9 we have found that when the behaviour of a cylindrical shell is dominated by longitudinal stretching and circumferential bending, respectively, it is convenient to discuss the various possible boundary conditions at the ends in terms of an analogous problem involving a beam on an elastic foundation. Here we shall use the same idea. It will be convenient to study first the problem of the beam on its elastic foundation, pure and simple, and then subsequently to consider the formulae which are necessary to make the transformation from this problem to the analogous problem concerning a cylindrical shell.

17.5.1 Free vibration of a beam on an elastic foundation

Consider a uniform beam whose mass per unit length and bending stiffness are m and B, respectively, and which is attached to a simple uniform elastic foundation of stiffness k per unit length: see fig 17.7. Let the coordinate x denote position along the beam and w the (small) transverse displacement. Under static conditions (cf. section 9.3) the governing equation of the beam is

$$p = Bw'''' + kw, \tag{17.52}$$

where p is an externally applied transverse load per unit length and $'$ denotes differentiation with respect to x. Now any problem of free vibrations may be

17.5 Boundary conditions

turned into one of statics by the use of d'Alembert's principle. If we suppose therefore that the displacement takes the form

$$w = \xi(x) \sin\omega t, \tag{17.53}$$

we have

$$\ddot{w} = -\omega^2 \xi(x) \sin\omega t. \tag{17.54}$$

Thus, putting $p = -m\ddot{w}$ in (17.52) and dividing throughout by $\sin\omega t$ we find the governing equation for the mode $\xi(x)$:

$$B\xi'''' + k\xi = m\omega^2 \xi$$

i.e.

$$\xi'''' + \left(\frac{k - m\omega^2}{B}\right)\xi = 0. \tag{17.55}$$

This equation (which is formally similar to (14.87)) must be solved in accordance with the prescribed boundary conditions; and when this has been done the corresponding value of natural frequency ω may be determined. In general there will be a number of different mode functions $\xi(x)$ which satisfy (17.55) and the given boundary conditions; and a definite frequency ω will be associated with each.

To fix ideas, consider first the case in which the beam is of length L and is simply-supported at the two ends. These conditions are clearly satisfied by

$$\xi = C \sin(\pi x/L), \tag{17.56}$$

where C is an arbitrary constant; and indeed this function evidently satisfies (17.55) provided

$$\pi^4/L^4 = (m\omega^2 - k)/B.$$

Therefore

$$\omega^2 = (\pi^4 B/mL^4) + (k/m). \tag{17.57}$$

Obviously there are also other modes of the form

$$\xi = C \sin(r\pi x/L), \tag{17.58}$$

Fig. 17.7. Variables used in the analysis of vibration of a uniform elastic beam attached to a uniform elastic foundation.

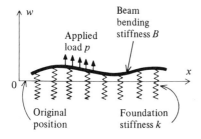

where r is an (integral) longitudinal wavenumber, each associated with a different frequency. But at present we are mainly concerned with the particular mode (17.56) which is clearly associated with the lowest value of ω.

There is a formal similarity between result (17.57) and (17.43), to which we shall return later. But a more striking feature is the fact that the first term on the right-hand side of (17.57) is exactly what it would be if we had considered an ordinary *free* beam, independent of any elastic foundation. Moreover, the second term corresponds precisely to the case of an isolated section of the beam oscillating simply as a mass on the spring provided by the associated elastic foundation. Thus we may write in this case

$$\omega^2 = \omega_b^2 + \omega_f^2, \tag{17.59}$$

where ω_b and ω_f are the natural frequency due to flexure of the *beam*, and the elastic distortion of the *foundation*, respectively. Note that for the 'beam' problem we retain the boundary conditions intact but remove the elastic foundation, while for the 'foundation' problem we consider an isolated piece, as before. Equation (17.59) holds not only for our specific boundary conditions of simple support, but also for general boundary conditions. The key to the situation is that the 'inertia' and 'spring' terms both come together in the governing equation (17.55); so that in solving the eigenvalue problem for a beam with particular boundary conditions we actually determine the value of $\omega^2 - k/m$; and since exactly the same eigenvalue problem has to be solved when $k = 0$, we have, in general, result (17.59). Note in particular that the contribution towards ω^2 from the foundation spring is completely dissociated from the boundary conditions.

It follows immediately that we can readily convert existing solutions for the vibration of a simple beam into solutions for the same beam, but now mounted on an elastic foundation. In this chapter we shall be concerned mostly with uniform beams on uniform foundations − corresponding to the vibration of uniform cylindrical shells − but in certain special circumstances the result also applies for non-uniform beams; see problem 17.4.

Fig. 17.8. Five sets of boundary conditions for a simple beam.

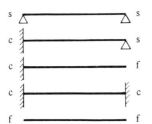

17.5 Boundary conditions

Fig. 17.8 shows a uniform beam with five different combinations of end-conditions. It is a straightforward matter to determine the modes and natural frequencies for each particular case, although in general the problem requires the solution of a characteristic equation by numerical means (see, e.g. Young, 1962, §61.4). Here we are primarily interested in natural frequencies. The most convenient way of presenting results for the different cases is in terms of an *equivalent simply-supported beam of length l* vibrating in its fundamental mode for which, by (17.57),

$$\omega = (\pi^2/l^2)(B/m)^{\frac{1}{2}}. \tag{17.60}$$

The frequencies for each beam are then determined by the relationship between the actual length L of the beam and the length l of the equivalent simply-supported beam. The most convenient form of this relationship is

$$l = L/(N + C_i), \tag{17.61}$$

where N is the (longitudinal) modenumber ($N = 1, 2, 3 \ldots$) and C_i is a constant depending on the boundary conditions for case i. Values of C_i, deduced from Young (1962, Table 6.1), are given in table 17.1.

The main point in this scheme of presentation of results is that with each set of boundary conditions and mode-number N there is associated an equivalent length of simply-supported beam vibrating in its fundamental mode. For a beam which is simply-supported at both ends, l is simply an integral fraction of L, so $C_1 = 0$ for all values of N. For the other cases the value of C_i is nonzero. But note that in each case the value of C_i is relatively insensitive to the value of N. Thus for a beam with one end clamped and the other simply-supported (case 2) $C_2 \approx 0.25$ for all values of N. Physically this means that the clamping of one end of a simply-supported beam is equivalent to shortening the beam by about $\frac{1}{4}$ of the half wavelength l. Clamping both ends (case 4) is equivalent to shortening the beam by about $\frac{1}{2}l$, and the effect of a sub-

Table 17.1. Values of C_i for use in (17.61) in the calculation of the length of an equivalent simply-supported beam.

Boundary conditions	Case	Value of N				
		1	2	3	4	5
(ss/ss)	1	$C_1 =$ 0	0	0	0	0
(clamped/ss)	2	$C_2 =$ 0.2498	0.2500	0.2500	0.2500	0.2500
(clamped/free)	3	$C_3 =$ −0.403	−0.5058	−0.4998	−0.5001	−0.5000
(clamped/clamped) (free/free)	4	$C_4 =$ 0.505	0.4997	0.5001	0.5000	0.5000

650   *Vibration of cylindrical shells*

sequent freeing of one end (case 3) is then equivalent to lengthening the beam by l.

Although this scheme of changes to the effective length in consequence of changes to the boundary conditions seems to be in accord with intuitive physical ideas, we should note that, paradoxically, the coefficients C – and hence also the natural frequencies – are precisely the same for clamped/clamped and free/free boundary conditions. The modeforms are, however, quite different; and moreover there is for the free/free case formally an additional natural frequency of *zero*, which corresponds physically to the unrestrained rigid-body motions which are possible when both ends are free. We shall return to this case later.

We have already stated that our main interest in the vibration of shells is the determination of the fundamental natural frequencies. In applying these results to shells, therefore, we shall be concerned mainly with the values of C_i for the case $N = 1$.

17.6   Natural frequencies for cylindrical shells having different boundary conditions

The developments of the preceding section make it particularly easy to transfer results from the beam problem to the analogous shell problem. The beam corresponds, as we have seen, to the behaviour of the S-surface of the shell in a 'long-wave' mode. The elastic foundation, corresponding to the effect of the B-surface in circumferential bending, affects the frequency but not the boundary conditions. Now the entire range of curves in fig. 17.6, corresponding to a shell 'held circular' at both ends, is derived in effect from the vibration of a beam with 'case 1' boundary conditions, supplemented by the elastic foundation, and interpreted for a range of values of n. The curve for each value of n corresponds to an interaction between lines BC and CF in fig. 17.5; and the line CF, corresponding to the contribution from circumferential bending, is directly analogous to the ω_f^2 term in (17.59). It follows that the effect of changing the boundary conditions of the shell is simply to alter the effective length of the shell in relation to its S-surface behaviour, without altering in any way the contribution from circumferential bending. Therefore the principal consequence of a change of boundary conditions is to shift every component curve in the festoons of fig. 17.6 by a uniform amount in the direction of the Ω-axis. This has been done in fig. 17.9. For all cases Ω is defined with respect to the actual overall length of the shell.

The first four sets of boundary conditions listed in table 17.1 produce a simple shift of the entire curve corresponding to the simply-supported boundary conditions. For the case of free/free boundaries, however, the situation

17.6 Frequencies for different boundary conditions

must be discussed separately. As shown in fig. 17.9, the mode having the lowest natural frequency is either the circumferential bending mode $n = 2$ with the generators remaining straight, or a simple 'beam' mode with the cross-sections remaining circular. The former is an inextensional mode which is only possible by virtue of the end-conditions; and it corresponds directly to rigid-body motions of the beam of section 17.5.1 on its elastic foundation. Thus, although the natural frequencies for the clamped/clamped and free/free beams respectively are in general the same, the extra frequency $\omega = 0$ for the free/free beam produces a radically different festoon in relation to the beam on elastic foundation and, by analogy, the cylindrical shell.

The correspondence of boundary conditions between the beam and the shell is just the same as in section 9.5. Thus a simple support condition for the beam corresponds to a shell which is 'held circular' but is otherwise unrestrained; a clamped beam corresponds to a shell which is held circular and also

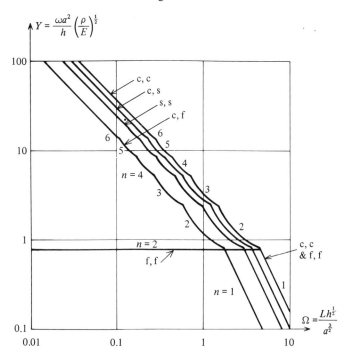

Fig. 17.9. The corrected festoon of fig. 17.6b together with corresponding curves for different boundary conditions of the shell, corresponding to the analogous 'beam' boundary conditions of fig. 17.8. Here L = overall length of shell.

prevented from axial movement, i.e. is for practical purposes fully clamped; and a free end of a beam corresponds to a free end of a shell.

The degree of shift corresponding to each set of boundary conditions is determined from the constants C_i for $N = 1$ in table 17.1. In many design situations the main problem is to determine the lowest natural frequency and the corresponding circumferential wavenumber. The curves of fig. 17.9 are useful for this purpose: given the dimensions of the shell, and hence the value of Ω, it is easy to see which mode has the lowest frequency, and what that frequency is.

Accuracy is often not of prime importance in preliminary design studies. In these circumstances it is sometimes useful to represent the curves of fig. 17.6b by two simple straight lines on the logarithmic plot. These are indicated by broken lines for a shell which is 'held circular' at both ends, and the corresponding formulae are

$$\left. \begin{array}{l} Y \approx 2.4/\Omega \\ n \approx 2.4/\Omega^{\frac{1}{2}} \end{array} \right\} \Omega < 2.9; \tag{17.62}$$

$$\left. \begin{array}{l} Y \approx 7.0/\Omega^2 \\ n = 1 \end{array} \right\} \Omega > 2.9. \tag{17.63}$$

The constants in these formulae have been rounded to two significant figures for obvious reasons. The value of n must of course be rounded to the nearest integer.

Inspection of fig. 17.6b reveals the degree of inaccuracy inherent in these simple formulae.

Formulae (17.62) and (17.63) apply to a cylindrical shell whose two ends are 'held circular'. Similar relations hold for the other boundary conditions which we have considered, and a generalised version of (17.62) and (17.63) may be presented as follows:

$$\left. \begin{array}{l} Y \approx K_1/\Omega \\ n \approx 1.57(K_1/\Omega)^{\frac{1}{2}} \end{array} \right\} \Omega < K_2; \tag{17.64}$$

$$\left. \begin{array}{l} Y \approx K_3/\Omega^2 \\ n = 1 \end{array} \right\} \Omega > K_2. \tag{17.65}$$

Values of the constants for the various sets of boundary conditions are given in table 17.2.

Expressions (17.64), (17.65) apply for the first four sets of boundary conditions, but for the fifth (17.64) must be replaced by

$$\left. \begin{array}{l} Y = 0.775 \\ n = 2 \end{array} \right\} \Omega < K_2. \tag{17.66}$$

The various constants are related to the constants given in table 17.1 for $N = 1$: in all cases, of course, the mode having the lowest natural frequency has the lowest wavenumber in the longitudinal direction.

Finally we remark that the formal similarities already noted between the present work and the buckling of cylindrical shells under external pressure extend to cases involving a variety of boundary conditions.

17.7 Concluding remarks

The calculation of natural frequencies of elastic systems is straightforward in principle, but the details can be tiresome. In this chapter we have kept the algebraic manipulation at a minimum level by considering a sequence of graded problems, each of which builds on the results of the previous one. A theme which recurs in the chapter is the usefulness of Rayleigh's principle; and it is safe to say that one of the benefits of this method is that it makes possible the derivation of relatively simple solutions covering a wide range of geometries.

A radically different approach to the same problem is also available, and has been used by many investigators. Love (1927, §334) gives a clear account of it in relation to the vibrations of a cylindrical shell. The basic idea is to set up the equations of motion in terms of the three displacement variables u, v and w; and then to assume a generalised mode of vibration and from this set up a 'frequency equation'. The frequencies are the roots of such an equation. As we remarked in section 17.1, Flügge was the first to solve a problem of this sort. He discovered that for any given mode (i.e. a pattern of nodal lines) there were three distinct frequencies. Two of these were associated with modes (Lamb's 'first class') in which the motion was essentially tangential, while the third involved both radial and tangential components of displacement. The third kind of mode was associated with much lower frequencies than the other two. This is not surprising, since the first two modes are analogous to in-plane modes of vibration of plates, for which the natural frequencies are obviously much higher than for flexural modes.

Table 17.2. Values of constants K_1, K_2 and K_3 for use in (17.64) to (17.66).

Case		K_1	K_2	K_3
1	(ss/ss)	2.4	2.9	6.9
2	(clamped/ss)	3.0	3.6	11
3	(clamped/free)	1.4	1.7	2.5
4	(clamped/clamped)	3.6	4.4	16
5	(free/free)		4.5	16

One of the problems inherent in any general method of analysis which includes all three components of displacement is that the method will yield, in principle, a large number of modes having high frequencies. This can cause difficulties in numerical work, since these additional modes can sometimes make a nuisance of themselves by intruding in an awkward way into the working. The methods of the present chapter eliminate such difficulties entirely by using only the displacement w (and in some cases v) to define the mode. This step certainly involves in principle the introduction of some approximations, but Rayleigh's principle enables us to see that the effects of these on the calculated frequencies are small.

It is appropriate to make a final remark about boundary conditions. One consequence of our scheme is to reduce Love's full eighth-order equation of the problem to a fourth-order one, and thereby to eliminate one-half of the boundary conditions of the problem. As we have seen in section 17.5.1, the effect of our various simplifications is to eliminate the boundary conditions associated with the 'B-surface', i.e. with the flexure of the surface. There are some circumstances in which this step is not valid. Essentially these involve shells for which the family of 'short-wave' solutions, which we have here discarded in favour of the 'long-wave' solutions, are in fact the more important group. Just as in chapter 8, the key to the conditions in which the short-wave solutions must be considered is that the value of the parameter $L/(at)^{\frac{1}{2}}$ is of order unity. For shells having this geometrical feature the results of the present chapter are not appropriate.

17.8 Problems

17.1 The frequency equation (17.20) for inextensional vibration of a ring of a cylindrical shell is based on the assumption that the mode of deformation involves only changes of curvature in the circumferential direction; that is, in the notation of section 2.3.3, $\kappa_y \neq 0$, $\kappa_x = 0$. Consider expression (2.33) for the strain energy of bending for a typical small element. Show that for a given value of κ_y, and $\kappa_{xy} = 0$, U_B is minimum when $\kappa_x = -\nu\kappa_y$, and that the strain energy is thereby reduced by a factor $(1 - \nu^2)$. Verify that this leads to an alteration in ω by the factor $(1 - \nu^2)^{\frac{1}{2}}$ *provided* the axial length of the ring is such that the change of cross-sectional shape in consequence of $\kappa_x \neq 0$ does not appreciably alter the expression for the kinetic energy of the ring, and that it is legitimate to put $\kappa_{xy} \approx 0$.

17.2 Write down the four simple expressions to which (17.39) reduces when each of the four terms, respectively, dominates. Determine the Y, Ω coordinates of points A and C in fig. 17.5. Verify that for both points $Y\Omega = \pi/(12)^{\frac{1}{4}}$, ir-

17.8 Problems

respective of the values of n and a/h; and hence that points A and C lie on a single, fixed line.

17.3 By examination of the coordinates of point A in fig. 17.5, show that if the data of fig. 17.5 were to be plotted with $\Omega a/h$ $(= l/(ah)^{\frac{1}{2}})$ as abscissa and Yh/a $(= \omega a(\rho/E)^{\frac{1}{2}})$ as ordinate, point A – and hence the 'short-wavelength' part of the curve – would be fixed irrespective of the values of n and a/h.

17.4 An initially straight linear-elastic beam is connected to a linear-elastic foundation. The flexural stiffness B, the mass per unit length m and the foundation spring constant k all vary with the axial coordinate x. Either by setting up the equation of motion, or by means of an energy argument, show that the contribution to the natural frequencies of the system from the elasticity of the foundation is additive in the sense of (17.59) if and only if $k(x) = Am(x)$, where A is a constant.

18

Shell structures and the theory of plasticity

18.1 Introduction

In most of the chapters of this book we have assumed that the material from which a shell is constructed behaves under stress in a linear-elastic manner. The materials which are used in structural engineering generally have a linear-elastic *range*, but behave inelastically when a certain level of stress is exceeded. Moreover at sufficiently high temperatures irreversible *creep* may be the most significant phenomenon.

It is obvious that there are some circumstances in which it is necessary for the designer to understand the behaviour of shells in the inelastic range. This subject is a large one, and in this chapter we shall give an introduction to part of it.

The aim of the present chapter is to give a glimpse, mainly through a few specific examples, of the ways in which the structural analyst may tackle problems connected with inelastic behaviour of shells. In general our plan will be to set up the simplest problems which illustrate various important points. But first it is necessary to discuss some general questions in connection with the scope of plastic theory, and the circumstances in which it is valid.

18.1.1 Plastic theory of structures

Engineering problems involving shell structures in which plasticity of the material plays an important part may be divided roughly into three categories, as follows.

(i) In many shell-manufacturing processes large-scale plastic deformation over the surface enables flat plates to be deformed into panels of spherical shells, complete torispherical pressure-vessel heads, or highly convoluted

18.1 Introduction

expansion bellows. In all of these and similar cases the material undergoes strains well into the plastic range, and there are also large overall changes in geometry during the process of deformation.

(ii) Some shell structures are subjected to loading which they resist by combinations of bending and stretching action over relatively small areas of the shell. Examples include ring-loads of the type shown in fig. 3.5a, and localised loads applied to spherical shells. In these cases the material can be strained locally well into the plastic range even though the overall distortion of the structure is relatively small.

(iii) In some *buckling* problems the behaviour of the structure can change drastically as soon as a relatively small region begins to *enter* the plastic range; and in fact it is relatively difficult to do experimental assays on the buckling of shell structures without encountering the harmful effects of plasticity.

These divisions are not water-tight. Thus, there is a class of problems lying between categories (i) and (ii) in which there are gross changes in geometry while relatively small parts of the shell are strained into the plastic range: examples include the collapse of submarine pipelines under external pressure (Palmer & Martin, 1975) and the crumpling of tubes in some energy-absorbing systems (Johnson & Reid, 1978). Also, lying somewhere in the region between categories (ii) and (iii) are problems in which geometry-change effects within the plastically-deforming zone produce unstable mechanical behaviour of the structure.

In the present chapter we shall be concerned almost entirely with problems of category (ii). The reason for exclusion of problems in category (i) is that we have developed (in chapter 6) techniques for the study of only relatively small-scale deformations of shells from their original configuration: and the problems of large-scale deformation are much more formidable. The reason for exclusion of problems in category (iii) is that a full study of buckling in the plastic range would form a relatively large addition to chapters 14 and 15: nevertheless, we have included a few simple observations on this topic in section 15.7.

The implication of remarks made under (ii) above to the effect that some structures can enter the plastic range of strain while the overall deformations remain relatively small may appear at first to be paradoxical. The best-known examples of this lie in the field of the plastic theory of beam- and frame-structures (Baker & Heyman, 1969; Neal, 1963). It is straightforward to demonstrate that a 'plastic hinge' in a simple beam made of material which does not strain-harden in the plastic range will develop plastic strains in the region of 1% when it rotates through an angle as small as 1 degree: see problem 18.1. It follows that if the elastic distortion of the remaining parts of such a beam is relatively small, the beam can be on the point of plastic collapse under load even though the overall deflection of the beam from its original form is relatively small. In order to satisfy these conditions the beam must in fact be of relatively 'stocky' proportions; i.e. the ratio of length to depth must not be too large (say of order 10, but depending on the strain at which the material reaches its elastic limit). In these circumstances the well-known 'simple' classical theory of plasticity, otherwise known as 'limit analysis', successfully predicts the loading conditions under which the beam will collapse plastically. This type of theory in relation to beam- and simple frame-structures is backed by ample experimental evidence.

Theory of this type is described as 'simple' and 'classical' because it ignores geometry-change effects in practically the same way as the linear theory of elastic structures does. In the present book, for example, we have deliberately ignored such geometry-change effects except in section 3.7.3 and chapters 14 to 16. One of the predictions of simple plastic theory is that during plastic collapse of the structure the load remains constant: the load reaches a well-defined 'limit'. This is illustrated by the central curve of the load-displacement relationship shown in fig. 18.1 for a structure whose loading is described by a single parameter or 'load factor'. Experiments on actual structures including beams, plates and shells which do not suffer large elastic deflections before the plastic range is reached show that the load may alternatively rise or fall with deflection in the plastic range, as illustrated by the other curves in fig. 18.1. Some 'rise' may be attributable to strain-hardening of the material in the plastic range, which is explicitly ignored in the simple plastic theory. But when this has been discounted the remaining rise or fall of the curve in the plastic range must be attributed to the 'geometry-change' effects which are also disregarded in the simple theory. The main argument in favour of limit

Fig. 18.1. Schematic load/deflection curves for structures made from elastic–plastic material. In the plastic range the curve may rise or fall or stay level, depending on circumstances.

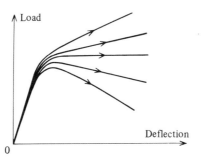

18.1 Introduction

analysis is that it does successfully predict the collapse loads of structures, except in those cases in which there are strongly negative geometry-change effects. In such cases the predicted collapse load may never be reached by the actual structure. We shall return to this point later.

One of the essential preconditions, mentioned above, for the applicability of simple plastic theory is that the structure itself should be sufficiently 'stocky'. At first this seems to be an unlikely condition ever to be met by a 'thin' shell structure. For example, an open-ended thin cylindrical shell would have to undergo substantial geometry-changes in the course of inextensional deformation before the elastic range of the material was exhausted; and thus simple plastic theory would clearly be inappropriate (see problem 18.2). On the other hand there are cases, as we have seen, in which a shell resists the application of load by means of stresses set up over a small localised region of extent $(rt)^{\frac{1}{2}}$, where r is a characteristic radius and t is the thickness. In such cases the key question concerns the 'stockiness' of this small region of the shell; and this can easily be of order 10 if the radius/thickness ratio is of the order of 100 or less. Experiments by Demir & Drucker (1963) on ring-loaded cylindrical shells, and by Dinno & Gill (1965) and Calladine & Goodall (1969) on simple pressure-vessel intersections justify the use of plastic theory in situations of this sort.

Most of this chapter will be concerned with the application of simple plastic theory to relatively simple problems involving shell structures. The theory is straightforward on account of the use of an idealised non-hardening 'perfectly plastic' material, and the explicit exclusion of geometry-change effects. In particular, it is possible to prove two general theorems, known as the upper-bound and the lower-bound theorem, respectively. These are powerful tools in the hands of the engineer, since they give upper- and lower-bound estimates of the limit load by means of relatively simple calculations. We shall describe these theorems later on.

In section 18.7.3 we shall describe a particular use of the upper-bound theorem which is specially effective for certain kinds of problem involving the application of localised loads to shells. An attractive feature of the method is that it enables us to follow rather easily changes of geometry that occur during the plastic deformation, and thus to make in a simple way approximate assessments of the destabilising effects of geometry-change in problems of this type.

Questions of this sort fall strictly within the scope of a more complete theory of perfectly plastic solids which has been developed by Hill (1957), Onat (1960) and others in order to consider explicitly the consequences of geometry-change effects. Nevertheless the method of section 18.7.3 gives some useful insights into this difficult area.

18.1.2 Elastic and plastic philosophies of structural design

The development of plastic theory in the mid-twentieth century, in distinction to the well-established nineteenth-century elastic theory, brought a renewed interest in the philosophy of structural design. The elastic theory enables an engineer to calculate stress levels in a structure which is subjected to given loading; and it seems natural therefore to design a structure in order that certain prescribed stress levels shall nowhere be exceeded. In marked contrast, the plastic theory lays emphasis on the state of the structure at the point of plastic 'collapse', and determines in particular the *limit load* (or simple plastic collapse load) of the structure or, for multiple independent loads, the *limit load envelope* of the structure. This suggests that a structure should be designed to be at the point of collapse when the design loads are multiplied by a prescribed *load factor*. Moreover, on this view it may be argued that the elastic scheme is irrational. For it is well known that any statically indeterminate structure may have appreciable lack-of-fit stresses, etc., in its nominally unloaded state. Such stresses are explicitly *ignored* in the elastic method of design; but they can be proved to be *irrelevant* to the collapse state in the plastic method.

If plastic collapse were the only possible mode of failure of a structure, this argument would be conclusive. Unfortunately, the real world is not so simple. In particular, local failure by *fatigue* either in the elastic or in the plastic range is possible and indeed common, in structures which are subjected to repeated or cyclic loadings. In these circumstances the analysis of peak stress levels by means of elastic theory is appropriate. In many structures the situation is more complicated than this discussion would suggest, for the behaviour of a structure under cyclic loading may become entirely elastic after an initial phase in which limited plastic flow takes place without the development of a collapse state. So-called *shakedown* analysis of structures is often the most rational basis of design for a structure such as a pressure-vessel which must tolerate cyclic loading during its operational lifetime. Shakedown analysis involves both elastic and plastic calculations, and is beyond the scope of this book: but see Leckie (1965); Leckie & Payne (1966).

Limit analysis according to simple plastic theory does however have a key part to play in design of structures against *overload* conditions. Thus, in a manufacturing plant the failure of control gear may involve a pressure-vessel being subjected to a pressure considerably in excess of the upper limit of the design cycle. Furthermore, a vessel may be subjected to an externally applied blast loading in consequence of the mechanical failure of neighbouring equipment. In either case, the satisfactory performance of the plant depends upon the integrity of the vessel in a 'once-only' application of load. In these circumstances the simple plastic theory is the appropriate tool for analysis and design.

18.1 Introduction

It may be remarked at this point that an alternative approach is possible to the analysis of plasticity in structures, along entirely different lines from the simple plastic theory which we have described above. The basic idea behind this alternative approach is to envisage the performance of a sequence of calculations essentially by means of elastic theory, in which the load is increased step-by-step, and any plastic strain which may occur is treated computationally in terms of the ideas of 'initial' or 'thermal' strain. Calculations of this sort presuppose large-scale computations. An advantage is that the development of plastic zones, etc. can be traced, and geometry-changes taken into account if this is desired: see Marcal (1969). Undoubtedly there are circumstances in which this approach is useful. From the point of view of the present chapter, however, a distinct disadvantage of this scheme is that the calculations must necessarily be performed on *ad hoc* structural cases. Thus overall patterns of behaviour do not emerge clearly. Furthermore the general advantages of the powerful structural theorems of simple plastic theory are not used; and this is especially significant in relation to the design advantages afforded by one of the theorems of simple plastic theory, which will be described in the next section.

18.1.3 The general equations of plastic theory

The simple plastic theory of structures is based on a particularly stark idealisation of the mechanical properties of the material. In the simplest version of plastic theory an element of material is considered to be *rigid* if the state of stress lies within a prescribed *yield condition*, but to *flow* irreversibly at constant stress when the yield condition is satisfied. The yield condition is in general a closed convex surface enclosing the origin in a suitable multi-dimensional *stress space*; and the mode of (incremental) plastic flow is determined by a *normality rule*. For a full description see, e.g. Prager (1959) or Calladine (1969a). At every point in the structure the state of stress must lie at or within the *yield condition*. The state of stress throughout the body, together with the prescribed loading, must also satisfy the equations of statical *equilibrium*; and finally the strain increments derived from the flow rule must satisfy the relations of geometric *compatibility*, which also involve boundary conditions. The three 'master' equations of the problem thus correspond precisely to those of elastic theory except that Hooke's law of the elastic theory is now replaced by the appropriate description of the idealised plastic material, i.e. by the yield condition together with its associated flow rule.

The character of the solution of a problem in plasticity is markedly different from that according to elastic theory. This is a direct consequence of the sharp nonlinearity of the prescription of the mechanical properties of the material. In particular, the solution involves a definite 'collapse' or 'limit' load for the structure.

We have remarked already that an important feature of simple plastic theory is the existence of two powerful theorems, usually known as the lower-bound theorem and the upper-bound theorem, respectively. These enable the analyst to make rapid upper- and lower-estimates of the limit load of a given structure at the expense of relatively simple calculations. The two theorems are well known, both in the context of structures (e.g. Symonds, 1962) and also in relation to metal-forming processes (e.g. Johnson & Mellor, 1973). There is no need to prove them formally here. But we shall illustrate their use in subsequent sections by means of various simple examples.

The two theorems apply equally well whether we are analysing a structure in terms of stress components or in terms of stress resultants such as bending moment and shearing force, which may be regarded as generalised stresses. Thus it is natural to discuss the state of stress in a shell in terms of stress resultants N, M, Q, etc. In this case it is necessary to begin by setting up an appropriate 'yield condition' in terms of generalised stresses. This involves the analysis of an element of the shell as a 'sub-structure' which is subjected to external loads in the form of the stress-resultants. Nevertheless, it is occasionally more convenient to revert to a description in terms of stress itself, as we shall see in section 18.7.

Throughout this chapter we shall use Tresca's description of the yield condition of the material. This is based on the physical hypothesis that the yield-point will be reached in an element when the greatest shearing stress on any plane reaches a critical value: see, e.g., Symonds (1962).

In the following sections we shall consider first a more-or-less complete solution of a very simple shell problem in terms of simple plastic theory; then we shall analyse some more difficult problems by means of the lower-bound theorem; and, lastly, we shall illustrate the use of the upper-bound theorem on a problem which may be extended to include some geometry-change effects.

18.2 Cylindrical shell subjected to axisymmetric loading applied at an edge

Consider a long thin circular cylindrical shell whose edge is subjected to an axisymmetric loading Q, M as shown in fig. 18.2. The shell is made from material which may be idealised as rigid/perfectly plastic with yield stress σ_0 in pure tension, and which yields under more general stress systems according to Tresca's hypothesis.

Simple plastic theory enables us to work out combinations M, Q for which the shell is at the point of plastic collapse. We shall therefore aim to determine

18.2 Axisymmetric load applied at an edge

the locus of collapse states in a Cartesian Q, M space. For the sake of simplicity we shall assume initially that there is zero pressure difference across the wall of the shell, and zero axial tension in the cylinder; but we shall consider later on the way in which our solution would be modified in the presence of loadings of this sort.

It will be convenient to use capital letters to represent the variables and shell dimensions, etc., and to reserve lower-case letters for dimensionless variables which we shall define in due course.

The ingredients of the complete mathematical problem are the equilibrium equations, the compatibility relations and the appropriate yield condition/flow rule. In *elastic* theory it is often convenient to combine the various equations into a single governing equation (see, e.g. section 3.3). In plastic theory, on the other hand, it is generally more convenient to approach the solution *either* via the equilibrium equations and yield condition *or* via the compatibility relations and the flow rule. In the present problem it is easier to take the former route. Accordingly we must first describe the equilibrium equations and the yield condition.

The stress resultants and equilibrium equations are exactly the same as in the corresponding elastic problem of chapter 3: see fig. 3.3 for sign conventions. The two equations of equilibrium are:

$$\frac{dM_x}{dX} = Q_x, \tag{18.1a}$$

$$\frac{dQ_x}{dX} = \frac{N_\theta}{A}. \tag{18.1b}$$

Here X is the axial coordinate and A is the radius of the shell.

In chapter 3 the elastic response of a typical element was described in

Fig. 18.2. A long thin cylindrical shell which is loaded at its free edge by symmetrically applied bending moment and shearing force: dimensions and sign conventions.

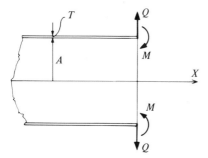

terms of a linear relation between the generalised stresses M_x, N_θ and the generalised strains κ_x, ϵ_θ. On the basis of arguments set out in section 2.2 for a *thin* shell, the generalised elastic law did not involve Q_x; and neither did it involve M_θ, since the latter was a 'reaction' which could be eliminated by means of the condition $\kappa_\theta = 0$.

In the present problem the mechanical properties of an element are specified by means of a yield-locus in a two-dimensional M_x, N_θ space, as shown in fig. 18.3. The shape of this is rather complicated, and consists of parabolic arcs and straight lines. A derivation is outlined in problem 18.3. Thus the locus is actually a cross-section of a three-dimensional yield-locus which must be used when $N_x \neq 0$: this is shown in fig. 18.4. For almost exactly the same reasons as in the corresponding elastic problem, Q_x and M_θ do not enter the specification of the mechanical properties of the element. By the symmetry of the present problem, $\dot{\kappa}_\theta = 0$, and M_θ is simply a self-equilibrating 'reaction'.

Corresponding to the yield-locus of fig. 18.3a is a *flow rule* which states that for a state of generalised stress corresponding to a point on the yield-locus, the outward-directed normal to the locus has the direction of the corresponding strain-increment vector $\dot{\kappa}_x$, $\dot{\epsilon}_\theta$ in a parallel Cartesian space, as shown. This is the well-known 'normality rule' of simple plastic theory (e.g. Symonds, 1962). Here we use the notation

$$\dot{\kappa} = \delta\kappa, \text{etc.}$$

Fig. 18.3. (*a*) Yield-locus in M_x, N_θ space when $N_x = 0$ for a shell element made from 'Tresca' perfectly plastic material. At any point on this locus the corresponding strain-increment vector is represented by an outward-directed normal. (*b*) The same locus plotted in terms of dimensionless variables, and showing the integration of (18.9).

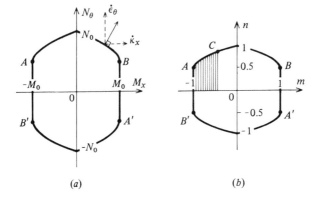

18.2 Axisymmetric load applied at an edge

to denote a small increment in a component of generalised strain. The components $\dot{\epsilon}_\theta, \dot{k}_x$ of the generalised strain increments *correspond* to the components N_θ, M_x of the generalised stress in the sense that the scalar product

$$N_\theta \dot{\epsilon}_\theta + M_x \dot{k}_x \tag{18.2}$$

represents the rate of dissipation of energy per unit area of shell.

In the theory of perfectly plastic solids the straining is quasi-static and time is not of the essence: but it is important nevertheless to distinguish between an *increment* of strain and the *total* strain (see, e.g., Prager, 1959). Note in particular that the flow rule does not determine the *magnitude* of the components $\dot{\epsilon}_\theta$ and $\dot{k}_x$, but only the *sense* of them and the ratio $\dot{k}_x/\dot{\epsilon}_\theta$.

In order to remind ourselves that in kinematic terms we are concerned with *increments* of displacement and strain, it is useful to use the term *velocity field* or, since we are concerned only with total displacements which are so small as to leave the overall geometry essentially unchanged, the better term *incipient* velocity field.

Consider the shell shown in fig. 18.2. In general, if the values of Q, M applied at the loaded edge are sufficient to cause plastic collapse, there must be a zone near the edge of the shell in which increments of plastic strain are taking place, and therefore in which elements of the shell are stressed to the yield-point. In this zone, therefore, the equilibrium equations (18.1) must be supplemented by the yield condition: and the arrangement is statically determinate in principle, since the yield condition furnishes a third relation between the three stress resultants M_x, Q_x and N_θ. The equations cannot be

Fig. 18.4. Three-dimensional yield-locus in M_x, N_θ, N_x space for an element made of 'Tresca' material when $\dot{k}_\theta = 0$: cf. fig. 18.3.

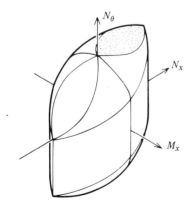

solved, however, until two boundary conditions have been specified. Now in an elastic version of this problem we may specify *arbitrary* values of the edge-loads M and Q, and solve the equations accordingly, cf. section 3.4. In the present plastic problem, on the other hand, we are only interested in those *particular* combinations of M and Q, to be determined, which produce a state of plastic collapse. Fortunately it is possible to fix the 'boundary' conditions at the interface between the plastically deforming part of the shell and the remaining rigid region as follows. Since the rigid region has $\dot\epsilon_\theta = \dot k_x = 0$, and $\dot\epsilon_\theta$ cannot undergo a step change across the interface, $\dot\epsilon_\theta$ must also be zero just inside the deforming zone. The normality rule then indicates that the material at the rigid-plastic boundary must be stressed at a point on one of the two straight segments AB', $A'B$ of the yield-locus. There is in fact a plastic hinge circle at the interface; and since $\dot\epsilon_\theta \neq 0$ further inside the plastically-deforming zone, the normality rule demands that the M_x, N_θ point corresponding to the interface must be at one of the four points A, B, A' or B' immediately adjacent to the curved parts of the locus. For the sake of definiteness let us suppose that the incipient velocity of the material immediately to the right of this boundary is *outward directed*; then in this case we may show that the appropriate point is A.

Another condition at the interface follows from considerations of statical equilibrium. Since $M_x = -M_0$ at the interface, as argued above, and since M_x cannot fall below this value on either side of the boundary without violating the yield condition, (18.1a) indicates that $Q_x = 0$ at the interface: Q_x is certainly continuous in this region in the absence of external loading.

At this juncture it is convenient to introduce dimensionless variables, and to drop some subscripts. First we define

$$n = N_\theta/\sigma_0 T, \quad m = 4M_x/\sigma_0 T^2, \tag{18.3}$$

so that the dimensionless yield-locus (n, m) passes through $(0, \pm 1)$ and $(\pm 1, 0)$: see fig. 18.3b. Then we define

$$x = X/(AT)^{\frac{1}{2}} \tag{18.4}$$

in general accordance with the previous experience of chapter 3, and lastly

$$q = 4Q_x A^{\frac{1}{2}}/\sigma_0 T^{\frac{3}{2}} \tag{18.5}$$

in such a way that the coefficients of the dimensionless equilibrium equations derived from (18.1) are purely numerical:

$$\frac{dm}{dx} = q, \tag{18.6a}$$

$$\frac{dq}{dx} = 4n. \tag{18.6b}$$

The conditions at the rigid-plastic boundary (say $x = 0$) are thus

$$m = -1, \quad n = 0.5, \quad q = 0. \tag{18.7}$$

There are various possible ways of solving the equations. Perhaps the most direct is to combine (18.6a) and (18.6b); then

$$q\frac{dq}{dx} = 4n\frac{dm}{dx}. \tag{18.8}$$

This may be integrated directly:

$$q_2^2 - q_1^2 = 8\int_1^2 n\,dm, \tag{18.9}$$

where 1, 2 represent any two locations within the plastically-deforming zone of the shell, and the integration is done around the perimeter of the yield-locus. Since $q = 0$ at the point corresponding to A we find that for a point C on the locus the above integration gives

$$q_C^2 = 8 \times \text{(shaded area in fig. 18.3}b). \tag{18.10}$$

The sign of q is positive, by (18.6b), since $n > 0$ in this region. This calculation is straightforward, and it produces the curved m, q locus which is plotted in fig. 18.5 for $q > 0$. When the m, n trajectory reaches point B, the calculation comes to a halt.

Essentially the calculation is a marching-out of the solution from a known interface condition into the deforming zone. We may determine the distance

Fig. 18.5. Collapse-load locus for the shell of fig. 18.2, obtained by using the yield-locus of fig. 18.3.

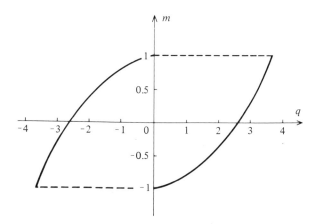

668   *Shell structures and the theory of plasticity*

δx corresponding to two adjacent points on the yield-locus by rearranging (18.6a) to give

$$\delta x = (\delta m)/q. \tag{18.11}$$

Thus, by means of numerical integration we may express the variables n, m, q as functions of x, as shown in fig. 18.6: here we have taken the origin of x at the interface between the rigid and the deforming regions. The above calculation presupposes that the shell extends in the positive x-direction at least as far as our integration. In fact we are at liberty to suppose that the shell ends at an arbitrary point within the plastically-deforming zone, and that the corresponding values of m, q are those which are applied externally. In other words, the curve $q > 0$ of fig. 18.5 is not only the trajectory of m, q along the plastically-deforming zone; it is also the locus of m, q which, when applied to the edge of the shell, will bring it to the point of plastic collapse.

The curve of fig. 18.5 in the region $q < 0$ corresponds to cases in which the incipient velocity is *inward directed*; in this case the integration proceeds from A' to B' around the lower edge of the locus of fig. 18.3b. The complete m, q locus is, of course, a closed figure. In fig. 18.5 the two curves have been joined by broken lines corresponding to solutions in which there is a plastic hinge circle at the very edge of the shell, but otherwise no plastically-deforming zone: the previous argument that $q = 0$ at the hinge does not apply when the hinge is at the very edge of the shell.

Fig. 18.6. Stress resultants as functions of x in the plastically deforming zone near the edge of the shell of fig. 18.2.

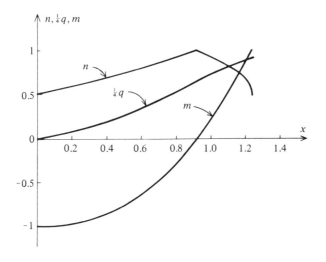

18.2 Axisymmetric load applied at an edge

This completes the solution of the problem in the plastically-deforming zone as far as the stress resultants are concerned. It is a straightforward matter now to solve the kinematic equations in order to determine the modeform of the collapse velocity field: see problem 18.4.

The solution is not strictly complete until it has been shown that there is a statically admissible set of stress resultants $n(x), m(x), q(x)$ in that part of the shell which has hitherto been assumed to be rigid, and which is therefore in the elastic range in the actual shell. These must not violate the yield condition. This part of the solution presents no difficulty, particularly if the shell is semi-infinite in the negative x-direction. The solution is not unique, and it is perhaps most convenient to 'borrow' a stress field from a previously obtained elastic analysis for this purpose. If the length of the shell is finite, however, there may be difficulty in satisfying boundary conditions at the remote end, unless the shell is sufficiently long: see section 3.8. Eason & Shield (1955) investigated the influence of free ends on the load-carrying capacity of relatively short shells.

Thus far the shell has been required to sustain no interior pressure. It is a straightforward matter to modify the calculation in order to take into account a uniform interior pressure P, provided N_x is still zero. First we define a dimensionless pressure p by

$$p = PA/\sigma_0 T; \qquad (18.12)$$

then we rewrite (18.6b) as

$$\frac{dq}{dx} = 4(n-p), \qquad (18.13)$$

and thus obtain in place of (18.9) the relation

$$q_2^2 - q_1^2 = 8\int_1^2 (n-p)dm, \qquad (18.14)$$

which is equally straightforward to evaluate.

If now we also add a uniform tension $N_x \neq 0$ to the shell, the required modification to the calculation goes deeper. In this case we must alter the yield-locus, for the curve in fig. 18.3 is based explicitly on the assumption that $N_x = 0$. Otherwise there is no change to the equations. The full locus N_x, N_θ, M_x is shown in fig. 18.4. As before, M_θ is a reaction, and the condition $\dot{\kappa}_\theta = 0$ has been imposed. The surface has a rather complicated form (Prager, 1959, §3.11), but it is possible to derive it relatively easily (Calladine, 1972c).

The above calculation illustrates a number of general points in connection with the analysis of structures made from rigid/perfectly plastic material.

These points include, among others, the static determinacy within a plastically-deforming zone; the special conditions which may occur at a rigid-plastic interface; the need for the use of trial-and-error in solving problems; and the nature of the conditions which must be fulfilled in the rigid zone.

In one important respect, however, the example is misleading as a guide to the analysis of shells having more complicated geometries. In general, a yield-locus such as that of fig. 18.3, which involves nonlinear curves, makes the resulting equations awkward. It is true that we negotiated this particular difficulty with ease in our simply cylindrical shell; but it is easy to produce counter-examples of more general shells (such as the one considered in section 18.5.1) for which our special method is not productive. For the purpose of analysis it is generally more convenient if the yield-locus is expressed in terms of *linear* functions of the generalised stresses. Since, of course, the yield-locus is always closed, these relations will be merely *piecewise* linear. Nevertheless, the problems associated with the transfer from one linear region to another are usually more straightforward to overcome than those which are associated with nonlinear functions. The most obvious example, of course, is Tresca's piecewise linear yield condition for an element subjected to independent principal stresses. In most problems the analysis may be done more expeditiously by means of Tresca's yield condition than by the use of von Mises' quadratic condition.

It may happen, of course, that the yield-locus for some particular problem in the appropriate generalised stress space is actually made up of nonlinear sectors (e.g. fig. 18.4), and that a piecewise linear version is in some sense incorrect. Fortunately, it is not difficult to circumvent difficulties of this sort. For example, it may be particularly convenient (as we shall see) to use a 'square' yield-locus n, m instead of the more complicated one shown in fig. 18.3. If this square *circumscribes* the correct locus (see fig. 18.7) then – by a general theorem of simple plastic theory (Drucker & Shield, 1959; Calladine, 1969a, §4.8) – the resulting m, q collapse load locus for the shell circumscribes the correct one. On the other hand, if the square *inscribes* the correct locus, then similarly the resulting m, q locus inscribes the correct one. But since in this case the two solutions for the 'square' yield-locus differ only by a simple scaling factor, we may obtain geometrically similar upper and lower bounds on the collapse locus rather easily. In the present case the linear dimensions of the inscribed and circumscribed loci are in the ratio 3/4, and consequently the resulting m, q locus is located to within reasonably close limits. The same considerations apply, of course, to the interpretation of results obtained for a Tresca material when it is known that the actual material obeys von Mises' yield condition: see Calladine, 1969a, §4.8.

18.2 Axisymmetric load applied at an edge

A piecewise linear yield condition may sometimes be obtained as a result of a physical simplification. Thus, if a solid element of a shell is replaced by a 'sandwich' element, the yield-locus of fig. 18.4 is transformed to a piecewise linear one. We shall use this physical idea in section 18.7.2. On the other hand, the form of a piecewise linear representation may be chosen, as below (fig. 18.7) for reasons of convenience, without any explicit physical basis.

18.2.1 Collapse of cylindrical shells, using a 'square' yield-locus

Consider the problem illustrated in fig. 18.8a, in which a long thin cylindrical shell is subjected to a uniform ring-load of intensity F per unit circumference: cf. fig. 3.5a. We wish to determine the value of F at which the point of collapse is reached, on the assumption that the yield-locus is 'square',

Fig. 18.7. Inscribed and circumscribed 'square' approximations to the yield-locus of fig. 18.3.

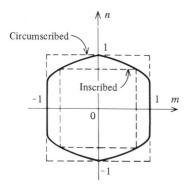

Fig. 18.8. (a) A long shell loaded by a uniform ring-load F, whose limiting value is to be determined for the yield-locus shown in (b).

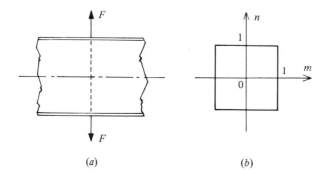

as shown in fig. 18.8b. It is clearly useful to define a dimensionless load intensity f on the same lines as q:

$$f = 4FA^{\frac{1}{2}}/\sigma_0 T^{\frac{3}{2}}. \qquad (18.15)$$

The value of f at collapse will be determined by a negative discontinuity in the value of q at the point of application of the load.

The equilibrium equations are just the same as before (see (18.6)):

$$\frac{dm}{dx} = q$$

$$\frac{dq}{dx} = 4n.$$

An immediate consequence of the adoption of a square yield-locus n, m is that the problem becomes directly analogous to that of a rigid/perfectly plastic beam supported on a rigid/perfectly plastic foundation, as shown in fig. 18.9a. The sides $m = \pm 1$ of the yield-locus correspond to the achievement of a plastic limit moment in the beam, while the sides $n = \pm 1$ correspond to yielding of the foundation.

The fact that the yield condition consists only of lines parallel to the two orthogonal axes indicates that the two effects are physically distinct and separable: hence the beam/foundation analogy.

It should be noted here that in general the idea of separating the stretching and bending effects in the shell into distinct S- and B-surfaces is not so useful in the plastic theory as in the elastic theory of uniform shells. The crux of the matter is that whereas the *elastic* law for the stretching and bending of a uniform element genuinely separates into two distinct parts (see chapter 2), the corresponding mechanical properties in the plastic range, as described in part by figs. 18.3 and 18.4 do not separate: instead, they involve in general a nonlinear *interaction* between the two effects. However, in the special case of the 'square' yield-locus of fig. 18.8b, the plastic properties of the S- and B-surfaces are effectively separated, and in consequence we may use the analogy of a beam and a foundation with unconnected mechanical properties.

It is easy to sketch a solution to the analogous beam problem in terms of the three-hinge mechanism shown in fig. 18.9b. The sense of rotation of the hinges is obvious, but the distance between them in the axial direction is as yet unknown. A sketch of the distribution of stress resultants in the region between the two outer hinges is also given in fig. 18.9c. Clearly $n = 1$ throughout, since the foundation is in a state of plastic yield. Therefore, by (18.6b) $q(x)$ has a constant positive slope in both zones; but it has a negative discontinuity at the centre on account of the concentrated external load there. Ac-

18.2 Axisymmetric load applied at an edge

cording to (18.6a) m is a quadratic function of x, which is limited by the yield condition to the range $-1 \leqslant m \leqslant 1$. The conditions at the two end hinges must be examined carefully. As they are hinges, the value of m must be exactly equal to -1. Now since there are no external loads in this region, q must be a continuous function of x (in contrast to the situation at the central hinge). If the value of q were nonzero at the hinge, dm/dx would also be nonzero and the condition $m \geqslant -1$ would be violated on one side of the hinge or the other; which is contrary to hypothesis. For this reason q must vanish at each of the two outer hinges: and consequently the variation of m is as shown.

It is now a simple matter to determine the separation, l, of the hinges. It is convenient to put the origin at the left-hand hinge, and to consider the portion of the beam between the left-hand and central hinges. Putting $n = 1$ in (18.6b), integrating, and setting $q(0) = 0$ we find

$$q = 4x. \tag{18.16}$$

Substituting this into (18.6a), integrating, and putting $m(0) = -1$ we have

$$m = 2x^2 - 1. \tag{18.17}$$

The location of the applied load is at the value of x for which $m = 1$, i.e.

Fig. 18.9. (a) A plastic beam on a plastic foundation, analogous to the problem of fig. 18.8. (b) Location of circumferential hinges, separated by an unknown distance l. (c) Distributions of stress resultants for the shell in the limiting state.

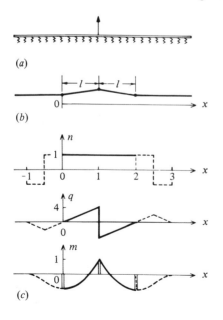

674   *Shell structures and the theory of plasticity*

$x = 1$. Under this load the value of q jumps from 4 to -4 (by symmetry), so $f = 8$.

In order to complete the solution we must verify that it is possible to satisfy the equilibrium equations in the non-deforming regions without violating the yield conditions. One of many possible ways of doing this is indicated in fig. 18.9c by broken lines. In the region immediately to the left of the left-hand hinge the stress resultants n, m and q are all brought to zero over a length $\Delta x = 1$. If there is a portion of shell of at least this length to the left of the hinge 0 (and a similar portion to the right of the third hinge) then our solution is correct. On the other hand, if the ends of the shell are free and the overall length of the shell is less than 4 units, then our solution does not satisfy all of the boundary conditions, and we must begin again: see problem 18.5.

In terms of the variables relating to the original shell, we thus find that if a concentrated ring-load is applied at a distance of at least $2(AT)^{\frac{1}{2}}$ from a free end, the collapse load is given by

$$F_0 = 2\sigma_0 T^{\frac{3}{2}}/A^{\frac{1}{2}}; \qquad (18.18)$$

the mechanism of collapse consists of three hinges set at spacings of $(AT)^{\frac{1}{2}}$, with the material between them in a state of hoopwise plastic flow.

In section 18.4 we shall consider some variations on the problem of the collapse of a long shell under a ring-load which is applied sufficiently far from the ends; but before we do so it will be useful to make some general remarks about our method of attack.

18.3   Upper- and lower-bound theorems

In the example described above we proceeded by solving the equilibrium equations subject to the constraints imposed first by the yield condition and second by any relevant boundary conditions. It is also possible to approach the same problem from the point of view of the kinematics of the mode of deformation of the system. Thus we may consider an incipient velocity field or 'mode of collapse' in which each generator deforms as shown in fig. 18.9b, with three concentrated hinges separating sections which remain straight: see fig. 18.10. Let the (unknown) spacing of the hinges be L in the axial direction, and let the incipient radial velocity at the central hinge be v. In such a velocity field, mechanical energy will be dissipated both in the rotation of the plastic hinges and also in the circumferential stretching of the material situated between the hinges; and on the other hand energy will be supplied by the radial line load F, since everywhere on the loaded circumference the point of application moves with velocity v. For this particular calculation it is perhaps more convenient to work with the dimensions etc. of the shell proper than with dimensionless quantities.

First let us calculate the energy dissipated in the circumferential hinges. The relative angular velocities of hinge rotation of the outer and inner hinges have magnitudes v/L and $2v/L$, respectively. The full plastic bending moment per unit length of circumference is M_0. Then the total rate of dissipation of energy in all four hinges of the postulated mechanism is $4M_0 v/L$ per unit circumference.

At the central hinge there is a circumferential strain-rate of v/A; and since the circumferential tensile-stress resultant at yield is equal to N_0, the rate of dissipation of energy in circumferential stretching is here equal to $N_0 v/A$ per unit area. For the simple mode which has been chosen, the mean rate of dissipation of energy is clearly one-half of this, and so over the entire length $2L$ the rate of dissipation of energy is equal to $N_0 vL/A$ per unit circumference.

Summing these components and equating to the energy input per unit circumference, we have

$$Fv = 4M_0 v/L + N_0 vL/A,$$

and thus

$$F = 4M_0/L + N_0 L/A. \qquad (18.19)$$

Here the value of F clearly depends on our choice of value of L. The right-hand side is a minimum with respect to L when

$$L^2 = 4M_0 A/N_0.$$

Putting N_0 and M_0 in terms of σ_0 and T, (see (18.3)) we have, simply

$$L = (AT)^{\frac{1}{2}}, \qquad (18.20)$$

and thus (cf. (18.18))

$$F = 2\sigma_0 T^{\frac{3}{2}}/A^{\frac{1}{2}}.$$

Fig. 18.10. Perspective sketch of the mode of collapse for the problem of Fig. 18.8.

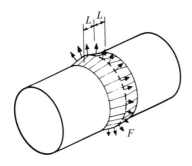

These results agree precisely with those obtained by the previous method.

The calculation illustrates the 'upper-bound' theorem in the simple plastic theory of structures. This theorem states that if the collapse load for a given structure is determined by means of an energy balance in relation to a hypothetical mode of collapse, then the calculation yields an answer which is an *upper bound* on the actual collapse load. If the hypothetical mode happens to coincide with a *correct* mode (i.e. one which satisfies *all* of the conditions of the problem), then the calculation produces the correct collapse load, as in the present example.

In section 18.7 we shall use the upper-bound method to study some problems for which it is particularly convenient: the kinematic aspects of these problems are more straightforward than the statical equilibrium aspects. Note, however, that the upper-bound method has the inherent disadvantage that if the range of hypothetical modes does not include a correct mode, then the resulting collapse loads necessarily exceed the actual collapse loads, and so provide an 'unsafe' basis for design. Thus, although the method possesses the obviously attractive feature that the state of affairs in any *rigid* regions does not enter the calculations, it nevertheless throws the onus on the analyst to make a good shot at defining the extent of any such regions.

A counterpart of this theorem in the simple plastic theory of structures is the so-called *lower-bound theorem*. In brief, this states that if a structure subjected to given applied loads has stress resultants which everywhere satisfy the equilibrium equations without violating the yield condition, then the structure will safely carry the applied loads.

Our earlier calculation may be adduced as an example of this theorem, provided it is interpreted a little differently from before. As far as the lower-bound theorem is concerned, all questions of *collapse mechanism* and *velocity field* are irrelevant: the theorem is concerned only with equilibrium stress distributions which do not violate the yield condition. In this connection we may view our previous calculation as an exercise in finding the maximum value of f consistent with the satisfaction of the equilibrium equations (18.6) and the yield condition ($-1 \leqslant m \leqslant 1, -1 \leqslant n \leqslant 1$). For these purposes we may ignore the fact that we obtained some clues to the proper form of the various stress resultants by imagining a mechanism of collapse. In section 18.5 we shall use the lower-bound theorem in circumstances where the full equations would pose formidable problems. Two attractions of the method are that it calculates collapse loads which are on the *safe* side of the actual collapse loads; and that it is particularly useful in *design* studies. Distributions of stress calculated according to elastic theory satisfy, in particular, the equations of equilibrium; consequently they may be used for the purposes of lower-bound plasticity calculations.

18.4 Cylindrical shell subjected to axisymmetric band-loading

In this section we consider the problem of a long cylindrical shell which is subjected to a uniform pressure P which acts over a region of length B in the axial direction, as shown in fig. 18.11a. Our aim is to determine the pressure required for collapse, as a function of the dimension B. The previous calculation considered in effect the case $B \to 0$; and for most purposes it will be more appropriate to consider the limiting value of the product PB in relation to the previously determined limit value of F.

Just as in section 18.2 we shall assume that $N_x = 0$ throughout; and we shall also use the 'square' yield-locus.

The dimensionless equilibrium equations are just the same as before, (18.6a), (18.13), except that now the unknown pressure P acts only over a sector of length B, or of length b in terms of the dimensionless coordinate x.

It is clear that provided the width B is sufficiently small, the general character of the function $m(x)$ will be the same as before (fig. 18.9), with the value of m passing from -1 to $+1$ and back to -1 in the plastically-deforming region. It is in fact most useful to concentrate attention on the form of the shear function $q(x)$, since its *slope* is related to n (and p) while the *integral* is related to m. A convenient scheme for visualising the general case is shown in fig. 18.11b. There are two curves on the diagram, marked $\int 4n dx$ and $\int 4p dx$, whose slopes are $4n$ and $4p$ respectively. The (algebraic) difference in ordinate between them represents q; and the area enclosed between them represents (by (18.6a)) the change in value of m between the corresponding points on the generator. In fig. 18.11b the curve $\int 4p dx$ has zero slope except in the loaded region, while the other curve has a constant slope of 4, corresponding to yield in hoopwise tension. We expect that there will be plastic hinges at the three locations of maximum/minimum m, which occur when

Fig. 18.11. (a) The same problem as in fig. 18.8, but with the load spread uniformly over a region of length B. (b) A convenient diagram for performing the relevant calculations.

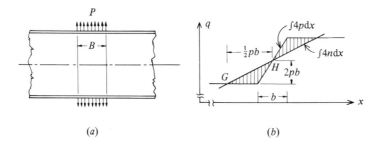

678 *Shell structures and the theory of plasticity*

$q = 0$, i.e. when two curves intersect. This requires that the shaded area between G and H should be equal to the total allowable 'throw' of 2 in the value of m between its extreme values in the yield-locus of fig. 18.8b.

The leading dimensions marked on fig. 18.11b are determined in terms of b on account of the slopes of the two lines; hence we find that our required condition on the enclosed area is given by

$$(p - 1)(\tfrac{1}{2}b)pb = 2. \tag{18.21}$$

Treating this as a quadratic equation in pb we find the following (positive) root:

$$pb = \tfrac{1}{2}b + [4 + (\tfrac{1}{2}b)^2]^{\tfrac{1}{2}}. \tag{18.22}$$

In the case $b = 0$ we replace bp by $\tfrac{1}{4}f$ (see (18.15)) and thus obtain, as before,

$$f = 8.$$

It is convenient to normalise the allowable value of pb with its value when $b \to 0$: thus we put

$$pb/pb)_{b=0} = \tfrac{1}{4}b + [1 + (\tfrac{1}{4}b)^2]^{\tfrac{1}{2}}. \tag{18.23}$$

This function is plotted in fig. 18.12: the load-carrying capacity pb increases more-or-less linearly with the width b of the loaded zone. From (18.23) we also obtain easily an expression for the pressure p:

$$2p = 1 + (1 + 16/b^2)^{\tfrac{1}{2}}. \tag{18.24}$$

Fig. 18.12. Distribution of dimensionless pressure p ($= PA/\sigma_0 T$) and 'total' load pb against b for the problem of Fig. 18.11.

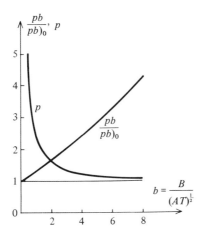

This is also plotted against b in fig. 18.12. It is, of course, not surprising that as $b \to \infty$, $p \to 1$, corresponding to the simple case of a long shell loaded by a uniform pressure along its entire length.

In the above analysis we have assumed implicitly that the shell extends a sufficient distance beyond the plastically-deforming zone in either direction for the boundary conditions at the ends of the shell to be immaterial: in these regions we may use precisely the same distribution of stress resultants as for the problem of the ring-load: see fig. 18.9.

It is straightforward to analyse the same problem by means of the upper-bound theorem: see problem 18.6.

Since we have used a *circumscribing* yield-locus for this problem, the remarks of section 18.2 apply to the solution.

18.5 Lower-bound analysis of axisymmetric pressure-vessels

Fig. 18.13a shows a pressure-vessel which has a spherical portion of radius R and a cylindrical portion of radius $\frac{1}{2}R$. The thickness T is uniform throughout, and in particular there is no thickening of the shell in the region of the junction. We wish to estimate the pressure at which the vessel experiences plastic collapse according to the simple plastic theory, as a fraction of the limit pressure for an unperforated spherical vessel of radius R and thickness T.

This problem is somewhat artificial, since no attempt has been made to 'reinforce' the junction. It represents, perhaps, the first stage of a design exercise. Furthermore the ratio of radii of the two parts of the vessel has been given a special value mainly for the sake of convenience.

Fig. 18.13. (*a*) A spherical pressure-vessel with a cylindrical branch. (*b*) Circumscribing version of the yield-locus of fig. 18.4, showing 'working faces' for two particular problems.

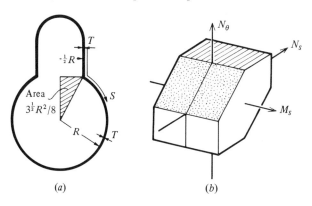

680 *Shell structures and the theory of plasticity*

It is clear that there is likely to be a 'local' state of plastic collapse in the region of the junction at a pressure which is only a fraction of that required to cause uniform plastic expansion of an unpierced spherical shell. Moreover, it is clear on intuitive grounds that the local collapse state will involve tensile yield in the hoopwise direction in the vicinity of the junction.

For this problem it is particularly convenient to use the simplified 'circumscribing' version of the appropriate yield-locus, as shown in fig. 18.13b, because it is clear that in the region of the junction the tensile-stress resultants in the circumferential (θ) and meridional (s) directions will lie in the order

$$N_\theta > N_s > 0. \tag{18.25}$$

Hence the yield condition is simply

$$N_\theta = N_0 = \sigma_0 T, \tag{18.26}$$

provided $|M_s|$ is not too large. Since this is a circumscribing' yield-locus, we must bear in mind the remarks of section 18.2 in the interpretation of the results.

The most convenient way of doing a lower-bound analysis is to adapt to the present configuration the scheme which was established in the preceding section.

The equations of equilibrium for an arbitrary shell of revolution under axisymmetric conditions may be taken directly from section 11.2.1. They are

$$N_s - N_s^* = \frac{U}{r_3}, \tag{18.27a}$$

$$N_\theta - N_\theta^* = \frac{dU}{dS}, \tag{18.27b}$$

$$\frac{dM_s}{dS} + \frac{M_s - M_\theta}{r_3} = \frac{U}{r_2}. \tag{18.27c}$$

Here, S (upper case) is the meridional contour length, and the radii $r_2(S), r_3(S)$ are defined in fig. 4.10. The quantities N_θ^*, N_s^* are found for a given loading by means of the membrane hypothesis: see section 4.5. The sign conventions for stress resultants N_s, N_θ, M_s are shown in fig. 11.3.

In the present problem both the geometry and the loading – a gauge pressure P – are particularly simple. Thus, in the cylindrical portion

$$\left. \begin{array}{l} r_2 = \tfrac{1}{2}R, \quad r_3 \to \infty, \\[4pt] N_\theta^* = \tfrac{1}{2}PR, \quad N_s^* = \tfrac{1}{4}PR; \end{array} \right\} \tag{18.28}$$

18.5 Lower-bound analysis of pressure-vessels

while in the spherical portion

$$r_2 = R, \quad r_3 = O(R),$$
$$N_\theta^* = N_s^* = \tfrac{1}{2}PR. \tag{18.29}$$

It seems clear intuitively that the meridional extent of the junction region will be small in comparison with r_3, and therefore that (18.27c) may be simplified without appreciable loss of accuracy by the removal of the second term on the left-hand side. We shall therefore work with (18.27b), i.e.

$$\frac{dU}{dS} = N_\theta - N_\theta^*$$

and

$$\frac{dM_s}{dS} = \frac{U}{R}\text{(sphere) or } \frac{2U}{R}\text{(cylinder)}, \tag{18.30}$$

and use (18.27a) at the end in order to check the magnitude of N_s.

It is convenient to use the diagram of fig. 18.14a, in which $U(S)$ is the dif-

Fig. 18.14. (a) Diagram analogous to fig. 18.11b for the problem of fig. 18.13. Line TRP has yet to be moved to its final position, in which $TSR = \tfrac{1}{2}RQP$. (b) Dimensionless limit pressure for the vessels of figs 18.13a and 18.15a. (c) Corresponding results for the 'torispherical' vessel of fig. 18.15b, showing a generalisation of the broken curve of (b) for arbitrary values of L/D, together with the effect of the inclusion of a toroidal 'knuckle'.

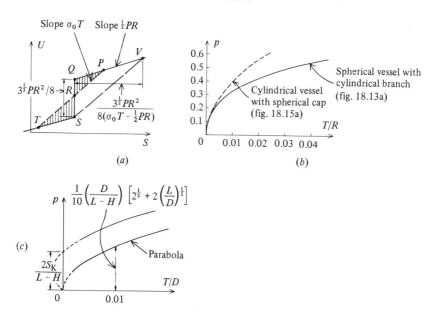

ference in ordinate between the two functions $\int N_\theta \, dS$ and $\int N_\theta^* \, dS$. The latter function has a positive jump at the junction equal in magnitude to the concentrated hoop tension which is required by the membrane hypothesis. For the present problem it is equal to $3^{\frac{1}{2}} PR^2/8$: see fig. 18.13a. The curve $\int N_\theta \, dS$ is sketched in with a constant slope equal to $\sigma_0 T$ on the assumption, already explained, that the yield condition allows full hoop tension in the vicinity of the junction. At points T, R and P on the diagram M_s has maximum/minimum values, and therefore (cf. section 18.4 and equation (18.30)) the shaded area TSR is equal to $\frac{1}{2}RM_0$, while that of RQP is equal to RM_0. In particular, the broken line must be positioned in such a way that the two areas are in the ratio $1/2$. This is straightforward, since the triangles are similar for this particular problem by virtue of our choice of radius-ratio. In the regions to the left of point T and to the right of point P it is necessary to propose an arrangement of stress resultants which does not violate the yield condition. This may be done without difficulty in the manner of fig. 18.9.

Let us define a dimensionless pressure p by

$$p = PR/2T\sigma_0. \tag{18.31}$$

We now have all the ingredients of the problem. The trigonometry of fig. 18.14a is straightforward, and we find that the resulting equation is

$$\frac{T}{R} = \frac{3}{8(3 + 2^{\frac{3}{2}})} \frac{p^2}{(1-p)} = \frac{0.064 \, p^2}{(1-p)}. \tag{18.32}$$

This relationship is plotted in fig. 18.14b. It is clear that p is sensitive to the ratio T/R, particularly for small values. If we wish to have $p > 0.5$, i.e. a limit pressure in excess of one half of the pressure-strength of the unpierced spherical vessel, we must have $T/R > 0.032$.

It is not difficult to extend this analysis to the case of a spherical closure to a cylindrical vessel, and indeed to the more general case of a 'torispherical closure' (fig. 18.15). There is only one new difficulty to be overcome in relation to these problems. In the vessel of fig. 18.13a it is clear that N_s and N_θ have the same sign throughout, and that the 'working face' of the circumscribed yield-locus is as shown in fig. 18.13b. For the torispherical end-closure of a cylindrical vessel we know from section 11.4.2 that under interior pressure N_s is likely to be positive throughout, while N_θ is likely to be strongly negative in the region of the junction. Therefore the 'working face' of the yield condition is likely to be the inclined plane

$$N_s - N_\theta = N_0 \tag{18.33}$$

18.5 Lower-bound analysis of pressure-vessels

rather than the plane $N_\theta = N_0$ used previously. The ensuing complications to the analysis may be overcome by combining (18.27a) and (18.27b) to give

$$\frac{dU}{dS} - \frac{U}{r_3} = (N_\theta - N_s) - (N_\theta^* - N_s^*). \tag{18.34}$$

Now the second term on the left-hand side is negligible in comparison with the first, so, by using the yield condition (18.33) we may simplify (18.34) to

$$\frac{dU}{dS} = -N_0 + N_s^* - N_\theta^*. \tag{18.35}$$

It is straightforward to evaluate $N_s^* - N_\theta^*$ as a function of position on the meridian, and to complete the calculation on the same lines as before. The details of the calculation are given by Calladine (1969b). In particular the following formula may be proved in relation to the vessel of fig. 18.15b:

$$\{p\,[(L-H)/D] - (S_K/D)\}^2 = (T/D)\,[2^{\frac{1}{2}} + 2\,(L/D)^{\frac{1}{2}}]^2 + (S_K/D)^2$$

where
$$\tag{18.36}$$

p is the pressure normalised with respect to the limit pressure for the cylindrical shell ($= pD/2\sigma_0 T$),
D is the diameter of the cylindrical shell,
L is the radius of the spherical cap,

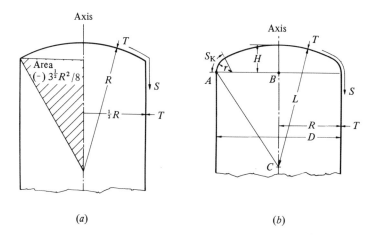

Fig. 18.15. (a) A cylindrical vessel with a spherical end-closure. (b) A vessel as in (a), but with more free dimensions, and including a toroidal 'knuckle'.

684   *Shell structures and the theory of plasticity*

T is the (uniform) thickness of the vessel

S_K is the contour length of the knuckle, measured along the meridian.

This curve is plotted in fig. 18.14*b* for the particular case $L = D$ and $S_K = 0$. The performance of the system is rather better than for a cylindrical branch in a spherical vessel having the same leading dimensions. In general, the presence of a knuckle enhances the value of p in almost direct proportion to S_K: see the graphical interpretation of (18.36) which is given in fig. 18.14*c*.

It must be recalled that the results displayed in figs. 18.12 and 18.14 have all been obtained by use of a *circumscribing* yield-locus.

18.5.1 *A shallow-shell example*

As a second example of the use of the lower-bound theorem we investigate a spherical pressure-vessel of radius A and thickness T which contains a small hole of radius L ($\ll A$) sealed by a simple rigid cover-plate as shown in fig. 18.16, which is assumed to apply a uniform shearing force at the edge of the hole in the direction of the radius of the sphere. We wish to find a lower-bound on the limit pressure for the system.

This problem is closely related to the analysis of small-radius cylindrical branches in spherical pressure-vessels (cf. section 18.5). In the vicinity of such a hole or branch it is obviously not reasonable to regard r_3 as large in comparison with the meridional extent of the plastic region, and so the method of analysis used in the previous section becomes inappropriate. However, for sufficiently small values of L/A it will be adequate to treat the problem as one in shallow-shell theory. For the sake of variety we shall use dimensionless stress-resultants for this problem. Also we shall use the 'correct' yield-locus as illustrated in fig. 18.4: in contrast to the results of section 18.5, the results as they stand will be true lower bounds. In fact, we need to consider only one

Fig. 18.16. A hole of radius L in a spherical shell of radius A and thickness T, which is closed by a simple 'cover-plate'.

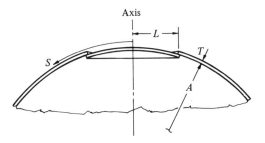

18.5 Lower-bound analysis of pressure-vessels

plane facet of this yield surface, defined by $n_\theta = 1$ and illustrated in fig. 18.17a. The derivation of the n_s, m_s locus in this plane is straightforward. A general scheme for determining the yield-locus in n, m space is to postulate an arbitrary distribution of stress components σ_s, σ_θ through the thickness in such a way that the Tresca yield condition is nowhere violated. Then the corresponding values of n_s, m_s etc. are calculated by integration; and by virtue of the lower-bound theorem they represent a 'safe' set of stress resultants.

In this particular case the task is straightforward: since $n_\theta = 1$, we must have $\sigma_\theta = \sigma_0$ throughout the entire thickness, and $\sigma_s = 0$. Note also that in consequence $m_\theta = 0$. But in the s-direction we may elect to have $\sigma_s = \sigma_0$ over only a fraction n_s of the thickness, and $\sigma_s = 0$ elsewhere, as shown in fig. 18.17b. The distribution of stress is uniform but different in each of the three

Fig. 18.17. (a) Yield-locus for the problem of fig. 18.16 (being essentially the shaded region of the surface shown in fig. 18.4), together with typical stress trajectories. (b) Hypothetical distribution of stress within an element of the shell, showing three distinct bands or layers of uniform stress, and corresponding to the area enclosed in (a). (c) Stress points for the three zones of (b) in relation to the two-dimensional 'Tresca' yield-locus. (d) Hypothetical distribution of stress in regions remote from the hole.

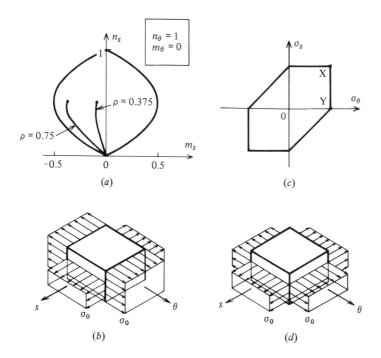

'bands' or 'layers' shown in the diagram. In the middle band the stress point is at X on the Tresca hexagon for plane stress (fig. 18.17c), while in the two outer bands the stress point is at Y. The location of the middle band within the thickness can clearly be altered at will. When it is located centrally it is clear that $m_s = 0$, but when it is moved off-centre $m_s \neq 0$. For a given value of n_s, m_s is proportional to the eccentricity of this band; and the largest permissible eccentricity corresponds to the band being hard up against either the upper or the lower surface of the element. It is straightforward to show that the limiting values of m_s are given by

$$m_s = \pm 2n_s(1 - n_s). \tag{18.37}$$

This defines the two parabolic arcs shown in fig. 18.17a, which together constitute a two-dimensional yield-locus when n_θ, $m_\theta = 1, 0$. Our task of analysis will involve satisfaction of the equilibrium equations of the problem while keeping n_s, m_s at any point within the yield-locus defined by (18.37).

The three equilibrium equations are exactly the same as (18.27), except that they may be simplified in connection with the assumption that the shell is 'shallow'.

We shall denote the meridional contour length by S; and for this problem locate the origin of S at the pole or apex. Thus, putting $r_3 \approx S$ in (18.27a and c) we obtain the approximate equations

$$N_s - N_s^* = U/S, \tag{18.38a}$$

$$\frac{dM_s}{dS} + \frac{M_s - M_\theta}{S} = \frac{U}{A}. \tag{18.38b}$$

We shall be mainly concerned with a zone in which $M_\theta = 0$, as explained above. Substituting for U from (18.38a) in (18.38b) and rearranging we find

$$\frac{d}{dS}(SM_s) = \frac{S^2}{A}(N_s - N_s^*). \tag{18.39}$$

The third equilibrium equation is exactly the same as (18.27b).

For a spherical shell under uniform interior pressure P we have

$$N_s^* = N_\theta^* = \tfrac{1}{2}PA. \tag{18.40}$$

Using these in (18.38a) and (18.27b) and eliminating U from the resulting equation we obtain the approximate expression

$$\frac{d}{dS}(SN_s) = N_\theta. \tag{18.41}$$

This equation is in fact precisely the same as for a symmetrically-loaded flat

18.5 Lower-bound analysis of pressure-vessels

plate in polar coordinates, when S is regarded as a radius. This illustrates yet again the decomposition of the equilibrium equations of shallow shells into those for S- and B-surfaces, respectively: cf. section 11.5. Next we define dimensionless variables, as follows:

$$n_i = N_i/\sigma_0 T, \quad m_i = 4M_i/\sigma_0 T^2, \quad \text{where } i = \theta, s;$$
$$s = S/L, \quad p = PA/2\sigma_0 T, \quad \rho = L/(AT)^{\frac{1}{2}}.$$
(18.42)

It follows from these definitions that

$$n_s{}^* = n_\theta{}^* = p.$$

Equilibrium equations (18.41) and (18.39) thus become

$$\frac{d}{ds}(sn_s) = n_\theta \qquad (18.43a)$$

$$\frac{d}{ds}(sm_s) = 4\rho^2 s^2(n_s - p). \qquad (18.43b)$$

Note that the geometrical hole-size parameter ρ appears only in the single equation (18.43b).

Suppose that a value of p has been given, and that we wish to construct a distribution of stress resultants which satisfies the equilibrium equations. Clearly $n_s = n_\theta = p$ satisfies (18.43a), but it does not satisfy the boundary condition

$$n_s(1) = 0 \qquad (18.44)$$

which is imposed by the nature of the cover-plate at the edge of the hole. In order to satisfy this condition we must have a region surrounding the hole (which we shall call zone I) in which $n_\theta > p$. Therefore, putting n_θ at its limit $n_\theta = 1$, we have

$$\frac{d}{ds}(sn_s) = 1,$$

from which we obtain, in conjunction with (18.44)

$$n_s = 1 - s^{-1}. \qquad (18.45)$$

Thus the value of n_s increases with s; and when it reaches $n_s = p$, (at $s = (1 - p)^{-1}$) we may change to the previous simple solution, $n_\theta = n_s = p$, which will hold throughout what we may call zone II. Thus a complete equilibrium solution for the tangential-stress resultants is:

$$n_\theta = 1, \quad n_s = 1 - s^{-1} \quad \text{for } 1 \leqslant s < (1 - p)^{-1}; \text{ zone I}$$
$$n_\theta = n_s = p \quad \text{for } s > (1 - p)^{-1}; \quad \text{zone II}.$$
(18.46)

This solution is shown in fig. 18.18 for the case $p = 0.5$. Note that there is a discontinuity in n_θ at the I/II boundary, but not in n_s: a discontinuity in n_s would violate the condition of equilibrium normal to the boundary plane, whereas a discontinuity in n_θ does not violate any equilibrium condition.

Now that we have a definite function $n_s(s)$ we may proceed to solve (18.43b) in zone I subject to the boundary condition $m_s(1) = 0$. We thus obtain

$$m_s = \tfrac{2}{3}\rho^2 \left[(2s^2 - 3s + s^{-1}) - 2p(s^2 - s^{-1})\right]. \tag{18.47}$$

This is also plotted in fig. 18.18, again for $p = 0.5$, in zone I. The curve has a minimum just before the I/II boundary is reached, since by (18.43b) sm_s has a minimum at the boundary itself. At the I/II boundary we have, from (18.47),

$$m_s = -2\rho^2 p^2 \left[(1 - \tfrac{2}{3}p)/(1-p)\right] = -m^*, \text{ say.} \tag{18.48}$$

We must, of course, specify the bending-stress resultants m_θ and m_s in zone II. The simplest plan is to put $m_s = m_\theta = -m^*$ in zone II, as shown: this clearly satisfies (18.38b), for when $M_\theta = M_s$ and $U = 0$, M_s is constant. Across the I/II boundary, equilibrium requires that M_s is continuous; but there may be a discontinuity in M_θ. Zone II begins in a 'shallow' part of the spherical shell, but as the distribution of stress resultants in it is so simple, we may envisage it extending into the entire remainder of the shell: the proposed stress resultants

$$n_\theta = n_s = p; \quad m_\theta = m_s = -m^* \tag{18.49}$$

satisfy both the 'shallow' and the 'deep' versions of the equilibrium equations exactly.

Fig. 18.18. Stress resultants plotted against meridional position for the problem of fig. 18.16, for a particular value of interior pressure. Stress resultants n_θ, n_s are independent of the hole-size parameter ρ.

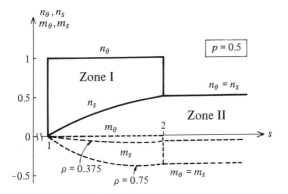

18.5 Lower-bound analysis of pressure-vessels

It may seem odd to be proposing a state of stress which requires uniform bending moment throughout a large region of the shell remote from the hole. It is unlikely that this will actually occur; but that question is not at issue here, since we simply need to satisfy the *equilibrium* equations for the purposes of applying the lower-bound theorem.

The state of stress in an element of zone II is shown schematically in fig. 18.17*d*. There is a 'band' of equal bi-axial stress within the element, sandwiched between two bands of zero stress. This scheme gives equal and non-zero values of m_θ and m_s when this band is off-centre. Clearly, if the yield condition is satisfied in zone I at the I/II boundary, then with the present distribution of stress resultants it is also satisfied throughout zone II.

So far we have ensured the satisfaction of the equilibrium equations without regard to the question of whether or not the yield condition is violated. For given values of ρ and p we may evaluate $n_s(s)$ and $m_s(s)$ and plot the corresponding trajectory onto the diagram of fig. 18.17*a*. This has been done in fig. 18.17*a* for the arbitrary choice of $p = 0.5$ and two values of ρ. And since each trajectory lies entirely within the yield-locus, the arrangement is 'safe'. Thus, by the lower-bound theorem, a shell having a hole for which $\rho = \frac{3}{4}$ can withstand at least $p = \frac{1}{2}$ without collapse. Now the sole consequence of changing the value of ρ is to increase $m_s(s)$ by a constant factor proportional to ρ^2, without changing $n_s(s)$. In other words an increase in ρ enlarges the n_s, m_s trajectory uniformly in the m_s-direction. Thus, in our previous example it is a simple matter in graphical terms to determine, for $p = \frac{1}{2}$, the value of ρ at which the m_s, n_s trajectory just touches the yield-locus. A graphical/numerical calculation shows that in this case $\rho = 0.84$; and this gives a point on the curve shown in fig. 18.19. The same procedure has been

Fig. 18.19. Lower-bound limit pressure p as a function of hole-size ρ for the problem of fig. 18.16.

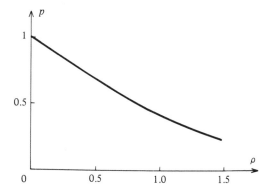

repeated for different values of p in order to determine the whole curve. The curve is 'safe' since it has been derived by means of an application of the lower-bound theorem. It is possible to obtain a somewhat higher curve, particularly in the region $\rho > 1$, by adopting alternative strategies; but this is beyond the scope of the present work.

It is clear from fig. 18.19 that even a hole of modest size can seriously reduce the pressure-strength of a spherical vessel.

18.6 Use of lower-bound theorem to design reinforcement for pressure-vessels

The calculation of the preceding section can be re-used to good effect in the design of reinforcing pads around holes in spherical vessels, in order to counter the weakening effect of a hole.

We discovered that in zone II our equilibrium-satisfying solution required only a fraction p of the thickness of the material to carry any stress: see fig. 18.17d. This immediately suggests that, within the context of the lower-bound theorem, we may imagine the *removal* of the unstressed material in zone II. In this way we end up with a modified vessel which now has thickness $T' = pT$ over most of its surface, but with a thickened pad of uniform thickness T surrounding the hole. This 'derived' shell constitutes a 'full strength' design for a vessel of radius A and thickness T'. Note that, since $m^* \neq 0$ in the original solution, the band of stressed material in fig. 18.17d is not central within the thickness. Therefore, in the 'derived' vessel the main shell and the thickened zone are connected eccentrically, as shown in fig. 18.20. The eccentricity of the connection at this point is in fact a crucial

Fig. 18.20. Geometry of an opening in a spherical shell with an eccentrically mounted reinforcing pad. In the shallow-shell region for which the diagram is relevant, the scale of the diagram in the axial direction is arbitrary.

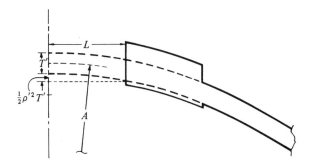

18.6 Lower-bound theorem and reinforcement of vessels

feature of the design. Note in particular that the bulk of the thickened zone surrounding the hole lies on the *outside* of the vessel. If the thickened region were to be connected to the shell without eccentricity, the pressure-strength would be considerably less.

In this connection it is convenient to define a thickening factor γ and an eccentricity factor η as follows. Let a shell of thickness T' be connected to a thicker shell, of thickness $\gamma T'$, in such a way that the radial separation of the two centre-surfaces is e, as shown in fig. 18.21c. Now in the special case in which the external surfaces connect *smoothly* on one side (fig. 18.21b), the value of e is equal to $\frac{1}{2}(\gamma - 1)T'$. Let us therefore define a dimensionless eccentricity factor η as follows:

$$\eta = 2e/(\gamma - 1)T'. \tag{18.50}$$

Thus the connection of fig. 18.21a has $\eta = 0$, while that of fig. 18.21b has $\eta = 1$. In particular, let us define the sense of η such that when the junction is flush on the *inside* surface, $\eta = +1$. In terms of our previous solution, from which the 'unwanted' portions of the shell in zone II were excised, we see that

$$\eta = |m^*/m_p|, \tag{18.51}$$

where m^* is $-m_s$ at the I/II junction, and m_p is the value of m_s on the yield-locus for $n_s = p$. Thus $m_s = \pm 2p(1 - p)$, and we obtain directly

$$\eta = \frac{L^2}{AT} \frac{p(1 - \frac{2}{3}p)}{(1 - p)^2}. \tag{18.52}$$

Fig. 18.21. Description of the 'eccentricity' η of a thickened pad.

(a) 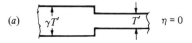 $\eta = 0$

(b) $\eta = 1$

(c) 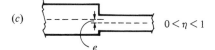 $0 < \eta < 1$

This formula needs to be rewritten for application to our 'derived' shell which has a basic-vessel thickness of T' and a thickened pad of thickness $\gamma T'$. After (18.42) we define a hole-size parameter ρ' as follows:

$$\rho' = L/(AT')^{\frac{1}{2}}. \tag{18.53}$$

Putting $p = \gamma^{-1}$ in (18.52) we thus obtain

$$\eta = \frac{\rho'^2(\gamma - \frac{2}{3})}{\gamma(\gamma - 1)^2}. \tag{18.54}$$

This is a general formula for the offset needed. In this notation the thickened zone extends from $S = L$ (the edge of the hole) to

$$S = L\gamma/(\gamma - 1). \tag{18.55}$$

Equations (18.54) and (18.55) can be used to design a thickened pad which will reinforce the hole to full pressure-strength by virtue of the lower-bound theorem. For a spherical vessel having given values of A, T' and L, the value of ρ' is known. When an arbitrary value of γ has been chosen, (18.55) gives the outer radius of the thickened region, and (18.54) gives the required offset at the outer edge. These formulae, however, pay no attention to the question of the violation of the yield condition. Fig. 18.17 is typical in the sense that when the values of p, ρ are chosen so that the n_s, m_s trajectory just touches the yield-locus, the end-point ($m_s = -m^*$) lies just inside the locus. On detailed investigation we find in general that as the value of ρ' is increased steadily, the yield condition is reached just before the offset η reaches the value of 1.

This point is clarified in fig. 18.22, which shows a plot of γ against ρ' for full pressure-strength reinforcement, with contours of constant η. Designs lying below the broken curve are invalid, since they violate the yield condition. Palmer (1969) was the first to point out the significance of positive offset in the design of effective reinforcement for holes in spherical pressure-vessels. He gave an analytical expression for the conditions under which the stress trajectory touches the yield condition (18.37), which has been plotted in fig. 18.22. It is for practical purposes indistinguishable from the line corresponding to $\eta = 1$.

18.6.1 A physical argument

The desirability of 'offsetting' a thickened pad reinforcement has been demonstrated in the preceding section. It is useful to reconsider the same point qualitatively by means of the following argument, which is based on 'gross' conditions of equilibrium.

18.6 Lower-bound theorem and reinforcement of vessels

Consider first a spherical shell of radius A and thickness T, which contains a hole of radius L and is closed by means of a simple cover-plate. It will be convenient to think of the axis of rotational symmetry as being vertical, with the hole at the top. The edge of the hole is strengthened by a compact bead of material which exerts a ring tension $T = \frac{1}{2}PAL$ when an interior pressure is applied. (Strictly the A in this expression should be replaced by $(A^2 - L^2)^{\frac{1}{2}}$: but the discrepancy is negligible for small holes). The remainder of the shell is in a state of pure membrane tension $N_\theta = N_s = \frac{1}{2}PA$ on account of this ideal compact reinforcement. Fig. 18.23 shows a portion of this shell which has been cut off by an arbitrary latitude circle and a meridional plane. We shall consider the conditions of equilibrium of this piece. It will be convenient to include within the system the pressurised fluid bounded by the meridional plane and the two planes normal to the axis which pass through the upper and lower latitude circles, respectively. In particular, the system does not include the cover-plate.

The system (shell and fluid) is subjected to various forces, which are shown schematically on the diagram of fig. 18.23a. First, there is uniform pressure P on each of the three plane surfaces. Second, there is a uniform tangential-stress resultant $N_s{}^*$ around the cut on the lower latitude circle and a uniform stress resultant $N_\theta{}^*$ on the meridional cuts. Third, there is a con-

Fig. 18.22. Design chart for full-strength reinforcement for a small hole in a spherical pressure-vessel. The thickening factor γ and eccentricity factor η are shown against the hole-size parameter $\rho' = L/(AT')^{\frac{1}{2}}$; cf. figs. 18.20 and 18.21.

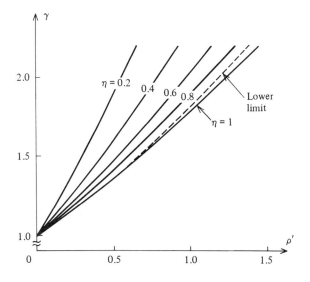

centrated tension T at each end of the diameter of the hole. Fourth, and last, is the nearly vertical uniformly distributed shearing force which is imposed by the cover-plate onto the edge of the hole. The whole arrangement is, of course, in equilibrium. Fig. 18.23b shows an elevation of the system, with each set of forces replaced by its resultant, except that the resultant of $N_s{}^*$ has been resolved into two components. It is not difficult to verify algebraically that the system is in equilibrium under these eight forces; but for present purposes we do not need to go through this calculation.

Suppose now that the compact ring of reinforcement at the edge of the hole is to be replaced by a general thickening of the shell in an annular region surrounding the hole. This change does not affect the plane boundaries of the fluid, and therefore the pressure resultants in fig. 18.23b are unchanged. The main change is that the force $2T$, which was formerly applied at the edge of the hole, is now lowered somewhat, to the level of the centroid of the cross-section of the annular region. But if the force $2T$ moves parallel to itself, and there are no other changes, the equilibrium condition for the system is at once violated; and it becomes necessary to provide around the lower edge of the shell a bending-stress resultant M_s. (Here we have rejected the possibility of moments M_θ on the meridonal cuts, since we expect 'full hoop tension' there.) The need for this bending moment, in order to preserve equilibrium, corresponds exactly to the analysis of the preceding section.

On the other hand, equilibrium can be preserved without the need for these bending-stress resultants if it can be arranged that the centroid of the cross-section of the added annular region stays at precisely the same level as

Fig. 18.23. Portion of a spherical shell in the vicinity of an opening, together with a section of the pressurised fluid. (a) Perspective view showing the forces acting on the system. (b) Elevation, showing resultant forces. In each diagram forces acting on the shell are shown by solid-headed arrows, while those acting on the fluid are shown by open-headed arrows.

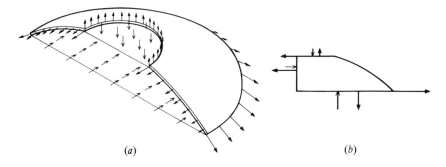

the original force $2T$. This can be achieved simply by thickening the shell more on the outside than on the inside. Again, this idea corresponds exactly to the previous analysis. It may seem strange that the required statical equivalence of the two forces can be obtained at the expense of such a small offset, which is typically only a small fraction of the thickness of the shell. Fig. 18.20 shows a scale drawing for the case $\rho = 1$, $\gamma = 2$, in which the various dimensions can be seen clearly. In any case the required offset η is proportional to ρ'^2, according to (18.54), when γ is fixed. This corresponds directly to the fact that the sagitta of the meridian at the edge of the hole (see fig. 18.20) is proportional to $T\rho'^2$: see appendix 8.

Broadly the same effects are found in the reinforcement of a junction between a spherical shell and a cylindrical branch having a relatively small radius. A branch can exert a pull on the shell at the same level as a compact ring, which is clearly an advantage; consequently it becomes easier to provide a reinforcing pad for a branch than for a simple hole of the same radius.

Another possible scheme is to make the reinforcing pad in the form of a shallow *cone* which touches the sphere at its outer edge: see Calladine (1966). This provides another way of 'raising' slightly the centroid of the cross-section of the added material.

The preceding section and the present one have illustrated the use of the lower-bound theorem of plastic theory in order to achieve an efficient design against plastic 'collapse' of the vessel. It turns out, moreover, that designs which are made in this way also have relatively low stress-concentration factors when they are analysed according to elastic theory: see Calladine (1966) and Palmer (1969). Thus, somewhat ironically, the methods of simple plastic theory, which afford a direct route of structural design, may be useful for the purposes of design in the elastic range in some circumstances.

18.7 An upper-bound method

As a tool for limit analysis of structures the upper-bound theorem of plastic theory is most attractive in situations where it is easy to envisage a plausible velocity field, but yet is relatively awkward to write down and solve the equilibrium equations. Perhaps the best-known example of this aspect of plastic theory in the structural field is the limit analysis of flat reinforced-concrete slabs by means of 'yield-line theory'. In such cases the computational advantages afforded by the method outweigh the disadvantages (i) that the method generally overestimates the strength of structures, and (ii) that in particular the use of an unrealistic velocity field may lead to serious error.

The upper-bound method of limit analysis has been applied successfully to shell structures by various authors: see, for example, Drucker & Shield

(1959), Hodge (1963), Gill (1964) and the review of Olszak & Sawczuk (1967, §6.3).

It turns out that although it is often a simple matter to analyse the kinematics of a proposed velocity field, the computation of the corresponding energy dissipation can be awkward, particularly if a complicated yield-locus, such as the one shown in fig. 18.4, is used. In general the function of the strain- and curvature-change increments which expresses the rate of dissipation of energy per unit area of shell surface is of the same order of complexity as the yield function itself. The contours of the dissipation function in terms of strain-rate resultants are geometrically similar; and they are in fact reciprocal surfaces to the yield surface with respect to a 'unit' sphere: see Calladine & Drucker (1962, Fig. 8). Thus, while a contour of the dissipation function corresponding to the circumscribing yield-locus of fig. 18.13b is a simple arrangement of plane facets, its counterpart in relation to the locus of fig. 18.4 is rather complicated and awkward.

In this section we shall present briefly a radically different approach to the computation of upper-bound limit loads for shells, which has been applied successfully to a number of difficult problems. The basic idea springs from the fact that the complications outlined above, which are associated with the complicated nature of the dissipation function, arise largely from the original decision to treat the actual shell as a *surface* endowed with certain mechanical properties which are expressed in terms of a yield function. If we consider for the moment the possibility of treating the shell as a *three-dimensional solid*, we find that the dissipation of energy per unit volume of material is a relatively simple function of the local strain-rates; and that although the integration of the dissipation of energy must now be performed over a volume rather than an area, the total computational effort required is much less.

This point may be made most clearly by means of an example concerning a circular plate of radius b and uniform thickness T, which is simply-supported on a concentric circle of radius a, and is subjected to a centrally applied transverse load F. This problem, together with several others, is discussed by Calladine (1968). The basic method is an extension of an idea by Haythornthwaite (1957). A diametral cross-section of the plate is shown in fig. 18.24. Let us consider a simple incipient velocity field in which the initially plane central surface of the plate is turned into a shallow cone. In the process each radial section rotates about its outer edge; and it will be simplest to imagine that each cross-section rotates as a rigid rectangle. Now the idea of a pivot 'at the outer edge' of the plate is in fact inadequate to specify exactly the motion of a plane area. We must specify the precise location of the *instantaneous centre* for the rigid, plane cross-section. By the kinematics of the

18.7 An upper-bound method

assumed support, this centre must lie on a line passing through the knife-edge of the hinge of the rolling pivot, and normal to the line of travel of the rollers. But this is merely one constraint, and we are thus free to choose the location of the instantaneous centre along this line. In fig. 18.24 we denote the instantaneous centre by I; and we shall attempt to fix its position within the thickness of the plate later on.

It is convenient to describe the incipient motion of the rigid cross-section with respect to a Cartesian coordinate system which is momentarily fixed in the plane of the cross-section, with the y-axis along the axis of rotational symmetry of the plate and the origin located so that the x-axis passes through I: see fig. 18.25. When the cross-section rotates with instantaneous angular velocity $\dot{\alpha}$ about I, a typical particle B has a linear velocity of magnitude $(IB)\dot{\alpha}$ in the direction perpendicular to IB.

Since the behaviour of every cross-section is the same, by hypothesis, the point B represents a circular hoop of material within the plate. The component of velocity in the y-direction corresponds to a rigid-body motion of the hoop; but the component in the x-direction involves an enlargement of the radius,

Fig. 18.24. Cross-section of simply-supported uniform plate with a central transverse load, showing a possible location of instantaneous centres I of rotation of cross-sections in a 'conical' incipient mode of deformation. The radii of the plate and its supports are b and a, respectively.

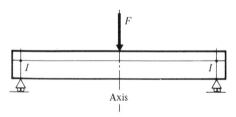

Fig. 18.25. Cross-section of the plate in its original position, ready to rotate about I as a rigid body through an incremental angle $\dot{\alpha}$.

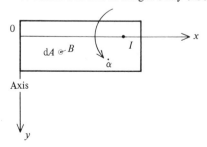

and hence a circumferential straining of the hoop. By elementary kinematics the component of velocity in the x-direction is $y\dot\alpha$, and therefore the circumferential strain-rate is given by

$$\dot\epsilon_\theta = y\dot\alpha/x. \tag{18.56}$$

The rate of dissipation of energy per unit volume of material in such a hoop is $\sigma_0 |\dot\epsilon_\theta|$. Associated with the point B is a small area dA of the cross-section, and consequently a hoop whose volume is $2\pi x dA$. Therefore the total rate of dissipation of energy in the hoop, $d\dot D$, is given by

$$d\dot D = \sigma_0 |y| (\dot\alpha/x) 2\pi x \, dA = 2\pi \sigma_0 \dot\alpha |y| dA. \tag{18.57}$$

Notice in particular that the radius x of the hoop does not appear in the final expression. The modulus signs here represent the fact that the energy dissipation rate is necessarily positive everywhere. The total rate of dissipation of energy in the entire plate is thus given by

$$\dot D = 2\pi\sigma_0 \dot\alpha \int |y| dA, \tag{18.58}$$

where the integration is over the entire (radial) cross-section.

The rate at which the external force F does work on the plate is equal to $Fa\dot\alpha$, where a is the radius at which the simple support is located. By equating this to D, we obtain an upper-bound on the limit load. Thus

$$F^{\mathrm{u}} = 2\pi\sigma_0 \dot\alpha J/a, \tag{18.59}$$

where

$$J = \int |y| \, dA. \tag{18.60}$$

Here the superscript indicates an upper bound. Now it is clear that the value of J depends on the initial location of the instantaneous centre I within the depth of the material; and therefore that we should aim to locate I in a way that minimises J. Fortunately, it is easy to show in general, for an area A of arbitrary shape, that the above integral, when calculated about an x-axis which is free to move parallel to itself, is always a minimum when the x-axis divides the area into *two equal parts*: see fig. 18.26 and problem 18.7. In the

Fig. 18.26. The location of the x-axis in order to minimise $\int |y| dA$ over the (arbitrary) area A divides the area into two equal parts.

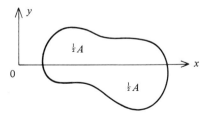

18.7 An upper-bound method

present case, therefore, we should fix I at mid-depth of the section. Thus we have

$$J = \tfrac{1}{4} bT^2, \tag{18.61}$$

and hence

$$F^u = \tfrac{1}{2} \pi \sigma_0 bT^2/a. \tag{18.62}$$

The superscript u remains, since we are still of course applying the upper-bound theorem: our minimisation merely gives the *best* upper bound from a family of calculations obtained by varying the location of I.

When $b = a$ we thus discover that F^u is independent of the radius of the plate; which is a well-known result from dimensional analysis. And indeed formula (18.62) agrees exactly with the upper-bound calculation as done in the usual way according to yield-line theory. Moreover, if we ignore the possibility of a purely local 'indentation' or 'punch-through' action under the concentrated load, the result agrees with a lower-bound analysis (see, e.g. Calladine, 1969a, Chapter 9) and therefore gives an exact limit load.

A particularly valuable feature of the method can be seen clearly in the example shown in fig. 18.27, which concerns a centrally loaded shallow *cone*. This cone may have been formed by plastic deformation of an originally flat plate under the action of a central force, or it may have been made originally in this form: in either case the analysis is the same. According to the present method we may use (18.59) as it stands, with the integral J merely evaluated afresh for the area marked in fig. 18.27. Just as in the earlier example, I is located to bisect the area, and the entire calculation is straightforward. On the other hand, this problem is quite beyond the scope of conventional yield-line analysis, for the smallest departure of the *plate* from a plane configuration renders yield-line theory inappropriate, and places the problem squarely in the field of *shell* structures.

In various possible applications of the method, it is sometimes desirable to

Fig. 18.27. Radial cross-section of a conical plate of radius a. The scale in the y-direction has been expanded for the sake of convenience.

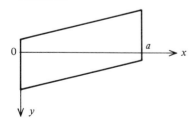

consider a mode which involves a 'conical' velocity field in the outer region of the plate/shell, while the central part executes a rigid-body motion. This central zone is therefore connected to the outer portion by a circumferential 'hinge circle'. The corresponding calculation is straightforward (see Calladine, 1968) and is most easily expressed by means of an equivalent 'area' calculation on the radial plane. Thus, in fig. 18.28a the hypothetical velocity field includes a circumferential hinge at radius c. The right-hand portion is the 'area' diagram corresponds to the conical velocity field for the outer part, while the inner portion corresponds to the energy dissipated at the circumferential hinge. By expressing the calculation in terms of a single area we may avail ourselves of the previous rule for the optimum location of the instantaneous centre: the level of I must bisect the total working area.

If the shallow conical shell is no longer simply-supported but is *clamped* at its outer edge, an outer circumferential hinge must be accounted for by means of an appropriate area as shown in fig. 18.28b. In particular, the effect of clamping the outer edge of a flat plate is thus to double the cross-sectional area over which J is evaluated, and hence to double the upper-bound estimate of the limit load.

In all of these cases the relevant area, although irregular, has straight boundaries, and the required integration can usually be done analytically. When the corresponding area has a more complicated shape, the calculation may have to be done numerically. A simple example would be the analysis of a circular plate whose thickness varies with radius.

Fig. 18.28. (a) 'Area' diagram for the conical plate of fig. 18.27 deforming in a mode in which a central portion of the plate, of radius c, descends as a rigid body. (b) The same, but for a plate which is built-in at its outer edge.

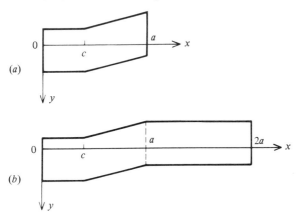

18.7 An upper-bound method

18.7.1 Application to a boss-loaded shell

A more complex example, which demonstrates convincingly the power of the method, is the problem illustrated in fig. 18.29. This has been studied in detail by Morris & Calladine (1969). It concerns an unpressurised spherical shell which is subjected to an inward-directed radial load P applied through a massive boss which may be regarded as rigid. The first aspect of the problem was to determine an upper-bound on the limit load P according to simple plastic theory in terms of a dimensionless parameter representing the size of the boss.

Experiments which were performed on shells of this sort (which had been machined from solid aluminium alloy) by Leckie & Penny (1968) showed an unstable snap-through behaviour in the plastic range under 'dead' loading; but these experimenters were subsequently able to follow the unstable path by means of a sufficiently stiff testing machine. The second aspect of Morris's work was to follow this unstable behaviour by performing a sequence of upper-bound analyses on a series of shells having progressively different geometry. The profile of each shell in the series was obtained from its predecessor by the addition of a suitable small multiple of the incipient velocity field which at the previous stage gave the minimum upper-bound load. At each stage the location of the outer hinge circle was regarded as a variable; and indeed in the later stages of the calculation the position of the inner hinge (originally adjacent to the boss) was also treated as a variable.

The step-by-step calculations agreed remarkably well with the experimental observations, thus providing a striking confirmation of the procedure in a problem for which a velocity field of a very simple kind was entirely adequate.

Fig. 18.29. Cross-section of a spherical shell which is loaded radially through a rigid boss.

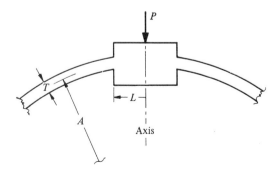

18.7.2 A calculation for a 'sandwich' shell

The calculations described above were perfectly straightforward except for the problem of performing the required integration over awkwardly shaped areas. Morris developed a simple general-purpose computer subroutine which could evaluate J over arbitrarily complicated shapes, and the optimum location of the two circumferential hinges at each stage was found numerically. For present purposes it is useful to describe a simplified version of these calculations. The simplification is achieved by the expedient of replacing conceptually the actual uniform shell by a 'sandwich' shell consisting of two thin sheets of material separated by a distance $\frac{1}{2}T$, and made of a material such that the product of the yield stress and thickness of each thin sheet is equal to $\frac{1}{2}\sigma_0 T$: in this way the sandwich mimics the original shell in both pure bi-axial tension and pure bi-axial bending. The material separating the two sheets is reckoned to be rigid in normal shear, but to offer no resistance to in-plane stress: this is a well-known idealisation (see Hodge, 1954).

Fig. 18.30a gives a meridional cross-section through the shell and the boss, showing both the original cross-sectional shape and also the location of the sandwich idealisation. The radii of the shell and boss are A and L, respectively.

Fig. 18.30. (a) As fig. 18.29, but showing a two-layer sandwich model for the shell. (b) Area diagram for a case in which the hypothetical position of the outer hinge is relatively close to the inner hinge; $\beta < \beta^*$. (c) As (b), but with the outer hinge at a larger radius; $\beta > \beta^*$.

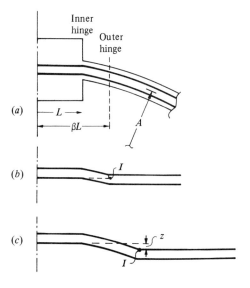

18.7 An upper-bound method

Consider an incipient velocity field which involves inner and outer circumferential hinges at radii L and βL, respectively, as shown in the diagram. The factor β will be regarded as a variable at present: we shall fix its value later in order to minimise the upper-bound value of P. The corresponding area diagram is shown in fig. 18.30b. It consists of two parallel lines each of whose 'area' in the calculation of J is to be reckoned as $\frac{1}{2}T$ per unit length. For a given hypothetical position of the outer hinge we must first locate I. There are two cases to consider, depending on the relative levels of the upper sheet at the outer hinge and the lower sheet at the inner hinge, which in turn depends on β. In the first case, shown in fig. 18.30b, the outer hinge in the upper sheet is higher than the inner hinge in the lower sheet; and the axis through I bisects the total area if it is situated *anywhere* along the short line through I. In spite of this ambiguity, the integral J has a constant value of $\frac{1}{2}\beta LT^2$. In the second case, shown in fig. 18.30c, the axis through I bisects the total area only if I is located in the upper sheet at the outer hinge: if it were higher than this, more than half of the total area would be below the level of I, whereas if it were lower than this more than half of the total area would be above the level of I. The corresponding value of J is rather more complicated to calculate than in the first case. However, it may readily be shown to exceed $\frac{1}{2}\beta LT^2$ by a quantity at least as large as zLT, where z is the dimension marked in fig. 18.30c. The calculation of z is straightforward, for in the context of shallow-shell theory we may regard the meridian as a parabolic arc. Thus we find that the difference in elevation of the upper sheet between the inner and outer hinges is equal to $L^2(\beta^2 - 1)/2A$, and so we obtain a general expression for z:

$$z/T = \tfrac{1}{2}\left[\rho^2(\beta^2 - 1) - 1\right]. \tag{18.63}$$

Here, for the sake of convenience, we have introduced the unifying geometrical parameter ρ, as before (see (18.42)). The second case occurs only when $z > 0$ and hence, by (18.63), when

$$\beta^2 > \beta*^2 = 1 + \rho^{-2}. \tag{18.64}$$

A plot of J against β is shown in fig. 18.31 for a boss having $\rho = 1$. For other values of ρ, $J(\beta)$ has the same general form: in particular there is always an abrupt change of slope at $\beta = \beta*$.

Having evaluated J as a function of the position β of the outer hinge, we now turn to the evaluation of the limit load itself. When the rotation-rate is $\dot{\alpha}$, the boss descends with velocity $(\beta - 1)L\dot{\alpha}$. Thus the upper bound on the load P is proportional to $J/(\beta - 1)$. Our next step is therefore to determine the value of β which minimises this quantity. It turns out that for all values of ρ the function $J/(\beta - 1)$ has its smallest value at $\beta = \beta*$. This result may be established by noting that in fig. 18.31 a line passing through the points $(\beta, J) = (1, 0)$ and $(\beta^*, J(\beta^*))$ lies below the curve $J(\beta)$ and touches it at $\beta = \beta^*$. This is true not only for the case $\rho = 1$, shown here, but also for all values of ρ. Hence in general the location of the outer hinge is determined directly by a very simple *rule*: the upper sheet at the outer hinge must be at the same level as the lower sheet at the inner hinge. It follows immediately that the minimum upper-bound on p is given in general by

$$P^u = \pi \sigma_0 T^2 \beta^* / (\beta^* - 1). \tag{18.65}$$

It is convenient to normalise this force with respect to the upper-bound point load, determined by the same method, for a flat plate which is clamped at the edges. Thus (see (18.62)) we define a dimensionless load f:

$$f^u = P^u / \pi \sigma_0 T^2 = \beta^* / (\beta^* - 1). \tag{18.66}$$

Hence, finally, from (18.64),

$$f^u = r/(r - \rho), \tag{18.67}$$

where $r = (\rho^2 + 1)^{\frac{1}{2}}$. This relationship is plotted in fig. 18.32, together with the corresponding results calculated numerically for a 'solid' (as distinct from a 'sandwich') shell. Also plotted are some of the experimental observations made by Leckie & Penny (1968) on the maximum loads reached in small-scale experiments on spherical vessels made from an aluminium alloy.

18.7.3 Calculation for changing geometry

So far, all of our calculations have been made in relation to shells in their original, undeformed calculation. Now when a shell deforms a little, according to the preferred incipient velocity field, its geometry clearly changes

Fig. 18.31. Integral J (see (18.60)) as a function of the position of the outer hinge.

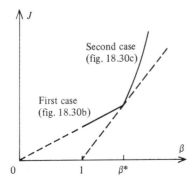

18.7 An upper-bound method

a little. Thus we must expect that the limit load also will change a little as the shell deforms. In principle this is true for any given structure. But in many cases the 'geometry-change effect' is a weak one, and the simple plastic theory (which explicitly ignores these effects) is a valid analytical tool. In the present problem the geometry-change effect is rather strong. In such cases the simple plastic theory is still a useful analytical tool *provided* that at each stage the calculations are performed with respect to the *current* geometrical form of the structure. In order to follow the correct sequence of geometrical configurations, it is necessary to have the *correct* velocity field at each stage. Thus our present calculation must be viewed with caution, since there is no guarantee that the optimum member of our family of hypothetical modes is actually the correct mode. For the problem in hand, however, it does seem clear that the optimised hypothetical mode is rather close to the correct mode. It should be noted here that the work of Allman & Gill (1968) on geometry-change effects in the plastic deformation of a cylindrical branch in a spherical pressure-vessel was done in exactly the same spirit.

The first step in the required sequence of calculations is to determine the upper-bound load for the shell whose meridian is formed by rotating the relevant portion of the meridian of the original shell: see fig. 18.30. The key to the situation is the two-fold observation (i) that when $\beta = \beta^*$ the value of J does not

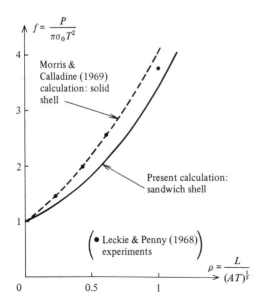

Fig. 18.32. Dimensionless limit load as a function of boss-size for the problem of fig. 18.29.

$f = \dfrac{P}{\pi \sigma_0 T^2}$

Morris & Calladine (1969) calculation: solid shell

Present calculation: sandwich shell

(• Leckie & Penny (1968) experiments)

$\rho = \dfrac{L}{(AT)^{\frac{1}{2}}}$

depend on the precise *form* of the meridional curve (see problem 18.8), and (ii) that the optimum value of β is given by the rule stated on p.704.

Hence it is possible to do a simple calculation of the change of f^u with deflection in the early stages of deformation. Suppose that, at some stage in the process, circumferential hinges are located at $x = L$ and βL. With the *original* geometry, the vertical separation of the two hinge circles is equal to $(\beta^2 - 1)L^2/2A$, as we have seen already. Now if the corresponding vertical separation in the deformed configuration is equal to $\frac{1}{2}T$ (according to our *rule*), the difference must be made up by the deflection Δ of the boss with respect to its original position. Thus we have

$$\Delta/T = \frac{1}{2}(\beta^2 - 1)\rho^2 - \frac{1}{2}. \tag{18.68}$$

The relationship (18.66) between f^u and β is unchanged, and hence we obtain

$$\beta = f/(f - 1), \tag{18.69}$$

where now we have dropped the superscript u. Substituting for β in (18.69) we find, finally,

$$\Delta/T = [(f - 1)^{-1} + \tfrac{1}{2}(f - 1)^{-2}]\rho^2 - \tfrac{1}{2}. \tag{18.70}$$

This relationship is plotted in fig. 18.33a for several values of ρ, for $\Delta/T < 1.5$. It is clear that the larger the initial value of f, the steeper the descent of the curve once deformation begins. In all cases the tangent to the curve at the axis $\Delta/T = 0$ intersects the line $f = 1$ at a value of Δ/T of the order of unity. In other words, the geometry-change effect is strong for this kind of structure.

Our simple rule for locating the outer hinge eventually ceases to be obviously valid when the configuration reaches the point at which the slope of the meridian adjacent to the boss changes sign. At this point the analysis becomes more complex, and eventually the preferred velocity field involves the movement of the inner hinge away from the boss. At this stage it is best to do the calculations numerically; and there is in fact no longer so much incentive to use a 'sandwich' idealisation of the shell. The results of detailed calculations by Morris & Calladine (1969) on a solid shell are shown in fig. 18.33b. These curves agree remarkably well with the corresponding experimental observations mentioned above. In particular, the geometry-change effects associated with *elastic* distortion of the shell before the plastic range is reached seem to be insignificant for this particular set of specimens.

By an extension of the same method, Morris & Calladine (1971) have investigated the response of a *cylindrical* shell to an inward-directed radial load applied through a boss. The calculations are more complex than those related to a spherical shell, since they involve a circumferential hinge having a non-

18.7 An upper-bound method

Fig. 18.33. Change of limit load as a function of displacement Δ of the boss. (*a*) Comparison of results obtained by present method for 'solid' and 'sandwich' shells. (*b*) Comparison between experiment (Leckie & Penny, 1968) and present theory (for 'solid' shells). (*c*) Corresponding theoretical curves for the loading of a cylindrical shell through a boss.

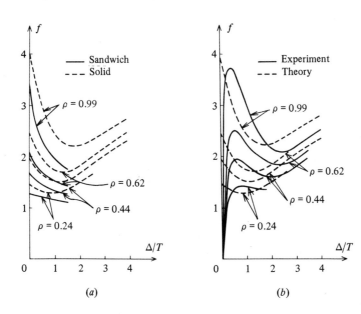
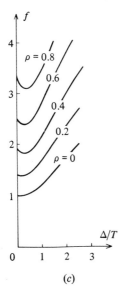

circular form. Some results are shown in fig. 18.33c. The conclusion is, broadly, that the *initial* limit load is almost exactly the same function of yield stress, radius and thickness as for a spherical shell having the same loading dimensions; but the subsequent behaviour as deflections grow is rather different and in particular is much more stable.

18.8   Discussion

The topics discussed in sections 18.4−18.7 illustrate the application of the methods of limit analysis to problems in shell structures. The same ideas may be used in the analysis and design of shells having a wide range of geometry. Except in section 18.7.3 we have been concerned with limit analysis of shells in their original configurations. It is usually not difficult to study the effect of geometry-changes on the limit loads, at least in a qualitative way, and to decide whether these effects are weak or strong, positive or negative. For example, the effect of changes in geometry on the limit pressure of torispherical pressure-vessel closures are straightforward to assess within the framework of the scheme described in section 18.5 for calculating the limit pressure of shells of revolution having arbitrary meridians.

The methods of section 18.7 may also be applied to problems involving detailed schemes of localised reinforcement.

18.9   Problems

18.1   When a localised plastic hinge is formed in a beam of depth t, it is found that the extent in the axial direction of the zone of plastically deforming material is approximately equal to t on the tensile and compressive sides of the beam. On the supposition that there is uniform plastic strain $\pm \epsilon$ in these regions, show that the corresponding rotation of the hinge is 2ϵ (radians); and thus that a rotation of 1 degree corresponds to a plastic strain of about $\pm 1\%$.

18.2   An open-ended thin-walled cylindrical shell of radius a and thickness t is initially stress-free. It is deformed inextensionally into an oval shape, so that the normal component of displacement is given by $w = w_0 \cos 2\theta$. Use (6.9e) to calculate the change of curvature κ in the circumferential direction, and hence to evaluate the peak circumferential 'bending strain' $\frac{1}{2}\kappa t$ in terms of w_0, a and t.

The shell is made of a material which has an elastic range of strain equal to 0.001. Show that the elastic limit is reached when $w_0/a \approx 7 \times 10^{-4} \, a/t$; and hence that displacements as large as $w_0/a = \frac{1}{10}$ occur within the elastic range provided $a/t > 150$.

18.3 In order to verify the yield-locus shown in fig. 18.3a in the first quadrant of M_x, N_θ space, consider the distribution of principal stress within a plate element of thickness T, as described in the table below. z is a through-thickness coordinate measured from the central plane, and the element is divided into three zones, of which the outer two have equal thickness $\frac{1}{2}(1-\alpha)T$: α is a variable, $0 \leq \alpha \leq 1$.

Zone		σ_x	σ_θ	σ_z
I	$\frac{1}{2}\alpha T < z \leq \frac{1}{2}T$	σ_0		0
II	$-\frac{1}{2}\alpha T < z < \frac{1}{2}\alpha T$	0	$0 \leq \sigma_\theta \leq \sigma_0$	0
III	$-\frac{1}{2}T \leq z < -\frac{1}{2}\alpha T$	$-\sigma_0$	0	0

First verify that Tresca's yield condition (cf. fig. 18.17c) is not violated in any of the three zones. Then confirm that $N_x = 0$, $M_x = M_0(1-\alpha^2)$ and $N_\theta \leq \frac{1}{2}N_0 \times (1+\alpha)$, where $M_0 = \frac{1}{4}\sigma_0 T^2$, $N_0 = \sigma_0 T$; and that this defines the yield-locus. Mark on the diagram points corresponding to $\alpha = 0, \frac{1}{2}$ and 1, respectively.

18.4 Suppose that stress resultants n, m, q have been determined as functions of x, and it is required to compute the corresponding incipient velocity field $\dot{w}(x)$.

First show that the flow rule for the curved portion of the yield-locus in the second quadrant of fig. 18.3a is given by

$$\dot{\epsilon}_\theta = -(2n-1)T\dot{k}_x;$$

and that in terms of a rotation-rate $\dot{\chi}$ defined by $\dot{\chi} = d\dot{w}/dx$ (where $x = X/(AT)^{\frac{1}{2}}$), this may be written

$$\dot{w} = (2n-1)\frac{d\dot{\chi}}{dx}.$$

Hence show that $\dot{w}$ may be calculated by an iterative scheme

$$\dot{w}_{i+1} \approx \dot{w}_i + \dot{\chi}_i \delta x,$$

where δx is a suitable small step size, and

$$\dot{\chi}_{i+1} = \dot{w}_{i+1}/[2n(x)-1].$$

By reading $n(x)$ from the curve of fig. 18.6 and starting with $\dot{\chi}(0) = C$, $\dot{w}(0) = 0$, perform a few cycles of the iteration. (Note that beyond the point in fig. 18.6 where dn/dx changes sign, the flow rule corresponding to the curve of the *first* quadrant of fig. 18.3a must be used.

710   *Shell structures and the theory of plasticity*

18.5 Consider the plastic collapse of a shell of overall length $3(AT)^{\frac{1}{2}}$, having free ends, which is loaded by a ring-load F, as in fig. 18.8a, at its central plane. Sketch a sequence of diagrams as in fig. 18.9, using a mechanism with a single central hinge and having points where $\dot{w} = 0$ at an unknown dimensionless distance α from each end. Set up a quadratic expression in terms of α for the value of the central bending moment m, and solve for the condition $m = 1$. Hence show that the collapse load is given by $f \approx 6.7$ (and $\alpha \approx 0.33$).

18.6 Use the upper-bound method to estimate the collapse pressure P in terms of the length B for the problem illustrated in fig. 18.11. Use the mode sketched in fig. 18.9b, set out the calculations as in section 18.3, and show that (18.22) is obtained when pb is minimised with respect to the disposable length.

Also consider separately a different mode in which the central hinge is replaced by two hinges, located at the edges of the pressurised band. Show that in this case the counterpart of (18.22) is $pb = 2 + b$. Plot this onto fig. 18.12.

18.7 Consider the integral J (see (18.60)) for a shape as in fig. 18.26 but with area A_1 above the x-axis and A_2 below it. Show that if the entire area is moved a small distance δ in the y-direction, the centroid of each part moves $\pm\delta$ in the y-direction (when terms of order δ^2 are neglected) and that consequently the value of J changes by $(A_1 - A_2)\delta$. Hence show that J is stationary with respect to this kind of movement of the x-axis when $A_1 = A_2$; and by doing a more refined calculation show that this condition minimises the value of J.

18.8 Consider the integral J (see (18.60)) for an area which consists of two separate parts. The upper part has area $\frac{1}{2}A$ and lies in the region $y > b > 0$, while the lower part has an equal area but lies in the region $y < -b$: thus the x-axis lies within an 'empty' band of width $2b$ which separates the two equal areas. Show that the value of J remains constant wherever the x-axis is moved (parallel to itself) in the band of width $2b$; and that this value is a true minimum.

Appendix 1

Theorems of structural mechanics

This appendix gives a brief sketch of various theorems in structural mechanics which are used in several parts of the book. These theorems apply to small deflections of elastic structures in the absence of buckling or other 'geometry-change' effects. In this appendix, for the sake of brevity and simplicity, they are described with reference to a simple plane pin-jointed truss which can be discussed in terms of a few, discrete, variables; but they can readily be translated into more general forms relevant to continuous structures (see appendix 2 on the idea of *corresponding* forces and displacements, etc.). The following description is restricted to frameworks whose members are made from weightless *linear*-elastic material, and which are stress-free in the initial configuration; but there is no difficulty in extending the scope of the theorems to include nonlinear elasticity and problems involving initial stress.

The description begins with the *principle of virtual work*, which makes a connection between the two distinct sets of conditions describing *statical equilibrium* and *geometric compatibility* of the various parts of the structure, respectively. This principle holds irrespective of the mechanical properties (or 'constitutive law') of the material from which the structure is made; and all of the various elastic theorems are derived directly from it by incorporation of the elastic material properties. (The theorems of the *plastic* theory of structures, which are used in chapter 18, are also derived directly from the principle of virtual work; but they are not proved here (see Calladine, 1969*a*).)

(I)   The principle of virtual work

The *statical* state of a general plane truss such as that shown in fig. A.1 is described by the external *forces* P_j which are applied to the joints, and the internal *tensions* T_i of the constituent bars. The forces and tensions satisfy the equations of statical equilibrium at the joints: in the diagram the tensions are depicted as imposing forces onto the joints. The geometric or *kinematic* state of the truss is described by reference to an initial or datum configuration: the joints have components of *displacement* u_j relative to this initial configuration, measured in each case in the direction of the corresponding external force: and the bars have extensions or *elongations* e_i relative to their lengths in the initial configuration. The elongations are related to the displacements by *geometrical* conditions, and are said to be *compatible* with them.

The *principle of virtual work* states that

$$\sum_j P_j u_j = \sum_i T_i e_i. \tag{A.1}$$

The summation on the left-hand side covers all joints which carry external force, and the summation on the right-hand side covers all of the bars in the structure. The power and usefulness of this equation lie in two aspects of it. The first is that the equation is *general*, i.e. equally valid for any arbitrary structure of this kind. The second is that the statically admissible set P_j, T_i is distinct from the kinematically admissible set u_j, e_i in the sense that the bar elongations e_i are not necessarily those which would occur under the tensions T_i on account of the elasticity, etc. of the bars; the *only* restriction binding u_j and e_i are those of geometric compatibility.

The proof of (A.1) is straightforward, and may be sketched as follows. Consider a structure which is carrying external forces P_j by means of bar tensions T_i. Now, keeping the joints fixed, remove the *bars* of the structure one by one, in each case supplying equal and opposed forces T_i to the two joints which were formerly linked by the bar carrying tension T_i. These forces lie along the line connecting the two joints. When every bar has been removed in this way, we are left with a complete set of isolated joints. Each joint is in equilibrium under any of the external forces P_j which act on it, together with the internal forces left behind when the bars were removed.

Now the condition of equilibrium of the forces acting on a typical joint is simply that the vector sum of all the forces is zero. If this condition is satis-

Fig. A.1.

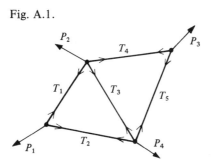

1 Theorems of structural mechanics

fied, it is certainly true that the (algebraic) sum of the *components* of the forces in any arbitrary prescribed direction are also zero. Equivalently, the sum of the scalar products of the forces with a vector of arbitrary magnitude in this same direction is zero. And indeed this last version may be interpreted as a statement that the sum of the *work* done by each of the forces acting is zero when the joint is given a *displacement* of arbitrary magnitude and direction, since the work done by a constant force when its point of application is given a displacement is equal to the scalar product of the force and displacement vectors.

Suppose now that each of the separate joints of the frame is given an arbitrary displacement, while the forces acting upon it remain constant. Since each joint is in a state of statical equilibrium under the forces acting on it, the sum of the work done by all of these forces is zero, as described above; and moreover the *global* sum of the work done by all of the forces acting on the entire system of joints is also zero.

Consider the composition of this global sum in two separate parts. Every external force appears just once, multiplied by the component of the displacement of the joint in the direction of the force. Hence we see that this part of the global sum constitutes precisely the left-hand side of (A.1). On the other hand, every internal tension of the original structure appears *twice* in the global sum, because each bar exerts opposing forces of equal magnitude on two joints. Taking these two work quantities together, we find that the tension T_i in a given member must be multiplied by the projection of the mutual approach of the two joints onto the original direction of the member connecting the joints. Now in general, when the joints move, the line connecting them rotates. If the angle of rotation is *small*, the combined work term described above is approximately equal to the product of the tension and the reduction of distance between the points. Hence, if these terms are all put onto the 'opposite side' of the equation we obtain in this global summation exactly the right-hand side of (A.1). The proof is completed by the remark that the separation of pairs of joints in this process of arbitrary displacement of the system of isolated joints is precisely equal to the extension which the corresponding bars connecting the joints would have undergone if these members had not been removed after all. And the condition that these members rotate through only small angles also ensures that the forces acting on the joints may indeed be regarded as constant when the joints move.

Note particularly that the set of joint displacements u_j and bar elongations e_i are *arbitrary* in the sense that they need only satisfy the conditions for geometric compatibility of the system. The restriction on the magnitude of the *rotation* of the members, which is a necessary part of the proof given above, is normally covered by the restriction that *all* of the displacements u_j are *small* in comparison with the lengths of the bars of which the structure is made. Equation (A.1) is not valid for *large* displacements unless u_j are such that the rotations of the members are negligibly small.

Equation (A.1) rests entirely on the fact that both forces and displacements are *vector* quantities. Observe that the properties of the *material* of which the bars are made do not enter the proof at any point.

(II)   **The theorem of minimum complementary energy**

This theorem is concerned with frameworks which are *statically indeterminate*, i.e. frameworks for which the equations of equilibrium are insufficient to determine the bar tensions and foundation reactions when arbitrary forces are applied to its joints. Any statically determinate frame can be made statically indeterminate by the insertion of a bar connecting two joints not previously linked. For present purposes it will be assumed that there is zero tension in every bar when the external forces are all zero: there is no initial stress or lack of fit. In order to determine all of the bar tensions in a statically indeterminate frame subjected to given external forces it is necessary to take account of the mechanical properties of the material from which the bars are made. Here we shall regard the bars as being *linear-elastic*; thus there is a linear relationship (Hooke's law) between the elongation e and the tension T in every bar:

$$e_i = \left(\frac{L_i}{A_i E_i}\right) T_i. \tag{A.2}$$

Here L_i, A_i and E_i are the length, cross-sectional area and Young's modulus for the bar in question. It will be assumed throughout that the elongation of a bar is small in comparison with its original length.

The theorem is concerned with a framework subjected to external forces P_j which actually cause joint displacements u_j. Now let T_i^* be *any* set of bar tensions in the structure in equilibrium with the applied loading P_j. The theorem states:

$$\sum_j \tfrac{1}{2} P_j u_j \leq \sum_i \tfrac{1}{2} T_i^{*2} (L_i/A_i E_i). \tag{A.3}$$

The equality sign holds only when T_i^* represents the *actual* set of bar tensions, which is not only in equilibrium with the applied loads but also is such that the corresponding elongations computed according to Hooke's law (A.2) are geometrically compatible with the joint displacements.

The quantity $\tfrac{1}{2} T^2 L/AE$ for a bar is known as the *complementary energy* of the bar. Inequality (A.3) may be interpreted as stating that the set of bar

1 Theorems of structural mechanics

tensions which also satisfies (via Hooke's law) the conditions of geometrical compatibility, is also the one which minimises the *total complementary energy* of the structure.

The idea of complementary energy emerges clearly in the proof of the theorem, which follows.

Let P_i, T_i, u_j, e_j represent the set of external forces, bar tensions, joint displacements and bar elongations which satisfy all of the equations of the problem; namely, equilibrium, compatibility and Hooke's law. Also let P_j, T_i^* represent the same set of external forces together with *any* set of bar tensions which merely satisfy the *equilibrium* equations of the joints. Let us now apply the equation of virtual work twice, using the two different statically admissible sets of forces and tensions but the same kinematically admissible set of displacements and elongations. Thus:

$$\left. \begin{array}{l} \sum_j P_j u_j = \sum_i T_i e_i, \\ \sum_j P_j u_j = \sum_i T_i^* e_i. \end{array} \right\} \quad (A.4)$$

By subtraction, therefore,

$$\sum_i (T_i^* - T_i) e_i = 0. \tag{A.5}$$

Consider the tension/elongation diagram for a typical member, shown in fig. A.2a; and define as the *complementary energy*, $C(T)$, the area shaded. Thus

$$C(T) = \int_0^T e \, dT = \tfrac{1}{2} T^2 L/AE. \tag{A.6}$$

We may now establish a simple inequality for the typical member by inspection of fig. A.2b:

$$C(T^*) - C(T) \geq (T^* - T)e. \tag{A.7}$$

Fig. A.2.

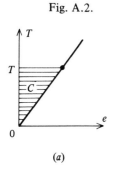
(a)

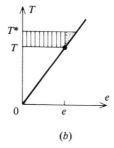
(b)

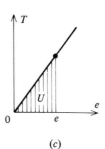
(c)

The quantity on the left-hand side represents the area of the shaded quadrilateral, in consequence of the above definition, while the quantity on the right-hand side represents the area of a rectangle which is smaller than this by the area of the small triangle. Note that T and e are the actual tension and elongation of the bar, related by Hooke's Law, while T^* is an arbitrary tension. Inequality (A.7) holds not only when $T^* > T$, as shown, but also when $T^* < T$: this may be verified directly by means of a separate diagram. The equality sign applies if and only if $T^* = T$.

Inequality (A.7) applies to each bar in turn. Consider now the sum of the inequalities for *all* of the members. By (A.5) the right-hand side is zero and thus

$$\sum_i C_i(T^*) \geqslant \sum_i C_i(T);$$

that is,

$$\sum_i \tfrac{1}{2} T_i^{*2}(L_i/A_i E_i) \geqslant \sum_i \tfrac{1}{2} T_i^2(L_i/A_i E_i). \tag{A.8}$$

Returning to (A.4), subsituting for e_i from (A.2), and multiplying by $\tfrac{1}{2}$ we find that the right-hand side of (A.8) is equal to the left-hand side of (A.3); and this establishes the required result.

The theorem has two main uses. The first depends on the fact that the actual total complementary energy of a statically indeterminate structure is *minimum* with respect to the redundant tensions. Thus the equations of compatibility may be derived in terms of the redundant tensions by a process of formal minimisation of the total complementary energy; and these may be solved in order to obtain the actual tensions in the loaded structure. The second use of (A.3) is to obtain an *estimate* of the total complementary energy of the structure by using guessed values of the redundant tensions. In this way an overestimate may be obtained of the flexibility of a structure which is subjected to a single load. The structure shown in fig. A.3 provides a simple example of this. The framework consists of six bars, each of cross-sectional area A and length l or $2^{\frac{1}{2}}l$. The material has Young's modulus E. We wish to *estimate* the displacement u of joint C due to the application of the external load P. Let us treat AC as the redundant bar, and let us guess that the tension in this bar is equal to $\tfrac{1}{2}P$. We can then find the tensions in the remaining bars by means of the joint-equilibrium equations; and these tensions will constitute a hypothetical equilibrium set T^*. Thus the tension in BD is $-\tfrac{1}{2}P$, and the tension in each of the four short bars is $P/2^{\frac{3}{2}}$. Consequently the total hypothetical complementary energy in the entire structure is equal to $\tfrac{1}{4}(1 + 2^{\frac{1}{2}}) \times$

1 Theorems of structural mechanics

$P^2 l/AE$, by summation of expressions like (A.6). Putting this into (A.3) we have

$$\tfrac{1}{2}Pu \leqslant \tfrac{1}{4}(1 + 2^{\tfrac{1}{2}})P^2 l/AE$$

and hence, finally

$$u \leqslant 1.207 \, Pl/AE, \qquad (A.9)$$

which is our estimate of u.

Instead of guessing the value of the tension in bar AC we could have regarded the tension in this member as a variable Q; and if we had then minimised the total complementary energy with respect to Q, we would have found the correct solution of the problem, namely

$$Q = P/2^{\tfrac{1}{2}}, \quad u = Pl/AE. \qquad (A.10)$$

Thus we confirm that the direct application of (A.3) for a hypothetical set of bar tensions overestimates the displacement of the joint.

(III) The theorem of minimum strain energy

This theorem is an almost exact counterpart of theorem II. It concerns a *kinematically indeterminate* structure to some of whose joints are applied a set of *displacements u_j*. It states that the displacements of the other joints, which are free to move, arrange themselves in such a way that a quantity known as the *total strain energy* is minimised. In applying this theorem to the structure of fig. A.3, for example, we must suppose that instead of a given *force* being applied to joint C, a displacement u is imposed on this joint instead. Then the theorem enables us to find an overestimate of the external force P which would be needed to provide this displacement. In some

Fig. A.3.

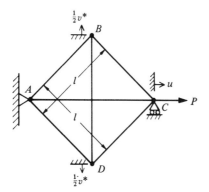

circumstances it is not possible to invert the boundary conditions of a problem in this way (consider, e.g. a truss loaded by external forces at two distinct points, for which the ratio of displacements at the corresponding points is not known); and it is then necessary to use the quite different theorem of minimum total potential energy: see below.

To prove the theorem, consider a structure which has displacements u_j imposed at some joints, displacements v_k at the remaining 'free' joints, bar elongations e_i, bar tensions T_i and external forces P_j at the joints where u_j are imposed. These quantities satisfy all of the static, kinematic and elastic equations of the problem. Consider also a hypothetical set of joint displacements and bar elongations u_j, v_k^* and e_i^* which satisfy merely the conditions of geometric compatibility. The displacements u_j are the same in both cases, since these components are prescribed in the present problem; but v_k^*, e_i^* are, in general, different from the corresponding 'exact' quantities.

Let us now apply the equation of virtual work twice, using the two different compatible sets of displacements and elongations but the same equilibrium set of forces and tensions. Thus:

$$\sum_j P_j u_j = \sum_i T_i e_i \qquad (A.11)$$

$$\sum_j P_j u_j = \sum_i T_i e_i^*.$$

By subtraction, therefore,

$$\sum_i T_i (e_i^* - e_i) = 0. \qquad (A.12)$$

Note that the displacements v_j, v_j^* do not enter these equations, since the 'free' components of joint displacement are associated with zero external force.

Consider the tension/elongation diagram for a typical member shown in fig. A.2c, and define as the *strain energy*, $U(e)$, the area shaded. Thus

$$U(e) = \int_0^e T \mathrm{d}e = \tfrac{1}{2} e^2 AE/L. \qquad (A.13)$$

Note that the strain energy of the bar is a function of its elongation; whereas the complementary energy of the bar (fig. A.2a) is a function of its *tension*. And although for a given bar in a given physical state U and C have numerically equal values (at least, for our *linear*-elastic material), nevertheless $U(e)$ and $C(T)$ are quite separate quantities, being functions of different variables.

The definition of $U(e)$ leads directly to the inequality

$$U(e^*) - U(e) \geqslant T(e^* - e) \qquad (A.14)$$

1 Theorems of structural mechanics

by considerations of area: cf. its counterpart (A.7). It follows from (A.12) that

$$\sum_i U_i(e^*) \geq \sum_i U_i(e);$$

and hence that

$$\tfrac{1}{2} P_j u_j \leq \sum_i \tfrac{1}{2} e_i^{*2} (A_i E_i/L_i). \tag{A.15}$$

This inequality is the required theorem.

For an example of the use of this theorem consider again the structure shown in fig. A.3. The displacement u is given. Let the horizontal displacement of B and D be $\tfrac{1}{2}u$ (an obvious guess), and let the vertical displacements be $\pm \tfrac{1}{2} v^*$, respectively, where v^*, is unknown. The elongation of each bar can be found in terms of u and v^* by means of a displacement diagram or otherwise; then the strain energy of each bar can be determined, and the total strain energy evaluated. If we guess that $v^* = \tfrac{1}{2}u$, we find that the total strain energy of the structure is $(5 \times 2^{\frac{1}{2}} + 1) AEu^2/16l$: hence we have, on using (A.15):

$$u \geq 0.99 \, Pl/AE. \tag{A.16}$$

The strain-energy theorem thus gives a lower bound on u, in contrast to the upper bound given by the complementary energy method: evidently the closeness of the bound to the correct value is due to the fact that the hypothetical value of v^* was fairly close to the exact value.

(IV)   **The theorem of minimum total potential energy**

This theorem concerns a structure which is subjected to given *forces* P_j, just as in theorem II, but it involves an energy quantity which is a function of the hypothetical *displacements* and *bar elongations* of the structure. The precise form of the energy quantities will emerge from the proof, as follows.

Let P_j, T_i, u_j, e_i represent the actual loads, tensions, displacements and elongations, respectively in the structure, and let u_j^*, e_i^* be a hypothetical set of displacements and elongations which satisfies merely the requirements of geometrical compatibility. By applying the equation of virtual work twice with the single statically admissible set and the two kinematically admissible sets, and subtracting the two equations, we obtain

$$\sum_j P_j (u_j^* - u_j) = \sum_i T_i (e_i^* - e_i)$$

as the counterpart of (A.5) and (A.12). Using (A.14) we find

$$\sum_i U_i(e) - \sum_j P_j u_j \leq \sum_i U_i(e^*) - \sum_j P_j u_j^*,$$

and hence, finally

$$-\sum_j \tfrac{1}{2} P_j u_j \leq \sum_i \tfrac{1}{2} e_i^{*2} (A_i E_i/L_i) - \sum_j P_j u_j^*. \tag{A.17}$$

This is the required inequality. The left-hand side is essentially the same as the left-hand side of (A.3) and (A.15). But the right-hand side now involves *two* quantities: the first is the total strain energy of the system, just as in theorem III, and the second may be described as the *potential energy of the loads*. Note that the potential energy term is proportional to displacement, while the strain-energy term is proportional to the square of the bar elongation. It follows that in the evaluation of the right-hand side for any hypothetical mode u_j^*, e_i^*, it is best to multiply by a scalar λ, say, and to minimise the resulting quadratic right-hand side of (A.17) with respect to λ.

Remarks

Theorem IV is not used in the present book. It has been included for the sake of completeness, in the sense that theorems II and IV both refer to a structure subjected to the same boundary conditions, whereas theorems II and III do not. A fourth theorem concerning elastic structures may be proved, but is not stated here: it concerns a structure subjected to given *displacements* (as in theorem III), which is analysed in relation to hypothetical tensions and external forces.

The definition of *complementary energy*, *strain energy* and *potential energy* emerge naturally in the above derivations as the quantities for which the various key inequalities can be set up. The definitions of $C(T)$ and $U(e)$ as integrals are valid for nonlinear elastic materials; and indeed they may be adapted to cover situations of initial stress, etc. in which the curves do not necessarily pass through the origin: the inequalities (A.7), (A.14) still hold in these circumstances provided the bars are *stable* in the sense that the slope dT/de is always positive. (Otherwise the area of the crucial 'triangles' would not necessarily be positive, and the various inequalities would not be rigorous.) The theorems also presuppose, of course, that the material is elastic (i.e. *reversible*); otherwise the various energy quantities cannot be defined in the first place.

Appendix 2

'Corresponding' load and deflection variables

A central notion in the concept of virtual work (appendix 1) is that both the external *force* and *displacement* quantities and the internal *tension* and *elongation* quantities are related to each other in the sense that the product of corresponding variables represents a quantity of work. If a single force P acts at a joint, the 'corresponding' measure of displacement of the joint is the component of the displacement in the (positive) direction of the line of action of the force. More generally, if the *components* of a force are specified, say U, V, W, in mutually perpendicular directions, the 'corresponding' displacements are the components of displacement u, v, w in the same directions; and the appropriate (scalar) work product is simply $Uu + Vv + Ww$.

We are not, however, limited to discussion of loads on structures in terms of force as such. A structure may be loaded by a *couple*, for which the corresponding displacement is an angle of rotation (measured in radians); or a *pressure*, for which the corresponding displacement is a 'swept volume'; or a uniform *line load*, for which the corresponding displacement is a 'swept area'.

In relation to internal variables we saw in appendix 1 that we must multiply the *tension* in a bar by the *elongation* in order to obtain the appropriate work quantity. For a uniform bar of length L and cross-sectional area A, precisely the same quantity would be obtained by evaluating $\sigma \epsilon V$, where $\sigma = T/A$ is the tensile stress, $\epsilon = e/L$ is the tensile strain and $V = AL$ is the volume of the bar. And indeed in a case of non-uniform stress and strain the appropriate calculation would be the integral $\int_V \sigma \epsilon dV$. This calculation is only valid for a state of uniaxial stress in a bar or body; but there is no difficulty in widening the scope of such a calculation to more general states of stress by considering the various components of stress. In the case of plate or shell structures it is often more convenient to deal in terms of stress resultants. Thus for a plate in plane stress the appropriate virtual work quantity would be, with respect to an x, y coordinate system in the plane of the plate,

$$\int_A (N_x \epsilon_x + N_y \epsilon_y + N_{xy} \gamma_{xy}) \, \mathrm{d}A$$

over the surface area of the plate: see chapter 2 for definitions of the various quantities. Again, in relation to the bending of a plate, the appropriate internal virtual work is

$$\int_A (M_x \kappa_x + M_y \kappa_y + 2M_{xy} \kappa_{xy}) \, \mathrm{d}A.$$

In these expressions the quantities $N_x, \ldots M_x, \ldots$ are the components of *generalised stress*, while $\epsilon_x \ldots, \kappa_x \ldots$ are the corresponding components of *generalised strain*. The concept of generalised stress and strain in relation to the internal quantities, together with generalised force and displacement as described above in relation to the external quantities, facilitates the discussion of the *general* properties of structural systems. Many aspects of structural mechanics are conducted in terms of generalised variables: tension, bending moment, shearing force and twisting moment are obvious examples of generalised stress in relation to prismatic structural elements; and the corresponding generalised strains are easy to deduce.

The principle of virtual work and all of the structural theorems in appendix 1 may be written in terms of whatever generalised forces, displacements, stresses and strains are most convenient for a given structure and its loading system; provided of course that the restrictions of 'small displacements' apply.

Appendix 3

Rayleigh's principle

This is an important energy principle in the fields of vibration and classical buckling of linear-elastic structures, which allows natural frequencies of vibration and critical buckling loads to be estimated by means of energy calculations with respect to hypothetical assumed modes. It was not included in appendix 1 because it cannot be derived directly from the principle of virtual work. Here we shall describe the principle in relation to natural frequencies of vibration; but we could have used the example of classical buckling equally well (cf. sections 17.2.2 and 14.7.2). The key idea, due to Rayleigh, is that in the small-amplitude vibration of linear-elastic structures any possible mode of deformation may be considered as a summation of *normal modes*, which are all mutually *orthogonal*. This orthogonality can be seen most clearly in a discrete system having a finite number, say n, degrees of freedom. The modes corresponding to the natural frequencies of the system are orthogonal since they are the eigenvectors of a real, symmetric matrix; and the symmetry of the matrix is a direct consequence of Maxwell's reciprocal theorem, which is only true for linear systems. We shall allocate subscripts $1 - n$ to the normal modes in order of increasing natural frequency.

Suppose that when the system is vibrating in the fundamental mode with a suitably defined amplitude a_1, the peak strain energy in the system is equal to $a_1^2 B_1$, where $B_1 > 0$. If the corresponding frequency is ω_1, (i.e. the *lowest natural frequency*), the peak kinetic energy is equal to $\omega_1^2 a_1^2 B_1$: see section 17.2.2. Similarly, for vibration of amplitude a_i in the ith mode at frequency ω_i, the corresponding peak energies are $a_i^2 B_i$ and $\omega_i^2 a_i^2 B_i$, respectively, with $B_i > 0$.

Consider now the calculation of frequency for an arbitrary hypothetical mode by means of an energy method. According to Rayleigh's analysis this mode may be considered as a linear combination of the n normal modes, with coefficients $a_1 \ldots a_n$, respectively; and since all of the modes are orthogonal (appendix 4) both the peak total strain energy and the peak total kinetic energy are found by summing the terms corresponding to the separate normal-mode components. The square of the natural frequency of this hypothetical mode is equal to the ratio of these two total energies; and thus we have

$$\omega^2 = \frac{a_1^2 B_1 \omega_1^2 + a_2^2 B_2 \omega_2^2 + \ldots + a_n^2 B_n \omega_n^2}{a_1^2 B_1 + a_2^2 B_2 + \ldots + a_n^2 B_n}. \tag{A.18}$$

Now if the hypothetical mode happens to coincide with one of the normal modes (say the mth), only one of the coefficients a_i is nonzero, and (A.18) simply yields $\omega^2 = \omega_m^2$. But in general this will not be the case, and the hypothetical mode will include contributions from most of the fundamental modes. Equation (A.18) states in effect that ω^2 is a weighted mean of $\omega_1^2 \ldots \omega_n^2$, with all of the weighting coefficients positive. It follows immediately that the *lowest possible* value of ω^2 is ω_1^2; and hence that the frequency computed by Rayleigh's method for an *arbitrary* hypothetical mode cannot possibly lie below the lowest natural frequency of the system. Likewise the computed value of ω^2 cannot possibly exceed ω_n^2. These are the only definite statements which can be made about the frequency ω computed in this way for an arbitrary hypothetical mode. It seems clear, however, from the form of (A.18) that if a particular hypothetical mode is rather similar to the mth mode without being exactly equal to it, then both the numerator and the denominator of the right-hand side of (A.18) will be strongly dominated by one term, and the calculated value of ω^2 will be close to ω_m^2. Throughout this discussion it has been assumed that the hypothetical modes satisfy adequately the boundary conditions of the problem.

Appendix 4

Orthogonal functions

A function $f(x)$ is said to be orthogonal to the function $g(x)$ in the region $0 < x < l$ if

$$\int_0^l f(x)\,g(x)\,\mathrm{d}x = 0. \tag{A.19}$$

Thus, if $f(x)$ and $g(x)$ represent two distinct orthogonal modes of displacement,

$$\int_0^l (f(x) + g(x))^2\,\mathrm{d}x = \int_0^l (f(x))^2\,\mathrm{d}x + \int_0^l (g(x))^2\,\mathrm{d}x.$$

It follows that, in this case, any energy quantity which involves an integration of the square of the displacements may be calculated for the combined mode $f(x) + g(x)$ by adding the energies which would be found for the two modes separately. It is easy to show in particular that the functions $\sin(m\pi x/l)$ and $\sin(n\pi x/l)$, where n and m are positive integers, are orthogonal unless $m = n$; that $\cos(m\pi x/l)$ and $\cos(n\pi x/l)$ are also orthogonal unless $m = n$; and that $\sin(m\pi x/l)$ and $\cos(n\pi x/l)$ are always orthogonal.

Appendix 5

Force-like and stress-like loads

In the theory of structures an important distinction is made between *generalised forces* and *generalised stresses*. Generalised forces – such as force and couple – are applied *externally* to structures, while generalised stresses – such as tension and bending moment – are *internal* variables, representing the state of stress of the material which the structure is made. Although the units of (e.g.) *couple* and *bending moment* are the same, these two quantities are quite different in character; and indeed this requires that different kinds of sign convention are used.

Notwithstanding this important general distinction between these two kinds of quantity, it is sometimes convenient to describe the external load acting on a structure with reference to the generalised stress which it produces in the structure at the point of application. Examples may be seen in figs. 7.4, 8.9, 8.10 and 8.11, where it is convenient to use well-defined generalised stresses as apparent edge-loads instead of defining a separate set of external force variables. The only practical disadvantage of this arrangement is that

Fig. A.4.

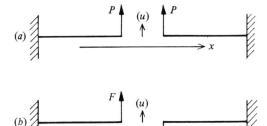

5 Force-like and stress-like loads

the relationship between the edge-loading and the deformation which it causes will require the insertion of a negative sign if it is used for a shell which extends in the opposite direction of x. The point can be illustrated sufficiently with respect to the two equal cantilevers shown in fig. A.4, which are mounted in opposite directions with respect to the x-coordinate. The relationship between the edge-*load* P and the corresponding edge-*displacement* u(fig. A.4a) is precisely the same for both cantilevers. On the other hand, if the loading is conceived, as in fig. A.4b in terms of the shearing force F which the loading produces at the tip (F is a generalised stress, with the sign convention shown in fig. 9.6b), the relationship between F and u has a different sign in the two cases.

Appendix 6

The 'static-geometric analogy'

If the various shallow-shell equations from chapter 8 which govern the behaviour of the S- and B-surfaces are assembled together, some remarkable formal analogies between them become obvious. Analogies of this kind were first pointed out in the 1940s by Lur'e and Goldenveiser (see Lur'e, 1961; Goldenveiser, 1961, §30), and they are known collectively as the 'static-geometric analogy'. They are peculiar to the theory of thin shells and have no counterpart in, e.g. the classical equations of three-dimensional elasticity.

These analogies emerge particularly clearly in the formulation of the equations of elastic shells in terms of the static and kinematic interaction of distinct 'stretching' and 'bending' surfaces. The following exposition follows closely that given by Calladine (1977b). Since it relates explicitly to shallow-shell equations, for which in particular the coordinates are aligned with the directions of principal curvature, the discussion cannot be regarded as complete. In fact the analogy holds when the equations are set up in terms of the most general curvilinear coordinate system; but it is usually regarded as being restricted to shells with zero surface loading (Naghdi, 1972, p.613). As will be seen, the introduction of *change of Gaussian curvature* (g) as a kinematic variable makes possible the extension of the analogy to shells loaded by *pressure* (p); and indeed these two variables turn out to be analogous in the present context.

The various equations may be collected together in two separate columns related to the S- and B-surfaces respectively, as follows.

$$\frac{N_x}{R_1} + \frac{N_y}{R_2} = p_S \quad \text{(S1)} \qquad \frac{\kappa_y}{R_1} + \frac{\kappa_x}{R_2} = g_B \quad \text{(B1)}$$

$$\left. \begin{array}{l} \dfrac{\partial N_x}{\partial x} + \dfrac{\partial N_{xy}}{\partial y} = 0 \\[6pt] \dfrac{\partial N_y}{\partial y} + \dfrac{\partial N_{xy}}{\partial x} = 0 \end{array} \right\} \quad \text{(S2)} \qquad \left. \begin{array}{l} \dfrac{\partial \kappa_y}{\partial x} - \dfrac{\partial \kappa_{xy}}{\partial y} = 0 \\[6pt] \dfrac{\partial \kappa_x}{\partial y} - \dfrac{\partial \kappa_{xy}}{\partial x} = 0 \end{array} \right\} \quad \text{(B2)}$$

6 The 'static-geometric analogy'

$$N_y = \frac{\partial^2 \phi}{\partial x^2}$$

$$N_{xy} = -\frac{\partial^2 \phi}{\partial x \partial y} \quad \text{(S3)}$$

$$N_x = \frac{\partial^2 \phi}{\partial y^2}$$

$$\kappa_x = -\frac{\partial^2 w}{\partial x^2}$$

$$\kappa_{xy} = -\frac{\partial^2 w}{\partial x \partial y} \quad \text{(B3)}$$

$$\kappa_y = -\frac{\partial^2 w}{\partial y^2}$$

$$\frac{\partial^2 \epsilon_y}{\partial x^2} - \frac{\partial^2 \gamma_{xy}}{\partial x \partial y} + \frac{\partial^2 \epsilon_x}{\partial y^2} = -g_S \quad \text{(S4)}$$

$$\frac{\partial^2 M_x}{\partial x^2} + \frac{2 \partial^2 M_{xy}}{\partial x \partial y} + \frac{\partial^2 M_y}{\partial y^2} = -p_B \quad \text{(B4)}$$

$$\epsilon_x = (N_x - \nu N_y)/Et$$
$$\epsilon_y = (N_y - \nu N_x)/Et \quad \text{(S5)}$$
$$\gamma_{xy} = 2(1+\nu)N_{xy}/Et$$

$$M_y = D(\kappa_y + \nu \kappa_x)$$
$$M_x = D(\kappa_x + \nu \kappa_y) \quad \text{(B5)}$$
$$M_{xy} = D(1-\nu)\kappa_{xy}$$

$$\Gamma^2 \phi = p_S \quad \text{(S6)} \qquad -\Gamma^2 w = g_B \quad \text{(B6)}$$

$$-(1/Et)\nabla^4 \phi = g_S \quad \text{(S7)} \qquad D\nabla^4 w = p_B. \quad \text{(B7)}$$

The provenance of these equations from chapters 2, 4, 6, 7 and 8 is given below. In some cases there are minor, but obvious, changes.

(S1) = (4.8), (S2) = (4.7) (with $q = 0$), (S3) = (7.36),
(S4) = (6.5), (S5) = (2.5), (S6) = (8.18), (S7) = (8.21),
(B1) = (6.6), (B3) = (8.8), (B4) = (8.7), (B5) = (2.28),
(B6) = (8.22), (B7) = (8.9).

Equation (B2) does not appear as such in the text: it follows an elimination of w from (B3).

In each case the equation on the left is formally analogous to the equation on the right. The variables which constitute the analogy may be set out thus:

$$\begin{aligned} N_x &\leftrightarrow \kappa_y & \epsilon_x &\leftrightarrow M_y & \phi &\leftrightarrow -w \\ N_y &\leftrightarrow \kappa_x & \epsilon_y &\leftrightarrow M_x & g_S &\leftrightarrow p_B \\ N_{xy} &\leftrightarrow -\kappa_{xy} & \gamma_{xy} &\leftrightarrow -2M_{xy} & p_S &\leftrightarrow g_B \end{aligned} \quad \text{(A.20)}$$

There is some minor untidiness over negative signs and constants; but this is largely a consequence of arbitrary definitions and sign conventions, and is of no fundamental significance.

Several remarkable features of the analogy may be listed, as follows.

(i) Force or stress quantities on one side are analogous to strain or displacement quantities on the other; and statical equilibrium equations on one side correspond to geometric compatibility equations on the other.

730   Appendices

(ii) The analogies are not between *corresponding* generalised stress and strain quantities (see appendix 2): e.g. N is analogous with κ and not with ϵ.

(iii) In the case of directionally subscripted quantities there is always an exchange of subscripts: e.g. $M_x \leftrightarrow \epsilon_y$; $N_x \leftrightarrow \kappa_y$.

(iv) The notion of separating the shell into two distinct surfaces brings out the analogy already mentioned between change of Gaussian curvature, g, and pressure, p: $g_S \leftrightarrow p_B$; $p_S \leftrightarrow g_B$.

The reasons behind the analogies may be explained by putting the equations in three groups, as follows:

(a) (S2), (S3) $\leftrightarrow$ (B2), (B3)
(b) (S1) $\leftrightarrow$ (B1)
(c) (S4) $\leftrightarrow$ (B4)

Analogies between the remaining equations follow directly by manipulation of the above. The first part (a) of the analogy is obvious from the fact that the relation between the Airy stress function ϕ and stress resultants N_x, etc., is just the same as that between displacement w and curvature changes κ_y, etc. In the plane-stress equations and plate-bending equations important variables are related by second derivatives: in one case force quantities, in the other displacement quantities. Note that the subscript exchange noted in (iii) above enters at this point.

The connection (b) between (S1) and (B1) can be seen with the help of fig. A.5 (cf. fig. 4.4a). Fig. A.5a shows the forces acting on the S-element on account of stress resultants N_x and N_y. Since the element is not twisted in the chosen x,y coordinate system (see section 8.2) the stress resultants on the

Fig. A.5.

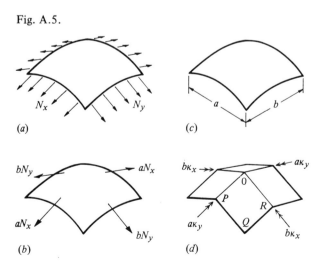

6 The 'static-geometric analogy'

various edges are statically equivalent to single forces, which are inclined to the xy plane as indicated in fig. A.5b. Equilibrium of the forces normal to the xy plane gives, for a sufficiently small element, the equilibrium equation (S1). (Stress resultants N_{xy} have not been shown, since they make no contribution to this equation.)

A comparable derivation of (B1) may be made if the doubly-curved shell element of fig. A.5c is first replaced by a polygonalised version, as shown in fig. A.5d. All of the curvature of the undeformed element has been concentrated in two creases, at which the hinge-angles are a/R_2 and b/R_1 respectively. The vertex has an angular defect which is easily calculated when these hinge-angles are small. However, we are concerned with the *change* in the angular defect when the hinge-angles are increased by $a\kappa_y$ and $b\kappa_x$ respectively. We can calculate this by imagining that we clamp one of the four faces, say 0PQR, make a slit on one edge, say 0P, then apply the small extra hinge rotation to each hinge separately, and finally find the resulting overlap of the cut edges 0P, which is the required change of angular defect. The calculation is facilitated by the fact that *small* angular rotations satisfy the laws of vector addition. We thus need to sum the four vectors shown in fig. A.5d. Each pair sums to a vector normal to the original xy plane, and the required sum is simply the sum of the four components in this normal direction. Since each of the original hinge directions makes a small angle with the xy plane the required sum is

$$ab(\kappa_x/R_2 + \kappa_y/R_1);$$

and as each of the planes makes but a small angle with the xy plane, this represents the change of angular defect to sufficient accuracy. The required result follows directly.

This method of derivation of equation (B1) makes its clear that the key to this part of the analogy is that both *forces* and *small angular rotations* are vectors.

The final stage (c) linking (S4) and (B4) is more complicated. Consider first a simpler problem concerning a beam, shown in fig. A.6a. Here the problem is to find the equilibrium equation relating the bending moment $M(x)$ in the beam to the transverse loading p per unit length. The simplest plan is to use virtual work, taking a virtual transverse displacement, shown in fig. A.6a, consisting of a shallow triangle of width $2h$ and height Δ. This gives an abrupt change in slope, or hinge rotation, at stations $x-h$, x and $x+h$ equal to Δ/h, $-2\Delta/h$ and Δ/h respectively; and the virtual work equation gives

$$(\Delta/h)(M_{x-h} - 2M_x + M_{x+h}) = -ph\Delta. \tag{A.21}$$

Here we have assumed that p is constant over the small interval $2h$ in x. In the limit as $h \to 0$ (and keeping $\Delta/h \ll 1$) we obtain

$$p = -\frac{d^2M}{dx^2}. \tag{A.22}$$

Note that by this choice of virtual displacement function we have avoided introducing shearing force to the equation.

Essentially the same method can be used to derive equilibrium equation (B4) for a flat plate, as follows. Consider a plate in the xy plane with a given distribution of bending and twisting moment. Take an arbitrary shallow pyramidal virtual normal displacement, as shown in fig. A.6b, centred on the point in question. If the plan of the pyramid is given and the height is Δ, it is a matter of simple geometry to calculate the (small) rotation θ_i of each of the radial and peripheral hinges. Applying virtual work and assuming that p_B is constant over the area in question we find

$$\tfrac{1}{3} p_S A \Delta = \sum_i M_i l_i \theta_i, \tag{A.23}$$

where

A is the area of the base of the pyramid,
l_i is the length of the hinge i,
M_i is the (mean) bending moment about the hinge.

It is easy to show that if the pyramid is given a square base of side $2h$ aligned with the x- and y-axes, (or indeed a rectangular base) application of (A.23) in the limit $h \to 0$ gives precisely equation (B4). It is of course necessary

Fig. A.6.

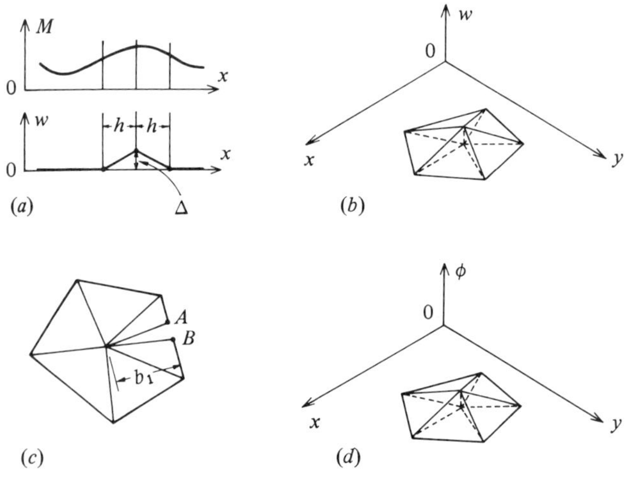

6 The 'static-geometric analogy'

to express (by means of Mohr's circle or otherwise) the bending moment on the diagonal hinges in terms of M_x, M_y, M_{xy}; but this is straightforward.

Consider now an analogous method for finding the change in Gaussian curvature at a point on the S-surface in terms of the strains in the surface. Fig. A.6c shows a flattened view of a small part of the undeformed polygonalised S-surface, consisting of the triangles surrounding a particular vertex. When the S-surface is strained there will be a consequent change of angular defect, which we wish to calculate. Fig. A.6d shows a stress function ϕ which we shall employ to find a local self-equilibrating virtual force field, which in turn we shall use in a virtual work calculation. It consists of a single shallow pyramid of exactly the same form as that previously used (see fig. A.6b): its base matches the triangular faces which surround the vertex. From the definition of ϕ (equation (S3)) we see that all stress resultants are zero where ϕ is a linear function of x and y; but corresponding to a discontinuity of slope θ_i in the function ϕ there is a line tension θ_i. Thus the pyramidal function ϕ corresponds to a self-equilibrating set of radial compressions and circumferential tensions lying along the plane projection of the edges of the triangles meeting at the vertex.

Now in order to determine the angular defect at a vertex of the polygonalised S-surface we first make a cut from the vertex and then flatten the faces meeting at the vertex. It is convenient to make this cut perpendicular to one of the peripheral edges, which we designate $i = 1$. The cut opens up to the required angular defect. At present we are interested in the *change* of this angular defect when the S-surface in the vicinity is given an arbitrary surface strain. We can calculate the change, say v, in the dimension AB (fig. A.6c) by employing virtual work, as follows, using the self-equilibrating force system described already. We have

$$v\theta_1 = \sum_i \epsilon_i l_i \theta_i, \qquad (A.24)$$

where ϵ_i is the (mean) tensile strain along line i. The geometry of the flattened vertex is of course not exactly the same as that of the plan of the shallow pyramid, on account of the angular defect. However, in the limit as the size of the triangular facets is reduced to zero the difference becomes negligible. Now the change in angular defect is equal to v/b_1, where b_1 is the dimension defined in fig. A.6c. The change in Gaussian curvature is equal to this divided by the associated area, $\frac{1}{3}A$. Thus $v = \frac{1}{3}g_B b_1 A$, and noting that $b_1 \theta_1 = \Delta$ we can rearrange (A.24) to give

$$\tfrac{1}{3}g_B A \Delta = \sum_i \epsilon_i l_i \theta_i. \qquad (A.25)$$

This equation is identical in form to (A.23), with M_i replaced by ϵ_i and p_S replaced by g_B. It follows that, just as (A.23) may be specialised to give (B4) so (A.25) may be specialised to give (S4). This analysis thus establishes the formal link between the two equations. It pivots on the analogy between w and ϕ, and the use of virtual work.

As we have already mentioned, the analogy is not confined to shallow shells. Thus we find in chapter 11 that there are formal analogies between the equilibrium equations (11.5) - (11.7) and the compatibility relations (11.10), (11.11). In this case the rotation χ is analogous to the shearing force U. The analogy also appears between equations (11.13) and (11.15), except that here there is a slight discrepancy in the form of terms enclosed { }. This small level inconsistency may be viewed in several ways. Thus, Simmonds (1975) has achieved a rigorous proof, by a fairly complicated argument, that omission of the offending terms is fully legitimate in the context of first-approximation theory: these terms are a manifestation of the small and unimportant ambiguities which can occur in 'first-order' derivation of the shell equations: see Koiter (1960). More recently Koiter (1980, §4) has shown that the Meissner-Reissner equations may be derived without the offending terms directly from the linearised general intrinsic equations of shells due to Koiter & Simmonds (1973). We may therefore use the equations as set out in chapter 11 safe in the knowledge that the apparent violation of the static-geometric analogy implied by (11.13) and (11.15) is ultimately of no significance.

The static-geometric analogy does not seem to have been particularly *useful* in the development of the theory of shell structures or the solution of practical problems since 1940. In part this has been due to the belief that it is valid only in the absence of surface loading. In this limited sense, however, the analogy enables us to see a correspondence between problems of statical analysis according to the membrane hypothesis by means of the Airy stress function (section 7.4) and problems involving inextensional deformation of shells (section 6.5); but both kinds of problem lie primarily in the realm of teaching exercises. The static-geometric analogy does, however, lead to some directly useful applications in certain *numerical* techniques for the solution of shell problems involving surface loading, as described by Pavlovic (1978).

Appendix 7

The area of a spherical polygon

The surface area of a simply-connected spherical polygon, drawn on the surface of a sphere of unit radius, depends only on the angles at its vertices. Each edge of such a polygon (or n-gon) is, by definition, an arc of a great circle, i.e. an arc of a circle whose plane includes the centre of the sphere. (A *plane* polygon, of course, has straight edges.) The formula for the surface area of a spherical n-gon may be developed as follows.

Consider first a spherical 2-gon, as shown in fig. A.7a. The two great semicircles which form its edges intersect at two diametrically opposite points on the sphere; for since the plane of every great circle passes through the centre of the sphere, the intersection of the two planes is a diameter of the sphere. Let the angle included between these two great semicircles at the vertices be α, as shown. It is clear from a view of the sphere along the line joining the two vertices that the surface area of the 2-gon is a fraction $\alpha/2\pi$ of the total surface area of the sphere; and thus that the total area of the 2-gon is 2α (since the total surface area of a sphere of unit radius is 4π).

Consider now a spherical triangle ABC whose vertices subtend angles α, β

Fig. A.7.

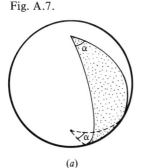
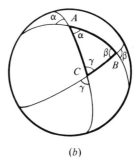

(a)   (b)

and γ, respectively, as shown in fig. A.7b. Each edge of the triangle is an arc of a great circle of the sphere. The three great circles are shown in fig. A.7b; and they intersect on the other side of the sphere to form a second, congruent spherical triangle whose vertices are diametrically opposite to A, B and C, respectively. Now imagine for the moment that the great circle passing through B and C is erased. The total area of the two spherical 2-gons having vertex angles α which are formed by the remaining two great circles is 4α. Similarly the surface area of the two 2-gons having vertex angles β which are formed by the two great circles passing through B is 4β; and similarly for the two 2-gons meeting at C. Now these six 2-gons together cover the entire surface of the sphere; but the area ABC is actually covered three times over, and so is its counterpart diametrically opposite. Therefore the total area of the six 2-gons is equal to the total area of the sphere plus four times the area of the spherical triangle ABC: consequently, if we designate Δ as the area of ABC we may write

$$4\alpha + 4\beta + 4\gamma = 4\pi + 4\Delta.$$

Hence

$$\Delta = \alpha + \beta + \gamma - \pi. \tag{A.26}$$

Now the sum of the interior angles of any *plane* triangle is π; therefore we may state that the area of a spherical triangle (on a sphere of unit radius) is equal to its *angular excess*; that is, the sum of the interior angles of the spherical triangle less the corresponding sum for a *plane* triangle.

It is now a straightforward matter to extend this result for a spherical triangle, or 3-gon, to the general case of a spherical polygon, or n-gon. Consider first a spherical 4-gon. It can be cut into two spherical triangles by a diagonal arc of a great circle; and it follows immediately that its area is equal to the sum of the interior angles of the 4-gon less the corresponding sum for a plane quadrilateral. Since any n-gon ($n > 3$) may be decomposed into $n-2$ triangles, we find that in general the area of a spherical n-gon is equal to its angular excess, namely the sum of its interior angles less the corresponding sum for a plane n-gon. (The area of a spherical 2-gon also satisfies this rule, since the area of a plane 2-gon is zero.)

Various simple examples may be used as checks of this result. Thus, a spherical triangle whose three vertices include right-angles covers one-eighth of the surface of a sphere; and a hemispherical cap may be considered as a spherical n-gon, with arbitrary $n \geq 2$, each of whose vertices has an interior angle of π.

Appendix 8

The 'sagitta' of an arc

The distance Δ marked in fig. A.8 between the mid-point of the arc and the chord is known as the *sagitta* of the arc. The name is latin for *arrow*, and refers to the configuration of a *bow*. Let the length of the chord be l, and let the arc be circular, of radius R. Then R may be determined approximately by means of the expression

$$R \approx l^2/8\Delta. \tag{A.27}$$

This formula is reasonably accurate provided the value of Δ/l is sufficiently small. It tends to underestimate the value of R; and the discrepancy, which tends to increase as $(\Delta/l)^2$, is around 4% when $\Delta/l = 0.1$.

The equation $\Delta = l^2/8R$ describes a *parabola* if R is regarded as a constant; and the error in formula (A.27) is due to the small difference in profile between shallow circular and parabolic arcs.

A somewhat more accurate version of the formula uses the *arc-length* λ instead of the chord l:

$$R \approx \lambda^2/8\Delta. \tag{A.28}$$

It tends to overestimate R, and the discrepancy is around 1.5% when $\Delta/\lambda = 0.1$.

Fig. A.8.

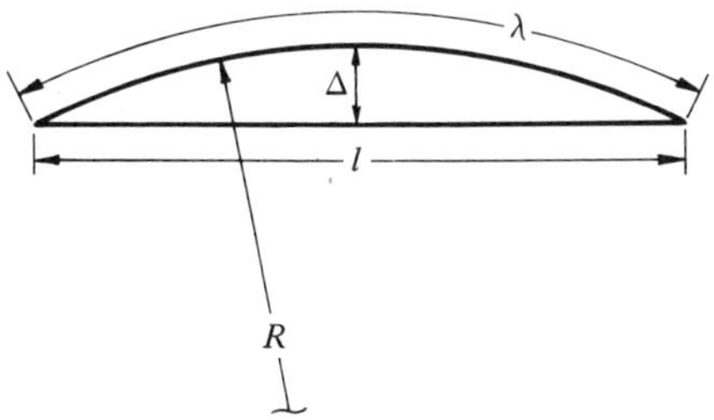

Appendix 9

Rigidity of polyhedral frames

The engineering design of space-frames is often facilitated by the use of an idealisation in which the actual structure is replaced conceptually by an assembly of rods and frictionless ball joints, or (as Maxwell put it) a collection of *lines* and *points*. If the idealised assembly is *rigid* when all of the bars or lines are inextensional – as distinct from being a *mechanism* – then the actual physical structure under consideration can be expected to carry loads applied at its joints primarily by means of tension and compression in its members. The next stage of the engineering calculation for such a structure is to perform a statical analysis of the tensions in the members, to invoke Hooke's law and then to compute the displacements of the assembly. But for the purposes of this appendix, we are concerned only with the question of the *rigidity* (or otherwise) of idealised frameworks made up from *inextensional* bars or lines.

This problem is one which attracts the attention of pure mathematicians. (Consideration of elasticity etc. would make the problem 'applied'.) These workers are inclined to think of the assembly of lines and points as their real structure, and any physical representation of the system by means of (e.g.) rubber connectors and wooden bars, or even structural steelwork, as conceptual idealisations of the reality under consideration. Here, then, we have a complete inversion of the engineer's view that the geometrical array of lines and points is a conceptual idealisation of the physical reality under consideration; the mathematician's is the *platonic* as opposed to the *aristotelian* view of nature.

Maxwell (1864), in developing his algebraic rule

$$b = 3j - 6 \qquad (A.29)$$

as the condition for a structure composed of b bars and j joints to be 'just stiff', was concerned to make a distinction between frameworks which were

9 Rigidity of polyhedral frames

statically determinate and those that were not, in the sense that the number of bar tensions was greater than the corresponding number of equations of statical equilibrium. The calculation of bar tensions in statically indeterminate frameworks necessitates the consideration of the mechanical properties of the material. Maxwell remarked in passing that this formula constituted a general rule, and that there would no doubt be exceptions to it.

Any framework whose bars lie along the edges of a *convex* simply-closed polyhedron with triangular faces (a convex 'deltahedron') is rigid: Euler's topological theorem (section 4.7.1) guarantees that Maxwell's rule is satisfied, and a theorem of Cauchy shows that convexity is sufficient for rigidity. A regular icosahedral framework (fig. A.9a) satisfies (A.29), is convex, and is therefore rigid. Convexity is not *necessary* for rigidity, however; a regular icosahedral frame which has been built with one vertex inverted (as in fig. A.9b) is certainly rigid. On the other hand, if the five bars meeting at one vertex of a regular icosahedron are steadily shortened but maintained at the same length as each other, there comes a limit to the length of these bars below which the apex cannot be connected. At this point the vertex is plane (fig. A.9c); the figure is no longer convex; and the figure is also no longer rigid, in the sense that this vertex may be moved a little out of the plane without incurring appreciable changes in length of the bars. This is often described as an 'infinitesimal mechanism'. According to Maxwell – who was primarily concerned with the structural performance of frameworks made from elastic bars – the stiffness of the frame against motions of this sort is 'of a low order'. On the other hand, it may be argued from the 'pure' point of view that the frame is actually rigid, in spite of the apparent mechanism in physical representations made of elastic substances: for if the bars or lines are *strictly* inextensional in the sense that no change of length, however small, is permissible, the vertex in question is immovable relative to the remainder of the assembly. Thus an 'infinitesimal mechanism' betrays to the engineer a potentially haz-

Fig. A.9.

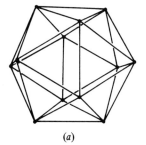
(a)
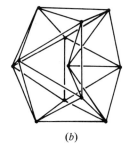
(b)
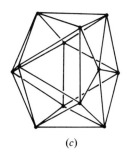
(c)

ardous feature of a structural design, while for our pure mathematician it does not constitute a mechanism at all.

The special case discussed above is one in which certain bars have the minimum length compatible with the assembly of the framework according to the given topological scheme. This aspect of such special cases appears to have been appreciated by Maxwell (see Calladine, 1978). Another class of special cases shares this same feature. Suppose we have a framework for which $b = 3j-7$. According to the usual analysis this assembly will constitute a *mechanism* having one degree of freedom. Imagine that one particular bar of the framework is now made steadily longer. The structure will distort in order to accommodate the increasing length of the member; but eventually a point will be reached at which the member cannot be extended without breaking the assembly. At this point it will be found in general that the assembly is no longer a mechanism, but is either *rigid* according to our mathematician or capable of distorting as an infinitesimal mechanism according to our engineer. Frameworks of this kind, with the value of b considerably less than $3j-6$ have been exploited by R. Buckminster Fuller, who calls them 'tensegrity' (= tension-integrity) structures (see Calladine, 1978). Their integrity ($\approx$ rigidity) depends on certain members being of minimum length; and in physical realisations it is necessary to arrange either for turnbuckles in some members, or for some members to be made of pre-stretched elastic threads. See Tarnai (1980) for a more thorough discussion of various anomalous frameworks.

Another kind of anomalous case involves a framework which satisfies Maxwell's rule and yet constitutes a genuine mechanism, i.e. not merely an infinitesimal one. The simplest example seems to be a 'chain' of six equal tetrahedra hinged together along opposite edges, and joined up to make a ring; $b = 30$ and $j = 12$, yet the outside parts can be rolled continuously, towards the interior of the 'torus'. It seems likely that the special features of this assembly are related to the fact that a torus, not being simply-connected, is *topologically* distinct from a simply-closed deltahedron.

The aims of this appendix have been threefold. Firstly, we have drawn attention to the existence of exceptions to Maxwell's rule for the construction of stiff frames. Secondly we have shown that the classification of exceptional cases depends to some extent on whether one regards the constituent members as completely rigid or slightly deformable. And thirdly, we have shown by means of various examples that the special cases are anomalous either by virtue of geometrical features (such as non-convexity, or maximum/minimum member lengths), or else on account of topological features: there is (as yet) no firm rule for spotting special cases.

Appendix 10

Fourier series

The following Fourier series are expressed in terms of θ in such a way that the period is 2π. They may be used directly in relation to functions having a period of 2π in polar coordinates; or they may be used for functions having a period $2\pi/n$ by means of the substitution $\theta = n\phi$, where $2\pi/n$ is the period of $f(\phi)$; or they may be interpreted in terms of a Cartesian coordinate x by means of the substitution $\theta = 2\pi x/T$, where T is the period of $f(x)$. In each case the origin has been chosen so that the waveform is an even function of θ.

Waveform   Series

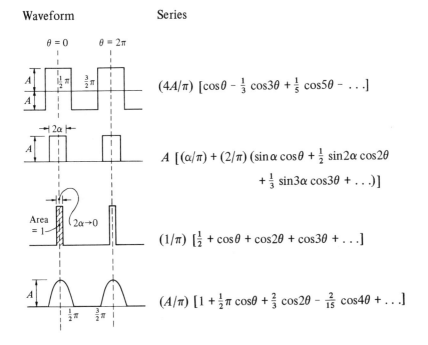

$(4A/\pi) [\cos\theta - \tfrac{1}{3} \cos 3\theta + \tfrac{1}{5} \cos 5\theta - \ldots]$

$A [(\alpha/\pi) + (2/\pi)(\sin\alpha \cos\theta + \tfrac{1}{2} \sin 2\alpha \cos 2\theta + \tfrac{1}{3} \sin 3\alpha \cos 3\theta + \ldots)]$

$(1/\pi) [\tfrac{1}{2} + \cos\theta + \cos 2\theta + \cos 3\theta + \ldots]$

$(A/\pi) [1 + \tfrac{1}{2}\pi \cos\theta + \tfrac{2}{3} \cos 2\theta - \tfrac{2}{15} \cos 4\theta + \ldots]$

Waveform	Series
(pulse train, width 2α, amplitude A, at $\theta=0, 2\pi$)	$(A/\pi)\,[(\sin\alpha - \alpha\cos\alpha) + (\alpha - \tfrac{1}{2}\sin 2\alpha)\cos\theta$ $+ (\sin\alpha + \tfrac{1}{3}\sin 3\alpha - \cos\alpha\sin 2\alpha)\cos 2\theta$ $+ (\tfrac{1}{2}\sin 2\alpha + \tfrac{1}{4}\sin 4\alpha - \tfrac{2}{3}\cos\alpha\sin 3\alpha)\cos 3\theta + \ldots]$
(triangular wave, amplitude $\pm A$, period 2π)	$(8A/\pi^2)\,[\cos\theta + \tfrac{1}{9}\cos 3\theta + \tfrac{1}{25}\cos 5\theta + \ldots]$

(After Howatson, Lund & Todd, 1972)

Appendix 11

Suggestions for further reading

Most of the references which I have cited are papers in journals, papers in proceedings of conferences, and sections of books. These journals, volumes of proceedings and books are all sources of further reading on the theory of shell structures. The following are some specific suggestions for further study.

The history of the subject is discussed from different viewpoints by Naghdi (1972) and Sechler (1974); and is also sketched by Flügge (1973, Bibliography).

The application of shell theory to practical problems in the aerospace field is described well by Babel, Christensen & Dixon (1974) and Bushnell (1981).

In his standard text on finite-element methods Zienkiewicz (1977) includes three chapters (13, 14, 16) on different types of finite-element calculation for shell structures.

A good example of the application of the membrane hypothesis to a shell of less simple form than those in chapter 4 is given by Martin & Scriven (1961).

Steele (1975) has written one of the few papers in the literature which uses change of Gaussian curvature as a variable for the description of distortion of surfaces (cf. chapter 6). His paper is concerned with the formation of a non-shallow shell (namely a cooking-pot) from a flat sheet by a process in which non-uniform surface stretching is imparted to the surface by beating.

For a discussion of non-symmetric behaviour of various non-cylindrical shells (cf. chapter 9) see Seide (1975).

Limit analysis (Chapter 18) was applied to the bending of curved pipes by Calladine (1974*b*). More recently Griffiths (1979) has used it to study the bending of curved pipes containing various slit-like defects. Ashton, McIntyre & Gill (1978) have applied plastic theory to the design of reinforcement of openings in pressure-vessels. The analysis of structures, including shells, which operate within the *creep* range of their material may be tackled in different ways: see, e.g. Hoff (1954*b*), Calladine (1972*b*) and Penny (1972).

Answers to selected problems

2.2 $(\sigma_x, \sigma_y, \tau_{xy}) = (12z/t^3)(M_x, M_y, M_{xy})$.

2.3 $1.5\nu(M_x + M_y)/Et = (\nu t/4E)(\sigma_x + \sigma_y)_{\text{surface}}$.

2.6 $\kappa_2 = M/E'R^2H$, where M is the bending moment per unit width of the beam.

2.8 $s = 1.5 M/Rt$ here gives $s = 1.0 M/a^2$.

3.6 $M_x = (\rho gal h/2 \times 3^{\frac{1}{2}})[1 - (ah)^{\frac{1}{2}}/3^{\frac{1}{4}}l]$. Peak bending stress/mean circumferential stress $= [3/(1 - \nu^2)]^{\frac{1}{2}}[1 - (ah)^{\frac{1}{2}}/3^{\frac{1}{4}}l]$. When $l/(ah)^{\frac{1}{2}} \gg 1$.

3.10 $M_{x0} = \frac{1}{4}\mu F(1 - \zeta + \frac{1}{3}\zeta^2 - \frac{1}{30}\zeta^4 \ldots)$.

3.13 $w_0, w_L = (0.95, 0.14)Fa^2/ELt$. $M_{x0}, M_{xL} = (0.125, -0.044)FL$.

3.19 No change of axial shortening.

3.23 Radial displacement $= (1 - \Phi) \times$ unconstrained displacement of an isolated hoop, cf. fig. 3.8. For low values of $l/(ah)^{\frac{1}{2}}$ there are significant longitudinal bending stresses.

4.1 Two semicircular ends, jointed tangentially by straights, like a toy railway track: this presupposes that alternate quarters of the *length* are loaded/unloaded, and that the load is normal to the current configuration.

4.4 $N_x = -(N_0 x/a)\cos\theta$; $N_\theta = 0$; $N_{x\theta} = N_0 \sin\theta$.

4.5 $N_x = -[f(l^2 - x^2)/a]\cos\theta$; $N_{x\theta} = -2fx\sin\theta$; $N_\theta = -fa\cos\theta$. $T = f(l^2 - x^2)$. At $x = 0$, $N_x = -(fl^2/a)\cos\theta$, $T = fl^2$.

4.6 $N_x = (n^2/2a)(l^2 - x^2)p_n\cos n\theta$; $N_\theta = ap_n\cos n\theta$; $N_{x\theta} = nxp_n\sin n\theta$ $(+C)$, including ends. Constant C represents a pure torque.

4.7 Upper half, $-\frac{1}{2}\pi < \theta < \frac{1}{2}\pi$: $N_x = N_\theta = N_{x\theta} = 0$.
Lower half, $\frac{1}{2}\pi < \theta < \frac{3}{2}\pi$: $N_x = -\frac{1}{2}\rho g(l^2 - x^2)\cos\theta$; $N_\theta = -\rho g a^2\cos\theta$; $N_{x\theta} = -\rho gax\sin\theta$. At $\theta = \pm\frac{1}{2}\pi$ a 'string' with tension $T = \frac{1}{2}\rho ga(l^2 - x^2)$, if $T = 0$ at $x = \pm l$: a line *compression*.

Answers to selected problems 745

4.8 $0 < x < h$: $N_x = -(n^2 p_n x^2/2a)\cos n\theta$, $N_\theta = ap_n \cos n\theta$; $N_{x\theta} = np_n x \sin n\theta$.
      $h < x < L$: $N_x = (n^2 p_n h/a)(\frac{1}{2}h - x)\cos n\theta$; $N_\theta = 0$; $N_{x\theta} = np_n h \sin n\theta$.

4.10 $N_{x\theta} = -lp_0 \Sigma n a_n \sin n\theta$; $N_x = (l^2 p_0/2a)\Sigma n^2 a_n \cos n\theta$.

4.16 $0 < \xi < 1$: $p = p_0 (1 - 3\xi^2 + 2\xi^3)$, $N_s/a = 0.05 p_0 (10 - 15\xi^2 + 8\xi^3)$,
      $N_\theta/a = 0.05 p_0 (10 - 45\xi^2 + 32\xi^3)$. $\xi > 1$: $p = 0$,
      $N_s/a = -N_\theta/a = 0.15 p_0/\xi^2$.

4.17 $N_\theta = -\frac{1}{2}fa(1 - 2r^2/a^2)$.

4.18 $N_s = -\frac{1}{2}fb[1 + (r/b)^2]^{\frac{1}{2}}$; $N_\theta = -\frac{1}{2}fb[1 + (r/b)^2]^{-\frac{1}{2}}$.

5.1 $Z = 2.5x^2 - 9xy + 0.5y^2$.

5.3 $\pm\theta$, where $\cos 2\theta = (c_1 + c_2)/(c_1 - c_2)$.

5.7 $+ = 70.435°$, $\Delta = 54.783°$; $46.957°$, $66.522°$, $66.522°$; 6 of the 9 triangles have edges of length $1, 1, 1.153$; the remaining 3 have edges, $1, 1, 0.920$. Alternatively, 3 triangles like the 6 above, with the remaining 6 having angles $58.696°, 60.652°, 60.652°$ (and edges of length $1.177, 1.177, 1.153$).

6.10 $w = w_n(x/l)\cos n\theta$; $v = -(w_n x/nl)\sin n\theta$ + constant;
      $u = -(aw_n/n^2 l)\cos n\theta$ + constant.

7.3   (b) $w = (q_0 L^4/\pi^4 EI)[\sin(\pi x/L) - (\pi x/L)(1 - x/L)]$, where transverse load intensity $q = q_0 \sin \pi x/L$, and EI is the flexural rigidity of the beam. (a) As above, but with quadratic expression in x omitted.

7.7   Largest radial displacements (and hence κ_θ) at $x/l \approx 0.42$.

7.8 $u = -(fL^3/3Eat)(1 - \xi^3)\cos\theta$; $v = A\sin\theta$; $w = -A\cos\theta$, where
      $A = (fL^4/Ea^2 t)(\frac{1}{4} - \frac{1}{3}\xi + \frac{1}{12}\xi^4)$: L is the length of the shell and $\xi = x/L$. θ is defined as in fig. 4.8.

7.9 $N_x^{\max} = n^2 p_n l^2/2a$, $w^{\max} = n^4 p_n l^4/8Ea^2 t$. $B = 1$ when $n^2 \approx 2.8 a^{\frac{3}{2}}/lt^{\frac{1}{2}}$ (taking $n^2 - 1 \approx n^2$). When end $x = l$ is held circular, $N_x^{\max}$ is reduced by a factor 4 while $w^{\max}$ is reduced by factor ≈ 23 ($w^{\max}$ at $x \approx 0.58 l$; so in expression for n^2, 2.8 is replaced by 6.8.

8.1 $(a^2/Et)(p_n/w_n) = [1 + (\lambda/\beta)^2]^{-2} + \frac{1}{12}\pi^4(\lambda^{-2} + \beta^{-2})^2$,
      where $\lambda = l/(ah)^{\frac{1}{2}}$, $\beta = b/(ah)^{\frac{1}{2}}$.

9.4 $w_0 B^{\frac{1}{4}} k^{\frac{3}{4}}/P = \frac{1}{3}\lambda^3/[1 + (11/140)\lambda^4]$, where $\lambda = Lk^{\frac{1}{4}}/B^{\frac{1}{4}}$. This follows the curve of fig. 9.7b closely up to $\lambda \approx 1.5$; is about 2.5% low at $\lambda = 2$ and about 5% low at $\lambda = 2.5$. Beyond this point the approximate curve falls as λ increases.

9.13   Mainly because the integration corresponds to the inverted V of the two asymptotes (cf. fig. 9.17) and does not allow for the smooth transition between them.

10.2 $q = (4q_0/\pi)\Sigma n^{-1}\sin n\theta$, where $\theta = \pi x/L$ and $n = 1, 3, 5, \ldots$.
      $M = (4L^2 q_0/\pi^3)\Sigma n^{-3}\sin n\theta$; $w = (4L^4 q_0/\pi^5)\Sigma n^{-5}\sin n\theta$.

11.2   At apex $2N_\theta/fa = -1$(sphere), -0.854 (paraboloid).

Difference = −0.146 (exact), −0.172 (by formula). At rim, $2N_\theta/fa = 0$ (sphere), −0.657 (paraboloid). Difference = 0.657 (exact), 0.586 (by formula). The agreement is worst at the rim, and very good in the region $r/a \approx 0.35$.

11.9 $N_\theta = -Et\alpha\Delta T$, $M_x = -Eth\alpha\Delta T/2 \times 3^{\frac{1}{2}}$.

14.3 (2, 1) preferred to (1, 1) when $-N_x > 0 > -N_y$ and $N_y/(-N_x) > \frac{3}{7}$. Similarly for (1, 2).

14.9 Equation (14.61) ($0.855 = (\pi/3)(2/3)^{\frac{1}{2}}$).

14.14 $I = 335t^4$, $706.25t^4$.

18.2 $\frac{1}{2}\kappa t = 1.5 w_0 t/a^2$.

References

Abbott, M. B. (1966). *An Introduction to the Method of Characteristics*. London: Thames and Hudson.

Akselrad, E. L. (1965). Refinement of the upper critical loading of pipe bending, taking account of the geometrical nonlinearity. (In Russian) *Izv. Akad. Nauk USSR, Otdelenie Teknicheskikh Nauk. Mech.* No. 4, 123–39.

Allman, D. J. & Gill, S. S. (1968). The effect of change of geometry on the limit pressure of a flush nozzle in a spherical pressure vessel. In *Engineering Plasticity*, ed. J. Heyman & F. A. Leckie, pp. 1–20. Cambridge University Press.

Arbocz, J. (1974). The effect of initial imperfections on shell stability. In *Thin Shell Structures*, ed. Y. C. Fung & E. E. Sechler, pp. 205–45. Englewood Cliffs, N. J.: Prentice-Hall, Inc.

Arbocz, J. & Babcock, C. D. (1969). The effect of general imperfections on the buckling of cylindrical shells. *Journal of Applied Mechanics*, 36, 28–38.

Arnold, R. N. & Warburton, G. B. (1953). The flexural vibrations of thin cylinders. *Proceedings of the Institution of Mechanical Engineers*, 167, 62–74.

Ashton, J. N., McIntyre, H. & Gill, S. S. (1978). A design procedure based on limit analysis for a pad-reinforced nozzle in a spherical pressure vessel. *International Journal of Mechanical Sciences*, 20, 747–57.

Ayrton, W. E. & Perry, J. (1886). On Struts. *The Engineer*, 62, 464–5; 513–15.

Babel, H. W., Christensen, R. H. & Dixon, H. H. (1974). Design, fracture control, fabrication and testing of pressurised space-vehicle structures. In *Thin Shell Structures*, ed. Y. C. Fung & E. E. Sechler, pp. 549–603. Englewood Cliffs, N. J.: Prentice-Hall Inc.

Baker, Lord & Heyman, J. (1969). *Plastic Design of Frames*. Cambridge University Press.

Baruch, M. & Singer, J. (1963). Effect of eccentricity of stiffeners on the general instability of stiffened cylindrical shells under hydrostatic pressure. *Journal of Mechanical Engineering Science*, 5, 23–7.

Basset, A. B. (1890). On the extension and flexure of cylindrical and spherical thin elastic shells. *Philosophical Transactions of the Royal Society of London, Series A*, 181, 433–80.

Batdorf, S. B. (1947). A simplified method of elastic-stability analysis for thin cylindrical shells. I: Donnell's equation. *N.A.C.A. Technical note 1341*. Washington: National Advisory Committee for Aeronautics.

Beskin, L. (1945). Bending of curved thin tubes. *Journal of Applied Mechanics*, 12, A1–A7.

Bickell, M. B. & Ruiz, C. (1967). *Pressure Vessel Design and Analysis*. London: Macmillan & Co. Ltd.

Bijlaard, P. P. (1955). Stresses from local loadings in cylindrical pressure vessels. *Transactions of the American Society of Mechanical Engineers*, 77, 805–14.

Blomfield, J. A. & Turner, C. E. (1972). Theory of thin elastic shells applied to pipe bends subject to bending and internal pressure. *Journal of Strain Anaylsis*, 7, 285–93.

Brazier, L. G. (1927). On the flexure of thin cylindrical shells and other 'thin' sections. *Proceedings of the Royal Society of London, Series A*, 116, 104–14.

British Standard 2654 (1973). *Vertical Steel Welded Storage Tanks with Butt-Welded Shells for the Petroleum Industry*. London: British Standards Institution.

British Standard 4485, part 4 (1975). *Specification for Water Cooling Towers. Part 4: Structural Design of Cooling Towers*. London: British Standards Institution.

Brush, D. O. & Almroth, B. O. (1975). *Buckling of Bars, Plates and Shells*. New York: McGraw-Hill Book Co.

Budiansky, B. (1969). Post-buckling behaviour of cylinders in torsion. In *Theory of Thin Shells, 2nd IUTAM Symposium*, ed. F. I. Niordson, pp. 212–32. Berlin: Springer-Verlag.

Budiansky, B. (1974). Theory of buckling and post-buckling of elastic structures. In *Advances in Applied Mechanics*, vol. 14, ed. C.-S. Yih, pp. 1–65. New York: Academic Press.

Budiansky, B. & Amazigo, J. C. (1968). Initial post-buckling behaviour of cylindrical shells under external pressure. *Journal of Mathematics and Physics*, 47, 223–35.

Budiansky, B. & Sanders, J. L. (1963). On the 'best' first-order linear shell theory. In *Progress in Applied Mechanics (the Prager Anniversary Volume)*, pp. 129–40. New York: Macmillan.

Bushnell, D. (1970). Analysis of buckling and vibration of ring-stiffened, segmented shells of revolution. *International Journal of Solids and Structures*, 6, 157–81.

Bushnell, D. (1973). Evaluation of various analytical models for buckling and vibration of stiffened shells. *A.I.A.A. Journal*, 11, 1283–91.

Bushnell, D. (1976). BOSOR5: Program for buckling of elastic-plastic complex shells of revolution, including large deflections and creep. *Computers and Structures*, 6, 221–39.

Bushnell, D. (1981). Buckling of shells – pitfall for designers. *A.I.A.A. Journal*, 19, 1183–226.

Bushnell, D. & Galletly, G. D. (1977). Stress and buckling of internally pressurised, elastic-plastic torispherical vessel heads: comparisons of test and theory. *Transactions of American Society of Mechanical Engineers, Series J, Journal of Pressure Vessel Technology*, 99, 39–53.

Calladine, C. R. (1966). On the design of reinforcement for openings and nozzles in thin spherical pressure vessels. *Journal of Mechanical Engineering Science*, 8, 1–14.

Calladine, C. R. (1968). Simple ideas in the large-deflection plastic theory of plates and slabs. In *Engineering Plasticity*, ed. J. Heyman & F. A. Leckie, pp. 93–127. Cambridge University Press.

Calladine, C. R. (1969a). *Engineering Plasticity*. Oxford: Pergamon Press.

Calladine, C. R. (1969b). Lower-bound analysis of symmetrically loaded shells of revolution. In *Pressure Vessel Technology (Proceedings of the First International Conference on Pressure Vessel Technology)*, vol. 1, ed. I. Berman, pp. 335–43. New York: American Society of Mechanical Engineers.

Calladine, C. R. (1972a). Structural consequences of small imperfections in elastic thin shells of revolution. *International Journal of Solids and Structures*, 8, 679–97.

Calladine, C. R. (1972b). Creep in torispherical pressure vessel heads. In *Creep in Structures 1970 (Proceedings of 2nd IUTAM Symposium)*, ed. J. Hult, pp. 247–68. Berlin: Springer-Verlag.

Calladine, C. R. (1972c). On the derivation of yield conditions for shells. *Journal of Applied Mechanics*, 39, 852–3.

Calladine, C. R. (1973a). A new finite-element method for analysing symmetrically loaded thin shells of revolution. *International Journal for Numerical Methods in Engineering*, 6, 475–87.

Calladine, C. R. (1973b). Inelastic buckling of columns: the effect of imperfections. *International Journal of Mechanical Sciences*, 15, 593–604.

Calladine, C. R. (1974a). Flexibility of axially symmetric bellows under axial loading. *International Journal of Mechanical Sciences*, 16, 843–53.

Calladine, C. R. (1974b). Limit analysis of curved tubes. *Journal of Mechanical Engineering Science*, 16, 85–7.

Calladine, C. R. (1977a). Thin-walled elastic shells analysed by a Rayleigh method. *International Journal of Solids and Structures*, 13, 515–30.

Calladine, C. R. (1977b). The static-geometric analogy in the equations of thin shell structures. *Mathematical Proceedings of the Cambridge Philosophical Society*, 82, 335–51.

Calladine, C. R. (1978). Buckminster Fuller's 'tensegrity' structures and Clerk Maxwell's rules for construction of stiff frames. *International Journal of Solids and Structures*, 14, 161–72.

Calladine, C. R. (1982). Natural frequencies of cooling-tower shells. *Journal of Sound and Vibration*, 82, 345–69.

Calladine, C. R. & Drucker, D. C. (1962). A bound method for creep analysis of structures: direct use of solutions in elasticity and plasticity. *Journal of Mechanical Engineering Science*, 4, 1–11.

Calladine, C. R. & Goodall, I. W. (1969). Plastic behaviour of thin cylindrical pressure vessels with circular cutouts and radial branches. *Journal of Mechanical Engineering Science*, 11, 351–63.

Calladine, C. R. & Paskaran, N. (1974). A re-appraisal of influence coefficients for the edges of thin elastic spherical shells subjected to symmetric loads. *Quarterly Journal of Mechanics and Applied Mathematics*, 27, 1–15.

Calladine, C. R. & Robinson, J. M. (1980). A simplified approach to the buckling of thin elastic shells. In *Theory of Shells (Proceedings of 3rd IUTAMS Symposium)* ed. W. T. Koiter & G. K. Mikhailov, pp. 173–96. Amsterdam: North-Holland Publishing Company.

C.E.G.B. (1965). *Report of the Committee of Inquiry into Collapse of Cooling Towers at Ferrybridge, Monday 1 November 1965*. London: Central Electricity Generating Board.

Clark, R. A. (1950). On the theory of thin elastic toroidal shells. *Journal of Mathematics and Physics*, 29, 146–78.

Clark, R. A. & Reissner, E. (1951). Bending of curved tubes. In *Advances in Applied Mechanics*, vol. 2, ed. R. von Mises & T. von Kármán, pp. 93–122. New York: Academic Press, Inc.

Cox, H.L. (1940). Stress analysis of thin metal construction. *Journal of the Royal Aeronautical Society*, 44, 231–272.

Crandall, S. H. & Dahl, N. C. (1957). The influence of pressure on the bending of curved tubes. *Proceedings of the 9th International Congress of Applied Mechanics (International Union of Theoretical and Applied Mechanics)*, vol. 6. pp. 101–11. Brussels: University of Brussels.

Crandall, S. H., Dahl, N. C. & Lardner, T. J. (1972). *An Introduction to the Mechanics of Solids* (2nd edn). New York: McGraw-Hill Book Company.

Croll, J. G. A. & Walker, A. C. (1972). *Elements of Structural Stability*. London: Macmillan.

D. A. St (1980). *Beulsicherheitsnachweise für Schalen*. Köln: Deutscher Ausschuss für Stahlbau.

Demir, H. H. & Drucker, D. C. (1963). An experimental study of cylindrical shells under ring loading. In *Progress in Applied Mechanics (the Prager Anniversary Volume)*, pp. 205–220. New York: Macmillan.

Dinno, K. S. & Gill, S. S. (1965). An experimental investigation into the plastic behaviour of flush nozzles in spherical pressure vessels. *International Journal of Mechanical Sciences*, 7, 817–39.

Donnell, L. H. (1933). Stability of thin-walled tubes under torsion. *N.A.C.A. Report 479*. Washington: National Advisory Committee for Aeronautics.

Donnell, L. H. (1976). *Beams, Plates and Shells*. New York: McGraw-Hill Book Co.

Donnell, L. H. & Wan, C. C. (1950). Effect of imperfections on buckling of thin cylinders and columns under axial compression. *Journal of Applied Mechanics*, 17, 73–83.

Drucker, D. C. & Shield, R. T. (1959). Limit analysis of symmetrically loaded thin shells of revolution. *Journal of Applied Mechanics*, 26, 61–8.

Eason, G. & Shield, R. T. (1955). The influence of free ends on the load-carrying capacities of cylindrical shells. *Journal of the Mechanics and Physics of Solids*, 4, 17–27.

Evensen, D. A. (1974). Nonlinear vibrations of circular cylindrical shells. In *Thin Shell Structures*, ed. Y. C. Fung & E. E. Sechler, pp. 133–55. Englewood Cliffs, N. J.: Prentice-Hall, Inc.

Fabian, O. (1977). Collapse of cylindrical elastic tubes under combined bending, pressure and axial loads. *International Journal of Solids and Structures*, 13, 1257–70.

Flügge, W. (1962). *Statik und Dynamik der Schalen* (3rd edn). Berlin: J. Springer.

Flügge, W. (1973). *Stresses in Shells* (2nd edn). Berlin: Springer-Verlag.

Flügge, W. & Elling, R. E. (1972). Singular solutions for shallow shells. *International Journal of Solids and Structures*, 8, 227–47.

Forsberg, K. (1964). Influence of boundary conditions on the modal characteristics of thin cylindrical shells. *Journal of the American Institute of Aeronautics and Astronautics*, 2, 2150–7.

Fung, Y. C. & Sechler, E. E. (1960). Instability of thin elastic shells. In *Structural Mechanics, Proceedings of the First Symposium on Naval Structural Mechanics*, ed. J. N. Goodier & N. J. Hoff, pp. 115–67. Oxford: Pergamon Press.

Galletly, G. D. (1959). Torispherical shells: a caution to designers. *Transactions of the American Society of Mechanical Engineers, Series B: Journal of Engineering for Industry*, 81, 51–66.

Galletly, G. D. (1960). Influence coefficients and pressure vessel analysis. *Transactions of the American Society of Mechanical Engineers, Series B: Journal of Engineering for Industry*, 82, 259–69.

Gauss, K. F. (1828). *Disquisitiones generales circa superficies curvas*. Göttingen. (English translation 1902, *General Investigation of Curved Surfaces* by J. C. Morehead & A. M. Hiltebeitel, Princeton; reprinted 1965 with introduction by R. Courant.

Hewlett, New York: Raven Press.)

Gellin, S. (1980). The plastic buckling of long cylindrical shells under pure bending. *International Journal of Solids and Structures*, 16, 397–407.

Gerard, G. & Becker, H. (1957). Handbook of structural stability, part 3: buckling of curved plates and shells. *N.A.C.A. Technical note 3783*. Washington: National Advisory Committee for Aeronautics.

Gibson, J. E. & Cooper, D. W. (1954). *The Design of Cylindrical Shell Roofs*. London: E. & F. N. Spon, Ltd.

Gill, S. S. (1964). The limit pressure for a flush cylindrical nozzle in a spherical pressure vessel. *International Journal of Mechanical Sciences*, 6, 105–15.

Goldenveiser, A. L. (1961). *Theory of Thin Shells*, translation ed. G. Herrmann. Oxford: Pergamon Press.

Griffiths, J. E. (1979). The effect of cracks on the limit load of pipe bends under in-plane bending: experimental study. *International Journal of Mechanical Sciences*, 21, 119–30.

Haythornthwaite, R. M. (1957). Beams with full end fixity. *Engineering*, 183, 110–12.

Hetényi, M. (1946). *Beams on Elastic Foundation*. Ann Arbor: University of Michigan Press.

Heyman, J. (1967). On shell solutions for masonry domes. *International Journal of Solids and Structures*, 3, 227–41.

Heyman, J. (1977). *Equilibrium of Shell Structures*. Oxford: Clarendon Press.

Hilbert, D. & Cohn-Vossen, S. (1952). *Geometry and the Imagination*, translator, P. Nemenyi. New York: Chelsea Publishing Company.

Hill, R. (1950). *The Mathematical Theory of Plasticity*. Oxford: Clarendon Press.

Hill, R. (1957). Stability of rigid–plastic solids.*Journal of the Mechanics and Physics of Solids*, 6, 1–8.

Hodge, P. G. (1954). The rigid–plastic analysis of symmetrically loaded cylindrical shells. *Journal of Applied Mechanics*, 21, 336–42.

Hodge, P. G. (1963). *Limit Analysis of Rotationally Symmetric Plates and Shells*. Englewood Cliffs, N. J.: Prentice-Hall.

Hoff, N. J. (1954a). Boundary-value problems of the thin-walled circular cylinder. *Journal of Applied Mechanics*, 21, 343–50.

Hoff, N. J. (1954b). Approximate analysis of structures in the presence of moderately large creep deformations. *Quarterly Journal of Applied Mathematics*, 12, 49–55.

Hoff, N. J. (1965). Low buckling stresses of axially compressed circular cylindrical shells of finite length. *Journal of Applied Mechanics*, 32, 533–41.

Holand, I. (1959). An application of Donnell's theory of circular cylindrical shells to the analysis of curved-edge disturbances. *Publications of the International Association of Bridge and Structural Engineering*, 19, 65–80.

Howatson, A. M., Lund, P. G. & Todd, J. D. (1972). *Engineering Tables and Data*, London: Chapman and Hall.

Hutchinson, J. W. (1965). Axial buckling of pressurised cylindrical shells. *A.I.A.A. Journal*, 3, 1461–6.

Hutchinson, J. W. (1967). Imperfection-sensitivity of externally pressurised spherical shells. *Journal of Applied Mechanics*, 34, 49–55.

Hutchinson, J. W. (1968). Buckling and initial postbuckling behaviour of oval cylindrical shells under axial compression. *Journal of Applied Mechanics*, 35, 66–72.

Hutchinson, J. W. (1974). Plastic buckling. In *Advances in Applied Mechanics*, vol. 14, ed. C.-S. Yih, pp. 67–144. New York: Academic Press.

Hutchinson, J. W. & Koiter, W. T. (1970). Postbuckling theory. *Applied Mechanics Reviews*, **23**, 1353–66.

Johnson, W. & Mellor, P. B. (1973). *Engineering Plasticity*. London: Van Nostrand Reinhold Co.

Johnson, W. & Reid, S. R. (1978). Metallic energy-dissipating systems. *Applied Mechanics Reviews*, **31**, 277–88.

Jolley, L. B. W. (1961). *Summation of Series* (2nd edn.). New York: Dover Publications, Inc.

Jones, N. (1966). On the design of pipe-bends. *Nuclear Engineering and Design*, **4**, 399–405.

Jordan, F. F. (1962). Stresses and deformations of the thin-walled pressurized torus. *Journal of Aerospace Sciences*, **29**, 213–25.

Kafka, P. G. & Dunn, M. B. (1956). Stiffness of curved circular tubes with internal pressure. *Journal of Applied Mechanics*, **23**, 247–54.

von Kármán, T., Dunn, L. G. & Tsien, H-S. (1940). The influence of curvature on the buckling characteristics of structures. *Journal of the Aeronautical Sciences*, **7**, 276–89.

von Kármán, T. & Tsien, H-S. (1941). The buckling of thin cylindrical shells under axial compression. *Journal of the Aeronautical Sciences*, **8**, 303–12.

M. W. Kellog Company (1956). *Design of Piping Systems* (2nd edn). New York: J. Wiley & Sons.

Kildegaard, A. (1969). Bending of a cylindrical shell subject to axial loading. In *Theory of Thin Shells (Proceedings of 2nd IUTAM Symposium)*, ed. F. I. Niordson. pp.301–15. Berlin: Springer-Verlag.

Koiter, W. T. (1945). The stability of elastic equilibrium. (In Dutch) Dissertation for the degree of Doctor in the Technical Sciences at the Technische Hooge School, Delft. Amsterdam: H. J. Paris. English translation by E. Riks (1970): Technical report AFFDL-TR-70-25, Air Force flight dynamics laboratory, Air Force Systems Command, Wright-Patterson Air Force Base, Ohio, USA.

Koiter, W. T. (1960). A consistent first approximation in the general theory of thin elastic shells. In *Theory of Thin Elastic Shells (Proceedings of lst IUTAM Symposium)*, ed. W. T. Koiter, pp. 12–33. Amsterdam: North-Holland Publishing Company.

Koiter, W. T. (1963a). Elastic stability and post-buckling behaviour. In *Proceedings of a Symposium on Non-Linear Problems*, ed. R. E. Langer, pp. 257–75. Madison, Wisconsin: University of Wisconsin Press.

Koiter, W. T. (1963b). The effect of axisymmetric imperfections on the buckling of cylindrical shells under axial compression. *Proceedings of Koninklijke Nederlandse Akademie van Wetenschappen*, **66**(*B*), 265–79.

Koiter, W. T. (1969a). Foundations and basic equations of shell theory: a survey of recent progress. In *Theory of Thin Shells (Proceedings of 2nd IUTAM Symposium)*, ed. F. I. Niordson, pp. 93–105. Berlin: Springer-Verlag.

Koiter, W. T. (1969b). The non-linear buckling problem of a complete spherical shell under uniform external pressure, I-IV. *Proceedings of Koninklijke Nederlandse Akademie van Wetenschappen*, **72**(*B*), 40–123.

Koiter, W. T. (1978). The influence of more-or-less localised short-wave imperfections on the buckling of circular cylindrical shells under axial compression (in a first approximation). In *Complex Analysis and its Applications (the I. N. Vekua Anniversary Volume)* pp. 242–4. Moscow: Nauka Publishing House.

Koiter, W. T. (1980). The intrinsic equations of shell theory, with some applications. In *Mechanics Today*, Vol. 5 *(the E. Reissner anniversary volume)*, ed. S. Nemat-Nasser, pp. 139–54. Oxford: Pergamon Press.

Koiter, W. T. & Simmonds, J. G. (1973). Foundations of shell theory. In *Theoretical and Applied Mechanics (Proceedings of 13th International Congress of Theoretical and Applied Mechanics)*, ed. E. Becker & G. K. Mikhailov, pp. 150–76. Berlin: Springer-Verlag.

Kraus, H. (1967). *Thin Elastic Shells*. New York: John Wiley & Sons, Inc.

Lamb, Sir H. (1890). On the determination of an elastic shell. *Proceedings of the London Mathematical Society*, 21, 119–46.

Lamb, Sir H. (1919). *An Elementary Course on Infinitesimal Calculus* (3rd edn). Cambridge University Press.

Leckie, F. A. (1965). Shakedown pressures for a flush cylindrical-spherical shell intersection. *Journal of Mechanical Engineering Science*, 7, 367–71.

Leckie, F. A. & Payne, D. J. (1966). Some observations on the design of spherical pressure vessels with flush cylindrical nozzles. *Proceedings of the Institution of Mechanical Engineers*, 180, 497–501.

Leckie, F. A. & Penny, R. K. (1963). Stress concentration factors for the stresses at nozzle intersections in pressure vessels. In *Welding Research Council Bulletin 90*, pp. 19–26. New York: Welding Research Council.

Leckie, F. A. & Penny, R. K. (1968). Plastic instability of a spherical shell. In *Engineering Plasticity*, ed. J. Heyman & F. A. Leckie, pp. 401–11. Cambridge University Press.

Leissa, A. W. (1973). *Vibration of Shells*, N.A.S.A. S.P. 288. Washington: National Aeronautics and Space Administration.

Love, A. E. H. (1888). On the small free vibrations and deformations of thin elastic shells. *Philosophical Transactions of the Royal Society of London, Series A*, 179, 491–546.

Love A. E. H. (1927). *A Treatise on the Mathematical Theory of Elasticity* (4th edn). Cambridge University Press.

Łukasiewicz, S. A. (1976). Introduction of concentrated loads in plates and shells. *Progress in Aerospace Science*, 17, 109–46.

Lur'e, A. I. (1961). On the static-geometric analogue of shell theory. In *Problems of Continuum Mechanics (the Muskhelisvili Anniversary Volume)*, ed. J. R. M. Radok, pp. 267–74. Philadelphia: Society for Industrial and Applied Mathematics.

Malik, Z., Morton, J. & Ruiz, C. (1980). Buckling under normal pressure of cylindrical shells with various end conditions. *Transactions of the American Society of Mechanical Engineers, series J, Journal of Pressure Vessel Technology*, 102, 107–10.

Marcal, P. V. (1969). Large-deflection analysis of elastic-plastic plates and shells. In *Pressure-Vessel Technology (Proceedings of the First International Conference on Pressure-Vessel Technology)*, vol. 1, ed. I. Berman, pp. 75–87. New York: American Society of Mechanical Engineers.

Marcal, P. V. & Turner, C. E. (1963). Numerical analysis of the elastic-plastic behaviour of axi-symmetrically loaded shells of revolution. *Journal of Mechanical Engineering Science*, 5, 232–7.

Martin, D. W. & Scriven, W. E. (1961). The calculation of membrane stresses in hyperbolic cooling towers. *Proceedings of the Institution of Civil Engineers*, 19, 503–14.

Maxwell, J. C. (1854). On the transformation of surfaces by bending. *Transactions of the Cambridge Philosophical Society*, 9, 455–469. (Reprinted, 1890, in *The Scientific Papers of J. C. Maxwell*, vol. 1, ed. W. D. Niven, pp. 81–114. Cambridge University Press.)

Maxwell, J. C. (1864). On the calculation of the equilibrium and stiffness of frames. *Philosophical Magazine*, 27, 294–99. (Reprinted, 1890, in *The Scientific Papers of J. C. Maxwell*, vol. 1, ed. W. D. Niven, pp. 598–604. Cambridge University Press.)

Morley, L. S. D. (1959). An improvement on Donnell's approximation for thin-walled circular cylinders. *Quarterly Journal of Mechanics and Applied Mathematics*, 12, 90–9.

Morris, A. J. & Calladine, C. R. (1969). The local strength of a thin spherical shell loaded radially through a rigid boss. In *Pressure Vessel Technology* (*Proceedings of the First International Conference on Pressure Vessel Technology*), vol. 1, ed. I. Berman, pp. 35–44. New York: American Society of Mechanical Engineers.

Morris, A. J. & Calladine, C. R. (1971). Simple upper-bound calculations for the indentation of cylindrical shells. *International Journal of Mechanical Sciences*, 13, 331–43.

Naghdi, P. M. (1963). Foundations of shell theory, In *Progress in Solid Mechanics*, vol. 4, ed. I. N. Sneddon & R. Hill, pp. 1–90. Amsterdam: North-Holland Publishing Co.

Naghdi, P. M. (1972). The theory of shells and plates. In *Handbuch der Physik*, vol. 6a.2, ed. C. Truesdell, pp. 425–640. Berlin: Springer-Verlag.

Neal, B. G. (1963). *The Plastic Methods of Structural Analysis* (2nd edn). London: Chapman & Hall, Ltd.

Novozhilov, V. V. (1964). *The Theory of Thin Shells*. Translation of 2nd Russian edn by P. G. Lowe, ed. J. R. M. Radok. Groningen: P. Noordhoff Ltd.

Olszak, W. & Sawczuk, A. (1967). *Inelastic Behaviour of Shells*. Groningen: P. Noordhoff, Ltd.

Olver, F. W. J. (1965). Bessel functions of integer order. In *Handbook of Mathematical Functions*, ed. M. Abramowitz & I. A. Stegun, pp. 355–433. New York: Dover Publications, Inc.

Onat, E. T. (1960). The influence of geometry changes on the load-deformation behaviour of plastic solids. In *Plasticity* (*Proceedings of the Second Symposium on Naval Structural Mechanics*), ed. E. H. Lee & P. S. Symonds, pp. 225–38. Oxford: Pergamon Press.

Palmer, A. C. (1969). A direct design technique for pressure-vessel intersections. In *Pressure Vessel Technology* (*Proceedings of the First International Conference on Pressure Vessel Technology*), vol. 1, ed. I. Berman, pp. 591–6. New York: American Society of Mechanical Engineers.

Palmer, A. C. & Martin, J. H. (1975). Buckle propagation in submarine pipelines. *Nature*, 254, 46–8.

Pardue, T. E. & Vigness, I. (1951). Properties of thin-walled curved tubes of short-bend radius. *Transactions of the American Society of Mechanical Engineers*, 73, 77–87.

Parkes, E. W. (1974). *Braced Frameworks* (2nd edn). Oxford: Pergamon Press.

Pavlovic, M. (1978). Numerical methods for the analysis of elastic thin shells. Dissertation submitted to the University of Cambridge for the degree of Doctor of Philosophy.

Penny, R. K. (1972). The creep of shells. In *Creep in Structures 1970* (*Proceedings of 2nd IUTAM Symposium*), ed. J. Hult, pp. 276–92. Berlin: Springer-Verlag.

Pflüger, A. (1961). *Elementary Statics of Shells* (translator, E. Galantay). New York: F. W. Dodge Corporation.

Prager, W. (1959). *Introduction to Plasticity*. Reading, Massachusetts: Addison-Wesley.
Ranjan, G. V. & Steele, C. R. (1975). Analysis of knuckle region between two smooth shells. *Journal of Applied Mechanics*, 42, 853-7.
Rayleigh, Lord (1881). On the infinitesimal bending of surfaces of revolution. *Proceedings of the London Mathematical Society*, 13, 4-16.
Rayleigh, Lord (1894). *The Theory of Sound* (2nd edn) vol. 1. London: Macmillan & Co.
Reddy, B. D. (1977). The elastic and plastic buckling of circular cylinders in bending. Dissertation submitted to the University of Cambridge for the Degree of Doctor of Philosophy.
Reddy, B. D. (1979). An experimental study of the plastic buckling of circular cylinders in pure bending. *International Journal of Solids and Structures*, 15, 669-83.
Reddy, B. D. & Calladine, C. R. (1978). Classical buckling of a thin-walled tube subjected to bending moment and internal pressure. *International Journal of Mechanical Sciences*, 20, 641-50.
Reissner, E. (1946). Stresses and small displacements of shallow spherical shells. I. *Journal of Mathematics and Physics*, 25, 80-5.
Reissner, E. (1959). On finite bending of pressurised tubes. *Journal of Applied Mechanics*, 26, 386-92.
Reissner, E. (1960). On some problems in shell theory. In *Structural Mechanics (Proceedings of the First Symposium on Naval Structural Mechanics)*, ed. J. N. Goodier & N. J. Hoff, pp. 74-114. Oxford: Pergamon Press.
Reissner, E. & Simmonds, J. G. (1966). Asymptotic solutions of boundary-value problems for elastic semi-infinite circular cylindrical shells. *Journal of Mathematics and Physics*, 45, 1-22.
Rodabaugh, E. C. & George, H. H. (1957). Effect of internal pressure on flexibility and stress-intensification factors of curved pipes or welding elbows. *Transactions of the American Society of Mechanical Engineers*, 79, 939-48.
Sanders, J. L. (1963). Non-linear theories for thin shells. *Quarterly of Applied Mathematics*, 21, 21-36.
Sanders, J. L. & Liepins, A. (1963). Toroidal membrane under internal pressure. *A.I.A.A. Journal*, 1, 2105-10.
Schorer, H. (1936). Line load action on thin cylindrical shells. *Transactions of the American Society of Civil Engineers*, 101, 767-802.
Sechler, E. E. (1974). The historical development of shell research and design. In *Thin Shell Structures*, ed. Y. C. Fung & E. E. Sechler, pp. 3-25. Englewood Cliffs, N. J.: Prentice-Hall, Inc.
Seide, P. (1975). *Small Elastic Deflections of Thin Shells*. Leyden: Noordhoff.
Seide, P. & Weingarten, V. I. (1961). On the buckling of circular cylinders under pure bending. *Journal of Applied Mechanics*, 28, 112-16.
Sewell, M. J. (1972), A survey of plastic buckling. In *Stability*, ed. H. Leipholz, pp. 85-197. Waterloo, Ontario: University of Waterloo Press.
Shanley, F. R. (1947). Inelastic column theory. *Journal of the Aeronautical Sciences*, 14, 261-7.
Shanley, F. R. (1967). *Mechanics of Materials*. New York: McGraw-Hill Book Co.
Simmonds, J. G. (1966). A set of simple, accurate equations for circular cylindrical elastic shells. *International Journal of Solids and Structures*, 2, 525-41.
Simmonds, J. G. (1975). Rigorous expunction of Poisson's ratio from the Reissner-Meissner equations. *International Journal of Solids and Structures*, 11, 1051-6.

Singer, J. (1969). The influence of stiffener geometry and spacing on the buckling of axially compressed cylindrical and conical shells. In *Theory of Thin Shells (Proceedings of 2nd IUTAM Symposium)*, ed. F. I. Niordson, pp. 234–63. Berlin: Springer-Verlag.

Singer, J., Baruch, M. & Harari, O. (1967). On the stability of eccentrically stiffened cylindrical shells under axial compression. *International Journal of Solids and Structures*, 3, 445–70.

Smith, R. T. & Ford, H. (1967). Experiments on pipelines and pipe bends subjected to three-dimensional loading. *Journal of Mechanical Engineering Science*, 9, 124–37.

Sobel, L. H. (1964). Effects of boundary conditions on the stability of cylinders subject to lateral and axial pressures. *A.I.A.A. Journal*, 2, 1437–40.

Southwell, R. V. (1913). On the collapse of tubes by external pressure. *Philosophical Magazine*, 25, 687–98.

Southwell, R. V. & Skan, S. W. (1924). On the stability under shearing forces of a flat elastic strip. *Proceedings of the Royal Society of London, Series A*, 105, 582–607.

Steele, C. R. (1974). Membrane solutions for shells with edge constraint. *Journal of the Engineering Mechanics Division, Proceedings of the American Society of Civil Engineers*, 100, 497–510.

Steele, C. R. (1975). Forming of thin shells. *Journal of Applied Mechanics*, 42, 884.

Stephens, W. B., Starnes, J. H. & Almroth, B. O. (1975). Collapse of long cylindrical shells under combined bending and pressure loads. *A.I.A.A. Journal*, 13, 20–4.

Symonds, P. S. (1951). Shakedown in continuous media. *Journal of Applied Mechanics*, 18, 85–9.

Symonds, P. S. (1962). Limit analysis. In *Handbook of Engineering Mechanics*, ed. W. Flügge, Chapter 49. New York: McGraw-Hill Book Company Inc.

Tarnai, T. (1980). Simultaneous static and kinematic indeterminacy of space trusses with cyclic symmetry. *International Journal of Solids and Structures*, 16, 347–59.

Tennyson, R. C. (1969). Buckling modes of circular cylindrical shells under axial compression. *A.I.A.A. Journal*, 7, 1481–7.

Tennyson, R. C. (1980). Interaction of cylindrical shell buckling experiments with theory. In *Theory of Shells (Proceedings of 3rd IUTAM Symposium on Shell Theory)*, ed. W. T. Koiter & G. K. Mikhailov, pp. 65–116. Amsterdam: North-Holland Publishing Company.

Thompson, J. M. T. (1962). The elastic instability of a complete spherical shell. *The Aeronautical Quarterly*, 13, 189–201.

Thompson, J. M. T. & Hunt, G. W. (1973). *A General Theory of Elastic Stability*. London: John Wiley & Sons.

Thomson, W. T. & Tait, P. G. (1879). *Treatise on Natural Philosophy*, vol. 1, part 1. Cambridge University Press.

Timoshenko, S. P. (1953). *History of Strength of Materials*. New York: McGraw-Hill Book Company.

Timoshenko, S. P. & Gere, J. M. (1961). *Theory of Elastic Stability* (2nd edn). New York: McGraw-Hill Book Company.

Timoshenko, S. P. & Goodier, J. N. (1970). *Theory of Elasticity* (3rd edn). New York: McGraw-Hill Book Company.

Timoshenko, S. P. & Woinowsky-Krieger, S. (1959). *Theory of Plates and Shells* (2nd edn). New York: McGraw-Hill Book Company.

Truesdell, C. (1945). The membrane theory of shells of revolution. *Transactions of the American Mathematical Society*, 58, 96–166.
Turner, C. E. & Ford, H. (1957). Stress and deflection studies of pipeline expansion bellows. *Proceedings of the Institution of Mechanical Engineers*, 171, 526–44.
van der Neut, A. (1947). *The general instability of stiffened cylindrical shells under axial compression*. Report S.314 (prepared for Reports and Transactions vol. 13). Amsterdam: Nationaal Luchtvaartlaboratorium.
Vlasov, V. Z. (1964). *General Theory of Shells and its Applications to Engineering* (translated from Russian version of 1949). NASA Technical Translation TTF-99. Washington: US National Aeronautics and Space Administration.
Warburton, G. B. (1976). *The Dynamical Behaviour of Structures* (2nd edn). Oxford: Pergamon Press.
Wells, A. A. (1950). A solution of beam-on-elastic foundation problems by means of a mechanical analogue. *Proceedings of the Institution of Mechanical Engineers*, 163, 307–10.
Windenberg, D. F. & Trilling, C. (1934). Collapse by instability of thin shells under external pressure. *Transactions of the American Society of Mechanical Engineers*, 56, 819–25.
Witt, P. J. (ed.) (1954). *Proceedings of a Symposium on Concrete Shell Roof Construction*. London: Cement and Concrete Association.
Wittrick, W. H. (1963). Interaction between membrane and edge stresses for thin cylinders under axially symmetrical loading. *Journal of the Royal Aeronautical Society*, 67, 172–4.
Yamaki, N. (1976). Experiments on the postbuckling behaviour of circular cylindrical shells under torsion. In *Buckling of Structures (Proceedings of IUTAM Symposium)*, ed. B. Budiansky, pp. 312–30. Berlin: Springer-Verlag.
Yamaki, N. & Kodama, S. (1980). Perturbation analyses for the postbuckling and imperfection sensitivity of circular cylindrical shells under torsion. In *Theory of Shells (Proceedings of 3rd IUTAM Symposium on Shell Theory)*, ed. W. T. Koiter & G. K. Mikhailov, pp. 635–67. Amsterdam: North-Holland Publishing Company.
Young, D. (1962). Vibration of continuous systems. In *Handbook of Engineering Mechanics*, ed. W. Flügge, Chapter 61. New York: McGraw-Hill Book Company.
Zick, L. P. & St Germain, A. R. (1963). Circumferential stresses in pressure vessels of revolution. *Transactions of the American Society of Mechanical Engineers, Series B: Journal of Engineering for Industry*, 85, 201–16.
Zienkiewicz, O. C. (1977). *The Finite-Element Method* (3rd edn). London: McGraw-Hill Book Co. (UK) Ltd.

Index

ABBOTT, M. B. 110
AIRY stress function 201, 324, 593, 730
AKSELRAD, E. L. 615
D'ALEMBERT, J. Le R. 110, 184, 631, 647
ALLMAN, D. J. 705
ALMROTH, B. O. 37, 476, 532, 550, 597, 620
AMAZIGO, J. C. 586
ANDERSON, R. J. 597, 613
angular defect in polygonalised surface 143, 185
angular excess in smooth surface 143
anisotropic shell 535
ARBOCZ, J. 555
ARGAND diagram 236
ARNOLD, R. N. 627, 642
ASHTON, J. N. 743
asymptotic analysis 423, 453
auxiliary sphere and circle 137
axial compression, buckling under 484
axisymmetric buckling mode 485, 541
axisymmetric loading 40, 96, 361, 413, 662
AYRTON, W. E. 475

B-surface 62, 221, 224, 227, 272, 310, 327, 403, 483, 565, 650, 672, 687, 728, 733
BABCOCK, C. D. 555
BABEL, H. W. 743
BAKER, LORD 657
band-loaded shell 57, 390, 677
BANTLIN, A. 429
barrel shell 105, 168
BARUCH, M. 536
BASSET, A. B. 11, 257

BATCH-HOPLEY formula 120
BATDORF, S. B. 492
beam analogy 208
beam-on-elastic-foundation 50, 240, 278, 463, 646
beam theory 277, 322
BECKER, H. 532
bellows 5, 413
bending mode of ring-vibration 630
bending strains 195, 424
bending- and stretching-strain energy 30
BESKIN, L. 444
BICKELL, M. B. 362
bifurcation of equilibrium paths 474, 573
BIJLAARD, P. P. 302
BLOMFIELD, J. A. 448, 461
BOSOR program 516
boundary conditions 51, 71, 112, 166, 249, 285, 293, 461, 512, 542, 614, 646, 650
boundary layer 11
Bourdon tube 437
BOUSSINESQ, J. 263
BRAZIER, L. G. 438, 595
BRITISH STANDARDS 120, 519
BRUSH, D. O. 37, 476, 532, 550
buckling 6, 473, 550, 609
BUDIANSKY, B. 36, 554, 586
BUSHNELL, D. 409, 516, 550, 743

CAUCHY, A. 5, 739
C. E. G. B. 638
change of Gaussian curvature 152, 175
characteristic length 47, 63, 280, 463
CHRISTENSEN, R. H. 743

Index

CLARK, R. A. 414, 429, 448, 453
classical buckling 473
closed box, closed surface 4, 125
CODAZZI, A. 178
COHN-VOSSEN, S. 114, 137
compatibility equations 45, 155, 230, 368, 474
competing buckling modes 493
complementary energy method 440, 443, 714
computer xiii, xvii
computer programs 516, 620
concentrated load applied to shell 33, 301, 307, 701
convexity 661, 739
COOPER, D. W. 318
correction of shallow-shell equations 242, 504, 644
corresponding variables 5, 51, 721
COX, H. L. 552
CRANDALL, S. H. 15, 448, 461
crease 165, 170
CROLL, J. G. A. 554
curvature-change 20
curvature–strain relations 156, 175
curved-beam effect 31
curved tube 429, 595
curvilinear coordinates xiii, 124, 171

DAHL, N. C. 15, 448, 461
D. A. St 516
DEMIR, H. H. 659
developable surface 127, 155
differential geometry 171
dimensionless length Ω 273, 501, 524, 614, 638
DINNO, K. S. 659
DIXON, H. H. 743
DONNELL, L. H. 218, 229, 283, 476, 521, 553, 593
DONNELL–MUSHTARI–VLASOV equations 232
doubly-periodic loading 191, 226, 488
droop of edge of shell roof 329
DRUCKER, D. C. 659, 670, 696
dual view of curvature of surface xv, 124, 156
DUNN, L. G. 551
DUNN, M. B. 461

EASON, G. 669
edge-beam to shell roof 338
edge-load coefficients for cylindrical shell 54, 69
'edge-string' of cylindrical shell 95, 119
effective meridian of shell of revolution 381
effective thickness of shell 48, 519
eigenvalue problem 475, 626
eighth-order equation 232, 654
elastic law for element of shell 19, 46, 368, 474
ELLING, R. E. 307
elliptic equation 111
elliptic or 'barrel' shell 105, 168, 240
energy methods 414, 440, 443, 524, 558, 603, 605, 714
envelope of family of curves 487, 501, 529, 643
equilibrium equations 44, 86, 105, 364, 449, 474
equilibrium of cap of shell 100, 376
equilibrium paths 475, 572
EULER, L. 114, 145, 475, 510, 739
EULER's topological theorem 114, 150
EVENSEN, D. A. 629
experiments on shells 455, 461, 467, 532, 555, 587, 597, 613, 627, 642, 659, 701
extensional vibration of a ring 630
extrinsic properties of a surface 126

FABIAN, O. 597, 613, 621
festoon of curves 487, 515, 640
finite-element calculations 72, 743
first- and second-class modes of vibration 631, 653
flanges, effect of 466
flat-plate closure to cylindrical shell 295
flat-plate region of behaviour 262, 291, 517
flexibility of bellows 413
flexibility factor for bellows 439
flexible membrane 436
flexural mode of vibration 630
flexural rigidity D of shell element 26
FLÜGGE, W. 83, 110, 222, 271, 277, 307, 318, 362, 367, 377, 380, 437, 476, 508, 627, 653, 743
FORD, H. 414, 448, 467
FORSBERG, K. 628
FOURIER series 59, 63, 271, 351, 449, 741
framework analogy for shell 113

760 *Index*

FULLER, R. B. 740
full-strength reinforcement of hole 693
fundamental mode of vibration 627
FUNG, Y. C. 532
funicular membrane in buckling analysis 563

GALLETLY, G. D. 362, 392, 409
GAUSSIAN curvature 83, 136, 193, 228, 324, 481, 563, 606, 728, 743
GAUSS, K. F. xv, 124, 175
GECKELER, J. W. 364, 370
GELLIN, S. 598
geodesic sphere 149
geometry-change effects in shells 595, 658, 704
geometry of general curved surface 125
GEORGE, H. H. 461
GERARD, G. 532
GERE, J. M. 475, 525
GIBSON, J. E. 318
GILL, S. S. 659, 696, 705, 743
GOLDENVEISER, A. L. 728
GOODALL, I. W. 659
GOODIER, J. N. 30, 263
gore of surface 147
GRIFFITHS, J. E. 743

HARARI, O. 536
HAYTHORNTHWAITE, R. M. 696
hemispherical bowl 627
HERTZ, H. R. 308
HETÉNYI, M. 69, 280, 313, 401, 464, 543
HEYMAN, J. 81, 377, 657
HILBERT, D. 114, 137
HILL, R. 659
HODGE, P. G. 696, 702
HOFF, N. J. 233, 543, 743
HOLAND, I. 233
HOOKE's law 9
horograph 137
HOWATSON, A. M. 742
HUNT, G. W. 554
HUTCHINSON, J. W. 550, 555, 585, 596, 609
hyperbolic equation 110
hyperbolic or 'waisted' shell 105, 169, 240
hypothesis xvi, 13, 80, 578, 676

imperfection-sensitive buckling 561
imperfect meridian of shell of revolution 374

improvement of shallow-shell equations 242, 504, 644
incompatible strains 191
inextensional deformation of surface 11, 145, 160, 168
inextensional distortion of meridian 415
inextensional mode of vibration 627
interactive buckling 554
interface pressure between B- and S-surfaces 62
internal pressure, stiffening effect of 456, 612
internal pressure, stresses due to 100, 361, 433
intrinsic properties of surface 126, 135

JOHNSON, W. 657, 662
JOLLEY, L. B. W. 64
JONES, N. 433, 448
JORDAN, F. F. 81, 436
junction problems in pressure-vessels 387, 679

KAFKA, P. G. 461
Von KÁRMÁN, T. 429, 551, 552, 593
M. W. KELLOG COMPANY 432, 455
KELVIN function 309
KILDEGAARD, A. 263
KIRCHHOFF, G. R. 257
KIRCHHOFF's hypothesis 13, 16, 24, 30, 258
KLUG, A. xviii
knuckle of pressure-vessel head 103
KODAMA, S. 591
KOITER, W. T. 12, 35, 479, 489, 539, 550, 555
KRAUS, H. 34, 362, 367

LAMB, SIR H. 11, 257, 631, 653
LAMÉ parameters 172, 179
LAPLACE operator 245
LARDNER, T. J. 15
Large-deflection analysis 621
LECKIE, F. A. 402, 660, 701
LEISSA, A. W. 628
LIEPINS, A. 436
limit load of plastic theory 662, 743
liquid-storage tanks 123, 518
load-sharing between shell and edge-beams 350
load-sharing between S- and B-surfaces 61, 226

local buckling in the flexure of tubes 609
local-buckling hypothesis in non-classical buckling 578
localised loading of shell 33, 301, 307, 701
long-wave region of behaviour 238, 250, 283, 291, 303, 465, 501, 638
LORENZ, R. 476
LOVE, A. E. H. xiv, 11, 16, 30, 34, 125, 171, 627, 630, 653
lower-bound theorem of plastic theory 659, 674, 679
low-frequency approximations 642
ŁUKASIEWICZ, S. A. 302
LUND, P. G. 742
LUR'E, A. I. 728

MALIK, Z. 516
map of shell behaviour 227, 238, 240, 288, 292, 297, 300, 304, 305, 307, 466, 491, 500, 502, 503, 515, 530, 539, 570, 619, 641, 651
MARCAL, P. V. 414, 661
MARTIN, D. W. 104, 120, 743
MARTIN, J. H. 657
MAXWELL, J. C. 51, 114, 143, 723, 738
McINTYRE, H. 743
MEISSNER, E. 369
MELLOR, P. B. 662
membrane, flexible 436
membrane hypothesis 80
meridian, imperfect 374
meridian of surface of revolution 96
Von MISES, R. 476, 670
MOHR circle 134
MORLEY, L. S. D. 244, 248
MORRIS, A. J. 701
MORTON, J. 516
most flexible mode of deformation 287, 305
MUSHTARI, K. M. 232

NAGHDI, P. M. 12, 728, 743
natural frequency of vibration 627
NEAL, B. G. 657
nearly-cylindrical shells 105, 168, 229
VAN DER NEUT, A. 537
NEWTON's laws 9
non-classical buckling 550
nonlinear behaviour of S-surface 581
nonlinear analysis of buckling 564
nonlinear effect of axial tension 66
nonlinear spring in analysis of buckling 556, 586

non-uniform shell 36, 535
normality condition of plastic theory 661
NOVOZHILOV, V. V. 12, 35, 116, 177, 239

OLSZAK, W. 696
OLVER, F. W. J. 309
ONAT, E. T. 659
optimum proportions for bellows 421
oil-storage tanks 123, 518, 547
open box, open surface 4, 125
orthogonal functions 725
orthotropic shells 36, 535
ovalisation of tube 443, 595

PALMER, A. C. 657, 692
PARDUE, T. E. 455, 467, 488
PARKES, E. W. 80, 114
PASKARAN, N. 371
PAVLOVIC, M. 734
PAYNE, D. J. 402, 660
PENNY, R. K. 402, 701, 743
PERRY, J. 475, 555
PFLÜGER, A. 373
'panel' flexibility and stiffness of cylindrical shell 194, 228, 485
pipe-bend 429
piping system 431
plane curves 128
plane string, equilibrium of 83
plastic beam on plastic foundation 673
plastic design of shells 660, 693
plastic instability and buckling 586, 705
plastic theory of structures 657
plate on elastic foundation 308
Platonic view of nature 738
pointed vertex in a surface 173
point load applied to shell 33, 301, 307, 701
POISSON's ratio 22, 261
polyhedral frame 738
polyhedron 114
PRAGER, W. 407, 661
pressure-vessels 100, 363, 679
prestress of shell 408
primary and secondary equilibrium paths 475
principal curvatures of surface 88, 129
principle of superposition 28
pulsating mode of vibration 630
pure mathematical viewpoint 738
pure twist of surface 24

quasi-inextensional deformation of shell 209

RANJAN, G. V. 392
RAYLEIGH, LORD 11, 171, 525, 627, 644, 653, 723
REDDY, B. D. 597, 609
REID, S. R. 657
reinforced-concrete shell 318
reinforced shell 535
REISSNER, E. 222, 250, 429, 448, 453
REISSNER, H. 369
rigidity of closed surface 5, 738
ring, elastic, 242
ring-loaded shell 54, 103, 671
ring-stiffened shell 535
ROBINSON, J. M. 578
RODABAUGH, E. C. 461
roots of Donnell's equation 233
RUIZ, C. 362, 516

S-surface 62, 221, 223, 227, 272, 310, 327, 403, 483, 564, 650, 672, 687, 728, 733
sagitta of arc 737
St GERMAIN, A. R. 101
St VENANT's principle 92, 258
SANDERS, J. L. 36, 177, 436
sandwich beam 15
sandwich shell 702
SAWCZUK, A. 696
SCHORER, H. 271
SCHWERIN, E. 476, 523
scientific method 6, 584
SCRIVEN, W. E. 104, 120, 743
SECHLER, E. E. 2, 532, 743
SEIDE, P. 597, 609, 743
self-weight loading 93, 326
semi-infinite cylindrical shell 41
semi-membrane theory 239
SEWELL, M. J. 585
shakedown 660
shallow-shell equations 156, 232, 269, 318, 398, 473, 550, 638, 684
SHANLEY, F. R. 475, 585
shear distortion 13
shear modulus 22
shell of revolution, symmetrically loaded 96, 179, 362, 679
shell roofs 318
SHIELD, R. T. 669, 670
short cylindrical shell under edge-loads 67

short-wave region of behaviour 238, 246, 254, 291, 303
SIMMONDS, J. G. 36, 248, 250
SINGER, J. 536
SKAN, S. W. 481, 530
SMITH, A. C. 289
SMITH, R. T. 448, 467
SOBEL, L. H. 517
solid angle subtended by portion, or vertex, of surface 136, 141
solid geometry 145
SOUTHWELL, R. V. 476, 481, 501, 530, 544, 614
spherical shell 307, 397, 679
spherical triangle, polygon 735
'square' mode of doubly-periodic displacement 494, 566, 589
stable equilibrium 559
STAGS program 620
STARNES, J. H. 597, 620
statical determinacy 199
statical equivalence of edge-loads 258
static–geometric analogy 231, 370, 728
STEELE, C. R. 392, 743
STEPHENS, W. B. 597, 620
stiffened shell 535
strain–curvature relations 156, 175
strain–displacement relations 157, 167, 181
strain energy method 29, 414, 524, 603, 605, 717
strain in bellows 424
stress-concentration factors 393, 455
stress reduction or moderation factors 57, 396
stress resultants 17
stringer-stiffened shell 535
SYMONDS, P. S. 407, 455, 662

TAIT, P. G. 137, 173
tangential loading of shell 197
TARNAI, T. 117, 740
TENNYSON, R. C. 555
THOMPSON, J. M. T. 554
THOMSON, W. T. (LORD KELVIN) 137, 173
through-thickness stress 33, 59
TIMOSHENKO, S. P. 31, 92, 257, 263, 279, 306, 308, 318, 367, 475, 525
TODD, J. D. 742
torispherical pressure-vessel head 102, 362, 390, 683

Index

toroidal surface 433
torsional buckling 591
total potential energy 558, 719
trapezoidal-edge effect 34
travelling wave 637
TRESCA's yield condition 392, 662, 670
trigonometrical identities 569
TRILLING, C. 501, 532
TRUESDELL, C. 82
TSIEN, H-S. 551, 552
TURNER, C. E. 414, 448
two-flanged beam 437, 598
two-surface model 61, 218, 272, 310, 327, 403, 477, 482, 563, 650, 728

unstable equilibrium 559, 595, 701
upper-bound theorem of plastic theory 659, 674, 695

VAN DER NEUT, A. 537
vibration of shells 627
VIGNESS, I. 448, 455, 467
virtual work 334, 417, 711

VLASOV, V. Z. 232, 239
'waisted' shell 105, 169, 240
WALKER, A. C. 554
WAN, C. C. 553
WARBURTON, G. B. 627, 642
WEINGARTEN, V. I. 597, 609
WELLS, A. A. 362
WINDENBERG, D. F. 501, 532
wind-loading on vertical cylinder 120
WINKLER foundation 279
WITT, P. J. 318
WITTRICK, W. H. 67
WOINOWSKY-KRIEGER, S. 306, 308, 318, 367

YAMAKI, N. 555, 591
yield condition 661
yield locus 664
YOUNG, D. 513, 630, 649
YOUNG's modulus 22

ZICK, L. P. 101
ZIENKIEWICZ, O. C. 743